Some Physical Constants

Quantity	Symbol	Value
Atomic mass unit	u	$1.660\ 538\ 782\ (83) \times 10^{-27}$ kg $931.494\ 028\ (23)$ MeV/c^2
Avogadro's number	N_A	$6.022\ 141\ 79\ (30) \times 10^{23}$ particles/mol
Bohr magneton	$\mu_B = \dfrac{e\hbar}{2m_e}$	$9.274\ 009\ 15\ (23) \times 10^{-24}$ J/T
Bohr radius	$a_0 = \dfrac{\hbar^2}{m_e e^2 k_e}$	$5.291\ 772\ 085\ 9\ (36) \times 10^{-11}$ m
Boltzmann's constant	$k_B = \dfrac{R}{N_A}$	$1.380\ 650\ 4\ (24) \times 10^{-23}$ J/K
Compton wavelength	$\lambda_C = \dfrac{h}{m_e c}$	$2.426\ 310\ 217\ 5\ (33) \times 10^{-12}$ m
Coulomb constant	$k_e = \dfrac{1}{4\pi\epsilon_0}$	$8.987\ 551\ 788 \ldots \times 10^9$ N·m^2/C^2 (exact)
Deuteron mass	m_d	$3.343\ 583\ 20\ (17) \times 10^{-27}$ kg $2.013\ 553\ 212\ 724\ (78)$ u
Electron mass	m_e	$9.109\ 382\ 15\ (45) \times 10^{-31}$ kg $5.485\ 799\ 094\ 3\ (23) \times 10^{-4}$ u $0.510\ 998\ 910\ (13)$ MeV/c^2
Electron volt	eV	$1.602\ 176\ 487\ (40) \times 10^{-19}$ J
Elementary charge	e	$1.602\ 176\ 487\ (40) \times 10^{-19}$ C
Gas constant	R	$8.314\ 472\ (15)$ J/mol·K
Gravitational constant	G	$6.674\ 28\ (67) \times 10^{-11}$ N·m^2/kg^2
Neutron mass	m_n	$1.674\ 927\ 211\ (84) \times 10^{-27}$ kg $1.008\ 664\ 915\ 97\ (43)$ u $939.565\ 346\ (23)$ MeV/c^2
Nuclear magneton	$\mu_n = \dfrac{e\hbar}{2m_p}$	$5.050\ 783\ 24\ (13) \times 10^{-27}$ J/T
Permeability of free space	μ_0	$4\pi \times 10^{-7}$ T·m/A (exact)
Permittivity of free space	$\epsilon_0 = \dfrac{1}{\mu_0 c^2}$	$8.854\ 187\ 817 \ldots \times 10^{-12}$ C^2/N·m^2 (exact)
Planck's constant	h	$6.626\ 068\ 96\ (33) \times 10^{-34}$ J·s
	$\hbar = \dfrac{h}{2\pi}$	$1.054\ 571\ 628\ (53) \times 10^{-34}$ J·s
Proton mass	m_p	$1.672\ 621\ 637\ (83) \times 10^{-27}$ kg $1.007\ 276\ 466\ 77\ (10)$ u $938.272\ 013\ (23)$ MeV/c^2
Rydberg constant	R_H	$1.097\ 373\ 156\ 852\ 7\ (73) \times 10^7$ m^{-1}
Speed of light in vacuum	c	$2.997\ 924\ 58 \times 10^8$ m/s (exact)

Note: These constants are the values recommended in 2006 by CODATA, based on a least-squares adjustment of data from different measurements. For a more complete list, see P. J. Mohr, B. N. Taylor, and D. B. Newell, "CODATA Recommended Values of the Fundamental Physical Constants: 2006." *Rev. Mod. Phys.* **80**:2, 633–730, 2008.

[a]The numbers in parentheses for the values represent the uncertainties of the last two digits.

Solar System Data

Body	Mass (kg)	Mean Radius (m)	Period (s)	Mean Distance from the Sun (m)
Mercury	3.30×10^{23}	2.44×10^6	7.60×10^6	5.79×10^{10}
Venus	4.87×10^{24}	6.05×10^6	1.94×10^7	1.08×10^{11}
Earth	5.97×10^{24}	6.37×10^6	3.156×10^7	1.496×10^{11}
Mars	6.42×10^{23}	3.39×10^6	5.94×10^7	2.28×10^{11}
Jupiter	1.90×10^{27}	6.99×10^7	3.74×10^8	7.78×10^{11}
Saturn	5.68×10^{26}	5.82×10^7	9.29×10^8	1.43×10^{12}
Uranus	8.68×10^{25}	2.54×10^7	2.65×10^9	2.87×10^{12}
Neptune	1.02×10^{26}	2.46×10^7	5.18×10^9	4.50×10^{12}
Pluto[a]	1.25×10^{22}	1.20×10^6	7.82×10^9	5.91×10^{12}
Moon	7.35×10^{22}	1.74×10^6	—	—
Sun	1.989×10^{30}	6.96×10^8	—	—

[a]In August 2006, the International Astronomical Union adopted a definition of a planet that separates Pluto from the other eight planets. Pluto is now defined as a "dwarf planet" (like the asteroid Ceres).

Physical Data Often Used

Average Earth–Moon distance	3.84×10^8 m
Average Earth–Sun distance	1.496×10^{11} m
Average radius of the Earth	6.37×10^6 m
Density of air (20°C and 1 atm)	1.20 kg/m^3
Density of air (0°C and 1 atm)	1.29 kg/m^3
Density of water (20°C and 1 atm)	1.00×10^3 kg/m^3
Free-fall acceleration	9.80 m/s^2
Mass of the Earth	5.97×10^{24} kg
Mass of the Moon	7.35×10^{22} kg
Mass of the Sun	1.99×10^{30} kg
Standard atmospheric pressure	1.013×10^5 Pa

Note: These values are the ones used in the text.

Some Prefixes for Powers of Ten

Power	Prefix	Abbreviation	Power	Prefix	Abbreviation
10^{-24}	yocto	y	10^1	deka	da
10^{-21}	zepto	z	10^2	hecto	h
10^{-18}	atto	a	10^3	kilo	k
10^{-15}	femto	f	10^6	mega	M
10^{-12}	pico	p	10^9	giga	G
10^{-9}	nano	n	10^{12}	tera	T
10^{-6}	micro	μ	10^{15}	peta	P
10^{-3}	milli	m	10^{18}	exa	E
10^{-2}	centi	c	10^{21}	zetta	Z
10^{-1}	deci	d	10^{24}	yotta	Y

FIFTH EDITION

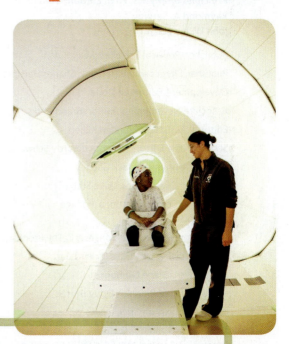

Principles of Physics

A CALCULUS-BASED TEXT

Volume 1

Raymond A. Serway
Emeritus, James Madison University

John W. Jewett, Jr.
Emeritus, California State Polytechnic University, Pomona

BROOKS/COLE
CENGAGE Learning

Australia • Brazil • Japan • Korea • Mexico • Singapore • Spain • United Kingdom • United States

BROOKS/COLE
CENGAGE Learning·

Principles of Physics, Fifth Edition, Volume 1
Raymond A. Serway
John W. Jewett, Jr.

Publisher, Physical Sciences: Mary Finch

Publisher, Physics and Astronomy: Charles Hartford

Development Editor: Ed Dodd

Associate Development Editor: Brandi Kirksey

Editorial Assistant: Brendan Killion

Senior Media Editor: Rebecca Berardy Schwartz

Marketing Manager: Jack Cooney

Marketing Coordinator: Julie Stefani

Marketing Communications Manager:
Darlene Macanan

Senior Content Project Manager: Cathy Brooks

Senior Art Director: Cate Rickard Barr

Print Buyer: Diane Gibbons

Manufacturing Planner: Doug Bertke

Rights Acquisition Specialist:
Shalice Shah-Caldwell

Production Service: MPS Limited, a Macmillan Company

Compositor: MPS Limited, a Macmillan Company

Text Designer: Brian Salisbury

Cover Designer: Brian Salisbury

Cover Images:

• Child prepares for proton therapy at the Roberts Proton Therapy Center in Philadelphia, PA: Copyright © 2011 Ed Cunicelli

• A full moon rises through an iridescent lenticular cloud at Mt. Rainier: © Caren Brinkema/ Science Faction/Getty Images

• Supernova remnant: NASA

• Infinity Bridge on the River Tees: © Paul Downing/ Flickr Select/Getty Images

For product information and technology assistance, contact us at
Cengage Learning Customer & Sales Support, 1-800-354-9706
For permission to use material from this text or product,
submit all requests online at **www.cengage.com/permissions.**
Further permissions questions can be emailed to
permissionrequest@cengage.com

Library of Congress Control Number: 2011937185

ISBN-13: 978-1-133-11027-9

ISBN-10: 1-133-11027-4

Brooks/Cole
20 Channel Center Street
Boston, MA 02210
USA

Cengage Learning is a leading provider of customized learning solutions with office locations around the globe, including Singapore, the United Kingdom, Australia, Mexico, Brazil and Japan. Locate your local office at **international.cengage.com/region**

Cengage Learning products are represented in Canada by Nelson Education, Ltd.

For your course and learning solutions, visit **www.cengage.com.**

Purchase any of our products at your local college store or at our preferred online store **www.cengagebrain.com.**

Instructors: Please visit **login.cengage.com** and log in to access instructor-specific resources.

We dedicate this book to our wives Elizabeth and Lisa and all our children and grandchildren for their loving understanding when we spent time on writing instead of being with them.

Printed in the United States of America
1 2 3 4 5 6 7 15 14 13 12 11

Welcome to your MCAT Test Preparation Guide

The **MCAT Test Preparation Guide** makes your copy of *Principles of Physics,* **Fifth Edition,** the most comprehensive MCAT study tool and classroom resource in introductory physics. The grid, which begins below and continues on the next two pages, outlines twelve concept-based **study courses** for the physics part of your MCAT exam. Use it to prepare for the MCAT, class tests, and your homework assignments.

Vectors

Skill Objectives: To calculate distance, calculate angles between vectors, calculate magnitudes, and to understand vectors.

Review Plan:

Distance and Angles: Chapter 1
- Section 1.6
- Active Figure 1.4
- Chapter Problem 33

Using Vectors: Chapter 1
- Sections 1.7–1.9
- Quick Quizzes 1.4–1.8
- Examples 1.6–1.8
- Active Figures 1.9, 1.16
- Chapter Problems 34, 35, 43, 44, 47, 51

Motion

Skill Objectives: To understand motion in two dimensions, to calculate speed and velocity, to calculate centripetal acceleration, and acceleration in free fall problems.

Review Plan:

Motion in 1 Dimension: Chapter 2
- Sections 2.1, 2.2, 2.4, 2.6, 2.7
- Quick Quizzes 2.3–2.6
- Examples 2.1, 2.2, 2.4–2.9
- Active Figure 2.12
- Chapter Problems 3, 5,13, 19, 21, 29, 31, 33

Motion in 2 Dimensions: Chapter 3
- Sections 3.1–3.3
- Quick Quizzes 3.2, 3.3
- Examples 3.1–3.4
- Active Figures 3.5, 3.7, 3.10
- Chapter Problems 1, 11, 13

Centripetal Acceleration: Chapter 3
- Sections 3.4, 3.5
- Quick Quizzes 3.4, 3.5
- Example 3.5
- Active Figure 3.14
- Chapter Problems 23, 31

Force

Skill Objectives: To know and understand Newton's laws, to calculate resultant forces and weight.

Review Plan:

Newton's Laws: Chapter 4
- Sections 4.1–4.6
- Quick Quizzes 4.1–4.6
- Example 4.1
- Chapter Problem 7

Resultant Forces: Chapter 4
- Section 4.7
- Quick Quiz 4.7
- Example 4.6
- Chapter Problems 29, 37

Gravity: Chapter 11
- Section 11.1
- Quick Quiz 11.1
- Chapter Problem 5

Equilibrium

Skill Objectives: To calculate momentum and impulse, center of gravity, and torque.

Review Plan:

Momentum: Chapter 8
- Section 8.1
- Quick Quiz 8.2
- Examples 8.2, 8.3

Impulse: Chapter 8
- Sections 8.2–8.4
- Quick Quizzes 8.3, 8.4
- Examples 8.4, 8.6
- Active Figures 8.8, 8.9
- Chapter Problems 5, 9, 11, 17, 21

Torque: Chapter 10
- Sections 10.5, 10.6
- Quick Quiz 10.7
- Example 10.8
- Chapter Problems 23, 30

Work

Skill Objectives: To calculate friction, work, kinetic energy, power, and potential energy.

Review Plan:

Friction: Chapter 5
- Section 5.1
- Quick Quizzes 5.1, 5.2

Work: Chapter 6
- Section 6.2
- Chapter Problems 3, 5

Kinetic Energy: Chapter 6
- Section 6.5
- Example 6.6

Power: Chapter 7
- Section 7.6
- Chapter Problem 29

Potential Energy: Chapters 6, 7
- Sections 6.6, 7.2
- Quick Quiz 6.6
Chapter 7
- Chapter Problem 3

Waves

Skill Objectives: To understand interference of waves, to calculate basic properties of waves, properties of springs, and properties of pendulums.

Review Plan:

Wave Properties: Chapters 12, 13
- Sections 12.1, 12.2, 13.1, 13.2
- Quick Quiz 13.1
- Examples 12.1, 13.2
- Active Figures 12.1, 12.2, 12.6, 12.8, 12.11
Chapter 13
- Problem 7

Pendulum: Chapter 12
- Sections 12.4, 12.5
- Quick Quizzes 12.5, 12.6
- Example 12.5
- Active Figure 12.13
- Chapter Problem 35

Interference: Chapter 14
- Sections 14.1, 14.2
- Quick Quiz 14.1
- Active Figures 14.1–14.3

Matter

Skill Objectives: To calculate density, pressure, specific gravity, and flow rates.

Review Plan:

Density: Chapters 1, 15
- Sections 1.1, 15.2

Pressure: Chapter 15
- Sections 15.1–15.4
- Quick Quizzes 15.1–15.4
- Examples 15.1, 15.3
- Chapter Problems 2, 11, 23, 27, 31

Flow Rates: Chapter 15
- Section 15.6
- Quick Quiz 15.5

Sound

Skill Objectives: To understand interference of waves, calculate properties of waves, the speed of sound, Doppler shifts, and intensity.

Review Plan:

Sound Properties: Chapters 13, 14
- Sections 13.2, 13.3, 13.6, 13.7, 14.3
- Quick Quizzes 13.2, 13.4, 13.7
- Example 14.3
- Active Figures 13.6, 13.7, 13.9, 13.22, 13.23
Chapter 13
- Problems 11, 15, 28, 34, 41, 45
Chapter 14
- Problem 27

Interference/Beats: Chapter 14
- Sections 14.1, 14.5
- Quick Quiz 14.6
- Active Figures 14.1–14.3, 14.12
- Chapter Problems 9, 45, 46

Light

Skill Objectives: To understand mirrors and lenses, to calculate the angles of reflection, to use the index of refraction, and to find focal lengths.

Review Plan:

Reflection: Chapter 25
- Sections 25.1–25.3
- Example 25.1
- Active Figure 25.5

Refraction: Chapter 25
- Sections 25.4, 25.5
- Quick Quizzes 25.2–25.5
- Example 25.2
- Chapter Problems 8, 16

Mirrors and Lenses: Chapter 26
- Sections 26.1–26.4
- Quick Quizzes 26.1–26.6
- Thinking Physics 26.2
- Examples 26.1–26.5
- Active Figures 26.2, 26.25
- Chapter Problems 27, 30, 33, 37

Electrostatics

Skill Objectives: To understand and calculate the electric field, the electrostatic force, and the electric potential.

Review Plan:

Coulomb's Law: Chapter 19
- Sections 19.2–19.4
- Quick Quiz 19.1–19.3
- Examples 19.1, 19.2
- Active Figure 19.7
- Chapter Problems 3, 9

Electric Field: Chapter 19
- Sections 19.5, 19.6
- Quick Quizzes 19.4, 19.5
- Active Figures 19.11, 19.20, 19.22

Potential: Chapter 20
- Sections 20.1–20.3
- Examples 20.1, 20.2
- Active Figure 20.8
- Chapter Problems 3, 5, 8, 11

Circuits

Skill Objectives: To understand and calculate current, resistance, voltage, and power, and to use circuit analysis.

Review Plan:

Ohm's Law: Chapter 21
- Sections 21.1, 21.2
- Quick Quizzes 21.1, 21.2
- Examples 21.1, 21.2
- Chapter Problem 8

Power and Energy: Chapter 21
- Section 21.5
- Quick Quiz 21.4
- Examples 21.4
- Active Figure 21.11
- Chapter Problems 21, 25, 31

Circuits: Chapter 21
- Sections 21.6–21.8
- Quick Quizzes 21.5–21.7
- Examples 21.6–21.8
- Active Figures 21.14, 21.15, 21.17
- Chapter Problems 31, 39, 47

Atoms

Skill Objectives: To understand decay processes and nuclear reactions and to calculate half-life.

Review Plan:

Atoms: Chapters, 11, 29
- Section 11.5
- Sections 29.1–29.6

Chapter 11
- Problems 37–43, 61

Decays: Chapter 30
- Sections 30.3, 30.4
- Quick Quizzes 30.3–30.6
- Examples 30.3–30.6
- Active Figures 30.8–30.11, 30.13, 30.14
- Chapter Problems 18, 23, 25

Nuclear Reactions: Chapter 30
- Section 30.5
- Active Figure 30.18
- Chapter Problems 32, 35

Contents

About the Authors

Raymond A. Serway received his doctorate at Illinois Institute of Technology and is Professor Emeritus at James Madison University. In 2011, he was awarded with an honorary doctorate degree from his alma mater, Utica College. He received the 1990 Madison Scholar Award at James Madison University, where he taught for 17 years. Dr. Serway began his teaching career at Clarkson University, where he conducted research and taught from 1967 to 1980. He was the recipient of the Distinguished Teaching Award at Clarkson University in 1977 and the Alumni Achievement Award from Utica College in 1985. As Guest Scientist at the IBM Research Laboratory in Zurich, Switzerland, he worked with K. Alex Müller, 1987 Nobel Prize recipient. Dr. Serway also was a visiting scientist at Argonne National Laboratory, where he collaborated with his mentor and friend, the late Dr. Sam Marshall. Dr. Serway is the coauthor of *College Physics,* ninth edition; *Physics for Scientists and Engineers,* eighth edition; *Essentials of College Physics; Modern Physics,* third edition; and the high school textbook *Physics,* published by Holt McDougal. In addition, Dr. Serway has published more than 40 research papers in the field of condensed matter physics and has given more than 60 presentations at professional meetings. Dr. Serway and his wife Elizabeth enjoy traveling, playing golf, fishing, gardening, singing in the church choir, and especially spending quality time with their four children, nine grandchildren, and a recent great grandson.

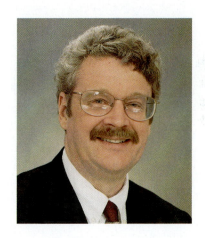

John W. Jewett, Jr. earned his undergraduate degree in physics at Drexel University and his doctorate at Ohio State University, specializing in optical and magnetic properties of condensed matter. Dr. Jewett began his academic career at Richard Stockton College of New Jersey, where he taught from 1974 to 1984. He is currently Emeritus Professor of Physics at California State Polytechnic University, Pomona. Throughout his teaching career, Dr. Jewett has been active in promoting effective physics education. In addition to receiving four National Science Foundation grants, he helped found and direct the Southern California Area Modern Physics Institute (SCAMPI) and Science IMPACT (Institute for Modern Pedagogy and Creative Teaching). Dr. Jewett's honors include the Stockton Merit Award at Richard Stockton College in 1980, selection as Outstanding Professor at California State Polytechnic University for 1991–1992, and the Excellence in Undergraduate Physics Teaching Award from the American Association of Physics Teachers (AAPT) in 1998. In 2010, he received an Alumni Lifetime Achievement Award from Drexel University in recognition of his contributions in physics education. He has given more than 100 presentations both domestically and abroad, including multiple presentations at national meetings of the AAPT. Dr. Jewett is the author of *The World of Physics: Mysteries, Magic, and Myth,* which provides many connections between physics and everyday experiences. In addition to his work as the coauthor for *Principles of Physics* he is also the coauthor on *Physics for Scientists and Engineers,* eighth edition, as well as *Global Issues,* a four-volume set of instruction manuals in integrated science for high school. Dr. Jewett enjoys playing keyboard with his all-physicist band, traveling, underwater photography, learning foreign languages, and collecting antique quack medical devices that can be used as demonstration apparatus in physics lectures. Most importantly, he relishes spending time with his wife Lisa and their children and grandchildren.

Preface

Principles of Physics is designed for a one-year introductory calculus-based physics course for engineering and science students and for premed students taking a rigorous physics course. This fifth edition contains many new pedagogical features—most notably, an integrated Web-based learning system and a structured problem-solving strategy that uses a modeling approach. Based on comments from users of the fourth edition and reviewers' suggestions, a major effort was made to improve organization, clarity of presentation, precision of language, and accuracy throughout.

This textbook was initially conceived because of well-known problems in teaching the introductory calculus-based physics course. The course content (and hence the size of textbooks) continues to grow, while the number of contact hours with students has either dropped or remained unchanged. Furthermore, traditional one-year courses cover little if any physics beyond the 19th century.

In preparing this textbook, we were motivated by the spreading interest in reforming the teaching and learning of physics through physics education research (PER). One effort in this direction was the Introductory University Physics Project (IUPP), sponsored by the American Association of Physics Teachers and the American Institute of Physics. The primary goals and guidelines of this project are to:

- Reduce course content following the "less may be more" theme;
- Incorporate contemporary physics naturally into the course;
- Organize the course in the context of one or more "story lines";
- Treat all students equitably.

Recognizing a need for a textbook that could meet these guidelines several years ago, we studied the various proposed IUPP models and the many reports from IUPP committees. Eventually, one of us (RAS) became actively involved in the review and planning of one specific model, initially developed at the U.S. Air Force Academy, entitled "A Particles Approach to Introductory Physics." An extended visit at the Academy was spent working with Colonel James Head and Lt. Col. Rolf Enger, the primary authors of the Particles model, and other members of that department. This most useful collaboration was the starting point of this project.

The other author (JWJ) became involved with the IUPP model called "Physics in Context," developed by John Rigden (American Institute of Physics), David Griffiths (Oregon State University), and Lawrence Coleman (University of Arkansas at Little Rock). This involvement led to National Science Foundation (NSF) grant support for the development of new contextual approaches and eventually to the contextual overlay that is used in this book and described in detail later in the Preface.

The combined IUPP approach in this book has the following features:

- It is an evolutionary approach (rather than a revolutionary approach), which should meet the current demands of the physics community.
- It deletes many topics in classical physics (such as alternating current circuits and optical instruments) and places less emphasis on rigid object motion, optics, and thermodynamics.
- Some topics in contemporary physics, such as fundamental forces, special relativity, energy quantization, and the Bohr model of the hydrogen atom, are introduced early in the textbook.
- A deliberate attempt is made to show the unity of physics and the global nature of physics principles.
- As a motivational tool, the textbook connects applications of physics principles to interesting biomedical situations, social issues, natural phenomena, and technological advances.

Other efforts to incorporate the results of physics education research have led to several of the features in this textbook described below. These include Quick Quizzes, Objective Questions, Pitfall Preventions, **What If?** features in worked examples, the use of energy bar charts, the modeling approach to problem solving, and the global energy approach introduced in Chapter 7.

Objectives

This introductory physics textbook has two main objectives: to provide the student with a clear and logical presentation of the basic concepts and principles of physics, and to strengthen an understanding of the concepts and principles through a broad range of interesting applications to the real world. To meet these objectives, we have emphasized sound physical arguments and problem-solving methodology. At the same time, we have attempted to motivate the student through practical examples that demonstrate the role of physics in other disciplines, including engineering, chemistry, and medicine.

Changes in the Fifth Edition

A number of changes and improvements have been made in the fifth edition of this text. Many of these are in response to recent findings in physics education research and to comments and suggestions provided by the reviewers of the manuscript and instructors using the first four editions. The following represent the major changes in the fifth edition:

New Contexts. The context overlay approach is described below under "Organization." The fifth edition introduces two new Contexts: for Chapter 15, "Heart Attacks," and for Chapters 22–23, "Magnetism in Medicine." Both of these new Contexts are aimed at applying the principles of physics to the biomedical field.

In the "Heart Attacks" Context, we study the flow of fluids through a pipe, as an analogy to the flow of blood through blood vessels in the human body. Various details of the blood flow are related to the dangers of cardiovascular disease. In addition, we discuss new developments in the study of blood flow and heart attacks using nanoparticles and computer imaging.

The "Magnetism in Medicine" Context explores the application of the principles of electromagnetism to diagnostic and therapeutic procedures in medicine. We begin by looking at historical uses of magnetism, including several "quack" medical devices. More modern applications include remote magnetic navigation in cardiac catheter ablation procedures for atrial fibrillation, transcranial magnetic stimulation for the treatment of depression, and magnetic resonance imaging as a diagnostic tool.

Worked Examples. All in-text worked examples have been recast and are now presented in a two-column format to better reinforce physical concepts. The left column shows textual information that describes the steps for solving the problem. The right column shows the mathematical manipulations and results of taking these steps. This layout facilitates matching the concept with its mathematical execution and helps students organize their work. The examples closely follow the General Problem-Solving Strategy introduced in Chapter 1 to reinforce effective problem-solving habits. In almost all cases, examples are solved symbolically until the end, when numerical values are substituted into the final symbolic result. This procedure allows students to analyze the symbolic result to see how the result depends on the parameters in the problem, or to take limits to test the final result for correctness. Most worked examples in the text may be assigned for homework in Enhanced WebAssign. A sample of a worked example can be found on the next page.

ENHANCED **WebAssign** Most worked examples are also available to be assigned as interactive examples in the Enhanced WebAssign homework management system.

Example 6.6 | A Block Pulled on a Frictionless Surface

A 6.0-kg block initially at rest is pulled to the right along a frictionless, horizontal surface by a constant horizontal force of 12 N. Find the block's speed after it has moved 3.0 m.

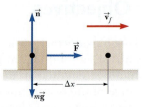

Figure 6.14 (Example 6.6) A block pulled to the right on a frictionless surface by a constant horizontal force.

Each solution has been written to closely follow the General Problem-Solving Strategy as outlined on pages 25–26 in Chapter 1, so as to reinforce good problem-solving habits.

SOLUTION

Conceptualize Figure 6.14 illustrates this situation. Imagine pulling a toy car across a table with a horizontal rubber band attached to the front of the car. The force is maintained constant by ensuring that the stretched rubber band always has the same length.

Categorize We could apply the equations of kinematics to determine the answer, but let us practice the energy approach. The block is the system, and three external forces act on the system. The normal force balances the gravitational force on the block, and neither of these vertically acting forces does work on the block because their points of application are horizontally displaced.

Analyze The net external force acting on the block is the horizontal 12-N force.

Use the work–kinetic energy theorem for the block, noting that its initial kinetic energy is zero:

$$W_{ext} = K_f - K_i = \tfrac{1}{2}mv_f^2 - 0 = \tfrac{1}{2}mv_f^2$$

Each step of the solution is detailed in a two-column format. The left column provides an explanation for each mathematical step in the right column, to better reinforce the physical concepts.

Solve for v_f and use Equation 6.1 for the work done on the block by $\vec{F}$:

$$v_f = \sqrt{\frac{2W_{ext}}{m}} = \sqrt{\frac{2F\Delta x}{m}}$$

Substitute numerical values:

$$v_f = \sqrt{\frac{2(12\ N)(3.0\ m)}{6.0\ kg}} = 3.5\ m/s$$

Finalize It would be useful for you to solve this problem again by modeling the block as a particle under a net force to find its acceleration and then as a particle under constant acceleration to find its final velocity.

What If? Suppose the magnitude of the force in this example is doubled to $F' = 2F$. The 6.0-kg block accelerates to 3.5 m/s due to this applied force while moving through a displacement $\Delta x'$. How does the displacement $\Delta x'$ compare with the original displacement Δx?

Answer If we pull harder, the block should accelerate to a given speed in a shorter distance, so we expect that $\Delta x' < \Delta x$. In both cases, the block experiences the same change in kinetic energy ΔK. Mathematically, from the work–kinetic energy theorem, we find that

$$W_{ext} = F'\Delta x' = \Delta K = F\Delta x$$
$$\Delta x' = \frac{F}{F'}\ \Delta x = \frac{F}{2F}\ \Delta x = \tfrac{1}{2}\Delta x$$

and the distance is shorter as suggested by our conceptual argument.

What If? statements appear in about 1/3 of the worked examples and offer a variation on the situation posed in the text of the example. For instance, this feature might explore the effects of changing the conditions of the situation, determine what happens when a quantity is taken to a particular limiting value, or question whether additional information can be determined about the problem situation. This feature encourages students to think about the results of the example and assists in conceptual understanding of the principles.

The final result is symbolic; numerical values are substituted into the final result.

Line-by-Line Revision of the Questions and Problems Set. For the Fifth Edition, the authors reviewed each question and problem and incorporated revisions designed to improve both readability and assignability. To make problems clearer to both students and instructors, this extensive process involved editing problems for clarity, editing for length, adding figures where appropriate, and introducing better problem architecture by breaking up problems into clearly defined parts.

Data from Enhanced WebAssign Used to Improve Questions and Problems. As part of the full-scale analysis and revision of the questions and problems sets, the authors utilized extensive user data gathered by WebAssign, from both instructors who assigned and students who worked on problems from previous editions of *Principles of Physics*. These data helped tremendously, indicating when the phrasing in problems could be clearer, thus providing guidance on how to revise problems so that they are more easily understandable for students and more easily assignable by instructors in Enhanced WebAssign. Finally, the data were used to ensure that the problems most often assigned were retained for this new edition. In each chapter's problems set, the top quartile of problems assigned in Enhanced WebAssign have blue-shaded problem numbers for easy identification, allowing professors to quickly and easily find the most popular problems assigned in Enhanced WebAssign.

To provide an idea of the types of improvements that were made to the problems, here is a problem from the fourth edition, followed by the problem as it now appears in the fifth edition, with explanations of how the problems were improved.

Problem from the Fourth Edition . . .

35. (a) Consider an extended object whose different portions have different elevations. Assume the free-fall acceleration is uniform over the object. Prove that the gravitational potential energy of the object–Earth system is given by $U_g = Mgy_{CM}$, where M is the total mass of the object and y_{CM} is the elevation of its center of mass above the chosen reference level. (b) Calculate the gravitational potential energy associated with a ramp constructed on level ground with stone with density 3 800 kg/m^3 and every-where 3.60 m wide (Fig. P8.35). In a side view, the ramp appears as a right triangle with height 15.7 m at the top end and base 64.8 m.

Figure P8.35

. . . As revised for the Fifth Edition:

37. Explorers in the jungle find an ancient monument in the shape of a large isosceles triangle as shown in Figure P8.37. The monument is made from tens of thousands of small stone blocks of density 3 800 kg/m^3. The monument is 15.7 m high and 64.8 m wide at its base and is everywhere 3.60 m thick from front to back. Before the monument was built many years ago, all the stone blocks lay on the ground. How much work did laborers do on the blocks to put them in position while building the entire monument? *Note*: The gravitational potential energy of an object–Earth system is given by $U_g = Mgy_{CM}$, where M is the total mass of the object and y_{CM} is the elevation of its center of mass above the chosen reference level.

> A storyline for the problem is provided.

> The requested quantity is made more personal by asking for work done by humans rather than asking for the gravitational potential energy.

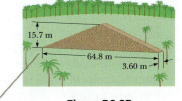

Figure P8.37

> The figure has been revised and dimensions added.

> The expression for the gravitational potential energy is provided, whereas it was requested to be proven in the original. This allows the problem to work better in Enhanced WebAssign.

Revised Questions Organization. We reorganized the end-of-chapter questions for this new edition. The previous edition's Questions section is now divided into two sections: Objective Questions and Conceptual Questions.

Objective Questions are multiple-choice, true/false, ranking, or other multiple guess-type questions. Some require calculations designed to facilitate students' familiarity with the equations, the variables used, the concepts the variables represent, and the relationships between the concepts. Others are more conceptual in nature and are designed to encourage conceptual thinking. Objective Questions are also written with the personal response system user in mind, and most of the questions could easily be used in these systems.

Conceptual Questions are more traditional short-answer and essay-type questions that require students to think conceptually about a physical situation.

Problems. The end-of-chapter problems are more numerous in this edition and more varied (in all, over 2 200 problems are given throughout the text). For the convenience of both the student and the instructor, about two-thirds of the problems are keyed to specific sections of the chapter, including Context Connection sections. The remaining problems, labeled "Additional Problems," are not keyed to specific sections. The

BIO icon identifies problems dealing with applications to the life sciences and medicine. Answers to odd-numbered problems are provided at the end of the book. For ease of identification, the problem numbers for straightforward problems are printed in **black**; intermediate-level problem numbers are printed in **blue**; and those of challenging problems are printed in **red**.

New Types of Problems. We have introduced four new problem types for this edition:

Q|C Quantitative/Conceptual problems contain parts that ask students to think both quantitatively and conceptually. An example of a Quantitative/Conceptual problem appears here:

The problem is identified with a **Q|C** icon.

55. **Q|C** A horizontal spring attached to a wall has a force constant of $k = 850$ N/m. A block of mass $m = 1.00$ kg is attached to the spring and rests on a frictionless, horizontal surface as in Figure P7.55. (a) The block is pulled to a position $x_i = 6.00$ cm from equilibrium and released. Find the elastic potential energy stored in the spring when the block is 6.00 cm from equilibrium and when the block passes through equilibrium. (b) Find the speed of the block as it passes through the equilibrium point. (c) What is the speed of the block when it is at a position $x_i/2 = 3.00$ cm? (d) Why isn't the answer to part (c) half the answer to part (b)?

Parts (a)–(c) of the problem ask for quantitative calculations.

Part (d) asks a conceptual question about the situation.

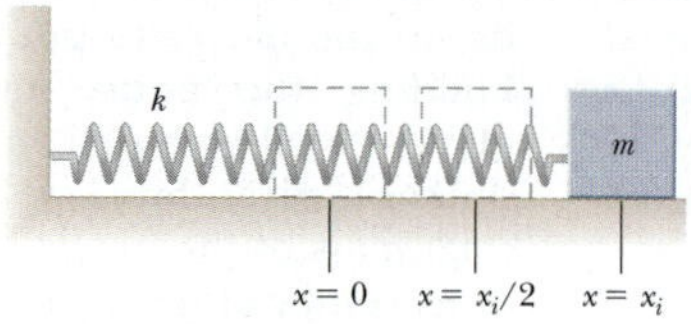

$x = 0$ $x = x_i/2$ $x = x_i$

Figure P7.55

S Symbolic problems ask students to solve a problem using only symbolic manipulation. A majority of survey respondents asked specifically for an increase in the number of symbolic problems found in the text because it better reflects the way instructors want their students to think when solving physics problems. An example of a Symbolic problem appears here:

The problem is identified with an **S** icon.

57. **S** **Review.** A uniform board of length L is sliding along a smooth, frictionless, horizontal plane as shown in Figure P7.57a (page 226). The board then slides across the boundary with a rough horizontal surface. The coefficient of kinetic friction between the board and the second surface is μ_k. (a) Find the acceleration of the board at the moment its front end has traveled a distance x beyond the boundary. (b) The board stops at the moment its back end reaches the boundary as shown in Figure P7.57b. Find the initial speed v of the board.

No numerical values appear in the problem statement.

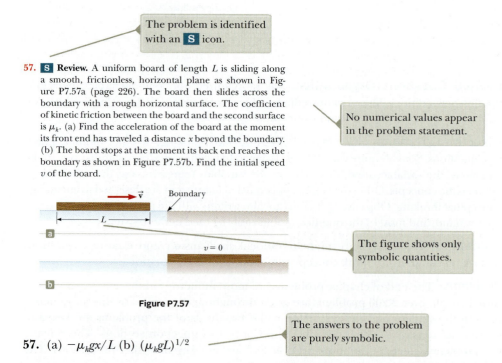

Boundary

L

a

$v = 0$

b

The figure shows only symbolic quantities.

Figure P7.57

The answers to the problem are purely symbolic.

57. (a) $-\mu_k gx/L$ (b) $(\mu_k gL)^{1/2}$

GP Guided Problems help students break problems into steps. A physics problem typically asks for one physical quantity in a given context. Often, however, several concepts must be used and a number of calculations are required to obtain that final answer. Many students are not accustomed to this level of complexity and often don't know where to start. A Guided Problem breaks a standard problem into smaller steps, enabling students to grasp all the concepts and strategies required to arrive at a correct solution. Unlike standard physics problems, guidance is often built into the problem statement. Guided Problems are reminiscent of how a student might interact with a professor in an office visit. These problems (there is one in every chapter of the text) help train students to break down complex problems into a series of simpler problems, an essential problem-solving skill. An example of a Guided Problem appears here:

The problem is identified with a **GP** icon.

28. **GP** A uniform beam resting on two pivots has a length $L = 6.00$ m and mass $M = 90.0$ kg. The pivot under the left end exerts a normal force n_1 on the beam, and the second pivot located a distance $\ell = 4.00$ m from the left end exerts a normal force n_2. A woman of mass $m = 55.0$ kg steps onto the left end of the beam and begins walking to the right as in Figure P10.28. The goal is to find the woman's position when the beam begins to tip. (a) What is the appropriate analysis model for the beam before it begins to tip? (b) Sketch a force diagram for the beam, labeling the gravitational and normal forces acting on the beam and placing the woman a distance x to the right of the first pivot, which is the origin. (c) Where is the woman when the normal force n_1 is the greatest? (d) What is n_1 when the beam is about to tip? (e) Use Equation 10.27 to find the value of n_2 when the beam is about to tip. (f) Using the result of part (d) and Equation 10.28, with torques computed around the second pivot, find the woman's position x when the beam is about to tip. (g) Check the answer to part (e) by computing torques around the first pivot point.

The goal of the problem is identified.

Analysis begins by identifying the appropriate analysis model.

Students are provided with suggestions for steps to solve the problem.

The calculation associated with the goal is requested.

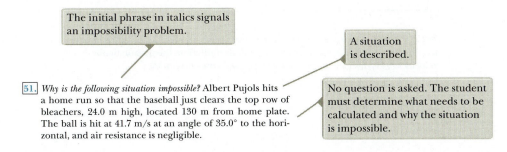

Figure P10.28

Impossibility Problems. Physics education research has focused heavily on the problem-solving skills of students. Although most problems in this text are structured in the form of providing data and asking for a result of computation, two problems in each chapter, on average, are structured as impossibility problems. They begin with the phrase *Why is the following situation impossible?* That is followed by the description of a situation. The striking aspect of these problems is that no question is asked of the students, other than that in the initial italics. The student must determine what questions need to be asked and what calculations need to be performed. Based on the results of these calculations, the student must determine why the situation described is not possible. This determination may require information from personal experience, common sense, Internet or print research, measurement, mathematical skills, knowledge of human norms, or scientific thinking.

These problems can be assigned to build critical thinking skills in students. They are also fun, having the aspect of physics "mysteries" to be solved by students individually or in groups. An example of an impossibility problem appears here:

The initial phrase in italics signals an impossibility problem.

A situation is described.

51. *Why is the following situation impossible?* Albert Pujols hits a home run so that the baseball just clears the top row of bleachers, 24.0 m high, located 130 m from home plate. The ball is hit at 41.7 m/s at an angle of 35.0° to the horizontal, and air resistance is negligible.

No question is asked. The student must determine what needs to be calculated and why the situation is impossible.

Increased Number of Paired Problems. Based on the positive feedback we received in a survey of the market, we have increased the number of paired problems in this edition. These problems are otherwise identical, one asking for a numerical solution and one asking for a symbolic derivation. There are now three pairs of these problems in most chapters, indicated by tan shading in the end-of-chapter problems set.

Thorough Revision of Artwork. Every piece of artwork in the Fifth Edition was revised in a new and modern style that helps express the physics principles at work in a clear and precise fashion. Every piece of art was also revised to make certain that the physical situations presented correspond exactly to the text discussion at hand.

Also added for this edition is a new feature: "focus pointers" that either point out important aspects of a figure or guide students through a process illustrated by the artwork or photo. This format helps those students who are more visual learners. Examples of figures with focus pointers appear below:

Figure 10.28 Two points on a rolling object take different paths through space.

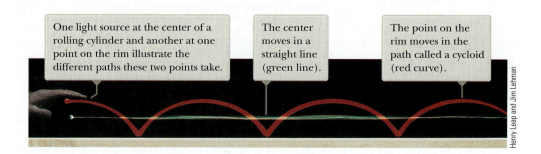

One light source at the center of a rolling cylinder and another at one point on the rim illustrate the different paths these two points take.

The center moves in a straight line (green line).

The point on the rim moves in the path called a cycloid (red curve).

Henry Leap and Jim Lehman

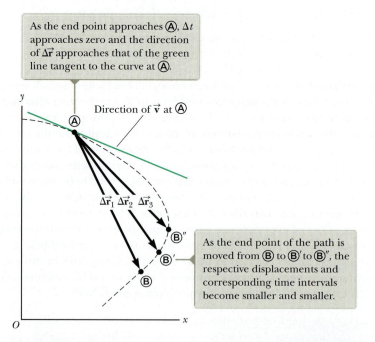

As the end point approaches Ⓐ, Δt approaches zero and the direction of $\Delta \vec{r}$ approaches that of the green line tangent to the curve at Ⓐ.

Direction of $\vec{v}$ at Ⓐ

$\Delta \vec{r}_1 \ \Delta \vec{r}_2 \ \Delta \vec{r}_3$

As the end point of the path is moved from Ⓑ to Ⓑ′ to Ⓑ″, the respective displacements and corresponding time intervals become smaller and smaller.

Figure 3.2 As a particle moves between two points, its average velocity is in the direction of the displacement vector $\Delta \vec{r}$. By definition, the instantaneous velocity at Ⓐ is directed along the line tangent to the curve at Ⓐ.

Expansion of the Analysis Model Approach. Students are faced with hundreds of problems during their physics courses. Instructors realize that a relatively small number of fundamental principles form the basis of these problems. When faced with a new problem, a physicist forms a *model* of the problem that can be solved in a simple way by identifying the fundamental principle that is applicable in the problem. For example, many problems involve conservation of energy, Newton's second law, or kinematic equations. Because the physicist has studied these principles extensively and understands the associated applications, he or she can apply this knowledge as a model for solving a new problem.

Although it would be ideal for students to follow this same process, most students have difficulty becoming familiar with the entire palette of fundamental principles that are available. It is easier for students to identify a *situation* rather than a fundamental principle. The *Analysis Model* approach we focus on in this revision lays out a standard set of situations that appear in most physics problems. These situations are based on an entity in one of four simplification models: particle, system, rigid object, and wave.

Once the simplification model is identified, the student thinks about what the entity is doing or how it interacts with its environment, which leads the student to identify a particular analysis model for the problem. For example, if an object is falling, the object is modeled as a particle. What it is doing is undergoing a constant acceleration due to gravity. The student has learned that this situation is described by the analysis model of a particle under constant acceleration. Furthermore, this model has a small number of equations associated with it for use in starting problems, the kinematic equations in Chapter 2. Therefore, an understanding of the situation has led to an analysis model, which then identifies a very small number of equations to start the problem, rather than the myriad equations that students see in the chapter. In this way, the use of analysis models leads the student to the fundamental principle the physicist would identify. As the student gains more experience, he or she will lean less on the analysis model approach and begin to identify fundamental principles directly, more like the physicist does. This approach is further reinforced in the end-of-chapter summary under the heading *Analysis Models for Problem Solving*.

Content Changes. The content and organization of the textbook are essentially the same as in the fourth edition. Several sections in various chapters have been streamlined, deleted, or combined with other sections to allow for a more balanced presentation. Chapters 6 and 7 have been completely reorganized to prepare students for a unified approach to energy that is used throughout the text. Updates have been added to reflect the current status of several areas of research and application of physics, including information on discoveries of new Kuiper belt objects (Chapter 11), comparisons of competing theories of pitch perception in humans (Chapter 14), progress in using grating light valves for optical applications (Chapter 27), new experiments to search for the cosmic background radiation (Chapter 28), developments in the search for evidence of a quark–gluon plasma (Chapter 31), and the status of the Large Hadron Collider (Chapter 31).

Organization

We have incorporated a "context overlay" scheme into the textbook, in response to the "Physics in Context" approach in the IUPP. This feature adds interesting applications of the material to real issues. We have developed this feature to be flexible; it is an "overlay" in the sense that the instructor who does not wish to follow the contextual approach can simply ignore the additional contextual features without sacrificing complete coverage of the existing material. We believe, though, that the benefits students will gain from this approach will be many.

The context overlay organization divides the text into nine sections, or "Contexts," after Chapter 1, as follows:

Context Number	Context	Physics Topics	Chapters
1	Alternative-Fuel Vehicles	Classical mechanics	2–7
2	Mission to Mars	Classical mechanics	8–11
3	Earthquakes	Vibrations and waves	12–14
4	Heart Attacks	Fluids	15
5	Global Warming	Thermodynamics	16–18
6	Lightning	Electricity	19–21
7	Magnetism in Medicine	Magnetism	22–23
8	Lasers	Optics	24–27
9	The Cosmic Connection	Modern physics	28–31

Each Context begins with an introductory section that provides some historical background or makes a connection between the topic of the Context and associated social issues. The introductory section ends with a "central question" that motivates study within the Context. The final section of each chapter is a "Context Connection," which discusses how the specific material in the chapter relates to the Context and to the central question. The final chapter in each Context is followed by a "Context Conclusion." Each Conclusion applies a combination of the principles learned in the various chapters of the Context to respond fully to the central question. Each chapter, as well as the Context Conclusions, includes problems related to the context material.

Text Features

Most instructors believe that the textbook selected for a course should be the student's primary guide for understanding and learning the subject matter. Furthermore, the textbook should be easily accessible and should be styled and written to facilitate instruction and learning. With these points in mind, we have included many pedagogical features, listed below, that are intended to enhance its usefulness to both students and instructors.

Problem Solving and Conceptual Understanding

General Problem-Solving Strategy. A general strategy outlined at the end of Chapter 1 (pages 25–26) provides students with a structured process for solving problems. In all remaining chapters, the strategy is employed explicitly in every example so that students learn how it is applied. Students are encouraged to follow this strategy when working end-of-chapter problems.

In most chapters, more specific strategies and suggestions are included for solving the types of problems featured in the end-of-chapter problems. This feature helps students identify the essential steps in solving problems and increases their skills as problem solvers.

Thinking Physics. We have included many Thinking Physics examples throughout each chapter. These questions relate the physics concepts to common experiences or extend the concepts beyond what is discussed in the textual material. Immediately following each of these questions is a "Reasoning" section that responds to the question. Ideally, the student will use these features to better understand physical concepts before being presented with quantitative examples and working homework problems.

MCAT Test Preparation Guide. Located at the front of the book, this guide outlines 12 concept-based study courses for the physics part of the MCAT exam. Students can use the guide to prepare for the MCAT exam, class tests, or homework assignments.

Active Figures. Many diagrams from the text have been animated to become Active Figures (identified in the figure legend), part of the Enhanced WebAssign online homework system. By viewing animations of phenomena and processes that cannot be fully represented on a static page, students greatly increase their conceptual understanding. In addition to viewing animations of the figures, students can see the outcome of changing variables, conduct suggested explorations of the principles involved in the figure, and take and receive feedback on quizzes related to the figure.

Quick Quizzes. Students are provided an opportunity to test their understanding of the physical concepts presented through Quick Quizzes. The questions require students to make decisions on the basis of sound reasoning, and some of the questions have been written to help students overcome common misconceptions. Quick Quizzes have been cast in an objective format, including multiple choice, true–false, and ranking. Answers to all Quick Quiz questions are found at the end of the text. Many instructors choose to use such questions in a "peer instruction" teaching style or with the use of personal response system "clickers," but they can be used in standard quiz format as well. An example of a Quick Quiz follows below.

QUICK QUIZ 6.5 A dart is inserted into a spring-loaded dart gun by pushing the spring in by a distance x. For the next loading, the spring is compressed a distance $2x$. How much faster does the second dart leave the gun compared with the first? (**a**) four times as fast (**b**) two times as fast (**c**) the same (**d**) half as fast (**e**) one-fourth as fast

Pitfall Preventions. Over 150 Pitfall Preventions (such as the one to the right) are provided to help students avoid common mistakes and misunderstandings. These features, which are placed in the margins of the text, address both common student misconceptions and situations in which students often follow unproductive paths.

Summaries. Each chapter contains a summary that reviews the important concepts and equations discussed in that chapter. New for the Fifth Edition is the Analysis Models for Problem Solving section of the Summary, which highlights the relevant analysis models presented in a given chapter.

Questions. As mentioned previously, the previous edition's Questions section is now divided into two sections: *Objective Questions* and *Conceptual Questions*. The instructor may select items to assign as homework or use in the classroom, possibly with "peer instruction" methods and possibly with personal response systems. More than seven hundred Objective and Conceptual Questions are included in this edition. Answers for selected questions are included in the *Student Solutions Manual/Study Guide*, and answers for all questions are found in the *Instructor's Solutions Manual*.

Problems. An extensive set of problems is included at the end of each chapter; in all, this edition contains over 2 200 problems. Answers for odd-numbered problems are provided at the end of the book. Full solutions for approximately 20% of the problems are included in the *Student Solutions Manual/Study Guide*, and solutions for all problems are found in the *Instructor's Solutions Manual*.

In addition to the new problem types mentioned previously, there are several other kinds of problems featured in this text:

- **Biomedical problems.** We added a number of problems related to biomedical situations in this edition (each indentified with a BIO icon), to highlight the relevance of physics principles to those students taking this course who are majoring in one of the life sciences.

Pitfall Prevention | 13.2
Two Kinds of Speed/Velocity
Do not confuse v, the speed of the wave as it propagates along the string, with v_y, the transverse velocity of a point on the string. The speed v is constant for a uniform medium, whereas v_y varies sinusoidally.

- **Paired Problems.** As an aid for students learning to solve problems symbolically, paired numerical and symbolic problems are included in all chapters of the text. Paired problems are identified by a common background screen.
- **Review problems.** Many chapters include review problems requiring the student to combine concepts covered in the chapter with those discussed in previous chapters. These problems (marked **Review**) reflect the cohesive nature of the principles in the text and verify that physics is not a scattered set of ideas. When facing a real-world issue such as global warming or nuclear weapons, it may be necessary to call on ideas in physics from several parts of a textbook such as this one.
- **"Fermi problems."** One or more problems in most chapters ask the student to reason in order-of-magnitude terms.
- **Design problems.** Several chapters contain problems that ask the student to determine design parameters for a practical device so that it can function as required.
- **Calculus-based problems.** Most chapters contain at least one problem applying ideas and methods from differential calculus and one problem using integral calculus.

The instructor's Web site, **www.cengage.com/physics/serway,** provides lists of all the various problem types, including problems most often assigned in Enhanced WebAssign, symbolic problems, quantitative/conceptual problems, Master It tutorials, Watch It solution videos, impossibility problems, paired problems, problems using calculus, problems encouraging or requiring computer use, problems with **What If?** parts, problems referred to in the chapter text, problems based on experimental data, order-of-magnitude problems, problems about biological applications, design problems, review problems, problems reflecting historical reasoning, and ranking questions.

Alternative Representations. We emphasize alternative representations of information, including mental, pictorial, graphical, tabular, and mathematical representations. Many problems are easier to solve if the information is presented in alternative ways, to reach the many different methods students use to learn.

Math Appendix. The math appendix (Appendix B), a valuable tool for students, shows the math tools in a physics context. This resource is ideal for students who need a quick review on topics such as algebra, trigonometry, and calculus.

Helpful Features

Style. To facilitate rapid comprehension, we have written the book in a clear, logical, and engaging style. We have chosen a writing style that is somewhat informal and relaxed so that students will find the text appealing and enjoyable to read. New terms are carefully defined, and we have avoided the use of jargon.

Important Definitions and Equations. Most important definitions are set in **boldface** or are set off from the paragraph in centered text for added emphasis and ease of review. Similarly, important equations are highlighted with a background screen to facilitate location.

Marginal Notes. Comments and notes appearing in the margin with a ▶ icon can be used to locate important statements, equations, and concepts in the text.

Pedagogical Use of Color. Readers should consult the **pedagogical color chart** (inside the front cover) for a listing of the color-coded symbols used in the text diagrams. This system is followed consistently throughout the text.

Mathematical Level. We have introduced calculus gradually, keeping in mind that students often take introductory courses in calculus and physics concurrently. Most steps are shown when basic equations are developed, and reference is often made to mathematical appendices near the end of the textbook. Although vectors are discussed in detail in

Chapter 1, vector products are introduced later in the text, where they are needed in physical applications. The dot product is introduced in Chapter 6, which addresses energy of a system; the cross product is introduced in Chapter 10, which deals with angular momentum.

Significant Figures. In both worked examples and end-of-chapter problems, significant figures have been handled with care. Most numerical examples are worked to either two or three significant figures, depending on the precision of the data provided. End-of-chapter problems regularly state data and answers to three-digit precision. When carrying out estimation calculations, we shall typically work with a single significant figure. (More discussion of significant figures can be found in Chapter 1, pages 10–12.)

Units. The international system of units (SI) is used throughout the text. The U.S. customary system of units is used only to a limited extent in the chapters on mechanics and thermodynamics.

Appendices and Endpapers. Several appendices are provided near the end of the textbook. Most of the appendix material represents a review of mathematical concepts and techniques used in the text, including scientific notation, algebra, geometry, trigonometry, differential calculus, and integral calculus. Reference to these appendices is made throughout the text. Most mathematical review sections in the appendices include worked examples and exercises with answers. In addition to the mathematical reviews, the appendices contain tables of physical data, conversion factors, and the SI units of physical quantities as well as a periodic table of the elements. Other useful information—fundamental constants and physical data, planetary data, a list of standard prefixes, mathematical symbols, the Greek alphabet, and standard abbreviations of units of measure—appears on the endpapers.

TextChoice Custom Options for *Principles of Physics*

Cengage Learning's digital library, TextChoice, enables you to build your custom version of Serway and Jewett's *Principles of Physics* from scratch. You may pick and choose the content you want to include in your text and even add your own original materials creating a unique, all-in-one learning solution. This all happens from the convenience of your desktop. Visit **www.textchoice.com** to start building your book today.

Cengage Learning offers the fastest and easiest way to create unique customized learning materials delivered the way you want. For more information about custom publishing options, visit **www.cengage.com/custom** or contact your local Cengage Learning representative.

Course Solutions That Fit Your Teaching Goals and Your Students' Learning Needs

Recent advances in educational technology have made homework management systems and audience response systems powerful and affordable tools to enhance the way you teach your course. Whether you offer a more traditional text-based course, are interested in using or are currently using an online homework management system such as Enhanced WebAssign, or are ready to turn your lecture into an interactive learning environment with JoinIn, you can be confident that the text's proven content provides the foundation for each and every component of our technology and ancillary package.

Homework Management Systems

Enhanced WebAssign for *Principles of Physics,* **Fifth Edition.** Exclusively from Cengage Learning, Enhanced WebAssign offers an extensive online program for physics to encourage the practice that's so critical for concept mastery. The meticulously crafted pedagogy and exercises in our proven texts become even more effective in Enhanced WebAssign. Enhanced WebAssign includes the Cengage YouBook, a highly customizable, interactive eBook. WebAssign includes:

- All of the quantitative end-of-chapter problems
- Selected problems enhanced with targeted feedback. An example of targeted feedback appears below:

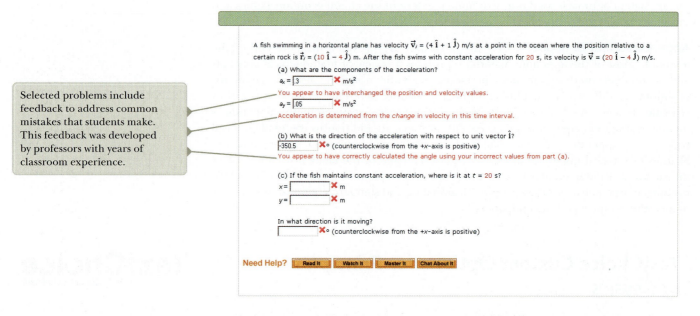

Selected problems include feedback to address common mistakes that students make. This feedback was developed by professors with years of classroom experience.

- Master It tutorials (indicated in the text by a **M** icon), to help students work through the problem one step at a time. An example of a Master It tutorial appears below:

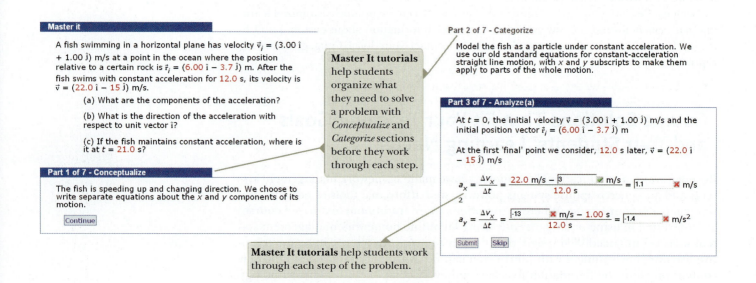

Master It tutorials help students organize what they need to solve a problem with *Conceptualize* and *Categorize* sections before they work through each step.

Master It tutorials help students work through each step of the problem.

- Watch It solution videos (indicated in the text by a **W** icon) that explain fundamental problem-solving strategies, to help students step through the problem. In addition, instructors can choose to include video hints of problem-solving strategies. A screen shot from a Watch It solution video appears below:

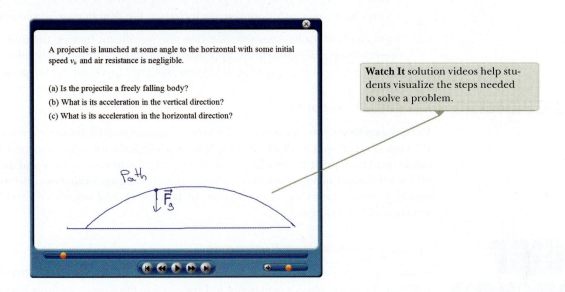

Watch It solution videos help students visualize the steps needed to solve a problem.

- Concept Checks
- Active Figure simulation tutorials
- PhET simulations
- Most worked examples, enhanced with hints and feedback, to help strengthen students' problem-solving skills
- Every Quick Quiz, giving your students ample opportunity to test their conceptual understanding
- The Cengage YouBook

WebAssign has a customizable and interactive eBook, the **Cengage YouBook**, that lets you tailor the textbook to fit your course and connect with your students. You can remove and rearrange chapters in the table of contents and tailor assigned readings that match your syllabus exactly. Powerful editing tools let you change as much as you'd like—or leave it just like it is. You can highlight key passages or add sticky notes to pages to comment on a concept in the reading, and then share any of these individual notes and highlights with your students, or keep them personal. You can also edit narrative content in the textbook by adding a text box or striking out text. With a handy link tool, you can drop in an icon at any point in the eBook that lets you link to your own lecture notes, audio summaries, video lectures, or other files on a personal website or anywhere on the web. A simple YouTube widget lets you easily find and embed videos from YouTube directly into eBook pages. There is a light discussion board that lets students and instructors find others in their class and start a chat session. The Cengage YouBook helps students go beyond just reading the textbook. Students can also highlight the text, add their own notes, and bookmark the text. Animations play right on the page at the point of learning so that they're not speed bumps to reading but true enhancements. Please visit **www.webassign .net/brookscole** to view an interactive demonstration of Enhanced WebAssign.

- Offered exclusively in WebAssign, **Quick Prep** for physics is algebra and trigonometry math remediation within the context of physics applications and principles. Quick Prep helps students succeed by using narratives illustrated throughout with video examples. The Master It tutorial problems allow students to assess and retune their understanding of the material. The Practice Problems that go along with each tutorial allow both the student and the instructor to test the student's understanding of the material.

Quick Prep includes the following features:

- 67 interactive tutorials
- 67 additional practice problems
- Thorough overview of each topic that includes video examples
- Taken before the semester begins or during the first few weeks of the course
- Can also be assigned alongside each chapter for "just in time" remediation

Topics include: units, scientific notation and significant figures; the motion of objects along a line; functions; approximation and graphing; probability and error; vectors, displacement, and velocity; spheres; force and vector projections.

CengageBrain.com

On **CengageBrain.com** students will be able to save up to 60% on their course materials through our full spectrum of options. Students will have the option to rent their textbooks, purchase print textbooks, e-textbooks, or individual e-chapters and audio books all for substantial savings over average retail prices. **CengageBrain.com** also includes access to Cengage Learning's broad range of homework and study tools and features a selection of free content.

Lecture Presentation Resources

PowerLecture with ExamView® and JoinIn for *Principles of Physics*, Fifth Edition. Bringing physics principles and concepts to life in your lectures has never been easier! The full-featured, two-volume **PowerLecture** Instructor's Resource DVD-ROM (Volume 1: Chapters 1–15; Volume 2: Chapters 16–31) provides everything you need for *Principles of Physics*, fifth edition. Key content includes the *Instructor's Solutions Manual* solutions, art and images from the text, pre-made chapter-specific PowerPoint lectures, ExamView test generator software with pre-loaded test questions, JoinIn response-system "clickers," Active Figures animations, and a physics movie library.

JoinIn. *Assessing to Learn in the Classroom* questions developed at the University of Massachusetts Amherst. This collection of 250 advanced conceptual questions has been tested in the classroom for more than ten years and takes peer learning to a new level. JoinIn helps you turn your lectures into an interactive learning environment that promotes conceptual understanding. Available exclusively for higher education from our partnership with Turning Technologies, JoinIn™ is the easiest way to turn your lecture hall into a personal, fully interactive experience for your students!

Assessment and Course Preparation Resources

A number of resources listed below will assist with your assessment and preparation processes.

Instructor's Solutions Manual by Vahe Peroomian (University of California at Los Angeles). Thoroughly revised for this edition, the *Instructor's Solutions Manual* contains complete worked solutions to all end-of-chapter problems in the textbook as well as answers to the even-numbered problems and all the questions. The solutions to problems new to the fifth edition are marked for easy identification. Volume 1 contains Chapters 1 through 15, and Volume 2 contains Chapters 16 through 31. Electronic files of the *Instructor's Solutions Manual* are available on the *PowerLecture*™ DVD-ROM.

Test Bank. The test bank is available on the two-volume *PowerLecture*™ DVD-ROM via the ExamView® test software. This two-volume test bank contains approximately 2 000 multiple-choice questions. Instructors may print and duplicate pages for distribution to students. Volume 1 contains Chapters 1 through 15, and Volume 2 contains Chapters 16 through 31. WebCT and Blackboard versions of the test bank are available on the instructor's companion site at **www.cengage/physics/serway.**

Instructor's Companion Web Site. Consult the instructor's site by pointing your browser to **www.cengage.com/physics/serway** for a problem correlation guide, PowerPoint lectures, and JoinIn audience response content. Instructors adopting the fifth edition of *Principles of Physics* may download these materials after securing the appropriate password from their local sales representative.

Supporting Materials for the Instructor

Supporting instructor materials are available to qualified adopters. Please consult your local Cengage Learning, Brooks/Cole representative for details. Visit **www.cengage.com/physics/serway** to:

- request a desk copy
- locate your local representative
- download electronic files of select support materials

Student Resources

Visit the *Principles of Physics* Web site at **www.cengagebrain.com/shop/ISBN/9781133104261** to see samples of select student supplements. Go to **CengageBrain.com** to purchase and access this product at Cengage Learning's preferred online store.

Student Solutions Manual/Study Guide by John R. Gordon, Vahe Peroomian, Raymond A. Serway, and John W. Jewett, Jr. This two-volume manual features detailed solutions to 20% of the end-of-chapter problems from the text. The manual also features a list of important equations, concepts, and notes from key sections of the text in addition to answers to selected end-of-chapter questions. Volume 1 contains Chapters 1 through 15, and Volume 2 contains Chapters 16 through 31.

Physics Laboratory Manual, Third Edition by David Loyd (Angelo State University) supplements the learning of basic physical principles while introducing laboratory procedures and equipment. Each chapter includes a prelaboratory assignment, objectives, an equipment list, the theory behind the experiment, experimental procedures, graphing exercises, and questions. A laboratory report form is included with each experiment so that the student can record data, calculations, and experimental results. Students are encouraged to apply statistical analysis to their data. A complete *Instructor's Manual* is also available to facilitate use of this lab manual.

Physics Laboratory Experiments, Seventh Edition by Jerry D. Wilson (Lander College) and Cecilia A. Hernández (American River College). This market-leading manual for the first-year physics laboratory course offers a wide range of class-tested experiments designed specifically for use in small to midsize lab programs. A series of integrated experiments emphasizes the use of computerized instrumentation and includes a set of "computer-assisted experiments" to allow students and instructors to gain experience with modern equipment. This option also enables instructors to determine the appropriate balance between traditional and computer-based experiments for their courses. By analyzing data through two different methods, students gain a greater understanding of the concepts behind the experiments. The seventh edition is updated with the latest information and techniques involving state-of-the-art equipment and a new Guided Learning feature addresses the growing interest in guided-inquiry pedagogy. Fourteen additional experiments are also available through custom printing.

Teaching Options

Although some topics found in traditional textbooks have been omitted from this textbook, instructors may find that the current text still contains more material than can be covered in a two-semester sequence. For this reason, we would like to offer the following suggestions. If you wish to place more emphasis on contemporary topics in physics, you should consider omitting parts or all of Chapters 15, 16, 17, 18, 24, 25, and 26. On the other hand, if you wish to follow a more traditional approach that places more emphasis on classical physics, you could omit Chapters 9, 11, 28, 29, 30, and 31. Either approach can be used without any loss in continuity. Other teaching options would fall somewhere between these two extremes by choosing to omit some or all of the following sections, which can be considered optional:

Acknowledgements

Prior to our work on this revision, we conducted two separate surveys of professors to gauge their textbook needs in the introductory calculus-based physics market. We were overwhelmed not only by the number of professors who wanted to take part in the survey but also by their insightful comments. Their feedback and suggestions helped shape the revision of this edition, and so we would like to thank the survey participants:

Anthony Abbott, *Willow-International Center;* Wagih Abdel Kader, *South Carolina State University;* Mikhail Agrest, *College of Charleston;* David Ailion, *University of Utah;* Rhett Allain, *Southeastern Louisiana University;* Bradley Antanaitis, *Lafayette College;* Vasudeva Rao Aravind, *Clarion University;* David Armstrong, *College of William & Mary;* Robert Arts, *University of Pikeville;* Robert Astalos, *Adams State College;* Charles Atchley, *Sauk Valley Community College;* Terry Austin, *LaGrange College;* Eric Ayars, *California State University, Chico;* Cristian Bahrim, *Lamar University;* Mirley Balasubramanya, *Texas A&M University-Corpus Christi;* Kenneth Balazovich, *University of Michigan;* David Balogh, *Fresno City College;* Alan Bardsley, *Salve Regina University;* Joseph Barranco, *San Francisco State University;* Perry Baskin, *Valdosta State University;* Celso Batalha, *Evergreen Valley College;* George Baxter, *Penn State Erie, The Behrend College;* Raymond Benge, *Tarrant County College;* Joel Berlinghieri, *The Citadel;* Matt Bigelow, *St. Cloud State University;* Raymond Bigliani, *Farmingdale State College, The State University of New York;* Mark S. Boley, *Western Illinois University;* Ken Bolland, *The Ohio State University;* Patrick Briggs, *The Citadel;* Douglas Brownell, *MiraCosta Community College;* William Bryant; Michael Butros, Victor *Valley College;* Gavin Buxton, *Robert Morris University;* Ralph Calhoun, *Northwest Florida State College;* Bruce Callen, *Drury University;* Bradley Carroll, *Weber State University;* Brian Carter, *Grossmont College;* Brian Carter, *San Diego City College;* Tom Carter, *College of DuPage;* Jennifer Cash, *South Carolina State University;* Cliff Castle, *Jefferson College;* Soumitra Chattopadhyay, *Georgia Highlands College;* Albert Chen, *Oklahoma Baptist University;* Li Chen, *Rhode Island College;* Norbert Chencinski, *College of Staten Island;* Kelvin Chu, *University of Vermont;* Darwin Church, *University of Cincinnati;* Michael Cohen, *Shippensburg University;* Robert Cohen, *East Stroudsburg University;* Stan Converse, *Wake Technical Community College;* S. Marie Cooper, *Immaculata University;* Susan Coppersmith, *University of Wisconsin-Madison;* Volker Crede, *Florida State University;* Demetra Czegan, *Seton Hill University;* Deborah Damcott, *Harper College;* Chris Davis, *University of Louisville;* Lynn Delker, *Diablo Valley College;* Todd Devore, *University of Alabama at Birmingham;* Alfonso Diaz Jimenez, *ADJOIN Research Center;* Susan DiFranzo, *Hudson Valley Community College;* Edward Dingler, *Southwest Virginia Community College;* Gregory Dolise, *Harrisburg Area Community College;* Sandra Doty, *Ohio University Lancaster;* Diana Driscoll, *Case Western Reserve University;* Mike Durren, *Lake Michigan College;* Ephraim Eisenberger, *Stevenson University;* Eleazer Ekwue, *College of Southern Maryland;* Terry Ellis, *Jacksonville University;* Mark Engebretson, *Augsburg College;* Tim Farris, *Volunteer State Community College;* Michael Fauerbach, *Florida Gulf Coast University;* Nail Fazleev, *University of Texas at Arlington;* Jason Ferguson, *Wichita State University;* Michael Ferralli, *Gannon University;* Chrisopher Fischer, *University of Kansas;* Kent Fisher, *Columbus State Community College;* Matthew Fleenor, *Roanoke College;* Richard Fleming, *Midwestern State University;* Terrence Flower, *St. Catherine University;* Marco Fornari, *Central Michigan University;* Juhan Frank, *Louisiana State University;* Allan Franklin, *University of Colorado;* Carl Frederickson, *University of Central Arkansas;* Rica French, *MiraCosta College;* Jerry Fuller, *Kilgore College;* Sambandamurthy Ganapathy, *University at Buffalo, The State University of New York;* Mark Gealy, *Concordia College;* Benjamin Geinstein, *University of California, San Diego;* Brian Geislinger, *Gadsden State Community College;* Dan Gibson, *Denison University;* Svetlana Gladycheva, *University of Washington;* Paul Goains, *San Jacinto College;* John Goehl, *Barry University;* John Goff, *Lynchburg College;* Charles Goodman, *Pitt Community College;* Christopher Gould, *University of Southern California;* Michael Graf, *Boston College;* Morris Greenwood, *San Jacinto College;* Allan Greer, *Gonzaga University;* Elena Gregg, *Oral Roberts University;* Alec Habig, *University of Minnesota Duluth;* Robert Hallock, *University of Massachusetts Amherst;* Dean Hamden, *Montclair State University;*

Carlos Handy, *Texas Southern University;* Baher Hanna, *Owens Community College;* Wayne Hayes, *Greenville Technical College;* Charles Henderson, *Western Michigan University;* Paul Henriksen, *James Madison University;* Thomas Herring, *Western Nevada College;* Yoshinao Hirai, *Central Lakes College;* Dean Hirschi, *Washington State Community College;* Stanley Hirschi, *Central Michigan University;* Pei-Chun Ho, *California State University, Fresno;* Roy Hoffer, *Bloomsburg University;* Mikel Holcomb, *West Virginia University;* Dawn Hollenbeck, *Rochester Institute of Technology;* Zdeslav Hrepic, *Columbus State University;* James Hudgings, *Camden County College;* Yung Huh, *South Dakota State University;* David Ingram, *Ohio University;* M. Islam, *The State University of New York at Potsdam;* Howard Jackson, *University of Cincinnati;* Shawn Jackson, *University of Arizona;* Sa-Han Jang, *Franklin and Marshall College;* Tim Jenkins, *University of Oregon;* Erik Jensen, *Chemeketa Community College;* Corinna Jobe, *College of the Canyons;* Charles Johnson, *South Georgia College;* Adam Johnston, *Weber State University;* Edwin Jones, *University of South Carolina;* Rex Joyner, *Indiana Institute of Technology;* Michael Kaplan, *Simmons College;* David Kardelis, *Utah State University-College of Eastern Utah;* Debora Katz, *United States Naval Academy;* Edward Kelsey, *Monmouth University;* Leonard Khazan, *Camden County College;* Nikolaos Kidonakis, *Kennesaw State University;* Derrick Kiley, *University of California, Merced;* Wayne Kinnison, *Texas A&M University-Kingsville;* Terence Kite, *Pepperdine University;* Joseph Klarfeld, *Queens College, the City University of New York;* Mario Klaric, *Midlands Technical College;* Suja Kochat, *Palo Alto College;* Michael Korth, *University of Minnesota Morris;* Aneta Koynova, *University of Texas at San Antonio;* Tatiana Krentsel, *Binghamton University, State University of New York;* George Kuck, *California State University, Long Beach;* Fred Kuttner, *University of California, Santa Cruz;* John LaBrasca, *Clark College;* Dan Lawrence, *Northwest Nazarene University;* Lynne Lawson, *Providence College;* Geoffrey Lenters, *Grand Valley State University;* John Lestrade, *Mississippi State University;* Rudi Lindner, *University of Michigan;* Beth Lindsey, *Penn State Greater Allegheny;* Kehfei Liu, *University of Kentucky;* Zengqiang Liu, *St. Cloud State University;* James Lockhart, *San Francisco State University;* Jorge Lopez, *University of Texas at El Paso;* Donald Luttermoser, *East Tennessee State University;* Ntungwa Maasha, *College of Coastal Georgia;* Terrence Maher, *Fayetteville Technical Community College;* Rizwan Mahmood, *Slippery Rock University;* Kingshuk Majumdar, *Grand Valley State University;* Igor Makasyuk, *Stanford University;* Jay Mancini; David Mandelbaum, *DeVry University;* David Marasco, *Foothill College;* Jelena Maricic, *Drexel University;* Pete Maritato, *Suffolk Community College;* Collette Marsh, *Harper College;* Devon Mason, *Albright College;* Martin Mason, *Mt. San Antonio College;* Sylvio May, *North Dakota State University;* Richard McCorkle, *University of Rhode Island;* Jimmy McCoy, *Tarleton State University;* Laura McCullough, *University of Wisconsin-Stout;* Ralph McGrew, *Broome Community College;* Rahul Mehta, *University of Central Arkansas;* Albert Menard, *Saginaw Valley State University;* William Mendoza, *Florida State College at Jacksonville;* Michael Meyer, *Michigan Technological University;* Karie Meyers, *Pima Community College;* Stanley Micklavzina, *University of Oregon;* Sudipa Mitra-Kirtley, *Rose-Hulman Institute of Technology;* Bob Moltaji, *Northeastern Illinois University;* Tamar More, *University of Portland;* Muhammad Asim Mubeen, *University at Albany, State University of New York;* Carl Mungan, *United States Naval Academy;* Mim Nakarmi, *Brooklyn College;* Majid Noori, *Cumberland County College;* Irina Novikova, *College of William and Mary;* Richard Olenick, *University of Dallas;* Grant O'Rielly, *University of Massachusetts Dartmouth;* Nicola Orsini, *Marshall University;* Edward

Osagie, *Lane College;* Eric Page, *University of San Diego;* Bruce Palmquist, *Central Washington University;* Dr. James Pazun, *Pfeiffer University;* David Pengra, *University of Washington;* Dorn Peterson, *James Madison University;* Anna Petrova-Mayor, *California State University, Chico;* Chris Pettit, *Emporia State University;* Ronald Phaneuf, *University of Nevada, Reno;* Daniel Phillips, *Ohio University;* Mark Pickar, *Minnesota State University, Mankato;* Alberto Pinkas, *New Jersey City University;* Dale Pleticha, *Gordon College;* William Powell, *Texas Lutheran University;* Norris Preyer, *College of Charleston;* Steve Quon, *Ventura College;* Andrew Rader, *Indiana University-Purdue University Indianapolis;* Stanley Radford, *The College at Brockport, State University of New York;* Bayani Ramirez, *San Jacinto College South;* Michael Ramsdell, *Clarkson University;* Steven Rehse, *Wayne State University;* Michael Richmond, *Rochester Institute of Technology;* Shadow Robinson, *Millsaps College;* Stephen Roeder, *San Diego State University;* John Rollino, *Rutgers University, Newark;* Alvin Rosenthal, *Western Michigan University;* Richard Ross, *Rock Valley College;* Louis Rubbo, *Coastal Carolina University;* Nanjundiah Sadanand, *Central Connecticut State University;* Mik Sawicki, *John A. Logan College;* Joseph Scanio, *University of Cincinnati;* Beth Schaefer, *Georgian Court University;* Scott Schultz, *Delta College;* Leon Scott, *Borough of Manhattan Community College;* Naidu Seetala, *Grambling State University;* Dastgeer Shaikh, *The University of Alabama in Huntsville;* Nimmi Sharma, *Central Connecticut State University;* Peter Sheldon, *Randolph College;* Douglas Sherman, *San Jose State University;* Paresh Shettigar, *Hawkeye Community College;* Earl Skelton, *Georgetown University;* Steve Spicklemire, *University of Indianapolis;* Philip Spickler, *Bridgewater College;* Glenn Spiczak, *University of Wisconsin-River Falls;* Susan Spillane, *McHenry County College;* Phillip Sprunger, *Louisiana State University;* Sudha Srinivas, *Northeastern Illinois University;* Jason Stalnaker, *Oberlin College;* Lawrence Staunton, *Drake University;* Gay Stewart, *University of Arkansas;* Chris Stockdale, *Marquette University;* Glenn Stracher, *East Georgia College;* Igor Strakovsky, *The George Washington University;* Dave Stuva, *Creighton University;* K. V. Sudhakar, *Montana Tech of the University of Montana;* Edward Sunderhaus, *Cincinnati State Technical and Community College;* Richard Swanson, *Sandhills Community College;* Steven Sweeney, *King's College;* Douglas Szper, *Lakeland College;* Nacira Tache, *Santa Fe College;* Kerry Tanimoto, *Honolulu Community College;* Vahe Tatoian, *Mt. San Antonio College;* Greg Thompson, *Adrian College;* Kristen Thompson, *Loras College;* Ionel Tifrea, *California State University, Fullerton;* Ramon Tirado; Daniela Topasna, *Virginia Military Institute;* Jim Tressel, *Massasoit Community College;* Som Tyagi, *Drexel University;* Bob Tyndall, *Carteret Community College;* Toshiya Ueta, *University of Denver;* Trina Van Ausdal, *Salt Lake Community College;* Bira van Kolck, *University of Arizona;* Brian Vermillion, *University of Indianapolis;* Joan Vogtman, *Potomac State College;* Sergei Voloshin, *Wayne State University;* Judy Vondruska, *South Dakota State University;* William Waggoner, *San Antonio College;* Fa-chung (Fred) Wang, *Prairie View A&M University;* Michael Weber, *Brigham Young University-Hawaii;* Margaret Wessling, *Los Angeles Pierce College;* James White, *Juniata College;* Daniel Willey, *Allegheny College;* Suzanne Willis, *Northern Illinois University;* Keith Willson, *Geneva College;* Stephen Wimpenny, *University of California, Riverside;* Jeff Winger, *Mississippi State University;* Krista Wood, *University of Cincinnati;* Hai-Sheng Wu, *Minnesota State University, Mankato;* Qilin Wu, *College of Southern Nevada;* Scott Yost, *The Citadel;* Anne Young, *Rochester Institute of Technology;* Hashim Yousif, *University of Pittsburgh at Bradford;* Chen Zeng, *The George Washington University;* Michael Ziegler, *The Ohio State University*

We thank the following people for their suggestions and assistance during the preparation of earlier editions of this textbook:

Edward Adelson, *Ohio State University;* Anthony Aguirre, *University of California at Santa Cruz;* Yildirim M. Aktas, *University of North Carolina—Charlotte;* Alfonso M. Albano, *Bryn Mawr College;* Royal Albridge, *Vanderbilt University;* Subash Antani, *Edgewood College;* Michael Bass, *University of Central Florida;* Harry Bingham, *University of California, Berkeley;* Billy E. Bonner, *Rice University;* Anthony Buffa, *California Polytechnic State University,* San Luis Obispo; Richard Cardenas, *St. Mary's University;* James Carolan, *University of British Columbia;* Kapila Clara Castoldi, *Oakland University;* Ralph V. Chamberlin, *Arizona State University;* Christopher R. Church, *Miami University (Ohio);* Gary G. DeLeo, *Lehigh University;* Michael Dennin, *University of California, Irvine;* Alan J. DeWeerd, *Creighton University;* Madi Dogariu, *University of Central Florida;* Gordon Emslie, *University of Alabama at Huntsville;*

Donald Erbsloe, *United States Air Force Academy;* William Fairbank, *Colorado State University;* Marco Fatuzzo, *University of Arizona;* Philip Fraundorf, *University of Missouri- St. Louis;* Patrick Gleeson, *Delaware State University;* Christopher M. Gould, *University of Southern California;* James D. Gruber, *Harrisburg Area Community College;* John B. Gruber, *San Jose State University;* Todd Hann, *United States Military Academy;* Gail Hanson, *Indiana University;* Gerald Hart, *Moorhead State University;* Dieter H. Hartmann, *Clemson University;* Richard W. Henry, *Bucknell University;* Athula Herat, *Northern Kentucky University;* Laurent Hodges, *Iowa State University;* Michael J. Hones, *Villanova University;* Huan Z. Huang, *University of California at Los Angeles;* Joey Huston, *Michigan State University;* George Igo, *University of California at Los Angeles;* Herb Jaeger, *Miami University;* David Judd, *Broward Community College;* Thomas H. Keil, *Worcester Polytechnic Institute;* V. Gordon Lind, *Utah State University;* Edwin Lo; Michael J. Longo, *University of Michigan;* Rafael Lopez-Mobilia, *University of Texas at San Antonio;* Roger M. Mabe, *United States Naval Academy;* David Markowitz, *University of Connecticut;* Thomas P. Marvin, *Southern Oregon University;* Bruce Mason, *University of Oklahoma at Norman;* Martin S. Mason, *College of the Desert;* Wesley N. Mathews, Jr., *Georgetown University;* Ian S. McLean, *University of California at Los Angeles;* John W. McClory, *United States Military Academy;* L. C. McIntyre, Jr., *University of Arizona;* Alan S. Meltzer, *Rensselaer Polytechnic Institute;* Ken Mendelson, *Marquette University;* Roy Middleton, *University of Pennsylvania;* Allen Miller, *Syracuse University;* Clement J. Moses, *Utica College of Syracuse University;* John W. Norbury, *University of Wisconsin-Milwaukee;* Anthony Novaco, *Lafayette College;* Romulo Ochoa, *The College of New Jersey;* Melvyn Oremland, *Pace University;* Desmond Penny, *Southern Utah University;* Steven J. Pollock, *University of Colorado-Boulder;* Prabha Ramakrishnan, *North Carolina State University;* Rex D. Ramsier, *The University of Akron;* Ralf Rapp, *Texas A&M University;* Rogers Redding, *University of North Texas;* Charles R. Rhyner, *University of Wisconsin-Green Bay;* Perry Rice, *Miami University;* Dennis Rioux, *University of Wisconsin-Oshkosh;* Richard Rolleigh, *Hendrix College;* Janet E. Seger, *Creighton University;* Gregory D. Severn, *University of San Diego;* Satinder S. Sidhu, *Washington College;* Antony Simpson, *Dalhousie University;* Harold Slusher, *University of Texas at El Paso;* J. Clinton Sprott, *University of Wisconsin at Madison;* Shirvel Stanislaus, *Valparaiso University;* Randall Tagg, *University of Colorado at Denver;* Cecil Thompson, *University of Texas at Arlington;* Harry W. K. Tom, *University of California at Riverside;* Chris Vuille, *Embry–Riddle Aeronautical University;* Fiona Waterhouse, *University of California at Berkeley;* Robert Watkins, *University of Virginia;* James Whitmore, *Pennsylvania State University*

Principles of Physics, fifth edition, was carefully checked for accuracy by Grant Hart (Brigham Young University), James E. Rutledge (University of California at Irvine), and Som Tyagi (Drexel University).

We are indebted to the developers of the IUPP models "A Particles Approach to Introductory Physics" and "Physics in Context," upon which much of the pedagogical approach in this textbook is based.

Vahe Peroomian wrote the initial draft of the new context on Heart Attacks, and we are very grateful for his efforts. He provided further assistance by reviewing early drafts of questions and problems sets.

We are grateful to John R. Gordon and Vahe Peroomian for writing the *Student Solutions Manual and Study Guide,* and to Vahe Peroomian for preparing an excellent *Instructor's Solutions Manual.* During the development of this text, the authors benefited from many useful discussions with colleagues and other physics instructors, including Robert Bauman, William Beston, Don Chodrow, Jerry Faughn, John R. Gordon, Kevin Giovanetti, Dick Jacobs, Harvey Leff, John Mallinckrodt, Clem Moses, Dorn Peterson, Joseph Rudmin, and Gerald Taylor.

Special thanks and recognition go to the professional staff at the Brooks/Cole Publishing Company—in particular, Charles Hartford, Ed Dodd, Brandi Kirksey, Rebecca Berardy-Schwartz, Jack Cooney, Cathy Brooks, Cate Barr, and Brendan Killion—for their fine work during the development and production of this textbook. We recognize the skilled production service provided by Jill Traut and the staff at Macmillan Solutions and the dedicated photo research efforts of Josh Garvin of the Bill Smith Group.

Finally, we are deeply indebted to our wives and children for their love, support, and long-term sacrifices.

RAYMOND A. SERWAY
St. Petersburg, Florida

JOHN W. JEWETT, JR.
Anaheim, California

To the Student

It is appropriate to offer some words of advice that should be of benefit to you, the student. Before doing so, we assume you have read the Preface, which describes the various features of the text and support materials that will help you through the course.

How to Study

Instructors are often asked, "How should I study physics and prepare for examinations?" There is no simple answer to this question, but we can offer some suggestions based on our own experiences in learning and teaching over the years.

First and foremost, maintain a positive attitude toward the subject matter, keeping in mind that physics is the most fundamental of all natural sciences. Other science courses that follow will use the same physical principles, so it is important that you understand and are able to apply the various concepts and theories discussed in the text.

Concepts and Principles

It is essential that you understand the basic concepts and principles before attempting to solve assigned problems. You can best accomplish this goal by carefully reading the textbook before you attend your lecture on the covered material. When reading the text, you should jot down those points that are not clear to you. Also be sure to make a diligent attempt at answering the questions in the Quick Quizzes as you come to them in your reading. We have worked hard to prepare questions that help you judge for yourself how well you understand the material. Study the **What If?** features that appear in many of the worked examples carefully. They will help you extend your understanding beyond the simple act of arriving at a numerical result. The Pitfall Preventions will also help guide you away from common misunderstandings about physics. During class, take careful notes and ask questions about those ideas that are unclear to you. Keep in mind that few people are able to absorb the full meaning of scientific material after only one reading; several readings of the text and your notes may be necessary. Your lectures and laboratory work supplement the textbook and should clarify some of the more difficult material. You should minimize your memorization of material. Successful memorization of passages from the text, equations, and derivations does not necessarily indicate that you understand the material. Your understanding of the material will be enhanced through a combination of efficient study habits, discussions with other students and with instructors, and your ability to solve the problems presented in the textbook. Ask questions whenever you believe that clarification of a concept is necessary.

Study Schedule

It is important that you set up a regular study schedule, preferably a daily one. Make sure that you read the syllabus for the course and adhere to the schedule set by your instructor. The lectures will make much more sense if you read the corresponding text material *before* attending them. As a general rule, you should devote about two hours of study time for each hour you are in class. If you are having trouble with the course, seek the advice of the instructor or other students who have taken the course. You may find it necessary to seek further instruction from experienced students. Very often, instructors offer review sessions in addition to regular class periods. Avoid the practice of delaying study until a day or two before an exam. More often than not, this approach has disastrous results. Rather than undertake an all-night study session before a test, briefly review the basic concepts and equations, and then get a good night's rest. If you believe that you need additional help in understanding the concepts, in preparing for exams, or in problem solving, we suggest that you acquire a copy of the *Student Solutions Manual/Study Guide* that accompanies this textbook.

Visit the *Principles of Physics* Web site at **www.cengagebrain.com/shop/ISBN/ 9781133104261** to see samples of select student supplements. You can purchase any Cengage Learning product at your local college store or at our preferred online store **CengageBrain.com.**

Use the Features

You should make full use of the various features of the text discussed in the Preface. For example, marginal notes are useful for locating and describing important equations and concepts, and **boldface** indicates important definitions. Many useful tables are contained in the appendices, but most are incorporated in the text where they are most often referenced. Appendix B is a convenient review of mathematical tools used in the text.

Answers to Quick Quizzes and odd-numbered problems are given at the end of the textbook, and solutions to selected end-of-chapter questions and problems are provided in the *Student Solutions Manual/Study Guide*. The table of contents provides an overview of the entire text, and the index enables you to locate specific material quickly. Footnotes are sometimes used to supplement the text or to cite other references on the subject discussed.

After reading a chapter, you should be able to define any new quantities introduced in that chapter and discuss the principles and assumptions that were used to arrive at certain key relations. The chapter summaries and the review sections of the *Student Solutions Manual/Study Guide* should help you in this regard. In some cases, you may find it necessary to refer to the textbook's index to locate certain topics. You should be able to associate with each physical quantity the correct symbol used to represent that quantity and the unit in which the quantity is specified. Furthermore, you should be able to express each important equation in concise and accurate prose.

Problem Solving

R. P. Feynman, Nobel laureate in physics, once said, "You do not know anything until you have practiced." In keeping with this statement, we strongly advise you to develop the skills necessary to solve a wide range of problems. Your ability to solve problems will be one of the main tests of your knowledge of physics; therefore, you should try to solve as many problems as possible. It is essential that you understand basic concepts and principles before attempting to solve problems. It is good practice to try to find alternate solutions to the same problem. For example, you can solve problems in mechanics using Newton's laws, but very often an alternative method that draws on energy considerations is more direct. You should not deceive yourself into thinking that you understand a problem merely because you have seen it solved in class. You must be able to solve the problem and similar problems on your own.

The approach to solving problems should be carefully planned. A systematic plan is especially important when a problem involves several concepts. First, read the problem several times until you are confident you understand what is being asked. Look for any key words that will help you interpret the problem and perhaps allow you to make certain assumptions. Your ability to interpret a question properly is an integral part of problem solving. Second, you should acquire the habit of writing down the information given in a problem and those quantities that need to be found; for example, you might construct a table listing both the quantities given and the quantities to be found. This procedure is sometimes used in the worked examples of the textbook. Finally, after you have decided on the method you believe is appropriate for a given problem, proceed with your solution. The General Problem-Solving Strategy will guide you through complex problems. If you follow the steps of this procedure *(Conceptualize, Categorize, Analyze, Finalize),* you will find it easier to come up with a solution and gain more from your efforts. This strategy, located at the end of Chapter 1 (pages 25–26), is used in all worked examples in the remaining chapters so that you can learn how to apply it. Specific problem-solving strategies for certain types of situations are included in the text and appear with a special heading. These specific strategies follow the outline of the General Problem-Solving Strategy.

Often, students fail to recognize the limitations of certain equations or physical laws in a particular situation. It is very important that you understand and remember the assumptions that underlie a particular theory or formalism. For example, certain equations in kinematics apply only to a particle moving with constant acceleration. These equations are not valid for describing motion whose acceleration is not constant, such as the motion of an object connected to a spring or the motion of an object through a fluid. Study the Analysis Models for Problem Solving in the chapter summaries carefully so that you know how each model can be applied to a specific situation. The analysis models provide you with a logical structure for solving problems and help you develop your thinking skills to become more like those of a physicist. Use the analysis model approach to save you hours of looking for the correct equation and to make you a faster and more efficient problem solver.

Experiments

Physics is a science based on experimental observations. Therefore, we recommend that you try to supplement the text by performing various types of "hands-on" experiments either at home or in the laboratory. These experiments can be used to test ideas and models discussed in class or in the textbook. For example, the common Slinky toy is excellent for studying traveling waves, a ball swinging on the end of a long string can be used to investigate pendulum motion, various masses attached to the end of a vertical spring or rubber band can be used to determine its elastic nature, an old pair of polarized sunglasses and some discarded lenses and a magnifying glass are the components of various experiments in optics, and an approximate measure of the free-fall acceleration can be determined simply by measuring with a stopwatch the time interval required for a ball to drop from a known height. The list of such experiments is endless. When physical models are not available, be imaginative and try to develop models of your own.

New Media

If available, we strongly encourage you to use the **Enhanced WebAssign** product that is available with this textbook. It is far easier to understand physics if you see it in action, and the materials available in Enhanced WebAssign will enable you to become a part of that action.

It is our sincere hope that you will find physics an exciting and enjoyable experience and that you will benefit from this experience, regardless of your chosen profession. Welcome to the exciting world of physics!

The scientist does not study nature because it is useful; he studies it because he delights in it, and he delights in it because it is beautiful. If nature were not beautiful, it would not be worth knowing, and if nature were not worth knowing, life would not be worth living.

—Henri Poincaré

Life Science Applications and Problems

An Invitation to Physics

Stephen Inglis/Shutterstock.com

Stonehenge, in southern England, was built thousands of years ago. Various theories have been proposed about its function, including a burial ground, a healing site, and a place for ancestor worship. One of the more intriguing theories suggests that Stonehenge was an observatory, allowing for predictions of celestial events such as eclipses, solstices, and equinoxes.

Physics, the most fundamental physical science, is concerned with the basic principles of the universe. It is the foundation on which engineering, technology, and the other sciences—astronomy, biology, chemistry, and geology—are based. The beauty of physics lies in the simplicity of its fundamental theories and in the manner in which just a small number of basic concepts, equations, and assumptions can alter and expand our view of the world around us.

Classical physics, developed prior to 1900, includes the theories, concepts, laws, and experiments in classical mechanics, thermodynamics, electromagnetism, and optics. For example, Galileo Galilei (1564–1642) made significant contributions to classical mechanics through his work on the laws of motion with constant acceleration. In the same era, Johannes Kepler (1571–1630) used astronomical observations to develop empirical laws for the motions of planetary bodies.

The most important contributions to classical mechanics, however, were provided by Isaac Newton (1642–1727), who developed classical mechanics as a systematic theory and was one of the originators of calculus as a mathematical tool. Although major developments in classical physics continued in the 18th century, thermodynamics and electromagnetism were not developed until the latter part of the 19th century, principally because the apparatus for controlled experiments was either too crude or unavailable until then. Although many electric and magnetic phenomena had been studied earlier, the work of James Clerk Maxwell (1831–1879) provided a unified theory of electromagnetism. In this text, we shall treat the various disciplines of classical physics in separate sections; we will see, however, that the disciplines of mechanics and electromagnetism are basic to all the branches of physics.

A major revolution in physics, usually referred to as *modern physics,* began near the end of the 19th century. Modern physics developed mainly because many physical phenomena could not be explained by classical physics. The two most important developments in this modern era were the theories of relativity and quantum mechanics. Albert Einstein's theory of relativity

1

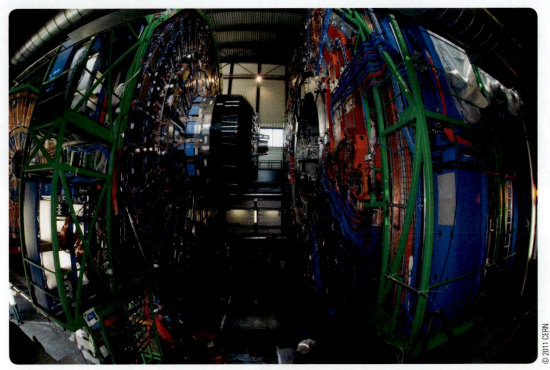

© 2011 CERN

The Compact Muon Solenoid (CMS) detector, part of the Large Hadron Collider operated by CERN (Conseil Européen pour la Recherche Nucléaire). The system is designed to detect and measure particles created in collisions of high-energy protons. Despite the word *compact* in the name, the detector is 15 meters in diameter. For a sense of scale, notice the worker in the blue helmet at the bottom of the photo as well as other workers in yellow helmets on the far side of the detector.

completely revolutionized the traditional concepts of space, time, and energy. This theory correctly describes the motion of objects moving at speeds comparable to the speed of light. The theory of relativity also shows that the speed of light is the upper limit of the speed of an object and that mass and energy are related. Quantum mechanics was formulated by a number of distinguished scientists to provide descriptions of physical phenomena at the atomic level.

Scientists continually work at improving our understanding of fundamental laws, and new discoveries are made every day. In many research areas, a great deal of overlap exists among physics, chemistry, and biology. Evidence for this overlap is seen in the names of some subspecialties in science: biophysics, biochemistry, chemical physics, biotechnology, and so on. Numerous technological advances in recent times are the result of the efforts of many scientists, engineers, and technicians. Some of the most notable developments in the latter half of the 20th century were (1) space missions to the Moon and other planets, (2) microcircuitry and high-speed computers, (3) sophisticated imaging techniques used in scientific research and medicine, and (4) several remarkable accomplishments in genetic engineering. The early years of the 21st century have seen additional developments. Materials such as carbon nanotubes are now experiencing a variety of new applications. The 2010 Nobel Prize in Physics was

awarded for experiments performed on graphene, a two-dimensional material formed from carbon atoms. Potential applications include incorporation into a variety of electrical components and biodevices such as those used in DNA sequencing. The impact of such developments and discoveries on society has indeed been great, and future discoveries and developments will very likely be exciting, challenging, and of great benefit to humanity.

To investigate the impact of physics on developments in our society, we will use a *contextual* approach to the study of the content in this textbook. The book is divided into nine *Contexts,* which relate the physics to social issues, natural phenomena, or medical/technological applications, as outlined here:

Chapters	Context
2–7	Alternative-Fuel Vehicles
8–11	Mission to Mars
12–14	Earthquakes
15	Heart Attacks
16–18	Global Warming
19–21	Lightning
22–23	Magnetism in Medicine
24–27	Lasers
28–31	The Cosmic Connection

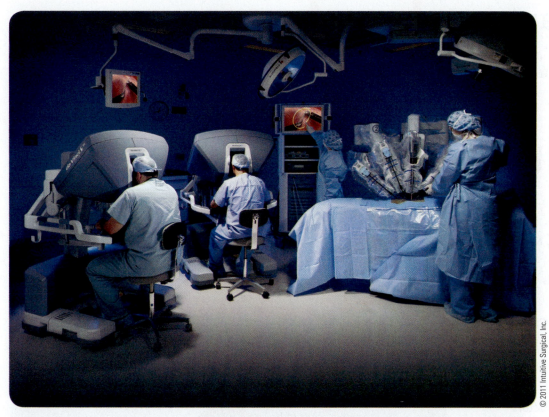

© 2011 Intuitive Surgical, Inc.

Physics is being used extensively today in the biomedical field. Shown here is the da Vinci Surgical System, a robotic device used to perform procedures such as prostatectomies, hysterectomies, mitral valve repairs, and coronary artery anastomosis. The surgeon sits at the console on the left and views a stereoscopic image of the surgery site. The movements of his hands are translated by a computer into movements of the robotic arms seen above the operating table at the right.

The Contexts provide a story line for each section of the text, which will help provide relevance and motivation for studying the material.

Each Context begins with a discussion of the topic, culminating in a *central question,* which forms the focus for the study of the physics in the Context. The final section of each chapter is a Context Connection, in which the material in the chapter is explored with the central question in mind. At the end of each Context, a Context Conclusion brings together all the principles necessary to respond as fully as possible to the central question.

In Chapter 1, we investigate some of the mathematical fundamentals and problem-solving strategies that we will use in our study of physics. The first Context, *Alternative-Fuel Vehicles,* is introduced just before Chapter 2; in this Context, the principles of classical mechanics are applied to the problem of designing, developing, producing, and marketing a vehicle that will help to reduce dependence on foreign oil and emit fewer harmful by-products into the atmosphere than current gasoline engines.

Introduction and Vectors

Chapter Outline

Raymond A. Serway

A signpost in Saint Petersburg, Florida, shows the distance and direction to several cities. Quantities that are defined by both a magnitude and a direction are called *vector quantities*.

The goal of physics is to provide a quantitative understanding of certain basic phenomena that occur in our Universe. Physics is a science based on experimental observations and mathematical analyses. The main objectives behind such experiments and analyses are to develop theories that explain the phenomenon being studied and to relate those theories to other established theories. Fortunately, it is possible to explain the behavior of various physical systems using relatively few fundamental laws. Analytical procedures require the expression of those laws in the language of mathematics, the tool that provides a bridge between theory and experiment. In this chapter, we shall discuss a few mathematical concepts and techniques that will be used throughout the text. In addition, we will outline an effective problem-solving strategy that should be adopted and used in your problem-solving activities throughout the text.

ENHANCED
WebAssign Interactive content from this and other chapters may be assigned online in Enhanced WebAssign.

1.1 | Standards of Length, Mass, and Time

To describe natural phenomena, we must make measurements associated with physical quantities, such as the length of an object. The laws of physics can be expressed as mathematical relationships among physical quantities that will be introduced and discussed throughout the book. In mechanics, the three fundamental quantities are length, mass, and time. All other quantities in mechanics can be expressed in terms of these three.

If we measure a certain quantity and wish to describe it to someone, a unit for the quantity must be specified and defined. For example, it would be meaningless for a visitor from another planet to talk to us about a length of 8.0 "glitches" if we did not know the meaning of the unit glitch. On the other hand, if someone familiar with our system of measurement reports that a wall is 2.0 meters high and our unit of length is defined to be 1.0 meter, we then know that the height of the wall is twice our fundamental unit of length. An international committee has agreed on a system of definitions and standards to describe fundamental physical quantities. It is called the **SI system** (Système International) of units. Its units of length, mass, and time are the meter, kilogram, and second, respectively.

Length

In A.D. 1120, King Henry I of England decreed that the standard of length in his country would be the yard and that the yard would be precisely equal to the distance from the tip of his nose to the end of his outstretched arm. Similarly, the original standard for the foot adopted by the French was the length of the royal foot of King Louis XIV. This standard prevailed until 1799, when the legal standard of length in France became the **meter,** defined as one ten-millionth of the distance from the equator to the North Pole.

Many other systems have been developed in addition to those just discussed, but the advantages of the French system have caused it to prevail in most countries and in scientific circles everywhere. Until 1960, the length of the meter was defined as the distance between two lines on a specific bar of platinum–iridium alloy stored under controlled conditions. This standard was abandoned for several reasons, a principal one being that the limited accuracy with which the separation between the lines can be determined does not meet the current requirements of science and technology. The definition of the meter was modified to be equal to 1 650 763.73 wavelengths of orange–red light emitted from a krypton-86 lamp. In October 1983, the meter was redefined to be **the distance traveled by light in a vacuum during a time interval of 1/299 792 458 second.** This value arises from the establishment of the speed of light in a vacuum as exactly 299 792 458 meters per second. We will use the standard scientific notation for numbers with more than three digits in which groups of three digits are separated by spaces rather than commas. Therefore, 1 650 763.73 and 299 792 458 in this paragraph are the same as the more popular American cultural notations of 1,650,763.73 and 299,792,458. Similarly, $\pi = 3.14159265$ is written as 3.141 592 65.

Mass

Mass represents a measure of the resistance of an object to changes in its motion. The SI unit of mass, the **kilogram,** is defined as **the mass of a specific platinum–iridium alloy cylinder kept at the International Bureau of Weights and Measures at Sèvres, France.** At this point, we should add a word of caution. Many beginning students of physics tend to confuse the physical quantities called *weight* and *mass*. For the present we shall not discuss the distinction between them; they will be clearly defined in later chapters. For now you should note that they are distinctly different quantities.

Time

Before 1967, the standard of time was defined in terms of the average length of a *mean solar day*. (A solar day is the time interval between successive appearances of the Sun at the highest point it reaches in the sky each day.) The basic unit of time, the **second,** was defined to be $(1/60)(1/60)(1/24) = 1/86\ 400$ of a mean solar day. In 1967, the second was redefined to take advantage of the great precision obtainable with a device known as an atomic clock (Fig. 1.1), which uses the characteristic frequency of the

Figure 1.1 A cesium fountain atomic clock. The clock will neither gain nor lose a second in 20 million years.

© 2005 Geoffrey Wheeler Photography

▶ Definition of the meter

▶ Definition of the kilogram

◀ **TABLE 1.1** | Approximate Values of Some Measured Lengths

	Length (m)
Distance from the Earth to the most remote quasar	1.4×10^{26}
Distance from the Earth to the most remote normal galaxies	9×10^{25}
Distance from the Earth to the nearest large galaxy (M 31, the Andromeda galaxy)	2×10^{22}
Distance from the Sun to the nearest star (Proxima Centauri)	4×10^{16}
One light-year	9.46×10^{15}
Mean orbit radius of the Earth	1.50×10^{11}
Mean distance from the Earth to the Moon	3.84×10^{8}
Distance from the equator to the North Pole	1.00×10^{7}
Mean radius of the Earth	6.37×10^{6}
Typical altitude (above the surface) of a satellite orbiting the Earth	2×10^{5}
Length of a football field	9.1×10^{1}
Length of this textbook	2.8×10^{-1}
Length of a housefly	5×10^{-3}
Size of smallest visible dust particles	$\sim 10^{-4}$
Size of cells of most living organisms	$\sim 10^{-5}$
Diameter of a hydrogen atom	$\sim 10^{-10}$
Diameter of a uranium nucleus	$\sim 10^{-14}$
Diameter of a proton	$\sim 10^{-15}$

◀ **TABLE 1.2** | Masses of Various Objects (Approximate Values)

	Mass (kg)
Visible Universe	$\sim 10^{52}$
Milky Way galaxy	$\sim 10^{42}$
Sun	1.99×10^{30}
Earth	5.98×10^{24}
Moon	7.36×10^{22}
Shark	$\sim 10^{3}$
Human	$\sim 10^{2}$
Frog	$\sim 10^{-1}$
Mosquito	$\sim 10^{-5}$
Bacterium	$\sim 10^{-15}$
Hydrogen atom	1.67×10^{-27}
Electron	9.11×10^{-31}

▶ Definition of the second

cesium-133 atom as the "reference clock." The second is now defined as **9 192 631 770 times the period of oscillation of radiation from the cesium atom.** It is possible today to purchase clocks and watches that receive radio signals from an atomic clock in Colorado, which the clock or watch uses to continuously reset itself to the correct time.

Approximate Values for Length, Mass, and Time

Approximate values of various lengths, masses, and time intervals are presented in Tables 1.1, 1.2, and 1.3, respectively. Note the wide range of values for these quantities.[1] You should study the tables and begin to generate an intuition for what is meant by a mass of 100 kilograms, for example, or by a time interval of 3.2×10^7 seconds.

Systems of units commonly used in science, commerce, manufacturing, and everyday life are (1) the *SI system,* in which the units of length, mass, and time are the meter (m), kilogram (kg), and second (s), respectively; and (2) the *U.S. customary system,* in which the units of length, mass, and time are the foot (ft), slug, and second, respectively. Throughout most of this text we shall use SI units because they are almost universally accepted in science and industry. We will make limited use of U.S. customary units in the study of classical mechanics.

Some of the most frequently used prefixes for the powers of ten and their abbreviations are listed in Table 1.4. For example, 10^{-3} m is equivalent to 1 millimeter (mm), and 10^3 m is 1 kilometer (km). Likewise, 1 kg is 10^3 grams (g), and 1 megavolt (MV) is 10^6 volts (V).

The variables length, time, and mass are examples of *fundamental quantities.* A much larger list of variables contains *derived quantities,* or quantities that can be expressed as a mathematical combination of fundamental quantities. Common examples are *area,* which is a product of two lengths, and *speed,* which is a ratio of a length to a time interval.

Another example of a derived quantity is **density**. The density ρ (Greek letter rho; a table of the letters in the Greek alphabet is provided at the back of the book) of any substance is defined as its *mass per unit volume:*

> **Pitfall Prevention** | **1.1**
> **Reasonable Values**
> Generating intuition about typical values of quantities when solving problems is important because you must think about your end result and determine if it seems reasonable. For example, if you are calculating the mass of a housefly and arrive at a value of 100 kg, this answer is *unreasonable* and there is an error somewhere.

▶ Definition of density

$$\rho \equiv \frac{m}{V}$$

1.1 ◀

---

[1] If you are unfamiliar with the use of powers of ten (scientific notation), you should review Appendix B.1.

TABLE 1.3 | Approximate Values of Some Time Intervals

	Time Interval (s)
Age of the Universe	4×10^{17}
Age of the Earth	1.3×10^{17}
Time interval since the fall of the Roman empire	5×10^{12}
Average age of a college student	6.3×10^{8}
One year	3.2×10^{7}
One day (time interval for one revolution of the Earth about its axis)	8.6×10^{4}
One class period	3.0×10^{3}
Time interval between normal heartbeats	8×10^{-1}
Period of audible sound waves	$\sim 10^{-3}$
Period of typical radio waves	$\sim 10^{-6}$
Period of vibration of an atom in a solid	$\sim 10^{-13}$
Period of visible light waves	$\sim 10^{-15}$
Duration of a nuclear collision	$\sim 10^{-22}$
Time interval for light to cross a proton	$\sim 10^{-24}$

TABLE 1.4 | Some Prefixes for Powers of Ten

Power	Prefix	Abbreviation
10^{-24}	yocto	y
10^{-21}	zepto	z
10^{-18}	atto	a
10^{-15}	femto	f
10^{-12}	pico	p
10^{-9}	nano	n
10^{-6}	micro	μ
10^{-3}	milli	m
10^{-2}	centi	c
10^{-1}	deci	d
10^{3}	kilo	k
10^{6}	mega	M
10^{9}	giga	G
10^{12}	tera	T
10^{15}	peta	P
10^{18}	exa	E
10^{21}	zetta	Z
10^{24}	yotta	Y

which is a ratio of mass to a product of three lengths. For example, aluminum has a density of $2.70 \times 10^3 \text{ kg/m}^3$, and lead has a density of $11.3 \times 10^3 \text{ kg/m}^3$. An extreme difference in density can be imagined by thinking about holding a 10-centimeter (cm) cube of Styrofoam in one hand and a 10-cm cube of lead in the other.

1.2 | Dimensional Analysis

In physics, the word *dimension* denotes the physical nature of a quantity. The distance between two points, for example, can be measured in feet, meters, or furlongs, which are all different ways of expressing the dimension of length.

The symbols used in this book to specify the dimensions[2] of length, mass, and time are L, M, and T, respectively. We shall often use square brackets [ ] to denote the dimensions of a physical quantity. For example, in this notation the dimensions of speed v are written $[v] = L/T$, and the dimensions of area A are $[A] = L^2$. The dimensions of area, volume, speed, and acceleration are listed in Table 1.5, along with their units in the two common systems. The dimensions of other quantities, such as force and energy, will be described as they are introduced in the text.

In many situations, you may be faced with having to derive or check a specific equation. Although you may have forgotten the details of the derivation, a useful and powerful procedure called **dimensional analysis** can be used as a consistency check, to assist in the derivation, or to check your final expression. Dimensional analysis makes use of the fact that dimensions can be treated as algebraic quantities. For example, quantities can be added or subtracted only if they have the same dimensions. Furthermore, the terms on both sides of an equation must have the same dimensions. By following these simple rules, you can use dimensional analysis to help determine

Pitfall Prevention | 1.2
Symbols for Quantities
Some quantities have a small number of symbols that represent them. For example, the symbol for time is almost always t. Other quantities might have various symbols depending on the usage. Length may be described with symbols such as x, y, and z (for position); r (for radius); a, b, and c (for the legs of a right triangle); ℓ (for the length of an object); d (for a distance); h (for a height); and so forth.

TABLE 1.5 | Dimensions and Units of Four Derived Quantities

Quantity	Area (A)	Volume (V)	Speed (v)	Acceleration (a)
Dimensions	L^2	L^3	L/T	L/T^2
SI units	m^2	m^3	m/s	m/s^2
U.S. customary units	ft^2	ft^3	ft/s	ft/s^2

[2] The *dimensions* of a variable will be symbolized by a capitalized, nonitalic letter, such as, in the case of length, L. The *symbol* for the variable itself will be italicized, such as L for the length of an object or t for time.

whether an expression has the correct form because the relationship can be correct only if the dimensions on the two sides of the equation are the same.

To illustrate this procedure, suppose you wish to derive an expression for the position x of a car at a time t if the car starts from rest at $t = 0$ and moves with constant acceleration a. In Chapter 2, we shall find that the correct expression for this special case is $x = \frac{1}{2}at^2$. Let us check the validity of this expression from a dimensional analysis approach.

The quantity x on the left side has the dimension of length. For the equation to be dimensionally correct, the quantity on the right side must also have the dimension of length. We can perform a dimensional check by substituting the basic dimensions for acceleration, L/T^2 (Table 1.5), and time, T, into the equation $x = \frac{1}{2}at^2$. That is, the dimensional form of the equation $x = \frac{1}{2}at^2$ can be written as

$$[x] = \frac{L}{T^2} T^2 = L$$

The dimensions of time cancel as shown, leaving the dimension of length, which is the correct dimension for the position x. Notice that the number $\frac{1}{2}$ in the equation has no units, so it does not enter into the dimensional analysis.

> **QUICK QUIZ** 1.1 **True or False:** Dimensional analysis can give you the numerical value of constants of proportionality that may appear in an algebraic expression.

Example 1.1 | **Analysis of an Equation**

Show that the expression $v = at$, where v represents speed, a acceleration, and t an instant of time, is dimensionally correct.

SOLUTION

Identify the dimensions of v from Table 1.5:

$$[v] = \frac{L}{T}$$

Identify the dimensions of a from Table 1.5 and multiply by the dimensions of t:

$$[at] = \frac{L}{T^2} T = \frac{L}{T}$$

Therefore, $v = at$ is dimensionally correct because we have the same dimensions on both sides. (If the expression were given as $v = at^2$, it would be dimensionally *incorrect*. Try it and see!)

1.3 | Conversion of Units

Pitfall Prevention | 1.3
Always Include Units
When performing calculations, make it a habit to include the units with every quantity and carry the units through the entire calculation. Avoid the temptation to drop the units during the calculation steps and then apply the expected unit to the number that results for an answer. By including the units in every step, you can detect errors if the units for the answer are incorrect.

Sometimes it is necessary to convert units from one system to another or to convert within a system, for example, from kilometers to meters. Equalities between SI and U.S. customary units of length are as follows:

1 mile (mi) = 1 609 m = 1.609 km   1 ft = 0.304 8 m = 30.48 cm
1 m = 39.37 in. = 3.281 ft   1 inch (in.) = 0.025 4 m = 2.54 cm

A more complete list of equalities can be found in Appendix A.

Units can be treated as algebraic quantities that can cancel each other. To perform a conversion, a quantity can be multiplied by a **conversion factor,** which is a fraction equal to 1, with numerator and denominator having different units, to provide the desired units in the final result. For example, suppose we wish to convert 15.0 in. to centimeters. Because 1 in. = 2.54 cm, we multiply by a conversion factor that is the appropriate ratio of these equal quantities and find that

$$15.0 \text{ in.} = (15.0 \text{ in.})\left(\frac{2.54 \text{ cm}}{1 \text{ in.}}\right) = 38.1 \text{ cm}$$

where the ratio in parentheses is equal to 1. Notice that we express 1 as 2.54 cm/1 in. (rather than 1 in./2.54 cm) so that the inch cancels with the unit in the original quantity. The remaining unit is the centimeter, which is our desired result.

QUICK QUIZ 1.2 The distance between two cities is 100 mi. What is the number of kilometers between the two cities? (**a**) smaller than 100 (**b**) larger than 100 (**c**) equal to 100

Example 1.2 | Is He Speeding?

On an interstate highway in a rural region of Wyoming, a car is traveling at a speed of 38.0 m/s. Is the driver exceeding the speed limit of 75.0 mi/h?

SOLUTION

Convert meters in the speed to miles:

$$(38.0 \text{ m/s})\left(\frac{1 \text{ mi}}{1\,609 \text{ m}}\right) = 2.36 \times 10^{-2} \text{ mi/s}$$

Convert seconds to hours: $(2.36 \times 10^{-2} \text{ mi/s})\left(\frac{60 \text{ s}}{1 \text{ min}}\right)\left(\frac{60 \text{ min}}{1 \text{ h}}\right) = 85.0 \text{ mi/h}$

The driver is indeed exceeding the speed limit and should slow down.

What If? What if the driver were from outside the United States and is familiar with speeds measured in kilometers per hour? What is the speed of the car in km/h?

Answer We can convert our final answer to the appropriate units:

$$(85.0 \text{ mi/h})\left(\frac{1.609 \text{ km}}{1 \text{ mi}}\right) = 137 \text{ km/h}$$

Figure 1.2 shows an automobile speedometer displaying speeds in both mi/h and km/h. Can you check the conversion we just performed using this photograph?

Figure 1.2 (Example 1.2) The speedometer of a vehicle that shows speeds in both miles per hour and kilometers per hour.

© Cengage Learning/Ed Dodd

1.4 | Order-of-Magnitude Calculations

Suppose someone asks you the number of bits of data on a typical musical compact disc. In response, it is not generally expected that you would provide the exact number but rather an estimate, which may be expressed in scientific notation. The estimate may be made even more approximate by expressing it as an order of *magnitude*, which is a power of ten determined as follows:

1. Express the number in scientific notation, with the multiplier of the power of ten between 1 and 10 and a unit.
2. If the multiplier is less than 3.162 (the square root of ten), the order of magnitude of the number is the power of ten in the scientific notation. If the multiplier is greater than 3.162, the order of magnitude is one larger than the power of ten in the scientific notation.

We use the symbol ~ for "is on the order of." Use the procedure above to verify the orders of magnitude for the following lengths:

$$0.008\,6 \text{ m} \sim 10^{-2} \text{ m} \qquad 0.002\,1 \text{ m} \sim 10^{-3} \text{ m} \qquad 720 \text{ m} \sim 10^{3} \text{ m}$$

Usually, when an order-of-magnitude estimate is made, the results are reliable to within about a factor of ten. If a quantity increases in value by three orders of magnitude, its value increases by a factor of about $10^{3} = 1\,000$.

Example 1.3 | The Number of Atoms in a Solid

Estimate the number of atoms in 1 cm^3 of a solid.

SOLUTION

From Table 1.1 we note that the diameter d of an atom is about 10^{-10} m. Let us assume that the atoms in the solid are spheres of this diameter. Then the volume of each sphere is about 10^{-30} m^3 (more precisely, volume $= 4\pi r^3/3 = \pi d^3/6$, where $r = d/2$). Therefore, because 1 cm$^3 = 10^{-6}$ m^3, the number of atoms in the solid is on the order of $10^{-6}/10^{-30} = 10^{24}$ atoms.

A more precise calculation would require additional knowledge that we could find in tables. Our estimate, however, agrees with the more precise calculation to within a factor of 10.

Example 1.4 | Breaths in a Lifetime

Estimate the number of breaths taken during an average human lifetime.

SOLUTION

We start by guessing that the typical human lifetime is about 70 years. Think about the average number of breaths that a person takes in 1 min. This number varies depending on whether the person is exercising, sleeping, angry, serene, and so forth. To the nearest order of magnitude, we shall choose 10 breaths per minute as our estimate. (This estimate is certainly closer to the true average value than an estimate of 1 breath per minute or 100 breaths per minute.)

Find the approximate number of minutes in a year:

$$1 \text{ yr} \left(\frac{400 \text{ days}}{1 \text{ yr}} \right) \left(\frac{25 \text{ h}}{1 \text{ day}} \right) \left(\frac{60 \text{ min}}{1 \text{ h}} \right) = 6 \times 10^5 \text{ min}$$

Find the approximate number of minutes in a 70-year lifetime:

$$\text{number of minutes} = (70 \text{ yr})(6 \times 10^5 \text{ min/yr})$$
$$= 4 \times 10^7 \text{ min}$$

Find the approximate number of breaths in a lifetime:

$$\text{number of breaths} = (10 \text{ breaths/min})(4 \times 10^7 \text{ min})$$
$$= \boxed{4 \times 10^8 \text{ breaths}}$$

Therefore, a person takes on the order of 10^9 breaths in a lifetime. Notice how much simpler it is in the first calculation above to multiply 400×25 than it is to work with the more accurate 365×24.

What If? What if the average lifetime were estimated as 80 years instead of 70? Would that change our final estimate?

Answer We could claim that $(80 \text{ yr})(6 \times 10^5 \text{ min/yr}) = 5 \times 10^7$ min, so our final estimate should be 5×10^8 breaths. This answer is still on the order of 10^9 breaths, so an order-of-magnitude estimate would be unchanged.

1.5 | Significant Figures

When certain quantities are measured, the measured values are known only to within the limits of the experimental uncertainty. The value of this uncertainty can depend on various factors, such as the quality of the apparatus, the skill of the experimenter, and the number of measurements performed. The number of **significant figures** in a measurement can be used to express something about the uncertainty. The number of significant figures is related to the number of numerical digits used to express the measurement, as we discuss below.

As an example of significant figures, suppose we are asked to measure the radius of a compact disc using a meterstick as a measuring instrument. Let us assume the accuracy to which we can measure the radius of the disc is ±0.1 cm. Because of the uncertainty of ±0.1 cm, if the radius is measured to be 6.0 cm, we can claim only that its radius lies somewhere between 5.9 cm and 6.1 cm. In this case, we say that the measured value of 6.0 cm has two significant figures. Note that *the significant figures include the first estimated digit.* Therefore, we could write the radius as (6.0 ± 0.1) cm.

Zeros may or may not be significant figures. Those used to position the decimal point in such numbers as 0.03 and 0.007 5 are not significant. Therefore, there are one and two significant figures, respectively, in these two values. When the zeros come after other digits, however, there is the possibility of misinterpretation. For example, suppose the mass of an object is given as 1 500 g. This value is ambiguous because we do not know whether the last two zeros are being used to locate the decimal point or whether they represent significant figures in the measurement. To remove this ambiguity, it is common to use scientific notation to indicate the number of significant figures. In this case, we would express the mass as 1.5×10^3 g if there are two significant figures in the measured value, 1.50×10^3 g if there are three significant figures, and 1.500×10^3 g if there are four. The same rule holds for numbers less than 1, so 2.3×10^{-4} has two significant figures (and therefore could be written 0.000 23) and 2.30×10^{-4} has three significant figures (also written as 0.000 230).

In problem solving, we often combine quantities mathematically through multiplication, division, addition, subtraction, and so forth. When doing so, you must make sure that the result has the appropriate number of significant figures. A good rule of thumb to use in determining the number of significant figures that can be claimed in a multiplication or a division is as follows:

When multiplying several quantities, the number of significant figures in the final answer is the same as the number of significant figures in the quantity having the smallest number of significant figures. The same rule applies to division.

Let's apply this rule to find the area of the compact disc whose radius we measured above. Using the equation for the area of a circle,

$$A = \pi r^2 = \pi (6.0 \text{ cm})^2 = 1.1 \times 10^2 \text{ cm}^2$$

If you perform this calculation on your calculator, you will likely see 113.097 335 5. It should be clear that you don't want to keep all of these digits, but you might be tempted to report the result as 113 cm². This result is not justified because it has three significant figures, whereas the radius only has two. Therefore, we must report the result with only two significant figures as shown above.

For addition and subtraction, you must consider the number of decimal places when you are determining how many significant figures to report:

When numbers are added or subtracted, the number of decimal places in the result should equal the smallest number of decimal places of any term in the sum or difference.

As an example of this rule, consider the sum

$$23.2 + 5.174 = 28.4$$

Notice that we do not report the answer as 28.374 because the lowest number of decimal places is one, for 23.2. Therefore, our answer must have only one decimal place.

The rules for addition and subtraction can often result in answers that have a different number of significant figures than the quantities with which you start. For example, consider these operations that satisfy the rule:

$$1.000\ 1 + 0.000\ 3 = 1.000\ 4$$

$$1.002 - 0.998 = 0.004$$

In the first example, the result has five significant figures even though one of the terms, 0.000 3, has only one significant figure. Similarly, in the second calculation, the result has only one significant figure even though the numbers being subtracted have four and three, respectively.

Pitfall Prevention | 1.4
Read Carefully
Notice that the rule for addition and subtraction is different from that for multiplication and division. For addition and subtraction, the important consideration is the number of *decimal places,* not the number of *significant figures.*

▶ Significant figure guidelines used in this book

In this book, most of the numerical examples and end-of-chapter problems will yield answers having three significant figures. When carrying out estimation calculations, we shall typically work with a single significant figure.

If the number of significant figures in the result of a calculation must be reduced, there is a general rule for rounding numbers: the last digit retained is increased by 1 if the last digit dropped is greater than 5. (For example, 1.346 becomes 1.35.) If the last digit dropped is less than 5, the last digit retained remains as it is. (For example, 1.343 becomes 1.34.) If the last digit dropped is equal to 5, the remaining digit should be rounded to the nearest even number. (This rule helps avoid accumulation of errors in long arithmetic processes.)

A technique for avoiding error accumulation is to delay the rounding of numbers in a long calculation until you have the final result. Wait until you are ready to copy the final answer from your calculator before rounding to the correct number of significant figures. In this book, we display numerical values rounded off to two or three significant figures. This occasionally makes some mathematical manipulations look odd or incorrect. For instance, looking ahead to Example 1.8 on page 21, you will see the operation −17.7 km + 34.6 km = 17.0 km. This looks like an incorrect subtraction, but that is only because we have rounded the numbers 17.7 km and 34.6 km for display. If all digits in these two intermediate numbers are retained and the rounding is only performed on the final number, the correct three-digit result of 17.0 km is obtained.

Example 1.5 | Installing a Carpet

A carpet is to be installed in a rectangular room whose length is measured to be 12.71 m and whose width is measured to be 3.46 m. Find the area of the room.

SOLUTION

If you multiply 12.71 m by 3.46 m on your calculator, you will see an answer of 43.976 6 m². How many of these numbers should you claim? Our rule of thumb for multiplication tells us that you can claim only the number of significant figures in your answer as are present in the measured quantity having the lowest number of significant figures. In this example, the lowest number of significant figures is three in 3.46 m, so we should express our final answer as 44.0 m².

◀ 1.6 | Coordinate Systems

Many aspects of physics deal in some way or another with locations in space. For example, the mathematical description of the motion of an object requires a method for specifying the object's position. Therefore, we first discuss how to describe the position of a point in space by means of coordinates in a graphical representation. A point on a line can be located with one coordinate, a point in a plane is located with two coordinates, and three coordinates are required to locate a point in space.

A coordinate system used to specify locations in space consists of

- A fixed reference point O, called the origin
- A set of specified axes or directions with an appropriate scale and labels on the axes
- Instructions that tell us how to label a point in space relative to the origin and axes

One convenient coordinate system that we will use frequently is the *Cartesian coordinate system*, sometimes called the *rectangular coordinate system*. Such a system in two dimensions is illustrated in Figure 1.3. An arbitrary point in this system is labeled with the coordinates (x, y). Positive x is taken to the right of the origin, and positive y is upward from the origin. Negative x is to the left of the origin, and negative y is downward from the origin. For example, the point P, which has coordinates $(5, 3)$, may be reached by going first 5 m to the right of the origin and then 3 m above the origin

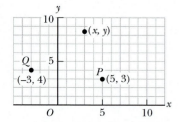

Figure 1.3 Designation of points in a Cartesian coordinate system. Each square in the xy plane is 1 m on a side. Every point is labeled with coordinates (x, y).

(*or* by going 3 m above the origin and then 5 m to the right). Similarly, the point Q has coordinates $(-3, 4)$, which correspond to going 3 m to the left of the origin and 4 m above the origin.

Sometimes it is more convenient to represent a point in a plane by its *plane polar coordinates* (r, θ), as in Active Figure 1.4a. In this coordinate system, r is the length of the line from the origin to the point, and θ is the angle between that line and a fixed axis, usually the positive x axis, with θ measured counterclockwise. From the right triangle in Active Figure 1.4b, we find that $\sin \theta = y/r$ and $\cos \theta = x/r$. (A review of trigonometric functions is given in Appendix B.4.) Therefore, starting with plane polar coordinates, one can obtain the Cartesian coordinates through the equations

$$x = r \cos \theta \qquad \text{1.2} \blacktriangleleft$$

$$y = r \sin \theta \qquad \text{1.3} \blacktriangleleft$$

Furthermore, if we know the Cartesian coordinates, the definitions of trigonometry tell us that

$$\tan \theta = \frac{y}{x} \qquad \text{1.4} \blacktriangleleft$$

and

$$r = \sqrt{x^2 + y^2} \qquad \text{1.5} \blacktriangleleft$$

You should note that these expressions relating the coordinates (x, y) to the coordinates (r, θ) apply only when θ is defined as in Active Figure 1.4a, where positive θ is an angle measured *counterclockwise* from the positive x axis. Other choices are made in navigation and astronomy. If the reference axis for the polar angle θ is chosen to be other than the positive x axis or if the sense of increasing θ is chosen differently, the corresponding expressions relating the two sets of coordinates will change.

◀1.7 | Vectors and Scalars

Each of the physical quantities that we shall encounter in this text can be placed in one of two categories, either a scalar or a vector. A **scalar** is a quantity that is completely specified by a positive or negative number with appropriate units. On the other hand, a **vector** is a physical quantity that must be specified by both magnitude and direction.

The number of grapes in a bunch (Fig. 1.5a) is an example of a scalar quantity. If you are told that there are 38 grapes in the bunch, this statement completely specifies

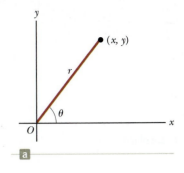

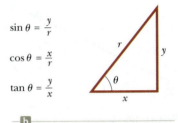

$$\sin \theta = \frac{y}{r}$$

$$\cos \theta = \frac{x}{r}$$

$$\tan \theta = \frac{y}{x}$$

Active Figure 1.4 (a) The plane polar coordinates of a point are represented by the distance r and the angle θ, where θ is measured in a counterclockwise direction from the positive x axis. (b) The right triangle used to relate (x, y) to (r, θ).

Mack Henley/Visuals Unlimited, Inc.

© Cengage Learning/George Semple

Figure 1.5 (a) The number of grapes in this bunch is one example of a scalar quantity. Can you think of other examples? (b) This helpful person pointing in the correct direction tells us to travel five blocks north to reach the courthouse. A vector is a physical quantity that is specified by both magnitude and direction.

the information; no specification of direction is required. Other examples of scalars are temperature, volume, mass, and time intervals. The rules of ordinary arithmetic are used to manipulate scalar quantities; they can be freely added and subtracted (assuming that they have the same units!), multiplied and divided.

Force is an example of a vector quantity. To describe the force on an object completely, we must specify both the direction of the applied force and the magnitude of the force.

Another simple example of a vector quantity is the **displacement** of a particle, defined as its *change in position*. The person in Figure 1.5b is pointing out the direction of your desired displacement vector if you would like to reach a destination such as the courthouse. She will also tell you the magnitude of the displacement along with the direction, for example, "5 blocks north."

Suppose a particle moves from some point Ⓐ to a point Ⓑ along a straight path, as in Figure 1.6. This displacement can be represented by drawing an arrow from Ⓐ to Ⓑ, where the arrowhead represents the direction of the displacement and the length of the arrow represents the magnitude of the displacement. If the particle travels along some other path from Ⓐ to Ⓑ, such as the broken line in Figure 1.6, its displacement is still the vector from Ⓐ to Ⓑ. The vector displacement along any indirect path from Ⓐ to Ⓑ is defined as being equivalent to the displacement represented by the direct path from Ⓐ to Ⓑ. The magnitude of the displacement is the shortest distance between the end points. Therefore, **the displacement of a particle is completely known if its initial and final coordinates are known.** The path need not be specified. In other words, the **displacement is independent of the path** if the end points of the path are fixed.

Note that the **distance** traveled by a particle is distinctly different from its displacement. The distance traveled (a scalar quantity) is the length of the path, which in general can be much greater than the magnitude of the displacement. In Figure 1.6, the length of the curved broken path is much larger than the magnitude of the solid black displacement vector.

If the particle moves along the x axis from position x_i to position x_f, as in Figure 1.7, its displacement is given by $x_f - x_i$. (The indices i and f refer to the initial and final values.) We use the Greek letter delta (Δ) to denote the *change* in a quantity. Therefore, we define the change in the position of the particle (the displacement) as

$$\Delta x \equiv x_f - x_i \qquad \textbf{1.6} \blacktriangleleft$$

From this definition we see that Δx is positive if x_f is greater than x_i and negative if x_f is less than x_i. For example, if a particle changes its position from $x_i = -5$ m to $x_f = 3$ m, its displacement is $\Delta x = +8$ m.

Many physical quantities in addition to displacement are vectors. They include velocity, acceleration, force, and momentum, all of which will be defined in later chapters. In this text, we will use boldface letters with an arrow over the letter, such as $\vec{\mathbf{A}}$, to represent vectors. Another common notation for vectors with which you should be familiar is a simple boldface character: **A**.

The magnitude of the vector $\vec{\mathbf{A}}$ is written with an italic letter A or, alternatively, $|\vec{\mathbf{A}}|$. The magnitude of a vector is always positive and carries the units of the quantity that the vector represents, such as meters for displacement or meters per second for velocity. Vectors combine according to special rules, which will be discussed in Sections 1.8 and 1.9.

▶ Displacement

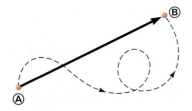

Figure 1.6 After a particle moves from Ⓐ to Ⓑ along an arbitrary path represented by the broken line, its displacement is a vector quantity shown by the arrow drawn from Ⓐ to Ⓑ.

▶ Distance

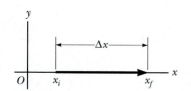

Figure 1.7 A particle moving along the x axis from x_i to x_f undergoes a displacement $\Delta x = x_f - x_i$.

QUICK QUIZ 1.3 Which of the following are vector quantities and which are scalar quantities? **(a)** your age **(b)** acceleration **(c)** velocity **(d)** speed **(e)** mass

> **THINKING PHYSICS 1.1**
>
> Consider your commute to work or school in the morning. Which is larger, the distance you travel or the magnitude of the displacement vector?
>
> **Reasoning** Unless you have a very unusual commute, the distance traveled *must* be larger than the magnitude of the displacement vector. The distance includes the results of all the twists and turns you make in following the roads from home to work or school. On the other hand, the magnitude of the displacement vector is the length of a straight line from your home to work or school. This length is often described informally as "the distance as the crow flies." The only way that the distance could be the same as the magnitude of the displacement vector is if your commute is a perfect straight line, which is highly unlikely! The distance could *never* be less than the magnitude of the displacement vector because the shortest distance between two points is a straight line. ◄

1.8 | Some Properties of Vectors

Equality of Two Vectors

Two vectors $\vec{A}$ and $\vec{B}$ are defined to be equal if they have the same units, the same magnitude, and the same direction. That is, $\vec{A} = \vec{B}$ only if $A = B$ *and* $\vec{A}$ and $\vec{B}$ point in the same direction. For example, all the vectors in Figure 1.8 are equal even though they have different starting points. This property allows us to translate a vector parallel to itself in a diagram without affecting the vector.

Figure 1.8 These four representations of vectors are equal because all four vectors have the same magnitude and point in the same direction.

Addition

The rules for vector sums are conveniently described using a graphical method. To add vector $\vec{B}$ to vector $\vec{A}$, first draw a diagram of vector $\vec{A}$ on graph paper, with its magnitude represented by a convenient scale, and then draw vector $\vec{B}$ to the same scale with its tail starting from the tip of $\vec{A}$, as in Active Figure 1.9a. The *resultant vector* $\vec{R} = \vec{A} + \vec{B}$ is the vector drawn from the tail of $\vec{A}$ to the tip of $\vec{B}$. The technique for adding two vectors is often called the "head-to-tail method."

When vectors are added, the sum is independent of the order of the addition. This independence can be seen for two vectors from the geometric construction in Active Figure 1.9b and is known as the **commutative law of addition:**

$$\vec{A} + \vec{B} = \vec{B} + \vec{A}$$ 1.7◄

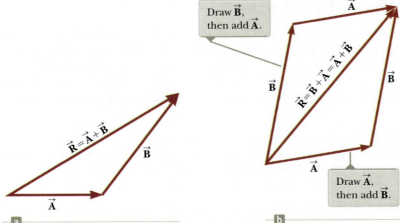

> **Pitfall Prevention | 1.6**
> **Vector Addition Versus Scalar Addition**
> Keep in mind that $\vec{A} + \vec{B} = \vec{C}$ is very different from $A + B = C$. The first equation is a vector sum, which must be handled carefully, such as with the graphical method described in Active Figure 1.9. The second equation is a simple algebraic addition of numbers that is handled with the normal rules of arithmetic.

Active Figure 1.9 (a) When vector $\vec{B}$ is added to vector $\vec{A}$, the resultant $\vec{R}$ is the vector that runs from the tail of $\vec{A}$ to the tip of $\vec{B}$. (b) This construction shows that $\vec{A} + \vec{B} = \vec{B} + \vec{A}$; vector addition is commutative.

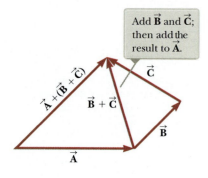

Add $\vec{B}$ and $\vec{C}$; then add the result to $\vec{A}$.

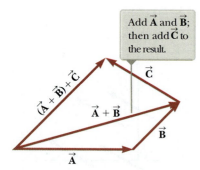

Add $\vec{A}$ and $\vec{B}$; then add $\vec{C}$ to the result.

Figure 1.10 Geometric constructions for verifying the associative law of addition.

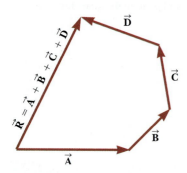

Figure 1.11 Geometric construction for summing four vectors. The resultant vector $\vec{R}$ closes the polygon and points from the tail of the first vector to the tip of the final vector.

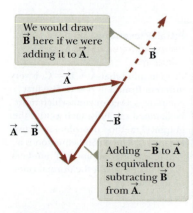

We would draw $\vec{B}$ here if we were adding it to $\vec{A}$.

Adding $-\vec{B}$ to $\vec{A}$ is equivalent to subtracting $\vec{B}$ from $\vec{A}$.

Figure 1.12 Subtracting vector $\vec{B}$ from vector $\vec{A}$. The vector $-\vec{B}$ is equal in magnitude to vector $\vec{B}$ and points in the opposite direction.

If three or more vectors are added, their sum is independent of the way in which they are grouped. A geometric demonstration of this property for three vectors is given in Figure 1.10. This property is called the **associative law of addition:**

$$\vec{A} + (\vec{B} + \vec{C}) = (\vec{A} + \vec{B}) + \vec{C} \qquad \textbf{1.8} \blacktriangleleft$$

Geometric constructions can also be used to add more than three vectors, as shown in Figure 1.11 for the case of four vectors. The resultant vector $\vec{R} = \vec{A} + \vec{B} + \vec{C} + \vec{D}$ is the vector that closes the polygon formed by the vectors being added. In other words, $\vec{R}$ is the vector drawn from the tail of the first vector to the tip of the last vector. Again, the order of the summation is unimportant.

In summary, **a vector quantity has both magnitude and direction and also obeys the laws of vector addition** as described in Active Figure 1.9 and Figures 1.10 and 1.11. When two or more vectors are added together, they must all have the same units and they must all be the same type of quantity. It would be meaningless to add a velocity vector (for example, 60 km/h to the east) to a displacement vector (for example, 200 km to the north) because these vectors represent different physical quantities. The same rule also applies to scalars. For example, it would be meaningless to add time intervals to temperatures.

Negative of a Vector

The negative of the vector $\vec{A}$ is defined as the vector that, when added to $\vec{A}$, gives zero for the vector sum. That is, $\vec{A} + (-\vec{A}) = 0$. The vectors $\vec{A}$ and $-\vec{A}$ have the same magnitude but point in opposite directions.

Subtraction of Vectors

The operation of vector subtraction makes use of the definition of the negative of a vector. We define the operation $\vec{A} - \vec{B}$ as vector $-\vec{B}$ added to vector $\vec{A}$:

$$\vec{A} - \vec{B} = \vec{A} + (-\vec{B}) \qquad \textbf{1.9} \blacktriangleleft$$

The geometric construction for subtracting two vectors is illustrated in Figure 1.12.

Multiplication of a Vector by a Scalar

If a vector $\vec{A}$ is multiplied by a positive scalar quantity s, the product $s\vec{A}$ is a vector that has the same direction as $\vec{A}$ and magnitude sA. If s is a negative scalar quantity, the vector $s\vec{A}$ is directed opposite to $\vec{A}$. For example, the vector $5\vec{A}$ is five times longer than $\vec{A}$ and has the same direction as $\vec{A}$; the vector $-\frac{1}{3}\vec{A}$ has one-third the magnitude of $\vec{A}$ and points in the direction opposite $\vec{A}$.

Multiplication of Two Vectors

Two vectors $\vec{A}$ and $\vec{B}$ can be multiplied in two different ways to produce either a scalar or a vector quantity. The **scalar product** (or dot product) $\vec{A} \cdot \vec{B}$ is a scalar quantity equal to $AB \cos \theta$, where θ is the angle between $\vec{A}$ and $\vec{B}$. The **vector product** (or cross product) $\vec{A} \times \vec{B}$ is a vector quantity whose magnitude is equal to $AB \sin \theta$. We shall discuss these products more fully in Chapters 6 and 10, where they are first used.

QUICK QUIZ 1.4 The magnitudes of two vectors $\vec{A}$ and $\vec{B}$ are $A = 12$ units and $B = 8$ units. Which pair of numbers represents the *largest* and *smallest* possible values for the magnitude of the resultant vector $\vec{R} = \vec{A} + \vec{B}$? **(a)** 14.4 units, 4 units **(b)** 12 units, 8 units **(c)** 20 units, 4 units **(d)** none of these answers

QUICK QUIZ 1.5 If vector $\vec{B}$ is added to vector $\vec{A}$, under what condition does the resultant vector $\vec{A} + \vec{B}$ have magnitude $A + B$? **(a)** $\vec{A}$ and $\vec{B}$ are parallel and in the same direction. **(b)** $\vec{A}$ and $\vec{B}$ are parallel and in opposite directions. **(c)** $\vec{A}$ and $\vec{B}$ are perpendicular.

1.9 | Components of a Vector and Unit Vectors

The graphical method of adding vectors is not recommended whenever high accuracy is required or in three-dimensional problems. In this section, we describe a method of adding vectors that makes use of the projections of vectors along coordinate axes. These projections are called the **components** of the vector or its **rectangular components.** Any vector can be completely described by its components.

Consider a vector $\vec{A}$ lying in the xy plane and making an arbitrary angle θ with the positive x axis as shown in Figure 1.13a. This vector can be expressed as the sum of two other *component vectors* $\vec{A}_x$, which is parallel to the x axis, and $\vec{A}_y$, which is parallel to the y axis. From Figure 1.13b, we see that the three vectors form a right triangle and that $\vec{A} = \vec{A}_x + \vec{A}_y$. We shall often refer to the "components of a vector $\vec{A}$," written A_x and A_y (without the boldface notation). The component A_x represents the projection of $\vec{A}$ along the x axis, and the component A_y represents the projection of $\vec{A}$ along the y axis. These components can be positive or negative. The component A_x is positive if the component vector $\vec{A}_x$ points in the positive x direction and is negative if $\vec{A}_x$ points in the negative x direction. A similar statement is made for the component A_y.

From Figure 1.13b and the definition of the sine and cosine of an angle, we see that $\cos\theta = A_x/A$ and $\sin\theta = A_y/A$. Hence, the components of $\vec{A}$ are given by

$$A_x = A\cos\theta \quad \text{and} \quad A_y = A\sin\theta \qquad \text{1.10}◄$$

The magnitudes of these components are the lengths of the two sides of a right triangle with a hypotenuse of length A. Therefore, the magnitude and direction of $\vec{A}$ are related to its components through the expressions

$$A = \sqrt{A_x^2 + A_y^2} \qquad \text{1.11}◄$$

▶ Magnitude of $\vec{A}$

$$\tan\theta = \frac{A_y}{A_x} \qquad \text{1.12}◄$$

▶ Direction of $\vec{A}$

To solve for θ, we can write $\theta = \tan^{-1}(A_y/A_x)$, which is read "$\theta$ equals the angle whose tangent is the ratio A_y/A_x." *Note that the signs of the components A_x and A_y depend on the angle θ.* For example, if $\theta = 120°$, A_x is negative and A_y is positive. If $\theta = 225°$, both A_x and A_y are negative. Figure 1.14 summarizes the signs of the components when $\vec{A}$ lies in the various quadrants.

If you choose reference axes or an angle other than those shown in Figure 1.13, the components of the vector must be modified accordingly. In many applications, it is more convenient to express the components of a vector in a coordinate system having axes that are not horizontal and vertical but are still perpendicular to each other.

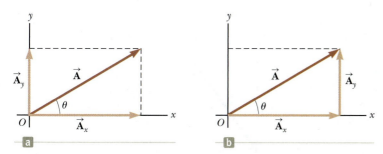

Figure 1.13 (a) A vector $\vec{A}$ lying in the xy plane can be represented by its component vectors $\vec{A}_x$ and $\vec{A}_y$. (b) The y component vector $\vec{A}_y$ can be moved to the right so that it adds to $\vec{A}_x$. The vector sum of the component vectors is $\vec{A}$. These three vectors form a right triangle.

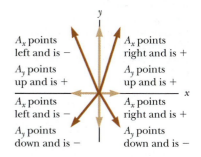

Figure 1.14 The signs of the components of a vector $\vec{A}$ depend on the quadrant in which the vector is located.

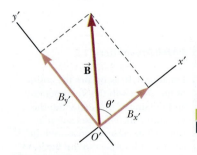

Figure 1.15 The component vectors of vector $\vec{\mathbf{B}}$ in a coordinate system that is tilted.

Suppose a vector $\vec{\mathbf{B}}$ makes an angle θ' with the x' axis defined in Figure 1.15. The components of $\vec{\mathbf{B}}$ along these axes are given by $B_{x'} = B \cos \theta'$ and $B_{y'} = B \sin \theta'$, as in Equation 1.10. The magnitude and direction of $\vec{\mathbf{B}}$ are obtained from expressions equivalent to Equations 1.11 and 1.12. Therefore, we can express the components of a vector in *any* coordinate system that is convenient for a particular situation.

▶ **QUICK QUIZ 1.6** Choose the correct response to make the sentence true: A component of a vector is (**a**) always, (**b**) never, or (**c**) sometimes larger than the magnitude of the vector.

Unit Vectors

Vector quantities often are expressed in terms of unit vectors. A **unit vector** is a dimensionless vector having a magnitude of exactly 1. Unit vectors are used to specify a given direction and have no other physical significance. We shall use the symbols $\hat{\mathbf{i}}$, $\hat{\mathbf{j}}$, and $\hat{\mathbf{k}}$ to represent unit vectors pointing in the x, y, and z directions, respectively. The "hat" over the letters is a common notation for a unit vector; for example, $\hat{\mathbf{i}}$ is called "i-hat." The unit vectors $\hat{\mathbf{i}}$, $\hat{\mathbf{j}}$, and $\hat{\mathbf{k}}$ form a set of mutually perpendicular vectors as shown in Active Figure 1.16a, where the magnitude of each unit vector equals 1; that is, $|\hat{\mathbf{i}}| = |\hat{\mathbf{j}}| = |\hat{\mathbf{k}}| = 1$.

Consider a vector $\vec{\mathbf{A}}$ lying in the xy plane, as in Active Figure 1.16b. The product of the component A_x and the unit vector $\hat{\mathbf{i}}$ is the component vector $\vec{\mathbf{A}}_x = A_x\hat{\mathbf{i}}$, which lies on the x axis and has magnitude A_x. Likewise, $A_y\hat{\mathbf{j}}$ is a component vector of magnitude A_y lying on the y axis. Therefore, the unit-vector notation for the vector $\vec{\mathbf{A}}$ is

$$\vec{\mathbf{A}} = A_x\hat{\mathbf{i}} + A_y\hat{\mathbf{j}} \qquad \textbf{1.13} ◀$$

Now suppose we wish to add vector $\vec{\mathbf{B}}$ to vector $\vec{\mathbf{A}}$, where $\vec{\mathbf{B}}$ has components B_x and B_y. The procedure for performing this sum is simply to add the x and y components separately. The resultant vector $\vec{\mathbf{R}} = \vec{\mathbf{A}} + \vec{\mathbf{B}}$ is therefore

$$\vec{\mathbf{R}} = (A_x + B_x)\hat{\mathbf{i}} + (A_y + B_y)\hat{\mathbf{j}} \qquad \textbf{1.14} ◀$$

From this equation, the components of the resultant vector are given by

$$R_x = A_x + B_x$$
$$R_y = A_y + B_y \qquad \textbf{1.15} ◀$$

Therefore, we see that in the component method of adding vectors, we add all the x components to find the x component of the resultant vector and use the same

Active Figure 1.16 (a) The unit vectors $\hat{\mathbf{i}}$, $\hat{\mathbf{j}}$, and $\hat{\mathbf{k}}$ are directed along the x, y, and z axes, respectively. (b) A vector $\vec{\mathbf{A}}$ lying in the xy plane has component vectors $A_x\hat{\mathbf{i}}$ and $A_y\hat{\mathbf{j}}$, where A_x and A_y are the components of $\vec{\mathbf{A}}$.

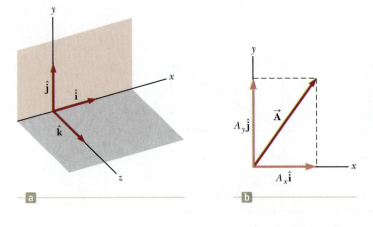

process for the y components. The procedure just described for adding two vectors $\vec{A}$ and $\vec{B}$ using the component method can be checked using a diagram like Figure 1.17.

The magnitude of $\vec{R}$ and the angle it makes with the x axis can be obtained from its components using the relationships

$$R = \sqrt{R_x^2 + R_y^2} = \sqrt{(A_x + B_x)^2 + (A_y + B_y)^2}$$ 1.16◄

$$\tan \theta = \frac{R_y}{R_x} = \frac{A_y + B_y}{A_x + B_x}$$ 1.17◄

The extension of these methods to three-dimensional vectors is straightforward. If $\vec{A}$ and $\vec{B}$ both have x, y, and z components, we express them in the form

$$\vec{A} = A_x \hat{i} + A_y \hat{j} + A_z \hat{k}$$

$$\vec{B} = B_x \hat{i} + B_y \hat{j} + B_z \hat{k}$$

The sum of $\vec{A}$ and $\vec{B}$ is

$$\vec{R} = \vec{A} + \vec{B} = (A_x + B_x)\hat{i} + (A_y + B_y)\hat{j} + (A_z + B_z)\hat{k}$$ 1.18◄

If a vector $\vec{R}$ has x, y, and z components, the magnitude of the vector is

$$R = \sqrt{R_x^2 + R_y^2 + R_z^2}$$

The angle θ_x that $\vec{R}$ makes with the x axis is given by

$$\cos \theta_x = \frac{R_x}{R}$$

with similar expressions for the angles with respect to the y and z axes.

The extension of our method to adding more than two vectors is also straightforward. For example, $\vec{A} + \vec{B} + \vec{C} = (A_x + B_x + C_x)\hat{i} + (A_y + B_y + C_y)\hat{j} + (A_z + B_z + C_z)\hat{k}$. Adding displacement vectors is relatively easy to visualize. We can also add other types of vectors, such as velocity, force, and electric field vectors, which we will do in later chapters.

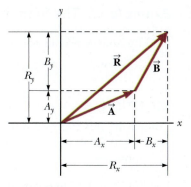

Figure 1.17 A geometric construction showing the relation between the components of the resultant $\vec{R}$ of two vectors and the individual components.

Pitfall Prevention | 1.8
Tangents on Calculators
Equation 1.17 involves the calculation of an angle by means of a tangent function. Generally, the inverse tangent function on calculators provides an angle between $-90°$ and $+90°$. As a consequence, if the vector you are studying lies in the second or third quadrant, the angle measured from the positive x axis will be the angle your calculator returns plus $180°$.

▶ **QUICK QUIZ 1.7** If at least one component of a vector is a positive number, the vector cannot **(a)** have any component that is negative, **(b)** be zero, **(c)** have three dimensions.

▶ **QUICK QUIZ 1.8** If $\vec{A} + \vec{B} = 0$, the corresponding components of the two vectors $\vec{A}$ and $\vec{B}$ must be **(a)** equal, **(b)** positive, **(c)** negative, **(d)** of opposite sign.

▶ **THINKING PHYSICS 1.2**

You may have asked someone directions to a destination in a city and been told something like, "Walk 3 blocks east and then 5 blocks south." If so, are you experienced with vector components?

Reasoning Yes, you are! Although you may not have thought of vector component language when you heard these directions, that is exactly what the directions represent. The perpendicular streets of the city reflect an xy coordinate system; we can assign the x axis to the east–west streets, and the y axis to the north–south streets. Therefore, the comment of the person giving you directions can be translated as, "Undergo a displacement vector that has an x component of $+3$ blocks and a y component of -5 blocks." You would arrive at the same destination by undergoing the y component first, followed by the x component, demonstrating the commutative law of addition. ◄

Example 1.6 | The Sum of Two Vectors

Find the sum of two displacement vectors $\vec{\mathbf{A}}$ and $\vec{\mathbf{B}}$ lying in the xy plane and given by

$$\vec{\mathbf{A}} = (2.0\hat{\mathbf{i}} + 2.0\,\hat{\mathbf{j}})\text{ m}\quad\text{and}\quad \vec{\mathbf{B}} = (2.0\hat{\mathbf{i}} - 4.0\,\hat{\mathbf{j}})\text{ m}$$

SOLUTION

Comparing this expression for $\vec{\mathbf{A}}$ with the general expression $\vec{\mathbf{A}} = A_x\hat{\mathbf{i}} + A_y\hat{\mathbf{j}} + A_z\hat{\mathbf{k}}$, we see that $A_x = 2.0$ m, $A_y = 2.0$ m, and $A_z = 0$. Likewise, $B_x = 2.0$ m, $B_y = -4.0$ m, and $B_z = 0$. We can use a two-dimensional approach because there are no z components.

Use Equation 1.14 to obtain the resultant vector $\vec{\mathbf{R}}$: $\vec{\mathbf{R}} = \vec{\mathbf{A}} + \vec{\mathbf{B}} = (2.0 + 2.0)\,\hat{\mathbf{i}}\text{ m} + (2.0 - 4.0)\,\hat{\mathbf{j}}\text{ m}$

Evaluate the components of $\vec{\mathbf{R}}$: $R_x = 4.0$ m $R_y = -2.0$ m

Use Equation 1.16 to find the magnitude of $\vec{\mathbf{R}}$: $R = \sqrt{R_x^2 + R_y^2} = \sqrt{(4.0\text{ m})^2 + (-2.0\text{ m})^2} = \sqrt{20}\text{ m} = \boxed{4.5\text{ m}}$

Find the direction of $\vec{\mathbf{R}}$ from Equation 1.17: $\tan\theta = \dfrac{R_y}{R_x} = \dfrac{-2.0\text{ m}}{4.0\text{ m}} = -0.50$

Your calculator likely gives the answer $-27°$ for $\theta = \tan^{-1}(-0.50)$. This answer is correct if we interpret it to mean 27° clockwise from the x axis. Our standard form has been to quote the angles measured counterclockwise from the $+x$ axis, and that angle for this vector is $\theta = \boxed{333.°}$

Example 1.7 | The Resultant Displacement

A particle undergoes three consecutive displacements: $\Delta\vec{\mathbf{r}}_1 = (15\hat{\mathbf{i}} + 30\,\hat{\mathbf{j}} + 12\hat{\mathbf{k}})$ cm, $\Delta\vec{\mathbf{r}}_2 = (23\hat{\mathbf{i}} - 14\,\hat{\mathbf{j}} - 5.0\hat{\mathbf{k}})$ cm, and $\Delta\vec{\mathbf{r}}_3 = (-13\hat{\mathbf{i}} + 15\,\hat{\mathbf{j}})$ cm. Find unit-vector notation for the resultant displacement and its magnitude.

SOLUTION

Although x is sufficient to locate a point in one dimension, we need a vector $\vec{\mathbf{r}}$ to locate a point in two or three dimensions. The notation $\Delta\vec{\mathbf{r}}$ is a generalization of the one-dimensional displacement Δx. Three-dimensional displacements are more difficult to conceptualize than those in two dimensions because the latter can be drawn on paper.

For this problem, let us imagine that you start with your pencil at the origin of a piece of graph paper on which you have drawn x and y axes. Move your pencil 15 cm to the right along the x axis, then 30 cm upward along the y axis, and then 12 cm *perpendicularly toward you away* from the graph paper. This procedure provides the displacement described by $\Delta\vec{\mathbf{r}}_1$. From this point, move your pencil 23 cm to the right parallel to the x axis, then 14 cm parallel to the graph paper in the $-y$ direction, and then 5.0 cm perpendicularly away from you toward the graph paper. You are now at the displacement from the origin described by $\Delta\vec{\mathbf{r}}_1 + \Delta\vec{\mathbf{r}}_2$. From this point, move your pencil 13 cm to the left in the $-x$ direction, and (finally!) 15 cm parallel to the graph paper along the y axis. Your final position is at a displacement $\Delta\vec{\mathbf{r}}_1 + \Delta\vec{\mathbf{r}}_2 + \Delta\vec{\mathbf{r}}_3$ from the origin.

To find the resultant displacement, add the three vectors:

$$\begin{aligned}\Delta\vec{\mathbf{r}} &= \Delta\vec{\mathbf{r}}_1 + \Delta\vec{\mathbf{r}}_2 + \Delta\vec{\mathbf{r}}_3\\ &= (15 + 23 - 13)\,\hat{\mathbf{i}}\text{ cm} + (30 - 14 + 15)\,\hat{\mathbf{j}}\text{ cm}\\ &\quad + (12 - 5.0 + 0)\hat{\mathbf{k}}\text{ cm}\\ &= \boxed{(25\hat{\mathbf{i}} + 31\,\hat{\mathbf{j}} + 7.0\hat{\mathbf{k}})\text{ cm}}\end{aligned}$$

Find the magnitude of the resultant vector:

$$\begin{aligned}R &= \sqrt{R_x^2 + R_y^2 + R_z^2}\\ &= \sqrt{(25\text{ cm})^2 + (31\text{ cm})^2 + (7.0\text{ cm})^2} = \boxed{40\text{ cm}}\end{aligned}$$

Example 1.8 | Taking a Hike

A hiker begins a trip by first walking 25.0 km southeast from her car. She stops and sets up her tent for the night. On the second day, she walks 40.0 km in a direction 60.0° north of east, at which point she discovers a forest ranger's tower.

(A) Determine the components of the hiker's displacement for each day.

SOLUTION

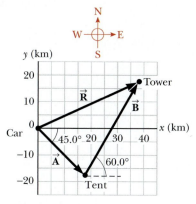

If we denote the displacement vectors on the first and second days by $\vec{A}$ and $\vec{B}$, respectively, and use the car as the origin of coordinates, we obtain the vectors shown in Figure 1.18. Drawing the resultant $\vec{R}$, we see that this problem is one we've solved before: an addition of two vectors.

Figure 1.18 (Example 1.8) The total displacement of the hiker is the vector $\vec{R} = \vec{A} + \vec{B}$.

Displacement $\vec{A}$ has a magnitude of 25.0 km and is directed 45.0° below the positive x axis.

Find the components of $\vec{A}$ using Equation 1.10:

$$A_x = A \cos(-45.0°) = (25.0 \text{ km})(0.707) = \boxed{17.7 \text{ km}}$$

$$A_y = A \sin(-45.0°) = (25.0 \text{ km})(-0.707) = \boxed{-17.7 \text{ km}}$$

The negative value of A_y indicates that the hiker walks in the negative y direction on the first day. The signs of A_x and A_y also are evident from Figure 1.18.

Find the components of $\vec{B}$ using Equation 1.10:

$$B_x = B \cos 60.0° = (40.0 \text{ km})(0.500) = \boxed{20.0 \text{ km}}$$

$$B_y = B \sin 60.0° = (40.0 \text{ km})(0.866) = \boxed{34.6 \text{ km}}$$

(B) Determine the components of the hiker's resultant displacement $\vec{R}$ for the trip. Find an expression for $\vec{R}$ in terms of unit vectors.

SOLUTION

Use Equation 1.15 to find the components of the resultant displacement $\vec{R} = \vec{A} + \vec{B}$:

$$R_x = A_x + B_x = 17.7 \text{ km} + 20.0 \text{ km} = \boxed{37.7 \text{ km}}$$

$$R_y = A_y + B_y = -17.7 \text{ km} + 34.6 \text{ km} = \boxed{17.0 \text{ km}}$$

Write the total displacement in unit-vector form:

$$\vec{R} = \boxed{(37.7\hat{\mathbf{i}} + 17.0\hat{\mathbf{j}}) \text{ km}}$$

Looking at the graphical representation in Figure 1.18, we estimate the position of the tower to be about (38 km, 17 km), which is consistent with the components of $\vec{R}$ in our result for the final position of the hiker. Also, both components of $\vec{R}$ are positive, putting the final position in the first quadrant of the coordinate system, which is also consistent with Figure 1.18.

What If? After reaching the tower, the hiker wishes to return to her car along a single straight line. What are the components of the vector representing this hike? What should the direction of the hike be?

Answer The desired vector $\vec{R}_{car}$ is the negative of vector $\vec{R}$:

$$\vec{R}_{car} = -\vec{R} = (-37.7\hat{\mathbf{i}} - 17.0\hat{\mathbf{j}}) \text{ km}$$

The direction is found by calculating the angle that the vector makes with the x axis:

$$\tan \theta = \frac{R_{car,y}}{R_{car,x}} = \frac{-17.0 \text{ km}}{-37.7 \text{ km}} = 0.450$$

which gives an angle of $\theta = 204.2°$, or 24.2° south of west.

⟨1.10 | Modeling, Alternative Representations, and Problem-Solving Strategy

Most courses in general physics require the student to learn the skills of problem solving, and examinations usually include problems that test such skills. This section describes some useful ideas that will enable you to enhance your understanding of physical concepts, increase your accuracy in solving problems, eliminate initial panic or lack of direction in approaching a problem, and organize your work.

One of the primary problem-solving methods in physics is to form an appropriate **model** of the problem. **A model is a simplified substitute for the real problem that allows us to solve the problem in a relatively simple way.** As long as the predictions of the model agree to our satisfaction with the actual behavior of the real system, the model is valid. If the predictions do not agree, the model must be refined or replaced with another model. The power of modeling is in its ability to reduce a wide variety of very complex problems to a limited number of classes of problems that can be approached in similar ways.

In science, a model is very different from, for example, an architect's scale model of a proposed building, which appears as a smaller version of what it represents. A scientific model is a theoretical construct and may have no visual similarity to the physical problem. A simple application of modeling is presented in Example 1.9, and we shall encounter many more examples of models as the text progresses.

Models are needed because the actual operation of the Universe is extremely complicated. Suppose, for example, we are asked to solve a problem about the Earth's motion around the Sun. The Earth is very complicated, with many processes occurring simultaneously. These processes include weather, seismic activity, and ocean movements as well as the multitude of processes involving human activity. Trying to maintain knowledge and understanding of all these processes is an impossible task.

The modeling approach recognizes that none of these processes affects the motion of the Earth around the Sun to a measurable degree. Therefore, these details are all ignored. In addition, as we shall find in Chapter 11, the size of the Earth does not affect the gravitational force between the Earth and the Sun; only the masses of the Earth and Sun and the distance between them determine this force. In a simplified model, the Earth is imagined to be a particle, an object with mass but zero size. This replacement of an extended object by a particle is called the **particle model,** which is used extensively in physics. By analyzing the motion of a particle with the mass of the Earth in orbit around the Sun, we find that the predictions of the particle's motion are in excellent agreement with the actual motion of the Earth.

The two primary conditions for using the particle model are as follows:

- The size of the actual object is of no consequence in the analysis of its motion.
- Any internal processes occurring in the object are of no consequence in the analysis of its motion.

Both of these conditions are in action in modeling the Earth as a particle. Its radius is not a factor in determining its motion, and internal processes such as thunderstorms, earthquakes, and manufacturing processes can be ignored.

Four categories of models used in this book will help us understand and solve physics problems. The first category is the **geometric model.** In this model, we form a geometric construction that represents the real situation. We then set aside the real problem and perform an analysis of the geometric construction. Consider a popular problem in elementary trigonometry, as in the following example.

Example **1.9** | Finding the Height of a Tree

You wish to find the height of a tree but cannot measure it directly. You stand 50.0 m from the tree and determine that a line of sight from the ground to the top of the tree makes an angle of 25.0° with the ground. How tall is the tree?

SOLUTION

Figure 1.19 shows the tree and a right triangle corresponding to the information in the problem superimposed over it. (We assume that the tree is exactly perpendicular to a perfectly flat ground.) In the triangle, we know the length of the horizontal leg and the angle between the hypotenuse and the horizontal leg. We can find the height of the tree by calculating the length of the vertical leg. We do so with the tangent function:

$$\tan \theta = \frac{\text{opposite side}}{\text{adjacent side}} = \frac{h}{50.0 \text{ m}}$$

$$h = (50.0 \text{ m}) \tan \theta = (50.0 \text{ m}) \tan 25.0° = \boxed{23.3 \text{ m}}$$

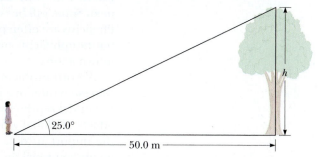

Figure 1.19 (Example 1.9) The height of a tree can be found by measuring the distance from the tree and the angle of sight to the top above the ground. This problem is a simple example of geometrically *modeling* the actual problem.

You may have solved a problem very similar to Example 1.9 but never thought about the notion of modeling. From the modeling approach, however, once we draw the triangle in Figure 1.19, the triangle is a geometric model of the real problem; it is a *substitute*. Until we reach the end of the problem, we do not imagine the problem to be about a *tree* but to be about a *triangle*. We use trigonometry to find the vertical leg of the triangle, leading to a value of 23.3 m. Because this leg *represents* the height of the tree, we can now return to the original problem and claim that the height of the tree is 23.3 m.

Other examples of geometric models include modeling the Earth as a perfect sphere, a pizza as a perfect disk, a meter stick as a long rod with no thickness, and an electric wire as a long, straight cylinder.

The particle model is an example of the second category of models, which we will call **simplification models.** In a simplification model, details that are not significant in determining the outcome of the problem are ignored. When we study rotation in Chapter 10, objects will be modeled as *rigid objects*. All the molecules in a rigid object maintain their exact positions with respect to one another. We adopt this simplification model because a spinning rock is much easier to analyze than a spinning block of gelatin, which is *not* a rigid object. Other simplification models will assume that quantities such as friction forces are negligible, remain constant, or are proportional to some power of the object's speed.

The third category is that of **analysis models,** which are general types of problems that we have solved before. An important technique in problem solving is to cast a new problem into a form similar to one we have already solved and which can be used as a model. As we shall see, there are about two dozen analysis models that can be used to solve most of the problems you will encounter. We will see our first analysis models in Chapter 2, where we will discuss them in more detail.

The fourth category of models is **structural models.** These models are generally used to understand the behavior of a system that is far different in scale from our macroscopic world—either much smaller or much larger—so that we cannot interact with it directly. As an example, the notion of a hydrogen atom as an electron in a circular orbit around a proton is a structural model of the atom. We will discuss this model and structural models in general in Chapter 11.

Intimately related to the notion of modeling is that of forming **alternative representations** of the problem. **A representation is a method of viewing or presenting the information related to the problem.** Scientists must be able to communicate complex ideas to individuals without scientific backgrounds. The best representation to use in conveying the information successfully will vary from one individual to the next. Some will be convinced by a well-drawn graph, and others will require a picture. Physicists are often persuaded to agree with a point of view by examining an equation, but nonphysicists may not be convinced by this mathematical representation of the information.

A word problem, such as those at the ends of the chapters in this book, is one representation of a problem. In the "real world" that you will enter after graduation, the initial representation of a problem may be just an existing situation, such as the effects of global warming or a patient in danger of dying. You may have to identify the important data and information, and then cast the situation yourself into an equivalent word problem!

Considering alternative representations can help you think about the information in the problem in several different ways to help you understand and solve it. Several types of representations can be of assistance in this endeavor:

Figure 1.20 A pictorial representation of a pop foul being hit by a baseball player.

- **Mental representation.** From the description of the problem, imagine a scene that describes what is happening in the word problem, then let time progress so that you understand the situation and can predict what changes will occur in the situation. This step is critical in approaching *every* problem.
- **Pictorial representation.** Drawing a picture of the situation described in the word problem can be of great assistance in understanding the problem. In Example 1.9, the pictorial representation in Figure 1.19 allows us to identify the triangle as a geometric model of the problem. In architecture, a blueprint is a pictorial representation of a proposed building.

Generally, a pictorial representation describes *what you would see* if you were observing the situation in the problem. For example, Figure 1.20 shows a pictorial representation of a baseball player hitting a short pop foul. Any coordinate axes included in your pictorial representation will be in two dimensions: x and y axes.

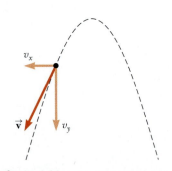

Figure 1.21 A simplified pictorial representation for the situation shown in Figure 1.20.

- **Simplified pictorial representation.** It is often useful to redraw the pictorial representation without complicating details by applying a simplification model. This process is similar to the discussion of the particle model described earlier. In a pictorial representation of the Earth in orbit around the Sun, you might draw the Earth and the Sun as spheres, with possibly some attempt to draw continents to identify which sphere is the Earth. In the simplified pictorial representation, the Earth and the Sun would be drawn simply as dots, representing particles. Figure 1.21 shows a simplified pictorial representation corresponding to the pictorial representation of the baseball trajectory in Figure 1.20. The notations v_x and v_y refer to the components of the velocity vector for the baseball. We shall use such simplified pictorial representations throughout the book.
- **Graphical representation.** In some problems, drawing a graph that describes the situation can be very helpful. In mechanics, for example, position–time graphs can be of great assistance. Similarly, in thermodynamics, pressure–volume graphs are essential to understanding. Figure 1.22 shows a graphical representation of the position as a function of time of a block on the end of a vertical spring as it oscillates up and down. Such a graph is helpful for understanding simple harmonic motion, which we study in Chapter 12.

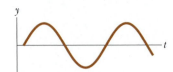

Figure 1.22 A graphical representation of the position as a function of time of a block hanging from a spring and oscillating.

A graphical representation is different from a pictorial representation, which is also a two-dimensional display of information but whose axes, if any, represent *length* coordinates. In a graphical representation, the axes may represent *any* two related variables. For example, a graphical representation may have axes for temperature and time. Therefore, in comparison to a pictorial representation, a graphical representation is generally *not* something you would see when observing the situation in the problem with your eyes.

- **Tabular representation.** It is sometimes helpful to organize the information in tabular form to help make it clearer. For example, some students find that making tables of known quantities and unknown quantities is helpful. The periodic table of the elements is an extremely useful tabular representation of information in chemistry and physics.
- **Mathematical representation.** The ultimate goal in solving a problem is often the mathematical representation. You want to move from the information contained in the word problem, through various representations of the problem that allow you to understand what is happening, to one or more equations that represent the situation in the problem and that can be solved mathematically for the desired result.

Besides what you might expect to learn about physics concepts, a very valuable skill you should acquire from your physics course is the ability to solve complicated problems. The way physicists approach complex situations and break them into manageable pieces is extremely useful. The following is a general problem-solving strategy to guide you through the steps. To help you remember the steps of the strategy, they are *Conceptualize*, *Categorize*, *Analyze*, and *Finalize*.

 GENERAL PROBLEM-SOLVING STRATEGY

Conceptualize

- The first things to do when approaching a problem are to *think about* and *understand* the situation. Study carefully any representations of the information (for example, diagrams, graphs, tables, or photographs) that accompany the problem. Imagine a movie, running in your mind, of what happens in the problem.
- If a pictorial representation is not provided, you should almost always make a quick drawing of the situation. Indicate any known values, perhaps in a table or directly on your sketch.
- Now focus on what algebraic or numerical information is given in the problem. Carefully read the problem statement, looking for key phrases such as "starts from rest" ($v_i = 0$) or "stops" ($v_f = 0$).
- Now focus on the expected result of solving the problem. Exactly what is the question asking? Will the final result be numerical or algebraic? Do you know what units to expect?
- Don't forget to incorporate information from your own experiences and common sense. What should a reasonable answer look like? For example, you wouldn't expect to calculate the speed of an automobile to be 5×10^6 m/s.

Categorize

- Once you have a good idea of what the problem is about, you need to *simplify* the problem. Remove the details that are not important to the solution. For example, model a moving object as a particle. If appropriate, ignore air resistance or friction between a sliding object and a surface.
- Once the problem is simplified, it is important to *categorize* the problem. Is it a simple *substitution problem* such that numbers can be substituted into an equation? If so, the problem is likely to be finished when this substitution is done. If not, you face what we call an *analysis problem:* the situation must be analyzed more deeply to reach a solution.
- If it is an analysis problem, it needs to be categorized further. Have you seen this type of problem before? Does it fall into the growing list of types of problems that you have solved previously? If so, identify any analysis

model(s) appropriate for the problem to prepare for the Analyze step below. Being able to classify a problem with an analysis model can make it much easier to lay out a plan to solve it. For example, if your simplification shows that the problem can be treated as a particle under constant acceleration and you have already solved such a problem (such as the examples we shall see in Section 2.6), the solution to the present problem follows a similar pattern.

Analyze

- Now you must analyze the problem and strive for a mathematical solution. Because you have already categorized the problem and identified an analysis model, it should not be too difficult to select relevant equations that apply to the type of situation in the problem. For example, if the problem involves a particle under constant acceleration (which we will study in Section 2.6), Equations 2.10 to 2.14 are relevant.
- Use algebra (and calculus, if necessary) to solve symbolically for the unknown variable in terms of what is given. Substitute in the appropriate numbers, calculate the result, and round it to the proper number of significant figures.

Finalize

- Examine your numerical answer. Does it have the correct units? Does it meet your expectations from your conceptualization of the problem? What about the algebraic form of the result? Does it make sense? Examine the variables in the problem to see whether the answer would change in a physically meaningful way if the variables were drastically increased or decreased or even became zero. Looking at limiting cases to see whether they yield expected values is a very useful way to make sure that you are obtaining reasonable results.
- Think about how this problem compared with others you have solved. How was it similar? In what critical ways did it differ? Why was this problem assigned? Can you figure out what you have learned by doing it? If it is a new category of problem, be sure you understand it so that you can use it as a model for solving similar problems in the future.

When solving complex problems, you may need to identify a series of subproblems and apply the problem-solving strategy to each. For simple problems, you probably don't need this strategy. When you are trying to solve a problem and you don't know what to do next, however, remember the steps in the strategy and use them as a guide.

In the rest of this book, we will label the *Conceptualize, Categorize, Analyze,* and *Finalize* steps explicitly in the worked examples. Many chapters in this book include a section labeled Problem-Solving Strategy that should help you through the rough spots. These sections are organized according to the General Problem-Solving Strategy outlined above and are tailored to the specific types of problems addressed in that chapter.

To clarify how this Strategy works, we repeat Example 1.8 on the next page with the particular steps of the Strategy identified.

When you **Conceptualize** a problem, try to understand the situation that is presented in the problem statement. Study carefully any representations of the information (for example, diagrams, graphs, tables, or photographs) that accompany the problem. Imagine a movie, running in your mind, of what happens in the problem.

Simplify the problem. Remove the details that are not important to the solution. Then **Categorize** the problem. Is it a simple substitution problem such that numbers can be substituted into an equation? If not, you face an analysis problem. In this case, identify the appropriate analysis model. (Analysis models will be introduced in Chapter 2.)

Now **Analyze** the problem. Select relevant equations from the analysis model. Solve symbolically for the unknown variable in terms of what is given. Substitute in the appropriate numbers, calculate the result, and round it to the proper number of significant figures.

Example 1.8 | Taking a Hike

A hiker begins a trip by first walking 25.0 km southeast from her car. She stops and sets up her tent for the night. On the second day, she walks 40.0 km in a direction 60.0° north of east, at which point she discovers a forest ranger's tower.

(A) Determine the components of the hiker's displacement for each day.

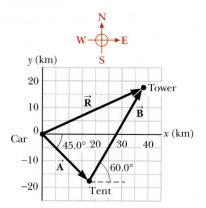

Figure 1.18 (Example 1.8) The total displacement of the hiker is the vector $\vec{R} = \vec{A} + \vec{B}$.

SOLUTION

Conceptualize We conceptualize the problem by drawing a sketch as in Figure 1.18. If we denote the displacement vectors on the first and second days by $\vec{A}$ and $\vec{B}$, respectively, and use the car as the origin of coordinates, we obtain the vectors shown in Figure 1.18.

Categorize Drawing the resultant $\vec{R}$, we can now categorize this problem as one we've solved before: an addition of two vectors. You should now have a hint of the power of categorization in that many new problems are very similar to problems we have already solved if we are careful to conceptualize them. Once we have drawn the displacement vectors and categorized the problem, this problem is no longer about a hiker, a walk, a car, a tent, or a tower. It is a problem about vector addition, one that we have already solved.

Analyze Displacement $\vec{A}$ has a magnitude of 25.0 km and is directed 45.0° below the positive x axis.

Find the components of $\vec{A}$ using Equation 1.10:

$$A_x = A\cos(-45.0°) = (25.0 \text{ km})(0.707) = \boxed{17.7 \text{ km}}$$

$$A_y = A\sin(-45.0°) = (25.0 \text{ km})(-0.707) = \boxed{-17.7 \text{ km}}$$

The negative value of A_y indicates that the hiker walks in the negative y direction on the first day. The signs of A_x and A_y also are evident from Figure 1.18.

Find the components of $\vec{B}$ using Equation 1.10:

$$B_x = B\cos 60.0° = (40.0 \text{ km})(0.500) = \boxed{20.0 \text{ km}}$$

$$B_y = B\sin 60.0° = (40.0 \text{ km})(0.866) = \boxed{34.6 \text{ km}}$$

....

(B) Determine the components of the hiker's resultant displacement $\vec{R}$ for the trip. Find an expression for $\vec{R}$ in terms of unit vectors.

SOLUTION

Use Equation 1.15 to find the components of the resultant displacement $\vec{R} = \vec{A} + \vec{B}$:

$$R_x = A_x + B_x = 17.7 \text{ km} + 20.0 \text{ km} = \boxed{37.7 \text{ km}}$$

$$R_y = A_y + B_y = -17.7 \text{ km} + 34.6 \text{ km} = \boxed{17.0 \text{ km}}$$

Write the total displacement in unit-vector form:

$$\vec{R} = \boxed{(37.7\hat{\mathbf{i}} + 17.0\hat{\mathbf{j}}) \text{ km}}$$

Finalize the problem. Examine the numerical answer. Does it have the correct units? Does it meet your expectations from your conceptualization of the problem? Does the answer make sense? What about the algebraic form of the result? Examine the variables in the problem to see whether the answer would change in a physically meaningful way if the variables were drastically increased or decreased or even became zero.

What If? questions will appear in many examples in the text, and offer a variation on the situation just explored. This feature encourages you to think about the results of the example and assists in conceptual understanding of the principles.

1.8 *cont.*

Finalize Looking at the graphical representation in Figure 1.18, we estimate the position of the tower to be about (38 km, 17 km), which is consistent with the components of $\vec{R}$ in our result for the final position of the hiker. Also, both components of $\vec{R}$ are positive, putting the final position in the first quadrant of the coordinate system, which is also consistent with Figure 1.18.

What If? After reaching the tower, the hiker wishes to return to her car along a single straight line. What are the components of the vector representing this hike? What should the direction of the hike be?

Answer The desired vector $\vec{R}_{car}$ is the negative of vector $\vec{R}$:

$$\vec{R}_{car} = -\vec{R} = (-37.7\hat{i} - 17.0\hat{j}) \text{ km}$$

The direction is found by calculating the angle that the vector makes with the x axis:

$$\tan \theta = \frac{R_{car,y}}{R_{car,x}} = \frac{-17.0 \text{ km}}{-37.7 \text{ km}} = 0.450$$

which gives an angle of $\theta = 204.2°$, or 24.2° south of west.

SUMMARY

Mechanical quantities can be expressed in terms of three fundamental quantities—**length, mass,** and **time**—which in the SI system have the units **meters** (m), **kilograms** (kg), and **seconds** (s), respectively. It is often useful to use the method of **dimensional analysis** to check equations and to assist in deriving expressions.

The **density** of a substance is defined as its mass per unit volume:

$$\rho \equiv \frac{m}{V} \qquad \text{1.1}$$

Vectors are quantities that have both magnitude and direction and obey the vector law of addition. **Scalars** are quantities that add algebraically.

Two vectors $\vec{A}$ and $\vec{B}$ can be added using the triangle method. In this method (see Active Fig. 1.9), the vector $\vec{R} = \vec{A} + \vec{B}$ runs from the tail of $\vec{A}$ to the tip of $\vec{B}$.

The x component A_x of the vector $\vec{A}$ is equal to its projection along the x axis of a coordinate system, where $A_x = A \cos \theta$ and where θ is the angle $\vec{A}$ makes with the x axis. Likewise, the y component A_y of $\vec{A}$ is its projection along the y axis, where $A_y = A \sin \theta$.

If a vector $\vec{A}$ has an x component equal to A_x and a y component equal to A_y, the vector can be expressed in unit-vector form as $\vec{A} = (A_x\hat{i} + A_y\hat{j})$. In this notation, $\hat{i}$ is a unit vector in the positive x direction and $\hat{j}$ is a unit vector in the positive y direction. Because $\hat{i}$ and $\hat{j}$ are unit vectors, $|\hat{i}| = |\hat{j}| = 1$. In three dimensions, a vector can be expressed as $\vec{A} = (A_x\hat{i} + A_y\hat{j} + A_z\hat{k})$, where $\hat{k}$ is a unit vector in the z direction.

The resultant of two or more vectors can be found by resolving all vectors into their x, y, and z components and adding their components:

$$\vec{R} = \vec{A} + \vec{B} = (A_x + B_x)\hat{i} + (A_y + B_y)\hat{j} + (A_z + B_z)\hat{k} \qquad \text{1.18}$$

Problem-solving skills and physical understanding can be improved by **modeling** the problem and by constructing **alternative representations** of the problem. Models helpful in solving problems include **geometric, simplification,** and **analysis models.** Scientists use **structural models** to understand systems larger or smaller in scale than those with which we normally have direct experience. Helpful representations include the **mental, pictorial, simplified pictorial, graphical, tabular,** and **mathematical representations.**

Complicated problems are best approached in an organized manner. Recall and apply the *Conceptualize, Categorize, Analyze,* and *Finalize* steps of the **General Problem-Solving Strategy** when you need them.

OBJECTIVE QUESTIONS

1. Answer each question yes or no. Must two quantities have the same dimensions (a) if you are adding them? (b) If you are multiplying them? (c) If you are subtracting them? (d) If you are dividing them? (e) If you are equating them?

2. The price of gasoline at a particular station is 1.5 euros per liter. An American student can use 33 euros to buy gasoline. Knowing that 4 quarts make a gallon and that 1 liter is close to 1 quart, she quickly reasons that she can buy how many gallons of gasoline? (a) less than 1 gallon (b) about 5 gallons (c) about 8 gallons (d) more than 10 gallons

3. Rank the following five quantities in order from the largest to the smallest. If two of the quantities are equal, give them equal rank in your list. (a) 0.032 kg (b) 15 g (c) 2.7×10^5 mg (d) 4.1×10^{-8} Gg (e) 2.7×10^8 μg

4. What is the y component of the vector $(3\hat{\mathbf{i}} - 8\hat{\mathbf{k}})$ m/s? (a) 3 m/s (b) −8 m/s (c) 0 (d) 8 m/s (e) none of those answers

5. Which of the following is the best estimate for the mass of all the people living on the Earth? (a) 2×10^8 kg (b) 1×10^9 kg (c) 2×10^{10} kg (d) 3×10^{11} kg (e) 4×10^{12} kg

6. What is the sum of the measured values 21.4 s + 15 s + 17.17 s + 4.003 s? (a) 57.573 s (b) 57.57 s (c) 57.6 s (d) 58 s (e) 60 s

7. One student uses a meterstick to measure the thickness of a textbook and obtains 4.3 cm ± 0.1 cm. Other students measure the thickness with vernier calipers and obtain four different measurements: (a) 4.32 cm ± 0.01 cm, (b) 4.31 cm ± 0.01 cm, (c) 4.24 cm ± 0.01 cm, and (d) 4.43 cm ± 0.01 cm. Which of these four measurements, if any, agree with that obtained by the first student?

8. Newton's second law of motion (Chapter 4) says that the mass of an object times its acceleration is equal to the net force on the object. Which of the following gives the correct units for force? (a) kg · m/s² (b) kg · m²/s² (c) kg/m · s² (d) kg · m²/s (e) none of those answers

9. What is the x component of the vector shown in Figure OQ1.9? (a) 3 cm (b) 6 cm (c) −4 cm (d) −6 cm (e) none of those answers

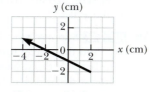

Figure OQ1.9 Objective Questions 9 and 10.

10. What is the y component of the vector shown in Figure OQ1.9? (a) 3 cm (b) 6 cm (c) −4 cm (d) −6 cm (e) none of those answers

11. Yes or no: Is each of the following quantities a vector? (a) force (b) temperature (c) the volume of water in a can (d) the ratings of a TV show (e) the height of a building (f) the velocity of a sports car (g) the age of the Universe

12. A vector lying in the xy plane has components of opposite sign. The vector must lie in which quadrant? (a) the first quadrant (b) the second quadrant (c) the third quadrant (d) the fourth quadrant (e) either the second or the fourth quadrant

13. Figure OQ1.13 shows two vectors $\vec{\mathbf{D}}_1$ and $\vec{\mathbf{D}}_2$. Which of the possibilities (a) through (d) is the vector $\vec{\mathbf{D}}_2 - 2\vec{\mathbf{D}}_1$, or (e) is it none of them?

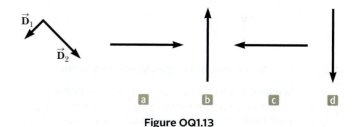

Figure OQ1.13

14. A vector points from the origin into the second quadrant of the xy plane. What can you conclude about its components? (a) Both components are positive. (b) The x component is positive, and the y component is negative. (c) The x component is negative, and the y component is positive. (d) Both components are negative. (e) More than one answer is possible.

15. What is the magnitude of the vector $(10\hat{\mathbf{i}} - 10\hat{\mathbf{k}})$ m/s? (a) 0 (b) 10 m/s (c) −10 m/s (d) 10 (e) 14.1 m/s

16. Vector $\vec{\mathbf{A}}$ lies in the xy plane. Both of its components will be negative if it points from the origin into which quadrant? (a) the first quadrant (b) the second quadrant (c) the third quadrant (d) the fourth quadrant (e) the second or fourth quadrants

CONCEPTUAL QUESTIONS

1. What natural phenomena could serve as alternative time standards?

2. Express the following quantities using the prefixes given in Table 1.4. (a) 3×10^{-4} m (b) 5×10^{-5} s (c) 72×10^2 g

3. Suppose the three fundamental standards of the metric system were length, *density*, and time rather than length, *mass*, and time. The standard of density in this system is to be defined as that of water. What considerations about water would you need to address to make sure that the standard of density is as accurate as possible?

4. If the component of vector $\vec{\mathbf{A}}$ along the direction of vector $\vec{\mathbf{B}}$ is zero, what can you conclude about the two vectors?

5. A book is moved once around the perimeter of a tabletop with the dimensions 1.0 m by 2.0 m. The book ends up at its initial position. (a) What is its displacement? (b) What is the distance traveled?

6. Can the magnitude of a vector have a negative value? Explain.

7. On a certain calculator, the inverse tangent function returns a value between $-90°$ and $+90°$. In what cases will this value correctly state the direction of a vector in the xy plane, by giving its angle measured counterclockwise from the positive x axis? In what cases will it be incorrect?

8. Is it possible to add a vector quantity to a scalar quantity? Explain.

▶ PROBLEMS

WebAssign The problems found in this chapter may be assigned online in Enhanced WebAssign.

1. denotes straightforward problem; **2.** denotes intermediate problem; **3.** denotes challenging problem

1. denotes full solution available in the *Student Solutions Manual/Study Guide*

1. denotes problems most often assigned in Enhanced WebAssign.

BIO denotes biomedical problem

GP denotes guided problem

M denotes Master It tutorial available in Enhanced WebAssign

Q|C denotes asking for quantitative and conceptual reasoning

S denotes symbolic reasoning problem

shaded denotes "paired problems" that develop reasoning with symbols and numerical values

W denotes Watch It video solution available in Enhanced WebAssign

Section 1.1 Standards of Length, Mass, and Time

Note: Consult the endpapers, appendices, and tables in the text whenever necessary in solving problems. For this chapter, Appendix B.3 and Table 15.1 may be particularly useful. Answers to odd-numbered problems appear in the back of the book.

1. Two spheres are cut from a certain uniform rock. One has radius 4.50 cm. The mass of the other is five times greater. Find its radius.

2. **Q|C** (a) Use information on the endpapers of this book to calculate the average density of the Earth. (b) Where does the value fit among those listed in Table 15.1 in Chapter 15? Look up the density of a typical surface rock like granite in another source and compare it with the density of the Earth.

3. An automobile company displays a die-cast model of its first car, made from 9.35 kg of iron. To celebrate its hundredth year in business, a worker will recast the model in gold from the original dies. What mass of gold is needed to make the new model?

4. **S** What mass of a material with density ρ is required to make a hollow spherical shell having inner radius r_1 and outer radius r_2?

Section 1.2 Dimensional Analysis

5. Which of the following equations are dimensionally correct? (a) $v_f = v_i + ax$ (b) $y = (2\ \text{m}) \cos(kx)$, where $k = 2\ \text{m}^{-1}$

6. **W** Figure P1.6 shows a *frustum of a cone*. Match each of the three expressions (a) $\pi(r_1 + r_2)[h^2 + (r_2 - r_1)^2]^{1/2}$, (b) $2\pi(r_1 + r_2)$, and (c) $\pi h(r_1^2 + r_1 r_2 + r_2^2)/3$ with the

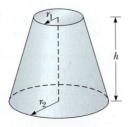

Figure P1.6

quantity it describes: (d) the total circumference of the flat circular faces, (e) the volume, or (f) the area of the curved surface.

7. The position of a particle moving under uniform acceleration is some function of time and the acceleration. Suppose we write this position as $x = ka^m t^n$, where k is a dimensionless constant. Show by dimensional analysis that this expression is satisfied if $m = 1$ and $n = 2$. Can this analysis give the value of k?

Section 1.3 Conversion of Units

8. A *section* of land has an area of 1 square mile and contains 640 acres. Determine the number of square meters in 1 acre.

9. **M** One gallon of paint (volume $= 3.78 \times 10^{-3}\ \text{m}^3$) covers an area of $25.0\ \text{m}^2$. What is the thickness of the fresh paint on the wall?

10. **BIO** **W** Suppose your hair grows at the rate 1/32 in. per day. Find the rate at which it grows in nanometers per second. Because the distance between atoms in a molecule is on the order of 0.1 nm, your answer suggests how rapidly layers of atoms are assembled in this protein synthesis.

11. An ore loader moves 1 200 tons/h from a mine to the surface. Convert this rate to pounds per second, using 1 ton = 2 000 lb.

12. The mass of the Sun is 1.99×10^{30} kg, and the mass of an atom of hydrogen, of which the Sun is mostly composed, is 1.67×10^{-27} kg. How many atoms are in the Sun?

13. **M** One cubic meter ($1.00\ \text{m}^3$) of aluminum has a mass of 2.70×10^3 kg, and the same volume of iron has a mass of 7.86×10^3 kg. Find the radius of a solid aluminum sphere that will balance a solid iron sphere of radius 2.00 cm on an equal-arm balance.

14. [S] Let ρ_{Al} represent the density of aluminum and ρ_{Fe} that of iron. Find the radius of a solid aluminum sphere that balances a solid iron sphere of radius r_{Fe} on an equal-arm balance.

15. [W] Assume it takes 7.00 min to fill a 30.0-gal gasoline tank. (a) Calculate the rate at which the tank is filled in gallons per second. (b) Calculate the rate at which the tank is filled in cubic meters per second. (c) Determine the time interval, in hours, required to fill a 1.00-m³ volume at the same rate. (1 U.S. gal = 231 in.³)

16. A hydrogen atom has a diameter of 1.06×10^{-10} m. The nucleus of the hydrogen atom has a diameter of approximately 2.40×10^{-15} m. (a) For a scale model, represent the diameter of the hydrogen atom by the playing length of an American football field (100 yards = 300 ft) and determine the diameter of the nucleus in millimeters. (b) Find the ratio of the volume of the hydrogen atom to the volume of its nucleus.

Section 1.4 Order-of-Magnitude Calculations

17. Find the order of magnitude of the number of table-tennis balls that would fit into a typical-size room (without being crushed).

18. (a) Compute the order of magnitude of the mass of a bathtub half full of water. (b) Compute the order of magnitude of the mass of a bathtub half full of copper coins.

19. To an order of magnitude, how many piano tuners reside in New York City? The physicist Enrico Fermi was famous for asking questions like this one on oral Ph.D. qualifying examinations.

20. An automobile tire is rated to last for 50 000 miles. To an order of magnitude, through how many revolutions will it turn over its lifetime?

Section 1.5 Significant Figures

21. The *tropical year*, the time interval from one vernal equinox to the next vernal equinox, is the basis for our calendar. It contains 365.242 199 days. Find the number of seconds in a tropical year.

22. [W] Carry out the arithmetic operations (a) the sum of the measured values 756, 37.2, 0.83, and 2; (b) the product $0.003\ 2 \times 356.3$; and (c) the product $5.620 \times \pi$.

23. [W] How many significant figures are in the following numbers? (a) 78.9 ± 0.2 (b) 3.788×10^9 (c) 2.46×10^{-6} (d) $0.005\ 3$

Note: Appendix B.8 on propagation of uncertainty may be useful in solving the next two problems.

24. The radius of a uniform solid sphere is measured to be (6.50 ± 0.20) cm, and its mass is measured to be (1.85 ± 0.02) kg. Determine the density of the sphere in kilograms per cubic meter and the uncertainty in the density.

25. A sidewalk is to be constructed around a swimming pool that measures (10.0 ± 0.1) m by (17.0 ± 0.1) m. If the sidewalk is to measure (1.00 ± 0.01) m wide by (9.0 ± 0.1) cm

thick, what volume of concrete is needed and what is the approximate uncertainty of this volume?

Note: The next four problems call upon mathematical skills that will be useful throughout the course.

26. [S] **Review.** From the set of equations

$$p = 3q$$

$$pr = qs$$

$$\tfrac{1}{2}pr^2 + \tfrac{1}{2}qs^2 = \tfrac{1}{2}qt^2$$

involving the unknowns p, q, r, s, and t, find the value of the ratio of t to r.

27. **Review.** A highway curve forms a section of a circle. A car goes around the curve as shown in the helicopter view of Figure P1.27. Its dashboard compass shows that the car is initially heading due east. After it travels $d = 840$ m, it is heading $\theta = 35.0°$ south of east. Find the radius of curvature of its path. *Suggestion:* You may find it useful to learn a geometric theorem stated in Appendix B.3.

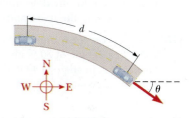

Figure P1.27

28. **Review.** Prove that one solution of the equation

$$2.00x^4 - 3.00x^3 + 5.00x = 70.0$$

is $x = -2.22$.

29. **Review.** Find every angle θ between 0 and 360° for which the ratio of $\sin \theta$ to $\cos \theta$ is -3.00.

Section 1.6 Coordinate Systems

30. Two points in the xy plane have Cartesian coordinates (2.00, -4.00) m and (-3.00, 3.00) m. Determine (a) the distance between these points and (b) their polar coordinates.

31. [W] The polar coordinates of a point are $r = 5.50$ m and $\theta = 240°$. What are the Cartesian coordinates of this point?

32. [S] Let the polar coordinates of the point (x, y) be (r, θ). Determine the polar coordinates for the points (a) $(-x, y)$, (b) $(-2x, -2y)$, and (c) $(3x, -3y)$.

33. [W] A fly lands on one wall of a room. The lower-left corner of the wall is selected as the origin of a two-dimensional Cartesian coordinate system. If the fly is located at the point having coordinates (2.00, 1.00) m, (a) how far is it from the origin? (b) What is its location in polar coordinates?

Section 1.7 Vectors and Scalars

Section 1.8 Some Properties of Vectors

34. [M] The displacement vectors $\vec{A}$ and $\vec{B}$ shown in Figure P1.34 have

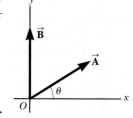

Figure P1.34 Problems 34 and 48.

magnitudes of 3.00 m. The direction of vector $\vec{A}$ is $\theta = 30.0°$. Find graphically (a) $\vec{A} + \vec{B}$, (b) $\vec{A} - \vec{B}$, (c) $\vec{B} - \vec{A}$, and (d) $\vec{A} - 2\vec{B}$. (Report all angles counterclockwise from the positive x axis.)

35. *Why is the following situation impossible?* A skater glides along a circular path. She defines a certain point on the circle as her origin. Later on, she passes through a point at which the distance she has traveled along the path from the origin is smaller than the magnitude of her displacement vector from the origin.

36. A plane flies from base camp to Lake A, 280 km away in the direction 20.0° north of east. After dropping off supplies, it flies to Lake B, which is 190 km at 30.0° west of north from Lake A. Graphically determine the distance and direction from Lake B to the base camp.

37. A roller-coaster car moves 200 ft horizontally and then rises 135 ft at an angle of 30.0° above the horizontal. It next travels 135 ft at an angle of 40.0° downward. What is its displacement from its starting point? Use graphical techniques.

Section 1.9 Components of a Vector and Unit Vectors

38. **W** Given the vectors $\vec{A} = 2.00\hat{i} + 6.00\hat{j}$ and $\vec{B} = 3.00\hat{i} - 2.00\hat{j}$, (a) draw the vector sum $\vec{C} = \vec{A} + \vec{B}$ and the vector difference $\vec{D} = \vec{A} - \vec{B}$. (b) Calculate $\vec{C}$ and $\vec{D}$, in terms of unit vectors. (c) Calculate $\vec{C}$ and $\vec{D}$ in terms of polar coordinates, with angles measured with respect to the positive x axis.

39. **Q|C** (a) Taking $\vec{A} = (6.00\hat{i} - 8.00\hat{j})$ units, $\vec{B} = (-8.00\hat{i} + 3.00\hat{j})$ units, and $\vec{C} = (26.0\hat{i} + 19.0\hat{j})$ units, determine a and b such that $a\vec{A} + b\vec{B} + \vec{C} = 0$. (b) A student has learned that a single equation cannot be solved to determine values for more than one unknown in it. How would you explain to him that both a and b can be determined from the single equation used in part (a)?

40. Find the horizontal and vertical components of the 100-m displacement of a superhero who flies from the top of a tall building following the path shown in Figure P1.40.

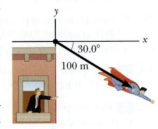

Figure P1.40

41. **W** A vector has an x component of −25.0 units and a y component of 40.0 units. Find the magnitude and direction of this vector.

42. Vector $\vec{B}$ has x, y, and z components of 4.00, 6.00, and 3.00 units, respectively. Calculate (a) the magnitude of $\vec{B}$ and (b) the angle that $\vec{B}$ makes with each coordinate axis.

43. **M** The vector $\vec{A}$ has x, y, and z components of 8.00, 12.0, and −4.00 units, respectively. (a) Write a vector expression for $\vec{A}$ in unit-vector notation. (b) Obtain a unit-vector expression for a vector $\vec{B}$ one-fourth the length of $\vec{A}$ pointing in the same direction as $\vec{A}$. (c) Obtain a unit-vector expression for a vector $\vec{C}$ three times the length of $\vec{A}$ pointing in the direction opposite the direction of $\vec{A}$.

44. Three displacement vectors of a croquet ball are shown in Figure P1.44, where $|\vec{A}| = 20.0$ units, $|\vec{B}| = 40.0$ units, and

$|\vec{C}| = 30.0$ units. Find (a) the resultant in unit-vector notation and (b) the magnitude and direction of the resultant displacement.

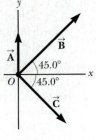

Figure P1.44

45. **M** A man pushing a mop across a floor causes it to undergo two displacements. The first has a magnitude of 150 cm and makes an angle of 120° with the positive x axis. The resultant displacement has a magnitude of 140 cm and is directed at an angle of 35.0° to the positive x axis. Find the magnitude and direction of the second displacement.

46. **W** Vector $\vec{A}$ has x and y components of −8.70 cm and 15.0 cm, respectively; vector $\vec{B}$ has x and y components of 13.2 cm and −6.60 cm, respectively. If $\vec{A} - \vec{B} + 3\vec{C} = 0$, what are the components of $\vec{C}$?

47. **M** Consider the two vectors $\vec{A} = 3\hat{i} - 2\hat{j}$ and $\vec{B} = -\hat{i} - 4\hat{j}$. Calculate (a) $\vec{A} + \vec{B}$, (b) $\vec{A} - \vec{B}$, (c) $|\vec{A} + \vec{B}|$, (d) $|\vec{A} - \vec{B}|$, and (e) the directions of $\vec{A} + \vec{B}$ and $\vec{A} - \vec{B}$.

48. Use the component method to add the vectors $\vec{A}$ and $\vec{B}$ shown in Figure P1.34. Express the resultant $\vec{A} + \vec{B}$ in unit-vector notation.

49. In an assembly operation illustrated in Figure P1.49, a robot moves an object first straight upward and then also to the east, around an arc forming one-quarter of a circle of radius 4.80 cm that lies in an east–west vertical plane. The robot then moves the object upward and to the north, through one-quarter of a circle of radius 3.70 cm that lies in a north–south vertical plane. Find (a) the magnitude of the total displacement of the object and (b) the angle the total displacement makes with the vertical.

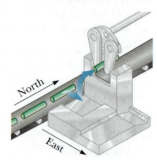

Figure P1.49

50. **W** Consider the three displacement vectors $\vec{A} = (3\hat{i} - 3\hat{j})$ m, $\vec{B} = (\hat{i} - 4\hat{j})$ m, and $\vec{C} = (2\hat{i} + 5\hat{j})$ m. Use the component method to determine (a) the magnitude and direction of the vector $\vec{D} = \vec{A} + \vec{B} + \vec{C}$ and (b) the magnitude and direction of $\vec{E} = -\vec{A} - \vec{B} + \vec{C}$.

51. **M** A person going for a walk follows the path shown in Figure P1.51. The total trip consists of four straight-line paths. At the end of the walk, what is the person's resultant displacement measured from the starting point?

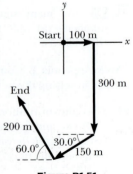

Figure P1.51

52. Express in unit-vector notation the following vectors, each of which has magnitude 17.0 cm. (a) Vector $\vec{E}$ is directed 27.0° counterclockwise from the positive x axis. (b) Vector $\vec{F}$ is directed 27.0° counterclockwise from the positive y axis. (c) Vector $\vec{G}$ is directed 27.0° clockwise from the negative y axis.

Section 1.10 Modeling, Alternative Representations, and Problem-Solving Strategy

53. A crystalline solid consists of atoms stacked up in a repeating lattice structure. Consider a crystal as shown in Figure P1.53a. The atoms reside at the corners of cubes of side $L = 0.200$ nm. One piece of evidence for the regular arrangement of atoms comes from the flat surfaces along which a crystal separates, or cleaves, when it is broken. Suppose this crystal cleaves along a face diagonal as shown in Figure P1.53b. Calculate the spacing d between two adjacent atomic planes that separate when the crystal cleaves.

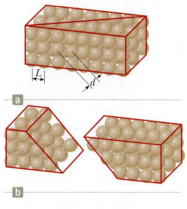

Figure P1.53

54. As she picks up her riders, a bus driver traverses four successive displacements represented by the expression

$$(-6.30 \text{ b})\hat{\mathbf{i}} - (4.00 \text{ b cos } 40°)\hat{\mathbf{i}} - (4.00 \text{ b sin } 40°)\hat{\mathbf{j}}$$
$$+ (3.00 \text{ b cos } 50°)\hat{\mathbf{i}} - (3.00 \text{ b sin } 50°)\hat{\mathbf{j}} - (5.00 \text{ b})\hat{\mathbf{j}}$$

Here b represents one city block, a convenient unit of distance of uniform size; $\hat{\mathbf{i}}$ is east; and $\hat{\mathbf{j}}$ is north. The displacements at 40° and 50° represent travel on roadways in the city that are at these angles to the main east–west and north–south streets. (a) Draw a map of the successive displacements. (b) What total distance did she travel? (c) Compute the magnitude and direction of her total displacement. The logical structure of this problem and of several problems in later chapters was suggested by Alan Van Heuvelen and David Maloney, *American Journal of Physics* **67**(3) 252–256, March 1999.

55. **W** A surveyor measures the distance across a straight river by the following method (Fig. P1.55). Starting directly across from a tree on the opposite bank, she walks $d = 100$ m along the riverbank to establish a baseline. Then she sights across to the tree. The angle from her baseline to the tree is $\theta = 35.0°$. How wide is the river?

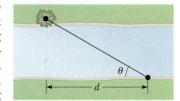

Figure P1.55

Additional Problems

56. The distance from the Sun to the nearest star is about 4×10^{16} m. The Milky Way galaxy (Fig. P1.56) is roughly a disk of diameter $\sim 10^{21}$ m and thickness $\sim 10^{19}$ m.

Figure P1.56 The Milky Way galaxy.

Richard Payne/NASA

Find the order of magnitude of the number of stars in the Milky Way. Assume the distance between the Sun and our nearest neighbor is typical.

57. In a situation in which data are known to three significant digits, we write 6.379 m = 6.38 m and 6.374 m = 6.37 m. When a number ends in 5, we arbitrarily choose to write 6.375 m = 6.38 m. We could equally well write 6.375 m = 6.37 m, "rounding down" instead of "rounding up," because we would change the number 6.375 by equal increments in both cases. Now consider an order-of-magnitude estimate, in which factors of change rather than increments are important. We write 500 m $\sim 10^3$ m because 500 differs from 100 by a factor of 5 while it differs from 1 000 by only a factor of 2. We write 437 m $\sim 10^3$ m and 305 m $\sim 10^2$ m. What distance differs from 100 m and from 1 000 m by equal factors so that we could equally well choose to represent its order of magnitude as $\sim 10^2$ m or as $\sim 10^3$ m?

58. **M** The consumption of natural gas by a company satisfies the empirical equation $V = 1.50t + 0.008\ 00t^2$, where V is the volume of gas in millions of cubic feet and t is the time in months. Express this equation in units of cubic feet and seconds. Assume a month is 30.0 days.

59. Vectors $\vec{\mathbf{A}}$ and $\vec{\mathbf{B}}$ have equal magnitudes of 5.00. The sum of $\vec{\mathbf{A}}$ and $\vec{\mathbf{B}}$ is the vector $6.00\hat{\mathbf{j}}$. Determine the angle between $\vec{\mathbf{A}}$ and $\vec{\mathbf{B}}$.

60. In physics, it is important to use mathematical approximations. (a) Demonstrate that for small angles ($<20°$)

$$\tan \alpha \approx \sin \alpha \approx \alpha = \frac{\pi \alpha'}{180°}$$

where α is in radians and α' is in degrees. (b) Use a calculator to find the largest angle for which $\tan \alpha$ may be approximated by α with an error less than 10.0%.

61. Two vectors $\vec{\mathbf{A}}$ and $\vec{\mathbf{B}}$ have precisely equal magnitudes. For the magnitude of $\vec{\mathbf{A}} + \vec{\mathbf{B}}$ to be 100 times larger than the magnitude of $\vec{\mathbf{A}} - \vec{\mathbf{B}}$, what must be the angle between them?

62. **S** Two vectors $\vec{\mathbf{A}}$ and $\vec{\mathbf{B}}$ have precisely equal magnitudes. For the magnitude of $\vec{\mathbf{A}} + \vec{\mathbf{B}}$ to be larger than the magnitude of $\vec{\mathbf{A}} - \vec{\mathbf{B}}$ by the factor n, what must be the angle between them?

63. There are nearly $\pi \times 10^7$ s in one year. Find the percentage error in this approximation, where "percentage error" is defined as

$$\text{percentage error} = \frac{|\text{assumed value} - \text{true value}|}{\text{true value}} \times 100\%$$

64. An air-traffic controller observes two aircraft on his radar screen. The first is at altitude 800 m, horizontal distance 19.2 km, and 25.0° south of west. The second aircraft is at altitude 1 100 m, horizontal distance 17.6 km, and 20.0° south of west. What is the distance between the two aircraft? (Place the x axis west, the y axis south, and the z axis vertical.)

65. A child loves to watch as you fill a transparent plastic bottle with shampoo (Fig P1.65, page 34). Every horizontal cross section of the bottle is circular, but the diameters of the circles have different values. You pour the brightly colored shampoo into the bottle at a constant rate of 16.5 cm³/s. At what rate is its level in the bottle rising (a) at a point where the diameter of the bottle is 6.30 cm and (b) at a point where the diameter is 1.35 cm?

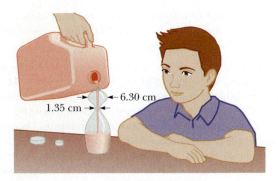

Figure P1.65

66. **BIO** One cubic centimeter of water has a mass of 1.00×10^{-3} kg. (a) Determine the mass of 1.00 m³ of water. (b) Biological substances are 98% water. Assume that they have the same density as water to estimate the masses of a cell that has a diameter of 1.00 μm, a human kidney, and a fly. Model the kidney as a sphere with a radius of 4.00 cm and the fly as a cylinder 4.00 mm long and 2.00 mm in diameter.

67. The helicopter view in Fig. P1.67 shows two people pulling on a stubborn mule. The person on the right pulls with a force $\vec{F}_1$ of magnitude 120 N and direction of $\theta_1 = 60.0°$. The person on the left pulls with a force $\vec{F}_2$ of magnitude 80.0 N and direction of $\theta_2 = 75.0°$. Find (a) the single force that is equivalent to the two forces shown and (b) the force that a third person would have to exert on the mule to make the resultant force equal to zero. The forces are measured in units of newtons (symbolized N).

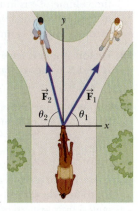

Figure P1.67

68. **GP** A woman wishing to know the height of a mountain measures the angle of elevation of a mountaintop as $12.0°$. After walking 1.00 km closer to the mountain on level ground, she finds the angle to be $14.0°$. (a) Draw a picture of the problem, neglecting the height of the woman's eyes above the ground. *Hint:* Use two triangles. (b) Using the symbol y to represent the mountain height and the symbol x to represent the woman's original distance from the mountain, label the picture. (c) Using the labeled picture, write two trigonometric equations relating the two selected variables. (d) Find the height y.

69. **Q|C** A pirate has buried his treasure on an island with five trees located at the points $(30.0$ m, -20.0 m$)$, $(60.0$ m, 80.0 m$)$, $(-10.0$ m, -10.0 m$)$, $(40.0$ m, -30.0 m$)$, and $(-70.0$ m, 60.0 m$)$, all measured relative to some origin, as shown in Figure P1.69. His ship's log instructs you to start at tree A and move toward tree B, but to cover only one-half the distance between A and B. Then move toward tree C, covering one-third the distance between your current location and C. Next move toward tree D, covering one-fourth the distance between where you are and D. Finally move toward tree E, covering one-fifth the distance between you and E, stop, and dig. (a) Assume you have correctly determined the order in which the pirate labeled the trees as A, B, C, D, and E as shown in the figure. What are the coordinates of the point where his treasure is buried? (b) **What If?** What if you do not really know the way the pirate labeled the trees?

What would happen to the answer if you rearranged the order of the trees, for instance, to B $(30$ m, -20 m$)$, A $(60$ m, 80 m$)$, E $(-10$ m, -10 m$)$, C $(40$ m, -30 m$)$, and D $(-70$ m, 60 m$)$? State reasoning to show that the answer does not depend on the order in which the trees are labeled.

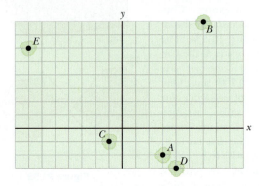

Figure P1.69

70. You stand in a flat meadow and observe two cows (Fig. P1.70). Cow A is due north of you and 15.0 m from your position. Cow B is 25.0 m from your position. From your point of view, the angle between cow A and cow B is $20.0°$, with cow B appearing to the right of cow A. (a) How far apart are cow A and cow B? (b) Consider the view seen by cow A. According to this cow, what is the angle between you and cow B? (c) Consider the view seen by cow B. According to this cow, what is the angle between you and cow A? *Hint:* What does the situation look like to a hummingbird hovering above the meadow? (d) Two stars in the sky appear to be $20.0°$ apart. Star A is 15.0 ly from the Earth, and star B, appearing to the right of star A, is 25.0 ly from the Earth. To an inhabitant of a planet orbiting star A, what is the angle in the sky between star B and our Sun?

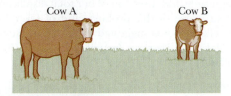

Figure P1.70 Your view of two cows in a meadow. Cow A is due north of you. You must rotate your eyes through an angle of $20.0°$ to look from cow A to cow B.

71. **S** A rectangular parallelepiped has dimensions a, b, and c as shown in Figure P1.71. (a) Obtain a vector expression for the face diagonal vector $\vec{R}_1$. (b) What is the magnitude of this vector? (c) Notice that $\vec{R}_1$, $c\hat{k}$, and $\vec{R}_2$ make a right triangle. Obtain a vector expression for the body diagonal vector $\vec{R}_2$.

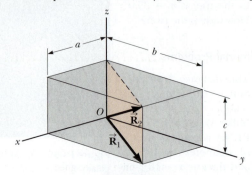

Figure P1.71

Alternative-Fuel Vehicles

The idea of self-propelled vehicles has been part of the human imagination for centuries. Leonardo da Vinci drew plans for a vehicle powered by a wound spring in 1478. This vehicle was never built although models have been constructed from his plans and appear in museums. Isaac Newton developed a vehicle in 1680 that operated by ejecting steam out the back, similar to a rocket engine. This invention did not develop into a useful device. Despite these and other attempts, self-propelled vehicles did not succeed; that is, they did not begin to replace the horse as a primary means of transportation until the 19th century.

The history of *successful* self-propelled vehicles begins in 1769 with the invention of a military tractor by Nicolas Joseph Cugnot in France. This vehicle, as well as Cugnot's follow-up vehicles, was powered by a steam engine. During the remainder of the 18th century and for most of the 19th century, additional steam-driven vehicles were developed in France, Great Britain, and the United States.

After the invention of the electric battery by Italian Alessandro Volta at the beginning of the 19th century and its further development over three decades came the invention of early electric vehicles in the 1830s. The development in 1859 of the storage battery, which could be recharged, provided significant impetus to the development of electric vehicles. By the early 20th century,

electric cars with a range of about 20 miles and a top speed of 15 miles per hour had been developed.

An internal combustion engine was designed but never built by Dutch physicist Christiaan Huygens in 1680. The invention of modern gasoline-powered internal combustion vehicles is generally credited to Gottlieb Daimler in 1885 and Karl Benz in 1886. Several earlier vehicles, dating back to 1807, however, used internal combustion engines operating on various fuels, including coal gas and primitive gasoline.

At the beginning of the 20th century, steam-powered, gasoline-powered, and electric cars shared the roadways in the United States. Electric cars did not possess the vibration, smell, and noise of gasoline-powered cars and did not suffer from the long start-up time intervals, up to 45 minutes, of steam-powered cars on cold mornings. Electric cars were especially preferred by women, who did not enjoy the difficult task of cranking a gasoline-powered car to start the engine. The limited range of electric cars was not a significant problem because the only roads that existed were in highly populated areas and cars were primarily used for short trips in town.

The end of electric cars in the early 20th century began with the following developments:

- 1901: A major discovery of crude oil in Texas reduced prices of gasoline to widely affordable levels.
- 1912: The electric starter for gasoline engines was invented, removing the physical task of cranking the engine.
- During the 1910s: Henry Ford successfully introduced mass production of internal combustion vehicles, resulting in a drop in the price of these vehicles to significantly less than that of an electric car.
- By the early 1920s: Roadways in the United States were of much better quality than in the previous decades, and now connected cities with each other, requiring vehicles with a longer range than when roadways existed only within city limits.

Because of these factors, the roadways were ruled by gasoline-powered cars almost exclusively by the 1920s. Gasoline, however, is a finite and short-lived commodity. We are approaching the end of our ability to use gasoline

Courtesy of The Exhibition Alliance, Inc., Hamilton, NY

Figure 1 A model of a spring-drive car designed by Leonardo da Vinci.

Figure 2 This magazine advertisement for an electric car is typical of this popular type of car in the early 20th century.

in transportation; some experts predict that diminishing supplies of crude oil will push the cost of gasoline to prohibitively high levels within two more decades. Furthermore, gasoline and diesel fuel result in serious

Figure 3 A bus running on natural gas operates in Port Huron, Michigan. Several cities such as Port Huron have established natural gas refueling centers so that a large percentage of their fleet can be operated with this fuel that is less expensive than diesel and emits fewer particulates into the atmosphere.

Figure 4 Modern electric cars can take advantage of an infrastructure set up in some localities to provide charging stations in parking lots.

tailpipe emissions that are harmful to the environment. As we look for a replacement for gasoline, we also want to pursue fuels that will be kinder to the atmosphere. Such fuels will help reduce the climate change effects of global warming, which we will study in Context 5.

What do the steam engine, the electric motor, and the internal combustion engine have in common? That is, what do they each extract from a source, be it a type of fuel or an electric battery? The answer to this question is *energy*. Regardless of the type of automobile, some source of energy must be provided. Energy is one of the physical concepts that we will investigate in this Context. A fuel such as gasoline contains energy due to its chemical composition and its ability to undergo a combustion process. The battery in an electric car also contains energy, again related to chemical composition, but in this case it is associated with an ability to produce an electric current.

One difficult social aspect of developing a new energy source for automobiles is that there must be a synchronized development of the new automobile along with the infrastructure for delivering the new source of energy. This aspect requires close cooperation between automotive corporations and energy manufacturers and suppliers. For example, electric cars cannot be used to travel long distances unless an infrastructure of charging stations develops in parallel with the development of electric cars.

As we draw near to the time when we run out of gasoline, our central question in this first Context is an important one for our future development:

What source besides gasoline can be used to provide energy for an automobile while reducing environmentally damaging emissions?

Motion in One Dimension

Chapter Outline

iStockphoto.com/technottechnotr

One of the physical quantities we will study in this chapter is the velocity of an object moving in a straight line. Downhill skiers can reach velocities with a magnitude greater than 100 km/h.

To begin our study of motion, it is important to be able to *describe* motion using the concepts of space and time without regard to the causes of the motion. This portion of mechanics is called *kinematics* (from the same root as the word *cinema*). In this chapter, we shall consider motion along a straight line, that is, one-dimensional motion. Chapter 3 extends our discussion to two-dimensional motion.

From everyday experience we recognize that motion represents continuous change in the position of an object. For example, if you are driving from your home to a destination, your position on the Earth's surface is changing.

The movement of an object through space (translation) may be accompanied by the rotation or vibration of the object. Such motions can be quite complex. It is often possible to simplify matters, however, by temporarily ignoring rotation and internal motions of the moving object. The result is the simplification model that we call the particle model, discussed in Chapter 1. In many situations, an object can be treated as a particle if the only motion being considered is translation through space. We will use the particle model extensively throughout this book.

⟨2.1⟩ Average Velocity

We begin our study of kinematics with the notion of average velocity. You may be familiar with a similar notion, average speed, from experiences with driving. If you drive your car 100 miles according to your odometer and it takes 2.0 hours to do so, your average speed is $(100 \text{ mi})/(2.0 \text{ h}) = 50 \text{ mi/h}$. For a particle moving through a distance d in a time interval Δt, the **average speed** v_{avg} is mathematically defined as

▶ Definition of average speed

$$v_{\text{avg}} \equiv \frac{d}{\Delta t}$$

2.1 ◀

Speed is not a vector, so there is no direction associated with average speed.

Average velocity may be a little less familiar to you due to its vector nature. Let us start by imagining the motion of a particle, which, through the particle model, can represent the motion of many types of objects. We shall restrict our study at this point to one-dimensional motion along the x axis.

The motion of a particle is completely specified if the position of the particle in space is known at all times. Consider a car moving back and forth along the x axis and imagine that we take data on the position of the car every 10 s. Active Figure 2.1a is a *pictorial* representation of this one-dimensional motion that shows the positions of the car at 10-s intervals. The six data points we have recorded are represented by the letters Ⓐ through Ⓕ. Table 2.1 is a *tabular* representation of the motion. It lists the data as entries for position at each time. The black dots in Active Figure 2.1b show a *graphical* representation of the motion. Such a plot is often called a **position–time graph.** The curved line in Active Figure 2.1b cannot be unambiguously drawn

Active Figure 2.1 A car moves back and forth along a straight line. Because we are interested only in the car's translational motion, we can model it as a particle. Several representations of the information about the motion of the car can be used. Table 2.1 is a tabular representation of the information. (a) A pictorial representation of the motion of the car. (b) A graphical representation, known as a position–time graph, of the car's motion in part (a). The average velocity $v_{x,\text{avg}}$ in the interval $t = 0$ to $t = 10$ s is obtained from the slope of the straight line connecting points Ⓐ and Ⓑ. (c) A velocity–time graph of the motion of the car in part (a).

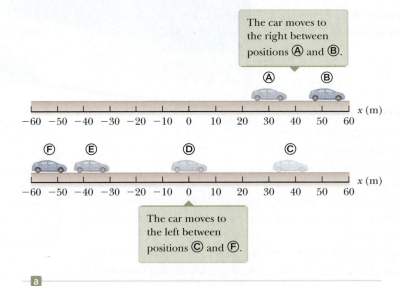

The car moves to the right between positions Ⓐ and Ⓑ.

The car moves to the left between positions Ⓒ and Ⓕ.

a

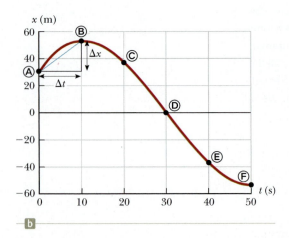

b

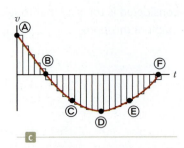

c

⟨TABLE 2.1⟩
Positions of the Car at Various Times

Position	t (s)	x (m)
Ⓐ	0	30
Ⓑ	10	52
Ⓒ	20	38
Ⓓ	30	0
Ⓔ	40	−37
Ⓕ	50	−53

through our six data points because we have no information about what happened between these points. The curved line is, however, a *possible* graphical representation of the position of the car at all instants of time during the 50 s.

If a particle is moving during a time interval $\Delta t = t_f - t_i$, the displacement of the particle is described as $\Delta \vec{\mathbf{x}} = \vec{\mathbf{x}}_f - \vec{\mathbf{x}}_i = (x_f - x_i)\hat{\mathbf{i}}$. (Recall from Chapter 1 that displacement is defined as the change in the position of the particle, which is equal to its final position value minus its initial position value.) Because we are considering only one-dimensional motion in this chapter, we shall drop the vector notation at this point and pick it up again in Chapter 3. The direction of a vector in this chapter will be indicated by means of a positive or negative sign.

The **average velocity** $v_{x,\text{avg}}$ of the particle is defined as the ratio of its displacement Δx to the time interval Δt during which the displacement takes place:

$$v_{x,\text{avg}} \equiv \frac{\Delta x}{\Delta t} = \frac{x_f - x_i}{t_f - t_i} \qquad \text{2.2} \blacktriangleleft$$

▶ Definition of average velocity

where the subscript x indicates motion along the x axis. From this definition we see that average velocity has the dimensions of length divided by time: meters per second in SI units and feet per second in U.S. customary units. The average velocity is *independent* of the path taken between the initial and final points. This independence is a major difference from the average speed discussed at the beginning of this section. The average velocity is independent of path because it is proportional to the displacement Δx, which depends only on the initial and final coordinates of the particle. Average speed (a scalar) is found by dividing the *distance* traveled by the time interval, whereas average velocity (a vector) is the *displacement* divided by the time interval. Therefore, average velocity gives us no details of the motion; rather, it only gives us the result of the motion. Finally, note that the average velocity in one dimension can be positive or negative, depending on the sign of the displacement. (The time interval Δt is always positive.) If the x coordinate of the particle increases during the time interval (i.e., if $x_f > x_i$), Δx is positive and $v_{x,\text{avg}}$ is positive, which corresponds to an average velocity in the positive x direction. On the other hand, if the coordinate decreases over time ($x_f < x_i$), Δx is negative; hence, $v_{x,\text{avg}}$ is negative, which corresponds to an average velocity in the negative x direction.

QUICK QUIZ 2.1 Under which of the following conditions is the magnitude of the average velocity of a particle moving in one dimension smaller than the average speed over some time interval? (**a**) A particle moves in the $+x$ direction without reversing. (**b**) A particle moves in the $-x$ direction without reversing. (**c**) A particle moves in the $+x$ direction and then reverses the direction of its motion. (**d**) There are no conditions for which it is true.

The average velocity can also be interpreted geometrically, as seen in the graphical representation in Active Figure 2.1b. A straight line can be drawn between any two points on the curve. Active Figure 2.1b shows such a line drawn between points Ⓐ and Ⓑ. Using a geometric model, this line forms the hypotenuse of a right triangle of height Δx and base Δt. The slope of the hypotenuse is the ratio $\Delta x/\Delta t$. Therefore, we see that the average velocity of the particle during the time interval t_i to t_f is equal to the slope of the straight line joining the initial and final points on the position–time graph. For example, the average velocity of the car between points Ⓐ and Ⓑ is $v_{x,\text{avg}} = (52 \text{ m} - 30 \text{ m})/(10 \text{ s} - 0) = 2.2 \text{ m/s}$.

We can also identify a geometric interpretation for the total displacement during the time interval. Active Figure 2.1c shows the velocity–time graphical representation of the motion in Active Figures 2.1a and 2.1b. The total time interval for the motion has been divided into small increments of duration Δt_n. During each of these increments, if we model the velocity as constant during the short increment, the displacement of the particle is given by $\Delta x_n = v_n \Delta t_n$.

Geometrically, the product on the right side of this expression represents the area of a thin rectangle associated with each time increment in Active Figure 2.1c;

Pitfall Prevention | 2.1
Average Speed and Average Velocity
The magnitude of the average velocity is *not* the average speed. Consider a particle moving from the origin to $x = 10$ m and then back to the origin in a time interval of 4.0 s. The magnitude of the average velocity is zero because the particle ends the time interval at the same position at which it started; the displacement is zero. The average speed, however, is the total distance divided by the time interval: 20 m/4.0 s = 5.0 m/s.

Pitfall Prevention | 2.2
Slopes of Graphs
In any graph of physical data, the slope represents the ratio of the change in the quantity represented on the vertical axis to the change in the quantity represented on the horizontal axis. Remember that a slope has units (unless both axes have the same units). The units of slope in Active Figures 2.1b and 2.2 (page 42) are meters per second, the units of velocity.

the height of the rectangle (measured from the time axis) is v_n, and the width is Δt_n. The total displacement of the particle will be the sum of the displacements during each of the increments:

$$\Delta x \approx \sum_n \Delta x_n = \sum_n v_n \Delta t_n$$

This sum is an approximation because we have modeled the velocity as constant in each increment, which is not the case. The term on the right represents the total area of all the thin rectangles. Now let us take the limit of this expression as the time increments shrink to zero, in which case the approximation becomes exact:

$$\Delta x = \lim_{\Delta t_n \to 0} \sum_n \Delta x_n = \lim_{\Delta t_n \to 0} \sum_n v_n \Delta t_n$$

In this limit, the sum of the areas of all the very thin rectangles becomes equal to the total area under the curve. Therefore, the displacement of a particle during the time interval t_i to t_f is equal to the area under the curve between the initial and final points on the velocity–time graph. We will make use of this geometric interpretation in Section 2.6.

Example 2.1 | Calculating the Average Velocity and Speed

Find the displacement, average velocity, and average speed of the car in Active Figure 2.1a between positions Ⓐ and Ⓕ.

SOLUTION

Conceptualize Consult the pictorial representation in Active Figure 2.1 to form a mental image of the car and its motion. Active Figure 2.1b shows a graphical representation of the motion in the form of a position–time graph for the particle.

Categorize We model the car as a particle. We will be substituting numerical values into definitions that we have seen, so this problem will be categorized as a substitution problem.

Analyze From the position–time graph given in Active Figure 2.1b, notice that $x_Ⓐ = 30$ m at $t_Ⓐ = 0$ s and that $x_Ⓕ = -53$ m at $t_Ⓕ = 50$ s.

Use Equation 1.6 to find the displacement of the car:

$$\Delta x = x_Ⓕ - x_Ⓐ = -53 \text{ m} - 30 \text{ m} = \boxed{-83 \text{ m}}$$

Use Equation 2.2 to find the car's average velocity:

$$v_{x,\text{avg}} = \frac{x_Ⓕ - x_Ⓐ}{t_Ⓕ - t_Ⓐ}$$

$$= \frac{-53 \text{ m} - 30 \text{ m}}{50 \text{ s} - 0 \text{ s}} = \frac{-83 \text{ m}}{50 \text{ s}} = \boxed{-1.7 \text{ m/s}}$$

We cannot unambiguously find the average speed of the car from the data in Table 2.1, because we do not have information about the positions of the car between the data points. If we adopt the assumption that the details of the car's position are described by the curve in Active Figure 2.1b, the distance traveled is 22 m (from Ⓐ to Ⓑ) plus 105 m (from Ⓑ to Ⓕ), for a total of 127 m.

Use Equation 2.1 to find the car's average speed:

$$v_{\text{avg}} = \frac{127 \text{ m}}{50 \text{ s}} = \boxed{2.5 \text{ m/s}}$$

Finalize The first result means that the car ends up 83 m in the negative direction (to the left, in this case) from where it started. This number has the correct units and is of the same order of magnitude as the supplied data. A quick look at Active Figure 2.1a indicates that it is the correct answer. The fact that the car ends up to the left of its initial position also makes it reasonable that the average velocity is negative.

Notice that the average speed is positive, as it must be. Suppose the red-brown curve in Active Figure 2.1b were different so that between 0 s and 10 s it went from Ⓐ up to 100 m and then came back down to Ⓑ. The average speed of the car would change because the distance is different, but the average velocity would not change.

Substitution problems generally do not have an extensive Analyze section other than the substitution of numbers into a given equation. Similarly, the Finalize step consists primarily of checking the units and making sure that the answer is reasonable. Therefore, for substitution problems after this one, we will not label Analyze or Finalize steps. We include those labels in this first example of a substitution problem just to demonstrate the process.

Example **2.2** | Motion of a Jogger

A jogger runs in a straight line, with an average velocity of magnitude 5.00 m/s for 4.00 min and then with an average velocity of magnitude 4.00 m/s for 3.00 min.

(A) What is the magnitude of the final displacement from her initial position?

SOLUTION

Conceptualize From your experience, imagine a runner running on a track. Notice that the runner runs more slowly on the average during the second time interval, as if he or she is tiring.

Categorize That this problem involves a jogger is not important; we model the jogger as a particle.

Analyze From the data for the two separate portions of the motion, find the displacement for each portion, using Equation 2.2:

$$v_{x,avg} = \frac{\Delta x}{\Delta t} \rightarrow \Delta x = v_{x,avg} \Delta t$$

$$\Delta x_{portion\ 1} = (5.00 \text{ m/s})(4.00 \text{ min})\left(\frac{60 \text{ s}}{1 \text{ min}}\right)$$

$$= 1.20 \times 10^3 \text{ m}$$

$$\Delta x_{portion\ 2} = (4.00 \text{ m/s})(3.00 \text{ min})\left(\frac{60 \text{ s}}{1 \text{ min}}\right)$$

$$= 7.20 \times 10^2 \text{ m}$$

We add these two displacements to find the total displacement of 1.92×10^3 m.

(B) What is the magnitude of her average velocity during this entire time interval of 7.00 min?

SOLUTION

Find the average velocity for the entire time interval using Equation 2.2:

$$v_{x,avg} = \frac{\Delta x}{\Delta t} = \frac{1.92 \times 10^3 \text{ m}}{7.00 \text{ min}}\left(\frac{1 \text{ min}}{60 \text{ s}}\right) = 4.57 \text{ m/s}$$

Finalize Notice that the average velocity is between the two velocities given in the problem, as expected, but is *not* the arithmetic mean of these two velocities.

2.2 | Instantaneous Velocity

Suppose you drive your car through a displacement of magnitude 40 miles and it takes exactly 1 hour to do so, from 1:00:00 P.M. to 2:00:00 P.M. Then the magnitude of your average velocity is 40 mi/h for the 1-h interval. How fast, though, were you going at the particular *instant* of time 1:20:00 P.M.? It is likely that your velocity varied during the trip, owing to hills, traffic lights, slow drivers ahead of you, and the like, so that there was not a single velocity maintained during the entire hour of travel. The velocity of a particle at any instant of time is called the *instantaneous velocity*.

Consider again the motion of the car shown in Active Figure 2.1a. Active Figure 2.2a (page 42) is the graphical representation again, with two blue lines representing average velocities over very different time intervals. One blue line represents the average velocity we calculated earlier over the interval from Ⓐ to Ⓑ. The second blue line represents the average velocity over the much longer interval Ⓐ to Ⓕ. How well does either of these represent the instantaneous velocity at point Ⓐ? In Active Figure 2.1a, the car begins to move to the right, which we identify as a positive velocity. The average velocity from Ⓐ to Ⓕ is *negative* (because the slope of the line from Ⓐ to Ⓕ

Active Figure 2.2 (a) Position–time graph for the motion of the car in Active Figure 2.1. (b) An enlargement of the upper left-hand corner of the graph.

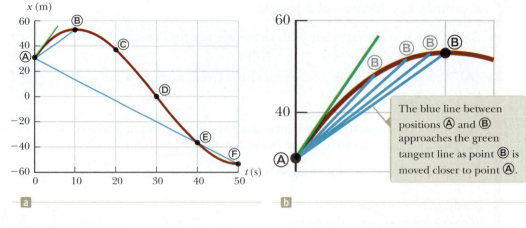

The blue line between positions Ⓐ and Ⓑ approaches the green tangent line as point Ⓑ is moved closer to point Ⓐ.

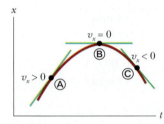

Figure 2.3 In the position–time graph shown, the velocity is positive at Ⓐ, where the slope of the tangent line is positive; the velocity is zero at Ⓑ, where the slope of the tangent line is zero; and the velocity is negative at Ⓒ, where the slope of the tangent line is negative.

▶ Definition of instantaneous velocity

is negative), so this velocity clearly is not an accurate representation of the instantaneous velocity at Ⓐ. The average velocity from interval Ⓐ to Ⓑ is *positive*, so this velocity at least has the right sign.

In Active Figure 2.2b, we show the result of drawing the lines representing the average velocity of the car as point Ⓑ is brought closer and closer to point Ⓐ. As that occurs, the slope of the blue line approaches that of the green line, which is the line drawn tangent to the curve at point Ⓐ. As Ⓑ approaches Ⓐ, the time interval that includes point Ⓐ becomes infinitesimally small. Therefore, the average velocity over this interval as the interval shrinks to zero can be interpreted as the instantaneous velocity at point Ⓐ. Furthermore, the slope of the line tangent to the curve at Ⓐ is the instantaneous velocity at the time $t_Ⓐ$. In other words, the **instantaneous velocity** v_x equals the limiting value of the ratio $\Delta x/\Delta t$ as Δt approaches zero:[1]

$$v_x \equiv \lim_{\Delta t \to 0} \frac{\Delta x}{\Delta t}$$

In calculus notation, this limit is called the *derivative* of x with respect to t, written dx/dt:

$$v_x \equiv \lim_{\Delta t \to 0} \frac{\Delta x}{\Delta t} = \frac{dx}{dt} \qquad \textbf{2.3} \blacktriangleleft$$

The instantaneous velocity can be positive, negative, or zero. When the slope of the position–time graph is positive, such as at point Ⓐ in Figure 2.3, v_x is positive. At point Ⓒ, v_x is negative because the slope is negative. Finally, the instantaneous velocity is zero at the peak Ⓑ (the turning point), where the slope is zero. From here on, we shall usually use the word *velocity* to designate instantaneous velocity.

The **instantaneous speed** of a particle is defined as the magnitude of the instantaneous velocity vector. Hence, by definition, *speed* can never be negative.

▶ **QUICK QUIZ 2.2** Are members of the highway patrol more interested in (a) your average speed or (b) your instantaneous speed as you drive?

If you are familiar with calculus, you should recognize that specific rules exist for taking the derivatives of functions. These rules, which are listed in Appendix B.6, enable us to evaluate derivatives quickly.

Suppose x is proportional to some power of t, such as

$$x = At^n$$

where A and n are constants. (This equation is a very common functional form.) The derivative of x with respect to t is

$$\frac{dx}{dt} = nAt^{n-1}$$

For example, if $x = 5t^3$, we see that $dx/dt = 3(5)t^{3-1} = 15t^2$.

---

[1] Note that the displacement Δx also approaches zero as Δt approaches zero. As Δx and Δt become smaller and smaller, however, the ratio $\Delta x/\Delta t$ approaches a value equal to the *true* slope of the line tangent to the x versus t curve.

> ## THINKING PHYSICS 2.1

Consider the following motions of an object in one dimension. (**a**) A ball is thrown directly upward, rises to its highest point, and falls back into the thrower's hand. (**b**) A race car starts from rest and speeds up to 100 m/s along a straight line. (**c**) A spacecraft on the way to another star drifts through empty space at constant velocity. Are there any instants of time in the motion of these objects at which the instantaneous velocity at the instant and the average velocity over the entire interval are the same? If so, identify the point(s).

Reasoning (**a**) The average velocity over the entire interval for the thrown ball is zero; the ball returns to the starting point at the end of the time interval. There is one point—at the top of the motion—at which the instantaneous velocity is zero. (**b**) The average velocity for the motion of the race car cannot be evaluated unambiguously with the information given, but its magnitude must be some value between 0 and 100 m/s. Because the magnitude of the instantaneous velocity of the car will have every value between 0 and 100 m/s at some time during the interval, there must be some instant at which the instantaneous velocity is equal to the average velocity over the entire interval. (**c**) Because the instantaneous velocity of the spacecraft is constant, its instantaneous velocity at *any* time and its average velocity over *any* time interval are the same. ◀

Example 2.3 | The Limiting Process

The position of a particle moving along the x axis varies in time according to the expression[2] $x = 3t^2$, where x is in meters and t is in seconds. Find the velocity in terms of t at any time.

SOLUTION

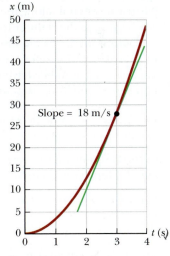

x (m)

Conceptualize The position–time graphical representation for this motion is shown in Figure 2.4. Before beginning the calculation, imagine the motion of the particle on the x axis. Does the particle ever reverse direction?

Categorize The entity in motion is already presented as a particle, so no simplification model is needed.

....

Analyze We can compute the velocity at any time t by using the definition of the instantaneous velocity.

If the initial coordinate of the particle at time t is $x_i = 3t^2$, find the coordinate at a later time $t + \Delta t$:

$$x_f = 3(t + \Delta t)^2 = 3[t^2 + 2t\,\Delta t + (\Delta t)^2]$$
$$= 3t^2 + 6t\,\Delta t + 3(\Delta t)^2$$

Find the displacement in the time interval Δt:

$$\Delta x = x_f - x_i = (3t^2 + 6t\,\Delta t + 3(\Delta t)^2) - (3t^2)$$
$$= 6t\,\Delta t + 3(\Delta t)^2$$

Find the average velocity in this time interval:

$$v_{x,\text{avg}} = \frac{\Delta x}{\Delta t} = \frac{6t\,\Delta t + 3(\Delta t)^2}{\Delta t} = 6t + 3\,\Delta t$$

To find the instantaneous velocity, take the limit of this expression as Δt approaches zero:

$$v_x = \lim_{\Delta t \to 0} \frac{\Delta x}{\Delta t} = 6t + 3(0) = \boxed{6t}$$

Figure 2.4 (Example 2.3) Position–time graph for a particle having an x coordinate that varies in time according to $x = 3t^2$. Note that the instantaneous velocity at $t = 3.0$ s is obtained from the slope of the green line tangent to the curve at this point.

continued

---

[2] Simply to make it easier to read, we write the equation as $x = 3t^2$ rather than as $x = (3.00 \text{ m/s}^2)t^{2.00}$. When an equation summarizes measurements, consider its coefficients to have as many significant digits as other data quoted in a problem. Also consider its coefficients to have the units required for dimensional consistency. When we start our clocks at $t = 0$, we usually do not mean to limit precision to a single digit. Consider any zero value in this book to have as many significant figures as you need.

2.3 *cont.*

Finalize Notice that this expression gives us the velocity at *any* general time *t*. It tells us that v_x is increasing linearly in time. It is then a straightforward matter to find the velocity at some specific time from the expression $v_x = 6t$ by substituting the value of the time. For example, at $t = 3.0$ s, the velocity is $v_x = 6(3) = 18$ m/s. Again, this answer can be checked from the slope at $t = 3.0$ s (the green line in Fig. 2.4).

We can also find v_x by taking the first derivative of *x* with respect to time, as in Equation 2.3. In this example, $x = 3t^2$, and we see that $v_x = dx/dt = 6t$, in agreement with our result from taking the limit explicitly.

Example 2.4 | Average and Instantaneous Velocity

A particle moves along the *x* axis. Its position varies with time according to the expression $x = -4t + 2t^2$, where *x* is in meters and *t* is in seconds. The position–time graph for this motion is shown in Figure 2.5a. Because the position of the particle is given by a mathematical function, the motion of the particle is completely known, unlike that of the car in Active Figure 2.1. Notice that the particle moves in the negative *x* direction for the first second of motion, is momentarily at rest at the moment $t = 1$ s, and moves in the positive *x* direction at times $t > 1$ s.

(A) Determine the displacement of the particle in the time intervals $t = 0$ to $t = 1$ s and $t = 1$ s to $t = 3$ s.

SOLUTION

Conceptualize From the graph in Figure 2.5a, form a mental representation of the particle's motion. Keep in mind that the particle does not move in a curved path in space such as that shown by the red-brown curve in the graphical representation. The particle moves only along the *x* axis in one dimension as shown in Figure 2.5b. At $t = 0$, is it moving to the right or to the left?

During the first time interval, the slope is negative and hence the average velocity is negative. Therefore, we know that the displacement between Ⓐ and Ⓑ must be a negative number having units of meters. Similarly, we expect the displacement between Ⓑ and Ⓓ to be positive.

Categorize We will be evaluating results from definitions given in the first two chapters, so we categorize this example as a substitution problem.

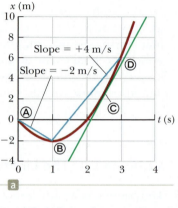

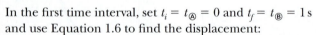

Figure 2.5 (Example 2.4) (a) Position–time graph for a particle having an *x* coordinate that varies in time according to the expression $x = -4t + 2t^2$. (b) The particle moves in one dimension along the *x* axis.

In the first time interval, set $t_i = t_Ⓐ = 0$ and $t_f = t_Ⓑ = 1$ s and use Equation 1.6 to find the displacement:

$$\Delta x_{Ⓐ \to Ⓑ} = x_f - x_i = x_Ⓑ - x_Ⓐ$$
$$= [-4(1) + 2(1)^2] - [-4(0) + 2(0)^2] = \boxed{-2 \text{ m}}$$

For the second time interval ($t = 1$ s to $t = 3$ s), set $t_i = t_Ⓑ = 1$ s and $t_f = t_Ⓓ = 3$ s:

$$\Delta x_{Ⓑ \to Ⓓ} = x_f - x_i = x_Ⓓ - x_Ⓑ$$
$$= [-4(3) + 2(3)^2] - [-4(1) + 2(1)^2] = \boxed{+8 \text{ m}}$$

These displacements can also be read directly from the position–time graph.

(B) Calculate the average velocity during these two time intervals.

SOLUTION

In the first time interval, use Equation 2.2 with $\Delta t = t_f - t_i = t_Ⓑ - t_Ⓐ = 1$ s:

$$v_{x,\text{avg}(Ⓐ \to Ⓑ)} = \frac{\Delta x_{Ⓐ \to Ⓑ}}{\Delta t} = \frac{-2 \text{ m}}{1 \text{ s}} = \boxed{-2 \text{ m/s}}$$

In the second time interval, $\Delta t = 2$ s:

$$v_{x,\text{avg}(Ⓑ \to Ⓓ)} = \frac{\Delta x_{Ⓑ \to Ⓓ}}{\Delta t} = \frac{8 \text{ m}}{2 \text{ s}} = \boxed{+4 \text{ m/s}}$$

These values are the same as the slopes of the blue lines joining these points in Figure 2.5a.

2.4 *cont.*

(C) Find the instantaneous velocity of the particle at $t = 2.5$ s.

SOLUTION

Measure the slope of the green line at $t = 2.5$ s (point ©) $v_x = \dfrac{10 \text{ m} - (-4 \text{ m})}{3.8 \text{ s} - 1.5 \text{ s}} = +6 \text{ m/s}$
in Figure 2.5a:

Notice that this instantaneous velocity is on the same order of magnitude as our previous results, that is, a few meters per second. Is that what you would have expected? Do you see any symmetry in the motion? For example, are there points at which the speed is the same? Is the velocity the same at these points?

2.3 | Analysis Model: Particle Under Constant Velocity

As mentioned in Section 1.10, the third category of models used in this book is that of *analysis models*. Such models help us analyze the situation in a physics problem and guide us toward the solution. **An analysis model is a problem we have solved before.** It is a description of either (1) the behavior of some physical entity or (2) the interaction between that entity and the environment. When you encounter a new problem, you should identify the fundamental details of the problem and attempt to recognize which, if any, of the types of problems you have already solved might be used as a model for the new problem.

This method is somewhat similar to the common practice in the legal profession of finding "legal precedents." If a previously resolved case can be found that is very similar legally to the present one, it is offered as a model and an argument is made in court to link them logically. The finding in the previous case can then be used to sway the finding in the present case. We will do something similar in physics. For a given problem, we search for a "physics precedent," a model with which we are already familiar and that can be applied to the present problem.

We shall generate analysis models based on four fundamental simplification models. The first simplification model is the particle model discussed in Chapter 1. We will look at a particle under various behaviors and environmental interactions. Further analysis models are introduced in later chapters based on simplification models of a *system*, a *rigid object*, and a *wave*. Once we have introduced these analysis models, we shall see that they appear over and over again later in the book in different situations.

When solving a problem, you should avoid browsing through the chapter looking for an equation that contains the unknown variable that is requested in the problem. In many cases, the equation you find may have nothing to do with the problem you are attempting to solve. It is *much* better to take this first step: **Identify the analysis model that is appropriate for the problem.** Think carefully about what is going on in the problem and match it to a situation you have seen before. What simplification model is appropriate for the entity involved in the problem? Is it a particle, a system, a rigid object, or a wave? Second, what is the entity doing or how is it interacting with its environment? For example, the analysis model in the title of this section indicates that we modeled the entity of interest as a particle. Furthermore, we determined that the particle is moving with constant velocity.

Once the analysis model is identified, there are a small number of equations from which to choose that are appropriate for that model. Therefore, **the model tells you which equation(s) to use for the mathematical representation.** In this section, we will learn what mathematical equations are associated with the particle under constant

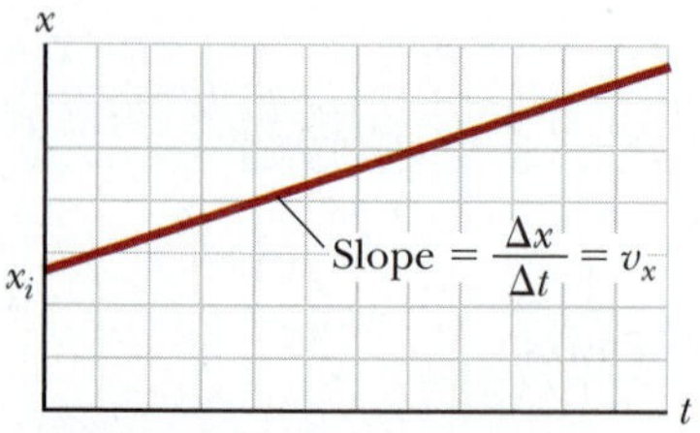

Figure 2.6 Position–time graph for a particle under constant velocity. The value of the constant velocity is the slope of the line.

velocity analysis model. In the future, when you identify the appropriate model in a problem as the particle under constant velocity, you will immediately know which equations to use to solve the problem.

Let us use Equation 2.2 to build our first analysis model. We imagine a particle moving with a constant velocity. The analysis model of a **particle under constant velocity** can be applied in *any* situation in which an entity that can be modeled as a particle is moving with constant velocity. This situation occurs frequently, so it is an important model.

If the velocity of a particle is constant, its instantaneous velocity at any instant during a time interval is the same as the average velocity over the interval, $v_x = v_{x,\text{avg}}$. Therefore, we start with Equation 2.2 to generate an equation to be used in the mathematical representation of this situation:

$$v_x = v_{x,\text{avg}} = \frac{\Delta x}{\Delta t} \qquad \text{2.4} \blacktriangleleft$$

Remembering that $\Delta x = x_f - x_i$, we see that $v_x = (x_f - x_i)/\Delta t$, or

$$x_f = x_i + v_x \Delta t$$

This equation tells us that the position of the particle is given by the sum of its original position x_i plus the displacement $v_x \Delta t$ that occurs during the time interval Δt. In practice, we usually choose the time at the beginning of the interval to be $t_i = 0$ and the time at the end of the interval to be $t_f = t$, so our equation becomes

▶ Position as a function of time for the particle under constant velocity model

$$\boxed{x_f = x_i + v_x t} \qquad \text{(for constant } v_x) \qquad \text{2.5} \blacktriangleleft$$

Equations 2.4 and 2.5 are the primary equations used in the model of a particle under constant velocity. They can be applied to particles or objects that can be modeled as particles. In the future, once you have identified a problem as requiring the particle under constant velocity model, either of these equations can be used to solve the problem.

Figure 2.6 is a graphical representation of the particle under constant velocity. On the position–time graph, the slope of the line representing the motion is constant and equal to the velocity. It is consistent with the mathematical representation, Equation 2.5, which is the equation of a straight line. The slope of the straight line is v_x and the y intercept is x_i in both representations.

Example 2.5 | Modeling a Runner as a Particle BIO

A kinesiologist is studying the biomechanics of the human body. (*Kinesiology* is the study of the movement of the human body. Notice the connection to the word *kinematics*.) She determines the velocity of an experimental subject while he runs along a straight line at a constant rate. The kinesiologist starts the stopwatch at the moment the runner passes a given point and stops it after the runner has passed another point 20 m away. The time interval indicated on the stopwatch is 4.0 s.

(A) What is the runner's velocity?

SOLUTION

Conceptualize You have probably watched track and field events at some point in your life, so it should be easy to conceptualize this situation.

Categorize We model the moving runner as a particle because the size of the runner and the movement of arms and legs are unnecessary details. Because the problem states that the subject runs at a constant rate, we can model him as a particle under constant velocity.

Analyze Having identified the model, we can use Equation 2.4 to find the constant velocity of the runner:

$$v_x = \frac{\Delta x}{\Delta t} = \frac{x_f - x_i}{\Delta t} = \frac{20 \text{ m} - 0}{4.0 \text{ s}} = \boxed{5.0 \text{ m/s}}$$

2.5 *cont.*

(B) If the runner continues his motion after the stopwatch is stopped, what is his position after 10 s has passed?

SOLUTION

Use Equation 2.5 and the velocity found in part (A) to find the position of the particle at time $t = 10$ s:

$$x_f = x_i + v_x t = 0 + (5.0 \text{ m/s})(10 \text{ s}) = \boxed{50 \text{ m}}$$

Finalize Is the result for part (A) a reasonable speed for a human? How does it compare to world-record speeds in 100-m and 200-m sprints? Notice that the value in part (B) is more than twice that of the 20-m position at which the stopwatch was stopped. Is this value consistent with the time of 10 s being more than twice the time of 4.0 s?

The mathematical manipulations for the particle under constant velocity stem from Equation 2.4 and its descendent, Equation 2.5. These equations can be used to solve for any variable in the equations that happens to be unknown if the other variables are known. For example, in part (B) of Example 2.5, we find the position when the velocity and the time are known. Similarly, if we know the velocity and the final position, we could use Equation 2.5 to find the time at which the runner is at this position. We shall present more examples of a particle under constant velocity in Chapter 3.

A particle under constant velocity moves with a constant speed along a straight line. Now consider a particle moving with a constant speed along a curved path. It can be represented with the **particle under constant speed model.** The primary equation for this model is Equation 2.1, with the average speed v_{avg} replaced by the constant speed v:

$$v \equiv \frac{d}{\Delta t} \qquad \qquad \textbf{2.6} \blacktriangleleft$$

As an example, imagine a particle moving at a constant speed in a circular path. If the speed is 5.00 m/s and the radius of the path is 10.0 m, we can calculate the time interval required to complete one trip around the circle:

$$v = \frac{d}{\Delta t} \rightarrow \Delta t = \frac{d}{v} = \frac{2\pi r}{v} = \frac{2\pi(10.0 \text{ m})}{5.00 \text{ m/s}} = 12.6 \text{ s}$$

2.4 | Acceleration

When the velocity of a particle changes with time, the particle is said to be *accelerating.* For example, the speed of a car increases when you "step on the gas," the car slows down when you apply the brakes, and it changes direction when you turn the wheel; these changes are all accelerations. We will need a precise definition of acceleration for our studies of motion.

Suppose a particle moving along the x axis has a velocity v_{xi} at time t_i and a velocity v_{xf} at time t_f. The **average acceleration** $a_{x,avg}$ of the particle in the time interval $\Delta t = t_f - t_i$ is defined as the ratio $\Delta v_x / \Delta t$, where $\Delta v_x = v_{xf} - v_{xi}$ is the *change* in velocity of the particle in this time interval:

$$a_{x,avg} \equiv \frac{v_{xf} - v_{xi}}{t_f - t_i} = \frac{\Delta v_x}{\Delta t} \qquad \qquad \textbf{2.7} \blacktriangleleft \qquad \blacktriangleright \text{ Definition of average acceleration}$$

Therefore, acceleration is a measure of how rapidly the velocity is changing. Acceleration is a vector quantity having dimensions of length divided by (time)^2, or L/T^2. Some of the common units of acceleration are meters per second per second (m/s^2) and feet per second per second (ft/s^2). For example, an acceleration of 2 m/s^2 means that the velocity changes by 2 m/s during each second of time that passes.

In some situations, the value of the average acceleration may be different for different time intervals. It is therefore useful to define the **instantaneous acceleration** as the limit of the average acceleration as Δt approaches zero, analogous to the definition of instantaneous velocity discussed in Section 2.2:

▶ Definition of instantaneous acceleration

$$a_x \equiv \lim_{\Delta t \to 0} \frac{\Delta v_x}{\Delta t} = \frac{dv_x}{dt} \qquad \textbf{2.8} \blacktriangleleft$$

That is, the instantaneous acceleration equals the derivative of the velocity with respect to time, which by definition is the slope of the velocity–time graph. Note that if a_x is positive, the acceleration is in the positive x direction, whereas negative a_x implies acceleration in the negative x direction. A negative acceleration does not necessarily mean that the particle is *moving* in the negative x direction, a point we shall address in more detail shortly. From now on, we use the term *acceleration* to mean instantaneous acceleration.

Because $v_x = dx/dt$, the acceleration can also be written

$$a_x = \frac{dv_x}{dt} = \frac{d}{dt}\left(\frac{dx}{dt}\right) = \frac{d^2x}{dt^2} \qquad \textbf{2.9} \blacktriangleleft$$

This equation shows that the acceleration equals the *second derivative* of the position with respect to time.

Figure 2.7 shows how the acceleration–time curve in a graphical representation can be derived from the velocity–time curve. In these diagrams, the acceleration of a particle at any time is simply the slope of the velocity–time graph at that time. Positive values of the acceleration correspond to those points (between 0 and $t_{\circledB}$) where the velocity in the positive x direction is increasing in magnitude (the particle is speeding up). The acceleration reaches a maximum at time $t_{\circledA}$, when the slope of the velocity–time graph is a maximum. The acceleration then goes to zero at time $t_{\circledB}$, when the velocity is a maximum (i.e., when the velocity is momentarily not changing and the slope of the v versus t graph is zero). Finally, the acceleration is negative when the velocity in the positive x direction is decreasing in magnitude (between $t_{\circledB}$ and $t_{\circledC}$).

> **Pitfall Prevention | 2.4**
> **Negative Acceleration**
> Keep in mind that *negative acceleration does not necessarily mean that an object is slowing down.* If the acceleration is negative and the velocity is negative, the object is speeding up!

> **Pitfall Prevention | 2.5**
> **Deceleration**
> The word *deceleration* has a common popular connotation as *slowing down.* When combined with the misconception in Pitfall Prevention 2.4 that negative acceleration means slowing down, the situation can be further confused by the use of the word *deceleration.* We will not use this word in this text.

QUICK QUIZ 2.3 Using Active Figure 2.8, match each v_x–t graph on the top with the a_x–t graph on the bottom that best describes the motion.

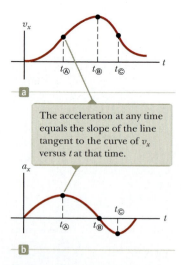

Figure 2.7 (a) The velocity–time graph for a particle moving along the x axis. (b) The instantaneous acceleration can be obtained from the velocity–time graph.

The acceleration at any time equals the slope of the line tangent to the curve of v_x versus t at that time.

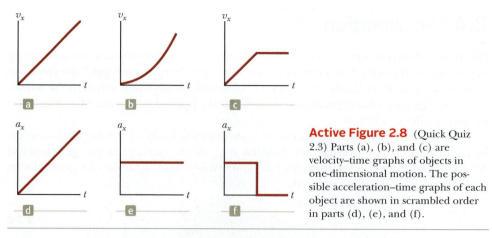

Active Figure 2.8 (Quick Quiz 2.3) Parts (a), (b), and (c) are velocity–time graphs of objects in one-dimensional motion. The possible acceleration–time graphs of each object are shown in scrambled order in parts (d), (e), and (f).

As an example of the computation of acceleration, consider the pictorial representation of a car's motion in Figure 2.9. In this case, the velocity of the car has changed from an initial value of 30 m/s to a final value of 15 m/s in a time interval of 2.0 s. The average acceleration during this time interval is

$$a_{x,\text{avg}} = \frac{15 \text{ m/s} - 30 \text{ m/s}}{2.0 \text{ s}} = -7.5 \text{ m/s}^2$$

The negative sign in this example indicates that the acceleration vector is in the negative x direction (to the left in Figure 2.9). For the case of motion in a straight line, the direction of the velocity of an object and the direction of its acceleration are related as follows. When the object's velocity and acceleration are in the same direction, the object is speeding up in that direction. On the other hand, when the object's velocity and acceleration are in opposite directions, the speed of the object decreases in time.

To help with this discussion of the signs of velocity and acceleration, let us take a peek ahead to Chapter 4, where we shall relate the acceleration of an object to the *force* on the object. We will save the details until that later discussion, but for now, let us borrow the notion that **the force on an object is proportional to the acceleration of the object:**

$$\vec{\mathbf{F}} \propto \vec{\mathbf{a}}$$

This proportionality indicates that acceleration is caused by force. What's more, as indicated by the vector notation in the proportionality, force and acceleration are in the same direction. Therefore, let us think about the signs of velocity and acceleration by forming a mental representation in which a force is applied to the object to cause the acceleration. Again consider the case in which the velocity and acceleration are in the same direction. This situation is equivalent to an object moving in a given direction and experiencing a force that pulls on it in the same direction. It is clear in this case that the object speeds up! If the velocity and acceleration are in opposite directions, the object moves one way and a force pulls in the opposite direction. In this case, the object slows down! It is very useful to equate the direction of the acceleration in these situations to the direction of a force because it is easier from our everyday experience to think about what effect a force will have on an object than to think only in terms of the direction of the acceleration.

> **QUICK QUIZ 2.4** If a car is traveling eastward and slowing down, what is the direction of the force on the car that causes it to slow down? **(a)** eastward **(b)** westward **(c)** neither of these directions

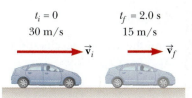

Figure 2.9 The velocity of the car decreases from 30 m/s to 15 m/s in a time interval of 2.0 s.

Example 2.6 | Average and Instantaneous Acceleration

The velocity of a particle moving along the x axis varies according to the expression $v_x = 40 - 5t^2$, where v_x is in meters per second and t is in seconds.

(A) Find the average acceleration in the time interval $t = 0$ to $t = 2.0$ s.

SOLUTION

Conceptualize Think about what the particle is doing from the mathematical representation. Is it moving at $t = 0$? In which direction? Does it speed up or slow down? Figure 2.10 is a v_x–t graph that was created from the velocity versus time expression given in the problem statement. Because the slope of the entire v_x–t curve is negative, we expect the acceleration to be negative.

Categorize While this problem does not involve an analysis model, it does involve taking a limit of a function, so it is a bit more sophisticated than a pure substitution problem.

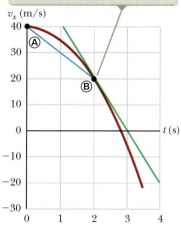

The acceleration at Ⓑ is equal to the slope of the green tangent line at $t = 2$ s, which is -20 m/s².

Figure 2.10 (Example 2.6) The velocity–time graph for a particle moving along the x axis according to the expression $v_x = 40 - 5t^2$.

....

Analyze Find the velocities at $t_i = t_{Ⓐ} = 0$ and $t_f = t_{Ⓑ} = 2.0$ s by substituting these values of t into the expression for the velocity:

$v_{xⒶ} = 40 - 5t_{Ⓐ}^2 = 40 - 5(0)^2 = +40$ m/s
$v_{xⒷ} = 40 - 5t_{Ⓑ}^2 = 40 - 5(2.0)^2 = +20$ m/s

continued

2.6 *cont.*

Find the average acceleration in the specified time interval $\Delta t = t_\circledB - t_\circledA = 2.0$ s:

$$a_{x,\text{avg}} = \frac{v_{xf} - v_{xi}}{t_f - t_i} = \frac{v_{x\circledB} - v_{x\circledA}}{t_\circledB - t_\circledA} = \frac{20 \text{ m/s} - 40 \text{ m/s}}{2.0 \text{ s} - 0 \text{ s}}$$

$$= \boxed{-10 \text{ m/s}^2}$$

Finalize The negative sign is consistent with our expectations: the average acceleration, represented by the slope of the blue line joining the initial and final points on the velocity–time graph, is negative.

(B) Determine the acceleration at $t = 2.0$ s.

SOLUTION

Analyze

Knowing that the initial velocity at any time t is $v_{xi} = 40 - 5t^2$, find the velocity at any later time $t + \Delta t$:

$$v_{xf} = 40 - 5(t + \Delta t)^2 = 40 - 5t^2 - 10t\,\Delta t - 5(\Delta t)^2$$

Find the change in velocity over the time interval Δt:

$$\Delta v_x = v_{xf} - v_{xi} = -10t\,\Delta t - 5(\Delta t)^2$$

To find the acceleration at any time t, divide this expression by Δt and take the limit of the result as Δt approaches zero:

$$a_x = \lim_{\Delta t \to 0} \frac{\Delta v_x}{\Delta t} = \lim_{\Delta t \to 0} (-10t - 5\Delta t) = -10t$$

Substitute $t = 2.0$ s:

$$a_x = (-10)(2.0) \text{ m/s}^2 = \boxed{-20 \text{ m/s}^2}$$

Finalize Because the velocity of the particle is positive and the acceleration is negative at this instant, the particle is slowing down.

   Notice that the answers to parts (A) and (B) are different. The average acceleration in part (A) is the slope of the blue line in Figure 2.10 connecting points Ⓐ and Ⓑ. The instantaneous acceleration in part (B) is the slope of the green line tangent to the curve at point Ⓑ. Notice also that the acceleration is *not* constant in this example. Situations involving constant acceleration are treated in Section 2.6.

⟨ **2.5** | Motion Diagrams

The concepts of velocity and acceleration are often confused with each other, but in fact they are quite different quantities. It is instructive to make use of the specialized pictorial representation called a **motion diagram** to describe the velocity and acceleration vectors while an object is in motion.

   A *stroboscopic photograph* of a moving object shows several images of the object taken as the strobe light flashes at a constant rate. Figure 2.1a is a motion diagram for the car studied in Section 2.1. Active Figure 2.11 represents three sets of strobe photographs of cars moving along a straight roadway in a single direction, from left to right. The time intervals between flashes of the stroboscope are equal in each part of the diagram. To distinguish between the two vector quantities, we use red arrows for velocity vectors and purple arrows for acceleration vectors in Active Figure 2.11. The vectors are sketched at several instants during the motion of the object. Let us describe the motion of the car in each diagram.

   In Active Figure 2.11a, the images of the car are equally spaced, and the car moves through the same displacement in each time interval. Therefore, the car moves with *constant positive velocity* and has *zero acceleration*. We could model the car as a particle and describe it using the particle under constant velocity analysis model.

   In Active Figure 2.11b, the images of the car become farther apart as time progresses. In this case, the velocity vector increases in time because the car's displacement between adjacent positions increases as time progresses. Therefore, the car is moving with a *positive velocity* and a *positive acceleration*. The velocity and acceleration are in the same direction. In terms of our earlier force discussion, imagine a force pulling on the car in the same direction it is moving: it speeds up.

   In Active Figure 2.11c, we interpret the car as slowing down as it moves to the right because its displacement between adjacent positions decreases as time progresses. In

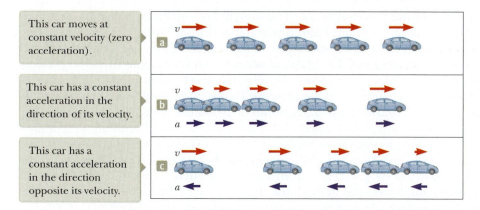

This car moves at constant velocity (zero acceleration).

a

This car has a constant acceleration in the direction of its velocity.

b

This car has a constant acceleration in the direction opposite its velocity.

c

Active Figure 2.11 Motion diagrams of a car moving along a straight roadway in a single direction. The velocity at each instant is indicated by a red arrow, and the constant acceleration is indicated by a purple arrow.

this case, the car moves initially to the right with a *positive velocity* and a *negative acceleration*. The velocity vector decreases in time and eventually reaches zero. (This type of motion is exhibited by a car that skids to a stop after its brakes are applied.) From this diagram we see that the acceleration and velocity vectors are *not* in the same direction. The velocity and acceleration are in opposite directions. In terms of our earlier force discussion, imagine a force pulling on the car opposite to the direction it is moving: it slows down.

The purple acceleration vectors in Active Figures 2.11b and 2.11c are all the same length. Therefore, these diagrams represent a motion with constant acceleration. This important type of motion is discussed in the next section.

QUICK QUIZ 2.5 Which of the following statements is true? **(a)** If a car is traveling eastward, its acceleration must be eastward. **(b)** If a car is slowing down, its acceleration must be negative. **(c)** A particle with constant acceleration can never stop and stay stopped.

2.6 | Analysis Model: Particle Under Constant Acceleration

If the acceleration of a particle varies in time, the motion may be complex and difficult to analyze. A very common and simple type of one-dimensional motion occurs when the acceleration is constant, such as for the motion of the cars in Active Figures 2.11b and 2.11c. In this case, the average acceleration over any time interval equals the instantaneous acceleration at any instant of time within the interval. Consequently, the velocity increases or decreases at the same rate throughout the motion. The **particle under constant acceleration** is a common analysis model that we can apply to appropriate problems. It is often used to model situations such as falling objects and braking cars.

If we replace $a_{x,\text{avg}}$ with the constant a_x in Equation 2.7, we find that

$$a_x = \frac{v_{xf} - v_{xi}}{t_f - t_i}$$

For convenience, let $t_i = 0$ and t_f be any arbitrary time t. With this notation, we can solve for v_{xf}:

$$v_{xf} = v_{xi} + a_x t \qquad \text{(for constant } a_x) \qquad \textbf{2.10} \blacktriangleleft$$

This expression enables us to predict the velocity at *any* time t if the initial velocity and constant acceleration are known. It is the first of four equations that can be used to solve problems using the particle under constant acceleration model. A graphical representation of position versus time for this motion is shown in Active Figure 2.12a. The velocity–time graph shown in Active Figure 2.12b is a straight line, the slope of

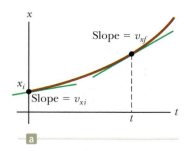

a

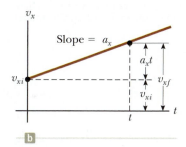

b

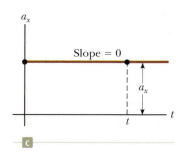

c

Active Figure 2.12 Graphical representations of a particle moving along the x axis with constant acceleration a_x. (a) The position–time graph, (b) the velocity–time graph, and (c) the acceleration–time graph.

▶ Velocity as a function of time for the particle under constant acceleration model

which is the constant acceleration a_x. The straight line on this graph is consistent with $a_x = dv_x/dt$ being a constant. From this graph and from Equation 2.10, we see that the velocity at any time t is the sum of the initial velocity v_{xi} and the change in velocity $a_x t$ due to the acceleration. The graph of acceleration versus time (Active Fig. 2.12c) is a straight line with a slope of zero because the acceleration is constant. If the acceleration were negative, the slope of Active Figure 2.12b would be negative and the horizontal line in Active Figure 2.12c would be below the time axis.

We can generate another equation for the particle under constant acceleration model by recalling a result from Section 2.1 that the displacement of a particle is the area under the curve on a velocity–time graph. Because the velocity varies linearly with time (see Active Fig. 2.12b), the area under the curve is the sum of a rectangular area (under the horizontal dashed line in Active Fig. 2.12b) and a triangular area (from the horizontal dashed line upward to the curve). Therefore,

$$\Delta x = v_{xi}\Delta t + \tfrac{1}{2}(v_{xf} - v_{xi})\Delta t$$

which can be simplified as follows:

$$\Delta x = (v_{xi} + \tfrac{1}{2}v_{xf} - \tfrac{1}{2}v_{xi})\Delta t = \tfrac{1}{2}(v_{xi} + v_{xf})\Delta t$$

In general, from Equation 2.2, the displacement for a time interval is

$$\Delta x = v_{x,\text{avg}}\Delta t$$

Comparing these last two equations, we find that the average velocity in any time interval is the arithmetic mean of the initial velocity v_{xi} and the final velocity v_{xf}:

$$v_{x,\text{avg}} = \tfrac{1}{2}(v_{xi} + v_{xf}) \qquad \text{(for constant } a_x) \qquad \textbf{2.11}\blacktriangleleft$$

◀ Average velocity for the particle under constant acceleration model

Remember that this expression is valid only when the acceleration is constant, that is, when the velocity varies linearly with time.

We now use Equations 2.2 and 2.11 to obtain the position as a function of time. Again we choose $t_i = 0$, at which time the initial position is x_i, which gives

$$\Delta x = v_{x,\text{avg}}\Delta t = \tfrac{1}{2}(v_{xi} + v_{xf})t$$

$$\boxed{x_f = x_i + \tfrac{1}{2}(v_{xi} + v_{xf})t} \qquad \text{(for constant } a_x) \qquad \textbf{2.12}\blacktriangleleft$$

◀ Position as a function of velocity and time for the particle under constant acceleration model

We can obtain another useful expression for the position by substituting Equation 2.10 for v_{xf} in Equation 2.12:

$$x_f = x_i + \tfrac{1}{2}[v_{xi} + (v_{xi} + a_x t)]t$$

$$\boxed{x_f = x_i + v_{xi}t + \tfrac{1}{2}a_x t^2} \qquad \text{(for constant } a_x) \qquad \textbf{2.13}\blacktriangleleft$$

◀ Position as a function of time for the particle under constant acceleration model

Note that the position at any time t is the sum of the initial position x_i, the displacement $v_{xi}t$ that would result if the velocity remained constant at the initial velocity, and the displacement $\tfrac{1}{2}a_x t^2$ because the particle is accelerating. Consider again the position–time graph for motion under constant acceleration shown in Active Figure 2.12a. The curve representing Equation 2.13 is a parabola, as shown by the t^2 dependence in the equation. The slope of the tangent to this curve at $t = 0$ equals the initial velocity v_{xi}, and the slope of the tangent line at any time t equals the velocity at that time.

Finally, we can obtain an expression that does not contain the time by substituting the value of t from Equation 2.10 into Equation 2.12, which gives

$$x_f = x_i + \tfrac{1}{2}(v_{xi} + v_{xf})\left(\frac{v_{xf} - v_{xi}}{a_x}\right) = x_i + \frac{v_{xf}^2 - v_{xi}^2}{2a_x}$$

$$\boxed{v_{xf}^2 = v_{xi}^2 + 2a_x(x_f - x_i)} \qquad \text{(for constant } a_x) \qquad \textbf{2.14}\blacktriangleleft$$

◀ Velocity as a function of position for the particle under constant acceleration model

TABLE 2.2 | Kinematic Equations for Motion of a Particle Under Constant Acceleration

Equation Number	Equation	Information Given by Equation
2.10	$v_{xf} = v_{xi} + a_x t$	Velocity as a function of time
2.12	$x_f = x_i + \frac{1}{2}(v_{xf} + v_{xi})t$	Position as a function of velocity and time
2.13	$x_f = x_i + v_{xi}t + \frac{1}{2}a_x t^2$	Position as a function of time
2.14	$v_{xf}^2 = v_{xi}^2 + 2a_x(x_f - x_i)$	Velocity as a function of position

Note: Motion is along the *x* axis. At $t = 0$, the position of the particle is x_i and its velocity is v_{xi}.

This expression is *not* an independent equation because it arises from combining Equations 2.10 and 2.12. It is useful, however, for those problems in which a value for the time is not involved.

If motion occurs in which the constant value of the acceleration is *zero*, Equations 2.10 and 2.13 become

$$\left. \begin{array}{l} v_{xf} = v_{xi} \\ x_f = x_i + v_{xi}t \end{array} \right\} \quad \text{when } a_x = 0$$

That is, when the acceleration is zero, the velocity remains constant and the position changes linearly with time. In this case, the particle under constant *acceleration* model reduces to the particle under constant *velocity* model.

Equations 2.10, 2.12, 2.13, and 2.14 are four **kinematic equations** that may be used to solve any problem in one-dimensional motion of a particle (or an object that can be modeled as a particle) under constant acceleration. If your analysis of a problem indicates that the particle under constant acceleration is the appropriate analysis model, select from these four equations to solve the problem. Keep in mind that these relationships were derived from the definitions of velocity and acceleration together with some simple algebraic manipulations and the requirement that the acceleration be constant. It is often convenient to choose the initial position of the particle as the origin of the motion so that $x_i = 0$ at $t = 0$. We will see cases, however, in which we must choose the value of x_i to be something other than zero.

The four kinematic equations for the particle under constant acceleration are listed in Table 2.2 for convenience. The choice of which kinematic equation or equations you should use in a given situation depends on what is known beforehand. Sometimes it is necessary to use two of these equations to solve for two unknowns, such as the position and velocity at some instant. You should recognize that the quantities that vary during the motion are velocity v_{xf}, position x_f, and time t. The other quantities—x_i, v_{xi}, and a_x—are *parameters* of the motion and remain constant.

> **PROBLEM-SOLVING STRATEGY:** Particle Under Constant Acceleration
>
> The following procedure is recommended for solving problems that involve an object undergoing a constant acceleration. As mentioned in Chapter 1, individual strategies such as this one will follow the outline of the General Problem-Solving Strategy from Chapter 1, with specific hints regarding the application of the general strategy to the material in the individual chapters.
>
> 1. **Conceptualize** Think about what is going on physically in the problem. Establish the mental representation.
>
> 2. **Categorize** Simplify the problem as much as possible. Confirm that the problem involves either a particle or an object that can be modeled as a particle and that it is moving with a constant acceleration. Construct an appropriate

pictorial representation, such as a motion diagram, or a graphical representation. Make sure all the units in the problem are consistent. That is, if positions are measured in meters, be sure that velocities have units of m/s and accelerations have units of m/s². Choose a coordinate system to be used throughout the problem.

3. **Analyze** Set up the mathematical representation. Choose an instant to call the "initial" time $t = 0$ and another to call the "final" time t. Let your choice be guided by what you know about the particle and what you want to know about it. The initial instant need not be when the particle starts to move, and the final instant will only rarely be when the particle stops moving. Identify all the quantities given in the problem and a separate list of those to be determined. A tabular representation of these quantities may be helpful to you. Select from the list of kinematic equations the one or ones that will enable you to determine the unknowns. Solve the equations.

4. **Finalize** Once you have determined your result, check to see if your answers are consistent with the mental and pictorial representations and that your results are realistic.

Example **2.7** | **Carrier Landing**

A jet lands on an aircraft carrier at a speed of 140 mi/h ($\approx$ 63 m/s).

(A) What is its acceleration (assumed constant) if it stops in 2.0 s due to an arresting cable that snags the jet and brings it to a stop?

SOLUTION

Conceptualize You might have seen movies or television shows in which a jet lands on an aircraft carrier and is brought to rest surprisingly fast by an arresting cable. A careful reading of the problem reveals that in addition to being given the initial speed of 63 m/s, we also know that the final speed is zero. We define our x axis as the direction of motion of the jet. Notice that we have no information about the change in position of the jet while it is slowing down.

Categorize Because the acceleration of the jet is assumed constant, we model it as a particle under constant acceleration.

Analyze
Equation 2.10 is the only equation in Table 2.2 that does not involve position, so we use it to find the acceleration of the jet, modeled as a particle:

$$a_x = \frac{v_{xf} - v_{xi}}{t} \approx \frac{0 - 63 \text{ m/s}}{2.0 \text{ s}}$$

$$= -32 \text{ m/s}^2$$

(B) If the jet touches down at position $x_i = 0$, what is its final position?

SOLUTION

Use Equation 2.12 to solve for the final position:

$$x_f = x_i + \tfrac{1}{2}(v_{xi} + v_{xf})t = 0 + \tfrac{1}{2}(63 \text{ m/s} + 0)(2.0 \text{ s}) = 63 \text{ m}$$

Finalize Given the size of aircraft carriers, a length of 63 m seems reasonable for stopping the jet. The idea of using arresting cables to slow down landing aircraft and enable them to land safely on ships originated at about the time of World War I. The cables are still a vital part of the operation of modern aircraft carriers.

What If? Suppose the jet lands on the deck of the aircraft carrier with a speed faster than 63 m/s but has the same acceleration due to the cable as that calculated in part (A). How will that change the answer to part (B)?

Answer If the jet is traveling faster at the beginning, it will stop farther away from its starting point, so the answer to part (B) should be larger. Mathematically, we see in Equation 2.12 that if v_{xi} is larger, then x_f will be larger.

Example 2.8 | Watch Out for the Speed Limit!

A car traveling at a constant speed of 45.0 m/s passes a trooper on a motorcycle hidden behind a billboard. One second after the speeding car passes the billboard, the trooper sets out from the billboard to catch the car, accelerating at a constant rate of 3.00 m/s². How long does it take her to overtake the car?

SOLUTION

Conceptualize A pictorial representation (Fig. 2.13) helps clarify the sequence of events.

Categorize The car is modeled as a particle under constant velocity, and the trooper is modeled as a particle under constant acceleration.

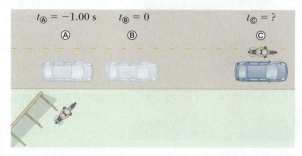

Figure 2.13 (Example 2.8) A speeding car passes a hidden trooper.

Analyze First, we write expressions for the position of each vehicle as a function of time. It is convenient to choose the position of the billboard as the origin and to set $t_{®} = 0$ as the time the trooper begins moving. At that instant, the car has already traveled a distance of 45.0 m from the billboard because it has traveled at a constant speed of $v_x = 45.0$ m/s for 1 s. Therefore, the initial position of the speeding car is $x_{®} = 45.0$ m.

Using the particle under constant velocity model, apply Equation 2.5 to give the car's position at any time t:

$$x_{car} = x_{®} + v_{x\,car}t$$

A quick check shows that at $t = 0$, this expression gives the car's correct initial position when the trooper begins to move: $x_{car} = x_{®} = 45.0$ m.

The trooper starts from rest at $t_{®} = 0$ and accelerates at $a_x = 3.00$ m/s² away from the origin. Use Equation 2.13 to give her position at any time t:

$$x_f = x_i + v_{xi}t + \tfrac{1}{2}a_x t^2$$
$$x_{trooper} = 0 + (0)t + \tfrac{1}{2}a_x t^2 = \tfrac{1}{2}a_x t^2$$

Set the positions of the car and trooper equal to represent the trooper overtaking the car at position ©:

$$x_{trooper} = x_{car}$$
$$\tfrac{1}{2}a_x t^2 = x_{®} + v_{x\,car}t$$

Rearrange to give a quadratic equation:

$$\tfrac{1}{2}a_x t^2 - v_{x\,car}t - x_{®} = 0$$

Solve the quadratic equation for the time at which the trooper catches the car (for help in solving quadratic equations, see Appendix B.2.):

$$t = \frac{v_{x\,car} \pm \sqrt{v_{x\,car}^2 + 2a_x x_{®}}}{a_x}$$

$$(1)\ t = \frac{v_{x\,car}}{a_x} \pm \sqrt{\frac{v_{x\,car}^2}{a_x^2} + \frac{2x_{®}}{a_x}}$$

Evaluate the solution, choosing the positive root because that is the only choice consistent with a time $t > 0$:

$$t = \frac{45.0\ \text{m/s}}{3.00\ \text{m/s}^2} + \sqrt{\frac{(45.0\ \text{m/s})^2}{(3.00\ \text{m/s}^2)^2} + \frac{2(45.0\ \text{m})}{3.00\ \text{m/s}^2}} = \boxed{31.0\ \text{s}}$$

Finalize Why didn't we choose $t = 0$ as the time at which the car passes the trooper? If we did so, we would not be able to use the particle under constant acceleration model for the trooper. Her acceleration would be zero for the first second and then 3.00 m/s² for the remaining time. By defining the time $t = 0$ as when the trooper begins moving, we can use the particle under constant acceleration model for her movement for all positive times.

What If? What if the trooper had a more powerful motorcycle with a larger acceleration? How would that change the time at which the trooper catches the car?

Answer If the motorcycle has a larger acceleration, the trooper should catch up to the car sooner, so the answer for the time should be less than 31 s. Because all terms on the right side of Equation (1) have the acceleration a_x in the denominator, we see symbolically that increasing the acceleration will decrease the time at which the trooper catches the car.

⟨2.7 | Freely Falling Objects

Galileo Galilei
Italian physicist and astronomer
(1564–1642)
Galileo formulated the laws that govern the motion of objects in free fall and made many other significant discoveries in physics and astronomy. Galileo publicly defended Nicolaus Copernicus's assertion that the Sun is at the center of the Universe (the heliocentric system). He published *Dialogue Concerning Two New World Systems* to support the Copernican model, a view that the Catholic Church declared to be heretical.

Georgios Kollidas/Shutterstock.com

Pitfall Prevention | 2.6
***g* and g**
Be sure not to confuse the italic symbol *g* for free-fall acceleration with the nonitalic symbol g used as the abbreviation for the unit gram.

It is well known that all objects, when dropped, fall toward the Earth with nearly constant acceleration. Legend has it that Galileo Galilei first discovered this fact by observing that two different weights dropped simultaneously from the Leaning Tower of Pisa hit the ground at approximately the same time. (Air resistance plays a role in the falling of an object, but for now we shall model falling objects as if they are falling through a vacuum; this is a simplification model.) Although there is some doubt that this particular experiment was actually carried out, it is well established that Galileo did perform many systematic experiments on objects moving on inclined planes. Through careful measurements of distances and time intervals, he was able to show that the displacement from an origin of an object starting from rest is proportional to the square of the time interval during which the object is in motion. This observation is consistent with one of the kinematic equations we derived for a particle under constant acceleration (Eq. 2.13, with $v_{xi} = 0$). Galileo's achievements in mechanics paved the way for Newton in his development of the laws of motion.

If a coin and a crumpled-up piece of paper are dropped simultaneously from the same height, there will be a small time difference between their arrivals at the floor. If this same experiment could be conducted in a good vacuum, however, where air friction is truly negligible, the paper and coin would fall with the same acceleration, regardless of the shape or weight of the paper, even if the paper were still flat. In the idealized case, where air resistance is ignored, such motion is referred to as *free-fall*. This point is illustrated very convincingly in Figure 2.14, which is a photograph of an apple and a feather falling in a vacuum. On August 2, 1971, such an experiment was conducted on the Moon by astronaut David Scott. He simultaneously released a geologist's hammer and a falcon's feather, and in unison they fell to the lunar surface. This demonstration surely would have pleased Galileo!

We shall denote the magnitude of the free-fall acceleration with the symbol *g*, representing a vector acceleration $\vec{g}$. At the surface of the Earth, *g* is approximately 9.80 m/s², or 980 cm/s², or 32 ft/s². Unless stated otherwise, we shall use the value 9.80 m/s² when doing calculations. Furthermore, we shall assume that the vector $\vec{g}$ is directed downward toward the center of the Earth.

When we use the expression *freely falling object*, we do not necessarily mean an object dropped from rest. A freely falling object is an object moving freely under the influence of gravity alone, regardless of its initial motion. Therefore, objects thrown upward or downward and those released from rest are all freely falling objects once they are

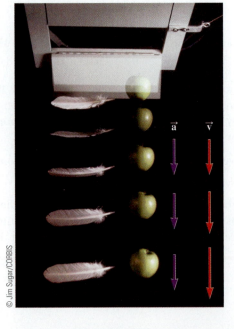

Figure 2.14 An apple and a feather, released from rest in a vacuum chamber, fall at the same rate, regardless of their masses. Ignoring air resistance, all objects fall to the Earth with the same acceleration of magnitude 9.80 m/s², as indicated by the purple arrows in this multiflash photograph. The velocity of the two objects increases linearly with time, as indicated by the series of red arrows.

© Jim Sugar/CORBIS

released! Because the value of g is constant as long as we are close to the surface of the Earth, we can model a freely falling object as a particle under constant acceleration.

In previous examples in this chapter, the particles were undergoing constant acceleration, as stated in the problem. Therefore, it may have been difficult to understand the need for modeling. We can now begin to see the need for modeling; we are *modeling* a real falling object with an analysis model. Notice that we are (1) ignoring air resistance and (2) assuming that the free-fall acceleration is constant. Therefore, the model of a particle under constant acceleration is a *replacement* for the real problem, which could be more complicated. If air resistance and any variation in g are small, however, the model should make predictions that agree closely with the real situation.

The equations developed in Section 2.6 for the particle under constant acceleration model can be applied to the falling object. The only necessary modification that we need to make in these equations for freely falling objects is to note that the motion is in the vertical direction, so we will use y instead of x, and that the acceleration is downward and of magnitude 9.80 m/s². Therefore, for a freely falling object we commonly take $a_y = -g = -9.80$ m/s², where the negative sign indicates that the acceleration of the object is downward. The choice of negative for the downward direction is arbitrary, but common.

QUICK QUIZ 2.6 A ball is thrown upward. While the ball is in free-fall, does its acceleration (a) increase, (b) decrease, (c) increase and then decrease, (d) decrease and then increase, or (e) remain constant?

> **Pitfall Prevention | 2.7**
> **Acceleration at the Top of the Motion**
> A common misconception is that the acceleration of a projectile at the top of its trajectory is zero. Although the velocity at the top of the motion of an object thrown upward momentarily goes to zero, *the acceleration is still that due to gravity* at this point. If the velocity and acceleration were both zero, the projectile would stay at the top.

> **Pitfall Prevention | 2.8**
> **The sign of g**
> Keep in mind that g is a *positive number*. It is tempting to substitute −9.80 m/s² for g, but resist the temptation. That the gravitational acceleration is downward is indicated explicitly by stating the acceleration as $a_y = -g$.

> **THINKING PHYSICS 2.2**
>
> A skydiver steps out of a stationary helicopter. A few seconds later, another skydiver steps out, so that both skydivers fall along the same vertical line. Ignore air resistance, so that both skydivers fall with the same acceleration, and model the skydivers as particles under constant acceleration. Does the vertical separation distance between them stay the same? Does the difference in their speeds stay the same?
>
> **Reasoning** At any given instant of time, the speeds of the skydivers are definitely different, because one had a head start over the other. In any time interval, however, each skydiver increases his or her speed by the same amount, because they have the same acceleration. Therefore, the difference in speeds remains the same. The first skydiver will always be moving with a higher speed than the second. In a given time interval, then, the first skydiver will have a larger displacement than the second. Therefore, the separation distance between them increases. ◄

Example 2.9 | Not a Bad Throw for a Rookie!

A stone thrown from the top of a building is given an initial velocity of 20.0 m/s straight upward. The stone is launched 50.0 m above the ground, and the stone just misses the edge of the roof on its way down as shown in Figure 2.15 (page 58).

(A) Using $t_Ⓐ = 0$ as the time the stone leaves the thrower's hand at position Ⓐ, determine the time at which the stone reaches its maximum height.

SOLUTION

Conceptualize You most likely have experience with dropping objects or throwing them upward and watching them fall, so this problem should describe a familiar experience. To simulate this situation, toss a small object upward and notice the time interval required for it to fall to the floor. Now imagine throwing that object upward from the roof of a building.

Recognize that the initial velocity is positive because the stone is launched upward. The velocity will change sign after the stone reaches its highest point, but the acceleration of the stone will *always* be downward.

Categorize Because the stone is in free fall, it is modeled as a particle under constant acceleration due to gravity.

continued

2.9 *cont.*

Analyze Choose an initial point just after the stone leaves the person's hand and a final point at the top of its flight.

Use Equation 2.10 to calculate the time at which the stone reaches its maximum height:

$$v_{yf} = v_{yi} + a_y t \quad \rightarrow \quad t = \frac{v_{yf} - v_{yi}}{a_y}$$

Substitute numerical values:

$$t = t_B = \frac{0 - 20.0 \text{ m/s}}{-9.80 \text{ m/s}^2} = \boxed{2.04 \text{ s}}$$

(B) Find the maximum height of the stone.

SOLUTION

As in part (A), choose the initial and final points at the beginning and the end of the upward flight.

Set $y_A = 0$ and substitute the time from part (A) into Equation 2.13 to find the maximum height:

$$y_{max} = y_B = y_A + v_{xA} t + \tfrac{1}{2} a_y t^2$$
$$y_B = 0 + (20.0 \text{ m/s})(2.04 \text{ s}) +$$
$$\tfrac{1}{2}(-9.80 \text{ m/s}^2)(2.04 \text{ s})^2 = \boxed{20.4 \text{ m}}$$

(C) Determine the velocity of the stone when it returns to the height from which it was thrown.

SOLUTION

Choose the initial point where the stone is launched and the final point when it passes this position coming down.

Substitute known values into Equation 2.14:

$$v_{yC}^2 = v_{yA}^2 + 2a_y(y_C - y_A)$$
$$v_{yC}^2 = (20.0 \text{ m/s})^2 + 2(-9.80 \text{ m/s}^2)(0 - 0) = 400 \text{ m}^2/\text{s}^2$$
$$v_{yC} = \boxed{-20.0 \text{ m/s}}$$

When taking the square root, we could choose either a positive or a negative root. We choose the negative root because we know that the stone is moving downward at point Ⓒ. The velocity of the stone when it arrives back at its original height is equal in magnitude to its initial velocity but is opposite in direction.

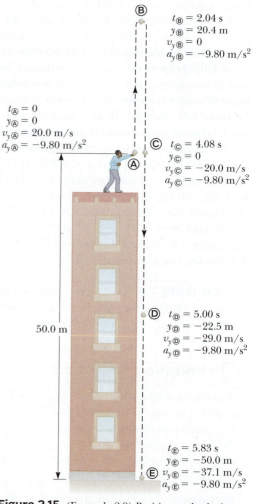

Figure 2.15 (Example 2.9) Position and velocity versus time for a freely falling stone thrown initially upward with a velocity $v_{yi} = 20.0$ m/s. Many of the quantities in the labels for points in the motion of the stone are calculated in the example. Can you verify the other values that are not?

(D) Find the velocity and position of the stone at $t = 5.00$ s.

SOLUTION

Choose the initial point just after the throw and the final point 5.00 s later.

Calculate the velocity at Ⓓ from Equation 2.10:

$$v_{yD} = v_{yA} + a_y t = 20.0 \text{ m/s} + (-9.80 \text{ m/s}^2)(5.00 \text{ s}) = \boxed{-29.0 \text{ m/s}}$$

Use Equation 2.13 to find the position of the stone at $t_D = 5.00$ s:

$$y_D = y_A + v_{yA} t + \tfrac{1}{2} a_y t^2$$
$$= 0 + (20.0 \text{ m/s})(5.00 \text{ s}) + \tfrac{1}{2}(-9.80 \text{ m/s}^2)(5.00 \text{ s})^2$$
$$= \boxed{-22.5 \text{ m}}$$

Finalize The choice of the time defined as $t = 0$ is arbitrary and up to you to select as the problem solver. As an example of this arbitrariness, choose $t = 0$ as the time at which the stone is at the highest point in its motion. Then solve parts (C) and (D) again using this new initial instant and notice that your answers are the same as those above.

What If? What if the throw were from 30.0 m above the ground instead of 50.0 m? Which answers in parts (A) to (D) would change?

Answer None of the answers would change. All the motion takes place in the air during the first 5.00 s. (Notice that even for a throw from 30.0 m, the stone is above the ground at $t = 5.00$ s.) Therefore, the height of the throw is not an issue. Mathematically, if we look back over our calculations, we see that we never entered the height of the throw into any equation.

◄2.8 | Context Connection: Acceleration Required by Consumers

We now have our first opportunity to address a Context in a closing section, as we will do in each remaining chapter. Our present Context is *Alternative-Fuel Vehicles*, and our central question is, *What source besides gasoline can be used to provide energy for an automobile while reducing environmentally damaging emissions?*

Consumers have been driving gasoline-powered vehicles for decades and have become used to a certain amount of acceleration. In addition, roadway features such as the lengths of freeway on-ramps have been designed with the expectation of a minimum acceleration required for a vehicle to merge with existing traffic. These experiences raise the question as to what kind of acceleration today's consumer would expect for an alternative-fuel vehicle that might replace a gasoline-powered vehicle. In turn, developers of alternative-fuel vehicles should strive for such an acceleration so as to satisfy consumer expectations and hope to generate a demand for the new vehicle.

If we consider published time intervals for accelerations from 0 to 60 mi/h for a number of automobile models, we find the data shown in the third column of Table 2.3 (page 60). The average acceleration of each vehicle is calculated from these data using Equation 2.7. It is clear from the upper part of this table (*Very expensive vehicles*) that acceleration upward of 20 mi/h · s is very expensive. The highest acceleration is 23.1 mi/h · s for the Bugatti Veyron 16.4 Super Sport and costs over two million dollars. A slightly lower acceleration can be had with the Shelby SuperCars Ultimate Aero for a bargain price of $654,000. The *Performance vehicles* between $44,000 and $102,000 show an average acceleration of 14.1 mi/h · s compared to 19.1 mi/h · s for the *Very expensive vehicles*. For the less affluent driver, the accelerations in the third section of the table (*Traditional vehicles*) have an average value of 6.9 mi/h · s. This number is typical of consumer-oriented gasoline-powered vehicles and provides an approximate standard for the acceleration desired in an alternative-fuel vehicle.

In the lower part of Table 2.3, we see data for five alternative vehicles. The average acceleration of these five cars is 6.2 mi/h · s, about 90% of the average value for the traditional vehicles. This acceleration is sufficiently large that it satisfies consumer demand for a car with "get-up-and-go." Figure 2.16 shows a graph of the cost of the vehicles in Table 2.3 versus acceleration. The graph clearly shows the skyrocketing cost of accelerations larger than 20 mi/h · s.

The Honda CR-Z, Honda Insight, and Toyota Prius are *hybrid vehicles*, which we will discuss further in the Context Conclusion. These vehicles combine a gasoline engine and an electric motor, both directly driving the wheels. The accelerations for these vehicles are among the lowest in the table. The disadvantage of the low acceleration is

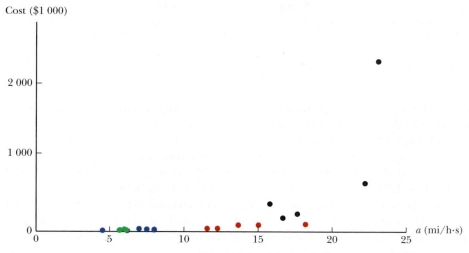

Figure 2.16 The cost to obtain a certain acceleration for alternative vehicles (in green), traditional vehicles (in blue), performance vehicles (in red), and very expensive vehicles (in black).

◤ **TABLE 2.3** | Accelerations of Various Vehicles, 0–60 mph

Automobile	Model Year	Time Interval, 0 to 60 mi/h (s)	Average Acceleration (mi/h · s)	Price
Very expensive vehicles:				
Bugatti Veyron 16.4 Super Sport	2011	2.60	23.1	$2,300,000
Lamborghini LP 570-4 Superleggera	2011	3.40	17.6	$240,000
Lexus LFA	2011	3.80	15.8	$375,000
Mercedes-Benz SLS AMG	2011	3.60	16.7	$186,000
Shelby SuperCars Ultimate Aero	2009	2.70	22.2	$654,000
Average		**3.22**	**19.1**	**$751,000**
Performance vehicles:				
Chevrolet Corvette ZR1	2010	3.30	18.2	$102,000
Dodge Viper SRT10	2010	4.00	15.0	$91,000
Jaguar XJL Supercharged	2011	4.40	13.6	$90,500
Acura TL SH-AWD	2009	5.20	11.5	$44,000
Dodge Challenger SRT8	2010	4.90	12.2	$45,000
Average		**4.36**	**14.1**	**$74,500**
Traditional vehicles:				
Buick Regal CXL Turbo	2011	7.50	8.0	$30,000
Chevrolet Tahoe 1500 LS (SUV)	2011	8.60	7.0	$40,000
Ford Fiesta SES	2010	9.70	6.2	$14,000
Hummer H3 (SUV)	2010	8.00	7.5	$34,000
Hyundai Sonata SE	2010	7.50	8.0	$25,000
Smart ForTwo	2010	13.30	4.5	$16,000
Average		**9.10**	**6.9**	**$26,500**
Alternative vehicles:				
Chevrolet Volt (hybrid)	2011	8.00	7.5	$41,000
Nissan Leaf (electric)	2011	10.00	6.0	$34,000
Honda CR-Z (hybrid)	2011	10.50	5.7	$25,000
Honda Insight (hybrid)	2010	10.60	5.7	$21,000
Toyota Prius (hybrid)	2010	9.80	6.1	$24,000
Average		**9.78**	**6.2**	**$29,000**

Note: Data given in this table as well as in similar tables in Chapters 3 through 6 were gathered from online sources such as road test reports and automobile manufacturer websites. Other data, such as the accelerations in this table, were calculated from the raw data.

offset by other factors. These vehicles obtain relatively high gas mileage, have very low emissions, and do not require recharging as does a pure electric vehicle.

The Chevrolet Volt and Nissan Leaf are vehicles driven only by electric motors. The Leaf is a pure electric vehicle: it has only batteries as an energy source. Once the batteries are depleted, the vehicle is inoperable, giving it an operating range of 73 miles (U.S. EPA) between charges. The Volt is a series hybrid (see the Context Conclusion for a discussion of types of hybrid vehicles): it has a gasoline engine, but the engine does not directly drive the wheels at normal speeds. The engine acts as a generator, charging the battery and allowing the vehicle to travel about 35 miles on electricity alone but over 350 miles between charges.

In comparison to the vehicles in Table 2.3, consider the acceleration of an even higher-level "performance vehicle," a typical drag racer, as shown in Figure 2.17.

Figure 2.17 In drag racing, acceleration is a highly desired quantity. In a distance of 1/4 mile, speeds of over 320 mi/h are reached, with the entire distance being covered in under 5 s.

Typical data show that such a vehicle covers a distance of 0.25 mi in 5.0 s, starting from rest. We can find the acceleration from Equation 2.13:

$$x_f = x_i + v_i t + \tfrac{1}{2}a_x t^2 = 0 + 0(t) + \tfrac{1}{2}(a_x)(t)^2 \rightarrow a_x = \frac{2x_f}{t^2}$$

$$a_x = \frac{2(0.25\text{ mi})}{(5.0\text{ s})^2} = 0.020\text{ mi/s}^2\left(\frac{3\,600\text{ s}}{1\text{ h}}\right) = 72\text{ mi/h}\cdot\text{s}$$

This value is much larger than any accelerations in the table, as would be expected. We can show that the acceleration due to gravity has the following value in units of mi/h · s:

$$g = 9.80\text{ m/s}^2 = 21.9\text{ mi/h}\cdot\text{s}$$

Therefore, the drag racer is moving horizontally with 3.3 times as much acceleration as it would move vertically if you pushed it off a cliff! (Of course, the horizontal acceleration can only be maintained for a very short time interval.)

As we investigate two-dimensional motion in the next chapter, we shall consider a different type of acceleration for vehicles, that associated with the vehicle turning in a sharp circle at high speed.

SUMMARY

The **average speed** of a particle during some time interval is equal to the ratio of the distance d traveled by the particle and the time interval Δt:

$$v_{avg} \equiv \frac{d}{\Delta t} \qquad \text{2.1}$$

The **average velocity** of a particle moving in one dimension during some time interval is equal to the ratio of the displacement Δx and the time interval Δt:

$$v_{x,avg} \equiv \frac{\Delta x}{\Delta t} \qquad \text{2.2}$$

The **instantaneous velocity** of a particle is defined as the limit of the ratio $\Delta x/\Delta t$ as Δt approaches zero:

$$v_x \equiv \lim_{\Delta t\to 0}\frac{\Delta x}{\Delta t} = \frac{dx}{dt} \qquad \text{2.3}$$

The **instantaneous speed** of a particle is defined as the magnitude of the instantaneous velocity vector.

The **average acceleration** of a particle moving in one dimension during some time interval is defined as the ratio of the change in its velocity Δv_x and the time interval Δt:

$$a_{x,avg} \equiv \frac{\Delta v_x}{\Delta t} \qquad \text{2.7}$$

The **instantaneous acceleration** is equal to the limit of the ratio $\Delta v_x/\Delta t$ as $\Delta t \to 0$. By definition, this limit equals the derivative of v_x with respect to t, or the time rate of change of the velocity:

$$a_x \equiv \lim_{\Delta t\to 0}\frac{\Delta v_x}{\Delta t} = \frac{dv_x}{dt} \qquad \text{2.8}$$

The slope of the tangent to the x versus t curve at any instant gives the instantaneous velocity of the particle.

The slope of the tangent to the v versus t curve gives the instantaneous acceleration of the particle.

An object falling freely experiences an acceleration directed toward the center of the Earth. If air friction is ignored and if the altitude of the motion is small compared with the Earth's radius, one can assume that the magnitude of the free-fall acceleration g is constant over the range of motion, where g is equal to 9.80 m/s², or 32 ft/s². Assuming y to be positive upward, the acceleration is given by $-g$, and the equations of kinematics for an object in free-fall are the same as those already given, with the substitutions $x\to y$ and $a_y \to -g$.

Analysis Models for Problem-Solving

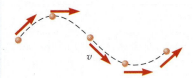

Particle Under Constant Velocity. If a particle moves in a straight line with a constant speed v_x, its constant velocity is given by

$$v_x = \frac{\Delta x}{\Delta t} \qquad \text{2.4} \blacktriangleleft$$

and its position is given by

$$x_f = x_i + v_x t \qquad \text{2.5} \blacktriangleleft$$

Particle Under Constant Speed. If a particle moves a distance d along a curved or straight path with a constant speed, its constant speed is given by

$$v = \frac{d}{\Delta t} \qquad \text{2.6} \blacktriangleleft$$

Particle Under Constant Acceleration. If a particle moves in a straight line with a constant acceleration a_x, its motion is described by the kinematic equations:

$$v_{xf} = v_{xi} + a_x t \qquad \text{2.10} \blacktriangleleft$$

$$v_{x,\text{avg}} = \frac{v_{xi} + v_{xf}}{2} \qquad \text{2.11} \blacktriangleleft$$

$$x_f = x_i + \tfrac{1}{2}(v_{xi} + v_{xf})t \qquad \text{2.12} \blacktriangleleft$$

$$x_f = x_i + v_{xi}t + \tfrac{1}{2}a_x t^2 \qquad \text{2.13} \blacktriangleleft$$

$$v_{xf}^2 = v_{xi}^2 + 2a_x(x_f - x_i) \qquad \text{2.14} \blacktriangleleft$$

OBJECTIVE QUESTIONS

1. One drop of oil falls straight down onto the road from the engine of a moving car every 5 s. Figure OQ2.1 shows the pattern of the drops left behind on the pavement. What is the average speed of the car over this section of its motion? (a) 20 m/s (b) 24 m/s (c) 30 m/s (d) 100 m/s (e) 120 m/s

← 600 m →

Figure OQ2.1

2. When applying the equations of kinematics for an object moving in one dimension, which of the following statements *must* be true? (a) The velocity of the object must remain constant. (b) The acceleration of the object must remain constant. (c) The velocity of the object must increase with time. (d) The position of the object must increase with time. (e) The velocity of the object must always be in the same direction as its acceleration.

3. A juggler throws a bowling pin straight up in the air. After the pin leaves his hand and while it is in the air, which statement is true? (a) The velocity of the pin is always in the same direction as its acceleration. (b) The velocity of the pin is never in the same direction as its acceleration. (c) The acceleration of the pin is zero. (d) The velocity of the pin is opposite its acceleration on the way up. (e) The velocity of the pin is in the same direction as its acceleration on the way up.

4. An arrow is shot straight up in the air at an initial speed of 15.0 m/s. After how much time is the arrow moving downward at a speed of 8.00 m/s? (a) 0.714 s (b) 1.24 s (c) 1.87 s (d) 2.35 s (e) 3.22 s

5. When the pilot reverses the propeller in a boat moving north, the boat moves with an acceleration directed south. Assume the acceleration of the boat remains constant in magnitude and direction. What happens to the boat? (a) It eventually stops and remains stopped. (b) It eventually stops and then speeds up in the forward direction. (c) It eventually stops and then speeds up in the reverse direction. (d) It never stops but loses speed more and more slowly forever. (e) It never stops but continues to speed up in the forward direction.

6. A pebble is dropped from rest from the top of a tall cliff and falls 4.9 m after 1.0 s has elapsed. How much farther does it drop in the next 2.0 s? (a) 9.8 m (b) 19.6 m (c) 39 m (d) 44 m (e) none of the above

7. A student at the top of a building of height h throws one ball upward with a speed of v_i and then throws a second ball downward with the same initial speed v_i. Just before it reaches the ground, is the final speed of the ball thrown upward (a) larger, (b) smaller, or (c) the same in magnitude, compared with the final speed of the ball thrown downward?

8. A rock is thrown downward from the top of a 40.0-m-tall tower with an initial speed of 12 m/s. Assuming negligible air resistance, what is the speed of the rock just before hitting the ground? (a) 28 m/s (b) 30 m/s (c) 56 m/s (d) 784 m/s (e) More information is needed.

9. As an object moves along the x axis, many measurements are made of its position, enough to generate a smooth, accurate graph of x versus t. Which of the following quantities for the object *cannot* be obtained from this graph *alone*? (a) the velocity at any instant (b) the acceleration at any instant (c) the displacement during some time interval (d) the average velocity during some time interval (e) the speed at any instant

10. You drop a ball from a window located on an upper floor of a building. It strikes the ground with speed v. You now repeat the drop, but your friend down on the ground throws another ball upward at the same speed v, releasing her ball at the same moment that you drop yours from the window. At some location, the balls pass each other. Is this location (a) *at* the halfway point between window and ground, (b) *above* this point, or (c) *below* this point?

11. A skateboarder starts from rest and moves down a hill with constant acceleration in a straight line, traveling for 6 s. In a second trial, he starts from rest and moves along the same straight line with the same acceleration for only 2 s. How does his displacement from his starting point in this second trial compare with that from the first trial? (a) one-third as large (b) three times larger (c) one-ninth as large (d) nine times larger (e) $1/\sqrt{3}$ times as large

12. A ball is thrown straight up in the air. For which situation are both the instantaneous velocity and the acceleration zero? (a) on the way up (b) at the top of its flight path (c) on the way down (d) halfway up and halfway down (e) none of the above

13. A hard rubber ball, not affected by air resistance in its motion, is tossed upward from shoulder height, falls to the sidewalk, rebounds to a smaller maximum height, and is caught on its way down again. This motion is represented in Figure OQ2.13, where the successive positions of the ball Ⓐ through Ⓔ are not equally spaced in time. At point Ⓓ the center of the ball is at its lowest point in the motion. The motion of the ball is along a straight, vertical line, but the diagram shows successive positions offset to the right to avoid overlapping. Choose the positive y direction to be upward. (a) Rank the situations Ⓐ through Ⓔ according to the speed of the ball $|v_y|$ at each point, with the largest speed first. (b) Rank the same situations according to the acceleration a_y of the ball at each point. (In both rankings, remember that zero is greater than a negative value. If two values are equal, show that they are equal in your ranking.)

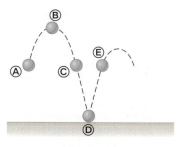

Figure OQ2.13

14. Each of the strobe photographs (a), (b), and (c) in Figure OQ2.14 was taken of a single disk moving toward the right, which we take as the positive direction. Within each photograph, the time interval between images is constant. (i) Which photograph shows motion with zero acceleration? (ii) Which photograph shows motion with positive acceleration? (iii) Which photograph shows motion with negative acceleration?

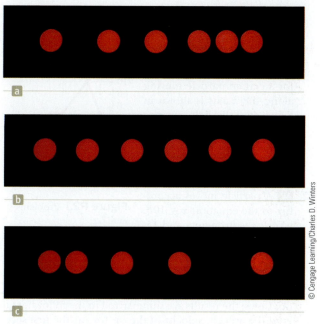

Figure OQ2.14

denotes answer available in *Student Solutions Manual/Study Guide*

CONCEPTUAL QUESTIONS

1. If a car is traveling eastward, can its acceleration be westward? Explain.

2. Try the following experiment away from traffic where you can do it safely. With the car you are driving moving slowly on a straight, level road, shift the transmission into neutral and let the car coast. At the moment the car comes to a complete stop, step hard on the brake and notice what you feel. Now repeat the same experiment on a fairly gentle, uphill slope. Explain the difference in what a person riding in the car feels in the two cases. (Brian Popp suggested the idea for this question.)

3. (a) Can the equations of kinematics (Eqs. 2.10–2.14) be used in a situation in which the acceleration varies in time? (b) Can they be used when the acceleration is zero?

4. (a) Can the velocity of an object at an instant of time be greater in magnitude than the average velocity over a time interval containing the instant? (b) Can it be less?

5. If the average velocity of an object is zero in some time interval, what can you say about the displacement of the object for that interval?

6. If the velocity of a particle is zero, can the particle's acceleration be zero? Explain.

7. If the velocity of a particle is nonzero, can the particle's acceleration be zero? Explain.

8. You throw a ball vertically upward so that it leaves the ground with velocity +5.00 m/s. (a) What is its velocity when it reaches its maximum altitude? (b) What is its acceleration at this point? (c) What is the velocity with which it returns to ground level? (d) What is its acceleration at this point?

9. Two cars are moving in the same direction in parallel lanes along a highway. At some instant, the velocity of car A exceeds the velocity of car B. Does that mean that the acceleration of car A is greater than that of car B? Explain.

PROBLEMS

WebAssign The problems found in this chapter may be assigned online in Enhanced WebAssign.

  1. denotes straightforward problem; **2.** denotes intermediate problem; **3.** denotes challenging problem

  1. denotes full solution available in the *Student Solutions Manual/ Study Guide*

  1. denotes problems most often assigned in Enhanced WebAssign.

  BIO denotes biomedical problem

  GP denotes guided problem

  M denotes Master It tutorial available in Enhanced WebAssign

  Q|C denotes asking for quantitative and conceptual reasoning

  S denotes symbolic reasoning problem

  shaded denotes "paired problems" that develop reasoning with symbols and numerical values

  W denotes Watch It video solution available in Enhanced WebAssign

Section 2.1 Average Velocity

1. **W** The position versus time for a certain particle moving along the *x* axis is shown in Figure P2.1. Find the average velocity in the time intervals (a) 0 to 2 s, (b) 0 to 4 s, (c) 2 s to 4 s, (d) 4 s to 7 s, and (e) 0 to 8 s.

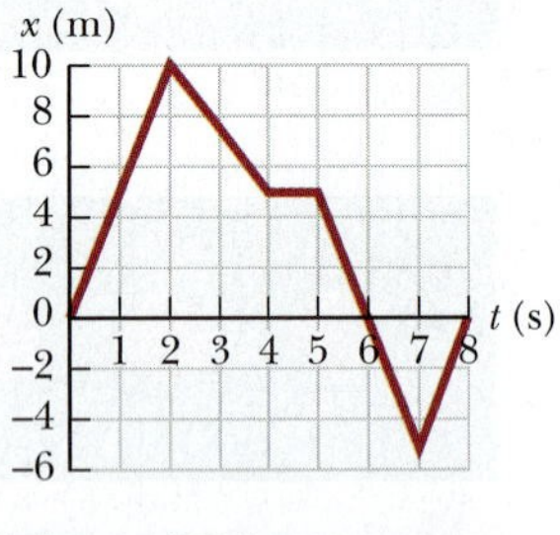

2. **W** A particle moves according to the equation $x = 10t^2$, where *x* is in meters and *t* is in seconds. (a) Find the average velocity for the time interval from 2.00 s to 3.00 s. (b) Find the average velocity for the time interval from 2.00 to 2.10 s.

x (m)

Figure P2.1 Problems 1 and 7.

3. The position of a pinewood derby car was observed at various times; the results are summarized in the following table. Find the average velocity of the car for (a) the first second, (b) the last 3 s, and (c) the entire period of observation.

t (s)	0	1.0	2.0	3.0	4.0	5.0
x (m)	0	2.3	9.2	20.7	36.8	57.5

4. **M** A person walks first at a constant speed of 5.00 m/s along a straight line from point Ⓐ to point Ⓑ and then back along the line from Ⓑ to Ⓐ at a constant speed of 3.00 m/s. (a) What is her average speed over the entire trip? (b) What is her average velocity over the entire trip?

Section 2.2 Instantaneous Velocity

5. A position–time graph for a particle moving along the *x* axis is shown in Figure P2.5. (a) Find the average velocity in the time interval $t = 1.50$ s to $t = 4.00$ s. (b) Determine the instantaneous velocity at $t = 2.00$ s by measuring the slope of the tangent line shown in the graph. (c) At what value of *t* is the velocity zero?

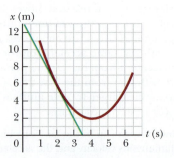

Figure P2.5

6. The position of a particle moving along the *x* axis varies in time according to the expression $x = 3t^2$, where *x* is in meters and *t* is in seconds. Evaluate its position (a) at $t = 3.00$ s and (b) at 3.00 s $+ \Delta t$. (c) Evaluate the limit of $\Delta x/\Delta t$ as Δt approaches zero to find the velocity at $t = 3.00$ s.

7. **W** Find the instantaneous velocity of the particle described in Figure P2.1 at the following times: (a) $t = 1.0$ s, (b) $t = 3.0$ s, (c) $t = 4.5$ s, and (d) $t = 7.5$ s.

8. (a) Use the data in Problem 2.3 to construct a smooth graph of position versus time. (b) By constructing tangents to the $x(t)$ curve, find the instantaneous velocity of the car at several instants. (c) Plot the instantaneous velocity versus time and, from this information, determine the average acceleration of the car. (d) What was the initial velocity of the car?

Section 2.3 Analysis Model: Particle Under Constant Velocity

9. A hare and a tortoise compete in a race over a straight course 1.00 km long. The tortoise crawls at a speed of 0.200 m/s toward the finish line. The hare runs at a speed of 8.00 m/s toward the finish line for 0.800 km and then stops to tease the slow-moving tortoise as the tortoise eventually passes by. The hare waits for a while after the tortoise passes and then runs toward the finish line again at 8.00 m/s. Both the hare and the tortoise cross the finish line at the exact same instant. Assume both animals, when moving, move steadily at their respective speeds. (a) How far is the tortoise from the finish line when the hare resumes the race? (b) For how long in time was the hare stationary?

Section 2.4 Acceleration

10. An object moves along the x axis according to the equation $x = 3.00t^2 - 2.00t + 3.00$, where x is in meters and t is in seconds. Determine (a) the average speed between $t = 2.00$ s and $t = 3.00$ s, (b) the instantaneous speed at $t = 2.00$ s and at $t = 3.00$ s, (c) the average acceleration between $t = 2.00$ s and $t = 3.00$ s, and (d) the instantaneous acceleration at $t = 2.00$ s and $t = 3.00$ s. (e) At what time is the object at rest?

11. **M** A particle moves along the x axis according to the equation $x = 2.00 + 3.00t - 1.00t^2$, where x is in meters and t is in seconds. At $t = 3.00$ s, find (a) the position of the particle, (b) its velocity, and (c) its acceleration.

12. A student drives a moped along a straight road as described by the velocity-versus-time graph in Figure P2.12. Sketch this graph in the middle of a sheet of graph paper. (a) Directly above your graph, sketch a graph of the position versus time, aligning the time coordinates of the two graphs. (b) Sketch a graph of the acceleration versus time directly below the velocity-versus-time graph, again aligning the time coordinates. On each graph, show the numerical values of x and a_x for all points of inflection. (c) What is the acceleration at $t = 6.00$ s? (d) Find the position (relative to the starting point) at $t = 6.00$ s. (e) What is the moped's final position at $t = 9.00$ s?

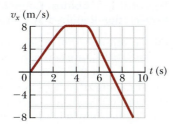

Figure P2.12

13. **W** A particle starts from rest and accelerates as shown in Figure P2.13. Determine (a) the particle's speed at $t = 10.0$ s and at $t = 20.0$ s, and (b) the distance traveled in the first 20.0 s.

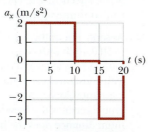

Figure P2.13

14. **W** A 50.0-g Super Ball traveling at 25.0 m/s bounces off a brick wall and rebounds at 22.0 m/s. A high-speed camera records this event. If the ball is in contact with the wall for 3.50 ms, what is the magnitude of the average acceleration of the ball during this time interval?

15. Figure P2.15 shows a graph of v_x versus t for the motion of a motorcyclist as he starts from rest and moves along the road in a straight line. (a) Find the average acceleration for the time interval $t = 0$ to $t = 6.00$ s. (b) Estimate the time at which the acceleration has its greatest positive value and the value of the acceleration at that instant. (c) When is the acceleration zero? (d) Estimate the maximum negative value of the acceleration and the time at which it occurs.

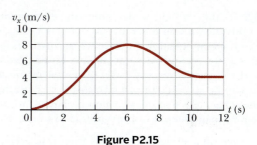

Figure P2.15

Section 2.5 Motion Diagrams

16. **Q|C** Draw motion diagrams for (a) an object moving to the right at constant speed, (b) an object moving to the right and speeding up at a constant rate, (c) an object moving to the right and slowing down at a constant rate, (d) an object moving to the left and speeding up at a constant rate, and (e) an object moving to the left and slowing down at a constant rate. (f) How would your drawings change if the changes in speed were not uniform, that is, if the speed were not changing at a constant rate?

Section 2.6 Analysis Model: Particle Under Constant Acceleration

17. **M** An object moving with uniform acceleration has a velocity of 12.0 cm/s in the positive x direction when its x coordinate is 3.00 cm. If its x coordinate 2.00 s later is −5.00 cm, what is its acceleration?

18. The minimum distance required to stop a car moving at 35.0 mi/h is 40.0 ft. What is the minimum stopping distance for the same car moving at 70.0 mi/h, assuming the same rate of acceleration?

19. **W** A truck covers 40.0 m in 8.50 s while smoothly slowing down to a final speed of 2.80 m/s. (a) Find its original speed. (b) Find its acceleration.

20. **Q|C M** In Example 2.7, we investigated a jet landing on an aircraft carrier. In a later maneuver, the jet comes in for a landing on solid ground with a speed of 100 m/s, and its acceleration can have a maximum magnitude of 5.00 m/s² as it comes to rest. (a) From the instant the jet touches the runway, what is the minimum time interval needed before it can come to rest? (b) Can this jet land at a small tropical island airport where the runway is 0.800 km long? (c) Explain your answer.

21. An electron in a cathode-ray tube accelerates uniformly from 2.00×10^4 m/s to 6.00×10^6 m/s over 1.50 cm. (a) In what time interval does the electron travel this 1.50 cm? (b) What is its acceleration?

22. A speedboat moving at 30.0 m/s approaches a no-wake buoy marker 100 m ahead. The pilot slows the boat with a constant acceleration of -3.50 m/s^2 by reducing the throttle. (a) How long does it take the boat to reach the buoy? (b) What is the velocity of the boat when it reaches the buoy?

23. **W** The driver of a car slams on the brakes when he sees a tree blocking the road. The car slows uniformly with an acceleration of -5.60 m/s^2 for 4.20 s, making straight skid marks 62.4 m long, all the way to the tree. With what speed does the car then strike the tree?

24. **S** In the particle under constant acceleration model, we identify the variables and parameters v_{xi}, v_{xf}, a_x, t, and $x_f - x_i$. Of the equations in Table 2.2, the first does not involve $x_f - x_i$, the second does not contain a_x, the third omits v_{xf}, and the last leaves out t. So, to complete the set, there should be an equation *not* involving v_{xi}. (a) Derive it from the others. (b) Use the equation in part (a) to solve Problem 23 in one step.

25. A truck on a straight road starts from rest, accelerating at 2.00 m/s^2 until it reaches a speed of 20.0 m/s. Then the truck travels for 20.0 s at constant speed until the brakes are applied, stopping the truck in a uniform manner in an additional 5.00 s. (a) How long is the truck in motion? (b) What is the average velocity of the truck for the motion described?

26. A particle moves along the x axis. Its position is given by the equation $x = 2 + 3t - 4t^2$, with x in meters and t in seconds. Determine (a) its position when it changes direction and (b) its velocity when it returns to the position it had at $t = 0$.

27. **GP** A speedboat travels in a straight line and increases in speed uniformly from $v_i = 20.0$ m/s to $v_f = 30.0$ m/s in a displacement Δx of 200 m. We wish to find the time interval required for the boat to move through this displacement. (a) Draw a coordinate system for this situation. (b) What analysis model is most appropriate for describing this situation? (c) From the analysis model, what equation is most appropriate for finding the acceleration of the speedboat? (d) Solve the equation selected in part (c) symbolically for the boat's acceleration in terms of v_i, v_f, and Δx. (e) Substitute numerical values to obtain the acceleration numerically. (f) Find the time interval mentioned above.

Section 2.7 Freely Falling Objects

Note: In all problems in this section, ignore the effects of air resistance.

28. In a classic clip on *America's Funniest Home Videos,* a sleeping cat rolls gently off the top of a warm TV set. Ignoring air resistance, calculate the position and velocity of the cat after (a) 0.100 s, (b) 0.200 s, and (c) 0.300 s.

29. *Why is the following situation impossible?* Emily challenges her friend David to catch a $1 bill as follows. She holds

Figure P2.29

the bill vertically as shown in Figure P2.29, with the center of the bill between but not touching David's index finger and thumb. Without warning, Emily releases the bill. David catches the bill without moving his hand downward. David's reaction time is equal to the average human reaction time.

30. **W** A baseball is hit so that it travels straight upward after being struck by the bat. A fan observes that it takes 3.00 s for the ball to reach its maximum height. Find (a) the ball's initial velocity and (b) the height it reaches.

31. A daring ranch hand sitting on a tree limb wishes to drop vertically onto a horse galloping under the tree. The constant speed of the horse is 10.0 m/s, and the distance from the limb to the level of the saddle is 3.00 m. (a) What must be the horizontal distance between the saddle and limb when the ranch hand makes his move? (b) For what time interval is he in the air?

32. It is possible to shoot an arrow at a speed as high as 100 m/s. (a) If friction can be ignored, how high would an arrow launched at this speed rise if shot straight up? (b) How long would the arrow be in the air?

33. **M** A student throws a set of keys vertically upward to her sorority sister, who is in a window 4.00 m above. The second student catches the keys 1.50 s later. (a) With what initial velocity were the keys thrown? (b) What was the velocity of the keys just before they were caught?

34. **S** At time $t = 0$, a student throws a set of keys vertically upward to her sorority sister, who is in a window at distance h above. The second student catches the keys at time t. (a) With what initial velocity were the keys thrown? (b) What was the velocity of the keys just before they were caught?

35. **W** A ball is thrown directly downward with an initial speed of 8.00 m/s from a height of 30.0 m. After what time interval does it strike the ground?

Section 2.8 Context Connection: Acceleration Required by Consumers

36. (a) Show that the largest and smallest average accelerations in Table 2.3 are correctly computed from the measured time intervals required for the cars to speed up from 0 to 60 mi/h. (b) Convert both of these accelerations to the standard SI unit. (c) Modeling each acceleration as constant, find the distance traveled by both cars as they speed up. (d) If an automobile were able to maintain an acceleration of magnitude $a = g = 9.80$ m/s^2 on a horizontal roadway, what time interval would be required to accelerate from zero to 60.0 mi/h?

37. A certain automobile manufacturer claims that its deluxe sports car will accelerate from rest to a speed of 42.0 m/s in 8.00 s. (a) Determine the average acceleration of the car. (b) Assume that the car moves with constant acceleration. Find the distance the car travels in the first 8.00 s. (c) What is the speed of the car 10.0 s after it begins its motion if it can continue to move with the same acceleration?

38. As soon as a traffic light turns green, a car speeds up from rest to 50.0 mi/h with constant acceleration 9.00 mi/h/s. In the adjoining bicycle lane, a cyclist speeds up from rest to 20.0 mi/h with constant acceleration 13.0 mi/h/s. Each vehicle maintains constant velocity after reaching its cruising speed. (a) For what time interval is the bicycle ahead of the car? (b) By what maximum distance does the bicycle lead the car?

© Cengage Learning/George Semple

Additional Problems

39. A steam catapult launches a jet aircraft from the aircraft carrier *John C. Stennis*, giving it a speed of 175 mi/h in 2.50 s. (a) Find the average acceleration of the plane. (b) Modeling the acceleration as constant, find the distance the plane moves in this time interval.

40. An object is at $x = 0$ at $t = 0$ and moves along the x axis according to the velocity–time graph in Figure P2.40. (a) What is the object's acceleration between 0 and 4.0 s? (b) What is the object's acceleration between 4.0 s and 9.0 s? (c) What is the object's acceleration between 13.0 s and 18.0 s? (d) At what time(s) is the object moving with the lowest speed? (e) At what time is the object farthest from $x = 0$? (f) What is the final position x of the object at $t = 18.0$ s? (g) Through what total distance has the object moved between $t = 0$ and $t = 18.0$ s?

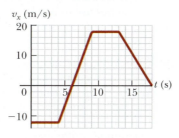

Figure P2.40

Note: The human body can undergo brief accelerations up to 15 times the free-fall acceleration g without injury or with only strained ligaments. Acceleration of long duration can do damage by preventing circulation of blood. Acceleration of larger magnitude can cause severe internal injuries, such as by tearing the aorta away from the heart. Problems 2.41 and 2.42 deal with variously large accelerations of the human body that you can compare with the $15g$ datum.

41. **BIO** **M** Colonel John P. Stapp, USAF, participated in studying whether a jet pilot could survive emergency ejection. On March 19, 1954, he rode a rocket-propelled sled that moved down a track at a speed of 632 mi/h. He and the sled were safely brought to rest in 1.40 s (Fig. P2.41). Determine (a) the negative acceleration he experienced and (b) the distance he traveled during this negative acceleration.

left, Courtesy U.S. Air Force; *right*, NASA/Photo Researchers Inc.

Figure P2.41 (*Left*) Col. John Stapp on the rocket sled. (*Right*) Col. Stapp's face is contorted by the stress of rapid negative acceleration.

42. **BIO** A woman is reported to have fallen 144 ft from the 17th floor of a building, landing on a metal ventilator box that she crushed to a depth of 18.0 in. She suffered only minor injuries. Ignoring air resistance, calculate (a) the speed of the woman just before she collided with the ventilator and (b) her average acceleration while in contact with the box. (c) Modeling her acceleration as constant, calculate the time interval it took to crush the box.

43. A ball starts from rest and accelerates at 0.500 m/s² while moving down an inclined plane 9.00 m long. When it reaches the bottom, the ball rolls up another plane, where it comes to rest after moving 15.0 m on that plane. (a) What is the speed of the ball at the bottom of the first plane? (b) During what time interval does the ball roll down the first plane? (c) What is the acceleration along the second plane? (d) What is the ball's speed 8.00 m along the second plane?

44. **Q|C** A glider of length ℓ moves through a stationary photogate on an air track. A photogate (Fig. P2.44) is a device that measures the time interval Δt_d during which the glider blocks a beam of infrared light passing across the photogate. The ratio $v_d = \ell/\Delta t_d$ is the average velocity of the glider over this part of its motion. Suppose the glider moves with constant acceleration. (a) Argue for or against the idea that v_d is equal to the instantaneous velocity of the glider when it is halfway through the photogate in space. (b) Argue for or against the idea that v_d is equal to the instantaneous velocity of the glider when it is halfway through the photogate in time.

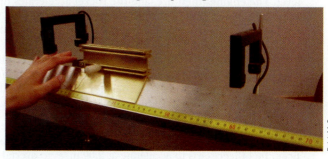

Ralph McGrew

Figure P2.44

45. **Review.** The biggest stuffed animal in the world is a snake 420 m long, constructed by Norwegian children. Suppose the snake is laid out in a park as shown in Figure P2.45, forming two straight sides of a 105° angle, with one side 240 m long. Olaf and Inge run a race they invent. Inge runs directly from the tail of the snake to its head, and Olaf starts from the same place at the same moment but runs along the snake. (a) If both children run steadily at 12.0 km/h, Inge reaches the head of the snake how much earlier than Olaf? (b) If Inge runs the race again at a constant speed of 12.0 km/h, at what constant speed must Olaf run to reach the end of the snake at the same time as Inge?

Figure P2.45

46. **Q|C** The Acela is an electric train on the Washington–New York–Boston run, carrying passengers at 170 mi/h. A velocity–time graph for the Acela is shown in Figure P2.46. (a) Describe the train's motion in each successive time interval. (b) Find the train's peak positive acceleration in the motion graphed. (c) Find the train's displacement in miles between $t = 0$ and $t = 200$ s.

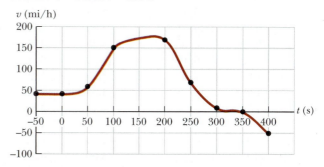

Figure P2.46 Velocity versus time graph for the Acela.

47. Liz rushes down onto a subway platform to find her train already departing. She stops and watches the cars go by. Each car is 8.60 m long. The first moves past her in 1.50 s and the second in 1.10 s. Find the constant acceleration of the train.

48. A commuter train travels between two downtown stations. Because the stations are only 1.00 km apart, the train never reaches its maximum possible cruising speed. During rush hour the engineer minimizes the time interval Δt between two stations by accelerating at a rate $a_1 = 0.100$ m/s² for a time interval Δt_1 and then immediately braking with acceleration $a_2 = -0.500$ m/s² for a time interval Δt_2. Find the minimum time interval of travel Δt and the time interval Δt_1.

49. A catapult launches a test rocket vertically upward from a well, giving the rocket an initial speed of 80.0 m/s at ground level. The engines then fire, and the rocket accelerates upward at 4.00 m/s² until it reaches an altitude of 1 000 m. At that point, its engines fail and the rocket goes into free fall, with an acceleration of −9.80 m/s². (a) For what time interval is the rocket in motion above the ground? (b) What is its maximum altitude? (c) What is its velocity just before it hits the ground? (You will need to consider the motion while the engine is operating and the free-fall motion separately.)

50. A motorist drives along a straight road at a constant speed of 15.0 m/s. Just as she passes a parked motorcycle police officer, the officer starts to accelerate at 2.00 m/s² to overtake her. Assuming that the officer maintains this acceleration, (a) determine the time interval required for the police officer to reach the motorist. Find (b) the speed and (c) the total displacement of the officer as he overtakes the motorist.

51. In a women's 100-m race, accelerating uniformly, Laura takes 2.00 s and Healan 3.00 s to attain their maximum speeds, which they each maintain for the rest of the race. They cross the finish line simultaneously, both setting a world record of 10.4 s. (a) What is the acceleration of each sprinter? (b) What are their respective maximum speeds? (c) Which sprinter is ahead at the 6.00-s mark, and by how much? (d) What is the maximum distance by which Healan is behind Laura, and at what time does that occur?

52. **Q|C** Astronauts on a distant planet toss a rock into the air. With the aid of a camera that takes pictures at a steady rate, they record the rock's height as a function of time as given in the following table. (a) Find the rock's average velocity in the time interval between each measurement and the next. (b) Using these average velocities to approximate instantaneous velocities at the midpoints of the time intervals, make a graph of velocity as a function of time. (c) Does the rock move with constant acceleration? If so, plot a straight line of best fit on the graph and calculate its slope to find the acceleration.

Time (s)	Height (m)	Time (s)	Height (m)
0.00	5.00	2.75	7.62
0.25	5.75	3.00	7.25
0.50	6.40	3.25	6.77
0.75	6.94	3.50	6.20
1.00	7.38	3.75	5.52
1.25	7.72	4.00	4.73
1.50	7.96	4.25	3.85
1.75	8.10	4.50	2.86
2.00	8.13	4.75	1.77
2.25	8.07	5.00	0.58
2.50	7.90		

53. **M** An inquisitive physics student and mountain climber climbs a 50.0-m-high cliff that overhangs a calm pool of water. He throws two stones vertically downward, 1.00 s apart, and observes that they cause a single splash. The first stone has an initial speed of 2.00 m/s. (a) How long after release of the first stone do the two stones hit the water? (b) What initial velocity must the second stone have if the two stones are to hit the water simultaneously? (c) What is the speed of each stone at the instant the two stones hit the water?

54. **Q|C** A hard rubber ball, released at chest height, falls to the pavement and bounces back to nearly the same height. When it is in contact with the pavement, the lower side of the ball is temporarily flattened. Suppose the maximum depth of the dent is on the order of 1 cm. Find the order of magnitude of the maximum acceleration of the ball while it is in contact with the pavement. State your assumptions, the quantities you estimate, and the values you estimate for them.

55. A man drops a rock into a well. (a) The man hears the sound of the splash 2.40 s after he releases the rock from rest. The speed of sound in air (at the ambient temperature) is 336 m/s. How far below the top of the well is the surface of the water? (b) **What If?** If the travel time for the sound is ignored, what percentage error is introduced when the depth of the well is calculated?

56. *Why is the following situation impossible?* A freight train is lumbering along at a constant speed of 16.0 m/s. Behind the freight train on the same track is a passenger train traveling in the same direction at 40.0 m/s. When the front of the passenger train is 58.5 m from the back of the freight train, the engineer on the passenger train recognizes the danger and hits the brakes of his train, causing the train to move with acceleration −3.00 m/s². Because of the engineer's action, the trains do not collide.

57. Two objects, A and B, are connected by a rigid rod that has a length L. The objects slide along perpendicular guide rails, as shown in Figure P2.57. If A slides to the left with a constant speed v, find the velocity of B when $\alpha = 60.0°$.

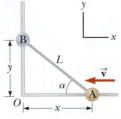

Figure P2.57

Chapter | 3

Motion in Two Dimensions

Chapter Outline

3.1 The Position, Velocity, and Acceleration Vectors

3.2 Two-Dimensional Motion with Constant Acceleration

3.3 Projectile Motion

3.4 Analysis Model: Particle in Uniform Circular Motion

3.5 Tangential and Radial Acceleration

3.6 Relative Velocity and Relative Acceleration

3.7 Context Connection: Lateral Acceleration of Automobiles

SUMMARY

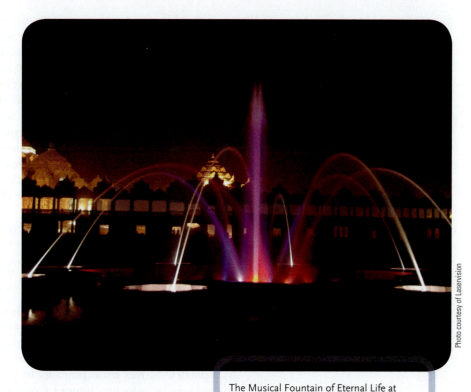

Photo courtesy of Laservision

The Musical Fountain of Eternal Life at the Swaminarayan Akshardham, a Hindu temple complex in New Delhi, India, presents a twelve-minute water, sound, and light show each evening. In this chapter, we will learn why the water arcs in the fountain have the shapes of parabolas.

In this chapter, we shall study the kinematics of an object that can be modeled as a particle moving in a plane. This motion is two dimensional. Some common examples of motion in a plane are the motions of satellites in orbit around the Earth, projectiles such as a thrown baseball, and the motion of electrons in uniform electric fields. We shall also study a particle in uniform circular motion and discuss various aspects of particles moving in curved paths.

3.1 | The Position, Velocity, and Acceleration Vectors

In Chapter 2, we found that the motion of a particle moving along a straight line such as the x axis is completely specified if its position is known as a function of time. Now let us extend this idea to motion in the xy plane. We will find equations for position and velocity that are the same as those in Chapter 2 except for their vector nature.

We begin by describing the position of a particle with a **position vector $\vec{r}$**, drawn from the origin of a coordinate system to the location of the particle in the xy plane, as in Figure 3.1 (page 70). At time t_i, the particle is at the point Ⓐ, and at some later time t_f, the particle is at Ⓑ, where the subscripts i and f refer to initial and final values. As the particle moves from Ⓐ to Ⓑ in the time interval $\Delta t = t_f - t_i$, the position vector changes from $\vec{r}_i$ to $\vec{r}_f$. As we learned in Chapter 2,

69

the displacement of a particle is the difference between its final position and its initial position:

$$\Delta \vec{r} \equiv \vec{r}_f - \vec{r}_i$$ 3.1 ◀

The direction of $\Delta \vec{r}$ is indicated in Figure 3.1.

The **average velocity** $\vec{v}_{avg}$ of the particle during the time interval Δt is defined as the ratio of the displacement to the time interval:

▶ Definition of average velocity

$$\vec{v}_{avg} \equiv \frac{\Delta \vec{r}}{\Delta t}$$ 3.2 ◀

Because displacement is a vector quantity and the time interval is a scalar quantity, we conclude that the average velocity is a *vector* quantity in the same direction as $\Delta \vec{r}$. Compare Equation 3.2 with its one-dimensional counterpart, Equation 2.2. The average velocity between points Ⓐ and Ⓑ is *independent of the path* between the two points. That is true because the average velocity is proportional to the displacement, which in turn depends only on the initial and final position vectors and not on the path taken between those two points. As with one-dimensional motion, if a particle starts its motion at some point and returns to this point via any path, its average velocity is zero for this trip because its displacement is zero.

Consider again the motion of a particle between two points in the *xy* plane, as shown in Figure 3.2. As the time intervals over which we observe the motion become smaller and smaller, the direction of the displacement approaches that of the line tangent to the path at the point Ⓐ.

The **instantaneous velocity** $\vec{v}$ is defined as the limit of the average velocity $\Delta \vec{r}/\Delta t$ as Δt approaches zero:

$$\vec{v} \equiv \lim_{\Delta t \to 0} \frac{\Delta \vec{r}}{\Delta t} = \frac{d\vec{r}}{dt}$$ 3.3 ◀

That is, the instantaneous velocity equals the derivative of the position vector with respect to time. The direction of the instantaneous velocity vector at any point in a particle's path is along a line that is tangent to the path at that point and in the direction of motion. The magnitude of the instantaneous velocity is called the *speed*.

As a particle moves from point Ⓐ to point Ⓑ along some path as in Figure 3.3, its instantaneous velocity changes from $\vec{v}_i$ at time t_i to $\vec{v}_f$ at time t_f. The **average**

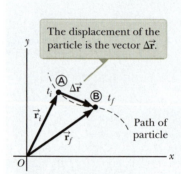

Figure 3.1 A particle moving in the *xy* plane is located with the position vector $\vec{r}$ drawn from the origin to the particle. The displacement of the particle as it moves from Ⓐ to Ⓑ in the time interval $\Delta t = t_f - t_i$ is equal to the vector $\Delta \vec{r} \equiv \vec{r}_f - \vec{r}_i$.

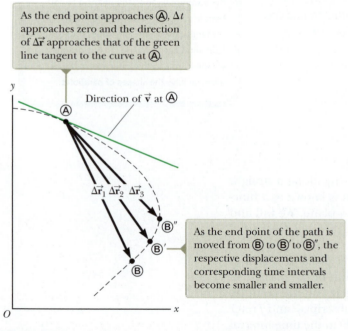

Figure 3.2 As a particle moves between two points, its average velocity is in the direction of the displacement vector $\Delta \vec{r}$. By definition, the instantaneous velocity at Ⓐ is directed along the line tangent to the curve at Ⓐ.

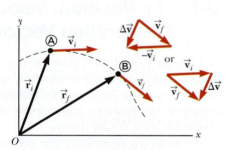

Figure 3.3 A particle moves from position Ⓐ to position Ⓑ. Its velocity vector changes from $\vec{v}_i$ at time t_i to $\vec{v}_f$ at time t_f. The vector addition diagrams at the upper right show two ways of determining the vector $\Delta \vec{v}$ from the initial and final velocities.

acceleration $\vec{\mathbf{a}}_{avg}$ of a particle over a time interval is defined as the ratio of the change in the instantaneous velocity $\Delta\vec{\mathbf{v}}$ to the time interval Δt:

$$\vec{\mathbf{a}}_{avg} \equiv \frac{\vec{\mathbf{v}}_f - \vec{\mathbf{v}}_i}{t_f - t_i} = \frac{\Delta\vec{\mathbf{v}}}{\Delta t} \qquad \textbf{3.4} \blacktriangleleft$$

▶ Definition of average acceleration

Because the average acceleration is the ratio of a vector quantity $\Delta\vec{\mathbf{v}}$ and a scalar quantity Δt, we conclude that $\vec{\mathbf{a}}_{avg}$ is a vector quantity in the same direction as $\Delta\vec{\mathbf{v}}$. Compare Equation 3.4 with the corresponding one-dimensional version, Equation 2.7. As indicated in Figure 3.3, the direction of $\Delta\vec{\mathbf{v}}$ is found by adding the vector $-\vec{\mathbf{v}}_i$ (the negative of $\vec{\mathbf{v}}_i$) to the vector $\vec{\mathbf{v}}_f$ because by definition $\Delta\vec{\mathbf{v}} = \vec{\mathbf{v}}_f - \vec{\mathbf{v}}_i$.

The **instantaneous acceleration** $\vec{\mathbf{a}}$ is defined as the limiting value of the ratio $\Delta\vec{\mathbf{v}}/\Delta t$ as Δt approaches zero:

$$\vec{\mathbf{a}} \equiv \lim_{\Delta t \to 0} \frac{\Delta\vec{\mathbf{v}}}{\Delta t} = \frac{d\vec{\mathbf{v}}}{dt} \qquad \textbf{3.5} \blacktriangleleft$$

▶ Definition of instantaneous acceleration

That is, the instantaneous acceleration equals the derivative of the velocity vector with respect to time. Compare Equation 3.5 with Equation 2.8.

It is important to recognize that various changes can occur that represent a particle undergoing an acceleration. First, the magnitude of the velocity vector (the speed) may change with time as in straight-line (one-dimensional) motion. Second, the direction of the velocity vector may change with time as its magnitude remains constant. Finally, both the magnitude and the direction of the velocity vector may change.

QUICK QUIZ 3.1 Consider the following controls in an automobile in motion: gas pedal, brake, steering wheel. What are the controls in this list that cause an acceleration of the car? (**a**) all three controls (**b**) the gas pedal and the brake (**c**) only the brake (**d**) only the gas pedal (**e**) only the steering wheel

Pitfall Prevention | 3.1
Vector Addition
Although the vector addition discussed in Chapter 1 involves *displacement* vectors, vector addition can be applied to *any* type of vector quantity. Figure 3.3, for example, shows the addition of *velocity* vectors using the graphical approach.

❮3.2 | Two-Dimensional Motion with Constant Acceleration

Let us consider two-dimensional motion during which the magnitude and direction of the acceleration remain unchanged. In this situation, we shall investigate motion as a two-dimensional version of the analysis in Section 2.6.

Before embarking on this investigation, we need to emphasize an important point regarding two-dimensional motion. Imagine an air hockey puck moving in a straight line along a perfectly level, friction-free surface of an air hockey table. Figure 3.4a shows a motion diagram from an overhead point of view of this puck. Recall that in Section 2.4 we related the acceleration of an object to a force on the object. Because there are no forces on the puck in the horizontal plane, it moves with constant velocity in the x direction. Now suppose you blow a puff of air on the puck as it passes your position, with the force from your puff of air *exactly* in the y direction. Because the force

The horizontal red vectors, representing the x component of the velocity, are the same length in both parts of the figure, which demonstrates that motion in two dimensions can be modeled as two independent motions in perpendicular directions.

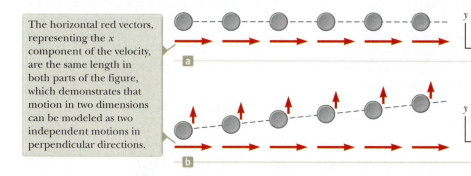

Figure 3.4 (a) A puck moves across a horizontal air hockey table at constant velocity in the x direction. (b) After a puff of air in the y direction is applied to the puck, the puck has gained a y component of velocity, but the x component is unaffected by the force in the perpendicular direction.

from this puff of air has no component in the x direction, it causes no acceleration in the x direction. It only causes a momentary acceleration in the y direction, causing the puck to have a constant y component of velocity once the force from the puff of air is removed. After your puff of air on the puck, its velocity component in the x direction is unchanged as shown in Figure 3.4b. The generalization of this simple experiment is that **motion in two dimensions can be modeled as two *independent* motions in each of the two perpendicular directions associated with the x and y axes. That is, any influence in the y direction does not affect the motion in the x direction and vice versa.**

The motion of a particle can be determined if its position vector $\vec{\mathbf{r}}$ is known at all times. The position vector for a particle moving in the xy plane can be written

$$\vec{\mathbf{r}} = x\hat{\mathbf{i}} + y\hat{\mathbf{j}} \qquad \textbf{3.6} \blacktriangleleft$$

where x, y, and $\vec{\mathbf{r}}$ change with time as the particle moves. If the position vector is known, the velocity of the particle can be obtained from Equations 3.3 and 3.6:

$$\vec{\mathbf{v}} = \frac{d\vec{\mathbf{r}}}{dt} = \frac{dx}{dt}\hat{\mathbf{i}} + \frac{dy}{dt}\hat{\mathbf{j}} = v_x\hat{\mathbf{i}} + v_y\hat{\mathbf{j}} \qquad \textbf{3.7} \blacktriangleleft$$

Because we are assuming that $\vec{\mathbf{a}}$ is constant in this discussion, its components a_x and a_y are also constants. Therefore, we can apply the equations of kinematics to the x and y components of the velocity vector separately. Substituting $v_x = v_{xf} = v_{xi} + a_x t$ and $v_y = v_{yf} = v_{yi} + a_y t$ into Equation 3.7 gives

$$\vec{\mathbf{v}}_f = (v_{xi} + a_x t)\hat{\mathbf{i}} + (v_{yi} + a_y t)\hat{\mathbf{j}}$$
$$= (v_{xi}\hat{\mathbf{i}} + v_{yi}\hat{\mathbf{j}}) + (a_x\hat{\mathbf{i}} + a_y\hat{\mathbf{j}})t$$

▶ Velocity vector as a function of time for a particle under constant acceleration

$$\boxed{\vec{\mathbf{v}}_f = \vec{\mathbf{v}}_i + \vec{\mathbf{a}}t} \qquad \textbf{3.8} \blacktriangleleft$$

This result states that the velocity $\vec{\mathbf{v}}_f$ of a particle at some time t equals the vector sum of its initial velocity $\vec{\mathbf{v}}_i$ and the additional velocity $\vec{\mathbf{a}}t$ acquired at time t as a result of its constant acceleration. This result is the same as Equation 2.10, except for its vector nature.

Similarly, from Equation 2.13 we know that the x and y coordinates of a particle moving with constant acceleration are

$$x_f = x_i + v_{xi}t + \tfrac{1}{2}a_x t^2 \quad \text{and} \quad y_f = y_i + v_{yi}t + \tfrac{1}{2}a_y t^2$$

Substituting these expressions into Equation 3.6 gives

$$\vec{\mathbf{r}}_f = (x_i + v_{xi}t + \tfrac{1}{2}a_x t^2)\hat{\mathbf{i}} + (y_i + v_{yi}t + \tfrac{1}{2}a_y t^2)\hat{\mathbf{j}}$$
$$= (x_i\hat{\mathbf{i}} + y_i\hat{\mathbf{j}}) + (v_{xi}\hat{\mathbf{i}} + v_{yi}\hat{\mathbf{j}})t + \tfrac{1}{2}(a_x\hat{\mathbf{i}} + a_y\hat{\mathbf{j}})t^2$$

▶ Position vector as a function of time for a particle under constant acceleration

$$\boxed{\vec{\mathbf{r}}_f = \vec{\mathbf{r}}_i + \vec{\mathbf{v}}_i t + \tfrac{1}{2}\vec{\mathbf{a}}t^2} \qquad \textbf{3.9} \blacktriangleleft$$

This equation implies that the final position vector $\vec{\mathbf{r}}_f$ is the vector sum of the initial position vector $\vec{\mathbf{r}}_i$ plus a displacement $\vec{\mathbf{v}}_i t$, arising from the initial velocity of the particle, and a displacement $\tfrac{1}{2}\vec{\mathbf{a}}t^2$, resulting from the uniform acceleration of the particle. It is the same as Equation 2.13 except for its vector nature.

Graphical representations of Equations 3.8 and 3.9 are shown in Active Figures 3.5a and 3.5b. Note from Active Figure 3.5b that $\vec{\mathbf{r}}_f$ is generally not along the direction of $\vec{\mathbf{r}}_i$, $\vec{\mathbf{v}}_i$, or $\vec{\mathbf{a}}$ because the relationship between these quantities is a vector expression. For the same reason, from Active Figure 3.5a we see that $\vec{\mathbf{v}}_f$ is generally not along the direction of $\vec{\mathbf{v}}_i$ or $\vec{\mathbf{a}}$. Finally, if we compare the two figures, we see that $\vec{\mathbf{v}}_f$ and $\vec{\mathbf{r}}_f$ are not in the same direction.

Because Equations 3.8 and 3.9 are *vector* expressions, we may also write their x and y component equations:

$$\vec{\mathbf{v}}_f = \vec{\mathbf{v}}_i + \vec{\mathbf{a}}t \;\rightarrow\; \begin{cases} v_{xf} = v_{xi} + a_x t \\ v_{yf} = v_{yi} + a_y t \end{cases}$$

$$\vec{\mathbf{r}}_f = \vec{\mathbf{r}}_i + \vec{\mathbf{v}}_i t + \tfrac{1}{2}\vec{\mathbf{a}}t^2 \;\rightarrow\; \begin{cases} x_f = x_i + v_{xi}t + \tfrac{1}{2}a_x t^2 \\ y_f = y_i + v_{yi}t + \tfrac{1}{2}a_y t^2 \end{cases}$$

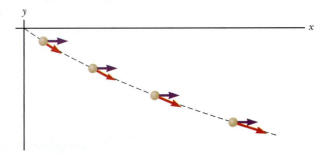

Active Figure 3.5 Vector representations and components of (a) the velocity and (b) the position of a particle moving with a constant acceleration $\vec{\mathbf{a}}$.

These components are illustrated in Active Figure 3.5. Consistent with our discussion related to Figure 3.4, two-dimensional motion having constant acceleration is equivalent to two *independent* motions in the x and y directions having constant accelerations a_x and a_y. Therefore, there is no new model for a particle under two-dimensional constant acceleration; the appropriate model is just the one-dimensional particle under constant acceleration applied twice, in the x and y directions separately!

Example 3.1 | Motion in a Plane

A particle moves in the xy plane, starting from the origin at $t = 0$ with an initial velocity having an x component of 20 m/s and a y component of -15 m/s. The particle experiences an acceleration in the x direction, given by $a_x = 4.0 \text{ m/s}^2$.

(A) Determine the total velocity vector at any time.

SOLUTION

Conceptualize The components of the initial velocity tell us that the particle starts by moving toward the right and downward. The x component of velocity starts at 20 m/s and increases by 4.0 m/s every second. The y component of velocity never changes from its initial value of -15 m/s. We sketch a motion diagram of the situation in Figure 3.6. Because the particle is accelerating in the $+x$ direction, its velocity component in this direction increases and the path curves as shown in the diagram. Notice that the spacing between successive images increases as time goes on because the speed is increasing. The placement of the acceleration and velocity vectors in Figure 3.6 helps us further conceptualize the situation.

Figure 3.6 (Example 3.1) Motion diagram for the particle.

Categorize Because the initial velocity has components in both the x and y directions, we categorize this problem as one involving a particle moving in two dimensions. Because the particle only has an x component of acceleration, we model it as a particle under constant acceleration in the x direction and a particle under constant velocity in the y direction.

Analyze To begin the mathematical analysis, we set $v_{xi} = 20$ m/s, $v_{yi} = -15$ m/s, $a_x = 4.0 \text{ m/s}^2$, and $a_y = 0$.

Use Equation 3.8 for the velocity vector:

$$\vec{\mathbf{v}}_f = \vec{\mathbf{v}}_i + \vec{\mathbf{a}}t = (v_{xi} + a_x t)\hat{\mathbf{i}} + (v_{yi} + a_y t)\hat{\mathbf{j}}$$

Substitute numerical values with the velocity in meters per second and the time in seconds:

$$\vec{\mathbf{v}}_f = [20 + (4.0)t]\hat{\mathbf{i}} + [-15 + (0)t]\hat{\mathbf{j}}$$

$$(1) \quad \vec{\mathbf{v}}_f = [(20 + 4.0t)\hat{\mathbf{i}} - 15\hat{\mathbf{j}}]$$

Finalize Notice that the x component of velocity increases in time while the y component remains constant; this result is consistent with our prediction.

continued

3.1 *cont.*

(B) Calculate the velocity and speed of the particle at $t = 5.0$ s and the angle the velocity vector makes with the x axis.

SOLUTION

Analyze

Evaluate the result from Equation (1) at $t = 5.0$ s:

$$\vec{\mathbf{v}}_f = [(20 + 4.0(5.0))\hat{\mathbf{i}} - 15\hat{\mathbf{j}}] = \boxed{(40\hat{\mathbf{i}} - 15\hat{\mathbf{j}})\,\text{m/s}}$$

Determine the angle θ that $\vec{\mathbf{v}}_f$ makes with the x axis at $t = 5.0$ s:

$$\theta = \tan^{-1}\left(\frac{v_{yf}}{v_{xf}}\right) = \tan^{-1}\left(\frac{-15\ \text{m/s}}{40\ \text{m/s}}\right) = \boxed{-21°}$$

Evaluate the speed of the particle as the magnitude of $\vec{\mathbf{v}}_f$:

$$v_f = |\vec{\mathbf{v}}_f| = \sqrt{v_{xf}{}^2 + v_{yf}{}^2} = \sqrt{(40)^2 + (-15)^2}\ \text{m/s} = \boxed{43\ \text{m/s}}$$

Finalize The negative sign for the angle θ indicates that the velocity vector is directed at an angle of 21° below the positive x axis. Notice that if we calculate v_i from the x and y components of $\vec{\mathbf{v}}_i$, we find that $v_f > v_i$. Is that consistent with our prediction?

(C) Determine the x and y coordinates of the particle at any time t and its position vector at this time.

SOLUTION

Analyze

Use the components of Equation 3.9 with $x_i = y_i = 0$ at $t = 0$ with x and y in meters and t in seconds:

$$x_f = v_{xi}t + \tfrac{1}{2}a_x t^2 = \boxed{20t + 2.0t^2}$$
$$y_f = v_{yi}t = \boxed{-15t}$$

Express the position vector of the particle at any time t:

$$\vec{\mathbf{r}}_f = x_f\hat{\mathbf{i}} + y_f\hat{\mathbf{j}} = \boxed{(20t + 2.0t^2)\,\hat{\mathbf{i}} - 15t\,\hat{\mathbf{j}}}$$

Finalize Let us now consider a limiting case for very large values of t.

What If? What if we wait a very long time and then observe the motion of the particle? How would we describe the motion of the particle for large values of the time?

Answer Looking at Figure 3.6, we see the path of the particle curving toward the x axis. There is no reason to assume this tendency will change, which suggests that the path will become more and more parallel to the x axis as time grows large. Mathematically, Equation (1) shows that the y component of the velocity remains constant while the x component grows linearly with t. Therefore, when t is very large, the x component of the velocity will be much larger than the y component, suggesting that the velocity vector becomes more and more parallel to the x axis. Both x_f and y_f continue to grow with time, although x_f grows much faster.

3.3 | Projectile Motion

Anyone who has observed a baseball in motion (or, for that matter, any object thrown into the air) has observed projectile motion. The ball moves in a curved path when thrown at some angle with respect to the Earth's surface. **Projectile motion** of an object is surprisingly simple to analyze if the following two assumptions are made when building a model for these types of problems: (1) the free-fall acceleration g is constant over the range of motion and is directed downward,[1] and (2) the effect of air resistance is negligible.[2] With these assumptions, the path of a projectile, called

---

[1]In effect, this approximation is equivalent to assuming that the Earth is flat within the range of motion considered and that the maximum height of the object is small compared to the radius of the Earth.

[2]This approximation is often *not* justified, especially at high velocities. In addition, the spin of a projectile, such as a baseball, can give rise to some very interesting effects associated with aerodynamic forces (for example, a curve ball thrown by a pitcher).

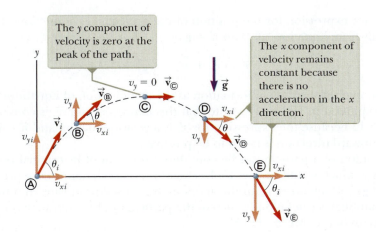

The y component of velocity is zero at the peak of the path.

The x component of velocity remains constant because there is no acceleration in the x direction.

Active Figure 3.7 The parabolic path of a projectile that leaves the origin (point Ⓐ) with a velocity $\vec{v}_i$. The velocity vector $\vec{v}$ changes with time in both magnitude and direction. This change is the result of acceleration $\vec{a} = \vec{g}$ in the negative y direction.

its *trajectory*, is *always* a parabola. **We shall use a simplification model based on these assumptions throughout this chapter.**

If we choose our reference frame such that the y direction is vertical and positive upward, $a_y = -g$ (as in one-dimensional free-fall) and $a_x = 0$ (because the only possible horizontal acceleration is due to air resistance, and it is ignored). Furthermore, let us assume that at $t = 0$, the projectile leaves the origin (point Ⓐ, $x_i = y_i = 0$) with speed v_i, as in Active Figure 3.7. If the vector $\vec{v}_i$ makes an angle θ_i with the horizontal, we can identify a right triangle in the diagram as a geometric model, and from the definitions of the cosine and sine functions we have

$$\cos \theta_i = \frac{v_{xi}}{v_i} \quad \text{and} \quad \sin \theta_i = \frac{v_{yi}}{v_i}$$

Therefore, the initial x and y components of velocity are

$$v_{xi} = v_i \cos \theta_i \quad \text{and} \quad v_{yi} = v_i \sin \theta_i$$

Substituting these expressions into Equations 3.8 and 3.9 with $a_x = 0$ and $a_y = -g$ gives the velocity components and position coordinates for the projectile at any time t:

$$v_{xf} = v_{xi} = v_i \cos \theta_i = \text{constant} \qquad \textbf{3.10} \blacktriangleleft$$

$$v_{yf} = v_{yi} - gt = v_i \sin \theta_i - gt \qquad \textbf{3.11} \blacktriangleleft$$

$$x_f = x_i + v_{xi}t = (v_i \cos \theta_i)t \qquad \textbf{3.12} \blacktriangleleft$$

$$y_f = y_i + v_{yi}t - \tfrac{1}{2}gt^2 = (v_i \sin \theta_i)t - \tfrac{1}{2}gt^2 \qquad \textbf{3.13} \blacktriangleleft$$

From Equation 3.10 we see that v_{xf} remains constant in time and is equal to v_{xi}; there is no horizontal component of acceleration. Therefore, we model the horizontal motion as that of a particle under constant velocity. For the y motion, note that the equations for v_{yf} and y_f are similar to Equations 2.10 and 2.13 for freely falling objects. Therefore, we can apply the model of a particle under constant acceleration to the y component. In fact, *all* the equations of kinematics developed in Chapter 2 are applicable to projectile motion.

If we solve for t in Equation 3.12 and substitute this expression for t into Equation 3.13, we find that

$$y_f = (\tan \theta_i) x_f - \left(\frac{g}{2 v_i^2 \cos^2 \theta_i} \right) x_f^2 \qquad \textbf{3.14} \blacktriangleleft$$

which is valid for angles in the range $0 < \theta_i < \pi/2$. This expression is of the form $y = ax - bx^2$, which is the equation of a parabola that passes through the origin. Therefore, we have proven that the trajectory of a projectile can be geometrically modeled as a parabola. The trajectory is *completely* specified if v_i and θ_i are known.

Pitfall Prevention | 3.2
Acceleration at the Highest Point
As discussed in Pitfall Prevention 2.7, many people claim that the acceleration of a projectile at the topmost point of its trajectory is zero. This mistake arises from confusion between zero vertical velocity and zero acceleration. If the projectile were to experience zero acceleration at the highest point, its velocity at that point would not change; rather, the projectile would move horizontally at constant speed from then on! That does not happen, however, because the acceleration is *not* zero anywhere along the trajectory.

A welder cuts holes through a heavy metal construction beam with a hot torch. The sparks generated in the process follow parabolic paths.

Lester Lefkowitz/Taxi/Getty Images

The vector expression for the position of the projectile as a function of time follows directly from Equation 3.9, with $\vec{a} = \vec{g}$:

$$\vec{r}_f = \vec{r}_i + \vec{v}_i t + \tfrac{1}{2}\vec{g}t^2$$

This equation gives the same information as the combination of Equations 3.12 and 3.13 and is plotted in Figure 3.8. Note that this expression for $\vec{r}_f$ is consistent with Equation 3.13 because the expression for $\vec{r}_f$ is a vector equation and $\vec{a} = \vec{g} = -g\hat{j}$ when the upward direction is taken to be positive.

The position of a particle can be considered the sum of its original position $\vec{r}_i$, the term $\vec{v}_i t$, which would be the displacement if no acceleration were present, and the term $\tfrac{1}{2}\vec{g}t^2$, which arises from the acceleration caused by gravity. In other words, if no gravitational acceleration occurred, the particle would continue to move along a straight path in the direction of $\vec{v}_i$.

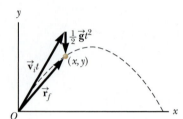

Figure 3.8 The position vector $\vec{r}_f$ of a projectile whose initial velocity at the origin is $\vec{v}_i$. The vector $\vec{v}_i t$ would be the position vector of the projectile if gravity were absent and the vector $\tfrac{1}{2}\vec{g}t^2$ is the particle's vertical displacement due to its downward gravitational acceleration.

▶ **QUICK QUIZ 3.2** **(i)** As a projectile thrown upward moves in its parabolic path (such as in Fig. 3.8), at what point along its path are the velocity and acceleration vectors for the projectile perpendicular to each other? **(a)** nowhere **(b)** the highest point **(c)** the launch point **(ii)** From the same choices, at what point are the velocity and acceleration vectors for the projectile parallel to each other?

Horizontal Range and Maximum Height of a Projectile

Let us assume that a projectile is launched over flat ground from the origin at $t = 0$ with a positive v_y component, as in Figure 3.9. This a common situation in sports, where baseballs, footballs, and golf balls often land at the same level from which they were launched.

There are two special points in this motion that are interesting to analyze: the peak point Ⓐ, which has Cartesian coordinates $(R/2, h)$, and the landing point Ⓑ, having coordinates $(R, 0)$. The distance R is called the *horizontal range* of the projectile, and h is its *maximum height*. Because of the symmetry of the trajectory, the projectile is at the maximum height h when its x position is half the range R. Let us find h and R in terms of v_i, θ_i, and g.

We can determine h by noting that at the peak $v_{y\text{Ⓐ}} = 0$. Therefore, Equation 3.11 can be used to determine the time $t_\text{Ⓐ}$ at which the projectile reaches the peak:

$$t_\text{Ⓐ} = \frac{v_i \sin \theta_i}{g}$$

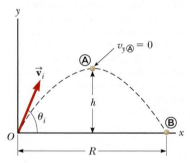

Figure 3.9 A projectile launched from the origin at $t = 0$ with an initial velocity $\vec{v}_i$. The maximum height of the projectile is h, and its horizontal range is R. At Ⓐ, the peak of the trajectory, the projectile has coordinates $(R/2, h)$.

Substituting this expression for $t_\text{Ⓐ}$ into Equation 3.13 and replacing y_f with h gives h in terms of v_i and θ_i:

$$h = (v_i \sin \theta_i)\frac{v_i \sin \theta_i}{g} - \tfrac{1}{2}g\left(\frac{v_i \sin \theta_i}{g}\right)^2$$

$$h = \frac{v_i^2 \sin^2 \theta_i}{2g} \qquad \text{3.15} \blacktriangleleft$$

Notice from the mathematical representation how you could increase the maximum height h: You could launch the projectile with a larger initial velocity, at a higher angle, or at a location with lower free-fall acceleration, such as on the Moon. Is that consistent with your mental representation of this situation?

The range R is the horizontal distance traveled in twice the time interval required to reach the peak. Equivalently, we are seeking the position of the projectile at a time $2t_\text{Ⓐ}$. Using Equation 3.12 and noting that $x_f = R$ at $t = 2t_\text{Ⓐ}$, we find that

$$R = (v_i \cos \theta_i)2t_\text{Ⓐ} = (v_i \cos \theta_i)\frac{2v_i \sin \theta_i}{g} = \frac{2v_i^2 \sin \theta_i \cos \theta_i}{g}$$

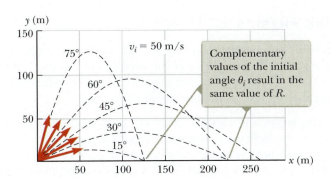

Because $\sin 2\theta = 2\sin\theta\cos\theta$, R can be written in the more compact form

$$R = \frac{v_i^2 \sin 2\theta_i}{g} \qquad \textbf{3.16} \blacktriangleleft$$

Notice from the mathematical expression how you could increase the range R: You could launch the projectile with a larger initial velocity or at a location with lower free-fall acceleration, such as on the Moon. Is that consistent with your mental representation of this situation?

The range also depends on the angle of the initial velocity vector. The maximum possible value of R from Equation 3.16 is given by $R_{max} = v_i^2/g$. This result follows from the maximum value of $\sin 2\theta_i$ being unity, which occurs when $2\theta_i = 90°$. Therefore, R is a maximum when $\theta_i = 45°$.

Active Figure 3.10 illustrates various trajectories for a projectile of a given initial speed. As you can see, the range is a maximum for $\theta_i = 45°$. In addition, for any θ_i other than 45°, a point with coordinates $(R, 0)$ can be reached by using either one of two complementary values of θ_i, such as 75° and 15°. Of course, the maximum height and the time of flight will be different for these two values of θ_i.

Pitfall Prevention | 3.3
The Height and Range Equations
Keep in mind that Equations 3.15 and 3.16 are useful for calculating h and R only for a symmetric path, as shown in Figure 3.9. If the path is not symmetric, *do not use these equations*. The general expressions given by Equations 3.10 through 3.13 are the *more important* results because they give the coordinates and velocity components of the projectile at any time t for any trajectory.

QUICK QUIZ 3.3 Rank the launch angles for the five paths in Active Figure 3.10 with respect to time of flight from the shortest time of flight to the longest.

PROBLEM-SOLVING STRATEGY: Projectile Motion

We suggest that you use the following approach when solving projectile motion problems:

1. **Conceptualize** Think about what is going on physically in the problem. Establish the mental representation by imagining the projectile moving along its trajectory.

2. **Categorize** Confirm that the problem involves a particle in free-fall and that air resistance is neglected. Select a coordinate system with x in the horizontal direction and y in the vertical direction.

3. **Analyze** If the initial velocity vector is given, resolve it into x and y components. Treat the horizontal motion and the vertical motion independently. Analyze the horizontal motion of the projectile with the particle under constant velocity model. Analyze the vertical motion of the projectile with the particle under constant acceleration model.

4. **Finalize** Once you have determined your result, check to see if your answers are consistent with the mental and pictorial representations and that your results are realistic.

> **THINKING PHYSICS 3.1**
>
> A home run is hit in a baseball game. The ball is hit from home plate into the stands along a parabolic path. What is the acceleration of the ball **(a)** while it is rising, **(b)** at the highest point of the trajectory, and **(c)** while it is descending after reaching the highest point? Ignore air resistance.
>
> **Reasoning** The answers to all three parts are the same: the acceleration is that due to gravity, $a_y = -9.80 \text{ m/s}^2$, because the gravitational force is pulling downward on the ball during the entire motion. During the rising part of the trajectory, the downward acceleration results in the decreasing positive values of the vertical component of the ball's velocity. During the falling part of the trajectory, the downward acceleration results in the increasing negative values of the vertical component of the velocity. ◀

Example 3.2 | That's Quite an Arm!

A stone is thrown from the top of a building upward at an angle of 30.0° to the horizontal with an initial speed of 20.0 m/s as shown in Figure 3.11. The height from which the stone is thrown is 45.0 m above the ground.

(A) How long does it take the stone to reach the ground?

SOLUTION

Conceptualize Study Figure 3.11, in which we have indicated the trajectory and various parameters of the motion of the stone.

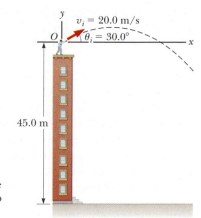

Figure 3.11
(Example 3.2) A stone is thrown from the top of a building.

Categorize We categorize this problem as a projectile motion problem. The stone is modeled as a particle under constant acceleration in the y direction and a particle under constant velocity in the x direction.

Analyze We have the information $x_i = y_i = 0$, $y_f = -45.0$ m, $a_y = -g$, and $v_i = 20.0$ m/s (the numerical value of y_f is negative because we have chosen the point of the throw as the origin).

Find the initial x and y components of the stone's velocity:

$$v_{xi} = v_i \cos \theta_i = (20.0 \text{ m/s}) \cos 30.0° = 17.3 \text{ m/s}$$
$$v_{yi} = v_i \sin \theta_i = (20.0 \text{ m/s}) \sin 30.0° = 10.0 \text{ m/s}$$

Express the vertical position of the stone from the vertical component of Equation 3.9:

$$y_f = y_i + v_{yi}t + \tfrac{1}{2}a_y t^2$$

Substitute numerical values:

$$-45.0 \text{ m} = 0 + (10.0 \text{ m/s})t + \tfrac{1}{2}(-9.80 \text{ m/s}^2)t^2$$

Solve the quadratic equation for t:

$$t = \boxed{4.22 \text{ s}}$$

(B) What is the speed of the stone just before it strikes the ground?

SOLUTION

Analyze Use the y component of Equation 3.8 to obtain the y component of the velocity of the stone just before it strikes the ground:

$$v_{yf} = v_{yi} + a_y t$$

Substitute numerical values, using $t = 4.22$ s:

$$v_{yf} = 10.0 \text{ m/s} + (-9.80 \text{ m/s}^2)(4.22 \text{ s}) = -31.3 \text{ m/s}$$

Use this component with the horizontal component $v_{xf} = v_{xi} = 17.3$ m/s to find the speed of the stone at $t = 4.22$ s:

$$v_f = \sqrt{v_{xf}^2 + v_{yf}^2} = \sqrt{(17.3 \text{ m/s})^2 + (-31.3 \text{ m/s})^2} = \boxed{35.8 \text{ m/s}}$$

3.2 *cont.*

Finalize Is it reasonable that the *y* component of the final velocity is negative? Is it reasonable that the final speed is larger than the initial speed of 20.0 m/s?

What If? What if a horizontal wind is blowing in the same direction as the stone is thrown and it causes the stone to have a horizontal acceleration component $a_x = 0.500$ m/s^2? Which part of this example, (A) or (B), will have a different answer?

Answer Recall that the motions in the *x* and *y* directions are independent. Therefore, the horizontal wind cannot affect the vertical motion. The vertical motion determines the time of the projectile in the air, so the answer to part (A) does not change. The wind causes the horizontal velocity component to increase with time, so the final speed will be larger in part (B). Taking $a_x = 0.500$ m/s^2, we find $v_{xf} = 19.4$ m/s and $v_f = 36.9$ m/s.

Example 3.3 | The End of the Ski Jump

A ski jumper leaves the ski track moving in the horizontal direction with a speed of 25.0 m/s as shown in Figure 3.12. The landing incline below her falls off with a slope of 35.0°. Where does she land on the incline?

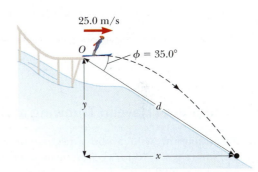

Figure 3.12 (Example 3.3) A ski jumper leaves the track moving in a horizontal direction.

SOLUTION

Conceptualize We can conceptualize this problem based on memories of observing winter Olympic ski competitions. We estimate the skier to be airborne for perhaps 4 s and to travel a distance of about 100 m horizontally. We should expect the value of *d*, the distance traveled along the incline, to be of the same order of magnitude.

Categorize We categorize the problem as one of a particle in projectile motion.

Analyze It is convenient to select the beginning of the jump as the origin. The initial velocity components are $v_{xi} = 25.0$ m/s and $v_{yi} = 0$. From the right triangle in Figure 3.12, we see that the jumper's *x* and *y* coordinates at the landing point are given by $x_f = d \cos \phi$ and $y_f = -d \sin \phi$.

Express the coordinates of the jumper as a function of time:

$$(1) \quad x_f = v_{xi} t$$

$$(2) \quad y_f = v_{yi} t + \tfrac{1}{2} a_y t^2 = -\tfrac{1}{2} g t^2$$

Substitute the values of x_f and y_f at the landing point:

$$(3) \quad d \cos \phi = v_{xi} t$$

$$(4) \quad -d \sin \phi = -\tfrac{1}{2} g t^2$$

Solve Equation (3) for *t* and substitute the result into Equation (4):

$$-d \sin \phi = -\tfrac{1}{2} g \left(\frac{d \cos \phi}{v_{xi}} \right)^2$$

Solve for *d*:

$$d = \frac{2 v_{xi}^2 \sin \phi}{g \cos^2 \phi} = \frac{2(25.0 \text{ m/s})^2 \sin 35.0°}{(9.80 \text{ m/s}^2) \cos^2 35.0°} = 109 \text{ m}$$

Evaluate the *x* and *y* coordinates of the point at which the skier lands:

$$x_f = d \cos \phi = (109 \text{ m}) \cos 35.0° = \boxed{89.3 \text{ m}}$$

$$y_f = -d \sin \phi = -(109 \text{ m}) \sin 35.0° = \boxed{-62.5 \text{ m}}$$

Finalize Let us compare these results with our expectations. We expected the horizontal distance to be on the order of 100 m, and our result of 89.3 m is indeed on this order of magnitude. It might be useful to calculate the time interval that the jumper is in the air and compare it with our estimate of about 4 s.

What If? Suppose everything in this example is the same except the ski jump is curved so that the jumper is projected upward at an angle from the end of the track. Is this design better in terms of maximizing the length of the jump?

Answer If the initial velocity has an upward component, the skier will be in the air longer and should therefore travel farther. Tilting the initial velocity vector upward, however, will reduce the horizontal component of the initial velocity.

continued

3.3 *cont.*

Therefore, angling the end of the ski track upward at a *large* angle may actually *reduce* the distance. Consider the extreme case: the skier is projected at 90° to the horizontal and simply goes up and comes back down at the end of the ski track! This argument suggests that there must be an optimal angle between 0° and 90° that represents a balance between making the flight time longer and the horizontal velocity component smaller.

Let us find this optimal angle mathematically. We modify Equations (1) through (4) in the following way, assuming the skier is projected at an angle θ with respect to the horizontal over a landing incline sloped with an arbitrary angle ϕ:

(1) and (3) $\rightarrow$ $x_f = (v_i \cos \theta)t = d \cos \phi$

(2) and (4) $\rightarrow$ $y_f = (v_i \sin \theta)t - \frac{1}{2}gt^2 = -d \sin \phi$

By eliminating the time t between these equations and using differentiation to maximize d in terms of θ, we arrive at the following equation for the angle θ that gives the maximum value of d:

$$\theta = 45° - \frac{\phi}{2}$$

For the slope angle in Figure 3.12, $\phi = 35.0°$; this equation results in an optimal launch angle of $\theta = 27.5°$. For a slope angle of $\phi = 0°$, which represents a horizontal plane, this equation gives an optimal launch angle of $\theta = 45°$, as we would expect (see Active Figure 3.10).

Example 3.4 | Javelin Throwing at the Olympics

An athlete throws a javelin a distance of 80.0 m at the Olympics held at the equator, where $g = 9.78$ m/s². Four years later the Olympics are held at the North Pole, where $g = 9.83$ m/s². Assuming that the thrower provides the javelin with exactly the same initial velocity as she did at the equator, how far does the javelin travel at the North Pole?

SOLUTION

A javelin can be thrown over a very long distance by a world class athlete.

Conceptualize In traveling between these two locations, we most likely would not feel any difference in the weight of an object. The increased gravity at the North Pole, however, will cause the javelin to return to the ground sooner and shorten its range compared to the throw at the equator.

Categorize In the absence of any information about how the javelin is affected by moving through the air, we adopt the free-fall model for the javelin. Track and field events are normally held on flat fields. Therefore, we surmise that the javelin returns to the same vertical position from which it was thrown and therefore that the trajectory is symmetric. These assumptions allow us to use Equations 3.15 and 3.16 to analyze the motion. The difference in range is due to the difference in the free-fall acceleration at the two locations.

....

Analyze To solve this problem, we will set up a ratio based on the range of the projectile being mathematically related to the acceleration due to gravity. This technique of solving by ratios is very powerful and should be studied and understood so that it can be applied in future problem solving.

Use Equation 3.16 to express the range of the particle at each of the two locations:

$$R_{\text{North Pole}} = \frac{v_i^2 \sin 2\theta_i}{g_{\text{North Pole}}}$$

$$R_{\text{equator}} = \frac{v_i^2 \sin 2\theta_i}{g_{\text{equator}}}$$

Divide the first equation by the second to establish a relationship between the ratio of the ranges and the ratio of the free-fall accelerations. Note that the problem states that the same initial velocity is provided to the javelin at both locations, so v_i and θ_i are the same in the numerator and denominator of the ratio:

$$\frac{R_{\text{North Pole}}}{R_{\text{equator}}} = \frac{\left(\dfrac{v_i^2 \sin 2\theta_i}{g_{\text{North Pole}}}\right)}{\left(\dfrac{v_i^2 \sin 2\theta_i}{g_{\text{equator}}}\right)} = \frac{g_{\text{equator}}}{g_{\text{North Pole}}}$$

3.4 *cont.*

Solve this equation for the range at the North Pole and substitute the numerical values:

$$R_{\text{North Pole}} = \frac{g_{\text{equator}}}{g_{\text{North Pole}}} R_{\text{equator}} = \frac{9.78 \text{ m/s}^2}{9.83 \text{ m/s}^2}(80.0 \text{ m})$$

$$= 79.6 \text{ m}$$

Finalize Notice one of the advantages of this powerful technique of setting up ratios; we do not need to know the magnitude (v_i) nor the direction (θ_i) of the initial velocity. As long as they are the same at both locations, they cancel in the ratio.

3.4 | Analysis Model: Particle in Uniform Circular Motion

Figure 3.13a shows a car moving in a circular path; we describe this motion by calling it **circular motion**. If the car is moving on this path with *constant speed v,* we call it **uniform circular motion**. Because it occurs so often, this type of motion is recognized as an analysis model called the **particle in uniform circular motion**. We discuss this model in this section.

It is often surprising to students to find that even though an object moves at a constant speed in a circular path, *it still has an acceleration.* To see why, consider the defining equation for average acceleration, $\vec{\mathbf{a}}_{\text{avg}} = \Delta\vec{\mathbf{v}}/\Delta t$ (Eq. 3.4). The acceleration depends on *the change in the velocity vector.* Because velocity is a vector quantity, an acceleration can be produced in two ways, as mentioned in Section 3.1: by a change in the *magnitude* of the velocity or by a change in the *direction* of the velocity. The latter situation is occurring for an object moving with constant speed in a circular path. The constant-magnitude velocity vector is always tangent to the path of the object and perpendicular to the radius of the circular path. Therefore, the velocity vector is constantly *changing.* We now show that the acceleration vector in uniform circular motion is always perpendicular to the path and always points toward the center of the circle.

Let us first argue conceptually that the acceleration must be perpendicular to the path followed by the particle. If not, there would be a component of the acceleration parallel to the path and therefore parallel to the velocity vector. Such an acceleration component would lead to a change in the speed of the object, which we model as a particle, along the path. This change, however, is inconsistent with our setup of the problem in which the particle moves with constant speed along the path. Therefore, for *uniform* circular motion, the acceleration vector can only have a component perpendicular to the path, which is toward the center of the circle.

Let us now find the magnitude of the acceleration of the particle. Consider the pictorial representation of the position and velocity vectors for the car modeled as a particle in Figure 3.13b. In addition, the figure shows the vector representing the change in position, $\Delta\vec{\mathbf{r}}$. The particle follows a circular path, part of which is shown by the dashed

Pitfall Prevention | 3.4
Acceleration of a Particle in Uniform Circular Motion
Remember that acceleration in physics is defined as a change in the *velocity,* not a change in the *speed* (contrary to the everyday interpretation). In circular motion, the velocity vector is changing in direction, so there is indeed an acceleration.

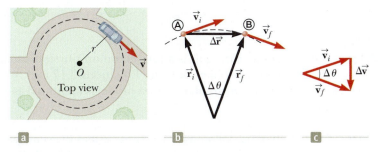

Figure 3.13 (a) A car moving along a circular path at constant speed is in uniform circular motion. (b) As the particle moves from Ⓐ to Ⓑ, its velocity vector changes from $\vec{\mathbf{v}}_i$ to $\vec{\mathbf{v}}_f$. (c) The construction for determining the direction of the change in velocity $\Delta\vec{\mathbf{v}}$, which is toward the center of the circle for small $\Delta\theta$.

curve. The particle is at Ⓐ at time t_i, and its velocity at that time is $\vec{\mathbf{v}}_i$; it is at Ⓑ at some later time t_f, and its velocity at that time is $\vec{\mathbf{v}}_f$. Let us also assume that $\vec{\mathbf{v}}_i$ and $\vec{\mathbf{v}}_f$ differ only in direction; their magnitudes are the same (i.e., $v_i = v_f = v$, because it is *uniform circular motion*). To calculate the acceleration of the particle, let us begin with the defining equation for average acceleration (Eq. 3.4):

$$\vec{\mathbf{a}}_{\text{avg}} = \frac{\vec{\mathbf{v}}_f - \vec{\mathbf{v}}_i}{t_f - t_i} = \frac{\Delta\vec{\mathbf{v}}}{\Delta t}$$

In Figure 3.13c, the velocity vectors in Figure 3.13b have been redrawn tail to tail. The vector $\Delta\vec{\mathbf{v}}$ connects the tips of the vectors, representing the vector addition, $\vec{\mathbf{v}}_f = \vec{\mathbf{v}}_i + \Delta\vec{\mathbf{v}}$. In Figures 3.13b and 3.13c, we can identify triangles that can serve as geometric models to help us analyze the motion. The angle $\Delta\theta$ between the two position vectors in Figure 3.13b is the same as the angle between the velocity vectors in Figure 3.13c because the velocity vector $\vec{\mathbf{v}}$ is always perpendicular to the position vector $\vec{\mathbf{r}}$. Therefore, the two triangles are *similar*. (Two triangles are similar if the angle between any two sides is the same for both triangles and if the ratio of the lengths of these sides is the same.) This similarity enables us to write a relationship between the lengths of the sides for the two triangles:

$$\frac{|\Delta\vec{\mathbf{v}}|}{v} = \frac{|\Delta\vec{\mathbf{r}}|}{r}$$

where $v = v_i = v_f$ and $r = r_i = r_f$. This equation can be solved for $|\Delta\vec{\mathbf{v}}|$ and the expression so obtained can be substituted into $\vec{\mathbf{a}}_{\text{avg}} = \Delta\vec{\mathbf{v}}/\Delta t$ (Eq. 3.4) to give the magnitude of the average acceleration over the time interval for the particle to move from Ⓐ to Ⓑ:

$$|\vec{\mathbf{a}}_{\text{avg}}| = \frac{v}{r}\frac{|\Delta\vec{\mathbf{r}}|}{\Delta t}$$

Now imagine that we bring points Ⓐ and Ⓑ in Figure 3.13b very close together. As Ⓐ and Ⓑ approach each other, Δt approaches zero and the ratio $|\Delta\vec{\mathbf{r}}|/\Delta t$ approaches the speed v. In addition, the average acceleration becomes the instantaneous acceleration at point Ⓐ. Hence, in the limit $\Delta t \to 0$, the magnitude of the acceleration is

► Magnitude of centripetal acceleration

$$a_c = \frac{v^2}{r}$$ **3.17** ◄

An acceleration of this nature is called a **centripetal acceleration** (*centripetal* means *center-seeking*). The subscript on the acceleration symbol reminds us that the acceleration is centripetal.

In many situations, it is convenient to describe the motion of a particle moving with constant speed in a circle of radius r in terms of the **period** T, which is defined as the time interval required for one complete revolution. In the time interval T, the particle moves a distance of $2\pi r$, which is equal to the circumference of the particle's circular path. Therefore, because its speed is equal to the circumference of the circular path divided by the period, or $v = 2\pi r/T$, it follows that

► Period of a particle in uniform circular motion

$$T = \frac{2\pi r}{v}$$ **3.18** ◄

The particle in uniform circular motion is a very common physical situation and is useful as an analysis model for problem solving. Equations 3.17 and 3.18 are to be used when the particle in uniform circular motion model is identified as appropriate for a given situation.

Pitfall Prevention | 3.5
Centripetal Acceleration Is Not Constant
The magnitude of the centripetal acceleration vector is constant for uniform circular motion, but *the centripetal acceleration vector is not constant.* It always points toward the center of the circle, so it continuously changes direction as the particle moves.

▶ **QUICK QUIZ 3.4** Which of the following correctly describes the centripetal acceleration vector for a particle moving in a circular path? **(a)** constant and always perpendicular to the velocity vector for the particle **(b)** constant and always parallel to the velocity vector for the particle **(c)** of constant magnitude and always perpendicular to the velocity vector for the particle **(d)** of constant magnitude and always parallel to the velocity vector for the particle

> ### THINKING PHYSICS 3.2
>
> An airplane travels from Los Angeles to Sydney, Australia. After cruising altitude is reached, the instruments on the plane indicate that the ground speed holds rock-steady at 700 km/h and that the heading of the airplane does not change. Is the velocity of the airplane constant during the flight?
>
> **Reasoning** The velocity is not constant because of the curvature of the Earth. Even though the speed does not change and the heading is always toward Sydney (is that actually true?), the airplane travels around a significant portion of the Earth's circumference. Therefore, the direction of the velocity vector does indeed change. We could extend this situation by imagining that the airplane passes over Sydney and continues (assuming it has enough fuel!) around the Earth until it arrives at Los Angeles again. It is impossible for an airplane to have a constant velocity (relative to the Universe, not to the Earth's surface) and return to its starting point. ◀

Example 3.5 | The Centripetal Acceleration of the Earth

What is the centripetal acceleration of the Earth as it moves in its orbit around the Sun?

SOLUTION

Conceptualize Think about a mental image of the Earth in a circular orbit around the Sun. We will model the Earth as a particle and approximate the Earth's orbit as circular (it's actually slightly elliptical, as we discuss in Chapter 11).

Categorize The Conceptualize step allows us to categorize this problem as one of a particle in uniform circular motion.

Analyze We do not know the orbital speed of the Earth to substitute into Equation 3.17. With the help of Equation 3.18, however, we can recast Equation 3.17 in terms of the period of the Earth's orbit, which we know is one year, and the radius of the Earth's orbit around the Sun, which is 1.496×10^{11} m.

Combine Equations 3.17 and 3.18:

$$a_c = \frac{v^2}{r} = \frac{\left(\dfrac{2\pi r}{T}\right)^2}{r} = \frac{4\pi^2 r}{T^2}$$

Substitute numerical values:

$$a_c = \frac{4\pi^2(1.496 \times 10^{11}\text{ m})}{(1\text{ yr})^2}\left(\frac{1\text{ yr}}{3.156 \times 10^7\text{ s}}\right)^2 = \boxed{5.93 \times 10^{-3}\text{ m/s}^2}$$

Finalize This acceleration is much smaller than the free-fall acceleration on the surface of the Earth. An important technique we learned here is replacing the speed v in Equation 3.17 in terms of the period T of the motion. In many problems, it is more likely that T is known rather than v.

3.5 | Tangential and Radial Acceleration

Let us consider a more general motion than that presented in Section 3.4. Consider a particle moving to the right along a curved path where the velocity changes both in direction *and* in magnitude, as described in Active Figure 3.14 (page 84). In this situation, the velocity vector is always tangent to the path; the acceleration vector $\vec{a}$, however, is at some angle to the path. At each instant, the particle can be modeled as if it were moving on a circular path. The radius of the circular path is the radius of curvature of the path at that instant. In the next instant, the particle is moving as if on a different circular path, with a different center and a different radius than the previous one. At each of three points Ⓐ, Ⓑ, and Ⓒ in Active Figure 3.14, we see the dashed circles that form geometric models of circular paths for the actual path at each point.

Active Figure 3.14 The motion of
a particle along an arbitrary curved
path lying in the *xy* plane. If the
velocity vector $\vec{v}$ (always tangent to
the path) changes in direction and
magnitude, the acceleration vector $\vec{a}$
has a tangential component a_t and a
radial component a_r.

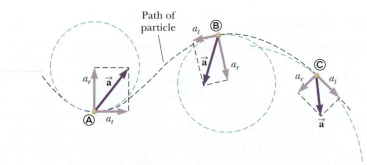

As the particle moves along the curved path in Active Figure 3.14, the direction of the total acceleration vector $\vec{a}$ changes from point to point. This vector can be resolved into two components based on an origin at the center of the model circle: a radial component a_r along the radius of the model circle and a tangential component a_t perpendicular to this radius. The *total* acceleration vector $\vec{a}$ can be written as the vector sum of the component vectors:

$$\vec{a} = \vec{a}_r + \vec{a}_t \qquad\qquad \textbf{3.19}◀$$

The tangential acceleration arises from the change in the speed of the particle and is given by

▶ Tangential acceleration

$$a_t = \frac{d|\vec{v}|}{dt} \qquad\qquad \textbf{3.20}◀$$

The radial acceleration is a result of the change in direction of the velocity vector and is given by

▶ Radial acceleration

$$a_r = -a_c = -\frac{v^2}{r}$$

where r is the radius of curvature of the path at the point in question, which is the radius of the model circle. We recognize the magnitude of the radial component of the acceleration as the centripetal acceleration discussed in Section 3.4. The negative sign indicates that the direction of the centripetal acceleration is toward the center of the model circle, opposite the direction of the radial unit vector $\hat{\mathbf{r}}$, which always points away from the center of the circle.

Because $\vec{a}_r$ and $\vec{a}_t$ are perpendicular component vectors of $\vec{a}$, it follows that $a = \sqrt{a_r^2 + a_t^2}$. At a given speed, a_r is large when the radius of curvature is small (as at points Ⓐ and Ⓑ in Active Fig. 3.14) and small when r is large (such as at point Ⓒ). The direction of $\vec{a}_t$ is either in the same direction as $\vec{v}$ (if v is increasing) or opposite $\vec{v}$ (if v is decreasing, as at point Ⓑ).

In the case of uniform circular motion, where v is constant, $a_t = 0$ and the acceleration is always radial, as described in Section 3.4. In other words, uniform circular motion is a special case of motion along a curved path. Furthermore, if the direction of $\vec{v}$ doesn't change, no radial acceleration occurs and the motion is one dimensional ($a_r = 0$, but a_t may not be zero).

QUICK QUIZ 3.5 A particle moves along a path, and its speed increases with time. **(i)** In which of the following cases are its acceleration and velocity vectors parallel? (a) when the path is circular. (b) when the path is straight. (c) when the path is a parabola. (d) never. **(ii)** From the same choices, in which case are its acceleration and velocity vectors perpendicular everywhere along the path?

❮3.6 | Relative Velocity and Relative Acceleration

In this section, we describe how observations made by different observers in different frames of reference are related to one another. A frame of reference can be described by a Cartesian coordinate system for which an observer is at rest with respect to the origin.

Let us conceptualize a sample situation in which there will be different observations for different observers. Consider the two observers A and B along the number line in Figure 3.15a. Observer A is located at the origin of a one-dimensional x_A axis, while observer B is at the position $x_A = -5$. We denote the position variable as x_A because observer A is at the origin of this axis. Both observers measure the position of point P, which is located at $x_A = +5$. Suppose observer B decides that he is located at the origin of an x_B axis as in Figure 3.15b. Notice that the two observers disagree on the value of the position of point P. Observer A claims point P is located at a position with a value of +5, whereas observer B claims it is located at a position with a value of +10. Both observers are correct, even though they make different measurements. Their measurements differ because they are making the measurement from different frames of reference.

Imagine now that observer B in Figure 3.15b is moving to the right along the x_B axis. Now the two measurements are even more different. Observer A claims point P remains at rest at a position with a value of +5, whereas observer B claims the position of P continuously changes with time, even passing him and moving behind him! Again, both observers are correct, with the difference in their measurements arising from their different frames of reference.

We explore this phenomenon further by considering two observers watching a man walking on a moving beltway at an airport in Figure 3.16. The woman standing on the moving beltway sees the man moving at a normal walking speed. The woman observing from the stationary floor sees the man moving with a higher speed because the beltway speed combines with his walking speed. Both observers look at the same man and arrive at different values for his speed. Both are correct; the difference in their measurements results from the relative velocity of their frames of reference.

In a more general situation, consider a particle located at point P in Figure 3.17 (page 86). Imagine that the motion of this particle is being described by two observers, observer A in a reference frame S_A fixed relative to the Earth and a second observer B in a reference frame S_B moving to the right relative to S_A (and therefore relative to the Earth) with a constant velocity $\vec{\mathbf{v}}_{BA}$. In this discussion of relative velocity, we use a double-subscript notation; the first subscript represents what is being observed, and the second represents who is doing the observing. Therefore, the notation $\vec{\mathbf{v}}_{BA}$ means the velocity of observer B (and the attached frame S_B) as measured by observer A. With this notation, observer B measures A to be moving to the left with a velocity $\vec{\mathbf{v}}_{AB} = -\vec{\mathbf{v}}_{BA}$. For purposes of this discussion, let us place each observer at her or his respective origin.

We define the time $t = 0$ as the instant at which the origins of the two reference frames coincide in space. Therefore, at time t, the origins of the reference frames will be separated by a distance $v_{BA}t$. We label the position P of the particle relative to observer A with the position vector $\vec{\mathbf{r}}_{PA}$ and that relative to observer B with the position vector $\vec{\mathbf{r}}_{PB}$, both at time t. From Figure 3.17, we see that the vectors $\vec{\mathbf{r}}_{PA}$ and $\vec{\mathbf{r}}_{PB}$ are related to each other through the expression

$$\vec{\mathbf{r}}_{PA} = \vec{\mathbf{r}}_{PB} + \vec{\mathbf{v}}_{BA}t \qquad \textbf{3.21} \blacktriangleleft$$

By differentiating Equation 3.21 with respect to time, noting that $\vec{\mathbf{v}}_{BA}$ is constant, we obtain

$$\frac{d\vec{\mathbf{r}}_{PA}}{dt} = \frac{d\vec{\mathbf{r}}_{PB}}{dt} + \vec{\mathbf{v}}_{BA}$$

$$\vec{\mathbf{u}}_{PA} = \vec{\mathbf{u}}_{PB} + \vec{\mathbf{v}}_{BA} \qquad \textbf{3.22} \blacktriangleleft \qquad \blacktriangleright \text{ Galilean velocity transformation}$$

where $\vec{\mathbf{u}}_{PA}$ is the velocity of the particle at P measured by observer A and $\vec{\mathbf{u}}_{PB}$ is its velocity measured by B. (We use the symbol $\vec{\mathbf{u}}$ for particle velocity rather than $\vec{\mathbf{v}}$, which we have already used for the relative velocity of two reference frames.) Equations 3.21 and 3.22 are known as **Galilean transformation equations.** They relate the position and velocity of a particle as measured by observers in relative motion. Notice the pattern of the subscripts in Equation 3.22. When relative velocities are

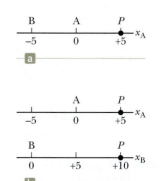

Figure 3.15 Different observers make different measurements. (a) Observer A is located at the origin, and Observer B is at a position of −5. Both observers measure the position of a particle at P. (b) If both observers see themselves at the origin of their own coordinate system, they disagree on the value of the position of the particle at P.

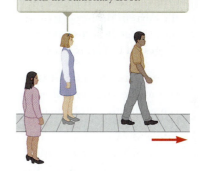

The woman standing on the beltway sees the man moving with a slower speed than does the other woman observing the man from the stationary floor.

Figure 3.16 Two observers measure the speed of a man walking on a moving beltway.

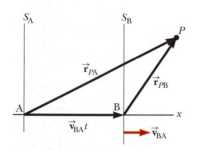

Figure 3.17 A particle located at *P* is described by two observers, one in the fixed frame of reference S_A and the other in the frame S_B, which moves to the right with a constant velocity $\vec{\mathbf{v}}_{BA}$. The vector $\vec{\mathbf{r}}_{PA}$ is the particle's position vector relative to S_A, and $\vec{\mathbf{r}}_{PB}$ is its position vector relative to S_B.

added, the inner subscripts (B) are the same and the outer ones (*P*, A) match the subscripts on the velocity on the left of the equation.

Although observers in two frames measure different velocities for the particle, they measure the *same acceleration* when $\vec{\mathbf{v}}_{BA}$ is constant. We can verify that by taking the time derivative of Equation 3.22:

$$\frac{d\vec{\mathbf{u}}_{PA}}{dt} = \frac{d\vec{\mathbf{u}}_{PB}}{dt} + \frac{d\vec{\mathbf{v}}_{BA}}{dt}$$

Because $\vec{\mathbf{v}}_{BA}$ is constant, $d\vec{\mathbf{v}}_{BA}/dt = 0$. Therefore, we conclude that $\vec{\mathbf{a}}_{PA} = \vec{\mathbf{a}}_{PB}$ because $\vec{\mathbf{a}}_{PA} = d\vec{\mathbf{u}}_{PA}/dt$ and $\vec{\mathbf{a}}_{PB} = d\vec{\mathbf{u}}_{PB}/dt$. That is, the acceleration of the particle measured by an observer in one frame of reference is the same as that measured by any other observer moving with constant velocity relative to the first frame.

Example 3.6 | A Boat Crossing a River

A boat crossing a wide river moves with a speed of 10.0 km/h relative to the water. The water in the river has a uniform speed of 5.00 km/h due east relative to the Earth.

(A) If the boat heads due north, determine the velocity of the boat relative to an observer standing on either bank.

SOLUTION

Conceptualize Imagine moving in a boat across a river while the current pushes you down the river. You will not be able to move directly across the river, but will end up downstream as suggested in Figure 3.18a.

Categorize Because of the combined velocities of you relative to the river and the river relative to the Earth, we can categorize this problem as one involving relative velocities.

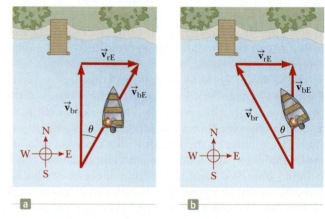

Figure 3.18 (Example 3.6) (a) A boat aims directly across a river and ends up downstream. (b) To move directly across the river, the boat must aim upstream.

Analyze We know $\vec{\mathbf{v}}_{br}$, the velocity of the *boat* relative to the *river*, and $\vec{\mathbf{v}}_{rE}$, the velocity of the *river* relative to the *Earth*. What we must find is $\vec{\mathbf{v}}_{bE}$, the velocity of the *boat* relative to the *Earth*. The relationship between these three quantities is $\vec{\mathbf{v}}_{bE} = \vec{\mathbf{v}}_{br} + \vec{\mathbf{v}}_{rE}$. The terms in the equation must be manipulated as vector quantities; the vectors are shown in Figure 3.18a. The quantity $\vec{\mathbf{v}}_{br}$ is due north; $\vec{\mathbf{v}}_{rE}$ is due east; and the vector sum of the two, $\vec{\mathbf{v}}_{bE}$, is at an angle θ as defined in Figure 3.18a.

Find the speed v_{bE} of the boat relative to the Earth using the Pythagorean theorem:

$$v_{bE} = \sqrt{v_{br}^2 + v_{rE}^2} = \sqrt{(10.0 \text{ km/h})^2 + (5.00 \text{ km/h})^2}$$
$$= 11.2 \text{ km/h}$$

Find the direction of $\vec{\mathbf{v}}_{bE}$:

$$\theta = \tan^{-1}\left(\frac{v_{rE}}{v_{br}}\right) = \tan^{-1}\left(\frac{5.00}{10.0}\right) = 26.6°$$

Finalize The boat is moving at a speed of 11.2 km/h in the direction 26.6° east of north relative to the Earth. Notice that the speed of 11.2 km/h is faster than your boat speed of 10.0 km/h. The current velocity adds to yours to give you a higher speed. Notice in Figure 3.18a that your resultant velocity is at an angle to the direction straight across the river, so you will end up downstream, as we predicted.

(B) If the boat travels with the same speed of 10.0 km/h relative to the river and is to travel due north as shown in Figure 3.18b, what should its heading be?

3.6 *cont.*

SOLUTION

Conceptualize/Categorize This question is an extension of part (A), so we have already conceptualized and categorized the problem. In this case, however, we must aim the boat upstream so as to go straight across the river.

Analyze The analysis now involves the new triangle shown in Figure 3.18b. As in part (A), we know $\vec{v}_{rE}$ and the magnitude of the vector $\vec{v}_{br}$, and we want $\vec{v}_{bE}$ to be directed across the river. Notice the difference between the triangle in Figure 3.18a and the one in Figure 3.18b: the hypotenuse in Figure 3.18b is no longer $\vec{v}_{bE}$.

Use the Pythagorean theorem to find v_{bE}:

$$v_{bE} = \sqrt{v_{br}^2 - v_{rE}^2} = \sqrt{(10.0 \text{ km/h})^2 - (5.00 \text{ km/h})^2} = 8.66 \text{ km/h}$$

Find the direction in which the boat is heading:

$$\theta = \tan^{-1}\left(\frac{v_{rE}}{v_{bE}}\right) = \tan^{-1}\left(\frac{5.00}{8.66}\right) = \boxed{30.0°}$$

Finalize The boat must head upstream so as to travel directly northward across the river. For the given situation, the boat must steer a course 30.0° west of north. For faster currents, the boat must be aimed upstream at larger angles.

What If? Imagine that the two boats in parts (A) and (B) are racing across the river. Which boat arrives at the opposite bank first?

Answer In part (A), the velocity of 10 km/h is aimed directly across the river. In part (B), the velocity that is directed across the river has a magnitude of only 8.66 km/h. Therefore, the boat in part (A) has a larger velocity component directly across the river and arrives first.

3.7 | Context Connection: Lateral Acceleration of Automobiles

An automobile does not travel in a straight line. It follows a two-dimensional path on a flat Earth surface and a three-dimensional path if there are hills and valleys. Let us restrict our thinking at this point to an automobile traveling in two dimensions on a flat roadway. During a turn, the automobile can be modeled as following an arc of a circular path at each point in its motion. Consequently, the automobile will have a centripetal acceleration.

A desired characteristic of automobiles is that they can negotiate a curve without rolling over. This characteristic depends on the centripetal acceleration. Imagine standing a book upright on a strip of sandpaper. If the sandpaper is moved slowly across the surface of a table with a very small acceleration, the book will stay upright. If the sandpaper is moved with a large acceleration, however, the book will fall over. That is what we would like to avoid in a car.

Imagine that instead of accelerating a book in one dimension we are centripetally accelerating a car in a circular path. The effect is the same. If there is too much centripetal acceleration, the car will "fall over" and will go into a sideways roll. The maximum possible centripetal acceleration that a car can exhibit without rolling over in a turn is called *lateral acceleration*. Two contributions to the lateral acceleration of a car are the height of the center of mass of the car above the ground and the side-to-side distance between the wheels. (We will study center of mass in Chapter 8.) The book in our demonstration has a relatively large ratio of the height of the center of mass to the width of the book upon which it is sitting, so it falls over relatively easily at low accelerations. An automobile has a much lower ratio of the height of the center of mass to the distance between the wheels. Therefore, it can withstand higher accelerations.

Consider the documented lateral acceleration of the vehicles from Table 2.3 listed in Table 3.1 (page 88). These values are given as multiples of *g*, the acceleration due to gravity. Notice that most of the very expensive vehicles and performance

TABLE 3.1 | Lateral Acceleration of Automobiles

Automobile	Lateral Acceleration (g)	Automobile	Lateral Acceleration (g)
Very expensive vehicles:		*Traditional vehicles:*	
Bugatti Veyron 16.4 Super Sport	1.40	Buick Regal CXL Turbo	0.85
Lamborghini LP 570-4 Superleggera	0.98	Chevrolet Tahoe 1500 LS (SUV)	0.70
Lexus LFA	1.04	Ford Fiesta SES	0.84
Mercedes-Benz SLS AMG	0.96	Hummer H3 (SUV)	0.66
Shelby SuperCars Ultimate Aero	1.05	Hyundai Sonata SE	0.85
Average	**1.09**	Smart ForTwo	0.72
		Average	**0.77**
Performance vehicles:		*Alternative vehicles:*	
Chevrolet Corvette ZR1	1.07	Chevrolet Volt (hybrid)	0.83
Dodge Viper SRT10	1.06	Nissan Leaf (electric)	0.79
Jaguar XJL Supercharged	0.88	Honda CR-Z (hybrid)	0.83
Acura TL SH-AWD	0.91	Honda Insight (hybrid)	0.74
Dodge Challenger SRT8	0.88	Toyota Prius (hybrid)	0.76
Average	**0.96**	**Average**	**0.79**

vehicles have a lateral acceleration close to that due to gravity and that the lateral acceleration of the Bugatti is 40% larger than that due to gravity. The Bugatti is a very stable vehicle!

In contrast, the lateral acceleration of nonperformance cars is lower because they generally are not designed to travel around turns at such a high speed as the performance cars. For example, the Buick Regal has a lateral acceleration of 0.85g. The two sport utility vehicles in the table have lateral accelerations lower than this value and such vehicles can have values as low as 0.62g. As a result, they are highly prone to rollovers in emergency maneuvers.

SUMMARY

If a particle moves with *constant* acceleration $\vec{a}$ and has velocity $\vec{v}_i$ and position $\vec{r}_i$ at $t = 0$, its velocity and position vectors at some later time t are

$$\vec{v}_f = \vec{v}_i + \vec{a}t \qquad \text{3.8} \blacktriangleleft$$

$$\vec{r}_f = \vec{r}_i + \vec{v}_i t + \tfrac{1}{2}\vec{a}t^2 \qquad \text{3.9} \blacktriangleleft$$

For two-dimensional motion in the xy plane under constant acceleration, these vector expressions are equivalent to two component expressions, one for the motion along x and one for the motion along y.

Projectile motion is a special case of two-dimensional motion under constant acceleration, where $a_x = 0$ and $a_y = -g$. In this case, the horizontal components of Equations 3.8 and 3.9 reduce to those of a particle under constant velocity:

$$v_{xf} = v_{xi} = \text{constant} \qquad \text{3.10} \blacktriangleleft$$

$$x_f = x_i + v_{xi}t \qquad \text{3.12} \blacktriangleleft$$

The vertical components of Equations 3.8 and 3.9 are those of a particle under constant acceleration:

$$v_{yf} = v_{yi} - gt \qquad \text{3.11} \blacktriangleleft$$

$$y_f = y_i + v_{yi}t - \tfrac{1}{2}gt^2 \qquad \text{3.13} \blacktriangleleft$$

where $v_{xi} = v_i \cos\theta_i$, $v_{yi} = v_i \sin\theta_i$, v_i is the initial speed of the projectile, and θ_i is the angle $\vec{v}_i$ makes with the positive x axis.

If a particle moves along a curved path in such a way that the magnitude and direction of $\vec{v}$ change in time, the particle has an acceleration vector that can be described by two components: (1) a radial component a_r arising from the change in direction of $\vec{v}$ and (2) a tangential component a_t arising from the change in magnitude of $\vec{v}$. The radial acceleration is called **centripetal acceleration,** and its direction is always toward the center of the circular path.

If an observer B is moving with velocity $\vec{\mathbf{v}}_{BA}$ with respect to observer A, their measurements of the velocity of a particle located at point P are related according to

$$\vec{\mathbf{u}}_{PA} = \vec{\mathbf{u}}_{PB} + \vec{\mathbf{v}}_{BA} \qquad 3.22 \blacktriangleleft$$

Equation 3.22 is the **Galilean transformation equation** for velocities and indicates that different observers will measure different velocities for the same particle.

Analysis Model for Problem Solving

Particle in Uniform Circular Motion If a particle moves in a circular path of radius r with a constant speed v, the magnitude of its centripetal acceleration is given by

$$a_c = \frac{v^2}{r} \qquad 3.17 \blacktriangleleft$$

and the **period** of the particle's motion is given by

$$T = \frac{2\pi r}{v} \qquad 3.18 \blacktriangleleft$$

OBJECTIVE QUESTIONS

☐ denotes answer available in *Student Solutions Manual/Study Guide*

1. In which of the following situations is the moving object appropriately modeled as a projectile? Choose all correct answers. (a) A shoe is tossed in an arbitrary direction. (b) A jet airplane crosses the sky with its engines thrusting the plane forward. (c) A rocket leaves the launch pad. (d) A rocket moves through the sky, at much less than the speed of sound, after its fuel has been used up. (e) A diver throws a stone under water.

2. A rubber stopper on the end of a string is swung steadily in a horizontal circle. In one trial, it moves at speed v in a circle of radius r. In a second trial, it moves at a higher speed $3v$ in a circle of radius $3r$. In this second trial, is its acceleration (a) the same as in the first trial, (b) three times larger, (c) one-third as large, (d) nine times larger, or (e) one-ninth as large?

3. Figure OQ3.3 shows a bird's-eye view of a car going around a high-way curve. As the car moves from point 1 to point 2, its speed doubles. Which of the vectors (a) through (e) shows the direction of the car's average acceleration between these two points?

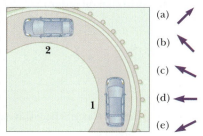

Figure OQ3.3

4. Entering his dorm room, a student tosses his book bag to the right and upward at an angle of 45° with the horizontal (Fig. OQ3.4). Air resistance does not affect the bag. The bag moves through point Ⓐ immediately after it leaves the student's hand, through point Ⓑ at the top of its flight, and through point Ⓒ immediately before it lands on the top

Figure OQ3.4

bunk bed. **(i)** Rank the following horizontal and vertical velocity components from the largest to the smallest. (a) $v_{Ⓐx}$ (b) $v_{Ⓐy}$ (c) $v_{Ⓑx}$ (d) $v_{Ⓑy}$ (e) $v_{Ⓒy}$. Note that zero is larger than a negative number. If two quantities are equal, show them as equal in your list. If any quantity is equal to zero, show that fact in your list. **(ii)** Similarly, rank the following acceleration components. (a) $a_{Ⓐx}$ (b) $a_{Ⓐy}$ (c) $a_{Ⓑx}$ (d) $a_{Ⓑy}$ (e) $a_{Ⓒy}$.

5. Does a car moving around a circular track with constant speed have (a) zero acceleration, (b) an acceleration in the direction of its velocity, (c) an acceleration directed away from the center of its path, (d) an acceleration directed toward the center of its path, or (e) an acceleration with a direction that cannot be determined from the given information?

6. An astronaut hits a golf ball on the Moon. Which of the following quantities, if any, remain constant as a ball travels through the vacuum there? (a) speed (b) acceleration (c) horizontal component of velocity (d) vertical component of velocity (e) velocity

7. A projectile is launched on the Earth with a certain initial velocity and moves without air resistance. Another projectile is launched with the same initial velocity on the Moon, where the acceleration due to gravity is one-sixth as large. How does the range of the projectile on the Moon compare with that of the projectile on the Earth? (a) It is one-sixth as large. (b) It is the same. (c) It is $\sqrt{6}$ times larger. (d) It is 6 times larger. (e) It is 36 times larger.

8. A baseball is thrown from the outfield toward the catcher. When the ball reaches its highest point, which statement is true? (a) Its velocity and its acceleration are both zero. (b) Its velocity is not zero, but its acceleration is zero. (c) Its velocity is perpendicular to its acceleration. (d) Its acceleration depends on the angle at which the ball was thrown. (e) None of statements (a) through (d) is true.

9. A student throws a heavy red ball horizontally from a balcony of a tall building with an initial speed v_i. At the same time, a second student drops a lighter blue ball from the balcony. Neglecting air resistance, which statement is true? (a) The blue ball reaches the ground first. (b) The balls reach the ground at the same instant.

(c) The red ball reaches the ground first. (d) Both balls hit the ground with the same speed. (e) None of statements (a) through (d) is true.

10. A sailor drops a wrench from the top of a sailboat's vertical mast while the boat is moving rapidly and steadily straight forward. Where will the wrench hit the deck? (a) ahead of the base of the mast (b) at the base of the mast (c) behind the base of the mast (d) on the windward side of the base of the mast (e) None of the choices (a) through (d) is true.

11. A set of keys on the end of a string is swung steadily in a horizontal circle. In one trial, it moves at speed v in a circle of radius r. In a second trial, it moves at a higher speed $4v$ in a circle of radius $4r$. In the second trial, how does the period of its motion compare with its period in the first trial? (a) It is the same as in the first trial. (b) It is 4 times larger. (c) It is one-fourth as large. (d) It is 16 times larger. (e) It is one-sixteenth as large.

12. A certain light truck can go around a curve having a radius of 150 m with a maximum speed of 32.0 m/s. To have the same acceleration, at what maximum speed can it go around a curve having a radius of 75.0 m? (a) 64 m/s (b) 45 m/s (c) 32 m/s (d) 23 m/s (e) 16 m/s

☐ denotes answer available in *Student Solutions Manual/Study Guide*

CONCEPTUAL QUESTIONS

1. Explain whether or not the following particles have an acceleration: (a) a particle moving in a straight line with constant speed and (b) a particle moving around a curve with constant speed.

2. Describe how a driver can steer a car traveling at constant speed so that (a) the acceleration is zero or (b) the magnitude of the acceleration remains constant.

3. If you know the position vectors of a particle at two points along its path and also know the time interval during which it moved from one point to the other, can you determine the particle's instantaneous velocity? Its average velocity? Explain.

4. Construct motion diagrams showing the velocity and acceleration of a projectile at several points along its path, assuming (a) the projectile is launched horizontally and (b) the projectile is launched at an angle θ with the horizontal.

5. A spacecraft drifts through space at a constant velocity. Suddenly, a gas leak in the side of the spacecraft gives it a

constant acceleration in a direction perpendicular to the initial velocity. The orientation of the spacecraft does not change, so the acceleration remains perpendicular to the original direction of the velocity. What is the shape of the path followed by the spacecraft in this situation?

6. An ice skater is executing a figure eight, consisting of two identically shaped, tangent circular paths. Throughout the first loop she increases her speed uniformly, and during the second loop she moves at a constant speed. Draw a motion diagram showing her velocity and acceleration vectors at several points along the path of motion.

7. A projectile is launched at some angle to the horizontal with some initial speed v_i, and air resistance is negligible. (a) Is the projectile a freely falling body? (b) What is its acceleration in the vertical direction? (c) What is its acceleration in the horizontal direction?

PROBLEMS

Section 3.1 The Position, Velocity, and Acceleration Vectors

1. A motorist drives south at 20.0 m/s for 3.00 min, then turns west and travels at 25.0 m/s for 2.00 min, and finally travels northwest at 30.0 m/s for 1.00 min. For this 6.00-min trip, find (a) the total vector displacement, (b) the average speed, and (c) the average velocity. Let the positive x axis point east.

2. Suppose the position vector for a particle is given as a function of time by $\vec{r}(t) = x(t)\hat{i} + y(t)\hat{j}$, with $x(t) = at + b$ and $y(t) = ct^2 + d$, where $a = 1.00$ m/s, $b = 1.00$ m, $c = 0.125$ m/s², and $d = 1.00$ m. (a) Calculate the average velocity during the time interval from $t = 2.00$ s to $t = 4.00$ s. (b) Determine the velocity and the speed at $t = 2.00$ s.

Section 3.2 Two-Dimensional Motion with Constant Acceleration

3. **W** A particle initially located at the origin has an acceleration of $\vec{a} = 3.00\hat{j}$ m/s² and an initial velocity of $\vec{v}_i = 5.00\hat{i}$ m/s. Find (a) the vector position of the particle at any time t, (b) the velocity of the particle at any time t, (c) the coordinates of the particle at $t = 2.00$ s, and (d) the speed of the particle at $t = 2.00$ s.

4. **BIO** It is not possible to see very small objects, such as viruses, using an ordinary light microscope. An electron microscope, however, can view such objects using an electron beam instead of a light beam. Electron microscopy has proved invaluable for investigations of viruses, cell membranes and sub-cellular structures, bacterial surfaces, visual receptors, chloroplasts, and the contractile properties of muscles. The "lenses" of an electron microscope consist of electric and magnetic fields that control the electron beam. As an example of the manipulation of an electron beam, consider an electron traveling away from the origin along the x axis in the xy plane with initial velocity $\vec{v}_i = v_i\hat{i}$. As it passes through the region $x = 0$ to $x = d$, the electron experiences acceleration $\vec{a} = a_x\hat{i} + a_y\hat{j}$, where a_x and a_y are constants. For the case $v_i = 1.80 \times 10^7$ m/s, $a_x = 8.00 \times 10^{14}$ m/s², and $a_y = 1.60 \times 10^{15}$ m/s², determine at $x = d = 0.010\,0$ m (a) the position of the electron, (b) the velocity of the electron, (c) the speed of the electron, and (d) the direction of travel of the electron (i.e., the angle between its velocity and the x axis).

5. **M** A fish swimming in a horizontal plane has velocity $\vec{v}_i = (4.00\,\hat{i} + 1.00\,\hat{j})$ m/s at a point in the ocean where the position relative to a certain rock is $\vec{r}_i = (10.0\,\hat{i} - 4.00\,\hat{j})$ m. After the fish swims with constant acceleration for 20.0 s, its velocity is $\vec{v} = (20.0\hat{i} - 5.00\hat{j})$ m/s. (a) What are the components of the acceleration of the fish? (b) What is the direction of its acceleration with respect to unit vector $\hat{i}$? (c) If the fish maintains constant acceleration, where is it at $t = 25.0$ s and in what direction is it moving?

6. At $t = 0$, a particle moving in the xy plane with constant acceleration has a velocity of $\vec{v}_i = (3.00\,\hat{i} - 2.00\,\hat{j})$ m/s and is at the origin. At $t = 3.00$ s, the particle's velocity is $\vec{v}_f = (9.00\hat{i} + 7.00\hat{j})$ m/s. Find (a) the acceleration of the particle and (b) its coordinates at any time t.

Section 3.3 Projectile Motion

Note: Ignore air resistance in all problems and take $g = 9.80$ m/s² at the Earth's surface.

7. **BIO** Mayan kings and many school sports teams are named for the puma, cougar, or mountain lion—*Felis concolor*—the best jumper among animals. It can jump to a height of 12.0 ft when leaving the ground at an angle of 45.0°. With what speed, in SI units, does it leave the ground to make this leap?

8. **BIO** The small archerfish (length 20 to 25 cm) lives in brackish waters of Southeast Asia from India to the Philippines. This aptly named creature captures its prey by shooting a stream of water drops at an insect, either flying or at rest. The bug falls into the water and the fish gobbles it up. The archerfish has high accuracy at distances of 1.2 m to 1.5 m, and it sometimes makes hits at distances up to 3.5 m. A groove in the roof of its mouth, along with a curled tongue, forms a tube that enables the fish to impart high velocity to the water in its mouth when it suddenly closes its gill flaps. Suppose the archerfish shoots at a target that is 2.00 m away, measured along a line at an angle of 30.0° above the horizontal. With what velocity must the water stream be launched if it is not to drop more than 3.00 cm vertically on its path to the target?

9. A cannon with a muzzle speed of 1 000 m/s is used to start an avalanche on a mountain slope. The target is 2 000 m from the cannon horizontally and 800 m above the cannon. At what angle, above the horizontal, should the cannon be fired?

10. An astronaut on a strange planet finds that she can jump a maximum horizontal distance of 15.0 m if her initial speed is 3.00 m/s. What is the free-fall acceleration on the planet?

11. **M** In a local bar, a customer slides an empty beer mug down the counter for a refill. The height of the counter is 1.22 m. The mug slides off the counter and strikes the floor 1.40 m from the base of the counter. (a) With what velocity did the mug leave the counter? (b) What was the direction of the mug's velocity just before it hit the floor?

12. **S** In a local bar, a customer slides an empty beer mug down the counter for a refill. The height of the counter is h. The mug slides off the counter and strikes the floor at distance d from the base of the counter. (a) With what velocity did the mug leave the counter? (b) What was the direction of the mug's velocity just before it hit the floor?

13. **M** A placekicker must kick a football from a point 36.0 m (about 40 yards) from the goal. Half the crowd hopes the ball will clear the crossbar, which is 3.05 m high. When kicked, the ball leaves the ground with a speed of 20.0 m/s at an angle of 53.0° to the horizontal. (a) By how much does the ball clear or fall short of clearing the crossbar? (b) Does the ball approach the crossbar while still rising or while falling?

14. **W** A ball is tossed from an upper-story window of a building. The ball is given an initial velocity of 8.00 m/s at an angle of 20.0° below the horizontal. It strikes the ground 3.00 s later. (a) How far horizontally from the base of the building does the ball strike the ground? (b) Find the height from which the ball was thrown. (c) How long does it take the ball to reach a point 10.0 m below the level of launching?

15. The speed of a projectile when it reaches its maximum height is one-half its speed when it is at half its maximum height. What is the initial projection angle of the projectile?

16. **S** A firefighter, a distance d from a burning building, directs a stream of water from a fire hose at angle θ_i above the horizontal as shown in Figure P3.16. If the initial speed of the stream is v_i, at what height h does the water strike the building?

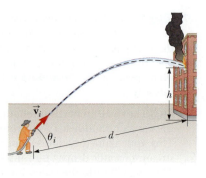

Figure P3.16

17. **W** A soccer player kicks a rock horizontally off a 40.0-m-high cliff into a pool of water. If the player hears the sound of the splash 3.00 s later, what was the initial speed given to the rock? Assume the speed of sound in air is 343 m/s.

18. A basketball star covers 2.80 m horizontally in a jump to dunk the ball (Fig. P3.18a). His motion through space can be modeled precisely as that of a particle at his *center of mass*, which we will define in Chapter 8. His center of mass is at elevation 1.02 m when he leaves the floor. It reaches a maximum height of 1.85 m above the floor and is at elevation 0.900 m when he touches down again. Determine (a) his time of flight (his "hang time"), (b) his horizontal and (c) vertical velocity components at the instant of takeoff, and (d) his takeoff angle. (e) For comparison, determine the hang time of a whitetail deer making a jump (Fig. P3.18b) with center-of-mass elevations $y_i = 1.20$ m, $y_{max} = 2.50$ m, and $y_f = 0.700$ m.

a, David Liam Kyle/NBAE/Getty Images;
b, B.G. Smith /Shutterstock.com

Figure P3.18

19. **GP** A student stands at the edge of a cliff and throws a stone horizontally over the edge with a speed of $v_i = 18.0$ m/s. The cliff is $h = 50.0$ m above a body of water as shown in Figure P3.19. (a) What are the coordinates of the initial position of the stone? (b) What are the components of the initial velocity of the stone? (c) What is the appropriate analysis model for the vertical motion of the stone? (d) What is the appropriate analysis model for the horizontal motion of the stone? (e) Write symbolic equations for the *x* and *y* components of the velocity of the stone as a function of time. (f) Write symbolic equations for the position of the stone as a function of time. (g) How long after being

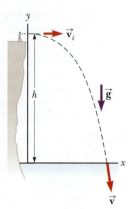

Figure P3.19

released does the stone strike the water below the cliff? (h) With what speed and angle of impact does the stone land?

20. The motion of a human body through space can be modeled as the motion of a particle at the body's center of mass, as we will study in Chapter 8. The components of the displacement of an athlete's center of mass from the beginning to the end of a certain jump are described by the equations

$$x_f = 0 + (11.2 \text{ m/s})(\cos 18.5°)t$$

$$0.360 \text{ m} = 0.840 \text{ m} + (11.2 \text{ m/s})(\sin 18.5°)t - \tfrac{1}{2}(9.80 \text{ m/s}^2)t^2$$

where t is in seconds and is the time at which the athlete ends the jump. Identify (a) the athlete's position and (b) his vector velocity at the takeoff point. (c) How far did he jump?

21. A playground is on the flat roof of a city school, 6.00 m above the street below (Fig. P3.21). The vertical wall of the building is $h = 7.00$ m high, forming a 1-m-high railing around the playground. A ball has fallen to the street below, and a passerby returns it by launching it at an angle of $\theta = 53.0°$ above the horizontal at a point $d = 24.0$ m from the base of the building wall. The ball takes 2.20 s to reach a point vertically above the wall. (a) Find the speed at which the ball was launched. (b) Find the vertical distance by which the ball clears the wall. (c) Find the horizontal distance from the wall to the point on the roof where the ball lands.

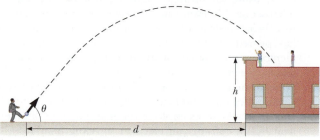

Figure P3.21

22. **S** A fireworks rocket explodes at height h, the peak of its vertical trajectory. It throws out burning fragments in all directions, but all at the same speed v. Pellets of solidified metal fall to the ground without air resistance. Find the smallest angle that the final velocity of an impacting fragment makes with the horizontal.

Section 3.4 Analysis Model: Particle in Uniform Circular Motion

23. The athlete shown in Figure P3.23 rotates a 1.00-kg discus along a circular path of radius 1.06 m. The maximum speed of the discus is 20.0 m/s. Determine the magnitude of the maximum radial acceleration of the discus.

ADRIAN DENNIS/AFP/Getty Images

Figure P3.23

24. In Example 3.5, we found the centripetal acceleration of the Earth as it revolves around the Sun. From information on the endpapers of this book, compute the centripetal acceleration of a point on the surface of the Earth at the equator caused by the rotation of the Earth about its axis.

25. As their booster rockets separate, Space Shuttle astronauts typically feel accelerations up to $3g$, where $g = 9.80$ m/s^2. In their training, astronauts ride in a device where they experience such an acceleration as a centripetal acceleration. Specifically, the astronaut is fastened securely at the end of a mechanical arm, which then turns at constant speed in a horizontal circle. Determine the rotation rate, in revolutions per second, required to give an astronaut a centripetal acceleration of $3.00g$ while in circular motion with radius 9.45 m.

26. Casting of molten metal is important in many industrial processes. *Centrifugal casting* is used for manufacturing pipes, bearings, and many other structures. A variety of sophisticated techniques have been invented, but the basic idea is as illustrated in Figure P3.26. A cylindrical enclosure is rotated rapidly and steadily about a horizontal axis. Molten metal is poured into the rotating cylinder and then cooled, forming the finished product. Turning the cylinder at a high rotation rate forces the solidifying metal strongly to the outside. Any bubbles are displaced toward the axis, so unwanted voids will not be present in the casting. Sometimes it is desirable to form a composite casting, such as for a bearing. Here a strong steel outer surface is poured and then inside it a lining of special low-friction metal. In some applications, a very strong metal is given a coating of corrosion-resistant metal. Centrifugal casting results in strong bonding between the layers.

Suppose a copper sleeve of inner radius 2.10 cm and outer radius 2.20 cm is to be cast. To eliminate bubbles and give high structural integrity, the centripetal acceleration of each bit of metal should be at least $100g$. What rate of rotation is required? State the answer in revolutions per minute.

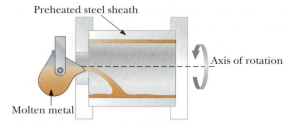

Figure P3.26

27. The astronaut orbiting the Earth in Figure P3.27 is preparing to dock with a Westar VI satellite. The satellite is in a circular orbit 600 km above the Earth's surface, where the free-fall acceleration is 8.21 m/s^2. Take the radius of the Earth as 6 400 km. Determine the speed of the satellite and the time interval required to complete one orbit around the Earth, which is the period of the satellite.

Figure P3.27

28. A tire 0.500 m in radius rotates at a constant rate of 200 rev/min. Find the speed and acceleration of a small stone lodged in the tread of the tire (on its outer edge).

Section 3.5 Tangential and Radial Acceleration

29. **M** A train slows down as it rounds a sharp horizontal turn, going from 90.0 km/h to 50.0 km/h in the 15.0 s it takes to round the bend. The radius of the curve is 150 m. Compute the acceleration at the moment the train speed reaches 50.0 km/h. Assume the train continues to slow down at this time at the same rate.

30. A point on a rotating turntable 20.0 cm from the center accelerates from rest to a final speed of 0.700 m/s in 1.75 s. At $t = 1.25$ s, find the magnitude and direction of (a) the radial acceleration, (b) the tangential acceleration, and (c) the total acceleration of the point.

31. **W** Figure P3.31 represents the total acceleration of a particle moving clockwise in a circle of radius 2.50 m at a certain instant of time. For that instant, find (a) the radial acceleration of the particle, (b) the speed of the particle, and (c) its tangential acceleration.

Figure P3.31

32. A ball swings counterclockwise in a vertical circle at the end of a rope 1.50 m long. When the ball is 36.9° past the lowest point on its way up, its total acceleration is $(-22.5\,\hat{\mathbf{i}} + 20.2\,\hat{\mathbf{j}})$ m/s^2. For that instant, (a) sketch a vector diagram showing the components of its acceleration, (b) determine the magnitude of its radial acceleration, and (c) determine the speed and velocity of the ball.

Section 3.6 Relative Velocity and Relative Acceleration

33. The pilot of an airplane notes that the compass indicates a heading due west. The airplane's speed relative to the air is 150 km/h. The air is moving in a wind at 30.0 km/h toward the north. Find the velocity of the airplane relative to the ground.

34. How long does it take an automobile traveling in the left lane at 60.0 km/h to pull alongside a car traveling in the right lane at 40.0 km/h if the cars' front bumpers are initially 100 m apart?

35. **M** **Q/C** A river has a steady speed of 0.500 m/s. A student swims upstream a distance of 1.00 km and swims back to the starting point. (a) If the student can swim at a speed of 1.20 m/s in still water, how long does the trip take? (b) How much time is required in still water for the same length swim? (c) Intuitively, why does the swim take longer when there is a current?

36. **Q/C** **S** A river flows with a steady speed v. A student swims upstream a distance d and then back to the starting point. The student can swim at speed c in still water. (a) In terms of d, v, and c, what time interval is required for the round trip? (b) What time interval would be required if the water were still? (c) Which time interval is larger? Explain whether it is always larger.

37. A car travels due east with a speed of 50.0 km/h. Raindrops are falling at a constant speed vertically with respect to the Earth. The traces of the rain on the side windows of the car make an angle of 60.0° with the vertical. Find the velocity of the rain with respect to (a) the car and (b) the Earth.

38. A Coast Guard cutter detects an unidentified ship at a distance of 20.0 km in the direction 15.0° east of north. The ship is traveling at 26.0 km/h on a course at 40.0° east of north. The Coast Guard wishes to send a speedboat to intercept and investigate the vessel. If the speedboat travels at 50.0 km/h, in what direction should it head? Express the direction as a compass bearing with respect to due north.

39. [M] A science student is riding on a flatcar of a train traveling along a straight, horizontal track at a constant speed of 10.0 m/s. The student throws a ball into the air along a path that he judges to make an initial angle of 60.0° with the horizontal and to be in line with the track. The student's professor, who is standing on the ground nearby, observes the ball to rise vertically. How high does she see the ball rise?

40. [S] Two swimmers, Chris and Sarah, start together at the same point on the bank of a wide stream that flows with a speed v. Both move at the same speed c (where $c > v$) relative to the water. Chris swims downstream a distance L and then upstream the same distance. Sarah swims so that her motion relative to the Earth is perpendicular to the banks of the stream. She swims the distance L and then back the same distance, with both swimmers returning to the starting point. In terms of L, c, and v, find the time intervals required (a) for Chris's round trip and (b) for Sarah's round trip. (c) Explain which swimmer returns first.

Section 3.7 Context Connection: Lateral Acceleration of Automobiles

41. A certain light truck can go around an unbanked curve having a radius of 150 m with a maximum speed of 32.0 m/s. With what maximum speed can it go around a curve having a radius of 75.0 m?

Additional Problems

42. A landscape architect is planning an artificial waterfall in a city park. Water flowing at 1.70 m/s will leave the end of a horizontal channel at the top of a vertical wall $h = 2.35$ m high, and from there it will fall into a pool (Fig. P3.42). (a) Will the space behind the waterfall be wide enough for a pedestrian walkway? (b) To sell her plan to the city council, the architect wants to build a model to standard scale, which is one-twelfth actual size. How fast should the water flow in the channel in the model?

Figure P3.42

43. *Why is the following situation impossible?* A normally proportioned adult walks briskly along a straight line in the $+x$ direction, standing straight up and holding his right arm vertical and next to his body so that the arm does not swing. His right hand holds a ball at his side a distance h above the floor. When the ball passes above a point marked as $x = 0$ on the horizontal floor, he opens his fingers to release the ball from rest relative to his hand. The ball strikes the ground for the first time at position $x = 7.00h$.

44. An astronaut on the surface of the Moon fires a cannon to launch an experiment package, which leaves the barrel moving horizontally. Assume the free-fall acceleration on

the Moon is one-sixth of that on the Earth. (a) What must the muzzle speed of the package be so that it travels completely around the Moon and returns to its original location? (b) What time interval does this trip around the Moon require?

45. *The "Vomit Comet."* In microgravity astronaut training and equipment testing, NASA flies a KC135A aircraft along a parabolic flight path. As shown in Figure P3.45, the aircraft climbs from 24 000 ft to 31 000 ft, where it enters a parabolic path with a velocity of 143 m/s nose high at 45.0° and exits with velocity 143 m/s at 45.0° nose low. During this portion of the flight, the aircraft and objects inside its padded cabin are in free fall; astronauts and equipment float freely as if there were no gravity. What are the aircraft's (a) speed and (b) altitude at the top of the maneuver? (c) What is the time interval spent in microgravity?

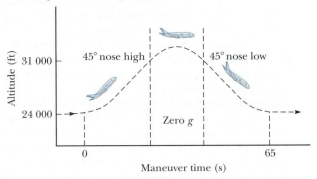

Figure P3.45

46. A projectile is fired up an incline (incline angle ϕ) with an initial speed v_i at an angle θ_i with respect to the horizontal ($\theta_i > \phi$) as shown in Figure P3.46. (a) Show that the projectile travels a distance d up the incline, where

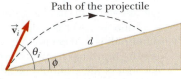

Figure P3.46

$$d = \frac{2v_i^2 \cos \theta_i \sin(\theta_i - \phi)}{g \cos^2 \phi}$$

(b) For what value of θ_i is d a maximum, and what is that maximum value?

47. A basketball player is standing on the floor 10.0 m from the basket as in Figure P3.47. The height of the basket is 3.05 m, and he shoots the ball at a 40.0° angle with the horizontal from a height of 2.00 m. (a) What is the acceleration of the basketball at the highest point in its trajectory? (b) At what speed must the player throw the basketball so that the ball goes through the hoop without striking the backboard?

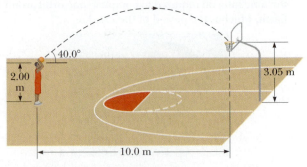

Figure P3.47

48. A truck loaded with cannonball watermelons stops suddenly to avoid running over the edge of a washed-out bridge (Fig. P3.48). The quick stop causes a number of melons to fly off the truck. One melon leaves the hood of the truck with an initial speed $v_i = 10.0$ m/s in the horizontal direction. A cross section of the bank has the shape of the bottom half of a parabola, with its vertex at the initial location of the projected watermelon and with the equation $y^2 = 16x$, where x and y are measured in meters. What are the x and y coordinates of the melon when it splatters on the bank?

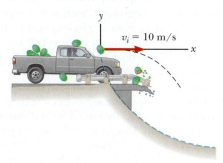

Figure P3.48

49. A ball on the end of a string is whirled around in a horizontal circle of radius 0.300 m. The plane of the circle is 1.20 m above the ground. The string breaks and the ball lands 2.00 m (horizontally) away from the point on the ground directly beneath the ball's location when the string breaks. Find the radial acceleration of the ball during its circular motion.

50. An outfielder throws a baseball to his catcher in an attempt to throw out a runner at home plate. The ball bounces once before reaching the catcher. Assume the angle at which the bounced ball leaves the ground is the same as the angle at which the outfielder threw it as shown in Figure P3.50, but that the ball's speed after the bounce is one-half of what it was before the bounce. (a) Assume the ball is always thrown with the same initial speed and ignore air resistance. At what angle θ should the fielder throw the ball to make it go the same distance D with one bounce (blue path) as a ball thrown upward at 45.0° with no bounce (green path)? (b) Determine the ratio of the time interval for the one-bounce throw to the flight time for the no-bounce throw.

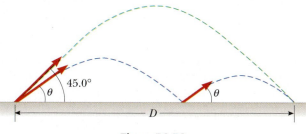

Figure P3.50

51. *Why is the following situation impossible?* Albert Pujols hits a home run so that the baseball just clears the top row of bleachers, 24.0 m high, located 130 m from home plate. The ball is hit at 41.7 m/s at an angle of 35.0° to the horizontal, and air resistance is negligible.

52. **Q|C** A skier leaves the ramp of a ski jump with a velocity of $v = 10.0$ m/s at $\theta = 15.0°$ above the horizontal as shown in Figure P3.52. The slope where she will land is inclined

downward at $\phi = 50.0°$, and air resistance is negligible. Find (a) the distance from the end of the ramp to where the jumper lands and (b) her velocity components just before the landing. (c) Explain how you think the results might be affected if air resistance were included.

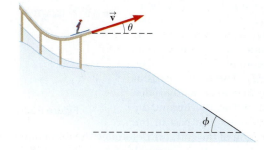

Figure P3.52

53. A World War II bomber flies horizontally over level terrain with a speed of 275 m/s relative to the ground and at an altitude of 3.00 km. The bombardier releases one bomb. (a) How far does the bomb travel horizontally between its release and its impact on the ground? Ignore the effects of air resistance. (b) The pilot maintains the plane's original course, altitude, and speed through a storm of flak. Where is the plane when the bomb hits the ground? (c) The bomb hits the target seen in the telescopic bombsight at the moment of the bomb's release. At what angle from the vertical was the bombsight set?

54. **S** A ball is thrown with an initial speed v_i at an angle θ_i with the horizontal. The horizontal range of the ball is R, and the ball reaches a maximum height $R/6$. In terms of R and g, find (a) the time interval during which the ball is in motion, (b) the ball's speed at the peak of its path, (c) the initial vertical component of its velocity, (d) its initial speed, and (e) the angle θ_i. (f) Suppose the ball is thrown at the same initial speed found in (d) but at the angle appropriate for reaching the greatest height that it can. Find this height. (g) Suppose the ball is thrown at the same initial speed but at the angle for greatest possible range. Find this maximum horizontal range.

55. **M** A car is parked on a steep incline, making an angle of 37.0° below the horizontal and overlooking the ocean, when its brakes fail and it begins to roll. Starting from rest at $t = 0$, the car rolls down the incline with a constant acceleration of 4.00 m/s², traveling 50.0 m to the edge of a vertical cliff. The cliff is 30.0 m above the ocean. Find (a) the speed of the car when it reaches the edge of the cliff, (b) the time interval elapsed when it arrives there, (c) the velocity of the car when it lands in the ocean, (d) the total time interval the car is in motion, and (e) the position of the car when it lands in the ocean, relative to the base of the cliff.

56. **S** A person standing at the top of a hemispherical rock of radius R kicks a ball (initially at rest on the top of the rock) to give it horizontal velocity $\vec{v}_i$ as shown in Figure P3.56. (a) What must be its minimum initial speed if the ball is

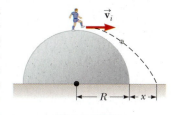

Figure P3.56

never to hit the rock after it is kicked? (b) With this initial speed, how far from the base of the rock does the ball hit the ground?

57. An aging coyote cannot run fast enough to catch a roadrunner. He purchases on eBay a set of jet-powered roller skates, which provide a constant horizontal acceleration of 15.0 m/s² (Fig. P3.57). The coyote starts at rest 70.0 m from the edge of a cliff at the instant the roadrunner zips past in the direction of the cliff. (a) Determine the minimum constant speed the roadrunner must have to reach the cliff before the coyote. At the edge of the cliff, the roadrunner escapes by making a sudden turn, while the coyote continues straight ahead. The coyote's skates remain horizontal and continue to operate while he is in flight, so his acceleration while in the air is $(15.0\,\hat{\mathbf{i}} - 9.80\,\hat{\mathbf{j}})$ m/s². (b) The cliff is 100 m above the flat floor of the desert. Determine how far from the base of the vertical cliff the coyote lands. (c) Determine the components of the coyote's impact velocity.

BEEP BEEP

Figure P3.57

58. **Q|C** Do not hurt yourself; do not strike your hand against anything. Within these limitations, describe what you do to give your hand a large acceleration. Compute an order-of-magnitude estimate of this acceleration, stating the quantities you measure or estimate and their values.

59. A catapult launches a rocket at an angle of 53.0° above the horizontal with an initial speed of 100 m/s. The rocket engine immediately starts a burn, and for 3.00 s the rocket moves along its initial line of motion with an acceleration of 30.0 m/s². Then its engine fails, and the rocket proceeds to move in free-fall. Find (a) the maximum altitude reached by the rocket, (b) its total time of flight, and (c) its horizontal range.

60. The water in a river flows uniformly at a constant speed of 2.50 m/s between parallel banks 80.0 m apart. You are to deliver a package directly across the river, but you can swim only at 1.50 m/s. (a) If you choose to minimize the time you spend in the water, in what direction should you head? (b) How far downstream will you be carried? (c) If you choose to minimize the distance downstream that the river carries you, in what direction should you head? (d) How far downstream will you be carried?

61. A fisherman sets out upstream on a river. His small boat, powered by an outboard motor, travels at a constant speed v in still water. The water flows at a lower constant speed v_w. The fisherman has traveled upstream for 2.00 km when his ice chest falls out of the boat. He notices that the chest is missing only after he has gone upstream for another 15.0 min. At that point, he turns around and heads back downstream, all the time traveling at the same speed relative to the water. He catches up with the floating ice chest just as he returns to his starting point. How fast is the river flowing? Solve this problem in two ways. (a) First, use the Earth as a reference frame. With respect to the Earth, the boat travels upstream at speed $v - v_w$ and downstream at $v + v_w$. (b) A second much simpler and more elegant solution is obtained by using the water as the reference frame. This approach has important applications in many more complicated problems; examples are calculating the motion of rockets and satellites and analyzing the scattering of subatomic particles from massive targets.

62. An enemy ship is on the east side of a mountain island as shown in Figure P3.62. The enemy ship has maneuvered to within 2 500 m of the 1 800-m-high mountain peak and can shoot projectiles with an initial speed of 250 m/s. If the western shoreline is horizontally 300 m from the peak, what are the distances from the western shore at which a ship can be safe from the bombardment of the enemy ship?

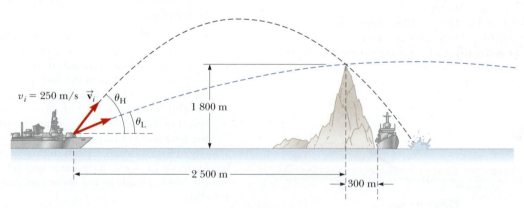

$v_i = 250$ m/s $\vec{\mathbf{v}}_i$ θ_H θ_L 1 800 m 2 500 m 300 m

Figure P3.62

The Laws of Motion

Chapter Outline

AL PARKER PHOTOGRAPHY/Shutterstock.com

In the preceding two chapters on kinematics, we *described* the motion of particles based on the definitions of position, velocity, and acceleration. Aside from our discussion of gravity for objects in free-fall, we did not address what factors might *influence* an object to move as it does. We would like to be able to answer general questions related to the influences on motion, such as "What mechanism causes changes in motion?" and "Why do some objects accelerate at higher rates than others?" In this first chapter on *dynamics,* we shall discuss the causes of the change in motion of particles using the concepts of force and mass. We will discuss the three fundamental laws of motion, which are based on experimental observations and were formulated about three centuries ago by Sir Isaac Newton.

By intuitively applying Newton's laws of motion, these two big horn sheep compete for dominance. They each exert forces against the Earth through muscular exertions of their legs, aided by the friction forces that keep them from slipping. The reaction forces of the Earth act back on the sheep and cause them to surge forward and butt heads. The goal is to force the opposing sheep out of equilibrium.

4.1 | The Concept of Force

As a result of everyday experiences, everyone has a basic understanding of the concept of force. When you push or pull an object, you exert a force on it. You exert a force when you throw or kick a ball. In these examples, the word *force* is associated with the result of muscular activity and with some change in the state of motion of an object. Forces do not always cause an object to move, however. For example, as you sit reading this book, the gravitational force acts on your body and yet you remain stationary. You can push on a heavy block of stone and yet fail to move it.

Figure 4.1 Some examples of forces applied to various objects. In each case, a force is exerted on the particle or object within the boxed area. The environment external to the boxed area provides this force.

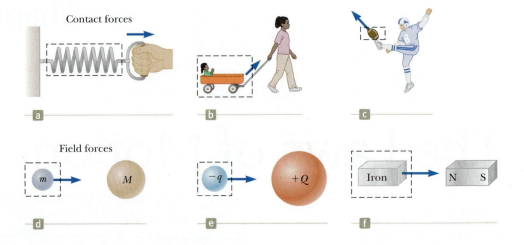

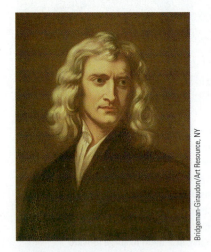

Bridgeman-Giraudon/Art Resource, NY

Isaac Newton
English Physicist and Mathematician (1642–1727)
Isaac Newton was one of the most brilliant scientists in history. Before the age of 30, he formulated the basic concepts and laws of mechanics, discovered the law of universal gravitation, and invented the mathematical methods of calculus. As a consequence of his theories, Newton was able to explain the motions of the planets, the ebb and flow of the tides, and many special features of the motions of the Moon and the Earth. He also interpreted many fundamental observations concerning the nature of light. His contributions to physical theories dominated scientific thought for two centuries and remain important today.

This chapter is concerned with the relation between the force on an object and the change in motion of that object. If you pull on a spring, as in Figure 4.1a, the spring stretches. If the spring is calibrated, the distance it stretches can be used to measure the strength of the force. If a child pulls on a wagon, as in Figure 4.1b, the wagon moves. When a football is kicked, as in Figure 4.1c, it is both deformed and set in motion. These examples all show the results of a class of forces called *contact forces*. That is, these forces represent the result of physical contact between two objects.

There exist other forces that do not involve physical contact between two objects. These forces, known as *field forces*, can act through empty space. The gravitational force between two objects that causes the free-fall acceleration described in Chapters 2 and 3 is an example of this type of force and is illustrated in Figure 4.1d. This gravitational force keeps objects bound to the Earth and gives rise to what we commonly call the *weight* of an object. The planets of our solar system are bound to the Sun under the action of gravitational forces. Another common example of a field force is the electric force that one electric charge exerts on another electric charge, as in Figure 4.1e. These charges might be an electron and a proton forming a hydrogen atom. A third example of a field force is the force that a bar magnet exerts on a piece of iron, as shown in Figure 4.1f.

The distinction between contact forces and field forces is not as sharp as you may have been led to believe by the preceding discussion. At the atomic level, all the forces classified as contact forces turn out to be caused by electric (field) forces similar in nature to the attractive electric force illustrated in Figure 4.1e. Nevertheless, in understanding macroscopic phenomena, it is convenient to use both classifications of forces.

We can use the linear deformation of a spring to measure force, as in the case of a common spring scale. Suppose a force is applied vertically to a spring that has a fixed upper end, as in Figure 4.2a. The spring can be calibrated by defining the unit force $\vec{\mathbf{F}}_1$ as the force that produces an elongation of 1.00 cm. If a force $\vec{\mathbf{F}}_2$, applied as in Figure 4.2b, produces an elongation of 2.00 cm, the magnitude of $\vec{\mathbf{F}}_2$ is 2.00 units. If the two forces $\vec{\mathbf{F}}_1$ and $\vec{\mathbf{F}}_2$ are applied simultaneously, as in Figure 4.2c, the elongation of the spring is 3.00 cm because the forces are applied in the same direction and their magnitudes add. If the two forces $\vec{\mathbf{F}}_1$ and $\vec{\mathbf{F}}_2$ are applied in perpendicular directions, as in Figure 4.2d, the elongation is $\sqrt{(1.00)^2 + (2.00)^2}$ cm $= \sqrt{5.00}$ cm $= 2.24$ cm. The single force $\vec{\mathbf{F}}$ that would produce this same elongation is the vector sum of $\vec{\mathbf{F}}_1$ and $\vec{\mathbf{F}}_2$, as described in Figure 4.2d. That is, $|\vec{\mathbf{F}}| = \sqrt{F_1^2 + F_2^2} = 2.24$ units, and its direction is $\theta = \tan^{-1}(-0.500) = -26.6°$. Because forces have been experimentally verified to behave as vectors, you *must* use the rules of vector addition to obtain the total force on an object.

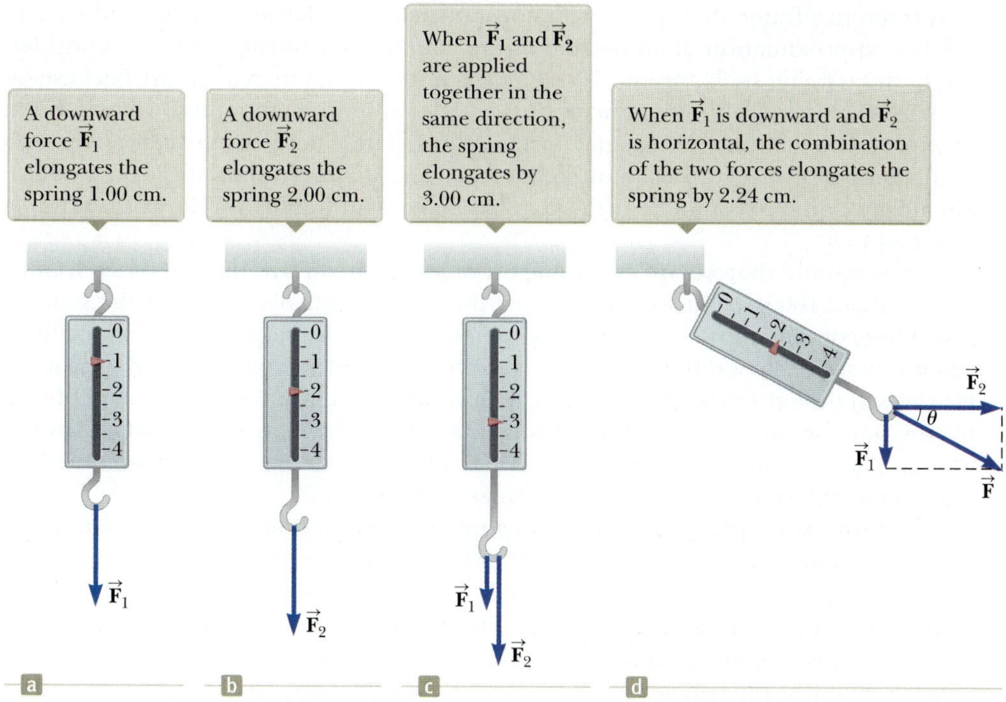

Figure 4.2 The vector nature of a force is tested with a spring scale.

A downward force $\vec{F}_1$ elongates the spring 1.00 cm.

A downward force $\vec{F}_2$ elongates the spring 2.00 cm.

When $\vec{F}_1$ and $\vec{F}_2$ are applied together in the same direction, the spring elongates by 3.00 cm.

When $\vec{F}_1$ is downward and $\vec{F}_2$ is horizontal, the combination of the two forces elongates the spring by 2.24 cm.

$\vec{F}_1$

$\vec{F}_2$

$\vec{F}_1$ $\vec{F}_2$

$\vec{F}_2$ θ $\vec{F}_1$ $\vec{F}$

a   b   c   d

4.2 | Newton's First Law

We begin our study of forces by imagining that you place a puck on a perfectly level air hockey table (Fig. 4.3). You expect that the puck will remain stationary when it is placed gently at rest on the table. Now imagine putting your air hockey table on a train moving with constant velocity. If the puck is placed on the table, the puck again remains where it is placed. If the train were to accelerate, however, the puck would start moving along the table, just as a set of papers on your dashboard slides onto the front seat of your car when you step on the gas.

As we saw in Section 3.6, a moving object can be observed from any number of reference frames. **Newton's first law of motion**, sometimes called the *law of inertia*, defines a special set of reference frames called *inertial frames*. This law can be stated as follows:

If an object does not interact with other objects, it is possible to identify a reference frame in which the object has zero acceleration.

Such a reference frame is called an **inertial frame of reference**. When the puck is on the air hockey table located on the ground, you are observing it from an inertial reference frame; there are no horizontal interactions of the puck with any other objects, and you observe it to have zero acceleration in the horizontal direction. When you are on the train moving at constant velocity, you are also observing the puck from an inertial reference frame. Any reference frame that moves with constant velocity relative to an inertial frame is itself an inertial frame. When the train accelerates, however, you are observing the puck from a **noninertial reference frame** because you and the train are accelerating relative to the inertial reference frame of the surface of the Earth. Although the puck appears to be accelerating according to your observations, we can identify a reference frame in which the puck has zero acceleration. For example, an observer standing outside the train on the ground sees the puck sliding relative to the table but always moving with the same velocity with respect to the ground as the train had before it started to accelerate (because there is almost no friction to "tie" the puck and the train together). Therefore, Newton's first law is still satisfied even though your observations say otherwise.

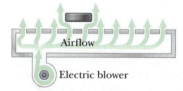

Airflow

Electric blower

Figure 4.3 On an air hockey table, air blown through holes in the surface allows the puck to move almost without friction. If the table is not accelerating, a puck placed on the table will remain at rest with respect to the table if there are no horizontal forces acting on it.

▶ Newton's first law

▶ Inertial frame of reference

A reference frame that moves with constant velocity relative to the distant stars is the best approximation of an inertial frame, and for our purposes we can consider the Earth as being such a frame. The Earth is not really an inertial frame because of its orbital motion around the Sun and its rotational motion about its own axis, both of which are related to centripetal accelerations. These accelerations, however, are small compared with g and can often be neglected. (This is a simplification model.) For this reason, we assume that the Earth is an inertial frame, as is any other frame attached to it.

Let us assume that we are observing an object from an inertial reference frame. Before about 1600, scientists believed that the natural state of matter was the state of rest. Observations showed that moving objects eventually stopped moving. Galileo was the first to take a different approach to motion and the natural state of matter. He devised thought experiments and concluded that it is not the nature of an object to stop once set in motion; rather, it is its nature to *resist changes in its motion*. In his words, "Any velocity once imparted to a moving body will be rigidly maintained as long as the external causes of retardation are removed."

Given our assumption of observations made from inertial reference frames, we can pose a more practical statement of Newton's first law of motion:

▶ Another statement of Newton's first law

> In the absence of external forces, when viewed from an inertial reference frame, an object at rest remains at rest and an object in motion continues in motion with a constant velocity (that is, with a constant speed in a straight line).

In simpler terms, we can say that **when no force acts on an object, the acceleration of the object is zero.** If nothing acts to change the object's motion, its velocity does not change. From the first law, we conclude that any *isolated object* (one that does not interact with its environment) is either at rest or moving with constant velocity. The tendency of an object to resist any attempt to change its velocity is called **inertia.**

Consider a spacecraft traveling in space, far removed from any planets or other matter. The spacecraft requires some propulsion system to change its velocity. If the propulsion system is turned off when the spacecraft reaches a velocity $\vec{\mathbf{v}}$, however, the spacecraft "coasts" in space with that velocity and the astronauts enjoy a "free ride" (i.e., no propulsion system is required to keep them moving at the velocity $\vec{\mathbf{v}}$).

Finally, recall our discussion in Chapter 2 about the proportionality between force and acceleration:

$$\vec{\mathbf{F}} \propto \vec{\mathbf{a}}$$

Newton's first law tells us that the velocity of an object remains constant if no force acts on an object; the object maintains its state of motion. The preceding proportionality tells us that if a force *does* act, a change does occur in the motion, measured by the acceleration. This notion will form the basis of Newton's second law, and we shall provide more details on this concept shortly.

Pitfall Prevention | 4.1
Newton's First Law
Newton's first law does *not* say what happens for an object with *zero net force*, that is, multiple forces that cancel; it says what happens *in the absence of external forces*. This subtle but important difference allows us to define force as that which causes a change in the motion. The description of an object under the effect of forces that balance is covered by Newton's second law.

QUICK QUIZ 4.1 Which of the following statements is most correct? (**a**) It is possible for an object to have motion in the absence of forces on the object. (**b**) It is possible to have forces on an object in the absence of motion of the object. (**c**) Neither statement (**a**) nor statement (**b**) is correct. (**d**) Both statements (**a**) and (**b**) are correct.

4.3 | Mass

Imagine playing catch with either a table-tennis ball or a bowling ball. Which ball is more likely to keep moving when you try to catch it? Which ball has the greater tendency to remain motionless when you try to throw it? The bowling ball is more resistant to changes in its velocity than the table-tennis ball. How can we quantify this concept?

▶ Definition of mass

Mass is that property of an object that specifies how much resistance an object exhibits to changes in its velocity, and as we learned in Section 1.1, the SI unit of mass is the kilogram. The greater the mass of an object, the less that object accelerates under the action of a given applied force.

To describe mass quantitatively, we begin by experimentally comparing the accelerations a given force produces on different objects. Suppose a force acting on an object of mass m_1 produces a change in motion of the object that we can quantify with the object's acceleration $\vec{a}_1$ and the *same force* acting on an object of mass m_2 produces an acceleration $\vec{a}_2$. The ratio of the two masses is defined as the *inverse* ratio of the magnitudes of the accelerations produced by the force:

$$\frac{m_1}{m_2} \equiv \frac{a_2}{a_1}$$

4.1 ◀

For example, if a given force acting on a 3-kg object produces an acceleration of 4 m/s^2, the same force applied to a 6-kg object produces an acceleration of 2 m/s^2. If one object has a known mass, the mass of the other object can be obtained from acceleration measurements.

Mass is an inherent property of an object and is independent of the object's surroundings and of the method used to measure it. Also, mass is a scalar quantity and therefore obeys the rules of ordinary arithmetic. That is, several masses can be combined in simple numerical fashion. For example, if you combine a 3-kg mass with a 5-kg mass, their total mass is 8 kg. We can verify this result experimentally by comparing the acceleration that a known force gives to several objects separately with the acceleration that the same force gives to the same objects combined as one unit.

Mass should not be confused with weight. Mass and weight are two different quantities. As we shall see later in this chapter, the weight of an object is equal to the magnitude of the gravitational force exerted on the object and varies with location. For example, a person who weighs 180 lb on the Earth weighs only about 30 lb on the Moon. On the other hand, the mass of an object is the same everywhere. An object having a mass of 2 kg on Earth also has a mass of 2 kg on the Moon.

▶ Mass and weight are different quantities

4.4 | Newton's Second Law

Newton's first law explains what happens to an object when no force acts on it: It either remains at rest or moves in a straight line with constant speed. This law allows us to define an inertial frame of reference. It also allows us to identify force as that which changes motion. Newton's second law answers the question of what happens to an object when it has one or more forces acting on it, based on our discussion of mass in the preceding section.

Imagine you are pushing a block of ice across a frictionless horizontal surface. When you exert some horizontal force $\vec{F}$, the block moves with some acceleration $\vec{a}$. Experiments show that if you apply a force twice as large to the same object, the acceleration doubles. If you increase the applied force to $3\vec{F}$, the original acceleration is tripled, and so on. From such observations, we conclude that the acceleration of an object is directly proportional to the net force acting on it. We alluded to this proportionality in our discussion of acceleration in Chapter 2. We also know from the preceding section that the magnitude of the acceleration of an object is inversely proportional to its mass: $|\mathbf{a}| \propto 1/m$.

These experimental observations are summarized in **Newton's second law:**

When viewed from an inertial reference frame, the acceleration of an object is directly proportional to the net force acting on it and inversely proportional to its mass.

▶ Newton's second law

We write this law as

$$\vec{a} \propto \frac{\sum \vec{F}}{m}$$

Pitfall Prevention | 4.2
Force Is the Cause of Changes in Motion
Be sure that you are clear on the role of force. Many times, students make the mistake of thinking that force is the cause of motion. An object can have motion in the absence of forces, as described in Newton's first law. Therefore, don't interpret force as the cause of *motion*. Be sure to understand that force is the cause of *changes* in motion.

where $\Sigma\vec{\mathbf{F}}$ is the **net force,** which is the vector sum of *all* forces acting on the object of mass m. If the object consists of a system of individual elements, the net force is the vector sum of all forces *external* to the system. Any *internal* forces—that is, forces between elements of the system—are not included because they do not affect the motion of the entire system. The net force is sometimes called the *resultant force,* the *sum of the forces,* the *total force,* or the *unbalanced force.*

Newton's second law in mathematical form is a statement of this relationship that makes the preceding proportionality an equality:[1]

▶ Mathematical representation of Newton's second law

$$\boxed{\sum \vec{\mathbf{F}} = m\vec{\mathbf{a}}}\qquad\qquad \textbf{4.2}\blacktriangleleft$$

Note that Equation 4.2 is a *vector* expression and hence is equivalent to the following three component equations:

▶ Newton's second law in component form

$$\sum F_x = ma_x \quad \sum F_y = ma_y \quad \sum F_z = ma_z \qquad \textbf{4.3}\blacktriangleleft$$

Newton's second law introduces us to a new analysis model, the particle under a net force. If a particle, or an object that can be modeled as a particle, is under the influence of a net force, Equation 4.2, the mathematical statement of Newton's second law, can be used to describe its motion. The acceleration is constant if the net force is constant. Therefore, the particle under a constant net force will have its motion described as a particle under constant acceleration. Of course, not all forces are constant, and when they are not, the particle cannot be modeled as one under constant acceleration. We shall investigate situations in this chapter and the next involving both constant and varying forces.

Pitfall Prevention | 4.3
$m\vec{\mathbf{a}}$ Is Not a Force
Equation 4.2 does *not* say that the product $m\vec{\mathbf{a}}$ is a force. All forces on an object are added vectorially to generate the net force on the left side of the equation. This net force is then equated to the product of the mass of the object and the acceleration that results from the net force. Do *not* include an "$m\vec{\mathbf{a}}$ force" in your analysis of the forces on an object.

QUICK QUIZ 4.2 An object experiences no acceleration. Which of the following *cannot* be true for the object? **(a)** A single force acts on the object. **(b)** No forces act on the object. **(c)** Forces act on the object, but the forces cancel.

QUICK QUIZ 4.3 You push an object, initially at rest, across a frictionless floor with a constant force for a time interval Δt, resulting in a final speed of v for the object. You then repeat the experiment, but with a force that is twice as large. What time interval is now required to reach the same final speed v? **(a)** $4\,\Delta t$ **(b)** $2\,\Delta t$ **(c)** Δt **(d)** $\Delta t/2$ **(e)** $\Delta t/4$

Unit of Force

The SI unit of force is the **newton,** which is defined as the force that, when acting on a 1-kg mass, produces an acceleration of $1\ \text{m/s}^2$.

From this definition and Newton's second law, we see that the newton can be expressed in terms of the fundamental units of mass, length, and time:

▶ Definition of the newton

$$1\ \text{N} \equiv 1\ \text{kg} \cdot \text{m/s}^2 \qquad\qquad \textbf{4.4}\blacktriangleleft$$

The units of mass, acceleration, and force are summarized in Table 4.1. Most of the calculations we shall make in our study of mechanics will be in SI units. Equalities between units in the SI and U.S. customary systems are given in Appendix A.

◀ TABLE 4.1 | Units of Mass, Acceleration, and Force

System of Units	Mass (M)	Acceleration (L/T²)	Force (ML/T²)
SI	kg	m/s²	N = kg · m/s²
U.S. customary	slug	ft/s²	lb = slug · ft/s²

---

[1]Equation 4.2 is valid only when the speed of the object is much less than the speed of light. We will treat the relativistic situation in Chapter 9.

> ### THINKING PHYSICS 4.1

In a train, the cars are connected by *couplers*. The couplers between the cars exert forces on the cars as the train is pulled by the locomotive in the front. Imagine that the train is speeding up in the forward direction. As you imagine moving from the locomotive to the last car, does the force exerted by the couplers *increase, decrease,* or *stay the same*? What if the engineer applies the brakes? How does the force vary from locomotive to last car in this case? (Assume that the only brakes applied are those on the engine.)

Reasoning The force *decreases* from the front of the train to the back. The coupler between the locomotive and the first car must apply enough force to accelerate all the remaining cars. As we move back along the train, each coupler is accelerating less mass behind it. The last coupler only has to accelerate the last car, so it exerts the smallest force. If the brakes are applied, the force decreases from front to back of the train also. The first coupler, at the back of the locomotive, must apply a large force to slow down all the remaining cars. The final coupler must only apply a force large enough to slow down the mass of the last car. ◀

Example 4.1 | An Accelerating Hockey Puck

A hockey puck having a mass of 0.30 kg slides on the frictionless, horizontal surface of an ice rink. Two hockey sticks strike the puck simultaneously, exerting the forces on the puck shown in Figure 4.4. The force $\vec{F}_1$ has a magnitude of 5.0 N, and the force $\vec{F}_2$ has a magnitude of 8.0 N. Determine both the magnitude and the direction of the puck's acceleration.

SOLUTION

Conceptualize Study Figure 4.4. Using your expertise in vector addition from Chapter 1, predict the approximate direction of the net force vector on the puck. The acceleration of the puck will be in the same direction.

Categorize Because we can determine a net force and we want an acceleration, this problem is categorized as one that may be solved using Newton's second law.

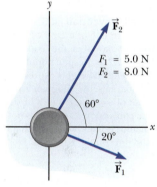

Figure 4.4 (Example 4.1) A hockey puck moving on a frictionless surface is subject to two forces $\vec{F}_1$ and $\vec{F}_2$.

$F_1 = 5.0$ N
$F_2 = 8.0$ N

Analyze Find the component of the net force acting on the puck in the *x* direction:

$$\sum F_x = F_{1x} + F_{2x} = F_1 \cos(-20°) + F_2 \cos 60°$$
$$= (5.0 \text{ N})(0.940) + (8.0 \text{ N})(0.500) = 8.7 \text{ N}$$

Find the component of the net force acting on the puck in the *y* direction:

$$\sum F_y = F_{1y} + F_{2y} = F_1 \sin(-20°) + F_2 \sin 60°$$
$$= (5.0 \text{ N})(-0.342) + (8.0 \text{ N})(0.866) = 5.2 \text{ N}$$

Use Newton's second law in component form (Eq. 4.3) to find the *x* and *y* components of the puck's acceleration:

$$a_x = \frac{\sum F_x}{m} = \frac{8.7 \text{ N}}{0.30 \text{ kg}} = 29 \text{ m/s}^2$$

$$a_y = \frac{\sum F_y}{m} = \frac{5.2 \text{ N}}{0.30 \text{ kg}} = 17 \text{ m/s}^2$$

Find the magnitude of the acceleration:

$$a = \sqrt{(29 \text{ m/s}^2)^2 + (17 \text{ m/s}^2)^2} = \boxed{34 \text{ m/s}^2}$$

Find the direction of the acceleration relative to the positive *x* axis:

$$\theta = \tan^{-1}\left(\frac{a_y}{a_x}\right) = \tan^{-1}\left(\frac{17}{29}\right) = \boxed{31°}$$

Finalize The vectors in Figure 4.4 can be added graphically to check the reasonableness of our answer. Because the acceleration vector is along the direction of the resultant force, a drawing showing the resultant force vector helps us check the validity of the answer. (Try it!)

continued

4.1 *cont.*

What If? Suppose three hockey sticks strike the puck simultaneously, with two of them exerting the forces shown in Figure 4.4. The result of the three forces is that the hockey puck shows *no* acceleration. What must be the components of the third force?

Answer If there is zero acceleration, the net force acting on the puck must be zero. Therefore, the three forces must cancel. We have found the components of the combination of the first two forces. The components of the third force must be of equal magnitude and opposite sign so that all the components add to zero. Therefore, $F_{3x} = -8.7$ N and $F_{3y} = -5.2$ N.

Pitfall Prevention | 4.4
"Weight of an Object"
We are familiar with the everyday phrase, the "weight of an object." Weight, however, is not an inherent property of an object; rather, it is a measure of the gravitational force between the object and the Earth (or other planet). Therefore, weight is a property of a *system* of items: the object and the Earth.

Pitfall Prevention | 4.5
Kilogram Is Not a Unit of Weight
You may have seen the "conversion" 1 kg = 2.2 lb. Despite popular statements of weights expressed in kilograms, the kilogram is not a unit of *weight*, it is a unit of *mass*. The conversion statement is not an equality; it is an *equivalence* that is valid only on the Earth's surface.

The life-support unit strapped to the back of astronaut Harrison Schmitt weighed 300 lb on the Earth and had a mass of 136 kg. During his training, a 50-lb mock-up with a mass of 23 kg was used. Although this strategy effectively simulated the reduced weight the unit would have on the Moon, it did not correctly mimic the unchanging mass. It was more difficult to accelerate the 136-kg unit (perhaps by jumping or twisting suddenly) on the Moon than it was to accelerate the 23-kg unit on the Earth.

4.5 | The Gravitational Force and Weight

We are well aware that all objects are attracted to the Earth. The force exerted by the Earth on an object is the **gravitational force** $\vec{\mathbf{F}}_g$. This force is directed toward the center of the Earth.[2] The magnitude of the gravitational force is called the **weight** F_g of the object.

We have seen in Chapters 2 and 3 that a freely falling object experiences an acceleration $\vec{\mathbf{g}}$ directed toward the center of the Earth. A freely falling object has only one force on it, the gravitational force, so the net force on the object in this situation is equal to the gravitational force:

$$\sum \vec{\mathbf{F}} = \vec{\mathbf{F}}_g$$

Because the acceleration of a freely falling object is equal to the free-fall acceleration $\vec{\mathbf{g}}$, it follows that

$$\sum \vec{\mathbf{F}} = m\vec{\mathbf{a}} \;\rightarrow\; \vec{\mathbf{F}}_g = m\vec{\mathbf{g}}$$

or, in magnitude,

$$F_g = mg \qquad\qquad \textbf{4.5} \blacktriangleleft$$

Because it depends on g, weight varies with location, as we mentioned in Section 4.3. Objects weigh less at higher altitudes than at sea level because g decreases with increasing distance from the center of the Earth. Hence, weight, unlike mass, is not an inherent property of an object. It is a property of the *system* of the object and the Earth. For example, if an object has a mass of 70 kg, its weight in a location where $g = 9.80$ m/s^2 is $mg = 686$ N. At the top of a mountain where $g = 9.76$ m/s^2, the object's weight would be 683 N. Therefore, if you want to lose weight without going on a diet, climb a mountain or weigh yourself at 30 000 ft during an airplane flight.

Because $F_g = mg$, we can compare the masses of two objects by measuring their weights with a spring scale. At a given location (so that g is fixed) the ratio of the weights of two objects equals the ratio of their masses.

Equation 4.5 quantifies the gravitational force on the object, but notice that this equation does not require the object to be moving. Even for a stationary object, or an object on which several forces act, Equation 4.5 can be used to calculate the magnitude of the gravitational force. This observation results in a subtle shift in the interpretation of m in the equation. The mass m in Equation 4.5 is playing the role of determining the strength of the gravitational attraction between the object and the Earth. This role is completely different from that previously described for mass, that of measuring the resistance to changes in motion in response to an external force. In that role, mass is also called **inertial mass.** We call m in Equation 4.5 the **gravitational mass.** Despite this quantity being different from inertial mass, it is one of the experimental conclusions in Newtonian dynamics that gravitational mass and inertial mass have the same value at the present level of experimental refinement.

---

[2]This statement represents a simplification model in that it ignores that the mass distribution of the Earth is not perfectly spherical.

4.6 | Newton's Third Law

Newton's third law conveys the notion that forces are always interactions between two objects:

If two objects interact, the force $\vec{F}_{12}$ exerted by object 1 on object 2 is equal in magnitude but opposite in direction to the force $\vec{F}_{21}$ exerted by object 2 on object 1:

$$\vec{F}_{12} = -\vec{F}_{21} \qquad \textbf{4.6} \blacktriangleleft$$

▶ Newton's third law

When it is important to designate forces as interactions between two objects, we will use this subscript notation, where $\vec{F}_{ab}$ means "the force exerted *by* a *on* b." The third law, illustrated in Figure 4.5a, is equivalent to stating that **forces always occur in pairs** or that a **single isolated force cannot exist.** The force that object 1 exerts on object 2 may be called the *action force,* and the force of object 2 on object 1 may be called the *reaction force.* In reality, either force can be labeled the action or reaction force. The action force is equal in magnitude to the reaction force and opposite in direction. In all cases, the action and reaction forces act on different objects and must be of the same type. For example, the force acting on a freely falling projectile is the gravitational force exerted by the Earth on the projectile $\vec{F}_g = \vec{F}_{Ep}$ (E = Earth, p = projectile), and the magnitude of this force is *mg*. The reaction to this force is the gravitational force exerted by the projectile on the Earth $\vec{F}_{pE} = -\vec{F}_{Ep}$. The reaction force $\vec{F}_{pE}$ must accelerate the Earth toward the projectile just as the action force $\vec{F}_{Ep}$ accelerates the projectile toward the Earth. Because the Earth has such a large mass, however, its acceleration as a result of this reaction force is negligibly small.

Another example of Newton's third law in action is shown in Figure 4.5b. The force $\vec{F}_{hn}$ exerted by the hammer on the nail (the action) is equal in magnitude and opposite the force $\vec{F}_{nh}$ exerted by the nail on the hammer (the reaction). This latter force stops the forward motion of the hammer when it strikes the nail.

Pitfall Prevention | 4.6
Newton's Third Law
Newton's third law is such an important and often misunderstood notion that it is repeated here in a Pitfall Prevention. In Newton's third law, action and reaction forces act on *different* objects. Two forces acting on the same object, even if they are equal in magnitude and opposite in direction, *cannot* be an action–reaction pair.

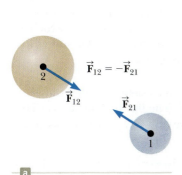

Figure 4.5 Newton's third law. (a) The force $\vec{F}_{12}$ exerted by object 1 on object 2 is equal in magnitude and opposite in direction to the force $\vec{F}_{21}$ exerted by object 2 on object 1. (b) The force $\vec{F}_{hn}$ exerted by the hammer on the nail is equal in magnitude and opposite in direction to the force $\vec{F}_{nh}$ exerted by the nail on the hammer.

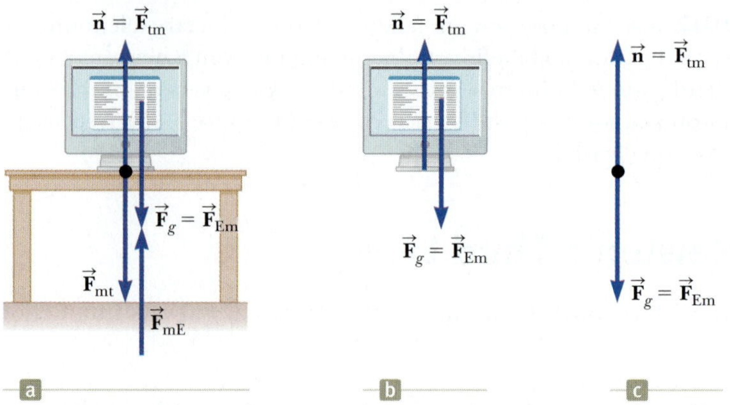

Figure 4.6 (a) When a computer monitor is at rest on a table, the forces acting on the monitor are the normal force $\vec{\mathbf{n}}$ and the gravitational force $\vec{\mathbf{F}}_g$. The reaction to $\vec{\mathbf{n}}$ is the force $\vec{\mathbf{F}}_{mt}$ exerted by the monitor on the table. The reaction to $\vec{\mathbf{F}}_g$ is the force $\vec{\mathbf{F}}_{mE}$ exerted by the monitor on the Earth. (b) A diagram showing the forces on the monitor. (c) A free-body diagram shows the monitor as a black dot with the forces acting on it.

▶ Normal force

> **Pitfall Prevention | 4.7**
> **n Does Not Always Equal mg**
> In the situation shown in Figure 4.6 and in many others, we find that $n = mg$ (the normal force has the same magnitude as the gravitational force). This result, however, is not generally true. If an object is on an incline, if there are applied forces with vertical components, or if there is a vertical acceleration of the system, then $n \neq mg$. *Always* apply Newton's second law to find the relationship between n and mg.

> **Pitfall Prevention | 4.8**
> **Free-Body Diagrams**
> The *most important* step in solving a problem using Newton's laws is to draw a proper simplified pictorial representation, the free-body diagram. Be sure to draw only those forces that act on the object you are isolating. Be sure to draw *all* forces acting on the object, including any field forces, such as the gravitational force.

The Earth exerts a gravitational force $\vec{\mathbf{F}}_g$ on any object. If the object is a computer monitor at rest on a table, as in the pictorial representation in Figure 4.6a, the reaction force to $\vec{\mathbf{F}}_g = \vec{\mathbf{F}}_{Em}$ is the force exerted by the monitor on the Earth $\vec{\mathbf{F}}_{mE} = -\vec{\mathbf{F}}_{Em}$. The monitor does not accelerate because it is held up by the table. The table exerts on the monitor an upward force $\vec{\mathbf{n}} = \vec{\mathbf{F}}_{tm}$, called the **normal force.**[3] This force prevents the monitor from falling through the table; it can have any value needed, up to the point at which the table breaks. From Newton's second law we see that, because the monitor has zero acceleration, it follows that $\Sigma\vec{\mathbf{F}} = \vec{\mathbf{n}} + \vec{\mathbf{F}}_g = 0$, or $n = mg$. The normal force balances the gravitational force on the monitor, so the net force on the monitor is zero. The reaction to **n** is the force exerted by the monitor downward on the table, $\vec{\mathbf{F}}_{mt} = -\vec{\mathbf{F}}_{tm}$.

Note that the forces acting on the monitor are $\vec{\mathbf{F}}_g$ and $\vec{\mathbf{n}}$, as shown in Figure 4.6b. The two reaction forces $\vec{\mathbf{F}}_{mE}$ and $\vec{\mathbf{F}}_{mt}$ are exerted by the monitor on the Earth and the table, respectively. Remember that the two forces in an action–reaction pair always act on two different objects.

Figure 4.6 illustrates an extremely important difference between a pictorial representation and a simplified pictorial representation for solving problems involving forces. Figure 4.6a shows many of the forces in the situation: those acting on the monitor, one acting on the table, and one acting on the Earth. Figure 4.6b, by contrast, shows only the forces on *one object*, the monitor, and is called a **force diagram,** or a *diagram showing the forces on the object.* The important simplified pictorial representation in Figure 4.6c is called a **free-body diagram.** In a free-body diagram, the particle model is used by representing the object as a dot and showing the forces that act on the object as being applied to the dot. When analyzing a particle under a net force, we are interested in the net force on one object, an object of mass m, which we will model as a particle. Therefore, a free-body diagram helps us isolate only those forces on the object and eliminate the other forces from our analysis.

◀ **QUICK QUIZ 4.5** (i) If a fly collides with the windshield of a fast-moving bus, which experiences an impact force with a larger magnitude? (a) The fly. (b) The bus. (c) The same force is experienced by both. (ii) Which experiences the greater acceleration? (a) The fly. (b) The bus. (c) The same acceleration is experienced by both.

◀ **QUICK QUIZ 4.6** Which of the following is the reaction force to the gravitational force acting on your body as you sit in your desk chair? (a) the normal force from the chair (b) the force you apply downward on the seat of the chair (c) neither of these forces

---

[3]The word *normal* is used because the direction of $\vec{\mathbf{n}}$ is always *perpendicular* to the surface.

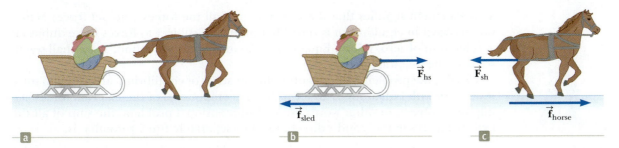

Figure 4.7 (Thinking Physics 4.2) (a) A horse pulls a sled through the snow. (b) The forces on the sled. (c) The forces on the horse.

> **THINKING PHYSICS 4.2**
>
> A horse pulls on a sled with a horizontal force, causing the sled to accelerate as in Figure 4.7a. Newton's third law says that the sled exerts a force of equal magnitude and opposite direction on the horse. In view of this situation, how can the sled accelerate? Don't these forces cancel?
>
> **Reasoning** When applying Newton's third law, it is important to remember that the forces involved act on different objects. Notice that the force exerted by the horse acts *on the sled*, whereas the force exerted by the sled acts *on the horse*. Because these forces act on different objects, they cannot cancel.
>
> The horizontal forces exerted on the *sled* alone are the forward force $\vec{\mathbf{F}}_{hs}$ exerted by the horse and the backward force of friction $\vec{\mathbf{f}}_{sled}$ between sled and surface (Fig. 4.7b). When $\vec{\mathbf{F}}_{hs}$ exceeds $\vec{\mathbf{f}}_{sled}$, the sled accelerates to the right.
>
> The horizontal forces exerted on the *horse* alone are the forward friction force $\vec{\mathbf{f}}_{horse}$ from the ground and the backward force $\vec{\mathbf{F}}_{sh}$ exerted by the sled (Fig. 4.7c). The resultant of these two forces causes the horse to accelerate. When $\vec{\mathbf{f}}_{horse}$ exceeds $\vec{\mathbf{F}}_{sh}$, the horse accelerates to the right. ◄

4.7 | Analysis Models Using Newton's Second Law

In this section, we discuss two analysis models for solving problems in which objects are either in equilibrium ($\vec{\mathbf{a}} = 0$) or are accelerating under the action of constant external forces. We shall assume that the objects behave as particles so that we need not worry about rotational motion or other complications. In this section, we also apply some additional simplification models. We ignore the effects of friction for those problems involving motion, which is equivalent to stating that the surfaces are *frictionless*. We usually ignore the masses of any ropes or strings involved. In this approximation, the magnitude of the force exerted at any point along a string is the same. In problem statements, the terms *light* and *of negligible mass* are used to indicate that a mass is to be ignored when you work the problem. These two terms are synonymous in this context.

Analysis Model: Particle in Equilibrium

Objects that are either at rest or moving with constant velocity are treated with the **particle in equilibrium** model. From Newton's second law with $\vec{\mathbf{a}} = 0$, this condition of equilibrium can be expressed as

$$\sum \vec{\mathbf{F}} = 0 \qquad\qquad \textbf{4.7} \blacktriangleleft$$

This statement signifies that the vector sum of all the forces (the net force) acting on an object in equilibrium is zero.[4] If a particle is subject to forces but exhibits an acceleration of zero, we use Equation 4.7 to analyze the situation, as we shall see in some of the following examples.

Usually, the problems we encounter in our study of equilibrium are easier to solve if we work with Equation 4.7 in terms of the components of the external forces acting on an object. In other words, in a two-dimensional problem, the sum of all the external forces in the x and y directions must separately equal zero; that is,

$$\sum F_x = 0 \qquad \sum F_y = 0 \qquad \text{4.8} \blacktriangleleft$$

The extension of Equations 4.8 to a three-dimensional situation can be made by adding a third component equation, $\sum F_z = 0$.

In a given situation, we may have balanced forces on an object in one direction but unbalanced forces in the other. Therefore, for a given problem, we may need to model the object as a particle in equilibrium for one component and a particle under a net force for the other.

QUICK QUIZ 4.7 Consider the two situations shown in Figure 4.8, in which no acceleration occurs. In both cases, all individuals pull with a force of magnitude F on a rope attached to a spring scale. Is the reading on the spring scale in part (i) of the figure (**a**) greater than, (**b**) less than, or (**c**) equal to the reading in part (ii)?

Figure 4.8 (Quick Quiz 4.7) (i) An individual pulls with a force of magnitude F on a spring scale attached to a wall. (ii) Two individuals pull with forces of magnitude F in opposite directions on a spring scale attached between two ropes.

Analysis Model: Particle Under a Net Force

If an object experiences an acceleration, its motion can be analyzed with the **particle under a net force** model. The appropriate equation for this model is Newton's second law, Equation 4.2:

$$\sum \vec{\mathbf{F}} = m\vec{\mathbf{a}} \qquad \text{4.2} \blacktriangleleft$$

Consider a crate being pulled to the right on a frictionless, horizontal floor as in Figure 4.9a. Of course, the floor directly under the boy must have friction; otherwise, his feet would simply slip when he tries to pull on the crate! Suppose you wish to find the acceleration of the crate and the force the floor exerts on it. The forces acting on the crate are illustrated in the free-body diagram in Figure 4.9b. Notice that the horizontal force $\vec{\mathbf{T}}$ being applied to the crate acts through the rope. The magnitude of $\vec{\mathbf{T}}$ is equal to the tension in the rope. In addition to the force $\vec{\mathbf{T}}$, the free-body diagram for the crate includes the gravitational force $\vec{\mathbf{F}}_g$ and the normal force $\vec{\mathbf{n}}$ exerted by the floor on the crate.

We can now apply Newton's second law in component form to the crate. The only force acting in the x direction is $\vec{\mathbf{T}}$. Applying $\sum F_x = ma_x$ to the horizontal motion gives

$$\sum F_x = T = ma_x \quad \text{or} \quad a_x = \frac{T}{m}$$

No acceleration occurs in the y direction because the crate moves only horizontally. Therefore, we use the particle in equilibrium model in the y direction. Applying the y component of Equation 4.7 yields

$$\sum F_y = n + (-F_g) = 0 \quad \text{or} \quad n = F_g$$

That is, the normal force has the same magnitude as the gravitational force but acts in the opposite direction.

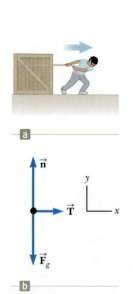

Figure 4.9 (a) A crate being pulled to the right on a frictionless floor. (b) The free-body diagram representing the external forces acting on the crate.

[4]This statement is only one condition of equilibrium for an object. An object moving through space is said to be in translational motion. If the object is spinning, it is said to be in rotational motion. A second condition of equilibrium is a statement of rotational equilibrium. This condition will be discussed in Chapter 10 when we discuss spinning objects. Equation 4.7 is sufficient for analyzing particle-like objects in translational motion, which are those of interest to us at this point.

If $\vec{\mathbf{T}}$ is a constant force, the acceleration $a_x = T/m$ also is constant. Hence, the crate is also modeled as a particle under constant acceleration in the x direction, and the equations of kinematics from Chapter 2 can be used to obtain the crate's position x and velocity v_x as functions of time.

> ## PROBLEM-SOLVING STRATEGY: Applying Newton's Laws
>
> The following procedure is recommended when dealing with problems involving Newton's laws.
>
> 1. **Conceptualize** Draw a simple, neat diagram of the system to help establish the mental representation. Establish convenient coordinate axes for each object in the system.
>
> 2. **Categorize** If an acceleration component for an object is zero, it is modeled as a particle in equilibrium in this direction and $\Sigma F = 0$. If not, the object is modeled as a particle under a net force in this direction and $\Sigma F = ma$.
>
> 3. **Analyze** Isolate the object whose motion is being analyzed. Draw a free-body diagram for this object. For systems containing more than one object, draw *separate* free-body diagrams for each object. *Do not* include in the free-body diagram forces exerted by the object on its surroundings.
>
>   Find the components of the forces along the coordinate axes. Apply Newton's second law, $\Sigma \vec{\mathbf{F}} = m\vec{\mathbf{a}}$, in component form. Check your dimensions to make sure all terms have units of force.
>
>   Solve the component equations for the unknowns. Remember that to obtain a complete solution, you generally must have as many independent equations as you have unknowns.
>
> 4. **Finalize** Make sure your results are consistent with the free-body diagram. Also check the predictions of your solutions for extreme values of the variables. By doing so, you can often detect errors in your results.

Example 4.2 | A Traffic Light at Rest

A traffic light weighing 122 N hangs from a cable tied to two other cables fastened to a support as in Figure 4.10a. The upper cables make angles of 37.0° and 53.0° with the horizontal. These upper cables are not as strong as the vertical cable and will break if the tension in them exceeds 100 N. Does the traffic light remain hanging in this situation, or will one of the cables break?

SOLUTION

Conceptualize Inspect the drawing in Figure 4.10a. Let us assume the cables do not break and nothing is moving.

Categorize If nothing is moving, no part of the system is accelerating. We can now model the light as a particle in equilibrium on which the net force is zero. Similarly, the net force on the knot (Fig. 4.10c) is zero.

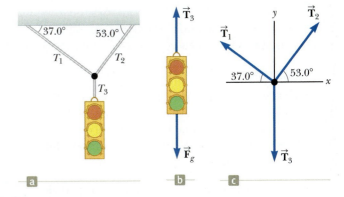

Figure 4.10 (Example 4.2) (a) A traffic light suspended by cables. (b) The forces acting on the traffic light. (c) The free-body diagram for the knot where the three cables are joined.

Analyze We construct a diagram of the forces acting on the traffic light, shown in Figure 4.10b, and a free-body diagram for the knot that holds the three cables together, shown in Figure 4.10c. This knot is a convenient object to choose because all the forces of interest act along lines passing through the knot.

Apply Equation 4.8 for the traffic light in the y direction:

$$\Sigma F_y = 0 \quad \rightarrow \quad T_3 - F_g = 0$$

$$T_3 = F_g = 122 \text{ N}$$

continued

4.2 *cont.*

Choose the coordinate axes as shown in Figure 4.10c and resolve the forces acting on the knot into their components:

Force	x Component	y Component
$\vec{T}_1$	$-T_1 \cos 37.0°$	$T_1 \sin 37.0°$
$\vec{T}_2$	$T_2 \cos 53.0°$	$T_2 \sin 53.0°$
$\vec{T}_3$	0	-122 N

Apply the particle in equilibrium model to the knot:

$$(1) \quad \sum F_x = -T_1 \cos 37.0° + T_2 \cos 53.0° = 0$$

$$(2) \quad \sum F_y = T_1 \sin 37.0° + T_2 \sin 53.0° + (-122 \text{ N}) = 0$$

Equation (1) shows that the horizontal components of $\vec{T}_1$ and $\vec{T}_2$ must be equal in magnitude, and Equation (2) shows that the sum of the vertical components of $\vec{T}_1$ and $\vec{T}_2$ must balance the downward force $\vec{T}_3$, which is equal in magnitude to the weight of the light.

Solve Equation (1) for T_2 in terms of T_1:

$$T_2 = T_1 \left(\frac{\cos 37.0°}{\cos 53.0°} \right) = 1.33 T_1$$

Substitute this value for T_2 into Equation (2):

$$T_1 \sin 37.0° + (1.33 T_1)(\sin 53.0°) - 122 \text{ N} = 0$$

$$T_1 = 73.4 \text{ N}$$

$$T_2 = 1.33 T_1 = 97.4 \text{ N}$$

Both values are less than 100 N (just barely for T_2), so the cables will not break.

Finalize Imagine changing some of the variables in the problem. What variables could be changed and what would their values have to be so that the cable would break? Suppose the two angles in Figure 4.10a are equal. What would be the relationship between T_1 and T_2?

Example 4.3 | The Runaway Car

A car of mass m is on an icy driveway inclined at an angle θ as in Figure 4.11a.

(A) Find the acceleration of the car, assuming the driveway is frictionless.

SOLUTION

Conceptualize Use Figure 4.11a to conceptualize the situation. From everyday experience, we know that a car on an icy incline will accelerate down the incline. (The same thing happens to a car on a hill with its brakes not set.)

Categorize We categorize the car as a particle under a net force because it accelerates. Furthermore, this example belongs to a very common category of problems in which an object moves under the influence of gravity on an inclined plane.

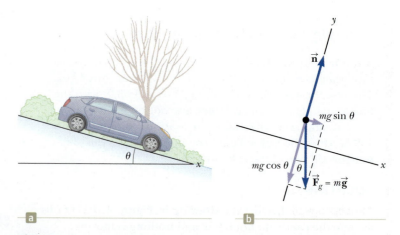

Figure 4.11 (Example 4.3) (a) A car on a frictionless incline. (b) The free-body diagram for the car. The black dot represents the position of the center of mass of the car. We will learn about center of mass in Chapter 8.

Analyze Figure 4.11b shows the free-body diagram for the car. The only forces acting on the car are the normal force $\vec{n}$ exerted by the inclined plane, which acts perpendicular to the plane, and the gravitational force $\vec{F}_g = m\vec{g}$, which acts vertically downward. For problems involving inclined planes, it is convenient to choose the coordinate axes with x along the incline and y perpendicular to it as in Figure 4.11b. With these axes, we represent the gravitational force by a component of magnitude $mg \sin \theta$ along the positive x axis and one of magnitude $mg \cos \theta$ along the negative y axis. Our choice of axes results in the car being modeled as a particle under a net force in the x direction and a particle in equilibrium in the y direction.

4.3 *cont.*

Apply these models to the car:

(1) $\sum F_x = mg \sin \theta = ma_x$

(2) $\sum F_y = n - mg \cos \theta = 0$

Solve Equation (1) for a_x:

(3) $a_x = \boxed{g \sin \theta}$

Finalize Note that the acceleration component a_x is independent of the mass of the car! It depends only on the angle of inclination and on g.

From Equation (2), we conclude that the component of $\vec{F}_g$ perpendicular to the incline is balanced by the normal force; that is, $n = mg \cos \theta$. This situation is a case in which the normal force is *not* equal in magnitude to the weight of the object (as discussed in Pitfall Prevention 4.7 on page 106).

It is possible, although inconvenient, to solve the problem with "standard" horizontal and vertical axes. You may want to try it, just for practice.

(B) Suppose the car is released from rest at the top of the incline and the distance from the car's front bumper to the bottom of the incline is d. How long does it take the front bumper to reach the bottom of the hill, and what is the car's speed as it arrives there?

SOLUTION

Conceptualize Imagine that the car is sliding down the hill and you use a stopwatch to measure the entire time interval until it reaches the bottom.

Categorize This part of the problem belongs to kinematics rather than to dynamics, and Equation (3) shows that the acceleration a_x is constant. Therefore, you should categorize the car in this part of the problem as a particle under constant acceleration.

Analyze Defining the initial position of the front bumper as $x_i = 0$ and its final position as $x_f = d$, and recognizing that $v_{xi} = 0$, apply Equation 2.13, $x_f = x_i + v_{xi}t + \frac{1}{2}a_x t^2$:

$$d = \tfrac{1}{2}a_x t^2$$

Solve for t:

(4) $t = \sqrt{\dfrac{2d}{a_x}} = \boxed{\sqrt{\dfrac{2d}{g \sin \theta}}}$

Use Equation 2.14, with $v_{xi} = 0$, to find the final velocity of the car:

$$v_{xf}^2 = 2a_x d$$

(5) $v_{xf} = \sqrt{2a_x d} = \boxed{\sqrt{2gd \sin \theta}}$

Finalize We see from Equations (4) and (5) that the time t at which the car reaches the bottom and its final speed v_{xf} are independent of the car's mass, as was its acceleration. Notice that we have combined techniques from Chapter 2 with new techniques from this chapter in this example. As we learn more techniques in later chapters, this process of combining analysis models and information from several parts of the book will occur more often. In these cases, use the General Problem-Solving Strategy to help you identify what analysis models you will need.

What If? What previously solved problem does this situation become if $\theta = 90°$?

Answer Imagine θ going to 90° in Figure 4.11. The inclined plane becomes vertical, and the car is an object in free fall! Equation (3) becomes

$$a_x = g \sin \theta = g \sin 90° = g$$

which is indeed the free-fall acceleration. (We find $a_x = g$ rather than $a_x = -g$ because we have chosen positive x to be downward in Fig. 4.11.) Notice also that the condition $n = mg \cos \theta$ gives us $n = mg \cos 90° = 0$. That is consistent with the car falling downward *next to* the vertical plane, in which case there is no contact force between the car and the plane.

Example 4.4 | The Atwood Machine

When two objects of unequal mass are hung vertically over a frictionless pulley of negligible mass as in Active Figure 4.12a, the arrangement is called an *Atwood machine*. The device is sometimes used in the laboratory to determine the value of *g*. Determine the magnitude of the acceleration of the two objects and the tension in the lightweight string.

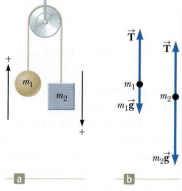

SOLUTION

Conceptualize Imagine the situation pictured in Active Figure 4.12a in action: as one object moves upward, the other object moves downward. Because the objects are connected by an inextensible string, their accelerations must be of equal magnitude.

Categorize The objects in the Atwood machine are subject to the gravitational force as well as to the forces exerted by the strings connected to them. Therefore, we can categorize this problem as one involving two particles under a net force.

Active Figure 4.12 (Example 4.4) The Atwood machine. (a) Two objects connected by a massless inextensible string over a frictionless pulley. (b) The free-body diagrams for the two objects.

Analyze The free-body diagrams for the two objects are shown in Active Figure 4.12b. Two forces act on each object: the upward force $\vec{T}$ exerted by the string and the downward gravitational force. In problems such as this one in which the pulley is modeled as massless and frictionless, the tension in the string on both sides of the pulley is the same. If the pulley has mass or is subject to friction, the tensions on either side are not the same and the situation requires techniques we will learn in Chapter 10.

We must be very careful with signs in problems such as this one. In Active Figure 4.12a, notice that if object 1 accelerates upward, object 2 accelerates downward. Therefore, for consistency with signs, if we define the upward direction as positive for object 1, we must define the downward direction as positive for object 2. With this sign convention, both objects accelerate in the same direction as defined by the choice of sign. Furthermore, according to this sign convention, the *y* component of the net force exerted on object 1 is $T - m_1 g$, and the *y* component of the net force exerted on object 2 is $m_2 g - T$.

Apply Newton's second law to object 1:	(1) $\sum F_y = T - m_1 g = m_1 a_y$
Apply Newton's second law to object 2:	(2) $\sum F_y = m_2 g - T = m_2 a_y$
Add Equation (2) to Equation (1), noticing that *T* cancels:	$- m_1 g + m_2 g = m_1 a_y + m_2 a_y$
Solve for the acceleration:	(3) $a_y = \left(\dfrac{m_2 - m_1}{m_1 + m_2} \right) g$
Substitute Equation (3) into Equation (1) to find *T*:	(4) $T = m_1 (g + a_y) = \left(\dfrac{2 m_1 m_2}{m_1 + m_2} \right) g$

Finalize The acceleration given by Equation (3) can be interpreted as the ratio of the magnitude of the unbalanced force on the system $(m_2 - m_1) g$ to the total mass of the system $(m_1 + m_2)$, as expected from Newton's second law. Notice that the sign of the acceleration depends on the relative masses of the two objects.

What If? Describe the motion of the system if the objects have equal masses, that is, $m_1 = m_2$.

Answer If we have the same mass on both sides, the system is balanced and should not accelerate. Mathematically, we see that if $m_1 = m_2$, Equation (3) gives us $a_y = 0$.

What If? What if one of the masses is much larger than the other: $m_1 \gg m_2$?

Answer In the case in which one mass is infinitely larger than the other, we can ignore the effect of the smaller mass. Therefore, the larger mass should simply fall as if the smaller mass were not there. We see that if $m_1 \gg m_2$, Equation (3) gives us $a_y = -g$.

Example 4.5 | One Block Pushes Another

Two blocks of masses m_1 and m_2, with $m_1 > m_2$, are placed in contact with each other on a frictionless, horizontal surface as in Active Figure 4.13a. A constant horizontal force $\vec{F}$ is applied to m_1 as shown.

(A) Find the magnitude of the acceleration of the system.

SOLUTION

Conceptualize Conceptualize the situation by using Active Figure 4.13a and realize that both blocks must experience the *same* acceleration because they are in contact with each other and remain in contact throughout the motion.

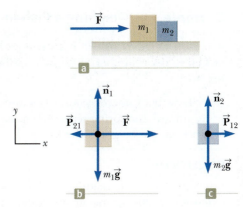

Active Figure 4.13 (Example 4.5) (a) A force is applied to a block of mass m_1, which pushes on a second block of mass m_2. (b) The forces acting on m_1. (c) The forces acting on m_2.

Categorize We categorize this problem as one involving a particle under a net force because a force is applied to a system of blocks and we are looking for the acceleration of the system.

Analyze First model the combination of two blocks as a single particle under a net force. Apply Newton's second law to the combination in the x direction to find the acceleration:

$$\sum F_x = F = (m_1 + m_2)a_x$$

$$(1)\ a_x = \frac{F}{m_1 + m_2}$$

Finalize The acceleration given by Equation (1) is the same as that of a single object of mass $m_1 + m_2$ and subject to the same force.

(B) Determine the magnitude of the contact force between the two blocks.

SOLUTION

Conceptualize The contact force is internal to the system of two blocks. Therefore, we cannot find this force by modeling the whole system (the two blocks) as a single particle.

Categorize Now consider each of the two blocks individually by categorizing each as a particle under a net force.

Analyze We construct a diagram of forces acting on the object for each block as shown in Active Figures 4.13b and 4.13c, where the contact force is denoted by $\vec{P}$. From Active Figure 4.13c, we see that the only horizontal force acting on m_2 is the contact force $\vec{P}_{12}$ (the force exerted by m_1 on m_2), which is directed to the right.

Apply Newton's second law to m_2:

$$(2)\ \sum F_x = P_{12} = m_2 a_x$$

Substitute the value of the acceleration a_x given by Equation (1) into Equation (2):

$$(3)\ P_{12} = m_2 a_x = \left(\frac{m_2}{m_1 + m_2}\right)F$$

Finalize This result shows that the contact force P_{12} is *less* than the applied force F. The force required to accelerate block 2 alone must be less than the force required to produce the same acceleration for the two-block system.

To finalize further, let us check this expression for P_{12} by considering the forces acting on m_1, shown in Active Figure 4.13b. The horizontal forces acting on m_1 are the applied force $\vec{F}$ to the right and the contact force $\vec{P}_{21}$ to the left (the force exerted by m_2 on m_1). From Newton's third law, $\vec{P}_{21}$ is the reaction force to $\vec{P}_{12}$, so $P_{21} = P_{12}$.

Apply Newton's second law to m_1:

$$(4)\ \sum F_x = F - P_{21} = F - P_{12} = m_1 a_x$$

Solve for P_{12} and substitute the value of a_x from Equation (1):

$$P_{12} = F - m_1 a_x = F - m_1\left(\frac{F}{m_1 + m_2}\right) = \left(\frac{m_2}{m_1 + m_2}\right)F$$

This result agrees with Equation (3), as it must.

What If? Imagine that the force $\vec{F}$ in Active Figure 4.13 is applied toward the left on the right-hand block of mass m_2. Is the magnitude of the force $\vec{P}_{12}$ the same as it was when the force was applied toward the right on m_1?

Answer When the force is applied toward the left on m_2, the contact force must accelerate m_1. In the original situation, the contact force accelerates m_2. Because $m_1 > m_2$, more force is required, so the magnitude of $\vec{P}_{12}$ is greater than in the original situation.

Example 4.6 | Weighing a Fish in an Elevator

A person weighs a fish of mass m on a spring scale attached to the ceiling of an elevator as illustrated in Figure 4.14.

(A) Show that if the elevator accelerates either upward or downward, the spring scale gives a reading that is different from the weight of the fish.

SOLUTION

Conceptualize The reading on the scale is related to the extension of the spring in the scale, which is related to the force on the end of the spring as in Figure 4.2. Imagine that the fish is hanging on a string attached to the end of the spring. In this case, the magnitude of the force exerted on the spring is equal to the tension T in the string. Therefore, we are looking for T. The force $\vec{T}$ pulls down on the string and pulls up on the fish.

Categorize We can categorize this problem by identifying the fish as a particle under a net force.

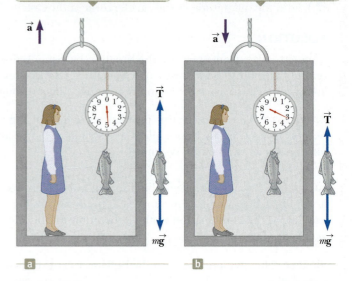

When the elevator accelerates upward, the spring scale reads a value greater than the weight of the fish.

When the elevator accelerates downward, the spring scale reads a value less than the weight of the fish.

Figure 4.14 (Example 4.6) A fish is weighed on a spring scale in an accelerating elevator car.

Analyze Inspect the diagrams of the forces acting on the fish in Figure 4.14 and notice that the external forces acting on the fish are the downward gravitational force $\vec{F}_g = m\vec{g}$ and the force $\vec{T}$ exerted by the string. If the elevator is either at rest or moving at constant velocity, the fish is a particle in equilibrium, so $\Sigma F_y = T - F_g = 0$ or $T = F_g = mg$. (Remember that the scalar mg is the weight of the fish.)

Now suppose the elevator is moving with an acceleration $\vec{a}$ relative to an observer standing outside the elevator in an inertial frame. The fish is now a particle under a net force.

Apply Newton's second law to the fish:

$$\Sigma F_y = T - mg = ma_y$$

Solve for T:

$$(1)\ T = ma_y + mg = mg\left(\frac{a_y}{g} + 1\right) = F_g\left(\frac{a_y}{g} + 1\right)$$

where we have chosen upward as the positive y direction. We conclude from Equation (1) that the scale reading T is greater than the fish's weight mg if $\vec{a}$ is upward, so a_y is positive (Fig. 4.14a), and that the reading is less than mg if $\vec{a}$ is downward, so a_y is negative (Fig. 4.14b).

(B) Evaluate the scale readings for a 40.0-N fish if the elevator moves with an acceleration $a_y = \pm 2.00$ m/s^2.

SOLUTION

Evaluate the scale reading from Equation (1) if $\vec{a}$ is upward:

$$T = (40.0\ \text{N})\left(\frac{2.00\ \text{m/s}^2}{9.80\ \text{m/s}^2} + 1\right) = 48.2\ \text{N}$$

Evaluate the scale reading from Equation (1) if $\vec{a}$ is downward:

$$T = (40.0\ \text{N})\left(\frac{-2.00\ \text{m/s}^2}{9.80\ \text{m/s}^2} + 1\right) = 31.8\ \text{N}$$

Finalize Take this advice: if you buy a fish in an elevator, make sure the fish is weighed while the elevator is either at rest or accelerating downward! Furthermore, notice that from the information given here, one cannot determine the direction of motion of the elevator.

What If? Suppose the elevator cable breaks and the elevator and its contents are in free fall. What happens to the reading on the scale?

Answer If the elevator falls freely, its acceleration is $a_y = -g$. We see from Equation (1) that the scale reading T is zero in this case; that is, the fish *appears* to be weightless.

4.8 | Context Connection: Forces on Automobiles

In the Context Connections of Chapters 2 and 3, we focused on two types of acceleration exhibited by a number of vehicles. In this chapter, we learned how the acceleration of an object is related to the force on the object. Let us apply this understanding to an investigation of the forces that are applied to automobiles when they are exhibiting their maximum acceleration in speeding up from rest to 60 mi/h.

The force that accelerates an automobile is the friction force from the ground. (We will study friction forces in detail in Chapter 5.) The engine applies a force to the wheels, attempting to rotate them so that the bottoms of the tires apply forces backward on the road surface. By Newton's third law, the road surface applies forces in the forward direction on the tires, causing the car to move forward. If we ignore air resistance, this force can be modeled as the net force on the automobile in the horizontal direction.

In Chapter 2, we investigated the 0 to 60 mi/h acceleration of a number of vehicles. Table 4.2 repeats this acceleration information and also shows the weight of the vehicle in pounds and the mass in kilograms. With both the acceleration and the mass, we can find the force driving the car forward, as shown in the last column of Table 4.2.

TABLE 4.2 | Driving Forces on Various Vehicles

Automobile	Model Year	Average Acceleration (mi/h·s)	Weight (lb)	Mass (kg)	Force (× 10³ N)
Very expensive vehicles:					
Bugatti Veyron 16.4 Super Sport	2011	23.1	4160	1887	19.5
Lamborghini LP 570-4 Superleggera	2011	17.6	2954	1340	10.5
Lexus LFA	2011	15.8	3580	1624	11.5
Mercedes-Benz SLS AMG	2011	16.7	3795	1721	12.8
Shelby SuperCars Ultimate Aero	2009	22.2	2750	1247	12.4
Average		**19.1**	**3448**	**1564**	**13.3**
Performance vehicles:					
Chevrolet Corvette ZR1	2010	18.2	3333	1512	12.3
Dodge Viper SRT10	2010	15.0	3460	1569	10.5
Jaguar XJL Supercharged	2011	13.6	4323	1961	11.9
Acura TL SH-AWD	2009	11.5	3860	1751	9.0
Dodge Challenger SRT8	2010	12.2	4140	1878	10.2
Average		**14.1**	**3823**	**1734**	**10.8**
Traditional vehicles:					
Buick Regal CXL Turbo	2011	8.0	3671	1665	6.0
Chevrolet Tahoe 1500 LS (SUV)	2011	7.0	5636	2556	8.0
Ford Fiesta SES	2010	6.2	2330	1057	2.9
Hummer H3 (SUV)	2010	7.5	4695	2130	7.1
Hyundai Sonata SE	2010	8.0	3340	1515	5.4
Smart ForTwo	2010	4.5	1825	828	1.7
Average		**6.9**	**3583**	**1625**	**5.2**
Alternative vehicles:					
Chevrolet Volt (hybrid)	2011	7.5	3500	1588	5.3
Nissan Leaf (electric)	2011	6.0	3500	1588	4.3
Honda CR-Z (hybrid)	2011	5.7	2637	1196	3.0
Honda Insight (hybrid)	2010	5.7	2723	1235	3.2
Toyota Prius (hybrid)	2010	6.1	3042	1380	3.8
Average		**6.2**	**3080**	**1397**	**3.9**

We can see some interesting results in Table 4.2. The forces in the very expensive and performance vehicle sections are all large compared with the forces in the other parts of the table. Furthermore, the average masses of very expensive and performance vehicles are less than 10% larger than those in the traditional vehicle portion of the table. Therefore, the large forces of the very expensive and performance vehicles translate into the very large accelerations exhibited by these vehicles. One standout in the very expensive vehicles is the Bugatti Veyron 16.4 Super Sport. It is the most massive car in the group, but the huge force generated on the tires results in its having the highest acceleration in the group. The second-highest acceleration in the group is the Shelby SuperCars Ultimate Aero. This vehicle has only 66% of the mass of the Bugatti, representing much less resistance to being accelerated. The force on the Shelby, however, is only 64% of that on the Bugatti, resulting in a lower acceleration of the Shelby, despite its smaller mass.

As expected, the forces exerted on the traditional vehicles are smaller than those of the very expensive and performance vehicles, corresponding to the smaller accelerations of this group. Notice, however, that the forces for the two SUVs are large. Because these two vehicles have accelerations that are somewhat similar to those of the other vehicles in this portion of the table, we can identify these large forces as being required to accelerate the larger mass of the SUVs.

Also as expected, the forces driving the alternative vehicles have the lowest average in the table. This finding is consistent with the accelerations of these vehicles being lower than those elsewhere in the table.

Another interesting entry in the table is the Smart ForTwo in the traditional vehicles. Its force is by far the smallest in the table, but its mass is also the smallest in the table. As a result, its acceleration is 4.5 mi/h · s, which, although not impressive, is enough to satisfy some consumers who are looking for other advantages offered by the Smart car, such as increased fuel efficiency.

SUMMARY

Newton's first law states that if an object does not interact with other objects, it is possible to identify a reference frame in which the object has zero acceleration. Therefore, if we observe an object from such a frame and no force is exerted on the object, an object at rest remains at rest and an object in uniform motion in a straight line maintains that motion.

Newton's first law defines an **inertial frame of reference,** which is a frame in which Newton's first law is valid.

Newton's second law states that the acceleration of an object is directly proportional to the net force acting on the object and inversely proportional to the object's mass.

Newton's third law states that if two objects interact, the force exerted by object 1 on object 2 is equal in magnitude but opposite in direction to the force exerted by object 2 on object 1. Therefore, an isolated force cannot exist in nature.

The **weight** of an object is equal to the product of its mass (a scalar quantity) and the magnitude of the free-fall acceleration, or

$$F_g = mg \qquad\qquad \text{4.5} \blacktriangleleft$$

Analysis Models for Problem Solving

Particle Under a Net Force If a particle of mass m experiences a nonzero net force, its acceleration is related to the net force by Newton's second law:

$$\sum \vec{\mathbf{F}} = m\vec{\mathbf{a}} \qquad\qquad \text{4.2} \blacktriangleleft$$

Particle in Equilibrium If a particle maintains a constant velocity (so that $\vec{\mathbf{a}} = 0$), which could include a velocity of zero, the forces on the particle balance and Newton's second law reduces to

$$\sum \vec{\mathbf{F}} = 0 \qquad\qquad \text{4.8} \blacktriangleleft$$

► OBJECTIVE QUESTIONS

1. The third graders are on one side of a schoolyard, and the fourth graders are on the other. They are throwing snowballs at each other. Between them, snowballs of various masses are moving with different velocities as shown in Figure OQ4.1. Rank the snowballs (a) through (e) according to the magnitude of the total force exerted on each one. Ignore air resistance. If two snowballs rank together, make that fact clear.

Figure OQ4.1

2. In Figure OQ4.2, a locomotive has broken through the wall of a train station. During the collision, what can be said about the force exerted by the locomotive on the wall? (a) The force exerted by the locomotive on the wall was larger than the force the wall could exert on the locomotive. (b) The force exerted by the locomotive on the wall was the same in magnitude as the force exerted by the wall on the locomotive. (c) The force exerted by the

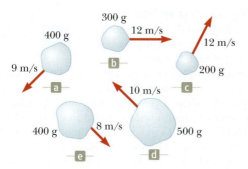

Figure OQ4.2

Studio Lévy and Sons

locomotive on the wall was less than the force exerted by the wall on the locomotive. (d) The wall cannot be said to "exert" a force; after all, it broke.

3. An experiment is performed on a puck on a level air hockey table, where friction is negligible. A constant horizontal force is applied to the puck, and the puck's acceleration is measured. Now the same puck is transported far into outer space, where both friction and gravity are negligible. The same constant force is applied to the puck (through a spring scale that stretches the same amount), and the puck's acceleration (relative to the distant stars) is measured. What is the puck's acceleration in outer space? (a) It is somewhat greater than its acceleration on the Earth. (b) It is the same as its acceleration on the Earth. (c) It is less than its acceleration on the Earth. (d) It is infinite because neither friction nor gravity constrains it. (e) It is very large because acceleration is inversely proportional to weight and the puck's weight is very small but not zero.

4. A truck loaded with sand accelerates along a highway. The driving force on the truck remains constant. What happens to the acceleration of the truck if its trailer leaks sand at a constant rate through a hole in its bottom? (a) It decreases at a steady rate. (b) It increases at a steady rate. (c) It increases and then decreases. (d) It decreases and then increases. (e) It remains constant.

5. If an object is in equilibrium, which of the following statements is *not* true? (a) The speed of the object remains constant. (b) The acceleration of the object is zero. (c) The net force acting on the object is zero. (d) The object must be at rest. (e) There are at least two forces acting on the object.

6. Two objects are connected by a string that passes over a frictionless pulley as in Active Figure 4.12a, where $m_1 < m_2$ and a_1 and a_2 are the magnitudes of the respective accelerations. Which mathematical statement is true regarding the magnitude of the acceleration a_2 of the mass m_2? (a) $a_2 < g$ (b) $a_2 > g$ (c) $a_2 = g$ (d) $a_2 < a_1$ (e) $a_2 > a_1$

► CONCEPTUAL QUESTIONS

1. A passenger sitting in the rear of a bus claims that she was injured when the driver slammed on the brakes, causing a suitcase to come flying toward her from the front of the bus. If you were the judge in this case, what disposition would you make? Why?

2. If a car is traveling due westward with a constant speed of 20 m/s, what is the resultant force acting on it?

3. A person holds a ball in her hand. (a) Identify all the external forces acting on the ball and the Newton's third-law reaction force to each one. (b) If the ball is dropped, what

force is exerted on it while it is falling? Identify the reaction force in this case. (Ignore air resistance.)

4. In the motion picture *It Happened One Night* (Columbia Pictures, 1934), Clark Gable is standing inside a stationary bus in front of Claudette Colbert, who is seated. The bus suddenly starts moving forward and Clark falls into Claudette's lap. Why did this happen?

5. If you hold a horizontal metal bar several centimeters above the ground and move it through grass, each leaf of grass bends out of the way. If you increase the speed

of the bar, each leaf of grass will bend more quickly. How then does a rotary power lawn mower manage to cut grass? How can it exert enough force on a leaf of grass to shear it off?

6. A spherical rubber balloon inflated with air is held stationary, with its opening, on the west side, pinched shut. (a) Describe the forces exerted by the air inside and outside the balloon on sections of the rubber. (b) After the balloon is released, it takes off toward the east, gaining speed rapidly. Explain this motion in terms of the forces now acting on the rubber. (c) Account for the motion of a skyrocket taking off from its launch pad.

7. A rubber ball is dropped onto the floor. What force causes the ball to bounce?

8. The mayor of a city reprimands some city employees because they will not remove the obvious sags from the cables that support the city traffic lights. What explanation can the employees give? How do you think the case will be settled in mediation?

9. Balancing carefully, three boys inch out onto a horizontal tree branch above a pond, each planning to dive in separately. The third boy in line notices that the branch is barely strong enough to support them. He decides to jump straight up and land back on the branch to break it, spilling all three into the pond. When he starts to carry out his plan, at what precise moment does the branch break? Explain. *Suggestion:* Pretend to be the third boy and imitate what he does in slow motion. If you are still unsure, stand on a bathroom scale and repeat the suggestion.

10. A child tosses a ball straight up. She says that the ball is moving away from her hand because the ball feels an upward "force of the throw" as well as the gravitational force. (a) Can the "force of the throw" exceed the gravitational force? How would the ball move if it did? (b) Can the "force of the throw" be equal in magnitude to the gravitational force? Explain. (c) What strength can accurately be attributed to the "force of the throw"? Explain. (d) Why does the ball move away from the child's hand?

11. A weightlifter stands on a bathroom scale. He pumps a barbell up and down. What happens to the reading on the scale as he does so? **What If?** What if he is strong enough to actually *throw* the barbell upward? How does the reading on the scale vary now?

12. When you push on a box with a 200-N force instead of a 50-N force, you can feel that you are making a greater effort. When a table exerts a 200-N normal force instead of one of smaller magnitude, is the table really doing anything differently?

13. Identify action–reaction pairs in the following situations: (a) a man takes a step (b) a snowball hits a girl in the back (c) a baseball player catches a ball (d) a gust of wind strikes a window

14. Give reasons for the answers to each of the following questions: (a) Can a normal force be horizontal? (b) Can a normal force be directed vertically downward? (c) Consider a tennis ball in contact with a stationary floor and with nothing else. Can the normal force be different in magnitude from the gravitational force exerted on the ball? (d) Can the force exerted by the floor on the ball be different in magnitude from the force the ball exerts on the floor?

15. An athlete grips a light rope that passes over a low-friction pulley attached to the ceiling of a gym. A sack of sand precisely equal in weight to the athlete is tied to the other end of the rope. Both the sand and the athlete are initially at rest. The athlete climbs the rope, sometimes speeding up and slowing down as he does so. What happens to the sack of sand? Explain.

16. In Figure CQ4.16, the light, taut, unstretchable cord B joins block 1 and the larger-mass block 2. Cord A exerts a force on block 1 to make it accelerate forward. (a) How does the magnitude of the force exerted by cord A on block 1 compare with the magnitude of the force exerted by cord B on block 2? Is it larger, smaller, or equal? (b) How does the acceleration of block 1 compare with the acceleration (if any) of block 2? (c) Does cord B exert a force on block 1? If so, is it forward or backward? Is it larger, smaller, or equal in magnitude to the force exerted by cord B on block 2?

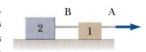

Figure CQ4.16

17. Twenty people participate in a tug-of-war. The two teams of ten people are so evenly matched that neither team wins. After the game they notice that a car is stuck in the mud. They attach the tug-of-war rope to the bumper of the car, and all the people pull on the rope. The heavy car has just moved a couple of decimeters when the rope breaks. Why did the rope break in this situation when it did not break when the same twenty people pulled on it in a tug-of-war?

18. Can an object exert a force on itself? Argue for your answer.

19. As shown in Figure CQ4.19, student A, a 55-kg girl, sits on one chair with metal runners, at rest on a classroom floor. Student B, an 80-kg boy, sits on an identical chair. Both students keep their feet off the floor. A rope runs from student A's hands around a light pulley and then over her shoulder to the hands of a teacher standing on the floor behind her. The low-friction axle of the pulley is attached to a second rope held by student B. All ropes run parallel to the chair runners. (a) If student A pulls on her end of the rope, will her chair or will B's chair slide on the floor? Explain why. (b) If instead the teacher pulls on his rope end, which chair slides? Why this one? (c) If student B pulls on his rope, which chair slides? Why? (d) Now the teacher ties his end of the rope to student A's chair. Student A pulls on the end of the rope in her hands. Which chair slides and why?

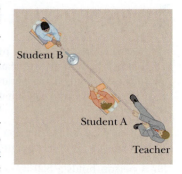

Figure CQ4.19

PROBLEMS

WebAssign The problems found in this chapter may be assigned online in Enhanced WebAssign.

1. denotes straightforward problem; **2.** denotes intermediate problem; **3.** denotes challenging problem

1. denotes full solution available in the *Student Solutions Manual/Study Guide*

1. denotes problems most often assigned in Enhanced WebAssign.

BIO denotes biomedical problem

GP denotes guided problem

M denotes Master It tutorial available in Enhanced WebAssign

Q|C denotes asking for quantitative and conceptual reasoning

S denotes symbolic reasoning problem

shaded denotes "paired problems" that develop reasoning with symbols and numerical values

W denotes Watch It video solution available in Enhanced WebAssign

Section 4.3 Mass

1. **W** A force $\vec{\mathbf{F}}$ applied to an object of mass m_1 produces an acceleration of 3.00 m/s². The same force applied to a second object of mass m_2 produces an acceleration of 1.00 m/s². (a) What is the value of the ratio m_1/m_2? (b) If m_1 and m_2 are combined into one object, find its acceleration under the action of the force $\vec{\mathbf{F}}$.

2. (a) A car with a mass of 850 kg is moving to the right with a constant speed of 1.44 m/s. What is the total force on the car? (b) What is the total force on the car if it is moving to the left?

Section 4.4 Newton's Second Law

3. **M** A toy rocket engine is securely fastened to a large puck that can glide with negligible friction over a horizontal surface, taken as the *xy* plane. The 4.00-kg puck has a velocity of $3.00\hat{\mathbf{i}}$ m/s at one instant. Eight seconds later, its velocity is $(8\hat{\mathbf{i}} + 10\hat{\mathbf{j}})$ m/s. Assuming the rocket engine exerts a constant horizontal force, find (a) the components of the force and (b) its magnitude.

4. Two forces, $\vec{\mathbf{F}}_1 = (-6\hat{\mathbf{i}} - 4\hat{\mathbf{j}})$ N and $\vec{\mathbf{F}}_2 = (-3\hat{\mathbf{i}} + 7\hat{\mathbf{j}})$ N, act on a particle of mass 2.00 kg that is initially at rest at coordinates $(-2.00$ m, $+4.00$ m$)$. (a) What are the components of the particle's velocity at $t = 10.0$ s? (b) In what direction is the particle moving at $t = 10.0$ s? (c) What displacement does the particle undergo during the first 10.0 s? (d) What are the coordinates of the particle at $t = 10.0$ s?

5. **W** A 3.00-kg object undergoes an acceleration given by $\vec{\mathbf{a}} = (2.00\hat{\mathbf{i}} + 5.00\hat{\mathbf{j}})$ m/s². Find (a) the resultant force acting on the object and (b) the magnitude of the resultant force.

6. **W** **Review.** Three forces acting on an object are given by $\vec{\mathbf{F}}_1 = (-2.00\hat{\mathbf{i}} + 2.00\hat{\mathbf{j}})$ N, $\vec{\mathbf{F}}_2 = (5.00\hat{\mathbf{i}} - 3.00\hat{\mathbf{j}})$ N, and $\vec{\mathbf{F}}_3 = (-45.0\hat{\mathbf{i}})$ N. The object experiences an acceleration of magnitude 3.75 m/s². (a) What is the direction of the acceleration? (b) What is the mass of the object? (c) If the object is initially at rest, what is its speed after 10.0 s? (d) What are the velocity components of the object after 10.0 s?

7. **M** Two forces $\vec{\mathbf{F}}_1$ and $\vec{\mathbf{F}}_2$ act on a 5.00-kg object. Taking $F_1 = 20.0$ N and $F_2 = 15.0$ N, find the accelerations of the object for the configurations of forces shown in parts (a) and (b) of Figure P4.7.

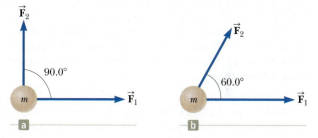

Figure P4.7

8. **W** A 3.00-kg object is moving in a plane, with its *x* and *y* coordinates given by $x = 5t^2 - 1$ and $y = 3t^3 + 2$, where *x* and *y* are in meters and *t* is in seconds. Find the magnitude of the net force acting on this object at $t = 2.00$ s.

Section 4.5 The Gravitational Force and Weight

9. A woman weighs 120 lb. Determine (a) her weight in newtons and (b) her mass in kilograms.

10. The gravitational force on a baseball is $-F_g\hat{\mathbf{j}}$. A pitcher throws the baseball with velocity $v\hat{\mathbf{i}}$ by uniformly accelerating it straight forward horizontally for a time interval $\Delta t = t - 0 = t$. If the ball starts from rest, (a) through what distance does it accelerate before its release? (b) What force does the pitcher exert on the ball?

11. **M** **Review.** An electron of mass 9.11×10^{-31} kg has an initial speed of 3.00×10^5 m/s. It travels in a straight line, and its speed increases to 7.00×10^5 m/s in a distance of 5.00 cm. Assuming its acceleration is constant, (a) determine the magnitude of the force exerted on the electron and (b) compare this force with the weight of the electron, which we ignored.

12. If a man weighs 900 N on the Earth, what would he weigh on Jupiter, where the free-fall acceleration is 25.9 m/s²?

13. The distinction between mass and weight was discovered after Jean Richer transported pendulum clocks from Paris, France, to Cayenne, French Guiana, in 1671. He found that they quite systematically ran slower in Cayenne than in Paris. The effect was reversed when the clocks returned to Paris. How much weight would a 90.0 kg person lose in traveling from Paris, where $g = 9.809\ 5$ m/s², to Cayenne, where $g = 9.780\ 8$ m/s²? (We will consider how the free-fall acceleration influences the period of a pendulum in Section 12.4.)

14. Besides the gravitational force, a 2.80-kg object is subjected to one other constant force. The object starts from rest and in 1.20 s experiences a displacement of $(4.20\hat{i} - 3.30\hat{j})$ m, where the direction of $\hat{j}$ is the upward vertical direction. Determine the other force.

Section 4.6 Newton's Third Law

15. The average speed of a nitrogen molecule in air is about 6.70×10^2 m/s, and its mass is 4.68×10^{-26} kg. (a) If it takes 3.00×10^{-13} s for a nitrogen molecule to hit a wall and rebound with the same speed but moving in the opposite direction, what is the average acceleration of the molecule during this time interval? (b) What average force does the molecule exert on the wall?

16. You stand on the seat of a chair and then hop off. (a) During the time interval you are in flight down to the floor, the Earth moves toward you with an acceleration of what order of magnitude? In your solution, explain your logic. Model the Earth as a perfectly solid object. (b) The Earth moves toward you through a distance of what order of magnitude?

17. A 15.0-lb block rests on the floor. (a) What force does the floor exert on the block? (b) A rope is tied to the block and is run vertically over a pulley. The other end is attached to a free-hanging 10.0-lb object. What now is the force exerted by the floor on the 15.0-lb block? (c) If the 10.0-lb object in part (b) is replaced with a 20.0-lb object, what is the force exerted by the floor on the 15.0-lb block?

Section 4.7 Analysis Models Using Newton's Second Law

18. W A block slides down a frictionless plane having an inclination of $\theta = 15.0°$. The block starts from rest at the top, and the length of the incline is 2.00 m. (a) Draw a free-body diagram of the block. Find (b) the acceleration of the block and (c) its speed when it reaches the bottom of the incline.

19. Two people pull as hard as they can on horizontal ropes attached to a boat that has a mass of 200 kg. If they pull in the same direction, the boat has an acceleration of 1.52 m/s² to the right. If they pull in opposite directions, the boat has an acceleration of 0.518 m/s² to the left. What is the magnitude of the force each person exerts on the boat? Disregard any other horizontal forces on the boat.

20. BIO A setup similar to the one shown in Figure P4.20 is often used in hospitals to support and apply a horizontal traction force to an injured leg. (a) Determine the force of tension in the rope supporting the leg. (b) What is the traction force exerted to the right on the leg?

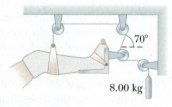

Figure P4.20

21. Review. Figure P4.21 shows a worker poling a boat—a very efficient mode of transportation—across a shallow lake. He pushes parallel to the length of the light pole, exerting a force of magnitude 240 N on the bottom of the lake. Assume the pole lies in the vertical plane containing the keel of the boat. At one moment, the pole makes an angle of 35.0° with the vertical and the water exerts a horizontal drag force of 47.5 N on the boat, opposite to its forward velocity of magnitude 0.857 m/s. The mass of the boat including its cargo and the worker is 370 kg. (a) The water exerts a buoyant force vertically upward on the boat. Find the magnitude of this force. (b) Model the forces as constant over a short interval of time to find the velocity of the boat 0.450 s after the moment described.

Figure P4.21

22. W The systems shown in Figure P4.22 are in equilibrium. If the spring scales are calibrated in newtons, what do they read? Ignore the masses of the pulleys and strings and assume the pulleys and the incline in Figure P4.22d are frictionless.

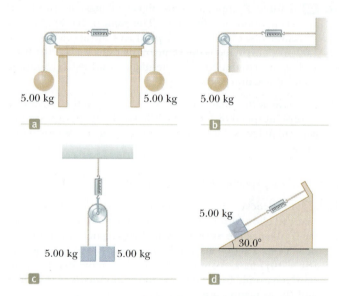

5.00 kg   5.00 kg   5.00 kg

a   b

5.00 kg

5.00 kg   5.00 kg   30.0°

c   d

Figure P4.22

23. W A bag of cement weighing 325 N hangs in equilibrium from three wires as suggested in Figure P4.23. Two of the wires make angles $\theta_1 = 60.0°$ and $\theta_2 = 40.0°$ with the horizontal. Assuming the system is in equilibrium, find the tensions T_1, T_2, and T_3 in the wires.

24. S A bag of cement whose weight is F_g hangs in equilibrium from three wires as shown in Figure P4.23. Two of the wires make angles θ_1 and θ_2 with the horizontal. Assuming the system is in equilibrium, show that the tension in the left-hand wire is

$$T_1 = \frac{F_g \cos \theta_2}{\sin(\theta_1 + \theta_2)}$$

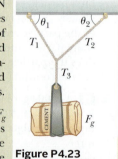

Figure P4.23
Problems 23 and 24.

25. M In Example 4.6, we investigated the apparent weight of a fish in an elevator. Now consider a 72.0-kg man standing on a spring scale in an elevator. Starting from rest, the elevator ascends, attaining its maximum speed of 1.20 m/s in 0.800 s.

It travels with this constant speed for the next 5.00 s. The elevator then undergoes a uniform acceleration in the negative *y* direction for 1.50 s and comes to rest. What does the spring scale register (a) before the elevator starts to move, (b) during the first 0.800 s, (c) while the elevator is traveling at constant speed, and (d) during the time interval it is slowing down?

26. Figure P4.26 shows loads hanging from the ceiling of an elevator that is moving at constant velocity. Find the tension in each of the three strands of cord supporting each load.

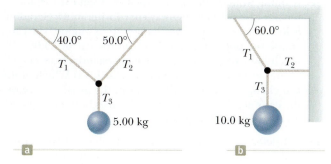

Figure P4.26

27. A simple accelerometer is constructed inside a car by suspending an object of mass *m* from a string of length *L* that is tied to the car's ceiling. As the car accelerates the string–object system makes a constant angle of θ with the vertical. (a) Assuming that the string mass is negligible compared with *m*, derive an expression for the car's acceleration in terms of θ and show that it is independent of the mass *m* and the length *L*. (b) Determine the acceleration of the car when $\theta = 23.0°$.

28. [W] An object of mass $m_1 = 5.00$ kg placed on a frictionless, horizontal table is connected to a string that passes over a pulley and then is fastened to a hanging object of mass $m_2 = 9.00$ kg as shown in Figure P4.28. (a) Draw free-body diagrams of both objects. Find (b) the magnitude of the acceleration of the objects and (c) the tension in the string.

Figure P4.28

29. An object of mass $m = 1.00$ kg is observed to have an acceleration $\vec{a}$ with a magnitude of 10.0 m/s² in a direction 60.0° east of north. Figure P4.29 shows a view of the object from above. The force $\vec{F}_2$ acting on the object has a magnitude of 5.00 N and is directed north. Determine the magnitude and direction of the one other horizontal force $\vec{F}_1$ acting on the object.

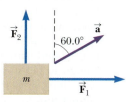

Figure P4.29

30. Two objects are connected by a light string that passes over a frictionless pulley as shown in Figure P4.30. Assume the incline is

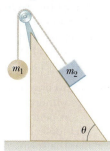

Figure P4.30

frictionless and take $m_1 = 2.00$ kg, $m_2 = 6.00$ kg, and $\theta = 55.0°$. (a) Draw free-body diagrams of both objects. Find (b) the magnitude of the acceleration of the objects, (c) the tension in the string, and (d) the speed of each object 2.00 s after it is released from rest.

31. A block is given an initial velocity of 5.00 m/s up a frictionless 20.0° incline (Fig. P4.31). How far up the incline does the block slide before coming to rest?

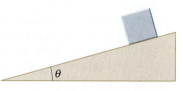

Figure P4.31

32. A car is stuck in the mud. A tow truck pulls on the car with the arrangement shown in Fig. P4.32. The tow cable is under a tension of 2 500 N and pulls downward and to the left on the pin at its upper end. The light pin is held in equilibrium by forces exerted by the two bars A and B. Each bar is a *strut;* that is, each is a bar whose weight is small compared to the forces it exerts and which exerts forces only through hinge pins at its ends. Each strut exerts a force directed parallel to its length. Determine the force of tension or compression in each strut. Proceed as follows. Make a guess as to which way (pushing or pulling) each force acts on the top pin. Draw a free-body diagram of the pin. Use the condition for equilibrium of the pin to translate the free-body diagram into equations. From the equations calculate the forces exerted by struts A and B. If you obtain a positive answer, you correctly guessed the direction of the force. A negative answer means that the direction should be reversed, but the absolute value correctly gives the magnitude of the force. If a strut pulls on a pin, it is in tension. If it pushes, the strut is in compression. Identify whether each strut is in tension or in compression.

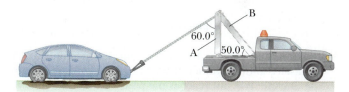

Figure P4.32

33. Two blocks, each of mass $m = 3.50$ kg, are hung from the ceiling of an elevator as in Figure P4.33. (a) If the elevator moves with an upward acceleration $\vec{a}$ of magnitude 1.60 m/s², find the tensions T_1 and T_2 in the upper and lower strings. (b) If the strings can withstand a maximum tension of 85.0 N, what maximum acceleration can the elevator have before a string breaks?

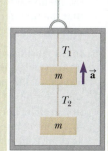

Figure P4.33
Problems 33 and 34.

34. [S] Two blocks, each of mass *m*, are hung from the ceiling of an elevator as in Figure P4.33. The elevator has an upward acceleration *a*. The strings have negligible mass. (a) Find the tensions T_1 and T_2 in the upper and lower strings in terms of *m*, *a*, and *g*. (b) Compare the two tensions and determine which string would break first if *a* is made sufficiently large. (c) What are the tensions if the cable supporting the elevator breaks?

35. In Figure P4.35, the man and the platform together weigh 950 N. The pulley can be modeled as frictionless. Determine how hard the man has to pull on the rope to lift himself steadily upward above the ground. (Or is it impossible? If so, explain why.)

Figure P4.35

36. Two objects with masses of 3.00 kg and 5.00 kg are connected by a light string that passes over a light frictionless pulley to form an Atwood machine as shown in Active Figure 4.12a. Determine (a) the tension in the string, (b) the acceleration of each object, and (c) the distance each object will move in the first second of motion if they start from rest.

37. **M** In the system shown in Figure P4.37, a horizontal force $\vec{F}_x$ acts on an object of mass $m_2 = 8.00$ kg. The horizontal surface is frictionless. Consider the acceleration of the sliding object as a function of F_x. (a) For what values of F_x does the object of mass $m_1 = 2.00$ kg accelerate upward? (b) For what values of F_x is the tension in the cord zero? (c) Plot the acceleration of the m_2 object versus F_x. Include values of F_x from -100 N to $+100$ N.

Figure P4.37

38. A frictionless plane is 10.0 m long and inclined at 35.0°. A sled starts at the bottom with an initial speed of 5.00 m/s up the incline. When it reaches the point at which it momentarily stops, a second sled is released from the top of this incline with an initial speed v_i. Both sleds reach the bottom of the incline at the same moment. (a) Determine the distance that the first sled traveled up the incline. (b) Determine the initial speed of the second sled.

39. In the Atwood machine discussed in Example 4.4 and shown in Active Figure 4.12a, $m_1 = 2.00$ kg and $m_2 = 7.00$ kg. The masses of the pulley and string are negligible by comparison. The pulley turns without friction, and the string does not stretch. The lighter object is released with a sharp push that sets it into motion at $v_i = 2.40$ m/s downward. (a) How far will m_1 descend below its initial level? (b) Find the velocity of m_1 after 1.80 s.

40. **S** An object of mass m_1 hangs from a string that passes over a very light fixed pulley P_1 as shown in Figure P4.40. The

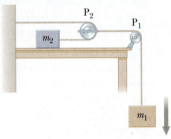

Figure P4.40

string connects to a second very light pulley P_2. A second string passes around this pulley with one end attached to a wall and the other to an object of mass m_2 on a frictionless, horizontal table. (a) If a_1 and a_2 are the accelerations of m_1 and m_2, respectively, what is the relation between these accelerations? Find expressions for (b) the tensions in the strings and (c) the accelerations a_1 and a_2 in terms of the masses m_1 and m_2, and g.

Section 4.8 Context Connection: Forces on Automobiles

41. A young woman buys an inexpensive used car for stock car racing. It can attain highway speed with an acceleration of 8.40 mi/h · s. By making changes to its engine, she can increase the net horizontal force on the car by 24.0%. With much less expense, she can remove material from the body of the car to decrease its mass by 24.0%. (a) Which of these two changes, if either, will result in the greater increase in the car's acceleration? (b) If she makes both changes, what acceleration can she attain?

42. A 1 000-kg car is pulling a 300-kg trailer. Together, the car and trailer move forward with an acceleration of 2.15 m/s². Ignore any force of air drag on the car and all frictional forces on the trailer. Determine (a) the net force on the car, (b) the net force on the trailer, (c) the force exerted by the trailer on the car, and (d) the resultant force exerted by the car on the road.

Additional Problems

43. **S** An object of mass M is held in place by an applied force $\vec{F}$ and a pulley system as shown in Figure P4.43. The pulleys are massless and frictionless. (a) Draw diagrams showing the forces on each pulley. Find (b) the tension in each section of rope, T_1, T_2, T_3, T_4, and T_5 and (c) the magnitude of $\vec{F}$.

44. Any device that allows you to increase the force you exert is a kind of *machine*. Some machines, such as the prybar or the inclined plane, are very simple. Some machines do not even look like machines. For example, your car is stuck in the mud and you can't pull hard enough to get it out. You do, however, have a long cable that you connect taut between your front bumper and the trunk of a stout tree. You now pull sideways on the cable at its midpoint, exerting a force f. Each half of the cable is displaced through a small angle θ from the straight line between the ends of the cable. (a) Deduce an expression for the force acting on the car. (b) Evaluate the cable tension for the case where $\theta = 7.00°$ and $f = 100$ N.

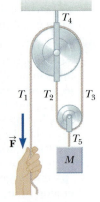

Figure P4.43

45. An inventive child named Nick wants to reach an apple in a tree without climbing the tree. Sitting in a chair connected to a rope that passes over a frictionless pulley (Fig. P4.45), Nick pulls on the loose end of the rope with such a force that the spring scale reads 250 N. Nick's true weight is 320 N, and the chair weighs 160 N. Nick's feet are not touching the ground. (a) Draw one pair of diagrams showing the

forces for Nick and the chair considered as separate systems and another diagram for Nick and the chair considered as one system. (b) Show that the acceleration of the system is *upward* and find its magnitude. (c) Find the force Nick exerts on the chair.

46. Q|C In the situation described in Problem 45 and Figure P4.45, the masses of the rope, spring balance, and pulley are negligible. Nick's feet are not touching the ground. (a) Assume Nick is momentarily at rest when he stops pulling down on the rope and passes the end of the rope to another child, of weight 440 N, who is standing on the ground next to him. The rope does not break. Describe the ensuing motion. (b) Instead, assume Nick is momentarily at rest when he ties the end of the rope to a strong hook projecting from the tree trunk. Explain why this action can make the rope break.

Figure P4.45
Problems 45 and 46

47. Two blocks of mass 3.50 kg and 8.00 kg are connected by a massless string that passes over a frictionless pulley (Fig. P4.47). The inclines are frictionless. Find (a) the magnitude of the acceleration of each block and (b) the tension in the string.

Figure P4.47

48. A 1.00-kg glider on a horizontal air track is pulled by a string at an angle θ. The taut string runs over a pulley and is attached to a hanging object of mass 0.500 kg as shown in Figure P4.48. (a) Show that the speed v_x of the glider and the speed v_y of the hanging object are related by $v_x = u v_y$, where $u = z(z^2 - h_0^2)^{-1/2}$. (b) The glider is released from rest. Show that at that instant the acceleration a_x of the glider and the acceleration a_y of the hanging object are related by $a_x = u a_y$. (c) Find the tension in the string at the instant the glider is released for $h_0 = 80.0$ cm and $\theta = 30.0°$.

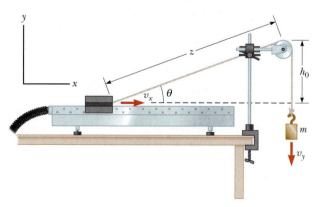

Figure P4.48

49. Q|C In Example 4.5, we pushed on two blocks on a table. Suppose three blocks are in contact with one another on a frictionless, horizontal surface as shown in Figure P4.49. A horizontal force $\vec{F}$ is applied to m_1. Take $m_1 = 2.00$ kg, $m_2 = 3.00$ kg, $m_3 = 4.00$ kg, and $F = 18.0$ N. (a) Draw a separate free-body diagram for each block. (b) Determine the acceleration of the blocks. (c) Find the *resultant* force on each block. (d) Find the magnitudes of the contact forces between the blocks. (e) You are working on a construction project. A coworker is nailing up plasterboard on one side of a light partition, and you are on the opposite side, providing "backing" by leaning against the wall with your back pushing on it. Every hammer blow makes your back sting. The supervisor helps you put a heavy block of wood between the wall and your back. Using the situation analyzed in parts (a) through (d) as a model, explain how this change works to make your job more comfortable.

Figure P4.49

50. S A mobile is formed by supporting four metal butterflies of equal mass m from a string of length L. The points of support are evenly spaced a distance ℓ apart as shown in Figure P4.50. The string forms an angle θ_1 with the ceiling at each endpoint. The center section of string is horizontal. (a) Find the tension in each section of string in terms of θ_1, m, and g. (b) In terms of θ_1, find the angle θ_2 that the sections of string between the outside butterflies and the inside butterflies form with the horizontal. (c) Show that the distance D between the endpoints of the string is

$$D = \frac{L}{5}\{2\cos\theta_1 + 2\cos[\tan^{-1}(\tfrac{1}{2}\tan\theta_1)] + 1\}$$

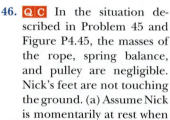

Figure P4.50

51. S What horizontal force must be applied to a large block of mass M shown in Figure P4.51 so that the tan blocks remain stationary relative to M? Assume all surfaces and the pulley are frictionless. Notice that the force exerted by the string accelerates m_2.

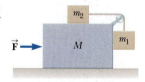

Figure P4.51 Problems 51 and 52.

52. S Initially, the system of objects shown in Figure P4.51 is held motionless. The pulley and all surfaces are frictionless. Let the force $\vec{F}$ be zero and assume that m_2 can move only vertically. At the instant after the system of objects is

released, find (a) the tension T in the string, (b) the acceleration of m_2, (c) the acceleration of M, and (d) the acceleration of m_1. (*Note:* The pulley accelerates along with the cart.)

53. **Review.** A block of mass $m = 2.00$ kg is released from rest at $h = 0.500$ m above the surface of a table, at the top of a $\theta = 30.0°$ incline as shown in Figure P4.53. The frictionless incline is fixed on a table of height $H = 2.00$ m. (a) Determine the acceleration of the block as it slides down the incline. (b) What is the velocity of the block as it leaves the incline? (c) How far from the table will the block hit the floor? (d) What time interval elapses between when the block is released and when it hits the floor? (e) Does the mass of the block affect any of the above calculations?

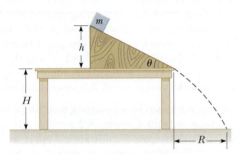

Figure P4.53 Problems 53 and 59

54. Q|C A student is asked to measure the acceleration of a glider on a frictionless, inclined plane, using an air track, a stopwatch, and a meterstick. The top of the track is measured to be 1.774 cm higher than the bottom of the track, and the length of the track is $d = 127.1$ cm. The cart is released from rest at the top of the incline, taken as $x = 0$, and its position x along the incline is measured as a function of time. For x values of 10.0 cm, 20.0 cm, 35.0 cm, 50.0 cm, 75.0 cm, and 100 cm, the measured times at which these positions are reached (averaged over five runs) are 1.02 s, 1.53 s, 2.01 s, 2.64 s, 3.30 s, and 3.75 s, respectively. (a) Construct a graph of x versus t^2, with a best-fit straight line to describe the data. (b) Determine the acceleration of the cart from the slope of this graph. (c) Explain how your answer to part (b) compares with the theoretical value you calculate using $a = g \sin \theta$ as derived in Example 4.3.

55. BIO If you jump from a desktop and land stiff-legged on a concrete floor, you run a significant risk that you will break a leg. To see how that happens, consider the average force stopping your body when you drop from rest from a height of 1.00 m and stop in a much shorter distance d. Your leg is

likely to break at the point where the cross-sectional area of the bone (the tibia) is smallest. This point is just above the ankle, where the cross-sectional area of one bone is about 1.60 cm². A bone will fracture when the compressive stress on it exceeds about 1.60×10^8 N/m². If you land on both legs, the maximum force that your ankles can safely exert on the rest of your body is then about

$$2(1.60 \times 10^8 \text{ N/m}^2)(1.60 \times 10^{-4} \text{ m}^2) = 5.12 \times 10^4 \text{ N}$$

Calculate the minimum stopping distance d that will not result in a broken leg if your mass is 60.0 kg. Don't try it! Bend your knees!

56. *Why is the following situation impossible?* On a single light vertical cable that does not stretch, a crane is lifting a 1207-kg Ferrari and, below it, a 1461-kg BMW Z8. Both cars move upward with speed 3.50 m/s and acceleration 1.25 m/s². The vertical cable has the same construction along its entire length and is not defective. Due to a tension exceeding the safety limit of the cable, it breaks just below the Ferrari.

57. M A car accelerates down a hill (Fig. P4.57), going from rest to 30.0 m/s in 6.00 s. A toy inside the car hangs by a string from the car's ceiling. The ball in the figure represents the toy, of mass 0.100 kg. The acceleration is such that the string remains perpendicular to the ceiling. Determine (a) the angle θ and (b) the tension in the string.

Figure P4.57

58. Q|C An 8.40-kg object slides down a fixed, frictionless, inclined plane. Use a computer to determine and tabulate (a) the normal force exerted on the object and (b) its acceleration for a series of incline angles (measured from the horizontal) ranging from 0° to 90° in 5° increments. (c) Plot a graph of the normal force and the acceleration as functions of the incline angle. (d) In the limiting cases of 0° and 90°, are your results consistent with the known behavior?

59. S In Figure P4.53, the incline has mass M and is fastened to the stationary horizontal tabletop. The block of mass m is placed near the bottom of the incline and is released with a quick push that sets it sliding upward. The block stops near the top of the incline as shown in the figure and then slides down again, always without friction. Find the force that the tabletop exerts on the incline throughout this motion in terms of m, M, g, and θ.

More Applications of Newton's Laws

Chapter Outline

Chris Graythen/Getty Images

In Chapter 4, we introduced Newton's laws of motion and applied them to situations in which we ignored friction. In this chapter, we shall expand our investigation to objects moving in the presence of friction, which will allow us to model situations more realistically. Such objects include those sliding on rough surfaces and those moving through viscous media such as liquids and air. We also apply Newton's laws to the dynamics of circular motion so that we can understand more about objects moving in circular paths under the influence of various types of forces.

Kyle Busch, driver of the #18 Snickers Toyota, leads Jeff Gordon, driver of the #24 Dupont Chevrolet, during the NASCAR Sprint Cup Series Kobalt Tools 500 at the Atlanta Motor Speedway on March 9, 2008, in Hampton, Georgia. The cars travel on a banked roadway to help them undergo circular motion on the turns.

5.1 | Forces of Friction

When an object moves either on a surface or through a viscous medium such as air or water, there is resistance to the motion because the object interacts with its surroundings. We call such resistance a **force of friction.** Forces of friction are very important in our everyday lives. They allow us to walk or run and are necessary for the motion of wheeled vehicles.

Imagine you are working in your garden and have filled a trash can with yard clippings. You then try to drag the trash can across the surface of your concrete patio as in Active Figure 5.1a (page 126). The patio surface is *real*, not an idealized, frictionless surface in a simplification model. If we apply an

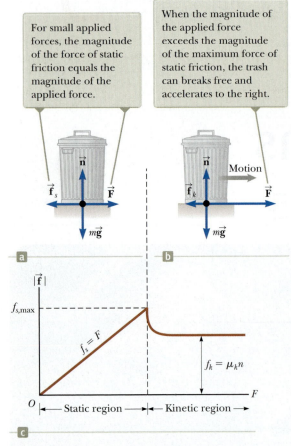

For small applied forces, the magnitude of the force of static friction equals the magnitude of the applied force.

When the magnitude of the applied force exceeds the magnitude of the maximum force of static friction, the trash can breaks free and accelerates to the right.

Active Figure 5.1 (a) and (b) When pulling on a trash can, the direction of the force of friction $\vec{f}$ between the can and a rough surface is opposite the direction of the applied force $\vec{F}$. (c) A graph of friction force versus applied force. Notice that $f_{s,max} > f_k$.

external horizontal force $\vec{F}$ to the trash can, acting to the right, the trash can remains stationary if $\vec{F}$ is small. The force that counteracts $\vec{F}$ and keeps the trash can from moving is applied at the base of the can by the surface and acts to the left. It is called the **force of static friction** $\vec{f}_s$. As long as the trash can is not moving, it is modeled as a particle in equilibrium and $f_s = F$. Therefore, if $\vec{F}$ is increased in magnitude, the magnitude of $\vec{f}_s$ also increases. Likewise, if $\vec{F}$ decreases, $\vec{f}_s$ also decreases.

Experiments show that the friction force arises from the nature of the two surfaces; because of their roughness, contact is made only at a few locations where peaks of the material touch. At these locations, the friction force arises in part because one peak physically blocks the motion of a peak from the opposing surface and in part from chemical bonding ("spot welds") of opposing peaks as they come into contact. Although the details of friction are quite complex at the atomic level, this force ultimately involves an electrical interaction between atoms or molecules.

If we increase the magnitude of $\vec{F}$, as in Active Figure 5.1b, the trash can eventually slips. When the trash can is on the verge of slipping, f_s is a maximum as shown in Active Figure 5.1c. If F exceeds $f_{s,max}$, the trash can moves and accelerates to the right. While the trash can is in motion, the friction force is less than $f_{s,max}$ (Active Fig. 5.1c). We call the friction force for an object in motion the **force of kinetic friction** $\vec{f}_k$. The net force $F - f_k$ in the x direction produces an acceleration to the right, according to Newton's second law. If we reduce the magnitude of $\vec{F}$ so that $F = f_k$, the acceleration is zero and the trash can moves to the right with constant speed. If the applied force is removed, the friction force acting to the left provides an acceleration of the trash can in the $-x$ direction and eventually brings it to rest.

Experimentally, one finds that, to a good approximation, both $f_{s,max}$ and f_k for an object on a surface are proportional to the normal force exerted by the surface on the object; therefore, we adopt a simplification model in which this approximation is assumed to be exact. The assumptions in this simplification model can be summarized as follows:

- The magnitude of the force of static friction between any two surfaces in contact can have the values

▶ Force of static friction

$$f_s \leq \mu_s n \qquad \textbf{5.1} ◀$$

where the dimensionless constant μ_s is called the **coefficient of static friction** and n is the magnitude of the normal force. The equality in Equation 5.1 holds when the surfaces are on the verge of slipping, that is, when $f_s = f_{s,max} \equiv \mu_s n$. This situation is called *impending motion*. The inequality holds when the component of the applied force parallel to the surfaces is less than this value.

- The magnitude of the force of kinetic friction acting between two surfaces is

▶ Force of kinetic friction

$$f_k = \mu_k n \qquad \textbf{5.2} ◀$$

where μ_k is the **coefficient of kinetic friction.** In our simplification model, this coefficient is independent of the relative speed of the surfaces.

- The values of μ_k and μ_s depend on the nature of the surfaces, but μ_k is generally less than μ_s. Table 5.1 lists some measured values.
- The direction of the friction force on an object is opposite to the actual motion (kinetic friction) or the impending motion (static friction) of the object relative to the surface with which it is in contact.

Pitfall Prevention | 5.1
The Equal Sign Is Used in Limited Situations
In Equation 5.1, the equal sign is used *only* when the surfaces are just about to break free and begin sliding. Do not fall into the common trap of using $f_s = \mu_s n$ in *any* static situation.

The approximate nature of Equations 5.1 and 5.2 is easily demonstrated by trying to arrange for an object to slide down an incline at constant speed. Especially at low speeds, the motion is likely to be characterized by alternate stick and slip episodes.

TABLE 5.1 | Coefficients of Friction

	μ_s	μ_k
Rubber on concrete	1.0	0.8
Steel on steel	0.74	0.57
Aluminium on steel	0.61	0.47
Glass on glass	0.94	0.4
Copper on steel	0.53	0.36
Wood on wood	0.25–0.5	0.2
Waxed wood on wet snow	0.14	0.1
Waxed wood on dry snow	—	0.04
Metal on metal (lubricated)	0.15	0.06
Teflon on Teflon	0.04	0.04
Ice on ice	0.1	0.03
Synovial joints in humans	0.01	0.003

Note: All values are approximate. In some cases, the coefficient of friction can exceed 1.0.

Pitfall Prevention | 5.2
The Direction of the Friction Force
Sometimes, an incorrect statement about the friction force between an object and a surface is made—"The friction force on an object is opposite to its motion or impending motion"—rather than the correct phrasing, "The friction force on an object is opposite to its motion or impending motion *relative to the surface.*" Think carefully about Quick Quiz 5.2.

The simplification model described in the bulleted list above has been developed so that we can solve problems involving friction in a relatively straightforward way.

Now that we have identified the characteristics of the friction force, we can include the friction force in the net force on an object in the model of a particle under a net force.

QUICK QUIZ 5.1 You press your physics textbook flat against a vertical wall with your hand, which applies a normal force perpendicular to the book. What is the direction of the friction force on the book due to the wall? (a) downward (b) upward (c) out from the wall (d) into the wall

QUICK QUIZ 5.2 A crate is located in the center of a flatbed truck. The truck accelerates to the east and the crate moves with it, not sliding at all. What is the direction of the friction force exerted by the truck on the crate? (a) It is to the west. (b) It is to the east. (c) No friction force exists because the crate is not sliding.

QUICK QUIZ 5.3 You are playing with your daughter in the snow. She sits on a sled and asks you to slide her across a flat, horizontal field. You have a choice of (a) pushing her from behind, by applying a force downward on her shoulders at 30° below the horizontal (Fig. 5.2a) or (b) attaching a rope to the front of the sled and pulling with a force at 30° above the horizontal (Fig 5.2b). Which would be easier for you and why?

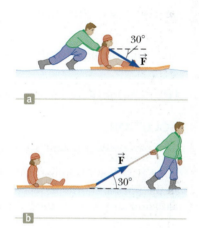

Figure 5.2 (Quick Quiz 5.3) A father tries to slide his daughter on a sled over snow by (a) pushing downward on her shoulders or (b) pulling upward on a rope attached to the sled. Which is easier?

Example 5.1 | The Sliding Hockey Puck

A hockey puck on a frozen pond is given an initial speed of 20.0 m/s.

(A) If the puck always remains on the ice and slides 115 m before coming to rest, determine the coefficient of kinetic friction between the puck and ice.

SOLUTION

Conceptualize Imagine that the puck in Figure 5.3 slides to the right and eventually comes to rest due to the force of kinetic friction.

Categorize The forces acting on the puck are identified in Figure 5.3, but the text of the problem provides kinematic variables. Therefore, we categorize the problem in two ways. First, it involves a particle under a net force: kinetic friction causes the puck to accelerate. Furthermore, because we model the force of kinetic friction as independent of speed, the acceleration of the puck is constant. So, we can also categorize this problem as one involving a particle under constant acceleration.

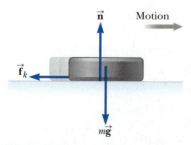

Figure 5.3 (Example 5.1) After the puck is given an initial velocity to the right, the only external forces acting on it are the gravitational force $m\vec{g}$, the normal force $\vec{n}$, and the force of kinetic friction $\vec{f}_k$.

continued

5.1 cont.

Analyze First, let's find the acceleration algebraically in terms of the coefficient of kinetic friction, using Newton's second law. Once we know the acceleration of the puck and the distance it travels, the equations of kinematics can be used to find the numerical value of the coefficient of kinetic friction. The diagram in Figure 5.3 shows the forces on the puck.

Apply the particle under a net force model in the x direction to the puck:

$$(1) \quad \sum F_x = -f_k = ma_x$$

Apply the particle in equilibrium model in the y direction to the puck:

$$(2) \quad \sum F_y = n - mg = 0$$

Substitute $n = mg$ from Equation (2) and $f_k = \mu_k n$ into Equation (1):

$$-\mu_k n = -\mu_k mg = ma_x$$
$$a_x = -\mu_k g$$

The negative sign means the acceleration is to the left in Figure 5.3. Because the velocity of the puck is to the right, the puck is slowing down. The acceleration is independent of the mass of the puck and is constant because we assume μ_k remains constant.

Apply the particle under constant acceleration model to the puck, using Equation 2.14, $v_{xf}^2 = v_{xi}^2 + 2a_x(x_f - x_i)$, with $x_i = 0$ and $v_f = 0$:

$$0 = v_{xi}^2 + 2a_x x_f = v_{xi}^2 - 2\mu_k g x_f$$

Solve for the coefficient of kinetic friction:

$$(3) \quad \mu_k = \frac{v_{xi}^2}{2g x_f}$$

Substitute the numerical values:

$$\mu_k = \frac{(20.0 \text{ m/s})^2}{2(9.80 \text{ m/s}^2)(115 \text{ m})} = \boxed{0.177}$$

(B) If the initial speed of the puck is halved, what would be the sliding distance?

SOLUTION

This part of the problem is a comparison problem and can be solved by a ratio technique such as that used in Example 3.4.

Solve Equation (3) in part (A) for the final position x_f of the puck and write it twice, once for the original situation and once for the halved initial velocity:

$$x_{f1} = \frac{v_{1xi}^2}{2\mu_k g}$$

$$x_{f2} = \frac{v_{2xi}^2}{2\mu_k g} = \frac{\left(\frac{1}{2}v_{1xi}\right)^2}{2\mu_k g} = \frac{1}{4}\frac{v_{1xi}^2}{2\mu_k g}$$

Divide the first equation by the second:

$$\frac{x_{f1}}{x_{f2}} = 4 \quad \rightarrow \quad x_{f2} = \boxed{\tfrac{1}{4}x_{f1}}$$

Finalize Notice in part (A) that μ_k is dimensionless, as it should be, and that it has a low value, consistent with an object sliding on ice. We learn in part (B) that halving the initial velocity of the puck reduces the sliding distance by 75%! Applying this idea to a sliding vehicle, we see that driving at low speeds on slippery roads is an important safety consideration.

Example **5.2** | Experimental Determination of μ_s and μ_k

The following is a simple method of measuring coefficients of friction. Suppose a block is placed on a rough surface inclined relative to the horizontal as shown in Active Figure 5.4. The incline angle is increased until the block starts to move. Show that you can obtain μ_s by measuring the critical angle θ_c at which this slipping just occurs.

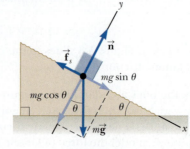

Active Figure 5.4 (Example 5.2) The external forces exerted on a block lying on a rough incline are the gravitational force $m\vec{g}$, the normal force $\vec{n}$, and the force of friction $\vec{f}_s$. For convenience, the gravitational force is resolved into a component $mg \sin \theta$ along the incline and a component $mg \cos \theta$ perpendicular to the incline.

SOLUTION

Conceptualize Consider Active Figure 5.4 and imagine that the block tends to slide down the incline due to the gravitational force. To simulate the situation, place a coin on this book's cover and tilt the book until the coin begins to slide.

5.2 *cont.*

Notice how this example differs from Example 4.3. When there is no friction on an incline, *any* angle of the incline will cause a stationary object to begin moving. When there is friction, however, there is no movement of the object for angles less than the critical angle.

Categorize The block is subject to various forces. Because we are raising the plane to the angle at which the block is just ready to begin to move but is not moving, we categorize the block as a particle in equilibrium.

Analyze The diagram in Active Figure 5.4 shows the forces on the block: the gravitational force $m\vec{g}$, the normal force $\vec{n}$, and the force of static friction $\vec{f}_s$. We choose x to be parallel to the plane and y perpendicular to it.

Apply Equation 4.7 to the block in both the x and y directions:

(1) $\sum F_x = mg \sin\theta - f_s = 0$

(2) $\sum F_y = n - mg \cos\theta = 0$

Substitute $mg = n/\cos\theta$ from Equation (2) into Equation (1):

(3) $f_s = mg \sin\theta = \left(\dfrac{n}{\cos\theta}\right) \sin\theta = n \tan\theta$

When the incline angle is increased until the block is on the verge of slipping, the force of static friction has reached its maximum value $\mu_s n$. The angle θ in this situation is the critical angle θ_c. Make these substitutions in Equation (3):

$\mu_s n = n \tan\theta_c$

$\mu_s = \tan\theta_c$

For example, if the block just slips at $\theta_c = 20.0°$, we find that $\mu_s = \tan 20.0° = 0.364$.

Finalize Once the block starts to move at $\theta \geq \theta_c$, it accelerates down the incline and the force of friction is $f_k = \mu_k n$. If θ is reduced to a value less than θ_c, however, it may be possible to find an angle θ'_c such that the block moves down the incline with constant speed as a particle in equilibrium again ($a_x = 0$). In this case, use Equations (1) and (2) with f_s replaced by f_k to find μ_k: $\mu_k = \tan\theta'_c$, where $\theta'_c < \theta_c$.

Example **5.3** | **Acceleration of Two Connected Objects When Friction Is Present**

A block of mass m_2 on a rough, horizontal surface is connected to a ball of mass m_1 by a lightweight cord over a lightweight, frictionless pulley as shown in Figure 5.5a. A force of magnitude F at an angle θ with the horizontal is applied to the block as shown, and the block slides to the right. The coefficient of kinetic friction between the block and surface is μ_k. Determine the magnitude of the acceleration of the two objects.

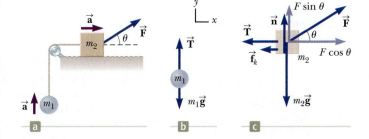

Figure 5.5 (Example 5.3) (a) The external force $\vec{F}$ applied as shown can cause the block to accelerate to the right. (b, c) Diagrams showing the forces on the two objects, assuming the block accelerates to the right and the ball accelerates upward.

SOLUTION

Conceptualize Imagine what happens as $\vec{F}$ is applied to the block. Assuming $\vec{F}$ is not large enough to lift the block, the block slides to the right and the ball rises.

Categorize We can identify forces and we want an acceleration, so we categorize this problem as one involving two particles under a net force, the ball and the block.

Analyze First draw force diagrams for the two objects as shown in Figures 5.5b and 5.5c. Notice that the string exerts a force of magnitude T on both objects. The applied force $\vec{F}$ has x and y components $F \cos\theta$ and $F \sin\theta$, respectively. Because the two objects are connected, we can equate the magnitudes of the x component of the acceleration of the block and the y component of the acceleration of the ball and call them both a. Let us assume the motion of the block is to the right.

continued

5.3 cont.

Apply the particle under a net force model to the block in the horizontal direction:	(1) $\sum F_x = F \cos \theta - f_k - T = m_2 a_x = m_2 a$
Because the block moves only horizontally, apply the particle in equilibrium model to the block in the vertical direction:	(2) $\sum F_y = n + F \sin \theta - m_2 g = 0$
Apply the particle under a net force model to the ball in the vertical direction:	(3) $\sum F_y = T - m_1 g = m_1 a_y = m_1 a$
Solve Equation (2) for n:	$n = m_2 g - F \sin \theta$
Substitute n into $f_k = \mu_k n$ from Equation 5.2:	(4) $f_k = \mu_k (m_2 g - F \sin \theta)$
Substitute Equation (4) and the value of T from Equation (3) into Equation (1):	$F \cos \theta - \mu_k (m_2 g - F \sin \theta) - m_1 (a + g) = m_2 a$
Solve for a:	(5) $a = \dfrac{F(\cos \theta + \mu_k \sin \theta) - (m_1 + \mu_k m_2)g}{m_1 + m_2}$

Finalize The acceleration of the block can be either to the right or to the left depending on the sign of the numerator in Equation (5). If the velocity is to the left, we must reverse the sign of f_k in Equation (1) because the force of kinetic friction must oppose the motion of the block relative to the surface. In this case, the value of a is the same as in Equation (5), with the two plus signs in the numerator changed to minus signs.

5.2 | Extending the Particle in Uniform Circular Motion Model

Solving problems involving friction is just one of many applications of Newton's second law. Let us now consider another common situation, associated with a particle in uniform circular motion. In Chapter 3, we found that a particle moving in a circular path of radius r with uniform speed v experiences a centripetal acceleration of magnitude

► Centripetal acceleration

$$a_c = \frac{v^2}{r}$$

The acceleration vector with this magnitude is directed toward the center of the circle and is *always* perpendicular to $\vec{v}$.

According to Newton's second law, if an acceleration occurs, a net force must be causing it. Because the acceleration is toward the center of the circle, the net force must be toward the center of the circle. Therefore, when a particle travels in a circular path, a force must be acting *inward* on the particle that causes the circular motion. We investigate the forces causing this type of acceleration in this section.

Consider a puck of mass m that is tied to a string of length r and is moving at constant speed in a horizontal, circular path as illustrated in Figure 5.6. Its weight is supported by a frictionless table, and the string is anchored to a peg at the center of the circular path of the puck. Why does the puck move in a circle? The natural tendency of the puck is to move in a straight-line path, according to Newton's first law; the string, however, prevents this motion along a straight line by exerting a radial force $\vec{F}_r$ on the puck to make it follow a circular path. This force, whose magnitude is the tension in the string, is directed along the length of the string toward the center of the circle as shown in Figure 5.6.

In this discussion, the tension in the string causes the circular motion of the puck. Other forces also cause objects to move in circular paths. For example, friction forces

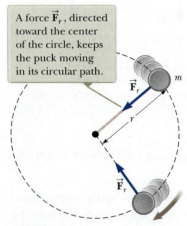

A force $\vec{F}_r$, directed toward the center of the circle, keeps the puck moving in its circular path.

Figure 5.6 An overhead view of a puck moving in a circular path in a horizontal plane.

cause automobiles to travel around curved roadways and the gravitational force causes a planet to orbit the Sun.

Regardless of the nature of the force acting on the particle in circular motion, we can apply Newton's second law to the particle along the radial direction:

$$\sum F = ma_c = m\frac{v^2}{r}$$ **5.3** ◀

In general, an object can move in a circular path under the influence of various types of forces, or a *combination* of forces, as we shall see in some of the examples that follow.

If the force acting on an object vanishes, the object no longer moves in its circular path; instead, it moves along a straight-line path tangent to the circle. This idea is illustrated in Active Figure 5.7 for the case of the puck moving in a circular path at the end of a string in a horizontal plane. If the string breaks at some instant, the puck moves along the straight-line path tangent to the circle at the position of the puck at this instant.

> **QUICK QUIZ 5.4** You are riding on a Ferris wheel (Fig. 5.8) that is rotating with constant speed. The car in which you are riding always maintains its correct upward orientation: it does not invert. **(i)** What is the direction of the normal force on you from the seat when you are at the top of the wheel? **(a)** upward **(b)** downward **(c)** impossible to determine. **(ii)** From the same choices, what is the direction of the net force on you when you are at the top of the wheel?

> ▶ **THINKING PHYSICS 5.1**
>
> The Copernican theory of the solar system is a structural model in which the planets are assumed to travel around the Sun in circular orbits. Historically, this theory was a break from the Ptolemaic theory, a structural model in which the Earth was at the center. When the Copernican theory was proposed, a natural question arose: What keeps the Earth and other planets moving in their paths around the Sun? An interesting response to this question comes from Richard Feynman: "In those days, one of the theories proposed was that the planets went around because behind them there were invisible angels, beating their wings and driving the planets forward. . . . It turns out that in order to keep the planets going around, the invisible angels must fly in a different direction."[1] What did Feynman mean by this statement?
>
> **Reasoning** The question asked by those at the time of Copernicus indicates that they did not have a proper understanding of inertia as described by Newton's first law. At that time in history, before Galileo and Newton, the interpretation was that *motion* was caused by force. This interpretation is different from our current understanding that *changes in motion* are caused by force. Therefore, it was natural for Copernicus's contemporaries to ask what force propelled a planet in its orbit. According to our current understanding, it is equally natural for us to realize that no force tangent to the orbit is necessary, that the motion simply continues owing to inertia.
>
> Therefore, in Feynman's imagery, the angels do not have to push the planet *from behind*. The angels must push *inward*, to provide the centripetal acceleration associated with the orbital motion of the planet. Of course, the angels are not real from a scientific point of view, but are a metaphor for the *gravitational force*. ◀

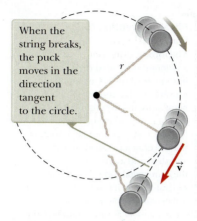

Active Figure 5.7 The string holding the puck in its circular path breaks.

When the string breaks, the puck moves in the direction tangent to the circle.

> **Pitfall Prevention | 5.3**
> **Direction of Travel When the String Is Cut**
> Study Active Figure 5.7 carefully. Many students have a misconception that the puck moves *radially* away from the center of the circle when the string is cut. The velocity of the puck is *tangent* to the circle. By Newton's first law, the puck simply continues to move in the direction that it is moving just as the force from the string disappears.

> **Pitfall Prevention | 5.4**
> **Centrifugal Force**
> The commonly heard phrase "centrifugal force" is described as a force pulling *outward* on an object moving in a circular path. If you are experiencing a "centrifugal force" on a rotating carnival ride, what is the other object with which you are interacting? You cannot identify another object because centrifugal force is a fictitious force.

Figure 5.8 (Quick Quiz 5.4) A Ferris wheel.

---

[1]R. P. Feynman, R. B. Leighton, and M. Sands, *The Feynman Lectures on Physics*, Vol. 1 (Reading, MA: Addison-Wesley, 1963), p. 7-2.

Pitfall Prevention | 5.5
Centripetal Force
The force causing centripetal acceleration is called *centripetal force* in some textbooks. Giving the force causing circular motion a name leads many students to consider it as a new *kind* of force rather than a new *role* for force. A common mistake is to draw the forces in a free-body diagram and then add another vector for the centripetal force. Yet it is not a separate force; it is one of our familiar forces *acting in the role of causing a circular motion*. For the motion of the Earth around the Sun, for example, the "centripetal force" is *gravity*. For a rock whirled on the end of a string, the "centripetal force" is the *tension* in the string. After this discussion, we shall no longer use the phrase *centripetal force*.

Robin Smith/GettyImages

The cars of a corkscrew roller coaster must travel in tight loops. The normal force from the track contributes to the centripetal acceleration. The gravitational force, because it remains constant in direction, is sometimes in the same direction as the normal force, but is sometimes in the opposite direction.

Example 5.4 | How Fast Can It Spin?

A puck of mass 0.500 kg is attached to the end of a cord 1.50 m long. The puck moves in a horizontal circle as shown in Figure 5.6. If the cord can withstand a maximum tension of 50.0 N, what is the maximum speed at which the puck can move before the cord breaks? Assume the string remains horizontal during the motion.

SOLUTION

Conceptualize It makes sense that the stronger the cord, the faster the puck can move before the cord breaks. Also, we expect a more massive puck to break the cord at a lower speed. (Imagine whirling a bowling ball on the cord!)

Categorize Because the puck moves in a circular path, we model it as a particle in uniform circular motion.

Analyze Incorporate the tension and the centripetal acceleration into Newton's second law as described by Equation 5.3:

$$T = m\frac{v^2}{r}$$

Solve for v:

$$(1) \quad v = \sqrt{\frac{Tr}{m}}$$

Find the maximum speed the puck can have, which corresponds to the maximum tension the string can withstand:

$$v_{max} = \sqrt{\frac{T_{max}r}{m}} = \sqrt{\frac{(50.0 \text{ N})(1.50 \text{ m})}{0.500 \text{ kg}}} = \boxed{12.2 \text{ m/s}}$$

Finalize Equation (1) shows that v increases with T and decreases with larger m, as we expected from our conceptualization of the problem.

What If? Suppose the puck moves in a circle of larger radius at the same speed v. Is the cord more likely or less likely to break?

Answer The larger radius means that the change in the direction of the velocity vector will be smaller in a given time interval. Therefore, the acceleration is smaller and the required tension in the string is smaller. As a result, the string is less likely to break when the puck travels in a circle of larger radius.

Example 5.5 | The Conical Pendulum

A small ball of mass m is suspended from a string of length L. The ball revolves with constant speed v in a horizontal circle of radius r as shown in Figure 5.9. (Because the string sweeps out the surface of a cone, the system is known as a *conical pendulum*.) Find an expression for v.

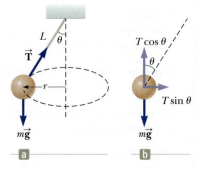

SOLUTION

Conceptualize Imagine the motion of the ball in Figure 5.9a and convince yourself that the string sweeps out a cone and that the ball moves in a horizontal circle.

Figure 5.9 (Example 5.5) (a) A conical pendulum. The path of the ball is a horizontal circle. (b) The forces acting on the ball.

Categorize The ball in Figure 5.9 does not accelerate vertically. Therefore, we model it as a particle in equilibrium in the vertical direction. It experiences a centripetal acceleration in the horizontal direction, so it is modeled as a particle in uniform circular motion in this direction.

Analyze Let θ represent the angle between the string and the vertical. In the diagram of forces acting on the ball in Figure 5.9b, the force $\vec{T}$ exerted by the string on the ball is resolved into a vertical component $T \cos \theta$ and a horizontal component $T \sin \theta$ acting toward the center of the circular path.

Apply the particle in equilibrium model in the vertical direction:

$$\sum F_y = T \cos \theta - mg = 0$$

$$(1)\ T \cos \theta = mg$$

Use Equation 5.3 from the particle in uniform circular motion model in the horizontal direction:

$$(2)\ \sum F_x = T \sin \theta = ma_c = \frac{mv^2}{r}$$

Divide Equation (2) by Equation (1) and use $\sin \theta / \cos \theta = \tan \theta$:

$$\tan \theta = \frac{v^2}{rg}$$

Solve for v:

$$v = \sqrt{rg \tan \theta}$$

Incorporate $r = L \sin \theta$ from the geometry in Figure 5.9a:

$$v = \sqrt{Lg \sin \theta \tan \theta}$$

Finalize Notice that the speed is independent of the mass of the ball. Consider what happens when θ goes to 90° so that the string is horizontal. Because the tangent of 90° is infinite, the speed v is infinite, which tells us the string cannot possibly be horizontal. If it were, there would be no vertical component of the force $\vec{T}$ to balance the gravitational force on the ball. That is why we mentioned in regard to Figure 5.6 that the puck's weight in the figure is supported by a frictionless table.

Example 5.6 | What Is the Maximum Speed of the Car?

A 1 500-kg car moving on a flat, horizontal road negotiates a curve as shown in Figure 5.10a (page 134). If the radius of the curve is 35.0 m and the coefficient of static friction between the tires and dry pavement is 0.523, find the maximum speed the car can have and still make the turn successfully.

SOLUTION

Conceptualize Imagine that the curved roadway is part of a large circle so that the car is moving in a circular path.

Categorize Based on the conceptualize step of the problem, we model the car as a particle in uniform circular motion in the horizontal direction. The car is not accelerating vertically, so it is modeled as a particle in equilibrium in the vertical direction.

Analyze Figure 5.10b shows the forces on the car. The force that enables the car to remain in its circular path is the force of static friction. (It is *static* because no slipping occurs at the point of contact between road and tires. If this force of static friction were zero—for example, if the car were on an icy road—the car would continue in a straight line and slide off the curved road.) The maximum speed v_{max} the car can have around the curve is the speed at which it is on the verge of skidding outward. At this point, the friction force has its maximum value $f_{s,max} = \mu_s n$.

continued

5.6 *cont.*

Apply Equation 5.3 in the radial direction for the maximum speed condition:

$$(1)\ f_{s,\text{max}} = \mu_s n = m\frac{v_{\text{max}}^2}{r}$$

Apply the particle in equilibrium model to the car in the vertical direction:

$$\sum F_y = 0 \rightarrow n - mg = 0 \rightarrow n = mg$$

Solve Equation (1) for the maximum speed and substitute for n:

$$(2)\ v_{\text{max}} = \sqrt{\frac{\mu_s n r}{m}} = \sqrt{\frac{\mu_s mgr}{m}} = \sqrt{\mu_s gr}$$

Substitute numerical values:

$$v_{\text{max}} = \sqrt{(0.523)(9.80\ \text{m/s}^2)(35.0\ \text{m})} = \boxed{13.4\ \text{m/s}}$$

Finalize This speed is equivalent to 30.0 mi/h. Therefore, if the speed limit on this roadway is higher than 30 mi/h, this roadway could benefit greatly from some banking, as in the next example! Notice that the maximum speed does not depend on the mass of the car, which is why curved highways do not need multiple speed limits to cover the various masses of vehicles using the road.

What If? Suppose a car travels this curve on a wet day and begins to skid on the curve when its speed reaches only 8.00 m/s. What can we say about the coefficient of static friction in this case?

Answer The coefficient of static friction between the tires and a wet road should be smaller than that between the tires and a dry road. This expectation is consistent with experience with driving because a skid is more likely on a wet road than a dry road.

To check our suspicion, we can solve Equation (2) for the coefficient of static friction:

$$\mu_s = \frac{v_{\text{max}}^2}{gr}$$

Substituting the numerical values gives

$$\mu_s = \frac{v_{\text{max}}^2}{gr} = \frac{(8.00\ \text{m/s})^2}{(9.80\ \text{m/s}^2)(35.0\ \text{m})} = 0.187$$

which is indeed smaller than the coefficient of 0.523 for the dry road.

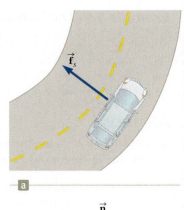

Figure 5.10 (Example 5.6) (a) The force of static friction directed toward the center of the curve keeps the car moving in a circular path. (b) The forces acting on the car.

Example 5.7 | The Banked Roadway

A civil engineer wishes to redesign the curved roadway in Example 5.6 in such a way that a car will not have to rely on friction to round the curve without skidding. In other words, a car moving at the designated speed can negotiate the curve even when the road is covered with ice. Such a road is usually *banked*, which means that the roadway is tilted toward the inside of the curve as seen in the opening photograph for this chapter. Suppose the designated speed for the ramp is to be 13.4 m/s (30.0 mi/h) and the radius of the curve is 35.0 m. At what angle should the curve be banked?

SOLUTION

Conceptualize The difference between this example and Example 5.6 is that the car is no longer moving on a flat roadway. Figure 5.11 shows the banked roadway, with the center of the circular path of the car far to the left of the figure. Notice that the horizontal component of the normal force participates in causing the car's centripetal acceleration.

Categorize As in Example 5.6, the car is modeled as a particle in equilibrium in the vertical direction and a particle in uniform circular motion in the horizontal direction.

Analyze On a level (unbanked) road, the force that causes the centripetal acceleration is the force of static friction between car and road as we saw in the preceding example. If the road is banked at an angle θ as in Figure 5.11, however, the normal force $\vec{\mathbf{n}}$ has a

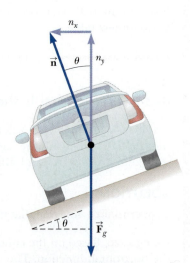

Figure 5.11 (Example 5.7) A car moves into the page and is rounding a curve on a road banked at an angle θ to the horizontal. When friction is neglected, the force that causes the centripetal acceleration and keeps the car moving in its circular path is the horizontal component of the normal force.

5.7 *cont.*

horizontal component toward the center of the curve. Because the ramp is to be designed so that the force of static friction is zero, only the component $n_x = n \sin \theta$ causes the centripetal acceleration.

Write Newton's second law for the car in the radial direction, which is the x direction:

$$(1) \quad \sum F_r = n \sin \theta = \frac{mv^2}{r}$$

Apply the particle in equilibrium model to the car in the vertical direction:

$$\sum F_y = n \cos \theta - mg = 0$$

$$(2) \quad n \cos \theta = mg$$

Divide Equation (1) by Equation (2):

$$(3) \quad \tan \theta = \frac{v^2}{rg}$$

Solve for the angle θ:

$$\theta = \tan^{-1} \left[\frac{(13.4 \text{ m/s})^2}{(35.0 \text{ m})(9.80 \text{ m/s}^2)} \right] = \boxed{27.6°}$$

Finalize Equation (3) shows that the banking angle is independent of the mass of the vehicle negotiating the curve. If a car rounds the curve at a speed less than 13.4 m/s, friction is needed to keep it from sliding down the bank (to the left in Fig. 5.11). A driver attempting to negotiate the curve at a speed greater than 13.4 m/s has to depend on friction to keep from sliding up the bank (to the right in Fig. 5.11).

What If? Imagine that this same roadway were built on Mars in the future to connect different colony centers. Could it be traveled at the same speed?

Answer The reduced gravitational force on Mars would mean that the car is not pressed as tightly to the roadway. The reduced normal force results in a smaller component of the normal force toward the center of the circle. This smaller component would not be sufficient to provide the centripetal acceleration associated with the original speed. The centripetal acceleration must be reduced, which can be done by reducing the speed v.

Mathematically, notice that Equation (3) shows that the speed v is proportional to the square root of g for a roadway of fixed radius r banked at a fixed angle θ. Therefore, if g is smaller, as it is on Mars, the speed v with which the roadway can be safely traveled is also smaller.

Example 5.8 | Riding the Ferris Wheel

A child of mass m rides on a Ferris wheel as shown in Figure 5.12a. The child moves in a vertical circle of radius 10.0 m at a constant speed of 3.00 m/s.

(A) Determine the force exerted by the seat on the child at the bottom of the ride. Express your answer in terms of the weight of the child, mg.

SOLUTION

Conceptualize Look carefully at Figure 5.12a. Based on experiences you may have had on a Ferris wheel or driving over small hills on a roadway, you would expect to feel lighter at the top of the path. Similarly, you would expect to feel heavier at the bottom of the path. At both the bottom of the path and the top, the normal and gravitational forces on the child act in *opposite* directions. The vector sum of these two forces gives a force of constant magnitude that keeps the child moving in a circular path at a constant speed. To yield net force vectors with the same magnitude, the normal force at the bottom must be greater than that at the top.

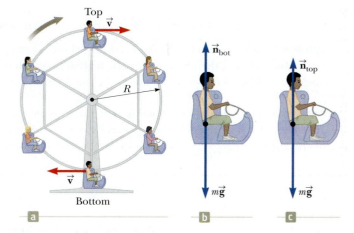

Figure 5.12 (Example 5.8) (a) A child rides on a Ferris wheel. (b) The forces acting on the child at the bottom of the path. (c) The forces acting on the child at the top of the path.

Categorize Because the speed of the child is constant, we can categorize this problem as one involving a particle (the child) in uniform circular motion, complicated by the gravitational force acting at all times on the child.

continued

5.8 *cont.*

Analyze We draw a diagram of forces acting on the child at the bottom of the ride as shown in Figure 5.12b. The only forces acting on him are the downward gravitational force $\vec{F}_g = m\vec{g}$ and the upward force $\vec{n}_{bot}$ exerted by the seat. The net upward force on the child that provides his centripetal acceleration has a magnitude $n_{bot} - mg$.

Apply Newton's second law to the child in the radial direction when he is at the bottom of the ride:

$$\sum F = n_{bot} - mg = m\frac{v^2}{r}$$

Solve for the force exerted by the seat on the child:

$$n_{bot} = mg + m\frac{v^2}{r} = mg\left(1 + \frac{v^2}{rg}\right)$$

Substitute the values given for the speed and radius:

$$n_{bot} = mg\left[1 + \frac{(3.00 \text{ m/s})^2}{(10.0 \text{ m})(9.80 \text{ m/s}^2)}\right]$$

$$= \boxed{1.09\, mg}$$

Hence, the magnitude of the force $\vec{n}_{bot}$ exerted by the seat on the child is *greater* than the weight of the child by a factor of 1.09. So, the child experiences an apparent weight that is greater than his true weight by a factor of 1.09.

(B) Determine the force exerted by the seat on the child at the top of the ride.

SOLUTION

Analyze The diagram of forces acting on the child at the top of the ride is shown in Figure 5.12c. The net downward force that provides the centripetal acceleration has a magnitude $mg - n_{top}$.

Apply Newton's second law to the child at this position:

$$\sum F = mg - n_{top} = m\frac{v^2}{r}$$

Solve for the force exerted by the seat on the child:

$$n_{top} = mg - m\frac{v^2}{r} = mg\left(1 - \frac{v^2}{rg}\right)$$

Substitute numerical values:

$$n_{top} = mg\left[1 - \frac{(3.00 \text{ m/s})^2}{(10.0 \text{ m})(9.80 \text{ m/s}^2)}\right]$$

$$= \boxed{0.908\, mg}$$

In this case, the magnitude of the force exerted by the seat on the child is *less* than his true weight by a factor of 0.908, and the child feels lighter.

Finalize The variations in the normal force are consistent with our prediction in the Conceptualize step of the problem.

What If? Suppose a defect in the Ferris wheel mechanism causes the speed of the child to increase to 10.0 m/s. What does the child experience at the top of the ride in this case?

Answer If the calculation above is performed with $v = 10.0$ m/s, the magnitude of the normal force at the top of the ride is negative, which is impossible. We interpret it to mean that the required centripetal acceleration of the child is larger than that due to gravity. As a result, the child will lose contact with the seat and will only stay in his circular path if there is a safety bar that provides a downward force on him to keep him in his seat. At the bottom of the ride, the normal force is $2.02\, mg$, which would be uncomfortable.

5.3 | Nonuniform Circular Motion

In Chapter 3, we found that if a particle moves with varying speed in a circular path, there is, in addition to the radial component of acceleration, a tangential component of magnitude dv/dt. Therefore, the net force acting on the particle must also have a radial and a tangential component as shown in Active Figure 5.13. That is, because the total acceleration is $\vec{a} = \vec{a}_r + \vec{a}_t$, the total force exerted on the particle is $\sum\vec{F} = \sum\vec{F}_r + \sum\vec{F}_t$. (We express the radial and tangential forces as net forces with the summation notation because each force could consist of multiple forces that combine.) The component vector $\Sigma\vec{F}_r$ is directed toward the center of the circle and is responsible for the centripetal acceleration. The component vector $\Sigma\vec{F}_t$ tangent to the circle is responsible for the tangential acceleration, which causes the speed of the particle to change with time.

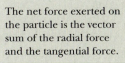

The net force exerted on the particle is the vector sum of the radial force and the tangential force.

QUICK QUIZ 5.5 Which of the following is *impossible* for a car moving in a circular path? Assume that the car is never at rest. (**a**) The car has tangential acceleration but no centripetal acceleration. (**b**) The car has centripetal acceleration but no tangential acceleration. (**c**) The car has both centripetal acceleration and tangential acceleration.

QUICK QUIZ 5.6 A bead slides freely along a curved wire lying on a horizontal surface at constant speed as shown by Figure 5.14. (**a**) Draw the vectors representing the force exerted by the wire on the bead at points Ⓐ, Ⓑ, and Ⓒ. (**b**) Suppose the bead in Figure 5.14 speeds up with constant tangential acceleration as it moves toward the right. Draw the vectors representing the force on the bead at points Ⓐ, Ⓑ, and Ⓒ.

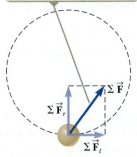

Active Figure 5.13 When the net force acting on a particle moving in a circular path has a tangential component vector $\Sigma\vec{F}_t$, its speed changes.

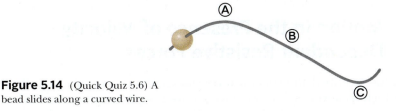

Figure 5.14 (Quick Quiz 5.6) A bead slides along a curved wire.

Example 5.9 | **Keep Your Eye on the Ball**

A small sphere of mass m is attached to the end of a cord of length R and set into motion in a *vertical* circle about a fixed point O as illustrated in Figure 5.15. Determine the tangential acceleration of the sphere and the tension in the cord at any instant when the speed of the sphere is v and the cord makes an angle θ with the vertical.

SOLUTION

Conceptualize Compare the motion of the sphere in Figure 5.15 with that of the child in Figure 5.12a associated with Example 5.8. Both objects travel in a circular path. Unlike the child in Example 5.8, however, the speed of the sphere is *not* uniform in this example because, at most points along the path, a tangential component of acceleration arises from the gravitational force exerted on the sphere.

Categorize We model the sphere as a particle under a net force and moving in a circular path, but it is not a particle in *uniform* circular motion. We need to use the techniques discussed in this section on nonuniform circular motion.

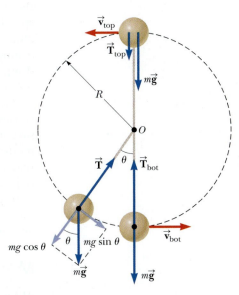

Figure 5.15 (Example 5.9) The forces acting on a sphere of mass m connected to a cord of length R and rotating in a vertical circle centered at O. Forces acting on the sphere are shown when the sphere is at the top and bottom of the circle and at an arbitrary location.

Analyze From the force diagram in Figure 5.15, we see that the only forces acting on the sphere are the gravitational force $\vec{F}_g = m\vec{g}$ exerted by the Earth and the force $\vec{T}$ exerted by the cord. We resolve $\vec{F}_g$ into a tangential component $mg \sin\theta$ and a radial component $mg \cos\theta$.

Apply Newton's second law to the sphere in the tangential direction:

$$\sum F_t = mg \sin\theta = ma_t$$
$$a_t = g \sin\theta$$

continued

5.9 *cont.*

Apply Newton's second law to the forces acting on the sphere in the radial direction, noting that both $\vec{\mathbf{T}}$ and $\vec{\mathbf{a}}_r$ are directed toward O:

$$\sum F_r = T - mg\cos\theta = \frac{mv^2}{R}$$

$$T = mg\left(\frac{v^2}{Rg} + \cos\theta\right)$$

Finalize Let us evaluate this result at the top and bottom of the circular path (Fig. 5.15):

$$T_{\text{top}} = mg\left(\frac{v_{\text{top}}^2}{Rg} - 1\right) \quad T_{\text{bot}} = mg\left(\frac{v_{\text{bot}}^2}{Rg} + 1\right)$$

These results have similar mathematical forms as those for the normal forces n_{top} and n_{bot} on the child in Example 5.8, which is consistent with the normal force on the child playing a similar physical role in Example 5.8 as the tension in the string plays in this example. Keep in mind, however, that the normal force $\vec{\mathbf{n}}$ on the child in Example 5.8 is always upward, whereas the force $\vec{\mathbf{T}}$ in this example changes direction because it must always point inward along the string. Also note that v in the expressions above varies for different positions of the sphere, as indicated by the subscripts, whereas v in Example 5.8 is constant.

5.4 | Motion in the Presence of Velocity-Dependent Resistive Forces

Earlier, we described the friction force between a moving object and the surface along which it moves. So far, we have ignored any interaction between the object and the *medium* through which it moves. Let us now consider the effect of a medium such as a liquid or gas. The medium exerts a **resistive force** $\vec{\mathbf{R}}$ on the object moving through it. You feel this force if you ride in a car at high speed with your hand out the window; the force you feel pushing your hand backward is the resistive force of the air rushing past the car. The magnitude of this force depends on the relative speed between the object and the medium, and the direction of $\vec{\mathbf{R}}$ on the object is always opposite the direction of the object's motion relative to the medium. Some examples are the air resistance associated with moving vehicles (sometimes called air drag), the force of the wind on the sails of a sailboat, and the viscous forces that act on objects sinking through a liquid.

Generally, the magnitude of the resistive force increases with increasing speed. The resistive force can have a complicated speed dependence. In the following discussions, we consider two simplification models that allow us to analyze these situations. The first model assumes that the resistive force is proportional to the velocity, which is approximately the case for objects that fall through a liquid with low speed and for very small objects, such as dust particles, that move through air. The second model treats situations for which we assume that the magnitude of the resistive force is proportional to the square of the speed of the object. Large objects, such as a skydiver moving through air in free-fall, experience such a force.

Model 1: Resistive Force Proportional to Object Velocity

At low speeds, the resistive force acting on an object that is moving through a viscous medium is effectively modeled as being proportional to the object's velocity. The mathematical representation of the resistive force can be expressed as

$$\vec{\mathbf{R}} = -b\vec{\mathbf{v}} \qquad \qquad \textbf{5.4}\blacktriangleleft$$

where $\vec{\mathbf{v}}$ is the velocity of the object relative to the medium and b is a constant that depends on the properties of the medium and on the shape and dimensions of the object. The negative sign represents that the resistive force is opposite the velocity of the object relative to the medium.

Consider a sphere of mass m released from rest in a liquid, as in Active Figure 5.16a. We assume that the only forces acting on the sphere are the resistive force $\vec{\mathbf{R}}$

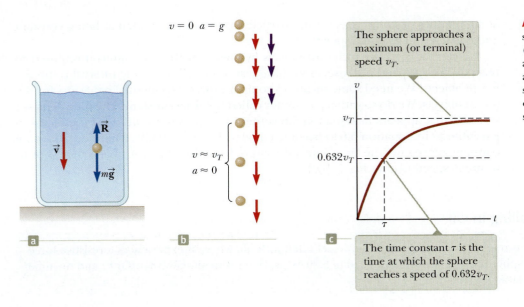

$v = 0$ $a = g$

The sphere approaches a maximum (or terminal) speed v_T.

$v \approx v_T$
$a \approx 0$

v_T

$0.632 v_T$

τ

The time constant τ is the time at which the sphere reaches a speed of $0.632 v_T$.

Active Figure 5.16 (a) A small sphere falling through a liquid. (b) A motion diagram of the sphere as it falls. Velocity vectors (red) and acceleration vectors (purple) are shown for each image after the first one. (c) A speed–time graph for the sphere.

and the weight $m\vec{\mathbf{g}}$, and we describe its motion using Newton's second law.[2] Considering the vertical motion and choosing the downward direction to be positive, we have

$$\sum F_y = ma_y \quad \rightarrow \quad mg - bv = m\frac{dv}{dt}$$

Dividing this equation by the mass m gives

$$\frac{dv}{dt} = g - \frac{b}{m}v \qquad\qquad \textbf{5.5} \blacktriangleleft$$

Equation 5.5 is called a *differential equation*; it includes both the speed v and the derivative of the speed. The methods of solving such an equation may not be familiar to you as yet. Note, however, that if we define $t = 0$ when $v = 0$, the resistive force is zero at this time and the acceleration dv/dt is simply g. As t increases, the speed increases, the resistive force increases, and the acceleration decreases. Therefore, this situation is one in which neither the velocity nor the acceleration of the particle is constant.

The acceleration becomes zero when the increasing resistive force eventually balances the weight. At this point, the object reaches its **terminal speed** v_T and from then on it continues to move with zero acceleration. The motion diagram in Active Figure 5.16b shows the sphere accelerating over the early part of its motion and then reaching terminal speed later on. After the object reaches terminal speed, its motion is that of a particle under constant velocity. The terminal speed can be obtained from Equation 5.5 by setting $a = dv/dt = 0$, which gives

$$mg - bv_T = 0 \quad \rightarrow \quad v_T = \frac{mg}{b}$$

The expression for v that satisfies Equation 5.5 with $v = 0$ at $t = 0$ is

$$v = \frac{mg}{b}(1 - e^{-bt/m}) = v_T(1 - e^{-t/\tau}) \qquad\qquad \textbf{5.6} \blacktriangleleft$$

where $v_T = mg/b$, $\tau = m/b$, and $e = 2.718\ 28$ is the base of the natural logarithm. This expression for v can be verified by substituting it back into Equation 5.5. (Try it!) This function is plotted in Active Figure 5.16c.

The mathematical representation of the motion (Eq. 5.6) indicates that the terminal speed is never reached because the exponential function is never exactly equal to zero. For all practical purposes, however, when the exponential function is very small

---

[2]A *buoyant* force also acts on any object surrounded by a fluid. This force is constant and equal to the weight of the displaced fluid, as will be discussed in Chapter 15. The effect of this force can be modeled by changing the apparent weight of the sphere by a constant factor, so we can ignore it here.

at large values of t, the speed of the particle can be approximated as being constant and equal to the terminal speed.

We cannot compare different objects by means of the time interval required to reach terminal speed because, as we have just discussed, this time interval is infinite for all objects! We need some means to compare these exponential behaviors for different objects. We do so with a parameter called the **time constant.** The time constant $\tau = m/b$ that appears in Equation 5.6 is the time interval required for the factor in parentheses in Equation 5.6 to become equal to $1 - e^{-1} = 0.632$. Therefore, the time constant represents the time interval required for the object to reach 63.2% of its terminal speed (Active Fig. 5.16c).

Example 5.10 | Sphere Falling in Oil

A small sphere of mass 2.00 g is released from rest in a large vessel filled with oil, where it experiences a resistive force proportional to its speed. The sphere reaches a terminal speed of 5.00 cm/s. Determine the time constant τ and the time at which the sphere reaches 90.0% of its terminal speed.

SOLUTION

Conceptualize With the help of Active Figure 5.16, imagine dropping the sphere into the oil and watching it sink to the bottom of the vessel. If you have some thick shampoo in a clear container, drop a marble in it and observe the motion of the marble.

Categorize We model the sphere as a particle under a net force, with one of the forces being a resistive force that depends on the speed of the sphere.

Analyze From $v_T = mg/b$, evaluate the coefficient b:

$$b = \frac{mg}{v_T} = \frac{(2.00 \text{ g})(980 \text{ cm/s}^2)}{5.00 \text{ cm/s}} = 392 \text{ g/s}$$

Evaluate the time constant τ:

$$\tau = \frac{m}{b} = \frac{2.00 \text{ g}}{392 \text{ g/s}} = \boxed{5.10 \times 10^{-3} \text{ s}}$$

Find the time t at which the sphere reaches a speed of $0.900v_T$ by setting $v = 0.900v_T$ in Equation 5.6 and solving for t:

$$0.900v_T = v_T(1 - e^{-t/\tau})$$

$$1 - e^{-t/\tau} = 0.900$$

$$e^{-t/\tau} = 0.100$$

$$-\frac{t}{\tau} = \ln (0.100) = -2.30$$

$$t = 2.30\tau = 2.30(5.10 \times 10^{-3} \text{ s}) = 11.7 \times 10^{-3} \text{ s}$$

$$= \boxed{11.7 \text{ ms}}$$

Finalize The sphere reaches 90.0% of its terminal speed in a very short time interval. You should have also seen this behavior if you performed the activity with the marble and the shampoo. Because of the short time interval required to reach terminal velocity, you may not have noticed the time interval at all. The marble may have appeared to immediately begin moving through the shampoo at a constant velocity.

Model 2: Resistive Force Proportional to Object Speed Squared

For large objects moving at high speeds through air, such as airplanes, skydivers, and baseballs, the magnitude of the resistive force is modeled as being proportional to the square of the speed:

$$R = \tfrac{1}{2}D\rho Av^2 \qquad \qquad \textbf{5.7}\blacktriangleleft$$

where ρ is the density of air, A is the cross-sectional area of the moving object measured in a plane perpendicular to its velocity, and D is a dimensionless empirical quantity called the *drag coefficient*. The drag coefficient has a value of about 0.5 for spherical objects moving through air but can be as high as 2 for irregularly shaped objects.

Consider an airplane in flight that experiences such a resistive force. Equation 5.7 shows that the force is proportional to the density of air and hence decreases with decreasing air density. Because air density decreases with increasing altitude, the resistive force on a jet airplane flying at a given speed will decrease with increasing altitude. Therefore, airplanes tend to fly at very high altitudes to take advantage of this reduced resistive force, which allows them to fly faster for a given engine thrust. Of course, this higher speed *increases* the resistive force, in proportion to the square of the speed, so a balance is struck between fuel economy and higher speed.

Now let us analyze the motion of a falling object subject to an upward air resistive force whose magnitude is given by Equation 5.7. Suppose an object of mass m is released from rest, as in Figure 5.17, from the position $y = 0$. The object experiences two external forces: the downward gravitational force $m\vec{g}$ and the upward resistive force $\vec{R}$. Hence, using Newton's second law,

$$\sum F = ma \quad \rightarrow \quad mg - \tfrac{1}{2}D\rho Av^2 = ma \qquad \text{5.8} \blacktriangleleft$$

Solving for a, we find that the object has a downward acceleration of magnitude

$$a = g - \left(\frac{D\rho A}{2m}\right)v^2 \qquad \text{5.9} \blacktriangleleft$$

Because $a = dv/dt$, Equation 5.9 is another differential equation that provides us with the speed as a function of time.

Again, we can calculate the terminal speed v_T because when the gravitational force is balanced by the resistive force, the net force is zero and therefore the acceleration is zero. Setting $a = 0$ in Equation 5.9 gives

$$g - \left(\frac{D\rho A}{2m}\right)v_T^2 = 0$$

$$v_T = \sqrt{\frac{2mg}{D\rho A}} \qquad \text{5.10} \blacktriangleleft$$

Table 5.2 lists the terminal speeds for several objects falling through air, all computed on the assumption that the drag coefficient is 0.5.

> ◣ **QUICK QUIZ 5.7** Consider a sky surfer falling through air, as in Figure 5.18, before reaching his terminal speed. As the speed of the sky surfer increases, the magnitude of his acceleration **(a)** remains constant, **(b)** decreases until it reaches a constant nonzero value, or **(c)** decreases until it reaches zero.

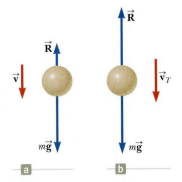

Figure 5.17 (a) An object falling through air experiences a resistive drag force $\vec{R}$ and a gravitational force $\vec{F}_g = m\vec{g}$. (b) The object reaches terminal speed when the net force acting on it is zero, that is, when $\vec{R} = -\vec{F}_g$, or $R = mg$. Before that occurs, the acceleration varies with speed according to Equation 5.9.

Figure 5.18 (Quick Quiz 5.7) A sky surfer takes advantage of the upward force of the air on his board.

© 2011 Oliver Furrer/Jupiterimages Corporation

> ◣ **TABLE 5.2** | Terminal Speeds for Various Objects Falling Through Air

Object	Mass (kg)	Cross-sectional Area (m²)	v_T (m/s)[a]
Skydiver	75	0.70	60
Baseball (radius 3.7 cm)	0.145	4.2×10^{-3}	33
Golf ball (radius 2.1 cm)	0.046	1.4×10^{-3}	32
Hailstone (radius 0.50 cm)	4.8×10^{-4}	7.9×10^{-5}	14
Raindrop (radius 0.20 cm)	3.4×10^{-5}	1.3×10^{-5}	9.0

[a]The drag coefficient D is assumed to be 0.5 in each case.

142   **CHAPTER 5** | More Applications of Newton's Laws

◄5.5 | The Fundamental Forces of Nature

We have described a variety of forces experienced in our everyday activities, such as the gravitational force acting on all objects at or near the Earth's surface and the force of friction as one surface slides over another. Newton's second law tells us how to relate the forces to the object's or particle's acceleration.

In addition to these familiar macroscopic forces in nature, forces also act in the atomic and subatomic world. For example, atomic forces within the atom are responsible for holding its constituents together and nuclear forces act on different parts of the nucleus to keep its parts from separating.

Until recently, physicists believed that there were four fundamental forces in nature: the gravitational force, the electromagnetic force, the strong force, and the weak force. We shall discuss these forces individually and then consider the current view of fundamental forces.

The Gravitational Force

The **gravitational force** is the mutual force of attraction between any two objects in the Universe. It is interesting and rather curious that although the gravitational force can be very strong between macroscopic objects, it is inherently the weakest of all the fundamental forces. For example, the gravitational force between the electron and proton in the hydrogen atom has a magnitude on the order of 10^{-46} N, whereas the electromagnetic force between these same two particles is on the order of 10^{-7} N.

In addition to his contributions to the understanding of motion, Newton studied gravity extensively. **Newton's law of universal gravitation** states that every particle in the Universe attracts every other particle with a force that is directly proportional to the product of the masses of the particles and inversely proportional to the square of the distance between them. If the particles have masses m_1 and m_2 and are separated by a distance r, as in Figure 5.19, the magnitude of the gravitational force is

▶ Newton's law of universal gravitation

$$F_g = G\frac{m_1 m_2}{r^2}$$

5.11◄

where $G = 6.674 \times 10^{-11}$ N·m²/kg² is the **universal gravitational constant.** More detail on the gravitational force will be provided in Chapter 11.

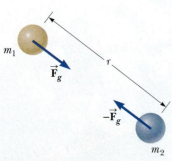

Figure 5.19 Two particles with masses m_1 and m_2 attract each other with a force of magnitude Gm_1m_2/r^2.

The Electromagnetic Force

The **electromagnetic force** is the force that binds atoms and molecules in compounds to form ordinary matter. It is much stronger than the gravitational force. The force that causes a rubbed comb to attract bits of paper and the force that a magnet exerts on an iron nail are electromagnetic forces. Essentially all forces at work in our macroscopic world, apart from the gravitational force, are manifestations of the electromagnetic force. For example, friction forces, contact forces, tension forces, and forces in elongated springs are consequences of electromagnetic forces between charged particles in proximity.

The electromagnetic force involves two types of particles: those with positive charge and those with negative charge. (More information on these two types of charge is provided in Chapter 19.) Unlike the gravitational force, which is always an attractive interaction, the electromagnetic force can be either attractive or repulsive, depending on the charges on the particles.

▶ Coulomb's law

Coulomb's law expresses the magnitude of the *electrostatic force*[3] F_e between two charged particles separated by a distance r:

$$F_e = k_e \frac{q_1 q_2}{r^2}$$

5.12◄

---

[3]The electrostatic force is the electromagnetic force between two electric charges that are at rest. If the charges are moving, magnetic forces are also present; these forces will be studied in Chapter 22.

where q_1 and q_2 are the charges on the two particles, measured in units called *coulombs* (C), and k_e (= 8.99 × 10^9 N·m²/C²) is the **Coulomb constant.** Note that the electrostatic force has the same mathematical form as Newton's law of universal gravitation (see Eq. 5.11), with charge playing the mathematical role of mass and the Coulomb constant being used in place of the universal gravitational constant. The electrostatic force is attractive if the two charges have opposite signs and is repulsive if the two charges have the same sign, as indicated in Figure 5.20.

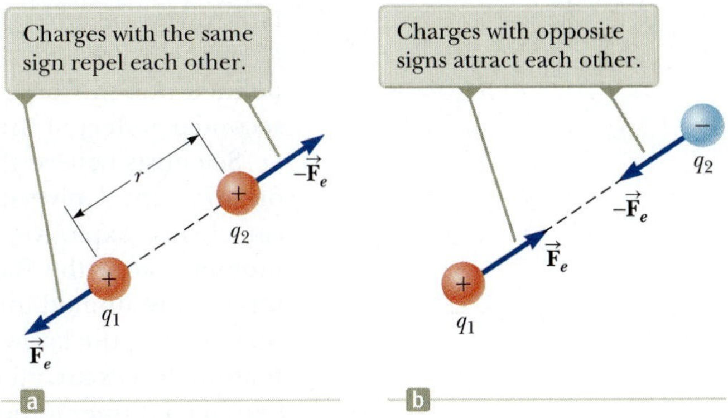

The smallest amount of isolated charge found in nature (so far) is the charge on an electron or proton. This fundamental unit of charge is given the symbol e and has the magnitude $e = 1.60 × 10^{-19}$ C. An electron has charge $-e$, whereas a proton has charge $+e$. Theories developed in the latter half of the 20th century propose that protons and neutrons are made up of smaller particles called **quarks,** which have charges of either $\frac{2}{3}e$ or $-\frac{1}{3}e$ (discussed further in Chapter 31). Although experimental evidence has been found for such particles inside nuclear matter, free quarks have never been detected.

Charges with the same sign repel each other.

Charges with opposite signs attract each other.

Figure 5.20 Two point charges separated by a distance r exert an electrostatic force on each other given by Coulomb's law.

The Strong Force

An atom, as we currently model it, consists of an extremely dense positively charged nucleus surrounded by a cloud of negatively charged electrons, with the electrons attracted to the nucleus by the electric force. All nuclei except those of hydrogen are combinations of positively charged protons and neutral neutrons (collectively called nucleons), yet why does the repulsive electrostatic force between the protons not cause nuclei to break apart? Clearly, there must be an attractive force that counteracts the strong electrostatic repulsive force and is responsible for the stability of nuclei. This force that binds the nucleons to form a nucleus is called the **nuclear force.** It is one manifestation of the **strong force,** which is the force between quarks, which we will discuss in Chapter 31. Unlike the gravitational and electromagnetic forces, which depend on distance in an inverse-square fashion, the nuclear force is extremely short range; its strength decreases very rapidly outside the nucleus and is negligible for separations greater than approximately 10^{-14} m.

The Weak Force

The **weak force** is a short-range force that tends to produce instability in certain nuclei. It was first observed in naturally occurring radioactive substances and was later found to play a key role in most radioactive decay reactions. The weak force is about 10^{34} times stronger than the gravitational force and about 10^3 times weaker than the electromagnetic force.

The Current View of Fundamental Forces

For years, physicists have searched for a simplification scheme that would reduce the number of fundamental forces needed to describe physical phenomena. In 1967, physicists predicted that the electromagnetic force and the weak force, originally thought to be independent of each other and both fundamental, are in fact manifestations of one force, now called the **electroweak** force. This prediction was confirmed experimentally in 1984. We shall discuss it more fully in Chapter 31.

We also now know that protons and neutrons are not fundamental particles; current models of protons and neutrons theorize that they are composed of simpler particles called quarks, as mentioned previously. The quark model has led to a modification of our understanding of the nuclear force. Scientists now define the strong force as the force that binds the quarks to one another in a nucleon (proton or neutron). This force is also referred to as a **color force,** in reference to a property of quarks called "color," which we shall investigate in Chapter 31. The previously defined nuclear force, the force that acts between nucleons, is now interpreted as a secondary effect of the strong force between the quarks.

Scientists believe that the fundamental forces of nature are closely related to the origin of the Universe. The Big Bang theory states that the Universe began with a cataclysmic explosion about 14 billion years ago. According to this theory, the first moments after the Big Bang saw such extremes of energy that all the fundamental forces were unified into one force. Physicists are continuing their search for connections among the known fundamental forces, connections that could eventually prove that the forces are all merely different forms of a single superforce. This fascinating search continues to be at the forefront of physics.

5.6 | Context Connection: Drag Coefficients of Automobiles

In the Context Connection of Chapter 4, we ignored air resistance and assumed that the driving force on the tires was the only force on the vehicle in the horizontal direction. Given our understanding of velocity-dependent forces from Section 5.4, we should understand now that air resistance could be a significant factor in the design of an automobile.

Table 5.3 shows the drag coefficients for the vehicles that we have investigated in previous chapters. Notice that the coefficients for the very expensive, performance, and traditional vehicles vary from 0.27 to 0.43, with the average coefficient in the three portions of the table almost the same. A look at the alternative vehicles shows that this parameter is the lowest on the average for all the vehicles, with the Chevrolet Volt and Toyota Prius having the lowest values in the entire table. The discontinued GM EV1, an electric car produced between 1996 and 1999, had a remarkable coefficient of just 0.19.

Designers of alternative-fuel vehicles try to squeeze every last mile of travel out of the energy that is stored in the vehicle in the form of fuel or an electric battery. A significant method of doing so is to reduce the force of air resistance so that the net force driving the car forward is as large as possible.

TABLE 5.3 | Drag Coefficients of Various Vehicles

Automobile		Drag Coefficient	Automobile		Drag Coefficient
Very expensive vehicles:			*Traditional vehicles:*		
Bugatti Veyron 16.4 Super Sport		0.36	Buick Regal CXL Turbo		0.27
Lamborghini LP 570-4 Superleggera		0.31	Chevrolet Tahoe 1500 LS (SUV)		0.42
Lexus LFA		0.31	Ford Fiesta SES		0.33
Mercedes-Benz SLS AMG		0.36	Hummer H3 (SUV)		0.43
Shelby SuperCars Ultimate Aero		0.36	Hyundai Sonata SE		0.32
	Average	**0.34**	Smart ForTwo		0.34
				Average	**0.35**
Performance vehicles:			*Alternative vehicles:*		
Chevrolet Corvette ZR1		0.28	Chevrolet Volt (hybrid)		0.26
Dodge Viper SRT10		0.40	Nissan Leaf (electric)		0.29
Jaguar XJL Supercharged		0.29	Honda CR-Z (hybrid)		0.30
Acura TL SH-AWD		0.29	Honda Insight (hybrid)		0.28
Dodge Challenger SRT8		0.35	Toyota Prius (hybrid)		0.25
	Average	**0.32**		**Average**	**0.28**

Figure 5.21 (a) The Chevrolet Corvette ZR1 has a streamlined shape that contributes to its low drag coefficient of 0.28. (b) The Hummer H3 is not streamlined like the Corvette and consequently has a much higher drag coefficient of 0.43.

A number of techniques can be used to reduce the drag coefficient. Two factors that help are a small frontal area and smooth curves from the front of the vehicle to the back. For example, the Chevrolet Corvette ZR1 shown in Figure 5.21a exhibits a streamlined shape that contributes to its low drag coefficient. As a comparison, consider a large, boxy vehicle, such as the Hummer H3 in Figure 5.21b. The drag coefficient for this vehicle is 0.43. (This is an improvement over the previous model, the H2, which had a coefficient of 0.57.) Another factor includes elimination or minimization of as many irregularities in the surfaces as possible, including door handles that project from the body, windshield wipers, wheel wells, and rough surfaces on headlamps and grills. An important consideration is the underside of the carriage. As air rushes beneath the car, there are many irregular surfaces associated with brakes, drive trains, suspension components, and so on. The drag coefficient can be made lower by assuring that the overall surface of the car's undercarriage is as smooth as possible.

SUMMARY

Forces of friction are complicated, but we design a simplification model for friction that allows us to analyze motion that includes the effects of friction. The **maximum force of static friction** $f_{s,\text{max}}$ between two surfaces is proportional to the normal force between the surfaces. This maximum force occurs when the surfaces are on the verge of slipping. In general, $f_s \leq \mu_s n$, where μ_s is the **coefficient of static friction** and n is the magnitude of the normal force. When an object slides over a rough surface, the **force of kinetic friction** $\vec{f}_k$ is opposite the direction of the velocity of the object relative to the surface and its magnitude is proportional to the magnitude of the normal force on the object. The magnitude is given by $f_k = \mu_k n$, where μ_k is the **coefficient of kinetic friction**. Usually, $\mu_k < \mu_s$.

An object moving through a liquid or gas experiences a **resistive force** that is velocity dependent. This resistive force, which is opposite the velocity of the object relative to the medium, generally increases with speed. The force depends on the object's shape and on the properties of the medium through which the object is moving. In the limiting case for a falling object, when the resistive force balances the weight ($a = 0$), the object reaches its **terminal speed.**

The fundamental forces existing in nature can be expressed as the following four: the gravitational force, the electromagnetic force, the strong force, and the weak force.

Analysis Model for Problem Solving

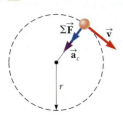

Particle in Uniform Circular Motion (Extension) With our new knowledge of forces, we can extend the model of a particle in uniform circular motion, first introduced in Chapter 3. Newton's second law applied to a particle moving in uniform circular motion states that the net force causing the particle to undergo a centripetal acceleration (Eq. 3.17) is related to the acceleration according to

$$\sum F = ma_c = m\frac{v^2}{r}$$

5.3

OBJECTIVE QUESTIONS

1. The driver of a speeding empty truck slams on the brakes and skids to a stop through a distance *d*. On a second trial, the truck carries a load that doubles its mass. What will now be the truck's "skidding distance"? (a) $4d$ (b) $2d$ (c) $\sqrt{2}d$ (d) d (e) $d/2$

2. The manager of a department store is pushing horizontally with a force of magnitude 200 N on a box of shirts. The box is sliding across the horizontal floor with a forward acceleration. Nothing else touches the box. What must be true about the magnitude of the force of kinetic friction acting on the box (choose one)? (a) It is greater than 200 N. (b) It is less than 200 N. (c) It is equal to 200 N. (d) None of those statements is necessarily true.

3. An object of mass *m* moves with acceleration $\vec{a}$ down a rough incline. Which of the following forces should appear in a free-body diagram of the object? Choose all correct answers. (a) the gravitational force exerted by the planet (b) $m\vec{a}$ in the direction of motion (c) the normal force exerted by the incline (d) the friction force exerted by the incline (e) the force exerted by the object on the incline

4. An office door is given a sharp push and swings open against a pneumatic device that slows the door down and then reverses its motion. At the moment the door is open the widest, (a) does the doorknob have a centripetal acceleration? (b) Does it have a tangential acceleration?

5. A crate remains stationary after it has been placed on a ramp inclined at an angle with the horizontal. Which of the following statements is or are correct about the magnitude of the friction force that acts on the crate? Choose all that are true. (a) It is larger than the weight of the crate. (b) It is equal to $\mu_s n$. (c) It is greater than the component of the gravitational force acting down the ramp. (d) It is equal to the component of the gravitational force acting down the ramp. (e) It is less than the component of the gravitational force acting down the ramp.

6. A pendulum consists of a small object called a bob hanging from a light cord of fixed length, with the top end of the cord fixed, as represented in Figure OQ5.6. The bob moves without friction, swinging equally high on both sides. It moves from its turning point *A* through point *B* and reaches its maximum speed at point *C*. (a) Of these points, is there a point where the bob has nonzero radial acceleration and zero tangential acceleration? If so, which point? What is the direction of its total acceleration at this point? (b) Of these points, is there a point where the bob has nonzero tangential acceleration and zero radial acceleration? If so, which point? What is the direction of its total acceleration at this point? (c) Is there a point where the bob has no acceleration? If so, which point? (d) Is there a point where the bob has both nonzero tangential and radial acceleration? If so, which point? What is the direction of its total acceleration at this point?

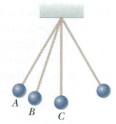

Figure OQ5.6

7. A door in a hospital has a pneumatic closer that pulls the door shut such that the doorknob moves with constant speed over most of its path. In this part of its motion,

(a) does the doorknob experience a centripetal acceleration? (b) Does it experience a tangential acceleration?

8. The driver of a speeding truck slams on the brakes and skids to a stop through a distance *d*. On another trial, the initial speed of the truck is half as large. What now will be the truck's skidding distance? (a) $2d$ (b) $\sqrt{2}d$ (c) d (d) $d/2$ (e) $d/4$

9. A child is practicing for a BMX race. His speed remains constant as he goes counterclockwise around a level track with two straight sections and two nearly semicircular sections as shown in the aerial view of Figure OQ5.9. (a) Rank the magnitudes of his acceleration at the points *A, B, C, D,* and *E* from largest to smallest. If his acceleration is the same size at two points, display that fact in your ranking. If his acceleration is zero, display that fact. (b) What are the directions of his velocity at points *A, B,* and *C*? For each point, choose one: north, south, east, west, or nonexistent. (c) What are the directions of his acceleration at points *A, B,* and *C*?

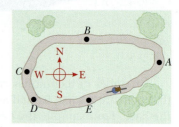

Figure OQ5.9

10. A large crate of mass *m* is placed on the flatbed of a truck but not tied down. As the truck accelerates forward with acceleration *a*, the crate remains at rest relative to the truck. What force causes the crate to accelerate? (a) the normal force (b) the gravitational force (c) the friction force (d) the *ma* force exerted by the crate (e) No force is required.

11. Before takeoff on an airplane, an inquisitive student on the plane dangles an iPod by its earphone wire. It hangs straight down as the plane is at rest waiting to take off. The plane then gains speed rapidly as it moves down the runway. (i) Relative to the student's hand, does the iPod (a) shift toward the front of the plane, (b) continue to hang straight down, or (c) shift toward the back of the plane? (ii) The speed of the plane increases at a constant rate over a time interval of several seconds. During this interval, does the angle the earphone wire makes with the vertical (a) increase, (b) stay constant, or (c) decrease?

12. Consider a skydiver who has stepped from a helicopter and is falling through air. Before she reaches terminal speed and long before she opens her parachute, does her speed (a) increase, (b) decrease, or (c) stay constant?

13. As a raindrop falls through the atmosphere, its speed initially changes as it falls toward the Earth. Before the raindrop reaches its terminal speed, does the magnitude of its acceleration (a) increase, (b) decrease, (c) stay constant at zero, (d) stay constant at 9.80 m/s², or (e) stay constant at some other value?

14. An object of mass *m* is sliding with speed v_i at some instant across a level tabletop, with which its coefficient of kinetic friction is μ. It then moves through a distance *d* and comes to rest. Which of the following equations for the speed v_i is reasonable? (a) $v_i = \sqrt{-2\mu mgd}$ (b) $v_i = \sqrt{2\mu mgd}$ (c) $v_i = \sqrt{-2\mu gd}$ (d) $v_i = \sqrt{2\mu gd}$ (e) $v_i = \sqrt{2\mu d}$

CONCEPTUAL QUESTIONS

☐ denotes answer available in *Student Solutions Manual/Study Guide*

1. A car is moving forward slowly and is speeding up. A student claims that "the car exerts a force on itself" or that "the car's engine exerts a force on the car." (a) Argue that this idea cannot be accurate and that friction exerted by the road is the propulsive force on the car. Make your evidence and reasoning as persuasive as possible. (b) Is it static or kinetic friction? *Suggestions:* Consider a road covered with light gravel. Consider a sharp print of the tire tread on an asphalt road, obtained by coating the tread with dust.

2. Your hands are wet, and the restroom towel dispenser is empty. What do you do to get drops of water off your hands? How does the motion of the drops exemplify one of Newton's laws? Which one?

3. Describe two examples in which the force of friction exerted on an object is in the direction of motion of the object.

4. Describe the path of a moving body in the event that (a) its acceleration is constant in magnitude at all times and perpendicular to the velocity, and (b) its acceleration is constant in magnitude at all times and parallel to the velocity.

5. An object executes circular motion with constant speed whenever a net force of constant magnitude acts perpendicular to the velocity. What happens to the speed if the force is not perpendicular to the velocity?

6. Consider a small raindrop and a large raindrop falling through the atmosphere. (a) Compare their terminal speeds. (b) What are their accelerations when they reach terminal speed?

7. Suppose you are driving a classic car. Why should you avoid slamming on your brakes when you want to stop in the shortest possible distance? (Many modern cars have anti-lock brakes that avoid this problem.)

8. A falling skydiver reaches terminal speed with her parachute closed. After the parachute is opened, what parameters change to decrease this terminal speed?

9. What forces cause (a) an automobile, (b) a propeller-driven airplane, and (c) a rowboat to move?

10. A pail of water can be whirled in a vertical path such that no water is spilled. Why does the water stay in the pail, even when the pail is above your head?

11. It has been suggested that rotating cylinders about 20 km in length and 8 km in diameter be placed in space and used as colonies. The purpose of the rotation is to simulate gravity for the inhabitants. Explain this concept for producing an effective imitation of gravity.

12. If someone told you that astronauts are weightless in orbit because they are beyond the pull of gravity, would you accept the statement? Explain.

13. **BIO** Why does a pilot tend to black out when pulling out of a steep dive?

PROBLEMS

ENHANCED Web**Assign** The problems found in this chapter may be assigned online in Enhanced WebAssign.

  1. denotes straightforward problem; **2.** denotes intermediate problem; **3.** denotes challenging problem

☐ **1.** denotes full solution available in the *Student Solutions Manual/Study Guide*

  1. denotes problems most often assigned in Enhanced WebAssign.

BIO denotes biomedical problem

GP denotes guided problem

M denotes Master It tutorial available in Enhanced WebAssign

Q|C denotes asking for quantitative and conceptual reasoning

S denotes symbolic reasoning problem

shaded denotes "paired problems" that develop reasoning with symbols and numerical values

W denotes Watch It video solution available in Enhanced WebAssign

Section 5.1 Forces of Friction

1. To determine the coefficients of friction between rubber and various surfaces, a student uses a rubber eraser and an incline. In one experiment, the eraser begins to slip down the incline when the angle of inclination is 36.0° and then moves down the incline with constant speed when the angle is reduced to 30.0°. From these data, determine the coefficients of static and kinetic friction for this experiment.

2. **Q|C** Before 1960, people believed that the maximum attainable coefficient of static friction for an automobile tire on a roadway was $\mu_s = 1$. Around 1962, three companies independently developed racing tires with coefficients of 1.6. This problem shows that tires have improved further since then. The shortest time interval in which a piston-engine car initially at rest has covered a distance of one-quarter mile is about 4.43 s. (a) Assume the car's rear wheels lift the front wheels off the pavement as shown in Figure P5.2. What minimum value of μ_s is necessary to achieve the record time? (b) Suppose the driver were able to increase his or her engine power, keeping other things equal. How would this change affect the elapsed time?

Figure P5.2

3. **W** A 25.0-kg block is initially at rest on a horizontal surface. A horizontal force of 75.0 N is required to set the block in motion, after which a horizontal force of 60.0 N is required to keep the block moving with constant speed. Find (a) the coefficient of static friction and (b) the coefficient of kinetic friction between the block and the surface.

4. **Review.** A car is traveling at 50.0 mi/h on a horizontal highway. (a) If the coefficient of static friction between road and tires on a rainy day is 0.100, what is the minimum distance in which the car will stop? (b) What is the stopping distance when the surface is dry and $\mu_s = 0.600$?

5. To meet a U.S. Postal Service requirement, employees' footwear must have a coefficient of static friction of 0.5 or more on a specified tile surface. A typical athletic shoe has a coefficient of static friction of 0.800. In an emergency, what is the minimum time interval in which a person starting from rest can move 3.00 m on the tile surface if she is wearing (a) footwear meeting the Postal Service minimum and (b) a typical athletic shoe?

6. **BIO** The person in Figure P5.6 weighs 170 lb. As seen from the front, each light crutch makes an angle of 22.0° with the vertical. Half of the person's weight is supported by the crutches. The other half is supported by the vertical forces of the ground on the person's feet. Assuming that the person is moving with constant velocity and the force exerted by the ground on the crutches acts along the crutches, determine (a) the smallest possible coefficient of friction between crutches and ground and (b) the magnitude of the compression force in each crutch.

22.0°   22.0°

Figure P5.6

7. **W** A 9.00-kg hanging object is connected by a light, inextensible cord over a light, frictionless pulley to a 5.00-kg block that is sliding on a flat table (Fig. P5.7). Taking the coefficient of kinetic friction as 0.200, find the tension in the string.

Figure P5.7

8. Consider a large truck carrying a heavy load, such as steel beams. A significant hazard for the driver is that the load may slide forward, crushing the cab, if the truck stops suddenly in an accident or even in braking. Assume, for example, that a 10 000-kg load sits on the flatbed of a 20 000-kg truck moving at 12.0 m/s. Assume that the load is not tied down to the truck, but has a coefficient of friction of 0.500 with the flatbed of the truck. (a) Calculate the minimum stopping distance for which the load will not slide forward relative to the truck. (b) Is any piece of data unnecessary for the solution?

9. **M** **Review.** A 3.00-kg block starts from rest at the top of a 30.0° incline and slides a distance of 2.00 m down the incline in 1.50 s. Find (a) the magnitude of the acceleration of the block, (b) the coefficient of kinetic friction between block and plane, (c) the friction force acting on the block, and (d) the speed of the block after it has slid 2.00 m.

10. **W** A woman at an airport is towing her 20.0-kg suitcase at constant speed by pulling on a strap at an angle θ above the horizontal (Fig. P5.10). She pulls on the strap with a 35.0-N force, and the friction force on the suitcase is 20.0 N. (a) Draw a free-body diagram of the suitcase. (b) What angle does the strap make with the horizontal? (c) What is the magnitude of the normal force that the ground exerts on the suitcase?

Figure P5.10

11. **Review.** One side of the roof of a house slopes up at 37.0°. A roofer kicks a round, flat rock that has been thrown onto the roof by a neighborhood child. The rock slides straight up the incline with an initial speed of 15.0 m/s. The coefficient of kinetic friction between the rock and the roof is 0.400. The rock slides 10.0 m up the roof to its peak. It crosses the ridge and goes into free fall, following a parabolic trajectory above the far side of the roof, with negligible air resistance. Determine the maximum height the rock reaches above the point where it was kicked.

12. **QC** A block of mass 3.00 kg is pushed up against a wall by a force $\vec{P}$ that makes an angle of $\theta = 50.0°$ with the horizontal as shown in Figure P5.12. The coefficient of static friction between the block and the wall is 0.250. (a) Determine the possible values for the magnitude of $\vec{P}$ that allow the block to remain stationary. (b) Describe what happens if $|\vec{P}|$ has a larger value and what happens if it is smaller. (c) Repeat parts (a) and (b), assuming the force makes an angle of $\theta = 13.0°$ with the horizontal.

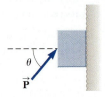

Figure P5.12

13. **M** Two blocks connected by a rope of negligible mass are being dragged by a horizontal force (Fig. P5.13). Suppose $F = 68.0$ N, $m_1 = 12.0$ kg, $m_2 = 18.0$ kg, and the coefficient of kinetic friction between each block and the surface is 0.100. (a) Draw a free-body diagram for each block. Determine (b) the acceleration of the system and (c) the tension T in the rope.

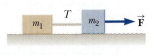

Figure P5.13

14. **QC** Three objects are connected on a table as shown in Figure P5.14. The coefficient of kinetic friction between the block of mass m_2 and the table is 0.350. The objects have masses of $m_1 = 4.00$ kg, $m_2 = 1.00$ kg, and $m_3 = 2.00$ kg, and the pulleys are frictionless. (a) Draw a free-body diagram of each object. (b) Determine the acceleration of each object, including its direction. (c) Determine the tensions in the two cords. **What If?** (d) If the tabletop were smooth, would the tensions increase, decrease, or remain the same? Explain.

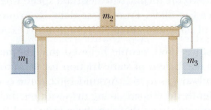

Figure P5.14

15. *Why is the following situation impossible?* Your 3.80-kg physics book is placed next to you on the horizontal seat of your car. The coefficient of static friction between the book and the seat is 0.650, and the coefficient of kinetic friction is 0.550. You are traveling forward at 72.0 km/h and brake to a stop with constant acceleration over a distance of 30.0 m. Your physics book remains on the seat rather than sliding forward onto the floor.

Section 5.2 Extending the Particle in Uniform Circular Motion Model

16. In the Bohr model of the hydrogen atom, an electron moves in a circular path around a proton. The speed of the electron is approximately 2.20×10^6 m/s. Find (a) the force acting on the electron as it revolves in a circular orbit of radius 0.530×10^{-10} m and (b) the centripetal acceleration of the electron.

17. **M** A light string can support a stationary hanging load of 25.0 kg before breaking. An object of mass $m = 3.00$ kg attached to the string rotates on a frictionless, horizontal table in a circle of radius $r = 0.800$ m, and the other end of the string is held fixed as in Figure P5.17. What range of speeds can the object have before the string breaks?

Figure P5.17

18. *Why is the following situation impossible?* The object of mass $m = 4.00$ kg in Figure P5.18 is attached to a vertical rod by two strings of length $\ell = 2.00$ m. The strings are attached to the rod at points a distance $d = 3.00$ m apart. The object rotates in a horizontal circle at a constant speed of $v = 3.00$ m/s, and the strings remain taut. The rod rotates along with the object so that the strings do not wrap onto the rod. **What If?** Could this situation be possible on another planet?

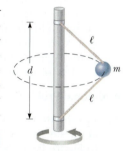

Figure P5.18

19. **W** A crate of eggs is located in the middle of the flatbed of a pickup truck as the truck negotiates a curve in the flat road. The curve may be regarded as an arc of a circle of radius 35.0 m. If the coefficient of static friction between crate and truck is 0.600, how fast can the truck be moving without the crate sliding?

20. Whenever two *Apollo* astronauts were on the surface of the Moon, a third astronaut orbited the Moon. Assume the orbit to be circular and 100 km above the surface of the Moon, where the acceleration due to gravity is 1.52 m/s². The radius of the Moon is 1.70×10^6 m. Determine (a) the astronaut's orbital speed and (b) the period of the orbit.

21. **W** Consider a conical pendulum (Fig. P5.21) with a bob of mass $m = 80.0$ kg on a string of length $L = 10.0$ m that makes an angle of $\theta = 5.00°$ with the vertical. Determine (a) the horizontal and vertical components of the force exerted by the wire on the pendulum and (b) the radial acceleration of the bob.

Figure P5.21

Section 5.3 Nonuniform Circular Motion

22. **Q|C** A roller coaster at the Six Flags Great America amusement park in Gurnee, Illinois, incorporates some clever design technology and some basic physics. Each vertical loop, instead of being circular, is shaped like a teardrop (Fig. P5.22). The cars ride on the inside of the loop at the top, and the speeds are fast enough to ensure the cars remain on the track. The biggest loop is 40.0 m high. Suppose the speed at the top of the loop is 13.0 m/s and the corresponding centripetal acceleration of the riders is 2g. (a) What is the radius of the arc of the teardrop at the top? (b) If the total mass of a car plus the riders is M, what force does the rail exert on the car at the top? (c) Suppose the roller coaster had a circular loop of radius 20.0 m. If the cars have the same speed, 13.0 m/s at the top, what is the centripetal acceleration of the riders at the top? (d) Comment on the normal force at the top in the situation described in part (c) and on the advantages of having teardrop-shaped loops.

Figure P5.22

23. Disturbed by speeding cars outside his workplace, Nobel laureate Arthur Holly Compton designed a speed bump (called the "Holly hump") and had it installed. Suppose a 1 800-kg car passes over a hump in a roadway that follows the arc of a circle of radius 20.4 m as shown in Figure P5.23. (a) If the car travels at 30.0 km/h, what force does the road exert on the car as the car passes the highest point of the hump? (b) **What If?** What is the maximum speed the car can have without losing contact with the road as it passes this highest point?

Figure P5.23 Problems 23 and 24.

24. **S** A car of mass m passes over a hump in a road that follows the arc of a circle of radius R as shown in Figure P5.23. (a) If the car travels at a speed v, what force does the road exert on the car as the car passes the highest point of the hump? (b) **What If?** What is the maximum speed the car can have without losing contact with the road as it passes this highest point?

25. An adventurous archeologist ($m = 85.0$ kg) tries to cross a river by swinging from a vine. The vine is 10.0 m long, and his speed at the bottom of the swing is 8.00 m/s. The archeologist doesn't know that the vine has a breaking strength of 1 000 N. Does he make it across the river without falling in?

26. **Q|C W** A pail of water is rotated in a vertical circle of radius 1.00 m. (a) What two external forces act on the water in the pail? (b) Which of the two forces is most important in causing the water to move in a circle? (c) What is the pail's minimum speed at the top of the circle if no water is to spill out? (d) Assume the pail with the speed found in part (c) were to suddenly disappear at the top of the circle. Describe the subsequent motion of the water. Would it differ from the motion of a projectile?

27. **M** A 40.0-kg child swings in a swing supported by two chains, each 3.00 m long. The tension in each chain at the lowest point is 350 N. Find (a) the child's speed at the lowest point and (b) the force exerted by the seat on the child at the lowest point. (Ignore the mass of the seat.)

28. **S** A child of mass m swings in a swing supported by two chains, each of length R. If the tension in each chain at the lowest point is T, find (a) the child's speed at the lowest point and (b) the force exerted by the seat on the child at the lowest point. (Ignore the mass of the seat.)

Section 5.4 Motion in the Presence of Velocity-Dependent Resistive Forces

29. **M** A small, spherical bead of mass 3.00 g is released from rest at $t = 0$ from a point under the surface of a viscous liquid. The terminal speed is observed to be $v_T = 2.00$ cm/s. Find (a) the value of the constant b that appears in Equation 5.4, (b) the time t at which the bead reaches $0.632v_T$, and (c) the value of the resistive force when the bead reaches terminal speed.

30. **Review.** (a) Estimate the terminal speed of a wooden sphere (density 0.830 g/cm³) falling through air, taking its radius as 8.00 cm and its drag coefficient as 0.500. (b) From what height would a freely falling object reach this speed in the absence of air resistance?

31. **W** A small piece of Styrofoam packing material is dropped from a height of 2.00 m above the ground. Until it reaches terminal speed, the magnitude of its acceleration is given by $a = g - Bv$. After falling 0.500 m, the Styrofoam effectively reaches terminal speed and then takes 5.00 s more to reach the ground. (a) What is the value of the constant B? (b) What is the acceleration at $t = 0$? (c) What is the acceleration when the speed is 0.150 m/s?

32. **S** Assume the resistive force acting on a speed skater is proportional to the square of the skater's speed v and is given by $f = -kmv^2$, where k is a constant and m is the skater's mass. The skater crosses the finish line of a straight-line race with speed v_i and then slows down by coasting on his skates. Show that the skater's speed at any time t after crossing the finish line is $v(t) = v_i/(1 + ktv_i)$.

33. A motorboat cuts its engine when its speed is 10.0 m/s and then coasts to rest. The equation describing the motion of the motorboat during this period is $v = v_i e^{-ct}$, where v is the speed at time t, v_i is the initial speed at $t = 0$, and c is a constant. At $t = 20.0$ s, the speed is 5.00 m/s. (a) Find the constant c. (b) What is the speed at $t = 40.0$ s? (c) Differentiate the expression for $v(t)$ and thus show that the acceleration of the boat is proportional to the speed at any time.

34. A 9.00-kg object starting from rest falls through a viscous medium and experiences a resistive force given by Equation 5.4. The object reaches one-half its terminal speed in 5.54 s. (a) Determine the terminal speed. (b) At what time is the speed of the object three-fourths the terminal speed? (c) How far has the object traveled in the first 5.54 s of motion?

Section 5.5 The Fundamental Forces of Nature

35. When a falling meteor is at a distance above the Earth's surface of 3.00 times the Earth's radius, what is its free-fall acceleration caused by the gravitational force exerted on it?

36. In a thundercloud, there may be electric charges of +40.0 C near the top of the cloud and −40.0 C near the bottom of the cloud. These charges are separated by 2.00 km. What is the electric force on the top charge?

37. Two identical isolated particles, each of mass 2.00 kg, are separated by a distance of 30.0 cm. What is the magnitude of the gravitational force exerted by one particle on the other?

38. Find the order of magnitude of the gravitational force that you exert on another person 2 m away. In your solution, state the quantities you measure or estimate and their values.

Section 5.6 Context Connection: Drag Coefficients of Automobiles

39. The mass of a sports car is 1 200 kg. The shape of the body is such that the aerodynamic drag coefficient is 0.250 and the frontal area is 2.20 m². Ignoring all other sources of friction, calculate the initial acceleration the car has if it has been traveling at 100 km/h and is now shifted into neutral and allowed to coast.

40. Consider a 1 300-kg car presenting front-end area 2.60 m² and having drag coefficient 0.340. It can achieve instantaneous acceleration 3.00 m/s² when its speed is 10.0 m/s. Ignore any force of rolling resistance. Assume that the only horizontal forces on the car are static friction forward exerted by the road on the drive wheels and resistance exerted by the surrounding air, with density 1.20 kg/m³. (a) Find the friction force exerted by the road. (b) Suppose the car body could be redesigned to have a drag coefficient of 0.200. If nothing else changes, what will be the car's acceleration? (c) Assume that the force exerted by the road remains constant. Then what maximum speed could the car attain with $D = 0.340$? (d) With $D = 0.200$?

Additional Problems

41. In a home laundry dryer, a cylindrical tub containing wet clothes is rotated steadily about a horizontal axis as shown in Figure P5.41. So that the clothes will dry uniformly, they are made to tumble. The rate of rotation of the smooth-walled tub is chosen so that a small piece of cloth will lose contact with the tub when the cloth is at an angle of $\theta = 68.0°$ above the horizontal. If the radius of the tub is $r = 0.330$ m, what rate of revolution is needed?

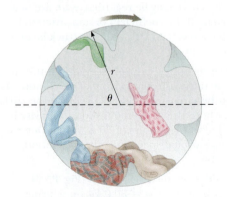

Figure P5.41

42. **S** A crate of weight F_g is pushed by a force $\vec{P}$ on a horizontal floor as shown in Figure P5.42. The coefficient of static friction is μ_s, and $\vec{P}$ is directed at angle θ below the horizontal.

(a) Show that the minimum value of P that will move the crate is given by

$$P = \frac{\mu_s F_g \sec\theta}{1 - \mu_s \tan\theta}$$

(b) Find the condition on θ in terms of μ_s for which motion of the crate is impossible for any value of P.

Figure P5.42

43. Consider the three connected objects shown in Figure P5.43. Assume first that the inclined plane is frictionless and that the system is in equilibrium. In terms of m, g, and θ, find (a) the mass M and (b) the tensions T_1

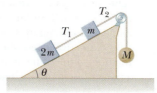

Figure P5.43

and T_2. Now assume that the value of M is double the value found in part (a). Find (c) the acceleration of each object and (d) the tensions T_1 and T_2. Next, assume that the coefficient of static friction between m and $2m$ and the inclined plane is μ_s and that the system is in equilibrium. Find (e) the maximum value of M and (f) the minimum value of M. (g) Compare the values of T_2 when M has its minimum and maximum values.

44. **S** A car rounds a banked curve as discussed in Example 5.7 and shown in Figure 5.11. The radius of curvature of the road is R, the banking angle is θ, and the coefficient of static friction is μ_s. (a) Determine the range of speeds the car can have without slipping up or down the road. (b) Find the minimum value for μ_s such that the minimum speed is zero.

45. The system shown in Figure P4.47 (Chapter 4) has an acceleration of magnitude 1.50 m/s^2. Assume that the coefficient of kinetic friction between block and incline is the same for both inclines. Find (a) the coefficient of kinetic friction and (b) the tension in the string.

46. An aluminum block of mass $m_1 = 2.00$ kg and a copper block of mass $m_2 = 6.00$ kg are connected by a light string over a frictionless pulley. They sit on a steel surface as shown in Figure P5.46, where $\theta = 30.0°$. (a) When they are

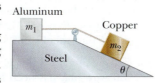

Figure P5.46

released from rest, will they start to move? If they do, determine (b) their acceleration and (c) the tension in the string. If they do not move, determine (d) the sum of the magnitudes of the forces of friction acting on the blocks.

47. **W** Figure P5.47 shows a photo of a swing ride at an amusement park. The structure consists of a horizontal, rotating, circular platform of diameter D from which seats of mass m are suspended at the end of massless chains of length d. When the system rotates at constant speed, the chains swing outward and make an angle θ with the vertical. Consider such

Figure P5.47

Stuart Gregory/Getty Images

a ride with the following parameters: $D = 8.00$ m, $d = 2.50$ m, $m = 10.0$ kg, and $\theta = 28.0°$. (a) What is the speed of each seat? (b) Draw a diagram of forces acting on a 40.0-kg child riding in a seat and (c) find the tension in the chain.

48. *Why is the following situation impossible?* A 1.30-kg toaster is not plugged in. The coefficient of static friction between the toaster and a horizontal countertop is 0.350. To make the toaster start moving, you carelessly pull on its electric cord. Unfortunately, the cord has become frayed from your previous similar actions and will break if the tension in the cord exceeds 4.00 N. By pulling on the cord at a particular angle, you successfully start the toaster moving without breaking the cord.

49. A space station, in the form of a wheel 120 m in diameter, rotates to provide an "artificial gravity" of 3.00 m/s^2 for persons who walk around on the inner wall of the outer rim. Find the rate of the wheel's rotation in revolutions per minute that will produce this effect.

50. A 5.00-kg block is placed on top of a 10.0-kg block (Fig. P5.50). A horizontal force of 45.0 N is applied to the 10-kg block, and the 5-kg block is tied to the wall. The coefficient of kinetic friction between all moving surfaces is 0.200. (a) Draw a free-body diagram for each block

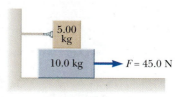

Figure P5.50

and identify the action-reaction forces between the blocks. (b) Determine the tension in the string and the magnitude of the acceleration of the 10-kg block.

51. **W** In Example 5.8, we investigated the forces a child experiences on a Ferris wheel. Assume the data in that example applies to this problem. What force (magnitude and direction) does the seat exert on a 40.0-kg child when the child is halfway between top and bottom?

52. A student builds and calibrates an accelerometer and uses it to determine the speed of her car around a certain unbanked highway curve. The accelerometer is a plumb bob with a protractor that she attaches to the roof of her car. A friend riding in the car with the student observes that the plumb bob hangs at an angle of 15.0° from the vertical when the car has a speed of 23.0 m/s. (a) What is the centripetal acceleration of the car rounding the curve? (b) What is the radius of the curve? (c) What is speed of the car if the plumb bob deflection is 9.00° while rounding the same curve?

53. **GP** **S** Two blocks of masses m_1 and m_2 are placed on a table in contact with each other as discussed in Example 4.5 and shown in Active Figure 4.13a. The coefficient of kinetic friction between the block of mass m_1 and the table is μ_1, and that between the block of mass m_2 and the table is μ_2. A horizontal force of magnitude F is applied to the block of mass m_1. We wish to find P, the magnitude of the contact force between the blocks. (a) Draw diagrams showing the forces for each block. (b) What is the net force on the system of two blocks? (c) What is the net force acting on m_1? (d) What is the net force acting on m_2? (e) Write Newton's second law in the x direction for each block. (f) Solve the two equations in two unknowns for the acceleration a of the blocks

in terms of the masses, the applied force *F*, the coefficients of friction, and *g*. (g) Find the magnitude *P* of the contact force between the blocks in terms of the same quantities.

54. *Why is the following situation impossible?* A mischievous child goes to an amusement park with his family. On one ride, after a severe scolding from his mother, he slips out of his seat and climbs to the top of the ride's structure, which is shaped like a cone with its axis vertical and its sloped sides making an angle of $\theta = 20.0°$ with the horizontal as shown in Figure P5.54. This part of the structure rotates about the vertical central axis when the ride operates. The child sits on the sloped surface at a point $d = 5.32$ m down the sloped side from the center of the cone and pouts. The coefficient of static friction between the boy and the cone is 0.700. The ride operator does not notice that the child has slipped away from his seat and so continues to operate the ride. As a result, the sitting, pouting boy rotates in a circular path at a speed of 3.75 m/s.

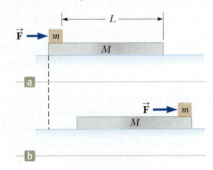

Figure P5.54

55. A block of mass $m = 2.00$ kg rests on the left edge of a block of mass $M = 8.00$ kg. The coefficient of kinetic friction between the two blocks is 0.300, and the surface on which the 8.00-kg block rests is frictionless. A constant horizontal force of magnitude $F = 10.0$ N is applied to the 2.00-kg block, setting it in motion as shown in Figure P5.55a. If the distance *L* that the leading edge of the smaller block travels on the larger block is 3.00 m, (a) in what time interval will the smaller block make it to the right side of the 8.00-kg block as shown in Figure P5.55b? (*Note:* Both blocks are set into motion when $\vec{F}$ is applied.) (b) How far does the 8.00-kg block move in the process?

Figure P5.55

56. **Q|C** **S** A puck of mass m_1 is tied to a string and allowed to revolve in a circle of radius *R* on a frictionless, horizontal table. The other end of the string passes through a small hole in the center of the table, and an object of mass m_2 is tied to it (Fig. P5.56). The suspended object remains in equilibrium while the puck on the tabletop revolves. Find

symbolic expressions for (a) the tension in the string, (b) the radial force acting on the puck, and (c) the speed of the puck. (d) Qualitatively describe what will happen in the motion of the puck if the value of m_2 is increased by placing a small additional load on the hanging load.

Figure P5.56

(e) Qualitatively describe what will happen in the motion of the puck if the value of m_2 is instead decreased by removing a part of the hanging load.

57. **M** A model airplane of mass 0.750 kg flies with a speed of 35.0 m/s in a horizontal circle at the end of a 60.0-m-long control wire as shown in Figure P5.57a. The forces exerted on the airplane are shown in Figure P5.57b: the tension in the control wire, the gravitational force, and aerodynamic lift that acts at $\theta = 20.0°$ inward from the vertical. Compute the tension in the wire, assuming it makes a constant angle of $\theta = 20.0°$ with the horizontal.

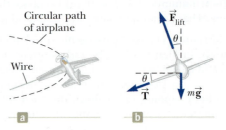

Figure P5.57

58. *Why is the following situation impossible?* A book sits on an inclined plane on the surface of the Earth. The angle of the plane with the horizontal is 60.0°. The coefficient of kinetic friction between the book and the plane of 0.300. At time $t = 0$, the book is released from rest. The book then slides through a distance of 1.00 m, measured along the plane, in a time interval of 0.483 s.

59. **Q|C** A single bead can slide with negligible friction on a stiff wire that has been bent into a circular loop of radius 15.0 cm as shown in Figure P5.59. The circle is always in a vertical plane and rotates steadily about its vertical diameter with a period of 0.450 s. The position of the bead is described by the angle θ that the radial line, from the center of the loop to the bead, makes with the vertical. (a) At what angle up from the bottom of the circle can the bead stay motionless relative to the turning circle? (b) **What If?** Repeat the problem, this time taking the period of the circle's rotation as 0.850 s. (c) Describe how the solution to part (b) is different from the solution to part (a). (d) For any period or loop size, is there always an angle at which the bead can stand still relative to the loop? (e) Are there ever more than two angles? Arnold Arons suggested the idea for this problem.

Figure P5.59

60. **Q|C** **S** An amusement park ride consists of a large vertical cylinder that spins about its axis fast enough that any person inside is held up against the wall when the floor drops away (Fig. P5.60). The coefficient of static

friction between person and wall is μ_s, and the radius of the cylinder is R. (a) Show that the maximum period of revolution necessary to keep the person from falling is $T = (4\pi^2 R\mu_s/g)^{1/2}$. (b) If the rate of revolution of the cylinder is made to be somewhat larger, what happens to the magnitude of each one of the forces acting on the person? What happens in the motion of the person? (c) If the rate of revolution of the cylinder is instead made to be somewhat smaller, what happens to the magnitude of each one of the forces acting on the person? What happens in the motion of the person?

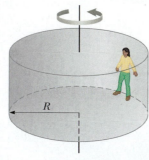

Figure P5.60

61. The expression $F = arv + br^2v^2$ gives the magnitude of the resistive force (in newtons) exerted on a sphere of radius r (in meters) by a stream of air moving at speed v (in meters per second), where a and b are constants with appropriate SI units. Their numerical values are $a = 3.10 \times 10^{-4}$ and $b = 0.870$. Using this expression, find the terminal speed for water droplets falling under their own weight in air, taking the following values for the drop radii: (a) 10.0 μm, (b) 100 μm, (c) 1.00 mm. For parts (a) and (c), you can obtain accurate answers without solving a quadratic equation by considering which of the two contributions to the air resistance is dominant and ignoring the lesser contribution.

62. Members of a skydiving club were given the following data to use in planning their jumps. In the table, d is the distance fallen from rest by a skydiver in a "free-fall stable spread position" versus the time of fall t. (a) Convert the distances in feet into meters. (b) Graph d (in meters) versus t. (c)

Determine the value of the terminal speed v_T by finding the slope of the straight portion of the curve. Use a least-squares fit to determine this slope.

t(s)	d(ft)	t(s)	d(ft)
0	0	11	1 309
1	16	12	1 483
2	62	13	1 657
3	138	14	1 831
4	242	15	2 005
5	366	16	2 179
6	504	17	2 353
7	652	18	2 527
8	808	19	2 701
9	971	20	2 875
10	1 138		

63. **M** Because the Earth rotates about its axis, a point on the equator experiences a centripetal acceleration of 0.033 7 m/s^2, whereas a point at the poles experiences no centripetal acceleration. If a person at the equator has a mass of 75 kg, calculate (a) the gravitational force (true weight) on the person and (b) the normal force (apparent weight) on the person. (c) Which force is greater? Assume the Earth is a uniform sphere and take $g = 9.800$ m/s^2.

64. **S** If a single constant force acts on an object that moves on a straight line, the object's velocity is a linear function of time. The equation $v = v_i + at$ gives its velocity v as a function of time, where a is its constant acceleration. What if velocity is instead a linear function of position? Assume that as a particular object moves through a resistive medium, its speed decreases as described by the equation $v = v_i - kx$, where k is a constant coefficient and x is the position of the object. Find the law describing the total force acting on this object.

Chapter 6

Energy of a System

Chapter Outline

Christopher Furlong/Getty Images

On a wind farm at the mouth of the River Mersey in Liverpool, England, the moving air does work on the blades of the windmills, causing the blades and the rotor of an electrical generator to rotate. Energy is transferred out of the system of the windmill by means of electricity.

The definitions of quantities such as position, velocity, acceleration, and force and associated principles such as Newton's second law have allowed us to solve a variety of problems. Some problems that could theoretically be solved with Newton's laws, however, are very difficult in practice, but they can be made much simpler with a different approach. Here and in the following chapters, we will investigate this new approach, which will include definitions of quantities that may not be familiar to you. Other quantities may sound familiar, but they may have more specific meanings in physics than in everyday life. We begin this discussion by exploring the notion of *energy*.

The concept of energy is one of the most important topics in science and engineering. In everyday life, we think of energy in terms of fuel for transportation and heating, electricity for lights and appliances, and foods for consumption. These ideas, however, do not truly define energy. They merely tell us that fuels are needed to do a job and that those fuels provide us with something we call energy.

Energy is present in the Universe in various forms. *Every* physical process that occurs in the Universe involves energy and energy transfers or transformations. Unfortunately, despite its extreme importance, energy cannot be easily defined. The variables in previous chapters were relatively concrete; we have everyday experience with velocities and forces, for example. Although we have *experiences* with energy, such as running out of gasoline or losing our electrical service following a violent storm, the *notion* of energy is more abstract.

The concept of energy can be applied to mechanical systems without resorting to Newton's laws. Furthermore, the energy approach allows us to understand thermal and electrical phenomena in later chapters of the book.

Our analysis models presented in earlier chapters were based on the motion of a *particle* or an object that could be modeled as a particle. We begin our new approach by focusing our attention on a *system* and analysis models based on the model of a system. These analysis models will be formally introduced in Chapter 7. In this chapter, we introduce systems and three ways to store energy in a system.

6.1 | Systems and Environments

In the system model, we focus our attention on a small portion of the Universe—the **system**—and ignore details of the rest of the Universe outside of the system. A critical skill in applying the system model to problems is *identifying the system.*

A valid system

- may be a single object or particle
- may be a collection of objects or particles
- may be a region of space (such as the interior of an automobile engine combustion cylinder)
- may vary with time in size and shape (such as a rubber ball, which deforms upon striking a wall)

> **Pitfall Prevention | 6.1**
> **Identify the System**
> The most important *first* step to take in solving a problem using the energy approach is to identify the appropriate system of interest.

Identifying the need for a system approach to solving a problem (as opposed to a particle approach) is part of the Categorize step in the General Problem-Solving Strategy outlined in Chapter 1. Identifying the particular system is a second part of this step.

No matter what the particular system is in a given problem, we identify a **system boundary,** an imaginary surface (not necessarily coinciding with a physical surface) that divides the Universe into the system and the **environment** surrounding the system.

As an example, imagine a force applied to an object in empty space. We can define the object as the system and its surface as the system boundary. The force applied to it is an influence on the system from the environment that acts across the system boundary. We will see how to analyze this situation from a system approach in a subsequent section of this chapter.

Another example was seen in Example 5.3, where the system can be defined as the combination of the ball, the block, and the cord. The influence from the environment includes the gravitational forces on the ball and the block, the normal and friction forces on the block, the force exerted by the pulley on the cord, and the applied force of magnitude F. The forces exerted by the cord on the ball and the block are internal to the system and therefore are not included as an influence from the environment.

There are a number of mechanisms by which a system can be influenced by its environment. The first one we shall investigate is *work*.

Figure 6.1 An eraser being pushed along a chalkboard tray by a force acting at different angles with respect to the horizontal direction.

6.2 | Work Done by a Constant Force

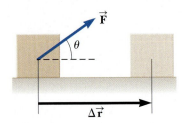

Figure 6.2 An object undergoes a displacement $\Delta\vec{\mathbf{r}}$ under the action of a constant force $\vec{\mathbf{F}}$.

► Work done by a constant force

Almost all the terms we have used thus far—velocity, acceleration, force, and so on—convey a similar meaning in physics as they do in everyday life. Now, however, we encounter a term whose meaning in physics is distinctly different from its everyday meaning: work.

To understand what work as an influence on a system means to the physicist, consider the situation illustrated in Figure 6.1. A force $\vec{\mathbf{F}}$ is applied to a chalkboard eraser, which we identify as the system, and the eraser slides along the tray. If we want to know how effective the force is in moving the eraser, we must consider not only the magnitude of the force but also its direction. Notice that the finger in Figure 6.1 applies forces in three different directions on the eraser. Assuming the magnitude of the applied force is the same in all three photographs, the push applied in Figure 6.1b does more to move the eraser than the push in Figure 6.1a. On the other hand, Figure 6.1c shows a situation in which the applied force does not move the eraser at all, regardless of how hard it is pushed (unless, of course, we apply a force so great that we break the chalkboard tray!). These results suggest that when analyzing forces to determine the influence they have on the system, we must consider the vector nature of forces. We must also consider the magnitude of the force. Moving a force of magnitude 2 N through a displacement represents a greater influence on the system than moving a force of magnitude 1 N through the same displacement. The magnitude of the displacement is also important. Moving the eraser 3 m along the tray represents a greater influence than moving it 2 cm if the same force is used in both cases.

Let us examine the situation in Figure 6.2, where the object (the system) undergoes a displacement along a straight line while acted on by a constant force of magnitude F that makes an angle θ with the direction of the displacement.

The **work** W done on a system by an agent exerting a constant force on the system is the product of the magnitude F of the force, the magnitude Δr of the displacement of the point of application of the force, and $\cos\theta$, where θ is the angle between the force and displacement vectors:

$$W \equiv F\,\Delta r \cos\theta \qquad\qquad \textbf{6.1} \blacktriangleleft$$

Notice in Equation 6.1 that work is a scalar, even though it is defined in terms of two vectors, a force $\vec{\mathbf{F}}$ and a displacement $\Delta\vec{\mathbf{r}}$. In Section 6.3, we explore how to combine two vectors to generate a scalar quantity.

Notice also that the displacement in Equation 6.1 is that of *the point of application of the force*. If the force is applied to a particle or a rigid object that can be modeled as a particle, this displacement is the same as that of the particle. For a deformable system, however, these displacements are not the same. For example, imagine pressing in on the sides of a balloon with both hands. The center of the balloon moves through zero displacement. The points of application of the forces from your hands on the sides of the balloon, however, do indeed move through a displacement as the balloon is compressed, and that is the displacement to be used in Equation 6.1. We will see other examples of deformable systems, such as springs and samples of gas contained in a vessel.

As an example of the distinction between the definition of work and our everyday understanding of the word, consider holding a heavy chair at arm's length for 3 min. At the end of this time interval, your tired arms may lead you to think you have done a considerable amount of work on the chair. According to our definition, however, you have done no work on it whatsoever. You exert a force to support the chair, but you do not move it. A force does no work on an object if the force does not move through a displacement. If $\Delta r = 0$, Equation 6.1 gives $W = 0$, which is the situation depicted in Figure 6.1c.

Equation 6.1 also shows that the work done by a force on a moving object is zero when the force applied is perpendicular to the displacement of its point of application. That is, if $\theta = 90°$, then $W = 0$ because $\cos 90° = 0$. For example, in Figure 6.3, the work done by the normal force on the object and the work done by the gravitational force on the object are both zero because both forces are perpendicular to the displacement and have zero components along an axis in the direction of $\Delta \vec{r}$.

The sign of the work also depends on the direction of $\vec{F}$ relative to $\Delta \vec{r}$. The work done by the applied force on a system is positive when the projection of $\vec{F}$ onto $\Delta \vec{r}$ is in the same direction as the displacement. For example, when an object is lifted, the work done by the applied force on the object is positive because the direction of that force is upward, in the same direction as the displacement of its point of application. When the projection of $\vec{F}$ onto $\Delta \vec{r}$ is in the direction opposite the displacement, W is negative. For example, as an object is lifted, the work done by the gravitational force on the object is negative. The factor $\cos \theta$ in the definition of W (Eq. 6.1) automatically takes care of the sign.

If an applied force $\vec{F}$ is in the same direction as the displacement $\Delta \vec{r}$, then $\theta = 0$ and $\cos 0 = 1$. In this case, Equation 6.1 gives

$$W = F \Delta r$$

The units of work are those of force multiplied by those of length. Therefore, the SI unit of work is the **newton · meter** ($N \cdot m = kg \cdot m^2/s^2$). This combination of units is used so frequently that it has been given a name of its own, the **joule** (J).

An important consideration for a system approach to problems is that **work is an energy transfer.** If W is the work done on a system and W is positive, energy is transferred *to* the system; if W is negative, energy is transferred *from* the system. Therefore, if a system interacts with its environment, this interaction can be described as a transfer of energy across the system boundary. The result is a change in the energy stored in the system. We will learn about the first type of energy storage in Section 6.5, after we investigate more aspects of work.

QUICK QUIZ 6.1 The gravitational force exerted by the Sun on the Earth holds the Earth in an orbit around the Sun. Let us assume that the orbit is perfectly circular. The work done by this gravitational force during a short time interval in which the Earth moves through a displacement in its orbital path is (a) zero (b) positive (c) negative (d) impossible to determine

QUICK QUIZ 6.2 Figure 6.4 shows four situations in which a force is applied to an object. In all four cases, the force has the same magnitude, and the displacement of the object is to the right and of the same magnitude. Rank the situations in order of the work done by the force on the object, from most positive to most negative.

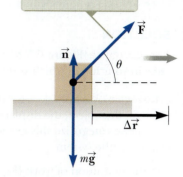

Figure 6.3 An object is displaced on a frictionless, horizontal surface. The normal force $\vec{n}$ and the gravitational force $m\vec{g}$ do no work on the object.

Pitfall Prevention | 6.3
Cause of the Displacement
We can calculate the work done by a force on an object, but that force is *not* necessarily the cause of the object's displacement. For example, if you lift an object, (negative) work is done on the object by the gravitational force, although gravity is not the cause of the object moving upward!

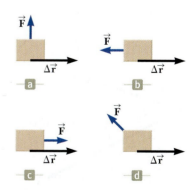

Figure 6.4 (Quick Quiz 6.2) A block is pulled by a force in four different directions. In each case, the displacement of the block is to the right and of the same magnitude.

Example 6.1 | Mr. Clean

A man cleaning a floor pulls a vacuum cleaner with a force of magnitude $F = 50.0$ N at an angle of 30.0° with the horizontal (Fig. 6.5). Calculate the work done by the force on the vacuum cleaner as the vacuum cleaner is displaced 3.00 m to the right.

SOLUTION

Conceptualize Figure 6.5 helps conceptualize the situation. Think about an experience in your life in which you pulled an object across the floor with a rope or cord.

Categorize We are asked for the work done on an object by a force and are given the force on the object, the displacement of the object, and the angle between the two vectors, so we categorize this example as a substitution problem. We identify the vacuum cleaner as the system.

Figure 6.5 (Example 6.1) A vacuum cleaner being pulled at an angle of 30.0° from the horizontal.

Use the definition of work (Eq. 6.1):

$$W = F\Delta r \cos\theta = (50.0 \text{ N})(3.00 \text{ m})(\cos 30.0°)$$
$$= \boxed{130 \text{ J}}$$

Notice in this situation that the normal force $\vec{n}$ and the gravitational $\vec{F}_g = m\vec{g}$ do no work on the vacuum cleaner because these forces are perpendicular to the displacements of their points of application. Furthermore, there was no mention of whether there was friction between the vacuum cleaner and the floor. The presence or absence of friction is not important when calculating the work done by the applied force. In addition, this work does not depend on whether the vacuum moved at constant velocity or if it accelerated.

Pitfall Prevention | 6.4
Work Is a Scalar
Although Equation 6.3 defines the work in terms of two vectors, *work is a scalar;* there is no direction associated with it. *All* types of energy and energy transfer are scalars. This fact is a major advantage of the energy approach because we often don't need vector calculations!

▶ Scalar product of any two vectors $\vec{A}$ and $\vec{B}$

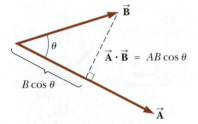

Figure 6.6 The scalar product $\vec{A} \cdot \vec{B}$ equals the magnitude of $\vec{A}$ multiplied by $B \cos\theta$, which is the projection of $\vec{B}$ onto $\vec{A}$.

6.3 | The Scalar Product of Two Vectors

Because of the way the force and displacement vectors are combined in Equation 6.1, it is helpful to use a convenient mathematical tool called the **scalar product** of two vectors. We write this scalar product of vectors $\vec{A}$ and $\vec{B}$ as $\vec{A} \cdot \vec{B}$. (Because of the dot symbol, the scalar product is often called the **dot product**.)

The scalar product of any two vectors $\vec{A}$ and $\vec{B}$ is defined as a scalar quantity equal to the product of the magnitudes of the two vectors and the cosine of the angle θ between them:

$$\vec{A} \cdot \vec{B} \equiv AB \cos\theta \qquad \textbf{6.2} \blacktriangleleft$$

As is the case with any multiplication, $\vec{A}$ and $\vec{B}$ need not have the same units.

By comparing this definition with Equation 6.1, we can express Equation 6.1 as a scalar product:

$$W = F\Delta r \cos\theta = \vec{F} \cdot \Delta\vec{r} \qquad \textbf{6.3} \blacktriangleleft$$

In other words, $\vec{F} \cdot \Delta\vec{r}$ is a shorthand notation for $F\Delta r \cos\theta$.

Before continuing with our discussion of work, let us investigate some properties of the dot product. Figure 6.6 shows two vectors $\vec{A}$ and $\vec{B}$ and the angle θ between them used in the definition of the dot product. In Figure 6.6, $B \cos\theta$ is the projection of $\vec{B}$ onto $\vec{A}$. Therefore, Equation 6.2 means that $\vec{A} \cdot \vec{B}$ is the product of the magnitude of $\vec{A}$ and the projection of $\vec{B}$ onto $\vec{A}$.[1]

From the right-hand side of Equation 6.2, we also see that the scalar product is **commutative.**[2] That is,

$$\vec{A} \cdot \vec{B} = \vec{B} \cdot \vec{A}$$

---

[1]This statement is equivalent to stating that $\vec{A} \cdot \vec{B}$ equals the product of the magnitude of $\vec{B}$ and the projection of $\vec{A}$ onto $\vec{B}$.

[2]In Chapter 10, you will see another way of combining vectors that proves useful in physics and is not commutative.

Finally, the scalar product obeys the **distributive law of multiplication,** so

$$\vec{A} \cdot (\vec{B} + \vec{C}) = \vec{A} \cdot \vec{B} + \vec{A} \cdot \vec{C}$$

The scalar product is simple to evaluate from Equation 6.2 when $\vec{A}$ is either perpendicular or parallel to $\vec{B}$. If $\vec{A}$ is perpendicular to $\vec{B}$ ($\theta = 90°$), then $\vec{A} \cdot \vec{B} = 0$. (The equality $\vec{A} \cdot \vec{B} = 0$ also holds in the more trivial case in which either $\vec{A}$ or $\vec{B}$ is zero.) If vector $\vec{A}$ is parallel to vector $\vec{B}$ and the two point in the same direction ($\theta = 0$), then $\vec{A} \cdot \vec{B} = AB$. If vector $\vec{A}$ is parallel to vector $\vec{B}$ but the two point in opposite directions ($\theta = 180°$), then $\vec{A} \cdot \vec{B} = -AB$. The scalar product is negative when $90° < \theta \le 180°$.

The unit vectors $\hat{i}$, $\hat{j}$, and $\hat{k}$, which were defined in Chapter 1, lie in the positive x, y, and z directions, respectively, of a right-handed coordinate system. Therefore, it follows from the definition of $\vec{A} \cdot \vec{B}$ that the scalar products of these unit vectors are

$$\hat{i} \cdot \hat{i} = \hat{j} \cdot \hat{j} = \hat{k} \cdot \hat{k} = 1 \qquad \textbf{6.4}◀ \qquad ▶ \text{ Scalar products of unit vectors}$$

$$\hat{i} \cdot \hat{j} = \hat{i} \cdot \hat{k} = \hat{j} \cdot \hat{k} = 0 \qquad \textbf{6.5}◀$$

According to Section 1.9, two vectors $\vec{A}$ and $\vec{B}$ can be expressed in unit-vector form as

$$\vec{A} = A_x\hat{i} + A_y\hat{j} + A_z\hat{k}$$

$$\vec{B} = B_x\hat{i} + B_y\hat{j} + B_z\hat{k}$$

Using these expressions for the vectors and the information given in Equations 6.4 and 6.5 shows that the scalar product of $\vec{A}$ and $\vec{B}$ reduces to

$$\vec{A} \cdot \vec{B} = A_xB_x + A_yB_y + A_zB_z \qquad \textbf{6.6}◀$$

(Details of the derivation are left for you in Problem 8 at the end of the chapter.) In the special case in which $\vec{A} = \vec{B}$, we see that

$$\vec{A} \cdot \vec{A} = A_x^2 + A_y^2 + A_z^2 = A^2$$

QUICK QUIZ 6.3 Which of the following statements is true about the relationship between the dot product of two vectors and the product of the magnitudes of the vectors? (a) $\vec{A} \cdot \vec{B}$ is larger than AB. (b) $\vec{A} \cdot \vec{B}$ is smaller than AB. (c) $\vec{A} \cdot \vec{B}$ could be larger or smaller than AB, depending on the angle between the vectors. (d) $\vec{A} \cdot \vec{B}$ could be equal to AB.

Example 6.2 | The Scalar Product

The vectors $\vec{A}$ and $\vec{B}$ are given by $\vec{A} = 2\hat{i} + 3\hat{j}$ and $\vec{B} = -\hat{i} + 2\hat{j}$.

(A) Determine the scalar product $\vec{A} \cdot \vec{B}$.

SOLUTION

Conceptualize There is no physical system to imagine here. Rather, it is purely a mathematical exercise involving two vectors.

Categorize Because we have a definition for the scalar product, we categorize this example as a substitution problem.

Substitute the specific vector expressions for $\vec{A}$ and $\vec{B}$:

$$\vec{A} \cdot \vec{B} = (2\hat{i} + 3\hat{j}) \cdot (-\hat{i} + 2\hat{j})$$
$$= -2\hat{i} \cdot \hat{i} + 2\hat{i} \cdot 2\hat{j} - 3\hat{j} \cdot \hat{i} + 3\hat{j} \cdot 2\hat{j}$$
$$= -2(1) + 4(0) - 3(0) + 6(1) = -2 + 6 = \boxed{4}$$

The same result is obtained when we use Equation 6.6 directly, where $A_x = 2$, $A_y = 3$, $B_x = -1$, and $B_y = 2$.

continued

6.2 cont.

(B) Find the angle θ between $\vec{\mathbf{A}}$ and $\vec{\mathbf{B}}$.

SOLUTION

Evaluate the magnitudes of $\vec{\mathbf{A}}$ and $\vec{\mathbf{B}}$ using the Pythagorean theorem:

$$A = \sqrt{A_x^2 + A_y^2} = \sqrt{(2)^2 + (3)^2} = \sqrt{13}$$

$$B = \sqrt{B_x^2 + B_y^2} = \sqrt{(-1)^2 + (2)^2} = \sqrt{5}$$

Use Equation 6.2 and the result from part (A) to find the angle:

$$\cos\theta = \frac{\vec{\mathbf{A}} \cdot \vec{\mathbf{B}}}{AB} = \frac{4}{\sqrt{13}\sqrt{5}} = \frac{4}{\sqrt{65}}$$

$$\theta = \cos^{-1}\frac{4}{\sqrt{65}} = \boxed{60.3°}$$

Example 6.3 | Work Done by a Constant Force

A particle moving in the xy plane undergoes a displacement given by $\Delta\vec{\mathbf{r}} = (2.0\hat{\mathbf{i}} + 3.0\hat{\mathbf{j}})$ m as a constant force $\vec{\mathbf{F}} = (5.0\hat{\mathbf{i}} + 2.0\hat{\mathbf{j}})$ N acts on the particle. Calculate the work done by $\vec{\mathbf{F}}$ on the particle.

SOLUTION

Conceptualize Although this example is a little more physical than the previous one in that it identifies a force and a displacement, it is similar in terms of its mathematical structure.

Categorize Because we are given force and displacement vectors and asked to find the work done by this force on the particle, we categorize this example as a substitution problem.

Substitute the expressions for $\vec{\mathbf{F}}$ and $\Delta\vec{\mathbf{r}}$ into Equation 6.3 and use Equations 6.4 and 6.5:

$$W = \vec{\mathbf{F}} \cdot \Delta\vec{\mathbf{r}} = [(5.0\hat{\mathbf{i}} + 2.0\hat{\mathbf{j}})\text{ N}] \cdot [(2.0\hat{\mathbf{i}} + 3.0\hat{\mathbf{j}})\text{ m}]$$

$$= (5.0\hat{\mathbf{i}} \cdot 2.0\hat{\mathbf{i}} + 5.0\hat{\mathbf{i}} \cdot 3.0\hat{\mathbf{j}} + 2.0\hat{\mathbf{j}} \cdot 2.0\hat{\mathbf{i}} + 2.0\hat{\mathbf{j}} \cdot 3.0\hat{\mathbf{j}})\text{ N} \cdot \text{m}$$

$$= [10 + 0 + 0 + 6]\text{ N} \cdot \text{m} = \boxed{16\text{ J}}$$

6.4 | Work Done by a Varying Force

Consider a particle being displaced along the x axis under the action of a force that varies with position. The particle is displaced in the direction of increasing x from $x = x_i$ to $x = x_f$. In such a situation, we cannot use $W = F\Delta r \cos\theta$ to calculate the work done by the force because this relationship applies only when $\vec{\mathbf{F}}$ is constant in magnitude and direction. If, however, we imagine that the particle undergoes a very small displacement Δx, shown in Figure 6.7a, the x component F_x of the force is approximately constant over this small interval; for this small displacement, we can approximate the work done on the particle by the force as

$$W \approx F_x \Delta x$$

which is the area of the shaded rectangle in Figure 6.7a. If we imagine the F_x versus x curve divided into a large number of such intervals, the total work done for the displacement from x_i to x_f is approximately equal to the sum of a large number of such terms:

$$W \approx \sum_{x_i}^{x_f} F_x \Delta x$$

If the size of the small displacements is allowed to approach zero, the number of terms in the sum increases without limit but the value of the sum approaches a definite value equal to the area bounded by the F_x curve and the x axis:

$$\lim_{\Delta x \to 0} \sum_{x_i}^{x_f} F_x \, \Delta x = \int_{x_i}^{x_f} F_x \, dx$$

Therefore, we can express the work done by F_x on the particle as it moves from x_i to x_f as

$$W = \int_{x_i}^{x_f} F_x \, dx \qquad \text{6.7} \blacktriangleleft$$

This equation reduces to Equation 6.1 when the component $F_x = F \cos \theta$ remains constant.

If more than one force acts on a system *and the system can be modeled as a particle,* the total work done on the system is just the work done by the net force. If we express the net force in the x direction as ΣF_x, the total work, or *net work,* done as the particle moves from x_i to x_f is

$$\sum W = W_{\text{ext}} = \int_{x_i}^{x_f} \left(\sum F_x \right) dx \quad \text{(particle)}$$

For the general case of a net force $\Sigma \vec{F}$ whose magnitude and direction may vary, we use the scalar product,

$$\sum W = W_{\text{ext}} = \int \left(\sum \vec{F} \right) \cdot d\vec{r} \quad \text{(particle)} \qquad \text{6.8} \blacktriangleleft$$

where the integral is calculated over the path that the particle takes through space. The subscript "ext" on work reminds us that the net work is done by an *external* agent on the system. We will use this notation in this chapter as a reminder and to differentiate this work from an *internal* work to be described shortly.

If the system cannot be modeled as a particle (for example, if the system is deformable), we cannot use Equation 6.8 because different forces on the system may move through different displacements. In this case, we must evaluate the work done by each force separately and then add the works algebraically to find the net work done on the system:

$$\sum W = W_{\text{ext}} = \sum_{\text{forces}} \left(\int \vec{F} \cdot d\vec{r} \right) \quad \text{(deformable system)}$$

Example 6.4 | Calculating Total Work Done from a Graph

A force acting on a particle varies with x as shown in Figure 6.8. Calculate the work done by the force on the particle as it moves from $x = 0$ to $x = 6.0$ m.

SOLUTION

Conceptualize Imagine a particle subject to the force in Figure 6.8. The force remains constant as the particle moves through the first 4.0 m and then decreases linearly to zero at 6.0 m. In terms of earlier discussions of motion, the particle could be modeled as a particle under constant acceleration for the first 4.0 m because the force is constant. Between 4.0 m and 6.0 m, however, the motion does not fit into one of our earlier analysis models because the acceleration of the particle is changing. If the particle starts from rest, its speed increases throughout the motion and the particle is always moving in the positive x direction. These details about its speed and direction are not necessary for the calculation of the work done, however.

Categorize Because the force varies during the entire motion of the particle, we must use the techniques for work done by varying forces. In this case, the graphical representation in Figure 6.8 can be used to evaluate the work done.

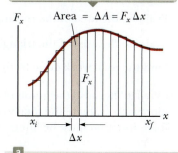
The total work done for the displacement from x_i to x_f is approximately equal to the sum of the areas of all the rectangles.

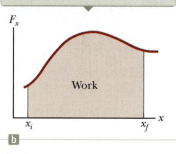
The work done by the component F_x of the varying force as the particle moves from x_i to x_f is *exactly* equal to the area under the curve.

Figure 6.7 (a) The work done on a particle by the force component F_x for the small displacement Δx is $F_x \Delta x$, which equals the area of the shaded rectangle. (b) The width Δx of each rectangle is shrunk to zero.

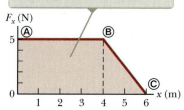
The net work done by this force is the area under the curve.

Figure 6.8 (Example 6.4) The force acting on a particle is constant for the first 4.0 m of motion and then decreases linearly with x from $x_{\circledB} = 4.0$ m to $x_{\circledC} = 6.0$ m.

continued

6.4 *cont.*

Analyze The work done by the force is equal to the area under the curve from $x_\text{Ⓐ} = 0$ to $x_\text{Ⓒ} = 6.0$ m. This area is equal to the area of the rectangular section from Ⓐ to Ⓑ plus the area of the triangular section from Ⓑ to Ⓒ.

Evaluate the area of the rectangle:	$W_{\text{Ⓐ to Ⓑ}} = (5.0 \text{ N})(4.0 \text{ m}) = 20 \text{ J}$
Evaluate the area of the triangle:	$W_{\text{Ⓑ to Ⓒ}} = \frac{1}{2}(5.0 \text{ N})(2.0 \text{ m}) = 5.0 \text{ J}$
Find the total work done by the force on the particle:	$W_{\text{Ⓐ to Ⓒ}} = W_{\text{Ⓐ to Ⓑ}} + W_{\text{Ⓑ to Ⓒ}} = 20 \text{ J} + 5.0 \text{ J} = \boxed{25 \text{ J}}$

Finalize Because the graph of the force consists of straight lines, we can use rules for finding the areas of simple geometric models to evaluate the total work done in this example. If a force does not vary linearly, such rules cannot be used and the force function must be integrated as in Equation 6.7 or 6.8.

Work Done by a Spring

A model of a common physical system on which the force varies with position is shown in Active Figure 6.9. The system is a block on a frictionless, horizontal surface and connected to a spring. For many springs, if the spring is either stretched or compressed a small distance from its unstretched (equilibrium) configuration, it exerts on the block a force that can be mathematically modeled as

▶ Spring force

$$F_s = -kx \qquad \text{6.9} \blacktriangleleft$$

where x is the position of the block relative to its equilibrium ($x = 0$) position and k is a positive constant called the **force constant** or the **spring constant** of the spring. In other words, the force required to stretch or compress a spring is proportional to the amount of stretch or compression x. This force law for springs is known as **Hooke's law.** The value of k is a measure of the *stiffness* of the spring. Stiff

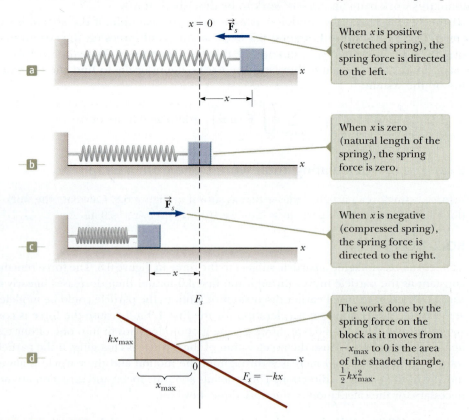

When x is positive (stretched spring), the spring force is directed to the left.

When x is zero (natural length of the spring), the spring force is zero.

When x is negative (compressed spring), the spring force is directed to the right.

The work done by the spring force on the block as it moves from $-x_\text{max}$ to 0 is the area of the shaded triangle, $\frac{1}{2}kx_\text{max}^2$.

Active Figure 6.9 The force exerted by a spring on a block varies with the block's position x relative to the equilibrium position $x = 0$. (a) x is positive. (b) x is zero. (c) x is negative. (d) Graph of F_s versus x for the block–spring system.

springs have large k values, and soft springs have small k values. As can be seen from Equation 6.9, the units of k are N/m.

The vector form of Equation 6.9 is

$$\vec{\mathbf{F}}_s = F_s\hat{\mathbf{i}} = -kx\hat{\mathbf{i}} \qquad \textbf{6.10}\blacktriangleleft$$

where we have chosen the x axis to lie along the direction the spring extends or compresses.

The negative sign in Equations 6.9 and 6.10 signifies that the force exerted by the spring is always directed *opposite* the displacement from equilibrium. When $x > 0$ as in Active Figure 6.9a so that the block is to the right of the equilibrium position, the spring force is directed to the left, in the negative x direction. When $x < 0$ as in Active Figure 6.9c, the block is to the left of equilibrium and the spring force is directed to the right, in the positive x direction. When $x = 0$ as in Active Figure 6.9b, the spring is unstretched and $F_s = 0$. Because the spring force always acts toward the equilibrium position ($x = 0$), it is sometimes called a *restoring force*.

If the spring is compressed until the block is at the point $-x_{max}$ and is then released, the block moves from $-x_{max}$ through zero to $+x_{max}$. It then reverses direction, returns to $-x_{max}$, and continues oscillating back and forth. We will study these oscillations in more detail in Chapter 12. For now, let's investigate the work done by the spring on the block over small portions of one oscillation.

Suppose the block is pushed to the left to a position $-x_{max}$ and is then released. We identify the block as our system and calculate the work W_s done by the spring force on the block as the block moves from $x_i = -x_{max}$ to $x_f = 0$. Applying Equation 6.8 and assuming the block may be modeled as a particle, we obtain

$$W_s = \int \vec{\mathbf{F}}_s \cdot d\vec{\mathbf{r}} = \int_{x_i}^{x_f} (-kx\hat{\mathbf{i}}) \cdot (dx\hat{\mathbf{i}}) = \int_{-x_{max}}^{0} (-kx)\, dx = \tfrac{1}{2}kx_{max}^2 \qquad \textbf{6.11}\blacktriangleleft$$

where we have used the integral $\int x^n\, dx = x^{n+1}/(n+1)$ with $n = 1$. The work done by the spring force is positive because the force is in the same direction as its displacement (both are to the right). Because the block arrives at $x = 0$ with some speed, it will continue moving until it reaches a position $+x_{max}$. The work done by the spring force on the block as it moves from $x_i = 0$ to $x_f = x_{max}$ is $W_s = -\tfrac{1}{2}kx_{max}^2$. The work is negative because for this part of the motion the spring force is to the left and its displacement is to the right. Therefore, the *net* work done by the spring force on the block as it moves from $x_i = -x_{max}$ to $x_f = x_{max}$ is *zero*.

Active Figure 6.9d is a plot of F_s versus x. The work calculated in Equation 6.11 is the area of the shaded triangle, corresponding to the displacement from $-x_{max}$ to 0. Because the triangle has base x_{max} and height kx_{max}, its area is $\tfrac{1}{2}kx_{max}^2$, agreeing with the work done by the spring as given by Equation 6.11.

If the block undergoes an arbitrary displacement from $x = x_i$ to $x = x_f$, the work done by the spring force on the block is

$$W_s = \int_{x_i}^{x_f} (-kx)\, dx = \tfrac{1}{2}kx_i^2 - \tfrac{1}{2}kx_f^2 \qquad \textbf{6.12}\blacktriangleleft$$

From Equation 6.12, we see that the work done by the spring force is zero for any motion that ends where it began ($x_i = x_f$). We shall make use of this important result in Chapter 7 when we describe the motion of this system in greater detail.

Equations 6.11 and 6.12 describe the work done by the spring on the block. Now let us consider the work done on the block by an *external agent* as the agent applies a force on the block and the block moves *very slowly* from $x_i = -x_{max}$ to $x_f = 0$ as in Figure 6.10. We can calculate this work by noting that at any value of the position, the *applied force* $\vec{\mathbf{F}}_{app}$ is equal in magnitude and opposite in direction to the spring force $\vec{\mathbf{F}}_s$, so $\vec{\mathbf{F}}_{app} = F_{app}\hat{\mathbf{i}} = -\vec{\mathbf{F}}_s = -(-kx\hat{\mathbf{i}}) = kx\hat{\mathbf{i}}$. Therefore, the work done by this applied force (the external agent) on the system of the block is

$$W_{ext} = \int \vec{\mathbf{F}}_{app} \cdot d\vec{\mathbf{r}} = \int_{x_i}^{x_f} (kx\hat{\mathbf{i}}) \cdot (dx\hat{\mathbf{i}}) = \int_{-x_{max}}^{0} kx\, dx = -\tfrac{1}{2}kx_{max}^2$$

If the process of moving the block is carried out very slowly, then $\vec{\mathbf{F}}_{app}$ is equal in magnitude and opposite in direction to $\vec{\mathbf{F}}_s$ at all times.

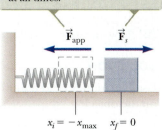

Figure 6.10 A block moves from $x_i = -x_{max}$ to $x_f = 0$ on a frictionless surface as a force $\vec{\mathbf{F}}_{app}$ is applied to the block.

This work is equal to the negative of the work done by the spring force for this displacement (Eq. 6.11). The work is negative because the external agent must push inward on the spring to prevent it from expanding, and this direction is opposite the direction of the displacement of the point of application of the force as the block moves from $-x_{max}$ to 0.

For an arbitrary displacement of the block, the work done on the system by the external agent is

$$W_{ext} = \int_{x_i}^{x_f} kx \, dx = \tfrac{1}{2}kx_f^2 - \tfrac{1}{2}kx_i^2$$

6.13 ◀

Notice that this equation is the negative of Equation 6.12.

▌**QUICK QUIZ 6.4** A dart is inserted into a spring-loaded dart gun by pushing the spring in by a distance x. For the next loading, the spring is compressed a distance $2x$. How much work is required to load the second dart compared with that required to load the first? (a) four times as much (b) two times as much (c) the same (d) half as much (e) one-fourth as much

---

Example 6.5 | Measuring *k* for a Spring

A common technique used to measure the force constant of a spring is demonstrated by the setup in Figure 6.11. The spring is hung vertically (Fig. 6.11a), and an object of mass m is attached to its lower end. Under the action of the "load" mg, the spring stretches a distance d from its equilibrium position (Fig. 6.11b).

(A) If a spring is stretched 2.0 cm by a suspended object having a mass of 0.55 kg, what is the force constant of the spring?

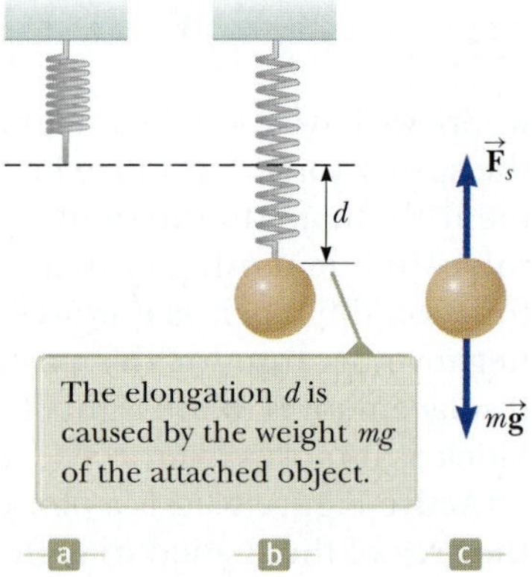

SOLUTION

Conceptualize Figure 6.11b shows what happens to the spring when the object is attached to it. Simulate this situation by hanging an object on a rubber band.

Categorize The object in Figure 6.11b is not accelerating, so it is modeled as a particle in equilibrium.

The elongation d is caused by the weight mg of the attached object.

Figure 6.11 (Example 6.5) Determining the force constant k of a spring.

Analyze Because the object is in equilibrium, the net force on it is zero and the upward spring force balances the downward gravitational force $m\vec{g}$ (Fig. 6.11c).

Apply the particle in equilibrium model to the object:

$$\vec{F}_s + m\vec{g} = 0 \ \rightarrow \ F_s - mg = 0 \ \rightarrow \ F_s = mg$$

Apply Hooke's law to give $F_s = kd$ and solve for k:

$$k = \frac{mg}{d} = \frac{(0.55 \text{ kg})(9.80 \text{ m/s}^2)}{2.0 \times 10^{-2} \text{ m}} = \boxed{2.7 \times 10^2 \text{ N/m}}$$

(B) How much work is done by the spring on the object as it stretches through this distance?

SOLUTION

Use Equation 6.12 to find the work done by the spring on the object:

$$W_s = 0 - \tfrac{1}{2}kd^2 = -\tfrac{1}{2}(2.7 \times 10^2 \text{ N/m})(2.0 \times 10^{-2} \text{ m})^2$$
$$= \boxed{-5.4 \times 10^{-2} \text{ J}}$$

Finalize As the object moves through the 2.0-cm distance, the gravitational force also does work on it. This work is positive because the gravitational force is downward and so is the displacement of the point of application of this force. Based on Equation 6.12 and the discussion afterward, would we expect the work done by the gravitational force to be $+5.4 \times 10^{-2}$ J? Let's find out.

6.5 *cont.*

Evaluate the work done by the gravitational force on the object:

$$W = \vec{\mathbf{F}} \cdot \Delta \vec{\mathbf{r}} = (mg)(d) \cos 0 = mgd$$

$$= (0.55 \text{ kg})(9.80 \text{ m/s}^2)(2.0 \times 10^{-2} \text{ m}) = 1.1 \times 10^{-1} \text{J}$$

If you expected the work done by gravity simply to be that done by the spring with a positive sign, you may be surprised by this result! To understand why that is not the case, we need to explore further, as we do in the next section.

6.5 | Kinetic Energy and the Work–Kinetic Energy Theorem

We have investigated work and identified it as a mechanism for transferring energy into a system. We have stated that work is an influence on a system from the environment, but we have not yet discussed the *result* of this influence on the system. One possible result of doing work on a system is that the system changes its speed. In this section, we investigate this situation and introduce our first type of energy that a system can possess, called *kinetic energy.*

Consider a system consisting of a single object. Figure 6.12 shows a block of mass m moving through a displacement directed to the right under the action of a net force $\Sigma\vec{\mathbf{F}}$, also directed to the right. We know from Newton's second law that the block moves with an acceleration $\vec{\mathbf{a}}$. If the block (and therefore the force) moves through a displacement $\Delta\vec{\mathbf{r}} = \Delta x \hat{\mathbf{i}} = (x_f - x_i)\hat{\mathbf{i}}$, the net work done on the block by the external net force $\Sigma\vec{\mathbf{F}}$ is

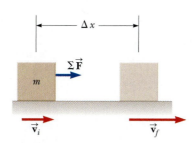

Figure 6.12 An object undergoing a displacement $\Delta\vec{\mathbf{r}} = \Delta x \hat{\mathbf{i}}$ and a change in velocity under the action of a constant net force $\Sigma\vec{\mathbf{F}}$.

$$W_{\text{ext}} = \int_{x_i}^{x_f} \Sigma F \, dx \qquad \qquad \textbf{6.14} \blacktriangleleft$$

Using Newton's second law, we substitute for the magnitude of the net force $\Sigma F = ma$ and then perform the following chain-rule manipulations on the integrand:

$$W_{\text{ext}} = \int_{x_i}^{x_f} ma \, dx = \int_{x_i}^{x_f} m \frac{dv}{dt} \, dx = \int_{x_i}^{x_f} m \frac{dv}{dx}\frac{dx}{dt} \, dx = \int_{v_i}^{v_f} mv \, dv$$

$$W_{\text{ext}} = \tfrac{1}{2}mv_f^2 - \tfrac{1}{2}mv_i^2 \qquad \qquad \textbf{6.15} \blacktriangleleft$$

where v_i is the speed of the block at $x = x_i$ and v_f is its speed at x_f.

Equation 6.15 was generated for the specific situation of one-dimensional motion, but it is a general result. It tells us that the work done by the net force on a particle of mass m is equal to the difference between the initial and final values of a quantity $\tfrac{1}{2}mv^2$. This quantity is so important that it has been given a special name, **kinetic energy:**

$$K \equiv \tfrac{1}{2}mv^2 \qquad \qquad \textbf{6.16} \blacktriangleleft \qquad \blacktriangleright \text{ Kinetic energy}$$

Kinetic energy represents the energy associated with the motion of the particle. Kinetic energy is a scalar quantity and has the same units as work. For example, a 2.0-kg object moving with a speed of 4.0 m/s has a kinetic energy of 16 J. Table 6.1 (page 166) lists the kinetic energies for various objects.

Equation 6.15 states that the work done on a particle by a net force $\Sigma\vec{\mathbf{F}}$ acting on it equals the change in kinetic energy of the particle. It is often convenient to write Equation 6.15 in the form

$$W_{\text{ext}} = K_f - K_i = \Delta K \qquad \qquad \textbf{6.17} \blacktriangleleft$$

Another way to write it is $K_f = K_i + W_{\text{ext}}$, which tells us that the final kinetic energy of an object is equal to its initial kinetic energy plus the change in energy due to the net work done on it.

◀ TABLE 6.1 | Kinetic Energies for Various Objects

Object	Mass (kg)	Speed (m/s)	Kinetic Energy (J)
Earth orbiting the Sun	5.97×10^{24}	2.98×10^4	2.65×10^{33}
Moon orbiting the Earth	7.35×10^{22}	1.02×10^3	3.82×10^{28}
Rocket moving at escape speed[a]	500	1.12×10^4	3.14×10^{10}
Automobile at 65 mi/h	2 000	29	8.4×10^5
Running athlete	70	10	3 500
Stone dropped from 10 m	1.0	14	98
Golf ball at terminal speed	0.046	44	45
Raindrop at terminal speed	3.5×10^{-5}	9.0	1.4×10^{-3}
Oxygen molecule in air	5.3×10^{-26}	500	6.6×10^{-21}

[a]Escape speed is the minimum speed an object must reach near the Earth's surface to move infinitely far away from the Earth.

We have generated Equation 6.17 by imagining doing work on a particle. We could also do work on a deformable system, in which parts of the system move with respect to one another. In this case, we also find that Equation 6.17 is valid as long as the net work is found by adding up the works done by each force and adding, as discussed earlier with regard to Equation 6.8.

Equation 6.17 is an important result known as the **work–kinetic energy theorem:**

▶ Work–kinetic energy theorem

> When work is done on a system and the only change in the system is in its speed, the net work done on the system equals the change in kinetic energy of the system.

The work–kinetic energy theorem indicates that the speed of a system *increases* if the net work done on it is *positive* because the final kinetic energy is greater than the initial kinetic energy. The speed *decreases* if the net work is *negative* because the final kinetic energy is less than the initial kinetic energy.

Because we have so far only investigated translational motion through space, we arrived at the work–kinetic energy theorem by analyzing situations involving translational motion. Another type of motion is *rotational motion,* in which an object spins about an axis. We will study this type of motion in Chapter 10. The work–kinetic energy theorem is also valid for systems that undergo a change in the rotational speed due to work done on the system. The windmill in the photograph at the beginning of this chapter is an example of work causing rotational motion.

The work–kinetic energy theorem will clarify a result seen earlier in this chapter that may have seemed odd. In Section 6.4, we arrived at a result of zero net work done when we let a spring push a block from $x_i = -x_{max}$ to $x_f = x_{max}$. Notice that because the speed of the block is continually changing, it may seem complicated to analyze this process. The quantity ΔK in the work–kinetic energy theorem, however, only refers to the initial and final points for the speeds; it does not depend on details of the path followed between these points. Therefore, because the speed is zero at both the initial and final points of the motion, the net work done on the block is zero. We will often see this concept of path independence in similar approaches to problems.

Let us also return to the mystery in the Finalize step at the end of Example 6.5. Why was the work done by gravity not just the value of the work done by the spring with a positive sign? Notice that the work done by gravity is larger than the magnitude of the work done by the spring. Therefore, the total work done by all forces on the object is positive. Imagine now how to create the situation in which the *only* forces on the object are the spring force and the gravitational force. You must support the object at the highest point and then remove your hand and let the object fall. If you do so, you know that when the object reaches a position 2.0 cm below your hand, it will be *moving,* which is consistent with Equation 6.17. Positive net work is

Pitfall Prevention | 6.5
Conditions for the Work–Kinetic Energy Theorem
The work–kinetic energy theorem is important but limited in its application; it is not a general principle. In many situations, other changes in the system occur besides its speed, and there are other interactions with the environment besides work. A more general principle involving energy is *conservation of energy* in Section 7.1.

Pitfall Prevention | 6.6
The Work–Kinetic Energy Theorem: Speed, Not Velocity
The work–kinetic energy theorem relates work to a change in the *speed* of a system, not a change in its velocity. For example, if an object is in uniform circular motion, its speed is constant. Even though its velocity is changing, no work is done on the object by the force causing the circular motion.

done on the object, and the result is that it has a kinetic energy as it passes through the 2.0-cm point.

The only way to prevent the object from having a kinetic energy after moving through 2.0 cm is to slowly lower it with your hand. Then, however, there is a third force doing work on the object, the normal force from your hand. If this work is calculated and added to that done by the spring force and the gravitational force, the net work done on the object is zero, which is consistent because the object is not moving at the 2.0-cm point.

Earlier, we indicated that work can be considered as a mechanism for transferring energy into a system. Equation 6.17 is a mathematical statement of this concept. When work W_{ext} is done on a system, the result is a transfer of energy across the boundary of the system. The result on the system, in the case of Equation 6.17, is a change ΔK in kinetic energy. In the next section, we investigate another type of energy that can be stored in a system as a result of doing work on the system.

QUICK QUIZ 6.5 A dart is inserted into a spring-loaded dart gun by pushing the spring in by a distance x. For the next loading, the spring is compressed a distance $2x$. How much faster does the second dart leave the gun compared with the first? (**a**) four times as fast (**b**) two times as fast (**c**) the same (**d**) half as fast (**e**) one-fourth as fast

---

THINKING PHYSICS 6.1

A man wishes to load a refrigerator onto a truck using a ramp at angle θ as shown in Figure 6.13. He claims that less work would be required to load the truck if the length L of the ramp were increased. Is his claim valid?

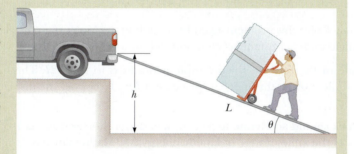

Figure 6.13 (Thinking Physics 6.1) A refrigerator attached to a frictionless, wheeled hand truck is moved up a ramp at constant speed.

Reasoning No. Suppose the refrigerator is wheeled on a hand truck up the ramp at constant speed. In this case, for the system of the refrigerator and the hand truck, $\Delta K = 0$. The normal force exerted by the ramp on the system is directed at 90° to the displacement of its point of application and so does no work on the system. Because $\Delta K = 0$, the work–kinetic energy theorem gives

$$W_{ext} = W_{by\ man} + W_{by\ gravity} = 0$$

The work done by the gravitational force equals the product of the weight mg of the system, the distance L through which the refrigerator is displaced, and $\cos(\theta + 90°)$. Therefore,

$$W_{by\ man} = -W_{by\ gravity} = -(mg)(L)[\cos(\theta + 90°)]$$

$$= mgL \sin\theta = mgh$$

where $h = L \sin\theta$ is the height of the ramp. Therefore, the man must do the same amount of work mgh on the system *regardless* of the length of the ramp. The work depends only on the height of the ramp. Although less force is required with a longer ramp, the point of application of that force moves through a greater displacement. ◀

Example 6.6 | A Block Pulled on a Frictionless Surface

A 6.0-kg block initially at rest is pulled to the right along a frictionless, horizontal surface by a constant horizontal force of 12 N. Find the block's speed after it has moved 3.0 m.

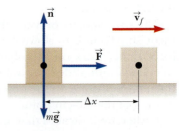

SOLUTION

Conceptualize Figure 6.14 illustrates this situation. Imagine pulling a toy car across a table with a horizontal rubber band attached to the front of the car. The force is maintained constant by ensuring that the stretched rubber band always has the same length.

Figure 6.14 (Example 6.6) A block pulled to the right on a frictionless surface by a constant horizontal force.

Categorize We could apply the equations of kinematics to determine the answer, but let us practice the energy approach. The block is the system, and three external forces act on the system. The normal force balances the gravitational force on the block, and neither of these vertically acting forces does work on the block because their points of application are horizontally displaced.

Analyze The net external force acting on the block is the horizontal 12-N force.

Use the work–kinetic energy theorem for the block, noting that its initial kinetic energy is zero:

$$W_{ext} = K_f - K_i = \tfrac{1}{2}mv_f^2 - 0 = \tfrac{1}{2}mv_f^2$$

Solve for v_f and use Equation 6.1 for the work done on the block by $\vec{F}$:

$$v_f = \sqrt{\frac{2W_{ext}}{m}} = \sqrt{\frac{2F\Delta x}{m}}$$

Substitute numerical values:

$$v_f = \sqrt{\frac{2(12\text{ N})(3.0\text{ m})}{6.0\text{ kg}}} = 3.5 \text{ m/s}$$

Finalize It would be useful for you to solve this problem again by modeling the block as a particle under a net force to find its acceleration and then as a particle under constant acceleration to find its final velocity.

What If? Suppose the magnitude of the force in this example is doubled to $F' = 2F$. The 6.0-kg block accelerates to 3.5 m/s due to this applied force while moving through a displacement $\Delta x'$. How does the displacement $\Delta x'$ compare with the original displacement Δx?

Answer If we pull harder, the block should accelerate to a given speed in a shorter distance, so we expect that $\Delta x' < \Delta x$. In both cases, the block experiences the same change in kinetic energy ΔK. Mathematically, from the work–kinetic energy theorem, we find that

$$W_{ext} = F'\Delta x' = \Delta K = F\Delta x$$

$$\Delta x' = \frac{F}{F'}\Delta x = \frac{F}{2F}\Delta x = \tfrac{1}{2}\Delta x$$

and the distance is shorter as suggested by our conceptual argument.

6.6 | Potential Energy of a System

So far in this chapter, we have defined a system in general, but have focused our attention primarily on single particles or objects under the influence of external forces. Let us now consider systems of two or more particles or objects interacting via a force that is *internal* to the system. The kinetic energy of such a system is the algebraic sum of the kinetic energies of all members of the system. There may be systems, however, in which one object is so massive that it can be modeled as stationary and its kinetic energy can be neglected. For example, if we consider a ball–Earth system as the ball falls to the Earth, the kinetic energy of the system can be considered as just the kinetic energy of the ball. The Earth moves so slowly in this process that we can ignore its kinetic energy. On the other hand, the kinetic energy of a system of two electrons must include the kinetic energies of both particles.

Imagine a system consisting of a book and the Earth, interacting via the gravitational force. We do some work on the system by lifting the book slowly from rest through a vertical displacement $\Delta \vec{\mathbf{r}} = (y_f - y_i)\hat{\mathbf{j}}$ as in Active Figure 6.15. According to our discussion of work as an energy transfer, this work done on the system must appear as an increase in energy of the system. The book is at rest before we perform the work and is at rest after we perform the work. Therefore, there is no change in the kinetic energy of the system.

Because the energy change of the system is not in the form of kinetic energy, it must appear as some other form of energy storage. After lifting the book, we could release it and let it fall back to the position y_i. Notice that the book (and therefore, the system) now has kinetic energy and that its source is in the work that was done in lifting the book. While the book was at the highest point, the system had the *potential* to possess kinetic energy, but it did not do so until the book was allowed to fall. Therefore, we call the energy storage mechanism before the book is released **potential energy.** We will find that the potential energy of a system can only be associated with specific types of forces acting between members of a system. The amount of potential energy in the system is determined by the *configuration* of the system. Moving members of the system to different positions or rotating them may change the configuration of the system and therefore its potential energy.

Let us now derive an expression for the potential energy associated with an object at a given location above the surface of the Earth. Consider an external agent lifting an object of mass m from an initial height y_i above the ground to a final height y_f as in Active Figure 6.15. We assume the lifting is done slowly, with no acceleration, so the applied force from the agent is equal in magnitude to the gravitational force on the object: the object is modeled as a particle in equilibrium moving at constant velocity. The work done by the external agent on the system (object and the Earth) as the object undergoes this upward displacement is given by the product of the upward applied force $\vec{\mathbf{F}}_{app}$ and the upward displacement of this force, $\Delta \vec{\mathbf{r}} = \Delta y \hat{\mathbf{j}}$:

$$W_{ext} = (\vec{\mathbf{F}}_{app}) \cdot \Delta \vec{\mathbf{r}} = (mg\hat{\mathbf{j}}) \cdot [(y_f - y_i)\hat{\mathbf{j}}] = mgy_f - mgy_i \qquad \textbf{6.18} \blacktriangleleft$$

This result is the net work done on the system because the applied force is the only force on the system from the environment. (Remember that the gravitational force is *internal* to the system.) Notice the similarity between Equation 6.18 and Equation 6.15. In each equation, the work done on a system equals a difference between the final and initial values of a quantity. In Equation 6.15, the work represents a transfer of energy into the system and the increase in energy of the system is kinetic in form. In Equation 6.18, the work represents a transfer of energy into the system and the system energy appears in a different form, which we have called potential energy.

Therefore, we can identify the quantity mgy as the **gravitational potential energy** U_g:

$$U_g \equiv mgy \qquad \textbf{6.19} \blacktriangleleft$$

The units of gravitational potential energy are joules, the same as the units of work and kinetic energy. Potential energy, like work and kinetic energy, is a scalar quantity. Notice that Equation 6.19 is valid only for objects near the surface of the Earth, where g is approximately constant.[3]

Using our definition of gravitational potential energy, Equation 6.18 can now be rewritten as

$$W_{ext} = \Delta U_g \qquad \textbf{6.20} \blacktriangleleft$$

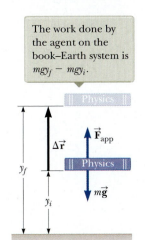

The work done by the agent on the book–Earth system is $mgy_f - mgy_i$.

Active Figure 6.15 An external agent lifts a book slowly from a height y_i to a height y_f.

Pitfall Prevention | 6.7
Potential Energy
The phrase *potential energy* does not refer to something that has the potential to become energy. Potential energy *is* energy.

Pitfall Prevention | 6.8
Potential Energy Belongs to a System
Potential energy is always associated with a *system* of two or more interacting objects. When a small object moves near the surface of the Earth under the influence of gravity, we may sometimes refer to the potential energy "associated with the object" rather than the more proper "associated with the system" because the Earth does not move significantly. We will not, however, refer to the potential energy "of the object" because this wording ignores the role of the Earth.

▶ Gravitational potential energy

---

[3]The assumption that g is constant is valid as long as the vertical displacement of the object is small compared with the Earth's radius.

which mathematically describes that the net external work done on the system in this situation appears as a change in the gravitational potential energy of the system.

Gravitational potential energy depends only on the vertical height of the object above the surface of the Earth. The same amount of work must be done on an object–Earth system whether the object is lifted vertically from the Earth or is pushed starting from the same point up a frictionless incline, ending up at the same height. We verified this statement for a specific situation of rolling a refrigerator up a ramp in Thinking Physics 6.1. This statement can be shown to be true in general by calculating the work done on an object by an agent moving the object through a displacement having both vertical and horizontal components:

$$W_{ext} = (\vec{F}_{app}) \cdot \Delta\vec{r} = (mg\hat{j}) \cdot [(x_f - x_i)\hat{i} + (y_f - y_i)\hat{j}] = mgy_f - mgy_i$$

where there is no term involving x in the final result because $\hat{j} \cdot \hat{i} = 0$.

In solving problems, you must choose a reference configuration for which the gravitational potential energy of the system is set equal to some reference value, which is normally zero. The choice of reference configuration is completely arbitrary because the important quantity is the *difference* in potential energy, and this difference is independent of the choice of reference configuration.

It is often convenient to choose as the reference configuration for zero gravitational potential energy the configuration in which an object is at the surface of the Earth, but this choice is not essential. Often, the statement of the problem suggests a convenient configuration to use.

QUICK QUIZ 6.6 Choose the correct answer. The gravitational potential energy of a system (a) is always positive (b) is always negative (c) can be negative or positive

Example 6.7 | The Proud Athlete and the Sore Toe

A trophy being shown off by a careless athlete slips from the athlete's hands and drops on his toe. Choosing floor level as the $y = 0$ point of your coordinate system, estimate the change in gravitational potential energy of the trophy–Earth system as the trophy falls. Repeat the calculation, using the top of the athlete's head as the origin of coordinates.

SOLUTION

Conceptualize The trophy changes its vertical position with respect to the surface of the Earth. Associated with this change in position is a change in the gravitational potential energy of the trophy–Earth system.

Categorize We evaluate a change in gravitational potential energy defined in this section, so we categorize this example as a substitution problem. Because there are no numbers, it is also an estimation problem.

The problem statement tells us that the reference configuration of the trophy–Earth system corresponding to zero potential energy is when the bottom of the trophy is at the floor. To find the change in potential energy for the system, we need to estimate a few values. Let's say the trophy has a mass of approximately 2 kg, and the top of a person's toe is about 0.03 m above the floor. Also, let's assume the trophy falls from a height of 0.5 m.

Calculate the gravitational potential energy of the trophy–Earth system just before the trophy is released:

$$U_i = mgy_i = (2 \text{ kg})(9.80 \text{ m/s}^2)(0.5 \text{ m}) = 9.80 \text{ J}$$

Calculate the gravitational potential energy of the trophy–Earth system when the trophy reaches the athlete's toe:

$$U_f = mgy_f = (2 \text{ kg})(9.80 \text{ m/s}^2)(0.03 \text{ m}) = 0.588 \text{ J}$$

Evaluate the change in gravitational potential energy of the trophy–Earth system:

$$\Delta U_g = 0.588 \text{ J} - 9.80 \text{ J} = -9.21 \text{ J}$$

We should probably keep only one digit because of the roughness of our estimates; therefore, we estimate that the change in gravitational potential energy is −9 J . The system had about 10 J of gravitational potential energy before the trophy began its fall and approximately 1 J of potential energy as the trophy reaches the top of the toe.

6.7 *cont.*

The second case presented indicates that the reference configuration of the system for zero potential energy is chosen to be when the trophy is at the athlete's head (even though the trophy is never at this position in its motion). We estimate this position to be 1.50 m above the floor.

Calculate the gravitational potential energy of the trophy–Earth system just before the trophy is released from its position 1 m below the athlete's head:

$$U_i = mgy_i = (2\ \text{kg})(9.80\ \text{m/s}^2)(-1\ \text{m}) = -19.6\ \text{J}$$

Calculate the gravitational potential energy of the trophy–Earth system when the trophy reaches the athlete's toe located 1.47 m below the athlete's head:

$$U_f = mgy_f = (2\ \text{kg})(9.80\ \text{m/s}^2)(-1.47\ \text{m}) = -28.8\ \text{J}$$

Evaluate the change in gravitational potential energy of the trophy–Earth system:

$$\Delta U_g = -28.8\ \text{J} - (-19.6\ \text{J}) = -9.2\ \text{J} \approx \boxed{-9\ \text{J}}$$

This value is the same as before, as it must be.

Elastic Potential Energy

Because members of a system can interact with one another by means of different types of forces, it is possible that there are different types of potential energy in a system. We are familiar with gravitational potential energy of a system in which members interact via the gravitational force. Let's explore a second type of potential energy that a system can possess.

Consider a system consisting of a block and a spring as shown in Active Figure 6.16 (page 172). In Section 6.4, we identified *only* the block as the system. Now we include both the block and the spring in the system and recognize that the spring force is the interaction between the two members of the system. The force that the spring exerts on the block is given by $F_s = -kx$ (Eq. 6.9). The work done by an external applied force F_{app} on a system consisting of a block connected to the spring is given by Equation 6.13:

$$W_{ext} = \tfrac{1}{2}kx_f^2 - \tfrac{1}{2}kx_i^2 \qquad \textbf{6.21}\blacktriangleleft$$

In this situation, the initial and final x coordinates of the block are measured from its equilibrium position, $x = 0$. Again (as in the gravitational case) we see that the work done on the system is equal to the difference between the initial and final values of an expression related to the system's configuration. The **elastic potential energy** function associated with the block–spring system is defined by

$$U_s \equiv \tfrac{1}{2}kx^2 \qquad \textbf{6.22}\blacktriangleleft \qquad \blacktriangleright \text{ Elastic potential energy}$$

The elastic potential energy of the system can be thought of as the energy stored in the deformed spring (one that is either compressed or stretched from its equilibrium position). The elastic potential energy stored in a spring is zero whenever the spring is undeformed ($x = 0$). Energy is stored in the spring only when the spring is either stretched or compressed. Because the elastic potential energy is proportional to x^2, we see that U_s is always positive in a deformed spring. Everyday examples of the storage of elastic potential energy can be found in old-style clocks or watches that operate from a wound-up spring and small wind-up toys for children.

Consider Active Figure 6.16, which shows a spring on a frictionless, horizontal surface. When a block is pushed against the spring by an external agent, the elastic potential energy and the total energy of the system increase as indicated in Figure 6.16b. When the spring is compressed a distance x_{max} (Active Fig. 6.16c), the elastic potential energy stored in the spring is $\tfrac{1}{2}kx_{max}^2$. When the block is released from rest, the spring exerts a force on the block and pushes the block to the right.

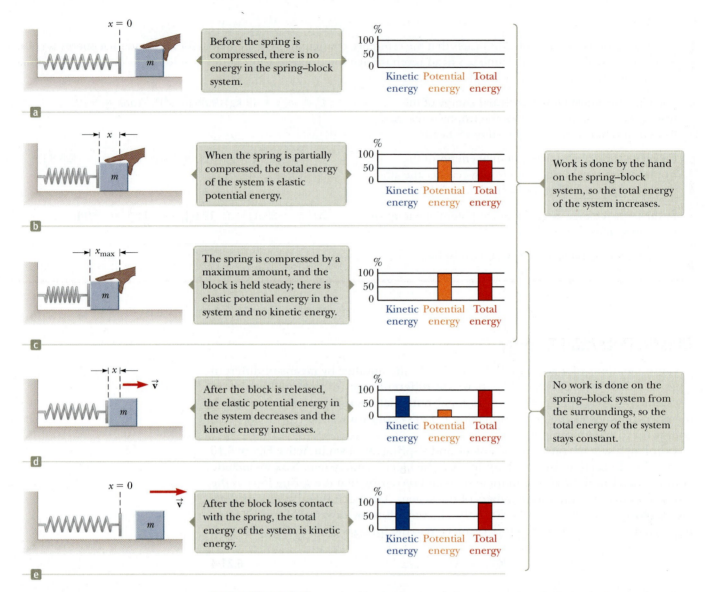

Before the spring is compressed, there is no energy in the spring–block system.

When the spring is partially compressed, the total energy of the system is elastic potential energy.

The spring is compressed by a maximum amount, and the block is held steady; there is elastic potential energy in the system and no kinetic energy.

After the block is released, the elastic potential energy in the system decreases and the kinetic energy increases.

After the block loses contact with the spring, the total energy of the system is kinetic energy.

Work is done by the hand on the spring–block system, so the total energy of the system increases.

No work is done on the spring–block system from the surroundings, so the total energy of the system stays constant.

Active Figure 6.16 A spring on a frictionless, horizontal surface is compressed a distance x_{max} when a block of mass m is pushed against it. The block is then released and the spring pushes it to the right, where the block eventually loses contact with the spring. Parts (a) through (e) show various instants in the process. Energy bar charts on the right of each part of the figure help keep track of the energy in the system.

The elastic potential energy of the system decreases, whereas the kinetic energy increases and the total energy remains fixed (Fig. 6.16d). When the spring returns to its original length, the stored elastic potential energy is completely transformed into kinetic energy of the block (Active Fig. 6.16e).

Energy Bar Charts

Active Figure 6.16 shows an important graphical representation of information related to energy of systems called an **energy bar chart.** The vertical axis represents the amount of energy of a given type in the system. The horizontal axis shows the types of energy in the system. The bar chart in Active Figure 6.16a shows that the system contains zero energy because the spring is relaxed and the block is not moving. Between Active Figure 6.16a and Active Figure 6.16c, the hand does work on the system, compressing the spring and storing elastic potential energy in the system. In Active Figure 6.16d, the block has been released and is moving to the right while still in contact with the spring. The height of the bar for the elastic potential energy of the system decreases, the kinetic energy bar increases, and the total energy bar remains fixed. In Active Figure 6.16e, the spring has returned to its relaxed length and the system now contains only kinetic energy associated with the moving block.

Energy bar charts can be a very useful representation for keeping track of the various types of energy in a system. For practice, try making energy bar charts for the book–Earth system in Active Figure 6.15 when the book is dropped from the higher position. Figure 6.17 associated with Quick Quiz 6.7 shows another system for which drawing an energy bar chart would be a good exercise. We will show energy bar charts in some figures in this chapter. Some Active Figures will not show a bar chart in the text but will include one in the animation in Enhanced WebAssign.

> **QUICK QUIZ 6.7** A ball is connected to a light spring suspended vertically as shown in Figure 6.17. When pulled downward from its equilibrium position and released, the ball oscillates up and down. **(i)** In the system of *the ball, the spring, and the Earth,* what forms of energy are there during the motion? (a) kinetic and elastic potential (b) kinetic and gravitational potential (c) kinetic, elastic potential, and gravitational potential (d) elastic potential and gravitational potential **(ii)** In the system of *the ball and the spring,* what forms of energy are there during the motion? Choose from the same possibilities (a) through (d).

6.7 | Conservative and Nonconservative Forces

We now introduce a third type of energy that a system can possess. Imagine that the book in Active Figure 6.18a has been accelerated by your hand and is now sliding to the right on the surface of a heavy table and slowing down due to the friction force. Suppose the *surface* is the system. Then the friction force from the sliding book does work on the surface. The force on the surface is to the right and the displacement of the point of application of the force is to the right because the book has moved to the right. The work done on the surface is positive, but the surface is not moving after the book has stopped. Positive work has been done on the surface, yet there is no increase in the surface's kinetic energy or the potential energy of any system.

From your everyday experience with sliding over surfaces with friction, you can probably guess that the surface will be *warmer* after the book slides over it. (Rub your hands together briskly to find out!) The work that was done on the surface has gone into warming the surface rather than increasing its speed or changing the configuration of a system. We call the energy associated with the temperature of a system its **internal energy,** symbolized E_{int}. (We will define internal energy more generally in Chapter 17.) In this case, the work done on the surface does indeed represent energy transferred into the system, but it appears in the system as internal energy rather than kinetic or potential energy.

Consider the book and the surface in Active Figure 6.18a together as a system. Initially, the system has kinetic energy because the book is moving. While the book is sliding, the internal energy of the system increases: the book and the surface are warmer than before. When the book stops, the kinetic energy has been completely transformed to internal energy. We can consider the work done by friction within the system—that is, between the book and the surface—as a *transformation mechanism* for energy. This work transforms the kinetic energy of the system into internal energy. Similarly, when a book falls straight down with no air resistance, the work done by the gravitational force within the book–Earth system transforms gravitational potential energy of the system to kinetic energy.

Active Figures 6.18b through 6.18d show energy bar charts for the situation in Active Figure 6.18a. In Active Figure 6.18b, the bar chart shows that the system

Figure 6.17 (Quick Quiz 6.7) A ball connected to a massless spring suspended vertically. What forms of potential energy are associated with the system when the ball is displaced downward?

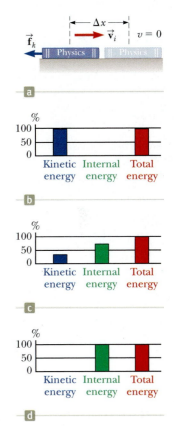

Active Figure 6.18 (a) A book sliding to the right on a horizontal surface slows down in the presence of a force of kinetic friction acting to the left. (b) An energy bar chart showing the energy in the system of the book and the surface at the initial instant of time. The energy of the system is all kinetic energy. (c) While the book is sliding, the kinetic energy of the system decreases as it is transformed to internal energy. (d) After the book has stopped, the energy of the system is all internal energy.

contains kinetic energy at the instant the book is released by your hand. We define the reference amount of internal energy in the system as zero at this instant. Active Figure 6.18c shows the kinetic energy transforming to internal energy as the book slows down due to the friction force. In Active Figure 6.18d, after the book has stopped sliding, the kinetic energy is zero and the system now contains only internal energy. Notice that the total energy bar in red has not changed during the process. The amount of internal energy in the system after the book has stopped is equal to the amount of kinetic energy in the system at the initial instant. This equality is described by an important principle called *conservation of energy*. We will explore this principle in Chapter 7.

Now consider in more detail an object moving downward near the surface of the Earth. The work done by the gravitational force on the object does not depend on whether it falls vertically or slides down a sloping incline with friction. All that matters is the change in the object's elevation. The energy transformation to internal energy due to friction on that incline, however, depends very much on the distance the object slides. The longer the incline, the more potential energy is transformed to internal energy. In other words, the path makes no difference when we consider the work done by the gravitational force, but it does make a difference when we consider the energy transformation due to friction forces. We can use this varying dependence on path to classify forces as either conservative or nonconservative. Of the two forces just mentioned, the gravitational force is conservative and the friction force is nonconservative.

Conservative Forces

Conservative forces have these two equivalent properties:

▶ Properties of conservative forces

1. The work done by a conservative force on a particle moving between any two points is independent of the path taken by the particle.
2. The work done by a conservative force on a particle moving through any closed path is zero. (A closed path is one for which the beginning point and the endpoint are identical.)

The gravitational force is one example of a conservative force; the force that an ideal spring exerts on any object attached to the spring is another. The work done by the gravitational force on an object moving between any two points near the Earth's surface is $W_g = -mg\hat{\mathbf{j}} \cdot [(y_f - y_i)\hat{\mathbf{j}}] = mgy_i - mgy_f$. From this equation, notice that W_g depends only on the initial and final y coordinates of the object and hence is independent of the path. Furthermore, W_g is zero when the object moves over any closed path (where $y_i = y_f$).

For the case of the object–spring system, the work W_s done by the spring force is given by $W_s = \frac{1}{2}kx_i^2 - \frac{1}{2}kx_f^2$ (Eq. 6.12). We see that the spring force is conservative because W_s depends only on the initial and final x coordinates of the object and is zero for any closed path.

We can associate a potential energy for a system with a force acting between members of the system, but we can do so only if the force is conservative. In general, the work W_{int} done by a conservative force on an object that is a member of a system as the system changes from one configuration to another is equal to the initial value of the potential energy of the system minus the final value:

$$W_{int} = U_i - U_f = -\Delta U \qquad \textbf{6.23} \blacktriangleleft$$

We use the subscript "int" in Equation 6.23 to remind us that the work we are discussing is done by one member of the system on another member and is therefore *internal* to the system. It is different from the work W_{ext} done *on* the system as a whole by an external agent. As an example, compare Equation 6.23 with the specific equation for the work done by the spring force (Eq. 6.12) as the extension of the spring changes.

Pitfall Prevention | 6.9
Similar Equation Warning
Compare Equation 6.23 with Equation 6.20. These equations are similar except for the negative sign, which is a common source of confusion. Equation 6.20 tells us that positive work done *by an outside agent* on a system causes an increase in the potential energy of the system (with no change in the kinetic or internal energy). Equation 6.23 states that work done *on a component of a system by a conservative force internal to the system* causes a decrease in the potential energy of the system.

Nonconservative Forces

A force is **nonconservative** if it does not satisfy properties 1 and 2 for conservative forces. We define the sum of the kinetic and potential energies of a system as the **mechanical energy** of the system:

$$E_{mech} \equiv K + U \qquad \qquad \textbf{6.24} \blacktriangleleft$$

where K includes the kinetic energy of all moving members of the system and U includes all types of potential energy in the system. For a book falling under the action of the gravitational force, the mechanical energy of the book–Earth system remains fixed; gravitational potential energy transforms to kinetic energy, and the total mechanical energy of the system remains constant. Nonconservative forces acting within a system, however, cause a *change* in the mechanical energy of the system. For example, for a book sent sliding on a horizontal surface that is not frictionless, the mechanical energy of the book–surface system is transformed to internal energy as we discussed earlier. Only part of the book's kinetic energy is transformed to internal energy in the book. The rest appears as internal energy in the surface. (When you trip and slide across a gymnasium floor, not only does the skin on your knees warm up, so does the floor!) Because the force of kinetic friction transforms the mechanical energy of a system into internal energy, it is a nonconservative force.

   As an example of the path dependence of the work for a nonconservative force, consider Figure 6.19. Suppose you displace a book between two points on a table. If the book is displaced in a straight line along the blue path between points Ⓐ and Ⓑ in Figure 6.19, you do a certain amount of work against the kinetic friction force to keep the book moving at a constant speed. Now, imagine that you push the book along the brown semicircular path in Figure 6.19. You perform more work against friction along this curved path than along the straight path because the curved path is longer. The work done on the book depends on the path, so the friction force *cannot* be conservative.

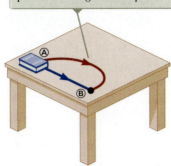

The work done in moving the book is greater along the brown path than along the blue path.

Figure 6.19 The work done against the force of kinetic friction depends on the path taken as the book is moved from Ⓐ to Ⓑ.

6.8 | Relationship Between Conservative Forces and Potential Energy

In the preceding section, we found that the work done on a member of a system by a conservative force between the members of the system does not depend on the path taken by the moving member. The work depends only on the initial and final coordinates. For such a system, we can define a **potential energy function** U such that the work done within the system by the conservative force equals the negative of the change in the potential energy of the system. Let us imagine a system of particles in which a conservative force $\vec{F}$ acts between the particles. Imagine also that the configuration of the system changes due to the motion of one particle along the x axis. The work done by the force $\vec{F}$ as the particle moves along the x axis is[4]

$$W_{int} = \int_{x_i}^{x_f} F_x\, dx = -\Delta U \qquad \qquad \textbf{6.25} \blacktriangleleft$$

where F_x is the component of $\vec{F}$ in the direction of the displacement. That is, the work done by a conservative force acting between members of a system equals the negative of the change in the potential energy of the system associated with that force when the system's configuration changes. We can also express Equation 6.25 as

$$\Delta U = U_f - U_i = -\int_{x_i}^{x_f} F_x\, dx \qquad \qquad \textbf{6.26} \blacktriangleleft$$

---

[4]For a general displacement, the work done in two or three dimensions also equals $-\Delta U$, where $U = U(x, y, z)$. We write this equation formally as $W_{int} = \int_i^f \vec{F} \cdot d\vec{r} = U_i - U_f$.

Therefore, ΔU is negative when F_x and dx are in the same direction, as when an object is lowered in a gravitational field or when a spring pushes an object toward equilibrium.

It is often convenient to establish some particular location x_i of one member of a system as representing a reference configuration and measure all potential energy differences with respect to it. We can then define the potential energy function as

$$U_f(x) = -\int_{x_i}^{x_f} F_x \, dx + U_i \qquad \textbf{6.27} \blacktriangleleft$$

The value of U_i is often taken to be zero for the reference configuration. It does not matter what value we assign to U_i because any nonzero value merely shifts $U_f(x)$ by a constant amount and only the *change* in potential energy is physically meaningful.

If the point of application of the force undergoes an infinitesimal displacement dx, we can express the infinitesimal change in the potential energy of the system dU as

$$dU = -F_x \, dx$$

Therefore, the conservative force is related to the potential energy function through the relationship[5]

$$\boxed{F_x = -\frac{dU}{dx}} \qquad \textbf{6.28} \blacktriangleleft$$

That is, the x component of a conservative force acting on a member within a system equals the negative derivative of the potential energy of the system with respect to x.

We can easily check Equation 6.28 for the two examples already discussed. In the case of the deformed spring, $U_s = \frac{1}{2}kx^2$; therefore,

$$F_s = -\frac{dU_s}{dx} = -\frac{d}{dx}\left(\frac{1}{2}kx^2\right) = -kx$$

which corresponds to the restoring force in the spring (Hooke's law). Because the gravitational potential energy function is $U_g = mgy$, it follows from Equation 6.28 that $F_g = -mg$ when we differentiate U_g with respect to y instead of x.

We now see that U is an important function because a conservative force can be derived from it. Furthermore, Equation 6.28 should clarify that adding a constant to the potential energy is unimportant because the derivative of a constant is zero.

> ◀ Relation of force between members of a system to the potential energy of the system

QUICK QUIZ 6.8 What does the slope of a graph of $U(x)$ versus x represent? (**a**) the magnitude of the force on the object (**b**) the negative of the magnitude of the force on the object (**c**) the x component of the force on the object (**d**) the negative of the x component of the force on the object

6.9 | Potential Energy for Gravitational and Electric Forces

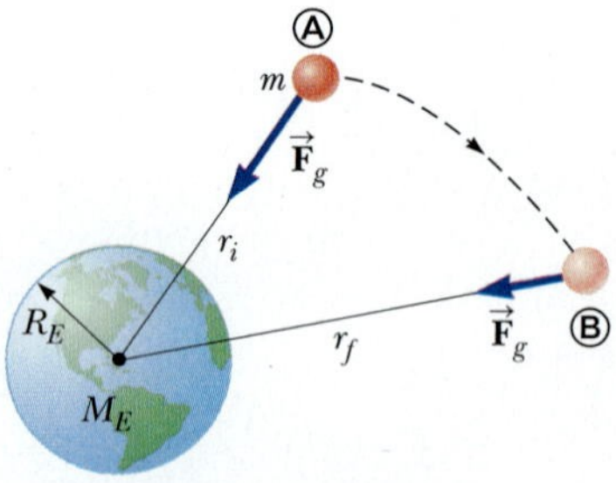

Earlier in this chapter we introduced the concept of gravitational potential energy, that is, the energy associated with a system of objects interacting via the gravitational force. We emphasized that the gravitational potential energy function, Equation 6.19, is valid only when the object of mass m is near the Earth's surface. We would like to find a more general expression for the gravitational potential energy that is valid for

Figure 6.20 As a particle of mass m moves from Ⓐ to Ⓑ above the Earth's surface, the potential energy of the particle–Earth system, given by Equation 6.31, changes because of the change in the particle–Earth separation distance r from r_i to r_f.

[5]In three dimensions, the expression is

$$\vec{F} = -\frac{\partial U}{\partial x}\hat{i} - \frac{\partial U}{\partial y}\hat{j} - \frac{\partial U}{\partial z}\hat{k}$$

where $(\partial U/\partial x)$ and so forth are partial derivatives. In the language of vector calculus, $\vec{F}$ equals the negative of the *gradient* of the scalar quantity $U(x, y, z)$.

all separation distances. Because the value of g varies with height, it follows that the general dependence of the potential energy function of the system on separation distance is more complicated than our simple expression, Equation 6.19.

Consider a particle of mass m moving between two points Ⓐ and Ⓑ above the Earth's surface as in Figure 6.20. The gravitational force on the particle due to the Earth, first introduced in Section 5.5, can be written in vector form as

$$\vec{F}_g = -G\frac{M_E m}{r^2}\hat{r}$$ **6.29**◀

where $\hat{r}$ is a unit vector directed from the Earth toward the particle and the negative sign indicates that the force is downward toward the Earth. This expression shows that the gravitational force depends on the radial coordinate r. Furthermore, the gravitational force is conservative. Equation 6.27 gives

$$U_f = -\int_{r_i}^{r_f} F(r)\,dr + U_i = GM_E m \int_{r_i}^{r_f}\frac{dr}{r^2} + U_i = GM_E m\left(-\frac{1}{r}\right)\Big|_{r_i}^{r_f} + U_i$$

or

$$U_f = -GM_E m\left(\frac{1}{r_f} - \frac{1}{r_i}\right) + U_i$$ **6.30**◀

As always, the choice of a reference configuration for the potential energy is completely arbitrary. It is customary to define the reference configuration as that for which the force is zero. Letting $U_i \to 0$ as $r_i \to \infty$, we obtain the important result

$$U_g = -G\frac{M_E m}{r}$$ **6.31**◀

for separation distances $r > R_E$, the radius of the Earth. Because of our choice of the reference configuration for zero potential energy, the function U_g is always negative (Fig. 6.21).

Although Equation 6.31 was derived for the particle–Earth system, it can be applied to *any* two particles. For *any pair* of particles of masses m_1 and m_2 separated by a distance r, the gravitational force of attraction is given by Equation 5.11 and the gravitational potential energy of the system of two particles is

$$U_g = -G\frac{m_1 m_2}{r}$$ **6.32**◀

This expression also applies to larger objects *if their mass distributions are spherically symmetric*, as first shown by Newton. In this case, r is measured between the centers of the spherical objects.

Equation 6.32 shows that the gravitational potential energy for any pair of particles varies as $1/r$ (whereas the force between them varies as $1/r^2$). Furthermore, the potential energy is *negative* because the force is attractive and we have chosen the potential energy to be zero when the particle separation is infinity. Because the force between the particles is attractive, we know that an external agent must do positive work to increase the separation between the two particles. The work done by the external agent produces an increase in the potential energy as the two particles are separated. That is, U_g becomes less negative as r increases.

We can extend this concept to three or more particles. In this case, the total potential energy of the system is the sum over all *pairs* of particles. Each pair contributes a term of the form given by Equation 6.32. For example, if the system contains three particles, as in Figure 6.22, we find that

$$U_{total} = U_{12} + U_{13} + U_{23} = -G\left(\frac{m_1 m_2}{r_{12}} + \frac{m_1 m_3}{r_{13}} + \frac{m_2 m_3}{r_{23}}\right)$$ **6.33**◀

The absolute value of U_{total} represents the work needed to separate all three particles by an infinite distance.

Pitfall Prevention | 6.10
What is r?
In Section 5.5, we discussed the gravitational force between two *particles*. In Equation 6.29, we present the gravitational force between a particle and an extended object, the Earth. We could also express the gravitational force between two extended objects, such as the Earth and the Sun. In these kinds of situations, remember that r is measured *between the centers* of the objects. Be sure *not* to measure r from the surface of the Earth.

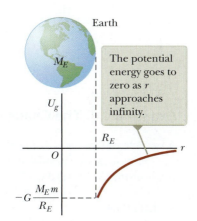

Figure 6.21 Graph of the gravitational potential energy U_g versus r for a particle above the Earth's surface.

Pitfall Prevention | 6.11
Gravitational Potential Energy
Be careful! Equation 6.32 looks similar to Equation 5.11 for the gravitational force, but there are two major differences. The gravitational force is a vector, whereas the gravitational potential energy is a scalar. The gravitational force varies as the *inverse square* of the separation distance, whereas the gravitational potential energy varies as the simple *inverse* of the separation distance.

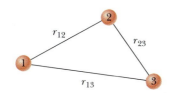

Figure 6.22 Three interacting particles.

> **THINKING PHYSICS 6.2**
>
> Why is the Sun hot?
>
> **Reasoning** The Sun was formed when a cloud of gas and dust coalesced, because of gravitational attraction, into a massive astronomical object. Let us define this cloud as our system and model the gas and dust as particles. Initially, the particles of the system were widely scattered, representing a large amount of gravitational potential energy. As the particles moved together to form the Sun, the gravitational potential energy of the system decreased. This potential energy was transformed to kinetic energy as the particles fell toward the center. As the speeds of the particles increased, many collisions occurred between particles, randomizing their motion and transforming the kinetic energy to internal energy, which represented an increase in temperature. As the particles came together, the temperature rose to a point at which nuclear reactions occurred. These reactions release huge amounts of energy that maintain the high temperature of the Sun. This process has occurred for every star in the Universe. ◀

Example 6.8 | The Change in Potential Energy

A particle of mass m is displaced through a small vertical distance Δy near the Earth's surface. Show that in this situation the general expression for the change in gravitational potential energy given by Equation 6.30 reduces to the familiar relationship $\Delta U = mg\,\Delta y$.

SOLUTION

Conceptualize Compare the two different situations for which we have developed expressions for gravitational potential energy: (1) a planet and an object that are far apart for which the energy expression is Equation 6.30 and (2) a small object at the surface of a planet for which the energy expression is Equation 6.19. We wish to show that these two expressions are equivalent.

Categorize This example is a substitution problem.

Combine the fractions in Equation 6.30:

$$(1) \quad \Delta U = -GM_E m\left(\frac{1}{r_f} - \frac{1}{r_i}\right) = GM_E m\left(\frac{r_f - r_i}{r_i r_f}\right)$$

Evaluate $r_f - r_i$ and $r_i r_f$ if both the initial and final positions of the particle are close to the Earth's surface:

$$r_f - r_i = \Delta y \qquad r_i r_f \approx R_E^2$$

Substitute these expressions into Equation (1):

$$\Delta U \approx \frac{GM_E m}{R_E^2}\,\Delta y$$

Use Equation 6.29 to express $GM_E m / R_E^2$ as the magnitude of the gravitational force F_g on an object of mass m at the Earth's surface:

$$\Delta U \approx F_g \Delta y$$

Use Equation 4.5 to express the gravitational force in terms of the acceleration due to gravity:

$$\Delta U \approx mg\Delta y$$

In Chapter 5, we discussed the electrostatic force between two point particles, which is given by Coulomb's law,

$$F_e = k_e \frac{q_1 q_2}{r^2} \qquad \textbf{6.34} \blacktriangleleft$$

Because this expression looks so similar to Newton's law of universal gravitation, we would expect that the generation of a potential energy function for this force would proceed in a similar way. That is indeed the case, and this procedure results in the **electric potential energy** function,

$$U_e = k_e \frac{q_1 q_2}{r}$$

6.35 ◀

As with the gravitational potential energy, the electric potential energy is defined as zero when the charges are infinitely far apart. Comparing this expression with that for the gravitational potential energy, we see the obvious differences in the constants and the use of charges instead of masses, but there is one more difference. The gravitational expression has a negative sign, but the electrical expression doesn't. For systems of objects that experience an attractive force, the potential energy decreases as the objects are brought closer together. Because we have defined zero potential energy at infinite separation, all real separations are finite and the energy must decrease from a value of zero. Therefore, all potential energies for systems of objects that attract must be negative. In the gravitational case, attraction is the only possibility. The constant, the masses, and the separation distance are all positive, so the negative sign must be included explicitly, as it is in Equation 6.32.

The electric force can be either attractive or repulsive. Attraction occurs between charges of opposite sign. Therefore, for the two charges in Equation 6.35, one is positive and one is negative if the force is attractive. The product of the charges provides the negative sign for the potential energy mathematically, and we do not need an explicit negative sign in the potential energy expression. In the case of charges with the same sign, either a product of two negative charges or two positive charges will be positive, leading to a positive potential energy. This conclusion is reasonable because to cause repelling particles to move together from infinite separation requires work to be done on the system, so the potential energy increases.

6.10 | Energy Diagrams and Equilibrium of a System

The motion of a system can often be understood qualitatively through a graph of its potential energy versus the position of a member of the system. Consider the potential energy function for a block–spring system, given by $U_s = \frac{1}{2}kx^2$. This function is plotted versus x in Active Figure 6.23a, where x is the position of the block. The force F_s exerted by the spring on the block is related to U_s through Equation 6.28:

$$F_s = -\frac{dU_s}{dx} = -kx$$

As we saw in Quick Quiz 6.8, the x component of the force is equal to the negative of the slope of the U-versus-x curve. When the block is placed at rest at the equilibrium position of the spring ($x = 0$), where $F_s = 0$, it will remain there unless some external force F_{ext} acts on it. If this external force stretches the spring from equilibrium, x is positive and the slope dU/dx is positive; therefore, the force F_s exerted by the spring is negative and the block accelerates back toward $x = 0$ when released. If the external force compresses the spring, x is negative and the slope is negative; therefore, F_s is positive and again the mass accelerates toward $x = 0$ upon release.

From this analysis, we conclude that the $x = 0$ position for a block–spring system is one of **stable equilibrium.** That is, any movement away from this position results in a force directed back toward $x = 0$. In general, configurations of a system in stable equilibrium correspond to those for which $U(x)$ for the system is a minimum.

If the block in Active Figure 6.23 is moved to an initial position x_{max} and then released from rest, its total energy initially is the potential energy $\frac{1}{2}kx_{max}^2$ stored in the spring. As the block starts to move, the system acquires kinetic energy and loses potential energy. The block oscillates (moves back and forth) between the two

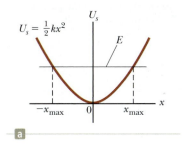

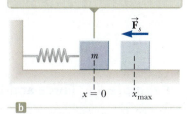

The restoring force exerted by the spring always acts toward $x = 0$, the position of stable equilibrium.

Active Figure 6.23 (a) Potential energy as a function of x for the frictionless block–spring system shown in (b). For a given energy E of the system, the block oscillates between the turning points, which have the coordinates $x = \pm x_{max}$.

Pitfall Prevention | 6.12
Energy Diagrams
A common mistake is to think that potential energy on the graph in an energy diagram represents the height of some object. For example, that is not the case in Active Figure 6.23, where the block is only moving horizontally.

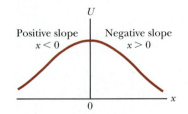

Figure 6.24 A plot of U versus x for a particle that has a position of unstable equilibrium located at $x = 0$. For any finite displacement of the particle, the force on the particle is directed away from $x = 0$.

points $x = -x_{max}$ and $x = +x_{max}$, called the *turning points*. In fact, because no energy is transformed to internal energy due to friction, the block oscillates between $-x_{max}$ and $+x_{max}$ forever. (We will discuss these oscillations further in Chapter 12.)

Another simple mechanical system with a configuration of stable equilibrium is a ball rolling about in the bottom of a bowl. Anytime the ball is displaced from its lowest position, it tends to return to that position when released.

Now consider a particle moving along the x axis under the influence of a conservative force F_x, where the U-versus-x curve is as shown in Figure 6.24. Once again, $F_x = 0$ at $x = 0$, and so the particle is in equilibrium at this point. This position, however, is one of **unstable equilibrium** for the following reason. Suppose the particle is displaced to the right ($x > 0$). Because the slope is negative for $x > 0$, $F_x = -dU/dx$ is positive and the particle accelerates away from $x = 0$. If instead the particle is at $x = 0$ and is displaced to the left ($x < 0$), the force is negative because the slope is positive for $x < 0$ and the particle again accelerates away from the equilibrium position. The position $x = 0$ in this situation is one of unstable equilibrium because for any displacement from this point, the force pushes the particle farther away from equilibrium and toward a position of lower potential energy. A pencil balanced on its point is in a position of unstable equilibrium. If the pencil is displaced slightly from its absolutely vertical position and is then released, it will surely fall over. In general, configurations of a system in unstable equilibrium correspond to those for which $U(x)$ for the system is a maximum.

Finally, a configuration called **neutral equilibrium** arises when U is constant over some region. Small displacements of an object from a position in this region produce neither restoring nor disrupting forces. A ball lying on a flat, horizontal surface is an example of an object in neutral equilibrium.

Example 6.9 | Force and Energy on an Atomic Scale

The potential energy associated with the force between two neutral atoms in a molecule can be modeled by the Lennard–Jones potential energy function:

$$U(x) = 4\epsilon\left[\left(\frac{\sigma}{x}\right)^{12} - \left(\frac{\sigma}{x}\right)^{6}\right]$$

where x is the separation of the atoms. The function $U(x)$ contains two parameters σ and ϵ that are determined from experiments. Sample values for the interaction between two atoms in a molecule are $\sigma = 0.263$ nm and $\epsilon = 1.51 \times 10^{-22}$ J. Using a spreadsheet or similar tool, graph this function and find the most likely distance between the two atoms.

SOLUTION

Conceptualize We identify the two atoms in the molecule as a system. Based on our understanding that stable molecules exist, we expect to find stable equilibrium when the two atoms are separated by some equilibrium distance.

Categorize Because a potential energy function exists, we categorize the force between the atoms as conservative. For a conservative force, Equation 6.28 describes the relationship between the force and the potential energy function.

....

Analyze Stable equilibrium exists for a separation distance at which the potential energy of the system of two atoms (the molecule) is a minimum.

Take the derivative of the function $U(x)$:

$$\frac{dU(x)}{dx} = 4\epsilon\frac{d}{dx}\left[\left(\frac{\sigma}{x}\right)^{12} - \left(\frac{\sigma}{x}\right)^{6}\right] = 4\epsilon\left[\frac{-12\sigma^{12}}{x^{13}} + \frac{6\sigma^{6}}{x^{7}}\right]$$

Minimize the function $U(x)$ by setting its derivative equal to zero:

$$4\epsilon\left[\frac{-12\sigma^{12}}{x_{eq}^{13}} + \frac{6\sigma^{6}}{x_{eq}^{7}}\right] = 0 \rightarrow x_{eq} = (2)^{1/6}\sigma$$

Evaluate x_{eq}, the equilibrium separation of the two atoms in the molecule:

$$x_{eq} = (2)^{1/6}(0.263 \text{ nm}) = \boxed{2.95 \times 10^{-10} \text{ m}}$$

6.9 *cont.*

We graph the Lennard–Jones function on both sides of this critical value to create our energy diagram as shown in Figure 6.25.

Finalize Notice that $U(x)$ is extremely large when the atoms are very close together, is a minimum when the atoms are at their critical separation, and then increases again as the atoms move apart. When $U(x)$ is a minimum, the atoms are in stable equilibrium, indicating that the most likely separation between them occurs at this point.

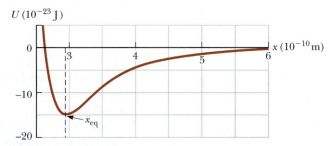

Figure 6.25 (Example 6.9) Potential energy curve associated with a molecule. The distance x is the separation between the two atoms making up the molecule.

6.11 | Context Connection: Potential Energy in Fuels

Fuel represents a storage mechanism for potential energy to be used to make a vehicle move. The standard fuel for automobiles for several decades has been *gasoline.* Gasoline is refined from crude oil that is present in the Earth. This oil represents the decay products of plant life that existed on the Earth, primarily from 100 to 600 million years ago. The source of energy in crude oil is hydrocarbons produced from molecules in the ancient plants.

The primary chemical reactions occurring in an internal combustion engine involve the oxidation of carbon and hydrogen:

$$C + O_2 \rightarrow CO_2$$

$$4H + O_2 \rightarrow 2H_2O$$

Both reactions release energy that is used to operate the automobile.

Notice the final products in these reactions. One is water, which is not harmful to the environment. Carbon dioxide, however, contributes to the greenhouse effect, which leads to global warming, which we will study in Context 5. The incomplete combustion of carbon and oxygen can form CO, carbon monoxide, which is a poisonous gas. Because air contains other elements besides oxygen, other harmful emission products, such as oxides of nitrogen, exist.

The amount of potential energy stored in a fuel and available from the fuel is typically called the *heat of combustion,* even though this term is a misuse of the word *heat.* For automotive gasoline, this value is about 44 MJ/kg. Because the efficiency of the engine is not 100%, only part of this energy eventually finds its way into kinetic energy of the car. We will study efficiencies of engines in Context 5.

Another common fuel is *diesel fuel.* The heat of combustion for diesel fuel is 42.5 MJ/kg, slightly lower than that for gasoline. Diesel engines, however, operate at a higher efficiency than gasoline engines, so they can extract a larger percentage of the available energy.

A number of additional fuels have been developed to operate internal combustion engines with minimal modifications. They are described briefly below.

Ethanol

Ethanol is the most widely used alternative fuel and is used for commercial fleet vehicles and increasingly for private vehicles. It is an alcohol made from such crops as corn, wheat, and barley. Because these crops can be grown, ethanol is renewable. The use of ethanol reduces carbon monoxide and carbon dioxide emissions compared with the use of normal gasoline.

Ethanol is mixed with gasoline to form the following mixtures:

E10: 10% ethanol, 90% gasoline

E85: 85% ethanol, 15% gasoline

The energy content of E85 is about 70% of that for gasoline, so the miles per gallon ratio will be lower than that for a vehicle powered by straight gasoline. On the other hand, the renewable nature of ethanol counteracts this disadvantage significantly.

You may have seen some automobiles labeled as "FLEXFUEL." These vehicles can operate from ethanol fuel over a continuous range from pure gasoline to E85. Whatever the mixture, the fuel is stored in a single fuel tank and sensors in the fuel system determine the amount of ethanol, automatically adjusting the fuel injection and spark timing accordingly.

Biodiesel

Biodiesel fuel is formed by a chemical reaction between alcohol and oils from field crops as well as vegetable oil, fat, and grease from commercial sources. Pacific Biodiesel in Hawaii makes biodiesel from used restaurant cooking oil, providing a usable fuel as well as diverting this used oil from landfills.

Biodiesel is available in the following forms:

B20: 20% biodiesel, 80% gasoline

B100: 100% biodiesel

B100 is nontoxic and biodegradable. The use of biodiesel reduces environmentally harmful tailpipe emissions significantly. Furthermore, tests have shown that the emission of cancer-causing particulate matter is reduced by 94% with the use of pure biodiesel.

The energy content of B100 is about 90% of that for conventional diesel. As with ethanol, the renewable nature of biodiesel counteracts this disadvantage significantly.

Natural Gas

Natural gas is a fossil fuel, originating from gas wells or as a by-product of the refining process for crude oil. It is primarily methane (CH_4), with smaller amounts of nitrogen, ethane, propane, and other gases. It burns cleanly and generates much lower amounts of harmful tailpipe emissions than gasoline. Natural gas vehicles are used in many fleets of buses, delivery trucks, and refuse haulers.

Although ethanol and biodiesel mixtures can be used in conventional engines with minimal modifications, a natural gas engine is much more heavily modified. In addition, the gas must be carried on board the vehicle in one of two ways that require higher-level technology than a simple fuel tank. One possibility is to liquefy the gas, requiring a well-insulated storage container to keep the gas at $-190°C$. The other possibility is to compress the gas to about 200 times atmospheric pressure and carry it in the vehicle in a high-pressure storage tank.

The energy content of natural gas is 48 MJ/kg, a bit higher than that for gasoline. Note that natural gas, like gasoline, is *not* a renewable source.

Propane

Propane is available commercially as liquefied petroleum gas, which is actually a mixture of propane, propylene, butane, and butylenes. It is a by-product of natural gas processing and refining of crude oil. Propane is the most widely accessible alternative fuel, with fueling facilities in all states of the United States.

Tailpipe emissions for propane-fueled vehicles are significantly lower than those for gasoline-powered vehicles. Tests show that carbon monoxide is reduced by 30% to 90%.

As with natural gas, high-pressure tanks are necessary to carry the fuel. In addition, propane is a nonrenewable resource. The energy content of propane is 46 MJ/kg, slightly higher than that of gasoline.

Electric Vehicles

In the Context introduction before Chapter 2, we discussed the electric cars that were on the roadways in the early part of the 20th century. As mentioned, these electric cars virtually disappeared around the 1920s due to several factors. One was that oil was plentiful during the 20th century and there was little incentive to operate vehicles on anything other than gasoline or diesel.

In the early 1970s, difficulties arose with regard to the availability of oil from the Middle East, leading to shortages at gas stations. At this time, interest arose anew in electric-powered vehicles. An early attempt to market a new electric vehicle was the Electrovette, an electric version of the Chevrolet Chevette.

Although the oil crisis eased somewhat, political instabilities in the Middle East created uncertainty in the availability of oil and interest in electric cars continued, albeit on a small scale. In the late 1980s, General Motors developed a prototype called the Impact, an electric car that could accelerate from 0 to 60 in 8 s and had a drag coefficient of 0.19, much lower than that of traditional cars. The Impact was the hit of the 1990 Los Angeles Auto Show. In the 1990s, the Impact became commercially available as the EV1. General Motors canceled the EV1 program in 2001 and recalled the vehicles. While a few of the EV1 vehicles were retained for museums, the vast majority of the vehicles were destroyed by being crushed.

Two major disadvantages of electric cars are the limited range, 70 to 100 mi, on a single charging of the batteries, and the several hours of time required to recharge the batteries. Despite these difficulties, new electric vehicles are available for the public, including the Nissan Leaf, discussed in Section 2.8, and the Tesla Roadster, a high-priced electric sports car that can accelerate from rest to 60 mi/h in 3.7 s. In addition, the Chevrolet Volt, also discussed in Section 2.8, is operated as an electric car over short trips. It solves the limited-range and charging-time problems for longer trips by incorporating a gasoline engine to charge the battery once the original charge has been depleted.

▶ SUMMARY

A **system** is most often a single particle, a collection of particles, or a region of space, and may vary in size and shape. A **system boundary** separates the system from the **environment.**

The **work** W done on a system by an agent exerting a constant force $\vec{F}$ on the system is the product of the magnitude Δr of the displacement of the point of application of the force and the component $F\cos\theta$ of the force along the direction of the displacement $\Delta\vec{r}$:

$$W \equiv F\Delta r\cos\theta \qquad \text{6.1} ◀$$

The **scalar product** (dot product) of two vectors $\vec{A}$ and $\vec{B}$ is defined by the relationship

$$\vec{A}\cdot\vec{B} \equiv AB\cos\theta \qquad \text{6.2} ◀$$

where the result is a scalar quantity and θ is the angle between the two vectors. The scalar product obeys the commutative and distributive laws.

If a varying force does work on a particle as the particle moves along the x axis from x_i to x_f, the work done by the force on the particle is given by

$$W = \int_{x_i}^{x_f} F_x\,dx \qquad \text{6.7} ◀$$

where F_x is the component of force in the x direction.

The **kinetic energy** of a particle of mass m moving with a speed v is

$$K \equiv \tfrac{1}{2}mv^2 \qquad \text{6.16} ◀$$

The **work–kinetic energy theorem** states that if work is done on a system by external forces and the only change in the system is in its speed,

$$W_{ext} = K_f - K_i = \Delta K = \tfrac{1}{2}mv_f^2 - \tfrac{1}{2}mv_i^2 \qquad \text{6.15, 6.17} ◀$$

If a particle of mass m is at a distance y above the Earth's surface, the **gravitational potential energy** of the particle–Earth system is

$$U_g \equiv mgy \qquad \textbf{6.19}\blacktriangleleft$$

The **elastic potential energy** stored in a spring of force constant k is

$$U_s \equiv \tfrac{1}{2}kx^2 \qquad \textbf{6.22}\blacktriangleleft$$

The **total mechanical energy of a system** is defined as the sum of the kinetic energy and the potential energy:

$$E_{\text{mech}} \equiv K + U \qquad \textbf{6.24}\blacktriangleleft$$

A force is **conservative** if the work it does on a particle that is a member of the system as the particle moves between two points is independent of the path the particle takes between the two points. Furthermore, a force is conservative if the work it does on a particle is zero when the particle moves through an arbitrary closed path and returns to its initial position. A force that does not meet these criteria is said to be **nonconservative.**

A **potential energy function** U can be associated only with a conservative force. If a conservative force $\vec{\textbf{F}}$ acts between members of a system while one member moves along the x axis from x_i to x_f, the change in the potential energy of the system equals the negative of the work done by that force:

$$U_f - U_i = -\int_{x_i}^{x_f} F_x\, dx \qquad \textbf{6.26}\blacktriangleleft$$

Systems can be in three types of equilibrium configurations when the net force on a member of the system is zero. Configurations of **stable equilibrium** correspond to those for which $U(x)$ is a minimum. Configurations of **unstable equilibrium** correspond to those for which $U(x)$ is a maximum. **Neutral equilibrium** arises when U is constant as a member of the system moves over some region.

▶ OBJECTIVE QUESTIONS |

1. Alex and John are loading identical cabinets onto a truck. Alex lifts his cabinet straight up from the ground to the bed of the truck, whereas John slides his cabinet up a rough ramp to the truck. Which statement is correct about the work done on the cabinet–Earth system? (a) Alex and John do the same amount of work. (b) Alex does more work than John. (c) John does more work than Alex. (d) None of those statements is necessarily true because the force of friction is unknown. (e) None of those statements is necessarily true because the angle of the incline is unknown.

2. Is the work required to be done by an external force on an object on a frictionless, horizontal surface to accelerate it from a speed v to a speed $2v$ (a) equal to the work required to accelerate the object from $v = 0$ to v, (b) twice the work required to accelerate the object from $v = 0$ to v, (c) three times the work required to accelerate the object from $v = 0$ to v, (d) four times the work required to accelerate the object from 0 to v, or (e) not known without knowledge of the acceleration?

3. A worker pushes a wheelbarrow with a horizontal force of 50 N on level ground over a distance of 5.0 m. If a friction force of 43 N acts on the wheelbarrow in a direction opposite that of the worker, what work is done on the wheelbarrow by the worker? (a) 250 J (b) 215 J (c) 35 J (d) 10 J (e) None of those answers is correct.

4. Mark and David are loading identical cement blocks onto David's pickup truck. Mark lifts his block straight up from the ground to the truck, whereas David slides his block up a ramp containing frictionless rollers. Which statement is true about the work done on the block–Earth system? (a) Mark does more work than David. (b) Mark and David do the same amount of work. (c) David does more work than Mark.

(d) None of those statements is necessarily true because the angle of the incline is unknown. (e) None of those statements is necessarily true because the mass of one block is not given.

5. Bullet 2 has twice the mass of bullet 1. Both are fired so that they have the same speed. If the kinetic energy of bullet 1 is K, is the kinetic energy of bullet 2 (a) $0.25K$, (b) $0.5K$, (c) $0.71K$, (d) K, or (e) $2K$?

6. As a simple pendulum swings back and forth, the forces acting on the suspended object are (a) the gravitational force, (b) the tension in the supporting cord, and (c) air resistance. (i) Which of these forces, if any, does no work on the pendulum at any time? (ii) Which of these forces does negative work on the pendulum at all times during its motion?

7. A block of mass m is dropped from the fourth floor of an office building and hits the sidewalk below at speed v. From what floor should the mass be dropped to double that impact speed? (a) the sixth floor (b) the eighth floor (c) the tenth floor (d) the twelfth floor (e) the sixteenth floor

8. If the net work done by external forces on a particle is zero, which of the following statements about the particle must be true? (a) Its velocity is zero. (b) Its velocity is decreased. (c) Its velocity is unchanged. (d) Its speed is unchanged. (e) More information is needed.

9. Let $\hat{\textbf{N}}$ represent the direction horizontally north, $\widehat{\textbf{NE}}$ represent northeast (halfway between north and east), and so on. Each direction specification can be thought of as a unit vector. Rank from the largest to the smallest the following dot products. Note that zero is larger than a negative number. If two quantities are equal, display that fact in your ranking. (a) $\hat{\textbf{N}} \cdot \hat{\textbf{N}}$ (b) $\hat{\textbf{N}} \cdot \widehat{\textbf{NE}}$ (c) $\hat{\textbf{N}} \cdot \hat{\textbf{S}}$ (d) $\hat{\textbf{N}} \cdot \hat{\textbf{E}}$ (e) $\widehat{\textbf{SE}} \cdot \hat{\textbf{S}}$

10. Figure OQ6.10 shows a light extended spring exerting a force F_s to the left on a block. (**i**) Does the block exert a force on the spring? Choose every correct answer. (a) No, it doesn't. (b) Yes, it does, to the left. (c) Yes, it does, to the right. (d) Yes, it does, and its magnitude is larger than F_s. (e) Yes, it does, and its magnitude is equal to F_s. (**ii**) Does the spring exert a force on the wall? Choose your answers from the same list (a) through (e).

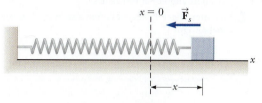

Figure OQ6.10

11. If the speed of a particle is doubled, what happens to its kinetic energy? (a) It becomes four times larger. (b) It becomes two times larger. (c) It becomes $\sqrt{2}$ times larger. (d) It is unchanged. (e) It becomes half as large.

12. A cart is set rolling across a level table, at the same speed on every trial. If it runs into a patch of sand, the cart exerts on the sand an average horizontal force of 6 N and travels a distance of 6 cm through the sand as it comes to a stop. If instead the cart runs into a patch of gravel on which the cart exerts an average horizontal force of 9 N, how far into the gravel will the cart roll before stopping? (a) 9 cm (b) 6 cm (c) 4 cm (d) 3 cm (e) none of those answers

13. A cart is set rolling across a level table, at the same speed on every trial. If it runs into a patch of sand, the cart exerts on the sand an average horizontal force of 6 N and travels a distance of 6 cm through the sand as it comes to a stop. If instead the cart runs into a patch of flour, it rolls an average of 18 cm before stopping. What is the average magnitude of the horizontal force the cart exerts on the flour? (a) 2 N (b) 3 N (c) 6 N (d) 18 N (e) none of those answers

14. A certain spring that obeys Hooke's law is stretched by an external agent. The work done in stretching the spring by 10 cm is 4 J. How much additional work is required to stretch the spring an additional 10 cm? (a) 2 J (b) 4 J (c) 8 J (d) 12 J (e) 16 J

15. (**i**) Rank the gravitational accelerations you would measure for the following falling objects: (a) a 2-kg object 5 cm above the floor, (b) a 2-kg object 120 cm above the floor, (c) a 3-kg object 120 cm above the floor, and (d) a 3-kg object 80 cm above the floor. List the one with the largest magnitude of acceleration first. If any are equal, show their equality in your list. (**ii**) Rank the gravitational forces on the same four objects, listing the one with the largest magnitude first. (**iii**) Rank the gravitational potential energies (of the object–Earth system) for the same four objects, largest first, taking $y = 0$ at the floor.

16. An ice cube has been given a push and slides without friction on a level table. Which is correct? (a) It is in stable equilibrium. (b) It is in unstable equilibrium. (c) It is in neutral equilibrium. (d) It is not in equilibrium.

CONCEPTUAL QUESTIONS

☐ denotes answer available in *Student Solutions Manual/Study Guide*

1. Discuss whether any work is being done by each of the following agents and, if so, whether the work is positive or negative. (a) a chicken scratching the ground (b) a person studying (c) a crane lifting a bucket of concrete (d) the gravitational force on the bucket in part (c) (e) the leg muscles of a person in the act of sitting down

2. Discuss the work done by a pitcher throwing a baseball. What is the approximate distance through which the force acts as the ball is thrown?

3. A certain uniform spring has spring constant k. Now the spring is cut in half. What is the relationship between k and the spring constant k' of each resulting smaller spring? Explain your reasoning.

4. (a) For what values of the angle θ between two vectors is their scalar product positive? (b) For what values of θ is their scalar product negative?

5. Can kinetic energy be negative? Explain.

6. Cite two examples in which a force is exerted on an object without doing any work on the object.

7. Does the kinetic energy of an object depend on the frame of reference in which its motion is measured? Provide an example to prove this point.

8. If only one external force acts on a particle, does it necessarily change the particle's (a) kinetic energy? (b) Its velocity?

9. Can a normal force do work? If not, why not? If so, give an example.

10. You are reshelving books in a library. You lift a book from the floor to the top shelf. The kinetic energy of the book on the floor was zero and the kinetic energy of the book on the top shelf is zero, so no change occurs in the kinetic energy, yet you did some work in lifting the book. Is the work–kinetic energy theorem violated? Explain.

11. A student has the idea that the total work done on an object is equal to its final kinetic energy. Is this idea true always, sometimes, or never? If it is sometimes true, under what circumstances? If it is always or never true, explain why.

12. Object 1 pushes on object 2 as the objects move together, like a bulldozer pushing a stone. Assume object 1 does 15.0 J of work on object 2. Does object 2 do work on object 1? Explain your answer. If possible, determine how much work and explain your reasoning.

▷ PROBLEMS

Section 6.2 Work Done by a Constant Force

1. In 1990, Walter Arfeuille of Belgium lifted a 281.5-kg object through a distance of 17.1 cm using only his teeth. (a) How much work was done on the object by Arfeuille in this lift, assuming the object was lifted at constant speed? (b) What total force was exerted on Arfeuille's teeth during the lift?

2. **W** A raindrop of mass 3.35×10^{-5} kg falls vertically at constant speed under the influence of gravity and air resistance. Model the drop as a particle. As it falls 100 m, what is the work done on the raindrop (a) by the gravitational force and (b) by air resistance?

3. **M** A block of mass $m = 2.50$ kg is pushed a distance $d = 2.20$ m along a frictionless, horizontal table by a constant applied force of magnitude $F = 16.0$ N directed at an angle $\theta = 25.0°$ below the horizontal as shown in Figure P6.3. Determine the work done on the block by (a) the applied force, (b) the normal force exerted by the table, (c) the gravitational force, and (d) the net force on the block.

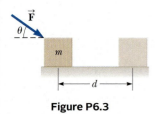

Figure P6.3

4. The record number of boat lifts, including the boat and its ten crew members, was achieved by Sami Heinonen and Juha Räsänen of Sweden in 2000. They lifted a total mass of 653.2 kg approximately 4 in. off the ground a total of 24 times. Estimate the total work done by the two men on the boat in this record lift, ignoring the negative work done by the men when they lowered the boat back to the ground.

5. **M** Spiderman, whose mass is 80.0 kg, is dangling on the free end of a 12.0-m-long rope, the other end of which is fixed to a tree limb above. By repeatedly bending at the waist, he is able to get the rope in motion, eventually getting it to swing enough that he can reach a ledge when the rope makes a 60.0° angle with the vertical. How much work was done by the gravitational force on Spiderman in this maneuver?

6. **Q|C** A shopper in a supermarket pushes a cart with a force of 35 N directed at an angle of 25° below the horizontal. The force is just sufficient to balance various friction forces, so the cart moves at constant speed. (a) Find the work done by the shopper on the cart as she moves down a 50.0-m-long aisle. (b) What is the net work done on the cart by all forces?

Why? (c) The shopper goes down the next aisle, pushing horizontally and maintaining the same speed as before. If the friction force doesn't change, would the shopper's applied force be larger, smaller, or the same? (d) What about the work done on the cart by the shopper?

Section 6.3 The Scalar Product of Two Vectors

7. Vector $\vec{\mathbf{A}}$ has a magnitude of 5.00 units, and vector $\vec{\mathbf{B}}$ has a magnitude of 9.00 units. The two vectors make an angle of 50.0° with each other. Find $\vec{\mathbf{A}} \cdot \vec{\mathbf{B}}$.

8. **S** For any two vectors $\vec{\mathbf{A}}$ and $\vec{\mathbf{B}}$, show that $\vec{\mathbf{A}} \cdot \vec{\mathbf{B}} = A_x B_x + A_y B_y + A_z B_z$. *Suggestions:* Write $\vec{\mathbf{A}}$ and $\vec{\mathbf{B}}$ in unit-vector form and use Equations 6.4 and 6.5.

Note: In Problems 9 through 12, calculate numerical answers to three significant figures as usual.

9. **M** A force $\vec{\mathbf{F}} = (6\hat{\mathbf{i}} - 2\hat{\mathbf{j}})$ N acts on a particle that undergoes a displacement $\Delta\vec{\mathbf{r}} = (3\hat{\mathbf{i}} + \hat{\mathbf{j}})$ m. Find (a) the work done by the force on the particle and (b) the angle between $\vec{\mathbf{F}}$ and $\Delta\vec{\mathbf{r}}$.

10. Find the scalar product of the vectors in Figure P6.10.

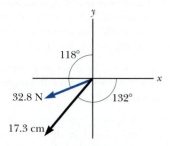

Figure P6.10

11. **W** For $\vec{\mathbf{A}} = 3\hat{\mathbf{i}} + \hat{\mathbf{j}} - \hat{\mathbf{k}}$, $\vec{\mathbf{B}} = -\hat{\mathbf{i}} + 2\hat{\mathbf{j}} + 5\hat{\mathbf{k}}$, and $\vec{\mathbf{C}} = 2\hat{\mathbf{j}} - 3\hat{\mathbf{k}}$, find $\vec{\mathbf{C}} \cdot (\vec{\mathbf{A}} - \vec{\mathbf{B}})$.

12. Using the definition of the scalar product, find the angles between (a) $\vec{\mathbf{A}} = 3\hat{\mathbf{i}} - 2\hat{\mathbf{j}}$ and $\vec{\mathbf{B}} = 4\hat{\mathbf{i}} - 4\hat{\mathbf{j}}$, (b) $\vec{\mathbf{A}} = -2\hat{\mathbf{i}} + 4\hat{\mathbf{j}}$ and $\vec{\mathbf{B}} = 3\hat{\mathbf{i}} - 4\hat{\mathbf{j}} + 2\hat{\mathbf{k}}$, and (c) $\vec{\mathbf{A}} = \hat{\mathbf{i}} - 2\hat{\mathbf{j}} + 2\hat{\mathbf{k}}$ and $\vec{\mathbf{B}} = 3\hat{\mathbf{j}} + 4\hat{\mathbf{k}}$.

13. Let $\vec{\mathbf{B}} = 5.00$ m at 60.0°. Let the vector $\vec{\mathbf{C}}$ have the same magnitude as $\vec{\mathbf{A}}$ and a direction angle greater than that of $\vec{\mathbf{A}}$ by 25.0°. Let $\vec{\mathbf{A}} \cdot \vec{\mathbf{B}} = 30.0$ m² and $\vec{\mathbf{B}} \cdot \vec{\mathbf{C}} = 35.0$ m². Find the magnitude and direction of $\vec{\mathbf{A}}$.

Section 6.4 Work Done by a Varying Force

14. [W] The force acting on a particle varies as shown in Figure P6.14. Find the work done by the force on the particle as it moves (a) from $x = 0$ to $x = 8.00$ m, (b) from $x = 8.00$ m to $x = 10.0$ m, and (c) from $x = 0$ to $x = 10.0$ m.

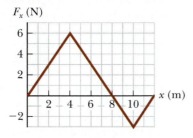

Figure P6.14

15. [W] A particle is subject to a force F_x that varies with position as shown in Figure P6.15. Find the work done by the force on the particle as it moves (a) from $x = 0$ to $x = 5.00$ m, (b) from $x = 5.00$ m to $x = 10.0$ m, and (c) from $x = 10.0$ m to $x = 15.0$ m. (d) What is the total work done by the force over the distance $x = 0$ to $x = 15.0$ m?

Figure P6.15 Problems 15 and 32.

16. An archer pulls her bowstring back 0.400 m by exerting a force that increases uniformly from zero to 230 N. (a) What is the equivalent spring constant of the bow? (b) How much work does the archer do on the string in drawing the bow?

17. [M] When a 4.00-kg object is hung vertically on a certain light spring that obeys Hooke's law, the spring stretches 2.50 cm. If the 4.00-kg object is removed, (a) how far will the spring stretch if a 1.50-kg block is hung on it? (b) How much work must an external agent do to stretch the same spring 4.00 cm from its unstretched position?

18. [S] A small particle of mass m is pulled to the top of a frictionless half-cylinder (of radius R) by a light cord that passes over the top of the cylinder as illustrated in Figure P6.18. (a) Assuming the particle moves at a constant speed, show that $F = mg \cos \theta$. *Note:* If the particle moves at constant speed, the component of its acceleration tangent to the cylinder must be zero at all times. (b) By directly integrating $W = \int \vec{\mathbf{F}} \cdot d\vec{\mathbf{r}}$, find the work done in moving the particle at constant speed from the bottom to the top of the half-cylinder.

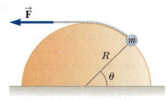

Figure P6.18

19. A light spring with spring constant 1 200 N/m is hung from an elevated support. From its lower end hangs a second light spring, which has spring constant 1 800 N/m. An object of mass 1.50 kg is hung at rest from the lower end of the second spring. (a) Find the total extension distance of the

pair of springs. (b) Find the effective spring constant of the pair of springs as a system. We describe these springs as *in series.*

20. [S] A light spring with spring constant k_1 is hung from an elevated support. From its lower end a second light spring is hung, which has spring constant k_2. An object of mass m is hung at rest from the lower end of the second spring. (a) Find the total extension distance of the pair of springs. (b) Find the effective spring constant of the pair of springs as a system.

21. In a control system, an accelerometer consists of a 4.70-g object sliding on a calibrated horizontal rail. A low-mass spring attaches the object to a flange at one end of the rail. Grease on the rail makes static friction negligible, but rapidly damps out vibrations of the sliding object. When subject to a steady acceleration of $0.800g$, the object should be at a location 0.500 cm away from its equilibrium position. Find the force constant of the spring required for the calibration to be correct.

22. [S] Express the units of the force constant of a spring in SI fundamental units.

23. A cafeteria tray dispenser supports a stack of trays on a shelf that hangs from four identical spiral springs under tension, one near each corner of the shelf. Each tray is rectangular, 45.3 cm by 35.6 cm, 0.450 cm thick, and with mass 580 g. (a) Demonstrate that the top tray in the stack can always be at the same height above the floor, however many trays are in the dispenser. (b) Find the spring constant each spring should have for the dispenser to function in this convenient way. (c) Is any piece of data unnecessary for this determination?

24. The force acting on a particle is $F_x = (8x - 16)$, where F is in newtons and x is in meters. (a) Make a plot of this force versus x from $x = 0$ to $x = 3.00$ m. (b) From your graph, find the net work done by this force on the particle as it moves from $x = 0$ to $x = 3.00$ m.

25. [W] A force $\vec{\mathbf{F}} = (4x\hat{\mathbf{i}} + 3y\hat{\mathbf{j}})$, where $\vec{\mathbf{F}}$ is in newtons and x and y are in meters, acts on an object as the object moves in the x direction from the origin to $x = 5.00$ m. Find the work $W = \int \vec{\mathbf{F}} \cdot d\vec{\mathbf{r}}$ done by the force on the object.

26. A light spring with force constant 3.85 N/m is compressed by 8.00 cm as it is held between a 0.250-kg block on the left and a 0.500-kg block on the right, both resting on a horizontal surface. The spring exerts a force on each block, tending to push the blocks apart. The blocks are simultaneously released from rest. Find the acceleration with which each block starts to move, given that the coefficient of kinetic friction between each block and the surface is (a) 0, (b) 0.100, and (c) 0.462.

27. A 6 000-kg freight car rolls along rails with negligible friction. The car is brought to rest by a combination of two coiled springs as illustrated in Figure P6.27 (page 188). Both springs are described by Hooke's law and have spring constants $k_1 = 1\,600$ N/m and $k_2 = 3\,400$ N/m. After the first spring compresses a distance of 30.0 cm, the second spring acts with the first to increase the force as additional compression occurs as shown in the graph. The car comes to rest 50.0 cm after first contacting the two-spring system. Find the car's initial speed.

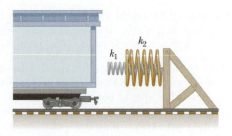

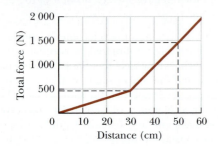

Figure P6.27

28. A 100-g bullet is fired from a rifle having a barrel 0.600 m long. Choose the origin to be at the location where the bullet begins to move. Then the force (in newtons) exerted by the expanding gas on the bullet is $15\,000 + 10\,000x - 25\,000x^2$, where x is in meters. (a) Determine the work done by the gas on the bullet as the bullet travels the length of the barrel. (b) **What If?** If the barrel is 1.00 m long, how much work is done, and (c) how does this value compare with the work calculated in part (a)?

29. Hooke's law describes a certain light spring of unstretched length 35.0 cm. When one end is attached to the top of a doorframe and a 7.50-kg object is hung from the other end, the length of the spring is 41.5 cm. (a) Find its spring constant. (b) The load and the spring are taken down. Two people pull in opposite directions on the ends of the spring, each with a force of 190 N. Find the length of the spring in this situation.

Section 6.5 Kinetic Energy and the Work–Kinetic Energy Theorem

30. **Q|C** A worker pushing a 35.0-kg wooden crate at a constant speed for 12.0 m along a wood floor does 350 J of work by applying a constant horizontal force of magnitude F on the crate. (a) Determine the value of F. (b) If the worker now applies a force greater than F, describe the subsequent motion of the crate. (c) Describe what would happen to the crate if the applied force is less than F.

31. **M** A 3.00-kg object has a velocity $(6.00\hat{\mathbf{i}} - 1.00\hat{\mathbf{j}})$ m/s. (a) What is its kinetic energy at this moment? (b) What is the net work done on the object if its velocity changes to $(8.00\hat{\mathbf{i}} + 4.00\hat{\mathbf{j}})$ m/s. (*Note*: From the definition of the dot product, $v^2 = \vec{\mathbf{v}} \cdot \vec{\mathbf{v}}$.)

32. **W** A 4.00-kg particle is subject to a net force that varies with position as shown in Figure P6.15. The particle starts from rest at $x = 0$. What is its speed at (a) $x = 5.00$ m, (b) $x = 10.0$ m, and (c) $x = 15.0$ m?

33. **W** A 0.600-kg particle has a speed of 2.00 m/s at point Ⓐ and kinetic energy of 7.50 J at point Ⓑ. What is (a) its kinetic energy at Ⓐ, (b) its speed at Ⓑ, and (c) the net work done on the particle by external forces as it moves from Ⓐ to Ⓑ?

34. **Review.** In an electron microscope, there is an electron gun that contains two charged metallic plates 2.80 cm apart. An electric force accelerates each electron in the beam from rest to 9.60% of the speed of light over this distance. (a) Determine the kinetic energy of the electron as it leaves the electron gun. Electrons carry this energy to a phosphorescent viewing screen where the microscope's image is formed, making it glow. For an electron passing between the plates in the electron gun, determine (b) the magnitude of the constant electric force acting on the electron, (c) the acceleration of the electron, and (d) the time interval the electron spends between the plates.

35. **M** A 2 100-kg pile driver is used to drive a steel I-beam into the ground. The pile driver falls 5.00 m before coming into contact with the top of the beam, and it drives the beam 12.0 cm farther into the ground before coming to rest. Using energy considerations, calculate the average force the beam exerts on the pile driver while the pile driver is brought to rest.

36. **Review.** A 7.80-g bullet moving at 575 m/s strikes the hand of a superhero, causing the hand to move 5.50 cm in the direction of the bullet's velocity before stopping. (a) Use work and energy considerations to find the average force that stops the bullet. (b) Assuming the force is constant, determine how much time elapses between the moment the bullet strikes the hand and the moment it stops moving.

37. **Review.** A 5.75-kg object passes through the origin at time $t = 0$ such that its x component of velocity is 5.00 m/s and its y component of velocity is -3.00 m/s. (a) What is the kinetic energy of the object at this time? (b) At a later time $t = 2.00$ s, the particle is located at $x = 8.50$ m and $y = 5.00$ m. What constant force acted on the object during this time interval? (c) What is the speed of the particle at $t = 2.00$ s?

38. **GP** **Q|C** **Review.** You can think of the work–kinetic energy theorem as a second theory of motion, parallel to Newton's laws in describing how outside influences affect the motion of an object. In this problem, solve parts (a), (b), and (c) separately from parts (d) and (e) so you can compare the predictions of the two theories. A 15.0-g bullet is accelerated from rest to a speed of 780 m/s in a rifle barrel of length 72.0 cm. (a) Find the kinetic energy of the bullet as it leaves the barrel. (b) Use the work–kinetic energy theorem to find the net work that is done on the bullet. (c) Use your result to part (b) to find the magnitude of the average net force that acted on the bullet while it was in the barrel. (d) Now model the bullet as a particle under constant acceleration. Find the constant acceleration of a bullet that starts from rest and gains a speed of 780 m/s over a distance of 72.0 cm. (e) Modeling the bullet as a particle under a net force, find the net force that acted on it during its acceleration. (f) What conclusion can you draw from comparing your results of parts (c) and (e)?

Section 6.6 Potential Energy of a System

39. A 0.20-kg stone is held 1.3 m above the top edge of a water well and then dropped into it. The well has a depth of 5.0 m. Relative to the configuration with the stone at the top edge of the well, what is the gravitational potential energy of the stone–Earth system (a) before the stone is released and (b) when it reaches the bottom of the well? (c) What is the

change in gravitational potential energy of the system from release to reaching the bottom of the well?

40. W A 400-N child is in a swing that is attached to a pair of ropes 2.00 m long. Find the gravitational potential energy of the child–Earth system relative to the child's lowest position when (a) the ropes are horizontal, (b) the ropes make a 30.0° angle with the vertical, and (c) the child is at the bottom of the circular arc.

41. A 1 000-kg roller coaster car is initially at the top of a rise, at point Ⓐ. It then moves 135 ft, at an angle of 40.0° below the horizontal, to a lower point Ⓑ. (a) Choose the car at point Ⓑ to be the zero configuration for gravitational potential energy of the roller coaster–Earth system. Find the potential energy of the system when the car is at points Ⓐ and Ⓑ, and the change in potential energy as the car moves between these points. (b) Repeat part (a), setting the zero configuration with the car at point Ⓐ.

Section 6.7 Conservative and Nonconservative Forces

42. M QC A 4.00-kg particle moves from the origin to position Ⓒ, having coordinates $x = 5.00$ m and $y = 5.00$ m (Fig. P6.42). One force on the particle is the gravitational force acting in the negative y direction. Using Equation 6.3, calculate the work done by the gravitational force on the particle as it goes from O to Ⓒ along (a) the purple path, (b) the red path, and (c) the blue path. (d) Your results should all be identical. Why?

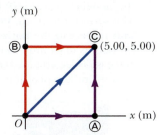

Figure P6.42 Problems 42 through 45.

43. M QC A force acting on a particle moving in the xy plane is given by $\vec{\mathbf{F}} = (2y\hat{\mathbf{i}} + x^2\hat{\mathbf{j}})$, where $\vec{\mathbf{F}}$ is in newtons and x and y are in meters. The particle moves from the origin to a final position having coordinates $x = 5.00$ m and $y = 5.00$ m as shown in Figure P6.42. Calculate the work done by $\vec{\mathbf{F}}$ on the particle as it moves along (a) the purple path, (b) the red path, and (c) the blue path. (d) Is $\vec{\mathbf{F}}$ conservative or nonconservative? (e) Explain your answer to part (d).

44. (a) Suppose a constant force acts on an object. The force does not vary with time or with the position or the velocity of the object. Start with the general definition for work done by a force

$$W = \int_i^f \vec{\mathbf{F}} \cdot d\vec{\mathbf{r}}$$

and show that the force is conservative. (b) As a special case, suppose the force $\vec{\mathbf{F}} = (3\hat{\mathbf{i}} + 4\hat{\mathbf{j}})$ N acts on a particle that moves from O to Ⓒ in Figure P6.42. Calculate the work done by $\vec{\mathbf{F}}$ on the particle as it moves along each one of the three paths shown in the figure and show that the work done along the three paths is identical.

45. QC An object moves in the xy plane in Figure P6.42 and experiences a friction force with constant magnitude 3.00 N, always acting in the direction opposite the object's velocity. Calculate the work that you must do to slide the object at constant speed against the friction force as the object

moves along (a) the purple path O to Ⓐ followed by a return purple path to O, (b) the purple path O to Ⓒ followed by a return blue path to O, and (c) the blue path O to Ⓒ followed by a return blue path to O. (d) Each of your three answers should be nonzero. What is the significance of this observation?

Section 6.8 Relationship Between Conservative Forces and Potential Energy

46. S The potential energy of a system of two particles separated by a distance r is given by $U(r) = A/r$, where A is a constant. Find the radial force $\vec{\mathbf{F}}_r$ that each particle exerts on the other.

47. M A single conservative force acts on a 5.00-kg particle within a system due to its interaction with the rest of the system. The equation $F_x = 2x + 4$ describes the force, where F_x is in newtons and x is in meters. As the particle moves along the x axis from $x = 1.00$ m to $x = 5.00$ m, calculate (a) the work done by this force on the particle, (b) the change in the potential energy of the system, and (c) the kinetic energy the particle has at $x = 5.00$ m if its speed is 3.00 m/s at $x = 1.00$ m.

48. A potential energy function for a system in which a two-dimensional force acts is of the form $U = 3x^3y - 7x$. Find the force that acts at the point (x, y).

49. *Why is the following situation impossible?* A librarian lifts a book from the ground to a high shelf, doing 20.0 J of work in the lifting process. As he turns his back, the book falls off the shelf back to the ground. The gravitational force from the Earth on the book does 20.0 J of work on the book while it falls. Because the work done was 20.0 J + 20.0 J = 40.0 J, the book hits the ground with 40.0 J of kinetic energy.

Section 6.9 Potential Energy for Gravitational and Electric Forces

50. How much energy is required to move a 1 000-kg object from the Earth's surface to an altitude twice the Earth's radius?

51. A satellite of the Earth has a mass of 100 kg and is at an altitude of 2.00×10^6 m. (a) What is the potential energy of the satellite–Earth system? (b) What is the magnitude of the gravitational force exerted by the Earth on the satellite? (c) What force does the satellite exert on the Earth?

52. A system consists of three particles, each of mass 5.00 g, located at the corners of an equilateral triangle with sides of 30.0 cm. (a) Calculate the potential energy describing the gravitational interactions internal to the system. (b) If the particles are released simultaneously, where will they collide?

53. At the Earth's surface, a projectile is launched straight up at a speed of 10.0 km/s. To what height will it rise? Ignore air resistance.

Section 6.10 Energy Diagrams and Equilibrium of a System

54. For the potential energy curve shown in Figure P6.54 (page 190), (a) determine whether the force F_x is positive, negative, or zero at the five points indicated. (b) Indicate points of stable, unstable, and neutral equilibrium. (c) Sketch the curve for F_x versus x from $x = 0$ to $x = 9.5$ m.

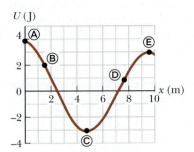

Figure P6.54

55. A right circular cone can theoretically be balanced on a horizontal surface in three different ways. Sketch these three equilibrium configurations and identify them as positions of stable, unstable, or neutral equilibrium.

Section 6.11 Context Connection: Potential Energy in Fuels

Note: Power will be defined in Section 7.6 as the rate of energy transfer, with a unit of a watt (W), equivalent to J/s. Therefore, a kilowatt-hour (kWh) is a unit of energy.

56. **Q|C** The power of sunlight reaching each square meter of the Earth's surface on a clear day in the tropics is close to 1 000 W. On a winter day in Manitoba, the power concentration of sunlight can be 100 W/m². Many human activities are described by a power per unit area on the order of 10^2 W/m² or less. (a) Consider, for example, a family of four paying $66 to the electric company every 30 days for 600 kWh of energy carried by electrical transmission to their house, which has floor dimensions of 13.0 m by 9.50 m. Compute the power per unit area used by the family. (b) Consider a car 2.10 m wide and 4.90 m long traveling at 55.0 mi/h using gasoline having "heat of combustion" 44.0 MJ/kg with fuel economy 25.0 mi/gal. One gallon of gasoline has a mass of 2.54 kg. Find the power per unit area used by the car. (c) Explain why direct use of solar energy is not practical for running a conventional automobile. (d) What are some uses of solar energy that are more practical?

57. In considering the energy supply for an automobile, the energy per unit mass of the energy source is an important parameter. As the chapter text points out, the "heat of combustion" or stored energy per mass is quite similar for gasoline, ethanol, diesel fuel, cooking oil, methane, and propane. For a broader perspective, compare the energy per mass in joules per kilogram for gasoline, lead-acid batteries, hydrogen, and hay. Rank the four in order of increasing energy density and state the factor of increase between each one and the next. Hydrogen has "heat of combustion" 142 MJ/kg. For wood, hay, and dry vegetable matter in general, this parameter is 17 MJ/kg. A fully charged 16.0-kg lead-acid battery can deliver power 1 200 W for 1.0 hr.

Additional Problems

58. **S** When an object is displaced by an amount x from stable equilibrium, a restoring force acts on it, tending to return the object to its equilibrium position. The magnitude of the restoring force can be a complicated function of x. In such

cases, we can generally imagine the force function $F(x)$ to be expressed as a power series in x as $F(x) = -(k_1x + k_2x^2 + k_3x^3 + \cdots)$. The first term here is just Hooke's law, which describes the force exerted by a simple spring for small displacements. For small excursions from equilibrium, we generally ignore the higher-order terms, but in some cases it may be desirable to keep the second term as well. If we model the restoring force as $F = -(k_1x + k_2x^2)$, how much work is done on an object in displacing it from $x = 0$ to $x = x_{max}$ by an applied force $-F$?

59. **Review.** A baseball outfielder throws a 0.150-kg baseball at a speed of 40.0 m/s and an initial angle of 30.0° to the horizontal. What is the kinetic energy of the baseball at the highest point of its trajectory?

60. *Why is the following situation impossible?* In a new casino, a supersized pinball machine is introduced. Casino advertising boasts that a professional basketball player can lie on top of the machine and his head and feet will not hang off the edge! The ball launcher in the machine sends metal balls up one side of the machine and then into play. The spring in the launcher (Fig. P6.60) has a force constant of 1.20 N/cm. The surface on which the ball moves is inclined $\theta = 10.0°$ with respect to the horizontal. The spring is initially compressed its maximum distance $d = 5.00$ cm. A ball of mass 100 g is projected into play by releasing the plunger. Casino visitors find the play of the giant machine quite exciting.

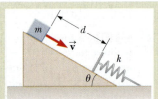

Figure P6.60

61. An inclined plane of angle $\theta = 20.0°$ has a spring of force constant $k = 500$ N/m fastened securely at the bottom so that the spring is parallel to the surface as shown in Figure P6.61. A block of mass $m = 2.50$ kg is placed on the plane at a distance $d = 0.300$ m from the spring.

Figure P6.61 Problems 61 and 62.

From this position, the block is projected downward toward the spring with speed $v = 0.750$ m/s. By what distance is the spring compressed when the block momentarily comes to rest?

62. **S** An inclined plane of angle θ has a spring of force constant k fastened securely at the bottom so that the spring is parallel to the surface. A block of mass m is placed on the plane at a distance d from the spring. From this position, the block is projected downward toward the spring with speed v as shown in Figure P6.61. By what distance is the spring compressed when the block momentarily comes to rest?

63. **Q|C** (a) Take $U = 5$ for a system with a particle at position $x = 0$ and calculate the potential energy of the system as a function of the particle position x. The force on the particle is given by $(8e^{-2x})\hat{i}$. (b) Explain whether the force is conservative or nonconservative and how you can tell.

64. The potential energy function for a system of particles is given by $U(x) = -x^3 + 2x^2 + 3x$, where x is the position of one particle in the system. (a) Determine the force F_x on the particle as a function of x. (b) For what values of x is the force equal to zero? (c) Plot $U(x)$ versus x and F_x versus x and indicate points of stable and unstable equilibrium.

65. **Q|C** **Review.** A light spring has unstressed length 15.5 cm. It is described by Hooke's law with spring constant 4.30 N/m. One end of the spring is held on a fixed vertical axle, and the other end is attached to a puck of mass m that can move without friction over a horizontal surface. The puck is set into motion in a circle with a period of 1.30 s. (a) Find the extension of the spring x as it depends on m. Evaluate x for (b) $m = 0.070\ 0$ kg, (c) $m = 0.140$ kg, (d) $m = 0.180$ kg, and (e) $m = 0.190$ kg. (f) Describe the pattern of variation of x as it depends on m.

66. The spring constant of an automotive suspension spring increases with increasing load due to a spring coil that is widest at the bottom, smoothly tapering to a smaller diameter near the top. The result is a softer ride on normal road surfaces from the wider coils, but the car does not bottom out on bumps because when the lower coils collapse, the stiffer coils near the top absorb the load. For such springs, the force exerted by the spring can be empirically found to be given by $F = ax^b$. For a tapered spiral spring that compresses 12.9 cm with a 1 000-N load and 31.5 cm with a 5 000-N load, (a) evaluate the constants a and b in the empirical equation for F and (b) find the work needed to compress the spring 25.0 cm.

67. **Q|C** **Review.** Two constant forces act on an object of mass $m = 5.00$ kg moving in the xy plane as shown in Figure P6.67. Force $\vec{F}_1$ is 25.0 N at 35.0°, and force $\vec{F}_2$ is 42.0 N at 150°. At time $t = 0$, the object is at the origin and has velocity $(4.00\hat{i} + 2.50\hat{j})$ m/s. (a) Express the two forces in unit-vector notation. Use unit-vector notation for your other answers. (b) Find the total force exerted on the object. (c) Find the object's acceleration. Now, considering the instant $t = 3.00$ s,

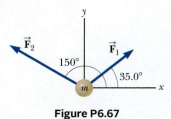

Figure P6.67

find (d) the object's velocity, (e) its position, (f) its kinetic energy from $\frac{1}{2}mv_f^2$, and (g) its kinetic energy from $\frac{1}{2}mv_i^2 + \sum \vec{F} \cdot \Delta\vec{r}$. (h) What conclusion can you draw by comparing the answers to parts (f) and (g)?

68. A particle of mass $m = 1.18$ kg is attached between two identical springs on a frictionless, horizontal tabletop. Both springs have spring constant k and are initially unstressed, and the particle is at $x = 0$. (a) The particle is pulled a distance x along a direction perpendicular to the initial configuration of the springs as shown in Figure P6.68. Show that the force exerted by the springs on the particle is

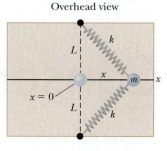

Figure P6.68

$$\vec{F} = -2kx\left(1 - \frac{L}{\sqrt{x^2 + L^2}}\right)\hat{i}$$

(b) Show that the potential energy of the system is

$$U(x) = kx^2 + 2kL\left(L - \sqrt{x^2 + L^2}\right)$$

(c) Make a plot of $U(x)$ versus x and identify all equilibrium points. Assume $L = 1.20$ m and $k = 40.0$ N/m. (d) If the particle is pulled 0.500 m to the right and then released, what is its speed when it reaches $x = 0$?

69. **Q|C** When different loads hang on a spring, the spring stretches to different lengths as shown in the following table. (a) Make a graph of the applied force versus the extension of the spring. (b) By least-squares fitting, determine the straight line that best fits the data. (c) To complete part (b), do you want to use all the data points, or should you ignore some of them? Explain. (d) From the slope of the best-fit line, find the spring constant k. (e) If the spring is extended to 105 mm, what force does it exert on the suspended object?

F (N)	2.0	4.0	6.0	8.0	10	12	14	16	18	20	22
L (mm)	15	32	49	64	79	98	112	126	149	175	190

Conservation of Energy

Chapter Outline

Jade Lee/Getty Images

Three youngsters enjoy the transformation of potential energy to kinetic energy on a waterslide. We can analyze processes such as these with the techniques developed in this chapter.

In Chapter 6, we introduced three methods for storing energy in a system: kinetic energy, associated with movement of members of the system; potential energy, determined by the configuration of the system; and internal energy, which is related to the temperature of the system.

We now consider analyzing physical situations using the energy approach for two types of systems: *nonisolated* and *isolated* systems. For nonisolated systems, we shall investigate ways that energy can cross the boundary of the system, resulting in a change in the system's total energy. This analysis leads to a critically important principle called *conservation of energy*. The conservation of energy principle extends well beyond physics and can be applied to biological organisms, technological systems, and engineering situations.

In isolated systems, energy does not cross the boundary of the system. For these systems, the total energy of the system is constant. If no nonconservative forces act within the system, we can use *conservation of mechanical energy* to solve a variety of problems.

Situations involving the transformation of mechanical energy to internal energy due to nonconservative forces require special handling. We investigate the procedures for these types of problems.

Finally, we recognize that energy can cross the boundary of a system at different rates. We describe the rate of energy transfer with the quantity *power*.

7.1 | Analysis Model: Nonisolated System (Energy)

As we have seen, an object, modeled as a particle, can be acted on by various forces, resulting in a change in its kinetic energy. If we choose the object as the system, this very simple situation is the first example of a *nonisolated system*, for which energy crosses the boundary of the system during some time interval due to an interaction with the environment. This scenario is common in physics problems. If a system does not interact with its environment, it is an isolated system, which we will study in Section 7.2.

The work–kinetic energy theorem from Chapter 6 is our first example of an energy equation appropriate for a nonisolated system. In the case of that theorem, the interaction of the system with its environment is the work done by the external force, and the quantity in the system that changes is the kinetic energy.

So far, we have seen only one way to transfer energy into a system: work. We mention below a few other ways to transfer energy into or out of a system. The details of these processes will be studied in other sections of the book. We illustrate mechanisms to transfer energy in Figure 7.1 and summarize them as follows.

Work, as we have learned in Chapter 6, is a method of transferring energy to a system by applying a force to the system such that the point of application of the force undergoes a displacement (Fig. 7.1a).

Mechanical waves (Chapters 13–14) are a means of transferring energy by allowing a disturbance to propagate through air or another medium. It is the method by which energy (which you detect as sound) leaves the system of your clock radio through the loudspeaker and enters your ears to stimulate the hearing process (Fig. 7.1b). Other examples of mechanical waves are seismic waves and ocean waves.

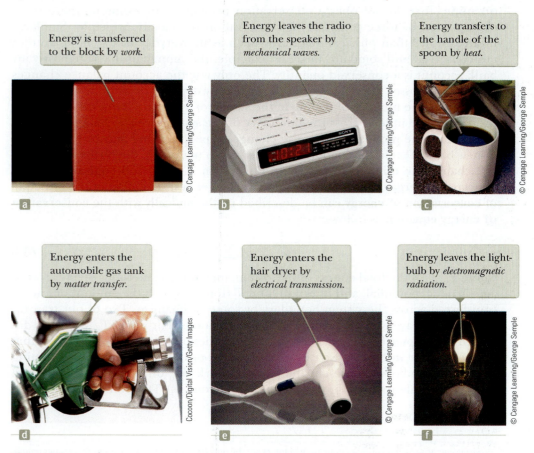

Energy is transferred to the block by *work.*

Energy leaves the radio from the speaker by *mechanical waves.*

Energy transfers to the handle of the spoon by *heat.*

Energy enters the automobile gas tank by *matter transfer.*

Energy enters the hair dryer by *electrical transmission.*

Energy leaves the light-bulb by *electromagnetic radiation.*

Figure 7.1 Energy transfer mechanisms. In each case, the system into which or from which energy is transferred is indicated.

Heat (Chapter 17) is a mechanism of energy transfer that is driven by a temperature difference between a system and its environment. For example, imagine dividing a metal spoon into two parts: the handle, which we identify as the system, and the portion submerged in a cup of coffee, which is part of the environment (Fig. 7.1c). The handle of the spoon becomes hot because fast-moving electrons and atoms in the submerged portion bump into slower ones in the nearby part of the handle. These particles move faster because of the collisions and bump into the next group of slow particles. Therefore, the internal energy of the spoon handle rises from energy transfer due to this collision process.

Matter transfer (Chapter 17) involves situations in which matter physically crosses the boundary of a system, carrying energy with it. Examples include filling your automobile tank with gasoline (Fig. 7.1d) and carrying energy to the rooms of your home by circulating warm air from the furnace, a process called *convection*.

Electrical transmission (Chapter 21) involves energy transfer into or out of a system by means of electric currents. It is how energy transfers into your hair dryer (Fig. 7.1e), stereo system, or any other electrical device.

Electromagnetic radiation (Chapter 24) refers to electromagnetic waves such as light (Fig. 7.1f), microwaves, and radio waves crossing the boundary of a system. Examples of this method of transfer include cooking a baked potato in your microwave oven and light energy traveling from the Sun to the Earth through space.[1]

A central feature of the energy approach is the notion that we can neither create nor destroy energy, that energy is always *conserved*. This feature has been tested in countless experiments, and no experiment has ever shown this statement to be incorrect. Therefore, **if the total amount of energy in a system changes, it can *only* be because energy has crossed the boundary of the system by a transfer mechanism such as one of the methods listed above.**

Energy is one of several quantities in physics that are conserved. We will see other conserved quantities in subsequent chapters. There are many physical quantities that do not obey a conservation principle. For example, there is no conservation of force principle or conservation of velocity principle. Similarly, in areas other than physical quantities, such as in everyday life, some quantities are conserved and some are not. For example, the money in the system of your bank account is a conserved quantity. The only way the account balance changes is if money crosses the boundary of the system by deposits or withdrawals. On the other hand, the number of people in the system of a country is not conserved. Although people indeed cross the boundary of the system, which changes the total population, the population can also change by people dying and by giving birth to new babies. Even if no people cross the system boundary, the births and deaths will change the number of people in the system. There is no equivalent in the concept of energy to dying or giving birth. The general statement of the principle of **conservation of energy** can be described mathematically with the **conservation of energy equation** as follows:

▶ Conservation of energy

$$\Delta E_{system} = \sum T \qquad \textbf{7.1} ◀$$

where E_{system} is the total energy of the system, including all methods of energy storage (kinetic, potential, and internal), and T (for *transfer*) is the amount of energy transferred across the system boundary by some mechanism. Two of our transfer mechanisms have well-established symbolic notations. For work, $T_{work} = W$ as discussed in Chapter 6, and for heat, $T_{heat} = Q$ as defined in Chapter 17. (Now that we are familiar with work, we can simplify the appearance of equations by letting the simple symbol W represent the external work W_{ext} on a system. For internal work, we will always use W_{int} to differentiate it from W.) The other four members

[1] Electromagnetic radiation and work done by field forces are the only energy transfer mechanisms that do not require molecules of the environment to be available at the system boundary. Therefore, systems surrounded by a vacuum (such as planets) can only exchange energy with the environment by means of these two possibilities.

of our list do not have established symbols, so we will call them T_{MW} (mechanical waves), T_{MT} (matter transfer), T_{ET} (electrical transmission), and T_{ER} (electromagnetic radiation).

The full expansion of Equation 7.1 is

$$\Delta K + \Delta U + \Delta E_{int} = W + Q + T_{MW} + T_{MT} + T_{ET} + T_{ER} \qquad \textbf{7.2} \blacktriangleleft$$

which is the primary mathematical representation of the energy version of the analysis model of the **nonisolated system.** (We will see other versions of the nonisolated system model, involving linear momentum and angular momentum, in later chapters.) In most cases, Equation 7.2 reduces to a much simpler one because some of the terms are zero. If, for a given system, all terms on the right side of the conservation of energy equation are zero, the system is an *isolated system,* which we study in the next section.

The conservation of energy equation is no more complicated in theory than the process of balancing your checking account statement. If your account is the system, the change in the account balance for a given month is the sum of all the transfers: deposits, withdrawals, fees, interest, and checks written. You may find it useful to think of energy as the *currency of nature!*

Suppose a force is applied to a nonisolated system and the point of application of the force moves through a displacement. Then suppose the only effect on the system is to change its speed. In this case, the only transfer mechanism is work (so that the right side of Eq. 7.2 reduces to just W) and the only kind of energy in the system that changes is the kinetic energy (so that ΔE_{system} reduces to just ΔK). Equation 7.2 then becomes

$$\Delta K = W$$

which is the work–kinetic energy theorem. This theorem is a special case of the more general principle of conservation of energy. We shall see several more special cases in future chapters.

> ◤ **QUICK QUIZ 7.1** By what transfer mechanisms does energy enter and leave (**a**) your television set? (**b**) Your gasoline-powered lawn mower? (**c**) Your hand-cranked pencil sharpener?

> ◤ **QUICK QUIZ 7.2** Consider a block sliding over a horizontal surface with friction. Ignore any sound the sliding might make. (**i**) If the system is the *block,* this system is (a) isolated (b) nonisolated (c) impossible to determine (**ii**) If the system is the *surface,* describe the system from the same set of choices. (**iii**) If the system is the *block and the surface,* describe the system from the same set of choices.

◤ 7.2 | Analysis Model: Isolated System (Energy)

In this section, we study another very common scenario in physics problems: a system is chosen such that no energy crosses the system boundary by any method. We begin by considering a gravitational situation. Think about the book–Earth system in Active Figure 6.15 in the preceding chapter. After we have lifted the book, there is gravitational potential energy stored in the system, which can be calculated from the work done by the external agent on the system, using $W = \Delta U_g$.

Let us now shift our focus to the work done *on the book alone* by the gravitational force (Fig. 7.2) as the book falls back to its original height. As the book falls from y_i to y_f, the work done by the gravitational force on the book is

$$W_{on\ book} = (m\vec{\mathbf{g}}) \cdot \Delta\vec{\mathbf{r}} = (-mg\hat{\mathbf{j}}) \cdot [(y_f - y_i)\hat{\mathbf{j}}] = mgy_i - mgy_f \qquad \textbf{7.3} \blacktriangleleft$$

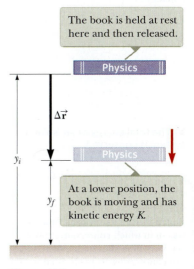

The book is held at rest here and then released.

At a lower position, the book is moving and has kinetic energy K.

Figure 7.2 A book is released from rest and falls due to work done by the gravitational force on the book.

From the work–kinetic energy theorem of Chapter 6, the work done on the book is equal to the change in the kinetic energy of the book:

$$W_{\text{on book}} = \Delta K_{\text{book}}$$

We can equate these two expressions for the work done on the book:

$$\Delta K_{\text{book}} = mgy_i - mgy_f \qquad \textbf{7.4} \blacktriangleleft$$

Let us now relate each side of this equation to the *system* of the book and the Earth. For the right-hand side of Equation 7.4,

$$mgy_i - mgy_f = -(mgy_f - mgy_i) = -\Delta U_g$$

where $U_g = mgy$ is the gravitational potential energy of the system. For the left-hand side of Equation 7.4, because the book is the only part of the system that is moving, we see that $\Delta K_{\text{book}} = \Delta K$, where K is the kinetic energy of the system. Therefore, with each side of Equation 7.4 replaced with its system equivalent, the equation becomes

$$\Delta K = -\Delta U_g \qquad \textbf{7.5} \blacktriangleleft$$

This equation can be manipulated to provide a very important general result for solving problems. First, we move the change in potential energy to the left side of the equation:

$$\Delta K + \Delta U_g = 0$$

The left side represents a sum of changes of the energy stored in the system. The right-hand side is zero because there are no transfers of energy across the boundary of the system; the book–Earth system is *isolated* from the environment. We developed this equation for a gravitational system, but it can be shown to be valid for a system with any type of potential energy. Therefore, for an isolated system,

$$\Delta K + \Delta U = 0 \qquad \textbf{7.6} \blacktriangleleft$$

We defined in Chapter 6 the sum of the kinetic and potential energies of a system as its mechanical energy:

▶ Mechanical energy of a system

$$E_{\text{mech}} \equiv K + U \qquad \textbf{7.7} \blacktriangleleft$$

where U represents the total of *all* types of potential energy. Because the system under consideration is isolated, Equations 7.6 and 7.7 tell us that the mechanical energy of the system is conserved:

▶ The mechanical energy of an isolated system with no nonconservative forces acting is conserved.

$$\Delta E_{\text{mech}} = 0 \qquad \textbf{7.8} \blacktriangleleft$$

Equation 7.8 is a statement of **conservation of mechanical energy** for an isolated system with no nonconservative forces acting. The mechanical energy in such a system is conserved: the sum of the kinetic and potential energies remains constant.

If there are nonconservative forces acting within the system, mechanical energy is transformed to internal energy as discussed in Section 6.7. If nonconservative forces act in an isolated system, the total energy of the system is conserved although the mechanical energy is not. In that case, we can express the conservation of energy of the system as

▶ The total energy of an isolated system is conserved.

$$\Delta E_{\text{system}} = 0 \qquad \textbf{7.9} \blacktriangleleft$$

where E_{system} includes all kinetic, potential, and internal energies. This equation is the most general statement of the energy version of the **isolated system** model. It is equivalent to Equation 7.2 with all terms on the right-hand side equal to zero.

Let us now write the changes in energy in Equation 7.6 explicitly:

$$(K_f - K_i) + (U_f - U_i) = 0$$

$$K_f + U_f = K_i + U_i \qquad \textbf{7.10} \blacktriangleleft$$

Pitfall Prevention | 7.2
Conditions on Equation 7.10
Equation 7.10 is only true for a system in which conservative forces act. We will see how to handle nonconservative forces in Sections 7.4 and 7.5.

For the gravitational situation of the falling book, Equation 7.10 can be written as

$$\tfrac{1}{2}mv_f^2 + mgy_f = \tfrac{1}{2}mv_i^2 + mgy_i$$

As the book falls to the Earth, the book–Earth system loses potential energy and gains kinetic energy such that the total of the two types of energy always remains constant.

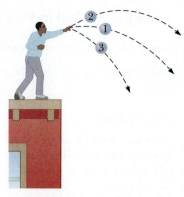

QUICK QUIZ 7.3 A rock of mass m is dropped to the ground from a height h. A second rock, with mass $2m$, is dropped from the same height. When the second rock strikes the ground, what is its kinetic energy? (**a**) twice that of the first rock (**b**) four times that of the first rock (**c**) the same as that of the first rock (**d**) half as much as that of the first rock (**e**) impossible to determine

QUICK QUIZ 7.4 Three identical balls are thrown from the top of a building, all with the same initial speed. As shown in Active Figure 7.3, the first is thrown horizontally, the second at some angle above the horizontal, and the third at some angle below the horizontal. Neglecting air resistance, rank the speeds of the balls at the instant each hits the ground.

Active Figure 7.3 (Quick Quiz 7.4) Three identical balls are thrown with the same initial speed from the top of a building.

> ### PROBLEM-SOLVING STRATEGY: Isolated Systems with No Nonconservative Forces: Conservation of Mechanical Energy
>
> Many problems in physics can be solved using the principle of conservation of energy for an isolated system. The following procedure should be used when you apply this principle:
>
> 1. **Conceptualize.** Study the physical situation carefully and form a mental representation of what is happening. As you become more proficient working energy problems, you will begin to be comfortable imagining the types of energy that are changing in the system.
>
> 2. **Categorize.** Define your system, which may consist of more than one object and may or may not include springs or other possibilities for storing potential energy. Determine if any energy transfers occur across the boundary of your system. If so, use the nonisolated system model, $\Delta E_{system} = \Sigma T$, from Section 7.1. If not, use the isolated system model, $\Delta E_{system} = 0$.
>
>   Determine whether any nonconservative forces are present within the system. If so, use the techniques of Sections 7.4 and 7.5. If not, use the principle of conservation of mechanical energy as outlined below.
>
> 3. **Analyze.** Choose configurations to represent the initial and final conditions of the system. For each object that changes elevation, select a reference position for the object that defines the zero configuration of gravitational potential energy for the system. For an object on a spring, the zero configuration for elastic potential energy is when the object is at its equilibrium position. If there is more than one conservative force, write an expression for the potential energy associated with each force.
>
>   Write the total initial mechanical energy E_i of the system for some configuration as the sum of the kinetic and potential energies associated with the configuration. Then write a similar expression for the total mechanical energy E_f of the system for the final configuration that is of interest. Because mechanical energy is *conserved*, equate the two total energies and solve for the quantity that is unknown.
>
> 4. **Finalize.** Make sure your results are consistent with your mental representation. Also make sure the values of your results are reasonable and consistent with connections to everyday experience.

Example **7.1** | **Ball in Free Fall**

A ball of mass m is dropped from a height h above the ground as shown in Active Figure 7.4.

(A) Neglecting air resistance, determine the speed of the ball when it is at a height y above the ground.

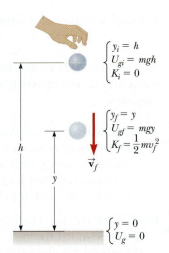

SOLUTION

Conceptualize Active Figure 7.4 and our everyday experience with falling objects allow us to conceptualize the situation. Although we can readily solve this problem with the techniques of Chapter 2, let us practice an energy approach.

Categorize We identify the system as the ball and the Earth. Because there is neither air resistance nor any other interaction between the system and the environment, the system is isolated and we use the isolated system model. The only force between members of the system is the gravitational force, which is conservative.

Active Figure 7.4 (Example 7.1) A ball is dropped from a height h above the ground. Initially, the total energy of the ball–Earth system is gravitational potential energy, equal to mgh relative to the ground. At the position y, the total energy is the sum of the kinetic and potential energies.

Analyze Because the system is isolated and there are no nonconservative forces acting within the system, we apply the principle of conservation of mechanical energy to the ball–Earth system. At the instant the ball is released, its kinetic energy is $K_i = 0$ and the gravitational potential energy of the system is $U_{gi} = mgh$. When the ball is at a position y above the ground, its kinetic energy is $K_f = \frac{1}{2}mv_f^2$ and the potential energy relative to the ground is $U_{gf} = mgy$.

Apply Equation 7.10:

$$K_f + U_{gf} = K_i + U_{gi}$$

$$\tfrac{1}{2}mv_f^2 + mgy = 0 + mgh$$

Solve for v_f:

$$v_f^2 = 2g(h - y) \quad \rightarrow \quad \boxed{v_f = \sqrt{2g(h - y)}}$$

The speed is always positive. If you had been asked to find the ball's velocity, you would use the negative value of the square root as the y component to indicate the downward motion.

(B) Determine the speed of the ball at y if at the instant of release it already has an initial upward speed v_i at the initial altitude h.

SOLUTION

Analyze In this case, the initial energy includes kinetic energy equal to $\frac{1}{2}mv_i^2$.

Apply Equation 7.10:

$$\tfrac{1}{2}mv_f^2 + mgy = \tfrac{1}{2}mv_i^2 + mgh$$

Solve for v_f:

$$v_f^2 = v_i^2 + 2g(h - y) \quad \rightarrow \quad \boxed{v_f = \sqrt{v_i^2 + 2g(h - y)}}$$

Finalize This result for the final speed is consistent with the expression $v_{yf}^2 = v_{yi}^2 - 2g(y_f - y_i)$ from the particle under constant acceleration model for a falling object, where $y_i = h$. Furthermore, this result is valid even if the initial velocity is at an angle to the horizontal (Quick Quiz 7.4) for two reasons: (1) the kinetic energy, a scalar, depends only on the magnitude of the velocity; and (2) the change in the gravitational potential energy of the system depends only on the change in position of the ball in the vertical direction.

What If? What if the initial velocity $\vec{v}_i$ in part (B) were downward? How would that affect the speed of the ball at position y?

Answer You might claim that throwing the ball downward would result in it having a higher speed at y than if you threw it upward. Conservation of mechanical energy, however, depends on kinetic and potential energies, which are scalars. Therefore, the direction of the initial velocity vector has no bearing on the final speed.

Example 7.2 | A Grand Entrance

You are designing an apparatus to support an actor of mass 65 kg who is to "fly" down to the stage during the performance of a play. You attach the actor's harness to a 130-kg sandbag by means of a lightweight steel cable running smoothly over two frictionless pulleys as in Figure 7.5a. You need 3.0 m of cable between the harness and the nearest pulley so that the pulley can be hidden behind a curtain. For the apparatus to work successfully, the sandbag must never lift above the floor as the actor swings from above the stage to the floor. Let us call the initial angle that the actor's cable makes with the vertical θ. What is the maximum value θ can have before the sandbag lifts off the floor?

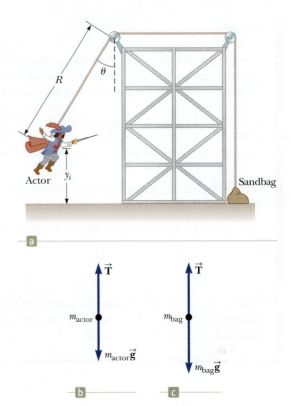

Figure 7.5 (Example 7.2) (a) An actor uses some clever staging to make his entrance. (b) The free-body diagram for the actor at the bottom of the circular path. (c) The free-body diagram for the sandbag if the normal force from the floor goes to zero.

SOLUTION

Conceptualize We must use several concepts to solve this problem. Imagine what happens as the actor approaches the bottom of the swing. At the bottom, the cable is vertical and must support his weight as well as provide centripetal acceleration of his body in the upward direction. At this point in his swing, the tension in the cable is the highest and the sandbag is most likely to lift off the floor.

Categorize Looking first at the swinging of the actor from the initial point to the lowest point, we model the actor and the Earth as an isolated system. We ignore air resistance, so there are no nonconservative forces acting. You might initially be tempted to model the system as nonisolated because of the interaction of the system with the cable, which is in the environment. The force applied to the actor by the cable, however, is always perpendicular to each element of the displacement of the actor and hence does no work. Therefore, in terms of energy transfers across the boundary, the system is isolated.

Analyze We first find the actor's speed as he arrives at the floor as a function of the initial angle θ and the radius R of the circular path through which he swings.

From the isolated system model, apply conservation of mechanical energy to the actor–Earth system:

$$K_f + U_f = K_i + U_i$$

Let y_i be the initial height of the actor above the floor and v_f be his speed at the instant before he lands. (Notice that $K_i = 0$ because the actor starts from rest and that $U_f = 0$ because we define the configuration of the actor at the floor as having a gravitational potential energy of zero.)

$$(1) \quad \tfrac{1}{2} m_{actor} v_f^2 + 0 = 0 + m_{actor} g y_i$$

From the geometry in Figure 7.5a, notice that $y_f = 0$, so $y_i = R - R\cos\theta = R(1 - \cos\theta)$. Use this relationship in Equation (1) and solve for v_f^2:

$$(2) \quad v_f^2 = 2gR(1 - \cos\theta)$$

Categorize Next, focus on the instant the actor is at the lowest point. Because the tension in the cable is transferred as a force applied to the sandbag, we model the actor at this instant as a particle under a net force. Because the actor moves along a circular arc, he experiences at the bottom of the swing a centripetal acceleration of v_f^2/r directed upward.

Analyze Apply Newton's second law from the particle under a net force model to the actor at the bottom of his path, using the free-body diagram in Figure 7.5b as a guide:

$$\sum F_y = T - m_{actor} g = m_{actor} \frac{v_f^2}{R}$$

$$(3) \quad T = m_{actor} g + m_{actor} \frac{v_f^2}{R}$$

continued

7.2 *cont.*

Categorize Finally, notice that the sandbag lifts off the floor when the upward force exerted on it by the cable exceeds the gravitational force acting on it; the normal force from the floor is zero when that happens. We do *not*, however, want the sandbag to lift off the floor. The sandbag must remain at rest, so we model it as a particle in equilibrium.

Analyze A force T of the magnitude given by Equation (3) is transmitted by the cable to the sandbag. If the sandbag remains at rest but is just ready to be lifted off the floor if any more force were applied by the cable, the normal force on it becomes zero and the particle in equilibrium model tells us that $T = m_{bag}g$ as in Figure 7.5c.

Substitute this condition and Equation (2) into Equation (3):

$$m_{bag}g = m_{actor}g + m_{actor}\frac{2gR(1 - \cos\theta)}{R}$$

Solve for $\cos\theta$ and substitute the given parameters:

$$\cos\theta = \frac{3m_{actor} - m_{bag}}{2m_{actor}} = \frac{3(65\text{ kg}) - 130\text{ kg}}{2(65\text{ kg})} = 0.50$$

$$\theta = \boxed{60°}$$

Finalize Here we had to combine several analysis models from different areas of our study. Notice that the length R of the cable from the actor's harness to the leftmost pulley did not appear in the final algebraic equation for $\cos\theta$. Therefore, the final answer is independent of R.

Example 7.3 | The Spring-Loaded Popgun

The launching mechanism of a popgun consists of a trigger-released spring (Active Fig. 7.6a). The spring is compressed to a position $y_Ⓐ$, and the trigger is fired. The projectile of mass m rises to a position $y_Ⓒ$ above the position at which it leaves the spring, indicated in Active Figure 7.6b as position $y_Ⓑ = 0$. Consider a firing of the gun for which $m = 35.0$ g, $y_Ⓐ = -0.120$ m, and $y_Ⓒ = 20.0$ m.

(A) Neglecting all resistive forces, determine the spring constant.

SOLUTION

Conceptualize Imagine the process illustrated in parts (a) and (b) of Active Figure 7.6. The projectile starts from rest, speeds up as the spring pushes upward on it, leaves the spring, and then slows down as the gravitational force pulls downward on it.

Categorize We identify the system as the projectile, the spring, and the Earth. We ignore both air resistance on the projectile and friction in the gun, so we model the system as isolated with no nonconservative forces acting.

Analyze Because the projectile starts from rest, its initial kinetic energy is zero. We choose the zero configuration for the gravitational potential energy of the system to be when the projectile leaves the spring. For this configuration, the elastic potential energy is also zero.

After the gun is fired, the projectile rises to a maximum height $y_Ⓒ$. The final kinetic energy of the projectile is zero.

From the isolated system model, write a conservation of mechanical energy equation for the system between points Ⓐ and Ⓒ:

$$K_Ⓒ + U_{gⒸ} + U_{sⒸ} = K_Ⓐ + U_{gⒶ} + U_{sⒶ}$$

Substitute for each energy:

$$0 + mgy_Ⓒ + 0 = 0 + mgy_Ⓐ + \tfrac{1}{2}kx^2$$

Solve for k:

$$k = \frac{2mg(y_Ⓒ - y_Ⓐ)}{x^2}$$

Substitute numerical values:

$$k = \frac{2(0.035\,0\text{ kg})(9.80\text{ m/s}^2)[20.0\text{ m} - (-0.120\text{ m})]}{(0.120\text{ m})^2} = \boxed{958\text{ N/m}}$$

7.3 *cont.*

Active Figure 7.6 (Example 7.3) A spring-loaded popgun (a) before firing and (b) when the spring extends to its relaxed length. (c) An energy bar chart for the popgun–projectile–Earth system before the popgun is loaded. The energy in the system is zero. (d) The popgun is loaded by means of an external agent doing work on the system to push the spring downward. Therefore the system is nonisolated during this process. After the popgun is loaded, elastic potential energy is stored in the spring and the gravitational potential energy of the system is lower because the projectile is below point Ⓑ. (e) As the projectile passes through point Ⓑ, all of the energy of the isolated system is kinetic. (f) When the projectile reaches point Ⓒ, all of the energy of the isolated system is gravitational potential.

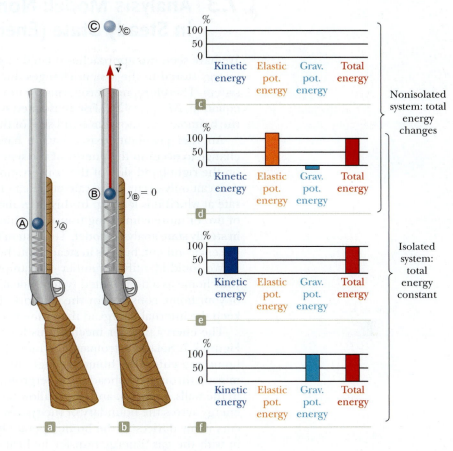

(B) Find the speed of the projectile as it moves through the equilibrium position Ⓑ of the spring as shown in Active Figure 7.6b.

SOLUTION

Analyze The energy of the system as the projectile moves through the equilibrium position of the spring includes only the kinetic energy of the projectile $\frac{1}{2}mv_\text{Ⓑ}^2$. Both types of potential energy are equal to zero for this configuration of the system.

Write a conservation of mechanical energy equation for the system between points Ⓐ and Ⓑ:

$$K_\text{Ⓑ} + U_{g\text{Ⓑ}} + U_{s\text{Ⓑ}} = K_\text{Ⓐ} + U_{g\text{Ⓐ}} + U_{s\text{Ⓐ}}$$

Substitute for each energy:

$$\tfrac{1}{2}mv_\text{Ⓑ}^2 + 0 + 0 = 0 + mgy_\text{Ⓐ} + \tfrac{1}{2}kx^2$$

Solve for $v_\text{Ⓑ}$:

$$v_\text{Ⓑ} = \sqrt{\frac{kx^2}{m} + 2gy_\text{Ⓐ}}$$

Substitute numerical values:

$$v_\text{Ⓑ} = \sqrt{\frac{(958\ \text{N/m})(0.120\ \text{m})^2}{(0.035\ 0\ \text{kg})} + 2(9.80\ \text{m/s}^2)(-0.120\ \text{m})} = \boxed{19.8\ \text{m/s}}$$

Finalize This example is the first one we have seen in which we must include two different types of potential energy. Notice in part (A) that we never needed to consider anything about the speed of the ball between points Ⓐ and Ⓒ, which is part of the power of the energy approach: changes in kinetic and potential energy depend only on the initial and final values, not on what happens between the configurations corresponding to these values.

⟨7.3 | Analysis Model: Nonisolated System in Steady State (Energy)

We have seen two approaches related to systems so far. In a nonisolated system, the energy stored in the system changes due to transfers across the boundaries of the system. Therefore, nonzero terms occur on both sides of the conservation of energy equation, $\Delta E_{system} = \Sigma T$. For an isolated system, no energy transfer takes place across the boundary, so the right-hand side of the equation is zero; that is, $\Delta E_{system} = 0$.

Another possibility exists that we have not yet addressed. It is possible for no change to occur in the energy of the system even though nonzero terms are present on the right-hand side of the conservation of energy equation, $0 = \Sigma T$. This situation can only occur if the rate at which energy is entering the system is equal to the rate at which it is leaving. In this case, the system is in steady state under the effects of two or more competing transfers, which we describe with the **nonisolated system in steady state** analysis model. The system is nonisolated because it is interacting with the environment, but it is in steady state because the system energy remains constant.

We could identify a number of examples of this type of situation. First, consider your home as a nonisolated system. Ideally, you would like to keep the temperature of your home constant for the comfort of the occupants. Therefore, your goal is to keep the internal energy in the home fixed.

The energy transfer mechanisms for the home are numerous, as we can see in Figure 7.7. Solar electromagnetic radiation is absorbed by the roof and walls of the home and enters the home through the windows. Energy enters by electrical transmission through overhead or underground wires to operate electrical devices. Leaks in the walls, windows, and doors allow warm or cold air to enter and leave, carrying energy across the boundary of the system by matter transfer. Matter transfer also occurs if any devices in the home operate from natural gas because energy is carried in with the gas. Energy transfer by heat occurs through the walls, windows, floor, and roof as a result of temperature differences between the inside and outside of the home. Therefore, we have a variety of transfers, but the energy in the home remains constant in the idealized case. In reality, the home is a system in *quasi-steady state* because some small temperature variations actually occur over a 24-h period, but we can imagine an idealized situation that conforms to the nonisolated system in steady-state model.

As a second example, consider the Earth and its atmosphere as a system. Because this system is located in the vacuum of space, the only possible types of energy

Figure 7.7 Energy enters and leaves a home by several mechanisms. The home can be modeled as a non-isolated system in steady state.

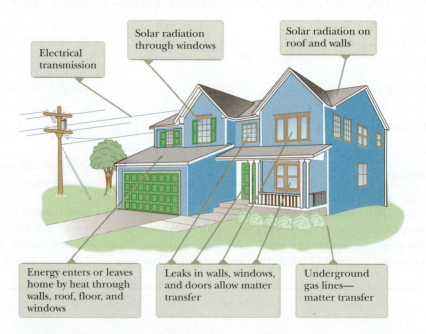

Electrical transmission

Solar radiation through windows

Solar radiation on roof and walls

Energy enters or leaves home by heat through walls, roof, floor, and windows

Leaks in walls, windows, and doors allow matter transfer

Underground gas lines— matter transfer

transfers are those that involve no contact between the system and external molecules in the environment. As mentioned in the footnote on page 194, only two types of transfer do not depend on contact with molecules: work done by field forces and electromagnetic radiation. The Earth–atmosphere system exchanges energy with the rest of the Universe only by means of electromagnetic radiation (ignoring work done by field forces and ignoring some small matter transfer as a result of cosmic ray particles and meteoroids entering the system and spacecraft leaving the system!). The primary input radiation is that from the Sun, and the output radiation is primarily infrared radiation emitted from the atmosphere and the ground. Ideally, these transfers are balanced so that the Earth maintains a constant temperature. In reality, however, the transfers are not *exactly* balanced, so the Earth is in quasi-steady state; measurements of the temperature show that it does appear to be changing. The change in temperature is very gradual and currently appears to be in the positive direction. This change is the essence of the social issue of global warming. (See Context 5, beginning on page 513.)

If we consider a time interval of several days, the human body can be modeled as another nonisolated system in steady state. If the body is at rest at the beginning and end of the time interval, there is no change in kinetic energy. Assuming that no major weight gain or loss occurs during this time interval, the amount of potential energy stored in the body as food in the stomach and fat remains constant on the average. If no fevers are experienced during this time interval, the internal energy of the body remains constant. Therefore, the change in the energy of the system is zero. Energy transfer methods during this time interval include work (you apply forces on objects that move), heat (your body is warmer than the surrounding air), matter transfer (breathing, eating), mechanical waves (you speak and hear), and electromagnetic radiation (you see, as well as absorb and emit radiation from your skin). Table 7.1 shows the amount of energy leaving the body by all methods during one hour of various activities.

TABLE 7.1 |
Energy Output for One Hour of Various Activities

Activity	Energy Output in One Hour (MJ)
Sleeping	0.27
Sitting at rest	0.42
Standing relaxed	0.44
Getting dressed	0.49
Typing	0.59
Walking on level ground (2.6 mi/h)	0.84
Painting a house	1.00
Bicycling on level ground (5.5 mi/h)	1.27
Shoveling snow	2.01
Swimming	2.09
Jogging (5.3 mi/h)	2.39
Rowing (20 strokes/min)	3.47
Walking up stairs	4.60

Adapted from L. Sherwood, *Fundamentals of Human Physiology*, 4th ed. (Belmont, CA: Brooks/Cole, 2012), p. 480

7.4 | Situations Involving Kinetic Friction

Consider again the book in Active Figure 6.18a sliding to the right on the surface of a heavy table and slowing down due to the friction force. Work is done by the friction force because there is a force and a displacement. Keep in mind, however, that our equations for work involve the displacement *of the point of application of the force*. A simple model of the friction force between the book and the surface is shown in Figure 7.8a. We have represented the entire friction force between the book and surface as being due to two identical teeth that have been spot-welded together.[2] One tooth projects upward from the surface, the other downward from the book, and they are welded at the points where they touch. The friction force acts at the junction of the two teeth. Imagine that the book slides a small distance d to the right as in Figure 7.8b. Because the teeth are modeled as identical, the junction of the teeth moves to the right by a distance $d/2$. Therefore, the displacement of the point of application of the friction force is $d/2$, but the displacement of the book is d!

In reality, the friction force is spread out over the entire contact area of an object sliding on a surface, so the force is not localized at a point. In addition, because the magnitudes of the friction forces at various points are constantly changing as individual spot welds occur, the surface and the book deform locally, and so on, the displacement of the point of application of the friction force is not at all the same as the displacement of the book. In fact, the displacement of the point of application of the friction force is not calculable and so neither is the work done by the friction force.

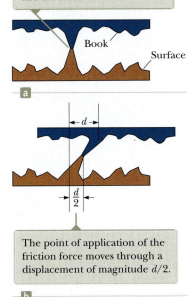

The entire friction force is modeled to be applied at the interface between two identical teeth projecting from the book and the surface.

The point of application of the friction force moves through a displacement of magnitude $d/2$.

Figure 7.8 (a) A simplified model of friction between a book and a surface. (b) The book is moved to the right by a distance d.

---

[2]Figure 7.8 and its discussion are inspired by a classic article on friction: B. A. Sherwood and W. H. Bernard, "Work and heat transfer in the presence of sliding friction," *American Journal of Physics*, **52**:1001, 1984.

The work–kinetic energy theorem is valid for a particle or an object that can be modeled as a particle. When a friction force acts, however, we cannot calculate the work done by friction. For such situations, Newton's second law is still valid for the system even though the work–kinetic energy theorem is not. The case of a non-deformable object like our book sliding on the surface[3] with friction can be handled in a relatively straightforward way.

Starting from a situation in which forces, including friction, are applied to the book, we can follow a similar procedure to that done in developing Equation 6.17. Let us start by writing Equation 6.8 for all forces other than friction:

$$\sum W_{\text{other forces}} = \int \left(\sum \vec{\mathbf{F}}_{\text{other forces}}\right) \cdot d\vec{\mathbf{r}} \qquad \text{7.11} \blacktriangleleft$$

The $d\vec{\mathbf{r}}$ in this equation is the displacement of the object because for forces other than friction, under the assumption that these forces do not deform the object, this displacement is the same as the displacement of the point of application of the forces. To each side of Equation 7.11 let us add the integral of the scalar product of the force of kinetic friction and $d\vec{\mathbf{r}}$. In doing so, we are not defining this quantity as work! We are simply saying that it is a quantity that can be calculated mathematically and will turn out to be useful to us in what follows.

$$\sum W_{\text{other forces}} + \int \vec{\mathbf{f}}_k \cdot d\vec{\mathbf{r}} = \int \left(\sum \vec{\mathbf{F}}_{\text{other forces}}\right) \cdot d\vec{\mathbf{r}} + \int \vec{\mathbf{f}}_k \cdot d\vec{\mathbf{r}}$$

$$= \int \left(\sum \vec{\mathbf{F}}_{\text{other forces}} + \vec{\mathbf{f}}_k\right) \cdot d\vec{\mathbf{r}}$$

The integrand on the right side of this equation is the net force $\sum \vec{\mathbf{F}}$, so

$$\sum W_{\text{other forces}} + \int \vec{\mathbf{f}}_k \cdot d\vec{\mathbf{r}} = \int \sum \vec{\mathbf{F}} \cdot d\vec{\mathbf{r}}$$

Incorporating Newton's second law $\sum \vec{\mathbf{F}} = m\vec{\mathbf{a}}$ gives

$$\sum W_{\text{other forces}} + \int \vec{\mathbf{f}}_k \cdot d\vec{\mathbf{r}} = \int m\vec{\mathbf{a}} \cdot d\vec{\mathbf{r}} = \int m\frac{d\vec{\mathbf{v}}}{dt} \cdot d\vec{\mathbf{r}} = \int_{t_i}^{t_f} m\frac{d\vec{\mathbf{v}}}{dt} \cdot \vec{\mathbf{v}}\, dt \qquad \text{7.12} \blacktriangleleft$$

where we have used Equation 3.3 to rewrite $d\vec{\mathbf{r}}$ as $\vec{\mathbf{v}}\, dt$. The scalar product obeys the product rule for differentiation (see Eq. B.30 in Appendix B.6), so the derivative of the scalar product of $\vec{\mathbf{v}}$ with itself can be written

$$\frac{d}{dt}(\vec{\mathbf{v}} \cdot \vec{\mathbf{v}}) = \frac{d\vec{\mathbf{v}}}{dt} \cdot \vec{\mathbf{v}} + \vec{\mathbf{v}} \cdot \frac{d\vec{\mathbf{v}}}{dt} = 2\frac{d\vec{\mathbf{v}}}{dt} \cdot \vec{\mathbf{v}}$$

where we have used the commutative property of the scalar product to justify the final expression in this equation. Consequently,

$$\frac{d\vec{\mathbf{v}}}{dt} \cdot \vec{\mathbf{v}} = \frac{1}{2}\frac{d}{dt}(\vec{\mathbf{v}} \cdot \vec{\mathbf{v}}) = \frac{1}{2}\frac{dv^2}{dt}$$

Substituting this result into Equation 7.12 gives

$$\sum W_{\text{other forces}} + \int \vec{\mathbf{f}}_k \cdot d\vec{\mathbf{r}} = \int_{t_i}^{t_f} m\left(\frac{1}{2}\frac{dv^2}{dt}\right) dt = \frac{1}{2}m\int_{v_i}^{v_f} d(v^2) = \frac{1}{2}mv_f^2 - \frac{1}{2}mv_i^2 = \Delta K$$

[3]The overall shape of the book remains the same, which is why we say it is nondeformable. On a microscopic level, however, there is deformation of the book's face as it slides over the surface.

Looking at the left side of this equation, notice that in the inertial frame of the surface, $\vec{\mathbf{f}}_k$ and $d\vec{\mathbf{r}}$ will be in opposite directions for every increment $d\vec{\mathbf{r}}$ of the path followed by the object. Therefore, $\vec{\mathbf{f}}_k \cdot d\vec{\mathbf{r}} = -f_k\, dr$. The previous expression now becomes

$$\sum W_{\text{other forces}} - \int f_k\, dr = \Delta K$$

In our model for friction, the magnitude of the kinetic friction force is constant, so f_k can be brought out of the integral. The remaining integral $\int dr$ is simply the sum of increments of length along the path, which is the total path length d. Therefore,

$$\sum W_{\text{other forces}} - f_k d = \Delta K \qquad \text{7.13} \blacktriangleleft$$

or

$$K_f = K_i - f_k d + \sum W_{\text{other forces}} \qquad \text{7.14} \blacktriangleleft$$

Equation 7.13 can be used when a friction force acts on an object. The change in kinetic energy is equal to the work done by all forces other than friction minus a term $f_k d$ associated with the friction force.

Considering the sliding book situation again, let's identify the larger system of the book *and* the surface as the book slows down under the influence of a friction force alone. There is no work done across the boundary of this system because the system does not interact with the environment. There are no other types of energy transfer occurring across the boundary of the system, assuming we ignore the inevitable sound the sliding book makes! In this case, Equation 7.2 becomes

$$\Delta E_{\text{system}} = \Delta K + \Delta E_{\text{int}} = 0$$

The change in kinetic energy of this book–surface system is the same as the change in kinetic energy of the book alone because the book is the only part of the system that is moving. Therefore, incorporating Equation 7.13 gives

$$-f_k d + \Delta E_{\text{int}} = 0$$

$$\Delta E_{\text{int}} = f_k d \qquad \text{7.15} \blacktriangleleft$$

▶ Change in internal energy due to a constant friction force within the system

The increase in internal energy of the system is therefore equal to the product of the friction force and the path length through which the block moves. In summary, a friction force transforms kinetic energy in a system to internal energy, and the increase in internal energy of the system is equal to its decrease in kinetic energy. Equation 7.13, with the help of Equation 7.15, can be written as

$$\sum W_{\text{other forces}} = W = \Delta K + \Delta E_{\text{int}}$$

which is a reduced form of Equation 7.2 and represents the nonisolated system model for a system within which a nonconservative force acts.

QUICK QUIZ 7.5 You are traveling along a freeway at 65 mi/h. Your car has kinetic energy. You suddenly skid to a stop because of congestion in traffic. Where is the kinetic energy your car once had? **(a)** It is all in internal energy in the road. **(b)** It is all in internal energy in the tires. **(c)** Some of it has transformed to internal energy and some of it transferred away by mechanical waves. **(d)** It is all transferred away from your car by various mechanisms.

> ## THINKING PHYSICS 7.1
>
> A car traveling at an initial speed v slides a distance d to a halt after its brakes lock. If the car's initial speed is instead $2v$ at the moment the brakes lock, estimate the distance it slides.
>
> **Reasoning** Let us assume the force of kinetic friction between the car and the road surface is constant and the same for both speeds. According to Equation 7.14, the friction force multiplied by the distance d is equal to the initial kinetic energy of the car (because $K_f = 0$ and there is no work done by other forces). If the speed is doubled, as it is in this example, the kinetic energy is quadrupled. For a given friction force, the distance traveled is four times as great when the initial speed is doubled, and so the estimated distance the car slides is $4d$. This result agrees with that in part (b) of Example 5.1, but it is determined by using energy techniques rather than force techniques. ◄

Example 7.4 | A Block Pulled on a Rough Surface

A 6.0-kg block initially at rest is pulled to the right along a horizontal surface by a constant horizontal force of 12 N.

(A) Find the speed of the block after it has moved 3.0 m if the surfaces in contact have a coefficient of kinetic friction of 0.15.

SOLUTION

Conceptualize This example is Example 6.6 (page 168), modified so that the surface is no longer frictionless. The rough surface applies a friction force on the block opposite to the applied force. As a result, we expect the speed to be lower than that found in Example 6.6.

Categorize The block is pulled by a force and the surface is rough, so we model the block–surface system as nonisolated with a nonconservative force acting.

Analyze Active Figure 7.9a illustrates this situation. Neither the normal force nor the gravitational force does work on the system because their points of application are displaced horizontally.

Active Figure 7.9 (Example 7.4) (a) A block pulled to the right on a rough surface by a constant horizontal force. (b) The applied force is at an angle θ to the horizontal.

Find the work done on the system by the applied force just as in Example 6.6:

$$\sum W_{\text{other forces}} = W_F = F\Delta x$$

Apply the particle in equilibrium model to the block in the vertical direction:

$$\sum F_y = 0 \;\rightarrow\; n - mg = 0 \;\rightarrow\; n = mg$$

Find the magnitude of the friction force:

$$f_k = \mu_k n = \mu_k mg = (0.15)(6.0\ \text{kg})(9.80\ \text{m/s}^2) = 8.82\ \text{N}$$

Find the final speed of the block from Equation 7.14:

$$\tfrac{1}{2}mv_f^2 = \tfrac{1}{2}mv_i^2 - f_k d + W_F$$

$$v_f = \sqrt{v_i^2 + \frac{2}{m}(-f_k d + F\Delta x)}$$

Substitute numerical values:

$$v_f = \sqrt{0 + \frac{2}{6.0\ \text{kg}}[-(8.82\ \text{N})(3.0\ \text{m}) + (12\ \text{N})(3.0\ \text{m})]} = \boxed{1.8\ \text{m/s}}$$

Finalize As expected, this value is less than the 3.5 m/s found in the case of the block sliding on a frictionless surface (see Example 6.6). The difference in kinetic energies between the block in Example 6.6 and the block in this example is equal to the increase in internal energy of the block–surface system in this example.

(B) Suppose the force $\vec{F}$ is applied at an angle θ as shown in Active Figure 7.9b. At what angle should the force be applied to achieve the largest possible speed after the block has moved 3.0 m to the right?

7.4 *cont.*

SOLUTION

Conceptualize You might guess that $\theta = 0$ would give the largest speed because the force would have the largest component possible in the direction parallel to the surface. Think about $\vec{F}$ applied at an arbitrary nonzero angle, however. Although the horizontal component of the force would be reduced, the vertical component of the force would reduce the normal force, in turn reducing the force of friction, which suggests that the speed could be maximized by pulling at an angle other than $\theta = 0$.

Categorize As in part (A), we model the block–surface system as nonisolated with a nonconservative force acting.

Analyze Find the work done by the applied force, noting that $\Delta x = d$ because the path followed by the block is a straight line:

$$\sum W_{\text{other forces}} = W_F = F \Delta x \cos \theta = Fd \cos \theta$$

Apply the particle in equilibrium model to the block in the vertical direction:

$$\sum F_y = n + F \sin \theta - mg = 0$$

Solve for n:

$$n = mg - F \sin \theta$$

Use Equation 7.14 to find the final kinetic energy for this situation:

$$K_f = K_i - f_k d + W_F$$
$$= 0 - \mu_k n d + Fd \cos \theta = -\mu_k (mg - F \sin \theta) d + Fd \cos \theta$$

Maximizing the speed is equivalent to maximizing the final kinetic energy. Consequently, differentiate K_f with respect to θ and set the result equal to zero:

$$\frac{dK_f}{d\theta} = -\mu_k(0 - F \cos \theta) d - Fd \sin \theta = 0$$
$$\mu_k \cos \theta - \sin \theta = 0$$
$$\tan \theta = \mu_k$$

Evaluate θ for $\mu_k = 0.15$:

$$\theta = \tan^{-1}(\mu_k) = \tan^{-1}(0.15) = \boxed{8.5°}$$

Finalize Notice that the angle at which the speed of the block is a maximum is indeed not $\theta = 0$. When the angle exceeds 8.5°, the horizontal component of the applied force is too small to be compensated by the reduced friction force and the speed of the block begins to decrease from its maximum value.

Example 7.5 | A Block–Spring System

A block of mass 1.6 kg is attached to a horizontal spring that has a force constant of 1 000 N/m as shown in Figure 7.10. The spring is compressed 2.0 cm and is then released from rest.

(A) Calculate the speed of the block as it passes through the equilibrium position $x = 0$ if the surface is frictionless.

SOLUTION

Conceptualize This situation has been discussed before, and it is easy to visualize the block being pushed to the right by the spring and moving with some speed at $x = 0$.

Categorize We identify the system as the block and model the block as a nonisolated system.

Analyze In this situation, the block starts with $v_i = 0$ at $x_i = -2.0$ cm, and we want to find v_f at $x_f = 0$.

Use Equation 6.11 to find the work done by the spring on the system with $x_{\text{max}} = x_i$:

$$\sum W_{\text{other forces}} = W_s = \tfrac{1}{2} k x_{\text{max}}^2$$

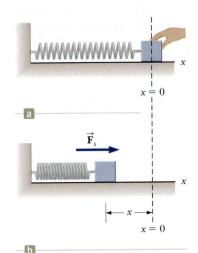

Figure 7.10 (Example 7.5) (a) A block attached to a spring is pushed inward from an initial position $x = 0$ by an external agent. (b) At position x, the block is released from rest and the spring pushes it to the right.

continued

7.5 *cont.*

Work is done on the block, and its speed changes. The conservation of energy equation, Equation 7.2, reduces to the work–kinetic energy theorem. Use that theorem to find the speed at $x = 0$:

$$W_s = \tfrac{1}{2}mv_f^2 - \tfrac{1}{2}mv_i^2$$

$$v_f = \sqrt{v_i^2 + \frac{2}{m}W_s} = \sqrt{v_i^2 + \frac{2}{m}\left(\tfrac{1}{2}kx_{max}^2\right)}$$

Substitute numerical values:

$$v_f = \sqrt{0 + \frac{2}{1.6 \text{ kg}}\left[\tfrac{1}{2}(1\,000 \text{ N/m})(0.020 \text{ m})^2\right]} = \boxed{0.50 \text{ m/s}}$$

Finalize Although this problem could have been solved in Chapter 6, it is presented here to provide contrast with the following part (B), which requires the techniques of this chapter.

(B) Calculate the speed of the block as it passes through the equilibrium position if a constant friction force of 4.0 N retards its motion from the moment it is released.

SOLUTION

Conceptualize The correct answer must be less than that found in part (A) because the friction force retards the motion.

Categorize We identify the system as the block and the surface. The system is nonisolated because of the work done by the spring, and there is a nonconservative force acting: the friction between the block and the surface.

Analyze Write Equation 7.14:

$$(1)\ K_f = K_i - f_k d + W_s$$

Substitute numerical values:

$$K_f = 0 - (4.0 \text{ N})(0.020 \text{ m}) + \tfrac{1}{2}(1\,000 \text{ N/m})(0.020 \text{ m})^2 = 0.12 \text{ J}$$

Write the definition of kinetic energy:

$$K_f = \tfrac{1}{2}mv_f^2$$

Solve for v_f and substitute numerical values:

$$v_f = \sqrt{\frac{2K_f}{m}} = \sqrt{\frac{2(0.12 \text{ J})}{1.6 \text{ kg}}} = \boxed{0.39 \text{ m/s}}$$

Finalize As expected, this value is less than the 0.50 m/s found in part (A).

What If? What if the friction force were increased to 10.0 N? What is the block's speed at $x = 0$?

Answer In this case, the value of $f_k d$ as the block moves to $x = 0$ is

$$f_k d = (10.0 \text{ N})(0.020 \text{ m}) = 0.20 \text{ J}$$

which is equal in magnitude to the kinetic energy at $x = 0$ for the frictionless case. (Verify it!) Therefore, all the kinetic energy has been transformed to internal energy by friction when the block arrives at $x = 0$, and its speed at this point is $v = 0$.

In this situation as well as that in part (B), the speed of the block reaches a maximum at some position other than $x = 0$. Problem 65 asks you to locate these positions.

7.5 | Changes in Mechanical Energy for Nonconservative Forces

Consider the book sliding across the surface in the preceding section. As the book moves through a distance d, the only force that does work on it is the force of kinetic friction. This force causes a change $-f_k d$ in the kinetic energy of the book as described by Equation 7.13.

Now, however, suppose the book is part of a system that also exhibits a change in potential energy. In this case, $-f_k d$ is the amount by which the mechanical energy of the system changes because of the force of kinetic friction. For example, if the book moves on an incline that is not frictionless, there is a change in both the kinetic energy and the gravitational potential energy of the book–Earth system. Consequently,

$$\Delta E_{\text{mech}} = \Delta K + \Delta U_g = -f_k d$$

In general, if a friction force acts within an isolated system,

$$\Delta E_{\text{mech}} = \Delta K + \Delta U = -f_k d \qquad \text{7.16} \blacktriangleleft$$

▶ Change in mechanical energy of a system due to friction within the system

where ΔU is the change in all forms of potential energy. Notice that Equation 7.16 reduces to Equation 7.10 if the friction force is zero.

If the system in which nonconservative forces act is nonisolated and the external influence on the system is by means of work, the generalization of Equation 7.13 is

$$\Delta E_{\text{mech}} = -f_k d + \sum W_{\text{other forces}} \qquad \text{7.17} \blacktriangleleft$$

Equation 7.17, with the help of Equations 7.7 and 7.15, can be written as

$$\sum W_{\text{other forces}} = W = \Delta K + \Delta U + \Delta E_{\text{int}}$$

This reduced form of Equation 7.2 represents the nonisolated system model for a system that possesses potential energy and within which a nonconservative force acts. In practice during problem solving, you do not need to use equations like Equation 7.15 or Equation 7.17. You can simply use Equation 7.2 and keep only those terms in the equation that correspond to the physical situation. See Example 7.8 for a sample of this approach.

> **PROBLEM-SOLVING STRATEGY: Systems with Nonconservative Forces**

The following procedure should be used when you face a problem involving a system in which nonconservative forces act:

1. **Conceptualize.** Study the physical situation carefully and form a mental representation of what is happening.

2. **Categorize.** Define your system, which may consist of more than one object. The system could include springs or other possibilities for storage of potential energy. Determine whether any nonconservative forces are present. If not, use the principle of conservation of mechanical energy as outlined in Section 7.2. If so, use the procedure discussed below.

    Determine if any work is done across the boundary of your system by forces other than friction. If so, use Equation 7.17 to analyze the problem. If not, use Equation 7.16.

3. **Analyze.** Choose configurations to represent the initial and final conditions of the system. For each object that changes elevation, select a reference position for the object that defines the zero configuration of gravitational potential energy for the system. For an object on a spring, the zero configuration for elastic potential energy is when the object is at its equilibrium position. If there is more than one conservative force, write an expression for the potential energy associated with each force.

    Use either Equation 7.16 or Equation 7.17 to establish a mathematical representation of the problem. Solve for the unknown.

4. **Finalize.** Make sure your results are consistent with your mental representation. Also make sure the values of your results are reasonable and consistent with connections to everyday experience.

Example 7.6 | Crate Sliding Down a Ramp

A 3.00-kg crate slides down a ramp. The ramp is 1.00 m in length and inclined at an angle of 30.0° as shown in Figure 7.11. The crate starts from rest at the top, experiences a constant friction force of magnitude 5.00 N, and continues to move a short distance on the horizontal floor after it leaves the ramp.

(A) Use energy methods to determine the speed of the crate at the bottom of the ramp.

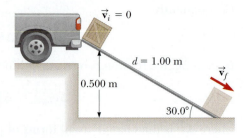

SOLUTION

Conceptualize Imagine the crate sliding down the ramp in Figure 7.11. The larger the friction force, the more slowly the crate will slide.

Figure 7.11 (Example 7.6) A crate slides down a ramp under the influence of gravity. The potential energy of the system decreases, whereas the kinetic energy increases.

Categorize We identify the crate, the surface, and the Earth as the system. The system is categorized as isolated with a nonconservative force acting.

Analyze Because $v_i = 0$, the initial kinetic energy of the system when the crate is at the top of the ramp is zero. If the y coordinate is measured from the bottom of the ramp (the final position of the crate, for which we choose the gravitational potential energy of the system to be zero) with the upward direction being positive, then $y_i = 0.500$ m.

Write the expression for the total mechanical energy of the system when the crate is at the top:

$$E_i = K_i + U_i = 0 + U_i = mgy_i$$

Write an expression for the final mechanical energy:

$$E_f = K_f + U_f = \tfrac{1}{2}mv_f^2 + 0 = \tfrac{1}{2}mv_f^2$$

Apply Equation 7.16:

$$\Delta E_{\text{mech}} = E_f - E_i = \tfrac{1}{2}mv_f^2 - mgy_i = -f_k d$$

Solve for v_f:

$$(1)\quad v_f = \sqrt{\frac{2}{m}(mgy_i - f_k d)}$$

Substitute numerical values:

$$v_f = \sqrt{\frac{2}{3.00\ \text{kg}}[(3.00\ \text{kg})(9.80\ \text{m/s}^2)(0.500\ \text{m}) - (5.00\ \text{N})(1.00\ \text{m})]} = \boxed{2.54\ \text{m/s}}$$

(B) How far does the crate slide on the horizontal floor if it continues to experience a friction force of magnitude 5.00 N?

SOLUTION

Analyze This part of the problem is handled in exactly the same way as part (A), but in this case we can consider the mechanical energy of the system to consist only of kinetic energy because the potential energy of the system remains fixed.

Write an expression for the mechanical energy of the system when the crate leaves the bottom of the ramp:

$$E_i = K_i = \tfrac{1}{2}mv_i^2$$

Apply Equation 7.16 with $E_f = 0$:

$$E_f - E_i = 0 - \tfrac{1}{2}mv_i^2 = -f_k d \quad \rightarrow \quad \tfrac{1}{2}mv_i^2 = f_k d$$

Solve for the distance d and substitute numerical values:

$$d = \frac{mv_i^2}{2f_k} = \frac{(3.00\ \text{kg})(2.54\ \text{m/s})^2}{2(5.00\ \text{N})} = \boxed{1.94\ \text{m}}$$

Finalize For comparison, you may want to calculate the speed of the crate at the bottom of the ramp in the case in which the ramp is frictionless. Also notice that the increase in internal energy of the system as the crate slides down the ramp is $f_k d = (5.00\ \text{N})(1.00\ \text{m}) = 5.00\ \text{J}$. This energy is shared between the crate and the surface, each of which is a bit warmer than before.

Also notice that the distance d the object slides on the horizontal surface is infinite if the surface is frictionless. Is that consistent with your conceptualization of the situation?

What If? A cautious worker decides that the speed of the crate when it arrives at the bottom of the ramp may be so large that its contents may be damaged. Therefore, he replaces the ramp with a longer one such that the new ramp makes an angle of 25.0° with the ground. Does this new ramp reduce the speed of the crate as it reaches the ground?

7.6 *cont.*

Answer Because the ramp is longer, the friction force acts over a longer distance and transforms more of the mechanical energy into internal energy. The result is a reduction in the kinetic energy of the crate, and we expect a lower speed as it reaches the ground.

Find the length d of the new ramp:

$$\sin 25.0° = \frac{0.500 \text{ m}}{d} \quad \rightarrow \quad d = \frac{0.500 \text{ m}}{\sin 25.0°} = 1.18 \text{ m}$$

Find v_f from Equation (1) in part (A):

$$v_f = \sqrt{\frac{2}{3.00 \text{ kg}}[(3.00 \text{ kg})(9.80 \text{ m/s}^2)(0.500 \text{ m}) - (5.00 \text{ N})(1.18 \text{ m})]} = 2.42 \text{ m/s}$$

The final speed is indeed lower than in the higher-angle case.

Example 7.7 | Block–Spring Collision

A block having a mass of 0.80 kg is given an initial velocity $v_Ⓐ = 1.2$ m/s to the right and collides with a spring whose mass is negligible and whose force constant is $k = 50$ N/m as shown in Figure 7.12.

(A) Assuming the surface to be frictionless, calculate the maximum compression of the spring after the collision.

SOLUTION

Conceptualize The various parts of Figure 7.12 help us imagine what the block will do in this situation. All motion takes place in a horizontal plane, so we do not need to consider changes in gravitational potential energy.

Categorize We identify the system to be the block and the spring. The block–spring system is isolated with no nonconservative forces acting.

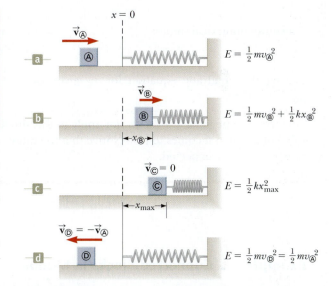

Figure 7.12 (Example 7.7) A block sliding on a frictionless, horizontal surface collides with a light spring. (a) Initially, the mechanical energy is all kinetic energy. (b) The mechanical energy is the sum of the kinetic energy of the block and the elastic potential energy in the spring. (c) The energy is entirely potential energy. (d) The energy is transformed back to the kinetic energy of the block. The total energy of the system remains constant throughout the motion.

Analyze Before the collision, when the block is at Ⓐ, it has kinetic energy and the spring is uncompressed, so the elastic potential energy stored in the system is zero. Therefore, the total mechanical energy of the system before the collision is just $\frac{1}{2}mv_Ⓐ^2$. After the collision, when the block is at Ⓒ, the spring is fully compressed; now the block is at rest and so has zero kinetic energy. The elastic potential energy stored in the system, however, has its maximum value $\frac{1}{2}kx^2 = \frac{1}{2}kx_{max}^2$, where the origin of coordinates $x = 0$ is chosen to be the equilibrium position of the spring and x_{max} is the maximum compression of the spring, which in this case happens to be $x_Ⓒ$. The total mechanical energy of the system is conserved because no nonconservative forces act on objects within the isolated system.

Write a conservation of mechanical energy equation:

$$K_Ⓒ + U_{sⒸ} = K_Ⓐ + U_{sⒶ}$$

$$0 + \frac{1}{2}kx_{max}^2 = \frac{1}{2}mv_Ⓐ^2 + 0$$

Solve for x_{max} and evaluate:

$$x_{max} = \sqrt{\frac{m}{k}}v_Ⓐ = \sqrt{\frac{0.80 \text{ kg}}{50 \text{ N/m}}}(1.2 \text{ m/s}) = \boxed{0.15 \text{ m}}$$

(B) Suppose a constant force of kinetic friction acts between the block and the surface, with $\mu_k = 0.50$. If the speed of the block at the moment it collides with the spring is $v_Ⓐ = 1.2$ m/s, what is the maximum compression $x_Ⓒ$ in the spring?

continued

7.7 *cont.*

SOLUTION

Conceptualize Because of the friction force, we expect the compression of the spring to be smaller than in part (A) because some of the block's kinetic energy is transformed to internal energy in the block and the surface.

Categorize We identify the system as the block, the surface, and the spring. This system is isolated but now involves a nonconservative force.

Analyze In this case, the mechanical energy $E_{mech} = K + U_s$ of the system is *not* conserved because a friction force acts on the block. From the particle in equilibrium model in the vertical direction, we see that $n = mg$.

Evaluate the magnitude of the friction force:

$$f_k = \mu_k n = \mu_k mg$$

Write the change in the mechanical energy of the system due to friction as the block is displaced from $x = 0$ to $x_{©}$:

$$\Delta E_{mech} = -f_k x_{©}$$

Substitute the initial and final energies:

$$\Delta E_{mech} = E_f - E_i = (0 + \tfrac{1}{2}kx_{©}^2) - (\tfrac{1}{2}mv_{Ⓐ}^2 + 0) = -f_k x_{©}$$

$$\tfrac{1}{2}kx_{©}^2 - \tfrac{1}{2}mv_{Ⓐ}^2 = -\mu_k mg x_{©}$$

Substitute numerical values:

$$\tfrac{1}{2}(50)x_{©}^2 - \tfrac{1}{2}(0.80)(1.2)^2 = -(0.50)(0.80)(9.80)x_{©}$$

$$25x_{©}^2 + 3.9x_{©} - 0.58 = 0$$

Solving the quadratic equation for $x_{©}$ gives $x_{©} = 0.093$ m and $x_{©} = -0.25$ m. The physically meaningful root is $x_{©} = \boxed{0.093 \text{ m.}}$

Finalize The negative root does not apply to this situation because the block must be to the right of the origin (positive value of x) when it comes to rest. Notice that the value of 0.093 m is less than the distance obtained in the frictionless case of part (A) as we expected.

Example 7.8 | Connected Blocks in Motion

Two blocks are connected by a light string that passes over a frictionless pulley as shown in Figure 7.13. The block of mass m_1 lies on a horizontal surface and is connected to a spring of force constant k. The system is released from rest when the spring is unstretched. If the hanging block of mass m_2 falls a distance h before coming to rest, calculate the coefficient of kinetic friction between the block of mass m_1 and the surface.

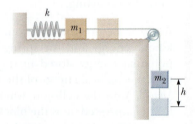

SOLUTION

Conceptualize The key word *rest* appears twice in the problem statement. This word suggests that the configurations of the system associated with rest are good candidates for the initial and final configurations because the kinetic energy of the system is zero for these configurations.

Categorize In this situation, the system consists of the two blocks, the spring, the surface, and the Earth. The system is isolated with a nonconservative force acting. We also model the sliding block as a particle in equilibrium in the vertical direction, leading to $n = m_1 g$.

Figure 7.13 (Example 7.8) As the hanging block moves from its highest elevation to its lowest, the system loses gravitational potential energy but gains elastic potential energy in the spring. Some mechanical energy is transformed to internal energy because of friction between the sliding block and the surface.

Analyze We need to consider two forms of potential energy for the system, gravitational and elastic: $\Delta U_g = U_{gf} - U_{gi}$ is the change in the system's gravitational potential energy, and $\Delta U_s = U_{sf} - U_{si}$ is the change in the system's elastic potential energy. The change in the gravitational potential energy of the system is associated with only the falling block because the vertical coordinate of the horizontally sliding block does not change. The initial and final kinetic energies of the system are zero, so $\Delta K = 0$.

7.8 *cont.*

For this example, let us start from Equation 7.2 to show how this approach would work in practice. Because the system is isolated, the entire right side of Equation 7.2 is zero. Based on the physical situation described in the problem, we see that there could be changes of kinetic energy, potential energy, and internal energy in the system. Write the corresponding reduction of Equation 7.2:

$$\Delta K + \Delta U + \Delta E_{\text{int}} = 0$$

Incorporate into this equation that $\Delta K = 0$ and that there are two types of potential energy:

$$(1)\ \Delta U_g + \Delta U_s + \Delta E_{\text{int}} = 0$$

Use Equation 7.15 to find the change in internal energy in the system due to friction between the horizontally sliding block and the surface, noticing that as the hanging block falls a distance h, the horizontally moving block moves the same distance h to the right:

$$(2)\ \Delta E_{\text{int}} = f_k h = (\mu_k n)\, h = \mu_k m_1 g h$$

Evaluate the change in gravitational potential energy of the system, choosing the configuration with the hanging block at the lowest position to represent zero potential energy:

$$(3)\ \Delta U_g = U_{gf} - U_{gi} = 0 - m_2 g h$$

Evaluate the change in the elastic potential energy of the system:

$$(4)\ \Delta U_s = U_{sf} - U_{si} = \tfrac{1}{2} k h^2 - 0$$

Substitute Equations (2), (3), and (4) into Equation (1):

$$-m_2 g h + \tfrac{1}{2} k h^2 + \mu_k m_1 g h = 0$$

Solve for μ_k:

$$\mu_k = \frac{m_2 g - \tfrac{1}{2} k h}{m_1 g}$$

Finalize This setup represents a method of measuring the coefficient of kinetic friction between an object and some surface. Notice that we do not need to remember which energy equation goes with which type of problem with this approach. You can always begin with Equation 7.2 and then tailor it to the physical situation. This process may include deleting terms, such as the kinetic energy term and all terms on the right-hand side in this example. It can also include expanding terms, such as rewriting ΔU due to two types of potential energy in this example.

THINKING PHYSICS 7.2

The energy bar charts in Figure 7.14 show three instants in the motion of the system in Figure 7.13 and described in Example 7.8. For each bar chart, identify the configuration of the system that corresponds to the chart.

Reasoning In Figure 7.14a, there is no kinetic energy in the system. Therefore, nothing in the system is moving. The bar chart shows that the system contains only gravitational potential energy and no internal energy yet, which corresponds to the configuration with the darker blocks in Figure 7.13 and represents the instant just after the system is released.

In Figure 7.14b, the system contains four types of energy. The height of the gravitational potential energy bar is at 50%, which tells us that the hanging block has moved halfway between its position corresponding to Figure 7.14a and the position defined as $y = 0$. Therefore, in this configuration, the hanging block is between the dark and light images of the hanging block in Figure 7.13. The system has gained kinetic energy because the blocks are

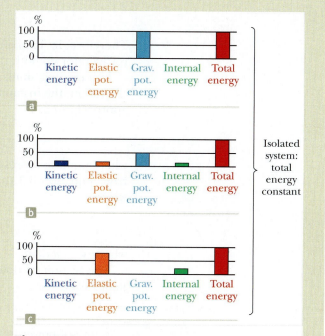

Figure 7.14 (Thinking Physics 7.2) Three energy bar charts are shown for the system in Figure 7.13.

moving, elastic potential energy because the spring is stretching, and internal energy because of friction between the block of mass m_1 and the surface.

In Figure 7.14c, the height of the gravitational potential energy bar is zero, telling us that the hanging block is at $y = 0$. In addition, the height of the kinetic energy bar is zero, indicating that the blocks have stopped moving momentarily. Therefore, the configuration of the system is that shown by the light images of the blocks in Figure 7.13. The height of the elastic potential energy bar is high because the spring is stretched its maximum amount. The height of the internal energy bar is higher than in Figure 7.14b because the block of mass m_1 has continued to slide over the surface. ◄

7.6 | Power

Consider Thinking Physics 6.1 again, which involved rolling a refrigerator up a ramp into a truck. Suppose the man is not convinced the work is the same regardless of the ramp's length and sets up a long ramp with a gentle rise. Although he does the same amount of work as someone using a shorter ramp, he takes longer to do the work because he has to move the refrigerator over a greater distance. Although the work done on both ramps is the same, there is *something* different about the tasks: the *time interval* during which the work is done.

The time rate of energy transfer is called the **instantaneous power** P and is defined as

► Definition of power

$$P \equiv \frac{dE}{dt}$$ 7.18 ◄

We will focus on work as the energy transfer method in this discussion, but keep in mind that the notion of power is valid for *any* means of energy transfer discussed in Section 7.1. If an external force is applied to an object (which we model as a particle) and if the work done by this force on the object in the time interval Δt is W, the **average power** during this interval is

$$P_{avg} = \frac{W}{\Delta t}$$

Therefore, in Thinking Physics 6.1, although the same work is done in rolling the refrigerator up both ramps, less power is required for the longer ramp.

In a manner similar to the way we approached the definition of velocity and acceleration, the instantaneous power is the limiting value of the average power as Δt approaches zero:

$$P = \lim_{\Delta t \to 0} \frac{W}{\Delta t} = \frac{dW}{dt}$$

where we have represented the infinitesimal value of the work done by dW. We find from Equation 6.3 that $dW = \vec{\mathbf{F}} \cdot d\vec{\mathbf{r}}$. Therefore, the instantaneous power can be written

$$P = \frac{dW}{dt} = \vec{\mathbf{F}} \cdot \frac{d\vec{\mathbf{r}}}{dt} = \vec{\mathbf{F}} \cdot \vec{\mathbf{v}}$$ 7.19 ◄

where $\vec{\mathbf{v}} = d\vec{\mathbf{r}}/dt$.

The SI unit of power is joules per second (J/s), also called the **watt** (W) after James Watt:

► The watt

$$1 \text{ W} = 1 \text{ J/s} = 1 \text{ kg} \cdot \text{m}^2/\text{s}^3$$

A unit of power in the U.S. customary system is the **horsepower** (hp):

$$1 \text{ hp} = 746 \text{ W}$$

A unit of energy (or work) can now be defined in terms of the unit of power. One **kilowatt-hour** (kWh) is the energy transferred in 1 h at the constant rate of 1 kW = 1 000 J/s. The amount of energy represented by 1 kWh is

$$1 \text{ kWh} = (10^3 \text{ W})(3\,600 \text{ s}) = 3.60 \times 10^6 \text{ J}$$

A kilowatt-hour is a unit of energy, not power. When you pay your electric bill, you are buying energy, and the amount of energy transferred by electrical transmission into a home during the period represented by the electric bill is usually expressed in kilowatt-hours. For example, your bill may state that you used 900 kWh of energy during a month and that you are being charged at the rate of 10¢ per kilowatt-hour. Your obligation is then $90 for this amount of energy. As another example, suppose an electric bulb is rated at 100 W. In 1.00 h of operation, it would have energy transferred to it by electrical transmission in the amount of (0.100 kW)(1.00 h) = 0.100 kWh = 3.60×10^5 J.

Example 7.9 | Power Delivered by an Elevator Motor

An elevator car (Fig. 7.15a) has a mass of 1 600 kg and is carrying passengers having a combined mass of 200 kg. A constant friction force of 4 000 N retards its motion.

(A) How much power must a motor deliver to lift the elevator car and its passengers at a constant speed of 3.00 m/s?

SOLUTION

Conceptualize The motor must supply the force of magnitude *T* that pulls the elevator car upward.

Categorize The friction force increases the power necessary to lift the elevator. The problem states that the speed of the elevator is constant, which tells us that *a* = 0. We model the elevator as a particle in equilibrium.

Analyze The free-body diagram in Figure 7.15b specifies the upward direction as positive. The *total* mass *M* of the system (car plus passengers) is equal to 1 800 kg.

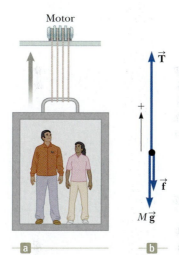

Figure 7.15 (Example 7.9) (a) The motor exerts an upward force $\vec{\mathbf{T}}$ on the elevator car. The magnitude of this force is the total tension *T* in the cables connecting the car and motor. The downward forces acting on the car are a friction force $\vec{\mathbf{f}}$ and the gravitational force $\vec{\mathbf{F}}_g = M\vec{\mathbf{g}}$. (b) The free-body diagram for the elevator car.

Using the particle in equilibrium model, apply Newton's second law to the car:

$$\sum F_y = T - f - Mg = 0$$

Solve for *T*:

$$T = f + Mg$$

Use Equation 7.19 and that $\vec{\mathbf{T}}$ is in the same direction as $\vec{\mathbf{v}}$ to find the power:

$$P = \vec{\mathbf{T}} \cdot \vec{\mathbf{v}} = Tv = (f + Mg)v$$

Substitute numerical values:

$$P = [(4\,000 \text{ N}) + (1\,800 \text{ kg})(9.80 \text{ m/s}^2)](3.00 \text{ m/s}) = \boxed{6.49 \times 10^4 \text{ W}}$$

(B) What power must the motor deliver at the instant the speed of the elevator is *v* if the motor is designed to provide the elevator car with an upward acceleration of 1.00 m/s²?

SOLUTION

Conceptualize In this case, the motor must supply the force of magnitude *T* that pulls the elevator car upward with an increasing speed. We expect that more power will be required to do that than in part (A) because the motor must now perform the additional task of accelerating the car.

Categorize In this case, we model the elevator car as a particle under a net force because it is accelerating.

Analyze Using the particle under a net force model, apply Newton's second law to the car:

$$\sum F_y = T - f - Mg = Ma$$

Solve for *T*:

$$T = M(a + g) + f$$

7.9 *cont.*

Use Equation 7.19 to obtain the required power:

$$P = Tv = [M(a + g) + f]v$$

Substitute numerical values:

$$P = [(1\ 800\ \text{kg})(1.00\ \text{m/s}^2 + 9.80\ \text{m/s}^2) + 4\ 000\ \text{N}]v$$

$$= (2.34 \times 10^4)v$$

where v is the instantaneous speed of the car in meters per second and P is in watts.

Finalize To compare with part (A), let $v = 3.00$ m/s, giving a power of

$$P = (2.34 \times 10^4\ \text{N})(3.00\ \text{m/s}) = 7.02 \times 10^4\ \text{W}$$

which is larger than the power found in part (A), as expected.

7.7 | Context Connection: Horsepower Ratings of Automobiles

As discussed in Section 4.8, an automobile moves because of Newton's third law. The engine attempts to rotate the wheels in such a direction as to push the Earth toward the back of the car because of the friction force between the wheels and the roadway. By Newton's third law, the Earth pushes in the opposite direction on the wheels, which is toward the front of the car. Because the Earth is much more massive than the car, the Earth remains stationary while the car moves forward.

This principle is the same one humans use for walking. By pushing your leg backward while your foot is on the ground, you apply a friction force backward on the surface of the Earth. By Newton's third law, the surface applies a forward friction force on you, which causes your body to move forward.

The strength of the friction force $\vec{f}$ exerted on a car by the roadway is related to the rate at which energy is transferred to the wheels to set them into rotation, which is the power of the engine:

$$P_{\text{avg}} = \frac{\Delta E}{\Delta t} = \frac{f\Delta x}{\Delta t} = fv \rightarrow P \leftrightarrow f$$

where the symbol $\leftrightarrow$ implies a relationship between the variables that is not necessarily an exact proportionality. In turn, the magnitude of the driving force is related to the acceleration of the car owing to Newton's second law:

$$f = ma \rightarrow f \propto a$$

Consequently, there should be a close relationship between the power rating of a vehicle and the possible acceleration of the vehicle:

$$P \leftrightarrow a$$

Let us see if this relationship exists for actual data. For automobiles, a common unit for power is the *horsepower* (hp), defined in Section 7.6. Table 7.2 shows the gasoline-powered automobiles we have studied in the preceding chapters. The third column provides the published horsepower rating of each vehicle. In Figure 7.16, we graph the acceleration values against the horsepower ratings for the vehicles. From this graph, we see a clear correlation between the acceleration and the horsepower as proposed above: as the horsepower rating goes up, the maximum possible acceleration increases. The two black data points at the far right of the graph lie below a line that could be drawn through the other data points. These two points represent the Bugatti Veyron 16.4 Super Sport and the Shelby SuperCars Ultimate Aero. These

TABLE 7.2 | **Horsepower Ratings and Accelerations of Various Vehicles**

Automobile	Average Acceleration (mi/h·s)	Horsepower Rating (hp)	Ratio HP to Acceleration (hp/mi/h·s)
Very expensive vehicles:			
Bugatti Veyron 16.4 Super Sport	23.1	1200	52
Lamborghini LP 570-4 Superleggera	17.6	570	32
Lexus LFA	15.8	560	35
Mercedes-Benz SLS AMG	16.7	563	34
Shelby SuperCars Ultimate Aero	22.2	1287	58
Average	**19.1**	**836**	**42.3**
Performance vehicles:			
Chevrolet Corvette ZR1	18.2	638	35
Dodge Viper SRT10	15.0	600	40
Jaguar XJL Supercharged	13.6	470	35
Acura TL SH-AWD	11.5	305	27
Dodge Challenger SRT8	12.2	425	35
Average	**14.1**	**488**	**34.2**
Traditional vehicles:			
Buick Regal CXL Turbo	8.0	220	28
Chevrolet Tahoe 1500 LS (SUV)	7.0	326	47
Ford Fiesta SES	6.2	120	19
Hummer H3 (SUV)	7.5	300	40
Hyundai Sonata SE	8.0	200	25
Smart ForTwo	4.5	70	16
Average	**6.9**	**206**	**29.0**
Alternative vehicles:			
Chevrolet Volt (hybrid)	7.5	74	10
Nissan Leaf (electric)	6.0	110	18
Honda CR-Z (hybrid)	5.7	122	21
Honda Insight (hybrid)	5.7	98	17
Toyota Prius (hybrid)	6.1	134	22
Average	**6.2**	**108**	**17.8**

are very high-powered vehicles, each sporting a horsepower rating of 1200 hp or more. The graph shows that this large increase in horsepower over the other vehicles results in a relatively modest increase in acceleration. This behavior is similar to that in Figure 2.16, in which a huge increase in cost is required for a relatively small increase in acceleration. Perhaps there is an upper limit of acceleration beyond which horsepower and money cannot take us.

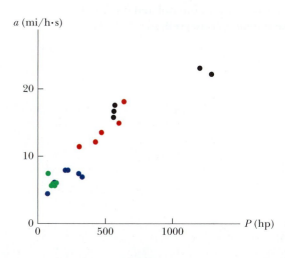

Figure 7.16 The acceleration as a function of horsepower rating for alternative vehicles (in green), traditional vehicles (in blue), performance vehicles (in red), and very expensive vehicles (in black).

SUMMARY

A **nonisolated system** is one for which energy crosses the boundary of the system. An **isolated system** is one for which no energy crosses the boundary of the system.

For a nonisolated system, we can equate the change in the total energy stored in the system to the sum of all the transfers of energy across the system boundary, which is a statement of **conservation of energy.** For an isolated system, the total energy is constant.

If a system is isolated and if no nonconservative forces are acting on objects inside the system, the total mechanical energy of the system is constant:

$$K_f + U_f = K_i + U_i \qquad \textbf{7.10}\blacktriangleleft$$

If nonconservative forces (such as friction) act between objects inside a system, mechanical energy is not conserved. In these situations, the difference between the total final mechanical energy and the total initial mechanical energy of the system equals the energy transformed to internal energy by the nonconservative forces.

If a friction force acts within an isolated system, the mechanical energy of the system is reduced and the appropriate equation to be applied is

$$\Delta E_{\text{mech}} = \Delta K + \Delta U = -f_k d \qquad \textbf{7.16}\blacktriangleleft$$

If a friction force acts within a nonisolated system, the appropriate equation to be applied is

$$\Delta E_{\text{mech}} = -f_k d + \sum W_{\text{other forces}} \qquad \textbf{7.17}\blacktriangleleft$$

The **instantaneous power** P is defined as the time rate of energy transfer:

$$P \equiv \frac{dE}{dt} \qquad \textbf{7.18}\blacktriangleleft$$

Analysis Models for Problem Solving

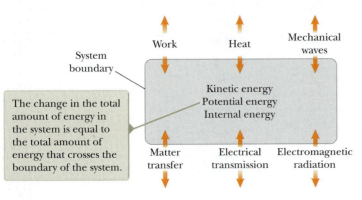

System boundary

The change in the total amount of energy in the system is equal to the total amount of energy that crosses the boundary of the system.

Work   Heat   Mechanical waves

Kinetic energy
Potential energy
Internal energy

Matter transfer   Electrical transmission   Electromagnetic radiation

Nonisolated System (Energy). The most general statement describing the behavior of a nonisolated system is the **conservation of energy equation:**

$$\Delta E_{\text{system}} = \sum T \qquad \textbf{7.1}\blacktriangleleft$$

Including the types of energy storage and energy transfer that we have discussed gives

$$\Delta K + \Delta U + \Delta E_{\text{int}} = \\ W + Q + T_{\text{MW}} + T_{\text{MT}} + T_{\text{ET}} + T_{\text{ER}} \qquad \textbf{7.2}\blacktriangleleft$$

For a specific problem, this equation is generally reduced to a smaller number of terms by eliminating the terms that are not appropriate to the situation.

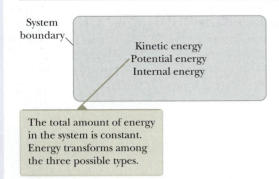

System boundary

Kinetic energy
Potential energy
Internal energy

The total amount of energy in the system is constant. Energy transforms among the three possible types.

Isolated System (Energy). The total energy of an isolated system is conserved, so

$$\Delta E_{\text{system}} = 0 \qquad \textbf{7.9}\blacktriangleleft$$

If no nonconservative forces act within the isolated system, the mechanical energy of the system is conserved, so

$$\Delta E_{\text{mech}} = 0 \qquad \textbf{7.8}\blacktriangleleft$$

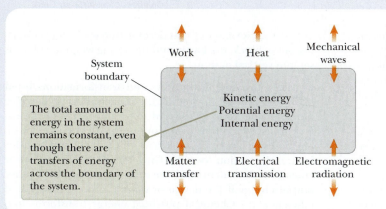

The total amount of energy in the system remains constant, even though there are transfers of energy across the boundary of the system.

Nonisolated System in Steady State (Energy). If there are energy inputs and outputs across the boundary of the system, but they are in balance, then the change in energy of the system is zero:

$$0 = \sum T$$

OBJECTIVE QUESTIONS

☐ denotes answer available in *Student Solutions Manual/Study Guide*

1. You hold a slingshot at arm's length, pull the light elastic band back to your chin, and release it to launch a pebble horizontally with speed 200 cm/s. With the same procedure, you fire a bean with speed 600 cm/s. What is the ratio of the mass of the bean to the mass of the pebble? (a) $\frac{1}{9}$ (b) $\frac{1}{3}$ (c) 1 (d) 3 (e) 9

2. An athlete jumping vertically on a trampoline leaves the surface with a velocity of 8.5 m/s upward. What maximum height does she reach? (a) 13 m (b) 2.3 m (c) 3.7 m (d) 0.27 m (e) The answer can't be determined because the mass of the athlete isn't given.

3. A pile driver drives posts into the ground by repeatedly dropping a heavy object on them. Assume the object is dropped from the same height each time. By what factor does the energy of the pile driver–Earth system change when the mass of the object being dropped is doubled? (a) $\frac{1}{2}$ (b) 1; the energy is the same (c) 2 (d) 4

4. Two children stand on a platform at the top of a curving slide next to a backyard swimming pool. At the same moment the smaller child hops off to jump straight down into the pool, the bigger child releases herself at the top of the frictionless slide. (i) Upon reaching the water, the kinetic energy of the smaller child compared with that of the larger child is (a) greater (b) less (c) equal. (ii) Upon reaching the water, the speed of the smaller child compared with that of the larger child is (a) greater (b) less (c) equal. (iii) During their motions from the platform to the water, the average acceleration of the smaller child compared with that of the larger child is (a) greater (b) less (c) equal.

5. Answer yes or no to each of the following questions. (a) Can an object–Earth system have kinetic energy and not gravitational potential energy? (b) Can it have gravitational potential energy and not kinetic energy? (c) Can it have both types of energy at the same moment? (d) Can it have neither?

6. A ball of clay falls freely to the hard floor. It does not bounce noticeably, and it very quickly comes to rest. What, then, has happened to the energy the ball had while it was falling? (a) It has been used up in producing the downward motion. (b) It has been transformed back into potential energy. (c) It has been transferred into the ball by heat. (d) It is in the ball and floor (and walls) as energy of invisible molecular motion. (e) Most of it went into sound.

7. What average power is generated by a 70.0-kg mountain climber who climbs a summit of height 325 m in 95.0 min? (a) 39.1 W (b) 54.6 W (c) 25.5 W (d) 67.0 W (e) 88.4 W

8. In a laboratory model of cars skidding to a stop, data are measured for four trials using two blocks. The blocks have identical masses but different coefficients of kinetic friction with a table: $\mu_k = 0.2$ and 0.8. Each block is launched with speed $v_i = 1$ m/s and slides across the level table as the block comes to rest. This process represents the first two trials. For the next two trials, the procedure is repeated but the blocks are launched with speed $v_i = 2$ m/s. Rank the four trials (a) through (d) according to the stopping distance from largest to smallest. If the stopping distance is the same in two cases, give them equal rank. (a) $v_i = 1$ m/s, $\mu_k = 0.2$ (b) $v_i = 1$ m/s, $\mu_k = 0.8$ (c) $v_i = 2$ m/s, $\mu_k = 0.2$ (d) $v_i = 2$ m/s, $\mu_k = 0.8$

9. At the bottom of an air track tilted at angle θ, a glider of mass m is given a push to make it coast a distance d up the slope as it slows down and stops. Then the glider comes back down the track to its starting point. Now the experiment is repeated with the same original speed but with a second identical glider set on top of the first. The airflow from the track is strong enough to support the stacked pair of gliders so that the combination moves over the track with negligible friction. Static friction holds the second glider stationary relative to the first glider throughout the motion. The coefficient of static friction between the two gliders is μ_s. What is the change in mechanical energy of the two-glider–Earth system in the up- and down-slope motion after the pair of gliders is released? Choose one. (a) $-2\mu_s mg$ (b) $-2mgd \cos \theta$ (c) $-2\mu_s mgd \cos \theta$ (d) 0 (e) $+2\mu_s mgd \cos \theta$

CONCEPTUAL QUESTIONS

1. One person drops a ball from the top of a building while another person at the bottom observes its motion. Will these two people agree (a) on the value of the gravitational potential energy of the ball–Earth system? (b) On the change in potential energy? (c) On the kinetic energy of the ball at some point in its motion?

2. In Chapter 6 the work–kinetic energy theorem, $W_{ext} = \Delta K$, was introduced. This equation states that work done on a system appears as a change in kinetic energy. It is a special-case equation, valid if there are no changes in any other type of energy such as potential or internal. Give two or three examples in which work is done on a system but the change in energy of the system is not a change in kinetic energy.

3. Does everything have energy? Give the reasoning for your answer.

4. Can a force of static friction do work? If not, why not? If so, give an example.

5. A bowling ball is suspended from the ceiling of a lecture hall by a strong cord. The ball is drawn away from its equilibrium position and released from rest at the tip of the demonstrator's nose as shown in Figure CQ7.5. The demonstrator remains stationary. (a) Explain why the ball does not strike her on its return swing. (b) Would this demonstrator be safe if the ball were given a push from its starting position at her nose?

Figure CQ7.5

6. You ride a bicycle. In what sense is your bicycle solar powered?

7. A block is connected to a spring that is suspended from the ceiling. Assuming air resistance is ignored, describe the energy transformations that occur within the system consisting of the block, the Earth, and the spring when the block is set into vertical motion.

8. Consider the energy transfers and transformations listed below in parts (a) through (e). For each part, (i) describe human-made devices designed to produce each of the energy transfers or transformations and, (ii) whenever possible, describe a natural process in which the energy transfer or transformation occurs. Give details to defend your choices, such as identifying the system and identifying other output energy if the device or natural process has limited efficiency. (a) Chemical potential energy transforms into internal energy. (b) Energy transferred by electrical transmission becomes gravitational potential energy. (c) Elastic potential energy transfers out of a system by heat. (d) Energy transferred by mechanical waves does work on a system. (e) Energy carried by electromagnetic waves becomes kinetic energy in a system.

9. In the general conservation of energy equation, state which terms predominate in describing each of the following devices and processes. For a process going on continuously, you may consider what happens in a 10-s time interval. State which terms in the equation represent original and final forms of energy, which would be inputs, and which outputs. (a) a slingshot firing a pebble (b) a fire burning (c) a portable radio operating (d) a car braking to a stop (e) the surface of the Sun shining visibly (f) a person jumping up onto a chair

10. A car salesperson claims that a 300-hp engine is a necessary option in a compact car, in place of the conventional 130-hp engine. Suppose you intend to drive the car within speed limits ($\leq$ 65 mi/h) on flat terrain. How would you counter this sales pitch?

PROBLEMS

Section 7.1 Analysis Model: Nonisolated System (Energy)

1. **S** For each of the following systems and time intervals, write the appropriate expanded version of Equation 7.2, the conservation of energy equation. (a) the heating coils in your toaster during the first five seconds after you turn the toaster on (b) your automobile from just before you fill it with gasoline until you pull away from the gas station at speed v (c) your body while you sit quietly and eat a peanut butter and jelly sandwich for lunch (d) your home during five minutes of a sunny afternoon while the temperature in the home remains fixed

2. **S** A ball of mass m falls from a height h to the floor. (a) Write the appropriate version of Equation 7.2 for the system of the ball and the Earth and use it to calculate the speed of the ball just before it strikes the Earth. (b) Write the appropriate version of Equation 7.2 for the system of the ball and use it to calculate the speed of the ball just before it strikes the Earth.

Section 7.2 Analysis Model: Isolated System (Energy)

3. **M** **Review.** A bead slides without friction around a loop-the-loop (Fig. P7.3). The bead is released from rest at a height $h = 3.50R$. (a) What is its speed at point Ⓐ? (b) How large is the normal force on the bead at point Ⓐ if its mass is 5.00 g?

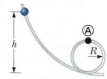

Figure P7.3

4. At 11:00 a.m. on September 7, 2001, more than one million British schoolchildren jumped up and down for one minute to simulate an earthquake. (a) Find the energy stored in the children's bodies that was converted into internal energy in the ground and their bodies and propagated into the ground by seismic waves during the experiment. Assume 1 050 000 children of average mass 36.0 kg jumped 12 times each, raising their centers of mass by 25.0 cm each time and briefly resting between one jump and the next. (b) Of the energy that propagated into the ground, most produced high-frequency "microtremor" vibrations that were rapidly damped and did not travel far. Assume 0.01% of the total energy was carried away by long-range seismic waves. The magnitude of an earthquake on the Richter scale is given by

$$M = \frac{\log E - 4.8}{1.5}$$

where E is the seismic wave energy in joules. According to this model, what was the magnitude of the demonstration quake?

5. **W** A block of mass 0.250 kg is placed on top of a light, vertical spring of force constant 5 000 N/m and pushed downward so that the spring is compressed by 0.100 m. After the block is released from rest, it travels upward and then leaves the spring. To what maximum height above the point of release does it rise?

6. **W** A block of mass $m = 5.00$ kg is released from point Ⓐ and slides on the frictionless track shown in Figure P7.6. Determine (a) the block's speed at points Ⓑ and Ⓒ and (b) the net work done by the gravitational force on the block as it moves from point Ⓐ to point Ⓒ.

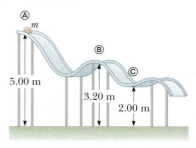

Figure P7.6

7. **M** Two objects are connected by a light string passing over a light, frictionless pulley as shown in Figure P7.7. The object of mass $m_1 = 5.00$ kg is released from rest at a height $h = 4.00$ m above the table. Using the isolated system model, (a) determine the speed of the object of mass $m_2 = 3.00$ kg just as the 5.00-kg object hits the table and (b) find the maximum height above the table to which the 3.00-kg object rises.

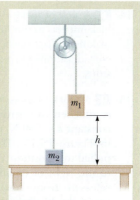

Figure P7.7 Problems 7 and 8.

8. **S** Two objects are connected by a light string passing over a light, frictionless pulley as shown in Figure P7.7. The object of mass m_1 is released from rest at height h above the table. Using the isolated system model, (a) determine the speed of m_2 just as m_1 hits the table and (b) find the maximum height above the table to which m_2 rises.

9. **S** **Review.** The system shown in Figure P7.9 consists of a light, inextensible cord; light, frictionless pulleys; and blocks of equal mass. Notice that block B is attached to one of the pulleys. The system is initially held at rest so that the blocks are at the same height above the ground. The blocks are then released. Find the speed of block A at the moment the vertical separation of the blocks is h.

Figure P7.9

10. **W** A 20.0-kg cannonball is fired from a cannon with muzzle speed of 1 000 m/s at an angle of 37.0° with the horizontal. A second ball is fired at an angle of 90.0°. Use the isolated system model to find (a) the maximum height reached by each ball and (b) the total mechanical energy of the ball–Earth system at the maximum height for each ball. Let $y = 0$ at the cannon.

11. A light, rigid rod is 77.0 cm long. Its top end is pivoted on a frictionless, horizontal axle. The rod hangs straight down at rest with a small, massive ball attached to its bottom end. You strike the ball, suddenly giving it a horizontal velocity so that it swings around in a full circle. What minimum speed at the bottom is required to make the ball go over the top of the circle?

Section 7.4 Situations Involving Kinetic Friction

12. **M** A crate of mass 10.0 kg is pulled up a rough incline with an initial speed of 1.50 m/s. The pulling force is 100 N parallel to the incline, which makes an angle of 20.0° with the horizontal. The coefficient of kinetic friction is 0.400, and the crate is pulled 5.00 m. (a) How much work is done by the gravitational force on the crate? (b) Determine the increase in internal energy of the crate–incline system owing to friction. (c) How much work is done by the 100-N force on the crate? (d) What is the change in kinetic energy of the crate? (e) What is the speed of the crate after being pulled 5.00 m?

13. A sled of mass m is given a kick on a frozen pond. The kick imparts to the sled an initial speed of 2.00 m/s. The coefficient of kinetic friction between sled and ice is 0.100. Use energy considerations to find the distance the sled moves before it stops.

14. **S** A sled of mass m is given a kick on a frozen pond. The kick imparts to the sled an initial speed of v. The coefficient of kinetic friction between sled and ice is μ_k. Use energy considerations to find the distance the sled moves before it stops.

15. **W** A block of mass $m = 2.00$ kg is attached to a spring of force constant $k = 500$ N/m as shown in Figure P7.15. The block is pulled to a position $x_i = 5.00$ cm to the right of equilibrium and released from rest. Find the speed the block has as it passes through equilibrium if (a) the horizontal surface is frictionless and (b) the coefficient of friction between block and surface is $\mu_k = 0.350$.

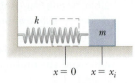

x = 0   x = x_i

Figure P7.15

16. A 40.0-kg box initially at rest is pushed 5.00 m along a rough, horizontal floor with a constant applied horizontal force of 130 N. The coefficient of friction between box and floor is 0.300. Find (a) the work done by the applied force, (b) the increase in internal energy in the box–floor system as a result of friction, (c) the work done by the normal force, (d) the work done by the gravitational force, (e) the change in kinetic energy of the box, and (f) the final speed of the box.

17. A smooth circular hoop with a radius of 0.500 m is placed flat on the floor. A 0.400-kg particle slides around the inside edge of the hoop. The particle is given an initial speed of 8.00 m/s. After one revolution, its speed has dropped to 6.00 m/s because of friction with the floor. (a) Find the energy transformed from mechanical to internal in the particle–hoop–floor system as a result of friction in one revolution. (b) What is the total number of revolutions the particle makes before stopping? Assume the friction force remains constant during the entire motion.

Section 7.5 Changes in Mechanical Energy for Nonconservative Forces

18. **Q|C** At time t_i, the kinetic energy of a particle is 30.0 J and the potential energy of the system to which it belongs is 10.0 J. At some later time t_f, the kinetic energy of the particle is 18.0 J. (a) If only conservative forces act on the particle, what are the potential energy and the total energy of the system at time t_f? (b) If the potential energy of the system at time t_f is 5.00 J, are any nonconservative forces acting on the particle? (c) Explain your answer to part (b).

19. A boy in a wheelchair (total mass 47.0 kg) has speed 1.40 m/s at the crest of a slope 2.60 m high and 12.4 m long. At the bottom of the slope his speed is 6.20 m/s. Assume air resistance and rolling resistance can be modeled as a constant friction force of 41.0 N. Find the work he did in pushing forward on his wheels during the downhill ride.

20. **Q|C** As shown in Figure P7.20, a green bead of mass 25 g slides along a straight wire. The length of the wire from point Ⓐ to point Ⓑ is 0.600 m, and point Ⓐ is 0.200 m higher than point Ⓑ. A constant friction force

Figure P7.20

of magnitude 0.025 0 N acts on the bead. (a) If the bead is released from rest at point Ⓐ, what is its speed at point Ⓑ? (b) A red bead of mass 25 g slides along a curved wire, subject to a friction force with the same constant magnitude as that on the green bead. If the green and red beads are released simultaneously from rest at point Ⓐ, which bead reaches point Ⓑ first? Explain.

21. **M** A 5.00-kg block is set into motion up an inclined plane with an initial speed of $v_i = 8.00$ m/s (Fig. P7.21). The block comes to rest after traveling $d = 3.00$ m along the plane, which is inclined at an angle of $\theta = 30.0°$ to the horizontal. For this motion, determine (a) the change in the block's kinetic energy, (b) the change in the potential energy of the block–Earth system, and (c) the friction force exerted on the block (assumed to be constant). (d) What is the coefficient of kinetic friction?

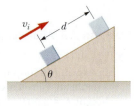

Figure P7.21

22. **W** The coefficient of friction between the block of mass $m_1 = 3.00$ kg and the surface in Figure P7.22 is $\mu_k = 0.400$. The system starts from rest. What is the speed of the ball of mass $m_2 = 5.00$ kg when it has fallen a distance $h = 1.50$ m?

Figure P7.22

23. **M** A 200-g block is pressed against a spring of force constant 1.40 kN/m until the block compresses the spring 10.0 cm. The spring rests at the bottom of a ramp inclined at 60.0° to the horizontal. Using energy considerations, determine how far up the incline the block moves from its initial position before it stops (a) if the ramp exerts no friction force on the block and (b) if the coefficient of kinetic friction is 0.400.

24. **Q|C** An 80.0-kg skydiver jumps out of a balloon at an altitude of 1 000 m and opens his parachute at an altitude of 200 m. (a) Assuming the total retarding force on the skydiver is constant at 50.0 N with the parachute closed and constant at 3 600 N with the parachute open, find the speed of the skydiver when he lands on the ground. (b) Do you think the skydiver will be injured? Explain. (c) At what height should the parachute be opened so that the final speed of the skydiver when he hits the ground is 5.00 m/s? (d) How realistic is the assumption that the total retarding force is constant? Explain.

25. **W** A toy cannon uses a spring to project a 5.30-g soft rubber ball. The spring is originally compressed by 5.00 cm and has a force constant of 8.00 N/m. When the cannon is fired, the ball moves 15.0 cm through the horizontal barrel of the cannon, and the barrel exerts a constant friction force of 0.032 0 N on the ball. (a) With what speed does the projectile leave the barrel of the cannon? (b) At what point does the ball have maximum speed? (c) What is this maximum speed?

26. A 1.50-kg object is held 1.20 m above a relaxed massless, vertical spring with a force constant of 320 N/m. The object is dropped onto the spring. (a) How far does the object compress the spring? (b) **What If?** Repeat part (a), but this time assume a constant air-resistance force of 0.700 N acts on the object during its motion. (c) **What If?** How far does the object compress the spring if the same experiment is performed on the Moon, where $g = 1.63$ m/s^2 and air resistance is neglected?

27. **GP** **Q|C** **S** A child of mass m starts from rest and slides without friction from a height h along a slide next to a pool (Fig. P7.27). She is launched from a height $h/5$ into the air over the pool. We wish to find the maximum height she reaches above the water in her projectile motion. (a) Is the child–Earth system isolated or nonisolated? Why? (b) Is there a nonconservative force acting within the system? (c) Define the configuration of the system when the child is at the water level as having zero gravitational potential energy. Express the total energy of the system when the child is at the top of the waterslide. (d) Express the total energy of the system when the child is at the launching point. (e) Express the total energy of the system when the child is at the highest point in her projectile motion. (f) From parts (c) and (d), determine her initial speed v_i at the launch point in terms of g and h. (g) From parts (d), (e), and (f), determine her maximum airborne height y_{max} in terms of h and the launch angle θ. (h) Would your answers be the same if the waterslide were not frictionless? Explain.

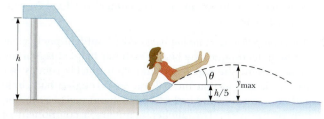

Figure P7.27

Section 7.6 Power

28. **Q|C** The electric motor of a model train accelerates the train from rest to 0.620 m/s in 21.0 ms. The total mass of the train is 875 g. (a) Find the minimum power delivered to the train by electrical transmission from the metal rails during the acceleration. (b) Why is it the minimum power?

29. **W** An 820-N Marine in basic training climbs a 12.0-m vertical rope at a constant speed in 8.00 s. What is his power output?

30. A certain rain cloud at an altitude of 1.75 km contains 3.20×10^7 kg of water vapor. How long would it take a 2.70-kW pump to raise the same amount of water from the Earth's surface to the cloud's position?

31. Make an order-of-magnitude estimate of the power a car engine contributes to speeding the car up to highway speed. In your solution, state the physical quantities you take as data and the values you measure or estimate for them. The mass of a vehicle is often given in the owner's manual.

32. Sewage at a certain pumping station is raised vertically by 5.49 m at the rate of 1 890 000 liters each day. The sewage, of density 1 050 kg/m^3, enters and leaves the pump at atmospheric pressure and through pipes of equal diameter. (a) Find the output mechanical power of the lift station. (b) Assume an electric motor continuously operating with average power 5.90 kW runs the pump. Find its efficiency.

33. When an automobile moves with constant speed down a highway, most of the power developed by the engine is used to compensate for the energy transformations due to friction forces exerted on the car by the air and the road. If the power developed by an engine is 175 hp, estimate the total friction force acting on the car when it is moving at a speed of 29 m/s. One horsepower equals 746 W.

34. An electric scooter has a battery capable of supplying 120 Wh of energy. If friction forces and other losses account for 60.0% of the energy usage, what altitude change can a rider achieve when driving in hilly terrain if the rider and scooter have a combined weight of 890 N?

35. An energy-efficient lightbulb, taking in 28.0 W of power, can produce the same level of brightness as a conventional lightbulb operating at power 100 W. The lifetime of the energy-efficient bulb is 10 000 h and its purchase price is $4.50, whereas the conventional bulb has a lifetime of 750 h and costs $0.42. Determine the total savings obtained by using one energy-efficient bulb over its lifetime as opposed to using conventional bulbs over the same time interval. Assume an energy cost of $0.200 per kilowatt-hour.

36. **S** An older-model car accelerates from 0 to speed v in a time interval of Δt. A newer, more powerful sports car accelerates from 0 to $2v$ in the same time period. Assuming the energy coming from the engine appears only as kinetic energy of the cars, compare the power of the two cars.

37. A 3.50-kN piano is lifted by three workers at constant speed to an apartment 25.0 m above the street using a pulley system fastened to the roof of the building. Each worker is able to deliver 165 W of power, and the pulley system is 75.0% efficient (so that 25.0% of the mechanical energy is transformed to other forms due to friction in the pulley). Neglecting the mass of the pulley, find the time required to lift the piano from the street to the apartment.

38. **BIO** Energy is conventionally measured in Calories as well as in joules. One Calorie in nutrition is one kilocalorie, defined as 1 kcal = 4 186 J. Metabolizing 1 g of fat can release 9.00 kcal. A student decides to try to lose weight by exercising. He plans to run up and down the stairs in a football stadium as fast as he can and as many times as necessary. To evaluate the program, suppose he runs up a flight of 80 steps, each 0.150 m high, in 65.0 s. For simplicity, ignore the energy he uses in coming down (which is small). Assume a typical efficiency for human muscles is 20.0%. This statement means that when your body converts 100 J from metabolizing fat, 20 J goes into doing mechanical work (here, climbing stairs). The remainder goes into extra internal energy. Assume the student's mass is 75.0 kg. (a) How many times must the student run the flight of stairs to lose 1.00 kg of fat? (b) What is his average power output, in watts and in horsepower, as he runs up the stairs? (c) Is this activity in itself a practical way to lose weight?

39. **BIO** For saving energy, bicycling and walking are far more efficient means of transportation than is travel by automobile. For example, when riding at 10.0 mi/h, a cyclist uses food energy at a rate of about 400 kcal/h above what he would use if merely sitting still. (In exercise physiology, power is often measured in kcal/h rather than in watts. Here 1 kcal = 1 nutritionist's Calorie = 4 186 J.) Walking at 3.00 mi/h requires about 220 kcal/h. It is interesting to

compare these values with the energy consumption required for travel by car. Gasoline yields about 1.30×10^8 J/gal. Find the fuel economy in equivalent miles per gallon for a person (a) walking and (b) bicycling.

40. A 650-kg elevator starts from rest. It moves upward for 3.00 s with constant acceleration until it reaches its cruising speed of 1.75 m/s. (a) What is the average power of the elevator motor during this time interval? (b) How does this power compare with the motor power when the elevator moves at its cruising speed?

41. A loaded ore car has a mass of 950 kg and rolls on rails with negligible friction. It starts from rest and is pulled up a mine shaft by a cable connected to a winch. The shaft is inclined at 30.0° above the horizontal. The car accelerates uniformly to a speed of 2.20 m/s in 12.0 s and then continues at constant speed. (a) What power must the winch motor provide when the car is moving at constant speed? (b) What maximum power must the winch motor provide? (c) What total energy has transferred out of the motor by work by the time the car moves off the end of the track, which is of length 1 250 m?

Section 7.7 Context Connection: Horsepower Ratings of Automobiles

42. Make an order-of-magnitude estimate of the output power a car engine contributes to speeding the car up to highway speed. For concreteness, consider your own car, if you use one, and make the calculation as precise as you wish. In your solution, state the physical quantities you take as data and the values you measure or estimate for them. The mass of the vehicle is given in the owner's manual. If you do not wish to estimate for a car, consider a bus or truck that you specify.

43. A certain automobile engine delivers 2.24×10^4 W (30.0 hp) to its wheels when moving at a constant speed of 27.0 m/s (≈ 60 mi/h). What is the resistive force acting on the automobile at that speed?

Additional Problems

44. **Q|C** Review. As shown in Figure P7.44, a light string that does not stretch changes from horizontal to vertical as it passes over the edge of a table. The string connects m_1, a 3.50-kg block originally at rest on the horizontal table at a height $h = 1.20$ m above the floor, to m_2, a hanging 1.90-kg block originally a distance $d = 0.900$ m above the floor. Neither the surface of the table nor its edge exerts a force of kinetic friction. The blocks start to move from rest. The sliding block m_1 is projected horizontally after reaching the edge of the table. The hanging block m_2 stops without bouncing when it strikes the floor. Consider the two blocks plus the Earth as the system. (a) Find the speed at which m_1 leaves the edge of the table. (b) Find the impact speed of m_1 on the floor. (c) What is the shortest length of the string so that it does not go taut

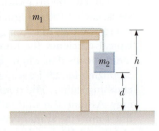

Figure P7.44

while m_1 is in flight? (d) Is the energy of the system when it is released from rest equal to the energy of the system just before m_1 strikes the ground? (e) Why or why not?

45. A small block of mass $m = 200$ g is released from rest at point Ⓐ along the horizontal diameter on the inside of a frictionless, hemispherical bowl of radius $R = 30.0$ cm (Fig. P7.45). Calculate (a) the gravitational potential energy of the block–Earth system when the block is at point Ⓐ relative to point Ⓑ, (b) the kinetic energy of the block at point Ⓑ, (c) its speed at point Ⓑ, and (d) its kinetic energy and the potential energy when the block is at point Ⓒ.

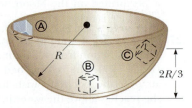

Figure P7.45 Problems 45 and 46.

46. **Q|C** **What If?** The block of mass $m = 200$ g described in Problem 45 (Fig. P7.45) is released from rest at point Ⓐ and the surface of the bowl is rough. The block's speed at point Ⓑ is 1.50 m/s. (a) What is its kinetic energy at point Ⓑ? (b) How much mechanical energy is transformed into internal energy as the block moves from point Ⓐ to point Ⓑ? (c) Is it possible to determine the coefficient of friction from these results in any simple manner? (d) Explain your answer to part (c).

47. *Why is the following situation impossible?* A softball pitcher has a strange technique: she begins with her hand at rest at the highest point she can reach and then quickly rotates her arm backward so that the ball moves through a half-circle path. She releases the ball when her hand reaches the bottom of the path. The pitcher maintains a component of force on the 0.180-kg ball of constant magnitude 12.0 N in the direction of motion around the complete path. As the ball arrives at the bottom of the path, it leaves her hand with a speed of 25.0 m/s.

48. A daredevil plans to bungee jump from a balloon 65.0 m above the ground. He will use a uniform elastic cord, tied to a harness around his body, to stop his fall at a point 10.0 m above the ground. Model his body as a particle and the cord as having negligible mass and obeying Hooke's law. In a preliminary test he finds that when hanging at rest from a 5.00-m length of the cord, his body weight stretches it by 1.50 m. He will drop from rest at the point where the top end of a longer section of the cord is attached to the stationary balloon. (a) What length of cord should he use? (b) What maximum acceleration will he experience?

49. A skateboarder with his board can be modeled as a particle of mass 76.0 kg, located at his center of mass (which we will study in Chapter 8. As shown in Figure P7.49, the skateboarder starts from rest in a crouching position at one lip of a half-pipe (point Ⓐ). The half-pipe is one half of a cylinder of radius 6.80 m with its axis horizontal. On his descent, the skateboarder moves without friction so that his center of mass moves through one quarter of a circle of radius 6.30 m. (a) Find his speed at the bottom of the half-pipe (point Ⓑ). (b) Immediately after passing point Ⓑ, he stands up and raises his arms, lifting his center of mass from 0.500 m to 0.950 m above the concrete (point Ⓒ). Next, the skateboarder glides upward with his center of mass moving in a quarter circle of radius 5.85 m. His body is horizontal

when he passes point Ⓓ, the far lip of the half-pipe. As he passes through point Ⓓ, the speed of the skateboarder is 5.14 m/s. How much chemical potential energy in the body of the skateboarder was converted to mechanical energy in the skateboarder–Earth system when he stood up at point Ⓑ? (c) How high above point Ⓓ does he rise? *Caution*: Do not try this stunt yourself without the required skill and protective equipment.

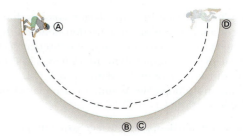

Figure P7.49

50. **S** **Review.** A boy starts at rest and slides down a frictionless slide as in Figure P7.50. The bottom of the track is a height h above the ground. The boy then leaves the track horizontally, striking the ground at a distance d as shown. Using energy methods, determine the initial height H of the boy above the ground in terms of h and d.

Figure P7.50

51. **M** A 4.00-kg particle moves along the x axis. Its position varies with time according to $x = t + 2.0t^3$, where x is in meters and t is in seconds. Find (a) the kinetic energy of the particle at any time t, (b) the acceleration of the particle and the force acting on it at time t, (c) the power being delivered to the particle at time t, and (d) the work done on the particle in the interval $t = 0$ to $t = 2.00$ s.

52. **Q|C** As the driver steps on the gas pedal, a car of mass 1 160 kg accelerates from rest. During the first few seconds of motion, the car's acceleration increases with time according to the expression

$$a = 1.16t - 0.210t^2 + 0.240t^3$$

where t is in seconds and a is in m/s². (a) What is the change in kinetic energy of the car during the interval from $t = 0$ to $t = 2.50$ s? (b) What is the minimum average power output of the engine over this time interval? (c) Why is the value in part (b) described as the *minimum* value?

53. Jonathan is riding a bicycle and encounters a hill of height 7.30 m. At the base of the hill, he is traveling at 6.00 m/s. When he reaches the top of the hill, he is traveling at 1.00 m/s. Jonathan and his bicycle together have a mass of 85.0 kg. Ignore friction in the bicycle mechanism and between the bicycle tires and the road. (a) What is the total external work done on the system of Jonathan and

the bicycle between the time he starts up the hill and the time he reaches the top? (b) What is the change in potential energy stored in Jonathan's body during this process? (c) How much work does Jonathan do on the bicycle pedals within the Jonathan–bicycle–Earth system during this process?

54. **S** Jonathan is riding a bicycle and encounters a hill of height h. At the base of the hill, he is traveling at a speed v_i. When he reaches the top of the hill, he is traveling at a speed v_f. Jonathan and his bicycle together have a mass m. Ignore friction in the bicycle mechanism and between the bicycle tires and the road. (a) What is the total external work done on the system of Jonathan and the bicycle between the time he starts up the hill and the time he reaches the top? (b) What is the change in potential energy stored in Jonathan's body during this process? (c) How much work does Jonathan do on the bicycle pedals within the Jonathan–bicycle–Earth system during this process?

55. **Q|C** A horizontal spring attached to a wall has a force constant of $k = 850$ N/m. A block of mass $m = 1.00$ kg is attached to the spring and rests on a frictionless, horizontal surface as in Figure P7.55. (a) The block is pulled to a position $x_i = 6.00$ cm from equilibrium and released. Find the elastic potential energy stored in the spring when the block is 6.00 cm from equilibrium and when the block passes through equilibrium. (b) Find the speed of the block as it passes through the equilibrium point. (c) What is the speed of the block when it is at a position $x_i/2 = 3.00$ cm? (d) Why isn't the answer to part (c) half the answer to part (b)?

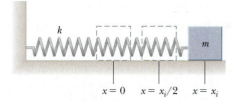

Figure P7.55

56. **Review.** *Why is the following situation impossible?* A new high-speed roller coaster is claimed to be so safe that the passengers do not need to wear seat belts or any other restraining device. The coaster is designed with a vertical circular section over which the coaster travels on the inside of the circle so that the passengers are upside down for a short time interval. The radius of the circular section is 12.0 m, and the coaster enters the bottom of the circular section at a speed of 22.0 m/s. Assume the coaster moves without friction on the track and model the coaster as a particle.

57. **S** **Review.** A uniform board of length L is sliding along a smooth, frictionless, horizontal plane as shown in Figure P7.57a (page 226). The board then slides across the boundary with a rough horizontal surface. The coefficient of kinetic friction between the board and the second surface is μ_k. (a) Find the acceleration of the board at the moment its front end has traveled a distance x beyond the boundary. (b) The board stops at the moment its back end reaches the boundary as shown in Figure P7.57b. Find the initial speed v of the board.

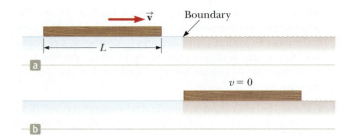

Figure P7.57

58. QC S As it plows a parking lot, a snowplow pushes
an ever-growing pile of snow
in front of it. Suppose a car
moving through the air is
similarly modeled as a cylinder of area A pushing a growing disk of air in front of it.
The originally stationary air
is set into motion at the constant speed v of the cylinder as
shown in Figure P7.58. In a time interval Δt, a new disk of air
of mass Δm must be moved a distance $v \Delta t$ and hence must
be given a kinetic energy $\frac{1}{2}(\Delta m) v^2$. Using this model, show
that the car's power loss owing to air resistance is $\frac{1}{2}\rho A v^3$
and that the resistive force acting on the car is $\frac{1}{2}\rho A v^2$, where
ρ is the density of air. Compare this result with the empirical
expression $\frac{1}{2}D\rho A v^2$ for the resistive force.

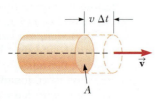

Figure P7.58

59. BIO While running, a person transforms about 0.600 J of
chemical energy to mechanical energy per step per kilogram of body mass. If a 60.0-kg runner transforms energy
at a rate of 70.0 W during a race, how fast is the person running? Assume that a running step is 1.50 m long.

60. Consider the popgun in Example 7.3. Suppose the projectile mass, compression distance, and spring constant remain
the same as given or calculated in the example. Suppose,
however, there is a friction force of magnitude 2.00 N acting
on the projectile as it rubs against the interior of the barrel.
The vertical length from point Ⓐ to the end of the barrel is
0.600 m. (a) After the spring is compressed and the popgun
fired, to what height does the projectile rise above point Ⓑ?
(b) Draw four energy bar charts for this situation, analogous
to those in Figures 7.6c-f.

61. A wind turbine on a wind farm turns in response to a force
of high-speed air resistance, $R = \frac{1}{2}D\rho A v^2$. The power available is $P = Rv = \frac{1}{2}D\rho\pi r^2 v^3$, where v is the wind speed and
we have assumed a circular face for the wind turbine of ra-
dius r. Take the drag coefficient as $D = 1.00$ and the density
of air from the front endpaper. For a wind turbine having
$r = 1.50$ m, calculate the power available with (a) $v =$
8.00 m/s and (b) $v = 24.0$ m/s. The power delivered to the
generator is limited by the efficiency of the system, about
25%. For comparison, a large American home uses about
2 kW of electric power.

62. BIO In bicycling for aerobic exercise, a woman wants her
heart rate to be between 136 and 166 beats per minute.
Assume that her heart rate is directly proportional to her
mechanical power output within the range relevant here.
Ignore all forces on the woman-plus-bicycle system except
for static friction forward on the drive wheel of the bicycle
and an air resistance force proportional to the square of her
speed. When her speed is 22.0 km/h, her heart rate is 90.0
beats per minute. In what range should her speed be so that
her heart rate will be in the range she wants?

63. BIO Make an order-of-magnitude estimate of your power
output as you climb stairs. In your solution, state the physical quantities you take as data and the values you measure
or estimate for them. Do you consider your peak power or
your sustainable power?

64. QC **What If?** Consider the roller coaster described in
Problem 56. Because of some friction between the coaster
and the track, the coaster enters the circular section at a
speed of 15.0 m/s rather than the 22.0 m/s in Problem 56.
Is this situation *more* or *less* dangerous for the passengers
than that in Problem 56? Assume the circular section is still
frictionless.

65. Consider the block–spring–surface system in part (B) of Example 7.5. (a) Using an energy approach, find the position
x of the block at which its speed is a maximum. (b) In the
What If? section of this example, we explored the effects of
an increased friction force of 10.0 N. At what position of the
block does its maximum speed occur in this situation?

66. **Review.** As a prank, someone has balanced a pumpkin at the
highest point of a grain silo. The silo is topped with a hemispherical cap that is frictionless when wet. The line from
the center of curvature of the cap to the pumpkin makes
an angle $\theta_i = 0°$ with the vertical. While you happen to be
standing nearby in the middle of a rainy night, a breath of
wind makes the pumpkin start sliding downward from rest.
It loses contact with the cap when the line from the center
of the hemisphere to the pumpkin makes a certain angle
with the vertical. What is this angle?

67. **Review.** The mass of a car is 1 500 kg. The shape of the
car's body is such that its aerodynamic drag coefficient is
$D = 0.330$ and its frontal area is 2.50 m². Assuming the drag
force is proportional to v^2 and ignoring other sources of
friction, calculate the power required to maintain a speed
of 100 km/h as the car climbs a long hill sloping at 3.20°.

68. W A 1.00-kg object slides
to the right on a surface
having a coefficient of kinetic friction 0.250 (Fig.
P7.68a). The object has
a speed of $v_i = 3.00$ m/s
when it makes contact
with a light spring (Fig.
P7.68b) that has a force
constant of 50.0 N/m.
The object comes to rest
after the spring has been
compressed a distance d
(Fig. P7.68c). The object
is then forced toward the
left by the spring (Fig.
P7.68d) and continues
to move in that direction
beyond the spring's un-
stretched position. Finally,
the object comes to rest a distance D to the left of the unstretched spring (Fig. P7.68e). Find (a) the distance of compression d, (b) the speed v at the unstretched position when

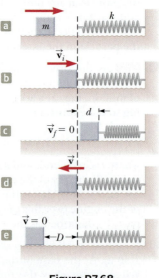

Figure P7.68

the object is moving to the left (Fig. P7.68d), and (c) the distance *D* where the object comes to rest.

69. A child's pogo stick (Fig. P7.69) stores energy in a spring with a force constant of 2.50×10^4 N/m. At position Ⓐ ($x_Ⓐ = -0.100$ m), the spring compression is a maximum and the child is momentarily at rest. At position Ⓑ ($x_Ⓑ = 0$), the spring is relaxed and the child is moving upward. At position Ⓒ, the child is again momentarily at rest at the top of the jump. The combined mass of child and pogo stick is 25.0 kg. Although the boy must lean forward to remain balanced, the angle is small, so let's assume the pogo stick is vertical. Also assume the boy does not bend his legs during the motion. (a) Calculate the total energy of the child–stick–Earth system, taking both gravitational and elastic potential energies as zero for $x = 0$. (b) Determine $x_Ⓒ$. (c) Calculate the speed of the child at $x = 0$. (d) Determine the value of x for which the kinetic energy of the system is a maximum. (e) Calculate the child's maximum upward speed.

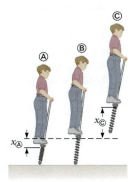

Figure P7.69

70. Ⓢ A block of mass *M* rests on a table. It is fastened to the lower end of a light, vertical spring. The upper end of the spring is fastened to a block of mass *m*. The upper block is pushed down by an additional force $3mg$, so the spring compression is $4mg/k$. In this configuration, the upper block is released from rest. The spring lifts the lower block off the table. In terms of *m*, what is the greatest possible value for *M*?

71. Ⓜ A 10.0-kg block is released from rest at point Ⓐ in Figure P7.71. The track is frictionless except for the portion between points Ⓑ and Ⓒ, which has a length of 6.00 m. The block travels down the track, hits a spring of force constant 2 250 N/m, and compresses the spring 0.300 m from its equilibrium position before coming to rest momentarily. Determine the coefficient of kinetic friction between the block and the rough surface between points Ⓑ and Ⓒ.

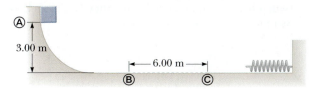

Figure P7.71

72. Consider the block–spring collision discussed in Example 7.7. (a) For the situation in part (B), in which the surface exerts a friction force on the block, show that the block never arrives back at $x = 0$. (b) What is the maximum value of the coefficient of friction that would allow the block to return to $x = 0$?

73. Ⓜ A block of mass $m_1 = 20.0$ kg is connected to a block of mass $m_2 = 30.0$ kg by a massless string that passes over a light, frictionless pulley. The 30.0-kg block is connected to a spring that has negligible mass and a force constant of $k = 250$ N/m as shown in Figure P7.73. The spring is unstretched when the system is as shown in the figure, and the incline is frictionless. The 20.0-kg block is pulled a distance $h = 20.0$ cm down the incline of angle $\theta = 40.0°$ (so that the 30.0-kg block is 40.0 cm above the floor) and released from rest. Find the speed of each block when the 30.0-kg block is 20.0 cm above the floor (that is, when the spring is unstretched).

Figure P7.73

74. **Review.** *Why is the following situation impossible?* An athlete tests her hand strength by having an assistant hang weights from her belt as she hangs onto a horizontal bar with her hands. When the weights hanging on her belt have increased to 80% of her body weight, her hands can no longer support her and she drops to the floor. Frustrated at not meeting her hand-strength goal, she decides to swing on a trapeze. The trapeze consists of a bar suspended by two parallel ropes, each of length ℓ, allowing performers to swing in a vertical circular arc (Fig. P7.74). The athlete holds the bar and steps off an elevated platform, starting from rest with the ropes at an angle $\theta_i = 60.0°$ with respect to the vertical. As she swings several times back and forth in a circular arc, she forgets her frustration related to the hand-strength test. Assume the size of the performer's body is small compared to the length ℓ and air resistance is negligible.

Figure P7.74

75. ⓆⒸ An airplane of mass 1.50×10^4 kg is in level flight, initially moving at 60.0 m/s. The resistive force exerted by air on the airplane has a magnitude of 4.0×10^4 N. By Newton's third law, if the engines exert a force on the exhaust gases to expel them out of the back of the engine, the exhaust gases exert a force on the engines in the direction of the airplane's travel. This force is called thrust, and the value of the thrust in this situation is 7.50×10^4 N. (a) Is the work done by the exhaust gases on the airplane during some time interval equal to the change in the airplane's kinetic energy? Explain. (b) Find the speed of the airplane after it has traveled 5.0×10^2 m.

76. Ⓢ A ball whirls around in a vertical circle at the end of a string. The other end of the string is fixed at the center of the circle. Assuming the total energy of the ball–Earth system remains constant, show that the tension in the string at the bottom is greater than the tension at the top by six times the ball's weight.

77. **Review.** In 1887 in Bridgeport, Connecticut, C. J. Belknap built the water slide shown in Figure P7.77 (page 228). A rider on a small sled, of total mass 80.0 kg, pushed off to start at the top of the slide (point Ⓐ) with a speed of 2.50 m/s.

The chute was 9.76 m high at the top and 54.3 m long. Along its length, 725 small wheels made friction negligible. Upon leaving the chute horizontally at its bottom end (point Ⓒ), the rider skimmed across the water of Long Island Sound for as much as 50 m, "skipping along like a flat pebble," before at last coming to rest and swimming ashore, pulling his sled after him. (a) Find the speed of the sled and rider at point Ⓒ. (b) Model the force of water friction as a constant retarding force acting on a particle. Find the magnitude of the friction force the water exerts on the sled. (c) Find the magnitude of the force the chute exerts on the sled at point Ⓑ. (d) At point Ⓒ, the chute is horizontal but curving in the vertical plane. Assume its radius of curvature is 20.0 m. Find the force the chute exerts on the sled at point Ⓒ.

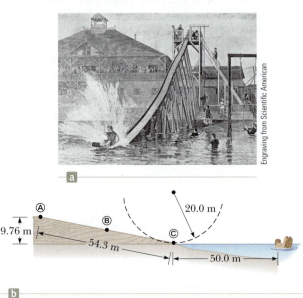

Engraving from Scientific American

a

b

Figure P7.77

78. Q|C Starting from rest, a 64.0-kg person bungee jumps from a tethered hot-air balloon 65.0 m above the ground. The bungee cord has negligible mass and unstretched length 25.8 m. One end is tied to the basket of the balloon and the other end to a harness around the person's body. The cord is modeled as a spring that obeys Hooke's law with a spring constant of 81.0 N/m, and the person's body is modeled as a particle. The hot-air balloon does not move. (a) Express the gravitational potential energy of the person–Earth system as a function of the person's variable height y above the ground. (b) Express the elastic potential energy of the cord as a function of y. (c) Express the total potential energy of the person–cord–Earth system as a function of y. (d) Plot a graph of the gravitational, elastic, and total potential energies as functions of y. (e) Assume air resistance is negligible. Determine the minimum height of the person above the ground during his plunge. (f) Does the potential energy graph show any equilibrium position or positions? If so, at what elevations? Are they stable or unstable? (g) Determine the jumper's maximum speed.

79. Q|C A block of mass 0.500 kg is pushed against a horizontal spring of negligible mass until the spring is compressed a distance x (Fig. P7.79). The force constant of the spring is 450 N/m. When it is released, the block travels along a frictionless, horizontal surface to point Ⓐ, the bottom of a

vertical circular track of radius $R = 1.00$ m, and continues to move up the track. The block's speed at the bottom of the track is $v_Ⓐ = 12.0$ m/s, and the block experiences an average friction force of 7.00 N while sliding up the track. (a) What is x? (b) If the block were to reach the top of the track, what would be its speed at that point? (c) Does the block actually reach the top of the track, or does it fall off before reaching the top?

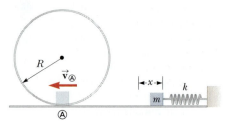

Figure P7.79

80. S A pendulum, comprising a light string of length L and a small sphere, swings in the vertical plane. The string hits a peg located a distance d below the point of suspension (Fig. P7.80). (a) Show that if the sphere is released from a height below that of the peg, it will return to this height after the string strikes the peg. (b) Show that if the pendulum is released from rest at the horizontal position ($\theta = 90°$) and is to swing in a complete circle centered on the peg, the minimum value of d must be $3L/5$.

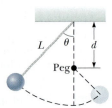

Figure P7.80

81. Jane, whose mass is 50.0 kg, needs to swing across a river (having width D) filled with person-eating crocodiles to save Tarzan from danger. She must swing into a wind exerting constant horizontal force $\vec{F}$, on a vine having length L and initially making an angle θ with the vertical (Fig. P7.81). Take $D = 50.0$ m, $F = 110$ N, $L = 40.0$ m, and $\theta = 50.0°$. (a) With what minimum speed must Jane begin her swing to just make it to the other side? (b) Once the rescue is complete, Tarzan and Jane must swing back across the river. With what minimum speed must they begin their swing? Assume Tarzan has a mass of 80.0 kg.

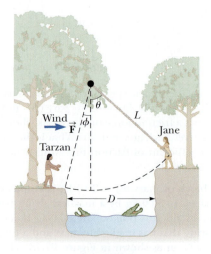

Figure P7.81

Problems 229

82. **S** A roller-coaster car shown in Figure P7.82 is released from rest from a height h and then moves freely with negligible friction. The roller-coaster track includes a circular loop of radius R in a vertical plane. (a) First suppose the car barely makes it around the loop; at the top of the loop, the riders are upside down and feel weightless. Find the required height h of the release point above the bottom of the loop in terms of R. (b) Now assume the release point is at or above the minimum required height. Show that the normal force on the car at the bottom of the loop exceeds the normal force at the top of the loop by six times the car's weight. The normal force on each rider follows the same rule. Such a large normal force is dangerous and very uncomfortable for the riders. Roller coasters are therefore not built with circular loops in vertical planes. Figure P5.22 (page 149) shows an actual design.

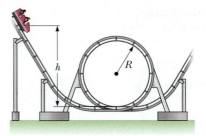

Figure P7.82

83. **BIO** In a needle biopsy, a narrow strip of tissue is extracted from a patient using a hollow needle. Rather than being pushed by hand, to ensure a clean cut the needle can be fired into the patient's body by a spring. Assume that the needle has mass 5.60 g, the light spring has force constant 375 N/m, and the spring is originally compressed 8.10 cm to project the needle horizontally without friction. After the needle leaves the spring, the tip of the needle moves through 2.40 cm of skin and soft tissue, which exerts on it a resistive force of 7.60 N. Next, the needle cuts 3.50 cm into an organ, which exerts on it a backward force of 9.20 N. Find (a) the maximum speed of the needle and (b) the speed at which a flange on the back end of the needle runs into a stop that is set to limit the penetration to 5.90 cm.

Present and Future Possibilities

N ow that we have explored some fundamental principles of classical mechanics, let us return to our central question for the *Alternative-Fuel Vehicles* Context:

> **What source besides gasoline can be used to provide energy for an automobile while reducing environmentally damaging emissions?**

Available Now—The Hybrid Electric Vehicle

As discussed in Section 6.11, a few purely electric vehicles are currently available, but they suffer from difficulties such as limited range and long charging times. Also available and more widely used by consumers are a growing number of **hybrid electric vehicles.** In these automobiles, a gasoline engine and an electric motor are combined to increase the fuel economy of the vehicle and reduce its emissions. Currently available models include the Toyota Prius and Honda Insight, which are originally designed hybrid vehicles, as well as other existing traditional gasoline-powered models that have been modified with a hybrid drive system.

Two major categories of hybrid vehicles are the **series hybrid** and the **parallel hybrid.** In a series hybrid, such as the Chevrolet Volt (Fig. 1) operating at low speeds, the gasoline engine does not provide propulsion energy to the transmission directly. The engine turns a generator, which in turn either charges the batteries or powers the electric motor. Only the electric motor is connected directly to the transmission to propel the car.

In a parallel hybrid, both the engine and the motor are connected to the transmission, so either one can provide propulsion energy for the car. The Honda Insight is a parallel hybrid. Both the engine and the motor provide power to the transmission, and the engine is running at all times while the car is moving. The goal of the development of this hybrid is maximum mileage, which is achieved through a number of design features. Because the engine is small, the Insight has lower emissions than a traditional gasoline-powered vehicle. Because the engine is running at all vehicle speeds, however, its emissions are not as low as those of the Toyota Prius.

Figure 2 shows the third-generation Toyota Prius, which is a series/parallel combination. At higher speeds, power to the wheels comes from both the gasoline engine and the electric motor. The vehicle has some aspects of a series hybrid, however, in that the electric motor alone accelerates the vehicle from rest until it is moving at a speed of about 15 mi/h (24 km/h). During this acceleration period, the engine is not running, so gasoline is not used and there is no emission. As a result, the average tailpipe

Figure 1 The Chevrolet Volt.

emissions are lower than those of the Insight. The Chevrolet Volt has the possibility of the lowest tailpipe emissions because, for repeated cycles of short trips alternating with recharging, the gasoline engine may not operate at all.

When a hybrid vehicle brakes, the motor acts as a generator and returns some of the kinetic energy of the vehicle back to the battery as electric potential energy. In a normal vehicle, this kinetic energy is not recoverable because it is transformed to internal energy in the brakes and roadway.

Gas mileage for hybrid vehicles is in the range of 40 to 55 mi/gal and emissions are far below those of a standard gasoline engine. A hybrid vehicle does not need to be charged like a purely electric vehicle. The battery that drives the electric motor is charged while the gasoline engine is running. Consequently, even though the hybrid vehicle has an electric motor like a pure electric vehicle, it can simply be filled at a gas station like a normal vehicle.

Hybrid electric vehicles are not strictly alternative-fuel vehicles because they use the same fuel as normal vehicles, gasoline. They do, however, represent an important step toward more efficient cars with lower emissions, and the increased mileage helps conserve crude oil.

Figure 2 The third-generation Toyota Prius.

In the Future—The Fuel Cell Vehicle

In an internal combustion engine, the chemical potential energy in the fuel is transformed to internal energy during an explosion initiated by a spark plug. The resulting expanding gases do work on pistons, directing energy to the wheels of the vehicle. In current development is the **fuel cell**, in which the conversion of the energy in the fuel to internal energy is not required. The fuel (hydrogen) is oxidized, and energy leaves the fuel cell by electrical transmission. The energy is used by an electric motor to drive the vehicle.

The advantages of this type of vehicle are many. There is no internal combustion engine to generate harmful emissions, so the vehicle is emission-free. Other than the energy used to power the vehicle, the only by-products are internal energy and water. The fuel is hydrogen, which is the most abundant element in the universe. The efficiency of a fuel cell is much higher than that of an internal combustion engine, so more of the potential energy in the fuel can be extracted.

That is all good news. The bad news is that fuel cell vehicles are only in the early prototype stage. Honda has a production model fuel-cell vehicle, the Honda FCX Clarity (Fig. 3). The Clarity is only available in the United States in southern California, where there are a few hydrogen filling stations, and only about 20 vehicles were operating there as of 2010. It will be many years before fuel cell vehicles are widely available to consumers. During these years, fuel cells must be perfected to operate in weather extremes, manufacturing infrastructure must be set up to supply the hydrogen, and a fueling infrastructure must be established to allow transfer of hydrogen into individual vehicles.

Problems

1. When a conventional car brakes to a stop, all (100%) its kinetic energy is converted into internal energy. None of this energy is available to get the car moving again. Consider a hybrid electric car of mass 1 300 kg moving at 22.0 m/s. (a) Calculate its kinetic energy. (b) The car uses its regenerative braking system to come to a stop at a red light. Assume that the motor-generator converts 70.0% of the car's kinetic energy into energy delivered to the battery by electrical transmission. The other 30.0% becomes internal energy. Compute the amount of energy charging up the battery. (c) Assume that the battery can give back 85.0% of the

Figure 3 The fuel filler inlet for hydrogen in the Honda FCX Clarity.

energy chemically stored in it. Compute the amount of this energy. The other 15.0% becomes internal energy. (d) When the light turns green, the car's motor-generator runs as a motor to convert 68.0% of the energy from the battery into kinetic energy of the car. Compute the amount of this energy and (e) the speed at which the car will be set moving with no other energy input. (f) Compute the overall efficiency of the braking-and-starting process. (g) Compute the net amount of internal energy produced.

2. In both a conventional car and a hybrid electric car, the gasoline engine is the original source of all the energy the car uses to push through the air and against rolling resistance of the road. In city traffic, a conventional gasoline engine must run at a wide variety of rotation rates and fuel inputs. That is, it must run at a wide variety of tachometer and throttle settings. It is almost never running at its maximum-efficiency point. In a hybrid electric car, on the other hand, the gasoline engine can run at maximum efficiency whenever it is on. A simple model can reveal the distinction numerically. Assume that the two cars both do 66.0 MJ of "useful" work in making the same trip to the drugstore. Let the conventional car run at 7.00% efficiency as it puts out useful energy 33.0 MJ and let it run at 30.0% efficiency as it puts out 33.0 MJ. Let the hybrid car run at 30.0% efficiency all the time. Compute (a) the required energy input for each car and (b) the overall efficiency of each.

Mission to Mars

In this Context, we shall investigate the physics necessary to send a spacecraft from Earth to Mars. If the two planets were sitting still in space, millions of kilometers apart, it would be a difficult enough proposition, but keep in mind that we are launching the spacecraft from a moving object, the Earth, and are aiming at a moving target, Mars. Furthermore, the spacecraft's motion is influenced by gravitational forces from the Earth, the Sun, and Mars as well as from any other massive objects in the vicinity. Despite these apparent difficulties, we can use the principles of physics to plan a successful mission.

In the 1970s, the Viking Project landed spacecraft on Mars to analyze the soil for signs of life. These tests were inconclusive. The United States returned to Mars in the 1990s with the Mars Global Surveyor, designed to perform careful mapping of the Martian surface, and Mars Pathfinder, which landed on Mars and deployed a roving robot to analyze rocks and soil. Not all trips have been successful. In 1999, Mars Polar Lander was launched to land near the polar ice cap and search for water. As it entered the Martian atmosphere, it sent its last data and was never heard from again. Mars Climate Orbiter was also lost in 1999 due to communication errors between the builder of the spacecraft and the mission control team.

In late 2003 and early 2004, arrivals of spacecraft at Mars were expected by three space agencies, the National Aeronautics and Space Administration (NASA) in the United States, the European Space Agency (ESA) in Europe, and the Japanese Aerospace Exploration Agency (JAXA) in Japan. The Japanese mission ended in failure when a stuck valve and electrical circuit problems affected a critical midcourse correction, resulting in the inability of the spacecraft, named *Nozomi*, to achieve an orbit around Mars. It passed about 1 000 km above the Martian surface on December 14, 2003, and then left the planet to continue its orbit around the Sun.

The European effort resulted in a successful injection of their *Mars Express* spacecraft into an orbit around Mars. A lander, named *Beagle 2*, descended to the surface. Unfortunately, no signals from the lander have

Figure 1 The Mars rover *Spirit* is tested in a clean room at the Jet Propulsion Laboratory in Pasadena, California.

NASA/JPL

Figure 2 An image from a camera on the Mars rover *Opportunity* shows a rock called the "Berry Bowl." The "berries" are sphere-like grains containing hematite, which scientists used to confirm the earlier presence of water on the surface. The circular area on the rock is the result of using the rover's rock abrasion tool to remove a layer of dust. In this way, a clean surface of the rock was available for spectral analysis by the rover's spectrometers.

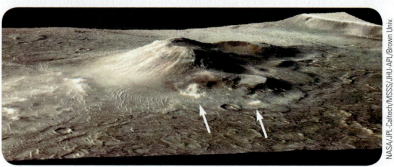

Figure 3 This volcanic cone on Mars has hydrothermal mineral deposits on the southern flanks and nearby terrains. Two of the largest deposits are marked by arrows, and the entire field of light-toned material on the left of the cone is hydrothermal deposits.

been detected and it is presumed lost. The *Mars Express* orbiter continues to send data and is equipped to perform scientific analyses from orbit.

The NASA effort was the most successful of the three missions, with the *Spirit* rover landing successfully on the surface of Mars on January 4, 2004. Its twin, *Opportunity*, also landed successfully, on January 24, 2004, on the opposite side of the planet from *Spirit*. Amazingly, *Opportunity* landed inside a crater, providing scientists with a wonderful opportunity to study the geology of an impact crater. Aside from a computer glitch that was successfully repaired, both rovers performed excellently and sent back very high-quality photographs of the Martian surface as well as large amounts of data including verification of water that once existed on the surface.

More recent observations in 2010 by NASA's Mars Reconnaissance Orbiter revealed a volcanic cone containing hydrothermal mineral deposits on the flanks of the cone. Researchers have identified one of the minerals as hydrated silica, and the new results suggest that in some regions, Mars may have supported microbial life. Excellent photographs near the North Pole of Mars were obtained using a camera mounted on the Orbiter as part of the High Resolution Imaging Science Experiment, HiRISE. The images show only small patches of ice at the surface whose structure is typical of icy permafrost that expands and contracts with changing seasons.

Many individuals dream of one day establishing colonies on Mars. This dream is far in the future; we are still learning

Figure 4 This HiRISE image shows some patches of surface ice near the North Pole of Mars.

much about Mars today and have yet taken only a handful of trips to the planet. Travel to Mars is still not an everyday occurrence, although we learn more from each mission. In this Context, we address the central question,

> **How can we undertake a successful transfer of a spacecraft from Earth to Mars?**

Chapter | 8

Momentum and Collisions

Chapter Outline

AP photos/Keystone/Regina Kuehne

Consider what happens when two cars collide as in the opening photograph for this chapter. Both cars change their motion from having a very large velocity to being at rest because of the collision. Because each car experiences a large change in velocity over a very short time interval, the average force on it is very large. By Newton's third law, each of the cars experiences a force of the same magnitude. By Newton's second law, the results of those forces on the motion of the car depends on the mass of the car.

One main objective of this chapter is to enable you to understand and analyze such events. As a first step, we shall introduce the concept of *momentum*, a term used to describe objects in motion. The concept of momentum leads us to a new conservation law and new analysis models incorporating momentum approaches for isolated and nonisolated systems. This conservation law is especially useful for treating problems that involve collisions between objects.

The concept of momentum allows the analysis of car collisions even without detailed knowledge of the forces involved. Such analysis can determine the relative velocity of the cars before the collision, and in addition aid engineers in designing safer vehicles. (The English translation of the text on the side of the trailer in the background is: "Pit stop for your vehicle.")

8.1 | Linear Momentum

In the preceding two chapters, we studied situations that are difficult to analyze with Newton's laws. We were able to solve problems involving these situations

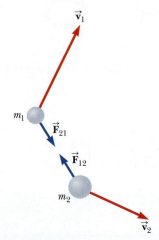

Figure 8.1 Two particles interact with each other. According to Newton's third law, we must have $\vec{\mathbf{F}}_{12} = -\vec{\mathbf{F}}_{21}$.

by applying a conservation principle, conservation of energy. Let us consider another situation and see if we can solve it with the models we have developed so far:

> A 60-kg archer stands at rest on frictionless ice and fires a 0.030-kg arrow horizontally at 85 m/s. With what velocity does the archer move across the ice after firing the arrow?

From Newton's third law, we know that the force that the bow exerts on the arrow is matched by a force in the opposite direction on the bow (and the archer). This force causes the archer to slide backward on the ice with the speed requested in the problem. We cannot determine this speed using motion models such as the particle under constant acceleration because we don't have any information about the acceleration of the archer. We cannot use force models such as the particle under a net force because we don't know anything about forces in this situation. Energy models are of no help because we know nothing about the work done in pulling the bowstring back or the elastic potential energy in the system related to the taut bowstring.

Despite our inability to solve the archer problem using models learned so far, this problem is very simple to solve if we introduce a new quantity that describes motion, *linear momentum*. To generate this new quantity, consider an isolated system of two particles (Fig. 8.1) with masses m_1 and m_2 moving with velocities $\vec{\mathbf{v}}_1$ and $\vec{\mathbf{v}}_2$ at an instant of time. Because the system is isolated, the only force on one particle is that from the other particle, and we can categorize this situation as one in which Newton's laws can be applied. If a force from particle 1 (for example, a gravitational force) acts on particle 2, there must be a second force—equal in magnitude but opposite in direction—that particle 2 exerts on particle 1. That is, the forces form a Newton's third law action–reaction pair so that $\vec{\mathbf{F}}_{12} = -\vec{\mathbf{F}}_{21}$. We can express this condition as a statement about the *system* of two particles as follows:

$$\vec{\mathbf{F}}_{21} + \vec{\mathbf{F}}_{12} = 0$$

Let us further analyze this situation by incorporating Newton's second law. At the instant shown in Figure 8.1, the interacting particles have accelerations corresponding to the forces on them. Therefore, replacing the force on each particle with $m\vec{\mathbf{a}}$ for the particle gives

$$m_1\vec{\mathbf{a}}_1 + m_2\vec{\mathbf{a}}_2 = 0$$

Now we replace the acceleration with its definition from Equation 3.5:

$$m_1\frac{d\vec{\mathbf{v}}_1}{dt} + m_2\frac{d\vec{\mathbf{v}}_2}{dt} = 0$$

If the masses m_1 and m_2 are constant, we can bring them inside the derivative operation, which gives

$$\frac{d(m_1\vec{\mathbf{v}}_1)}{dt} + \frac{d(m_2\vec{\mathbf{v}}_2)}{dt} = 0$$

$$\frac{d}{dt}(m_1\vec{\mathbf{v}}_1 + m_2\vec{\mathbf{v}}_2) = 0 \qquad \textbf{8.1}\blacktriangleleft$$

Notice that the derivative of the sum $m_1\vec{\mathbf{v}}_1 + m_2\vec{\mathbf{v}}_2$ with respect to time is zero. Consequently, this sum must be constant. We learn from this discussion that the quantity $m\vec{\mathbf{v}}$ for a particle is important in that the sum of the values of this quantity for the particles in an isolated system is conserved. We call this quantity *linear momentum*:

> The **linear momentum** $\vec{\mathbf{p}}$ of a particle or an object that can be modeled as a particle of mass m moving with a velocity $\vec{\mathbf{v}}$ is defined to be the product of the mass and velocity:[1]

▶ Definition of linear momentum of a particle

$$\vec{\mathbf{p}} \equiv m\vec{\mathbf{v}} \qquad \textbf{8.2}\blacktriangleleft$$

---

[1]This expression is nonrelativistic and is valid only when $v \ll c$, where c is the speed of light. In the next chapter, we discuss momentum for high-speed particles.

Linear momentum is a vector quantity because it equals the product of a scalar m and a vector $\vec{v}$. Its direction is along $\vec{v}$, it has dimensions ML/T, and its SI unit is kg·m/s.

If a particle is moving in an arbitrary direction in three-dimensional space, $\vec{p}$ has three components and Equation 8.2 is equivalent to the component equations

$$p_x = mv_x \quad p_y = mv_y \quad p_z = mv_z \qquad \textbf{8.3}\blacktriangleleft$$

As you can see from its definition, the concept of momentum provides a quantitative distinction between objects of different masses moving at the same velocity. For example, the momentum of a truck moving at 2 m/s is much greater in magnitude than that of a Ping-Pong ball moving at the same speed. Newton called the product $m\vec{v}$ the *quantity of motion*, perhaps a more graphic description than *momentum*, which comes from the Latin word for movement.

QUICK QUIZ 8.1 Two objects have equal kinetic energies. How do the magnitudes of their momenta compare? (a) $p_1 < p_2$ (b) $p_1 = p_2$ (c) $p_1 > p_2$ (d) not enough information to tell

QUICK QUIZ 8.2 Your physical education teacher throws a baseball to you at a certain speed and you catch it. The teacher is next going to throw you a medicine ball whose mass is ten times the mass of the baseball. You are given the following choices: You can have the medicine ball thrown with (a) the same speed as the baseball, (b) the same momentum, or (c) the same kinetic energy. Rank these choices from easiest to hardest to catch.

Let us use the particle model for an object in motion. By using Newton's second law of motion, we can relate the linear momentum of a particle to the net force acting on the particle. In Chapter 4, we learned that Newton's second law can be written as $\Sigma\vec{F} = m\vec{a}$. This form applies only when the mass of the particle remains constant, however. In situations where the mass is changing with time, one must use an alternative statement of Newton's second law: **The time rate of change of momentum of a particle is equal to the net force acting on the particle**, or

$$\Sigma\vec{F} = \frac{d\vec{p}}{dt} \qquad \textbf{8.4}\blacktriangleleft$$

▶ Newton's second law for a particle

If the mass of the particle is constant, the preceding equation reduces to our previous expression for Newton's second law:

$$\Sigma\vec{F} = \frac{d\vec{p}}{dt} = \frac{d(m\vec{v})}{dt} = m\frac{d\vec{v}}{dt} = m\vec{a}$$

It is difficult to imagine a particle whose mass is changing, but if we consider objects, a number of examples emerge. These examples include a rocket that is ejecting its fuel as it operates, a snowball rolling down a hill and picking up additional snow, and a watertight pickup truck whose bed is collecting water as it moves in the rain.

From Equation 8.4 we see that if the net force on an object is zero, the time derivative of the momentum is zero and therefore the momentum of the object must be constant. This conclusion should sound familiar because it is the model of a particle in equilibrium, expressed in terms of momentum. Of course, if the particle is *isolated* (that is, if it does not interact with its environment), no forces act on it and $\vec{p}$ remains unchanged, which is Newton's first law.

8.2 | Analysis Model: Isolated System (Momentum)

Using the definition of momentum, Equation 8.1 can be written as

$$\frac{d}{dt}(\vec{p}_1 + \vec{p}_2) = 0$$

Because the time derivative of the total system momentum $\vec{p}_{tot} = \vec{p}_1 + \vec{p}_2$ is *zero*, we conclude that the *total* momentum $\vec{p}_{tot}$ must remain constant:

$$\vec{p}_{tot} = \text{constant} \qquad \textbf{8.5}\blacktriangleleft$$

▶ Conservation of momentum for an isolated system

or, equivalently,

$$\vec{\mathbf{p}}_{1i} + \vec{\mathbf{p}}_{2i} = \vec{\mathbf{p}}_{1f} + \vec{\mathbf{p}}_{2f} \qquad \blacktriangleleft 8.6$$

where $\vec{\mathbf{p}}_{1i}$ and $\vec{\mathbf{p}}_{2i}$ are initial values and $\vec{\mathbf{p}}_{1f}$ and $\vec{\mathbf{p}}_{2f}$ are final values of the momentum during a period over which the particles interact. Equation 8.6 in component form states that the momentum components of the isolated system in the x, y, and z directions are all *independently constant;* that is,

$$\sum_{\text{system}} p_{ix} = \sum_{\text{system}} p_{fx} \qquad \sum_{\text{system}} p_{iy} = \sum_{\text{system}} p_{fy} \qquad \sum_{\text{system}} p_{iz} = \sum_{\text{system}} p_{fz} \qquad \blacktriangleleft 8.7$$

Equation 8.6 is the mathematical statement of a new analysis model, the **isolated system (momentum)**. It can be extended to any number of particles in an isolated system as we show in Section 8.7. We studied the energy version of the isolated system model in Chapter 7 and now we have a momentum version. In general, Equation 8.6 can be stated in words as follows:

> Whenever two or more particles in an isolated system interact, the total momentum of the system remains constant.

Notice that we have made no statement concerning the nature of the forces acting between members of the system. The only requirement is that the forces must be *internal* to the system. Therefore, momentum is conserved for an isolated system *regardless* of the nature of the internal forces, *even if the forces are nonconservative.*

Example 8.1 | Can We Really Ignore the Kinetic Energy of the Earth?

In Section 6.6, we claimed that we can ignore the kinetic energy of the Earth when considering the energy of a system consisting of the Earth and a dropped ball. Verify this claim.

SOLUTION

Conceptualize Imagine dropping a ball at the surface of the Earth. From your point of view, the ball falls while the Earth remains stationary. By Newton's third law, however, the Earth experiences an upward force and therefore an upward acceleration while the ball falls. In the calculation below, we will show that this motion is extremely small and can be ignored.

Categorize We identify the system as the ball and the Earth. We assume there are no forces on the system from outer space, so the system is isolated. Let's use the momentum version of the isolated system model.

Analyze We begin by setting up a ratio of the kinetic energy of the Earth to that of the ball. We identify v_E and v_b as the speeds of the Earth and the ball, respectively, after the ball has fallen through some distance.

Use the definition of kinetic energy to set up this ratio:

$$(1) \quad \frac{K_E}{K_b} = \frac{\frac{1}{2}m_E v_E^2}{\frac{1}{2}m_b v_b^2} = \left(\frac{m_E}{m_b}\right)\left(\frac{v_E}{v_b}\right)^2$$

Apply the isolated system (momentum) model: the initial momentum of the system is zero, so set the final momentum equal to zero:

$$p_i = p_f \quad \rightarrow \quad 0 = m_b v_b + m_E v_E$$

Solve the equation for the ratio of speeds:

$$\frac{v_E}{v_b} = -\frac{m_b}{m_E}$$

Substitute this expression for v_E/v_b in Equation (1):

$$\frac{K_E}{K_b} = \left(\frac{m_E}{m_b}\right)\left(-\frac{m_b}{m_E}\right)^2 = \frac{m_b}{m_E}$$

Substitute order-of-magnitude numbers for the masses:

$$\frac{K_E}{K_b} = \frac{m_b}{m_E} \sim \frac{1 \text{ kg}}{10^{25} \text{ kg}} \sim 10^{-25}$$

Finalize The kinetic energy of the Earth is a very small fraction of the kinetic energy of the ball, so we are justified in ignoring it in the kinetic energy of the system.

Example 8.2 | The Archer

Let us consider the situation proposed at the beginning of Section 8.1. A 60-kg archer stands at rest on frictionless ice and fires a 0.030-kg arrow horizontally at 85 m/s (Fig. 8.2). With what velocity does the archer move across the ice after firing the arrow?

SOLUTION

Conceptualize You may have conceptualized this problem already when it was introduced at the beginning of Section 8.1. Imagine the arrow being fired one way and the archer recoiling in the opposite direction.

Categorize As discussed in Section 8.1, we cannot solve this problem with models based on motion, force, or energy. Nonetheless, we *can* solve this problem very easily with an approach involving momentum.

Let us take the system to consist of the archer (including the bow) and the arrow. The system is not isolated because the gravitational force and the normal force from the ice act on the system. These forces, however, are vertical and perpendicular to the motion of the system. Therefore, there are no external forces in the horizontal direction, and we can apply the isolated system (momentum) model in terms of momentum components in this direction.

Figure 8.2 (Example 8.2) An archer fires an arrow horizontally to the right. Because he is standing on frictionless ice, he will begin to slide to the left across the ice.

Analyze The total horizontal momentum of the system before the arrow is fired is zero because nothing in the system is moving. Therefore, the total horizontal momentum of the system after the arrow is fired must also be zero. We choose the direction of firing of the arrow as the positive x direction. Identifying the archer as particle 1 and the arrow as particle 2, we have $m_1 = 60$ kg, $m_2 = 0.030$ kg, and $\vec{\mathbf{v}}_{2f} = 85\hat{\mathbf{i}}$ m/s.

Using the isolated system (momentum) model, set the final momentum of the system equal to the initial value of zero:

$$m_1\vec{\mathbf{v}}_{1f} + m_2\vec{\mathbf{v}}_{2f} = 0$$

Solve this equation for $\vec{\mathbf{v}}_{1f}$ and substitute numerical values:

$$\vec{\mathbf{v}}_{1f} = -\frac{m_2}{m_1}\vec{\mathbf{v}}_{2f} = -\left(\frac{0.030 \text{ kg}}{60 \text{ kg}}\right)(85\hat{\mathbf{i}} \text{ m/s}) = -0.042\hat{\mathbf{i}} \text{ m/s}$$

Finalize The negative sign for $\vec{\mathbf{v}}_{1f}$ indicates that the archer is moving to the left in Figure 8.2 after the arrow is fired, in the direction opposite the direction of motion of the arrow, in accordance with Newton's third law. Because the archer is much more massive than the arrow, his acceleration and consequent velocity are much smaller than the acceleration and velocity of the arrow. Notice that this problem sounds very simple, but we could not solve it with models based on motion, force, or energy. Our new momentum model, however, shows us that it not only *sounds* simple, it *is* simple!

What If? What if the arrow were fired in a direction that makes an angle θ with the horizontal? How will that change the recoil velocity of the archer?

Answer The recoil velocity should decrease in magnitude because only a component of the velocity of the arrow is in the x direction. Conservation of momentum in the x direction gives

$$m_1 v_{1f} + m_2 v_{2f} \cos \theta = 0$$

leading to

$$v_{1f} = -\frac{m_2}{m_1} v_{2f} \cos \theta$$

For $\theta = 0$, $\cos \theta = 1$ and the final velocity of the archer reduces to the value when the arrow is fired horizontally. For nonzero values of θ, the cosine function is less than 1 and the recoil velocity is less than the value calculated for $\theta = 0$. If $\theta = 90°$, then $\cos \theta = 0$ and $v_{1f} = 0$, so there is no recoil velocity. In this case, the archer is simply pushed downward harder against the ice as the arrow is fired.

Example 8.3 | Decay of the Kaon at Rest

One type of nuclear particle, called the *neutral kaon* (K^0), decays into a pair of other particles called *pions* (π^+ and π^-), which are oppositely charged but equal in mass, as in Figure 8.3. Assuming that the kaon is initially at rest, prove that the two pions must have momenta that are equal in magnitude and opposite in direction.

SOLUTION

Conceptualize Study Figure 8.3 carefully and imagine the kaon at rest decaying into two moving particles. Compare Figure 8.3 with Figure 8.2 and correlate the arrow and the archer with the individual pions.

Categorize Because the kaon does not interact with its surroundings, we model it as an isolated system. The system after the decay is the two pions.

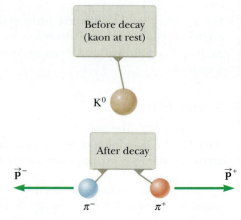

Figure 8.3 (Example 8.3) A kaon at rest decays into a pair of oppositely charged pions. The pions move apart with momenta of equal magnitudes but opposite directions.

Analyze Write an expression for decay of the kaon, represented in Figure 8.3:

$$K^0 \;\rightarrow\; \pi^+ + \pi^-$$

Let $\vec{\mathbf{p}}^+$ be the momentum of the positive pion and $\vec{\mathbf{p}}^-$ be the momentum of the negative pion after the decay, and find an expression for the final momentum $\vec{\mathbf{p}}_f$ of the isolated system of two pions:

$$\vec{\mathbf{p}}_f = \vec{\mathbf{p}}^+ + \vec{\mathbf{p}}^-$$

Because the kaon is at rest before the decay, we know that the initial system momentum $\vec{\mathbf{p}}_i = 0$. Furthermore, because the momentum of the isolated system is conserved, $\vec{\mathbf{p}}_i = \vec{\mathbf{p}}_f = 0$.

Incorporate this result in the previous equation:

$$0 = \vec{\mathbf{p}}^+ + \vec{\mathbf{p}}^- \;\rightarrow\; \vec{\mathbf{p}}^+ = -\vec{\mathbf{p}}^-$$

Finalize Therefore, we see that the two momentum vectors of the pions are equal in magnitude and opposite in direction.

8.3 | Analysis Model: Nonisolated System (Momentum)

As described by Equation 8.4, the momentum of a particle changes if a net force acts on the particle. Let us assume that a net force $\Sigma\vec{\mathbf{F}}$ acts on a particle and that this force may vary with time. According to Equation 8.4,

$$d\vec{\mathbf{p}} = \Sigma\vec{\mathbf{F}}\, dt \qquad\qquad \text{8.8} \blacktriangleleft$$

We can integrate this expression to find the change in the momentum of a particle during the time interval $\Delta t = t_f - t_i$. Integrating Equation 8.8 gives

$$\Delta\vec{\mathbf{p}} = \vec{\mathbf{p}}_f - \vec{\mathbf{p}}_i = \int_{t_i}^{t_f} \Sigma\vec{\mathbf{F}}\, dt \qquad\qquad \text{8.9} \blacktriangleleft$$

The integral of a force over the time interval during which it acts is called the **impulse** of the force. The impulse of the net force $\Sigma\vec{\mathbf{F}}$ is a vector defined by

▶ Impulse of a net force

$$\vec{\mathbf{I}} \equiv \int_{t_i}^{t_f} \Sigma\vec{\mathbf{F}}\, dt \qquad\qquad \text{8.10} \blacktriangleleft$$

From its definition, we see that impulse $\vec{\mathbf{I}}$ is a vector quantity having a magnitude equal to the area under the force–time curve as described in Figure 8.4a. It is assumed the force varies in time in the general manner shown in the figure and is nonzero in the time interval $\Delta t = t_f - t_i$. The direction of the impulse vector is the same as the direction of the change in momentum. Impulse has the dimensions of momentum, that is, ML/T. Impulse is *not* a property of a particle; rather, it is a measure of the degree to which an external force changes the particle's momentum.

Combining Equations 8.9 and 8.10 gives us an important statement known as the **impulse–momentum theorem:**

The change in the momentum of a particle is equal to the impulse of the net force acting on the particle:

$$\Delta\vec{\mathbf{p}} = \vec{\mathbf{I}} \qquad \text{8.11} \blacktriangleleft$$

◄ Impulse–momentum theorem for a particle

This statement is equivalent to Newton's second law. When we say that an impulse is given to a particle, we mean that momentum is transferred from an external agent to that particle. Equation 8.11 is identical in form to the conservation of energy equation, Equation 7.1, and its full expansion, Equation 7.2. Equation 8.11 is the most general statement of the principle of **conservation of momentum** and is called the **conservation of momentum equation.** In the case of a momentum approach, isolated systems tend to appear in problems more often than nonisolated systems, so, in practice, the conservation of momentum equation is often identified as the special case shown in Equation 8.6.

The left side of Equation 8.11 represents the change in the momentum of the system, which in this case is a single particle. The right side is a measure of how much momentum crosses the boundary of the system due to the net force being applied to the system. Equation 8.11 is the mathematical statement of a new analysis model, the **nonisolated system (momentum)** model. Although this equation is similar in form to Equation 7.1, there are several differences in its application to problems. First, Equation 8.11 is a vector equation, whereas Equation 7.1 is a scalar equation. Therefore, directions are important for Equation 8.11. Second, there is only one type of linear momentum and therefore only one way to store momentum in a system. In contrast, as we see from Equation 7.2, there are three ways to store energy in a system: kinetic, potential, and internal. Third, there is only one way to transfer momentum into a system: by the application of a force on the system over a time interval. Equation 7.2 shows six ways we have identified as transferring energy into a system. Therefore, there is no expansion of Equation 8.11 analogous to Equation 7.2.

Because the net force on a particle can generally vary in time as in Figure 8.4a, it is convenient to define a time-averaged net force $\left(\Sigma\vec{\mathbf{F}}\right)_{\text{avg}}$ given by

$$\left(\Sigma\vec{\mathbf{F}}\right)_{\text{avg}} \equiv \frac{1}{\Delta t}\int_{t_i}^{t_f}\Sigma\vec{\mathbf{F}}\,dt \qquad \text{8.12} \blacktriangleleft$$

where $\Delta t = t_f - t_i$. Therefore, we can express Equation 8.10 as

$$\vec{\mathbf{I}} = \left(\Sigma\vec{\mathbf{F}}\right)_{\text{avg}}\Delta t \qquad \text{8.13} \blacktriangleleft$$

The magnitude of this average net force, described in Figure 8.4b, can be thought of as the magnitude of the constant net force that would give the same impulse to the particle in the time interval Δt as the actual time-varying net force gives over this same interval.

In principle, if $\Sigma\vec{\mathbf{F}}$ is known as a function of time, the impulse can be calculated from Equation 8.10. The calculation becomes especially simple if the net force acting on the particle is constant. In this case, $\left(\Sigma\vec{\mathbf{F}}\right)_{\text{avg}}$ over a time interval is the same as the constant $\Sigma\vec{\mathbf{F}}$ at any instant within the interval, and Equation 8.13 becomes

$$\vec{\mathbf{I}} = \Sigma\vec{\mathbf{F}}\Delta t \qquad \text{8.14} \blacktriangleleft$$

In many physical situations, we shall use what is called the **impulse approximation,** in which we assume that one of the forces exerted on a particle acts for a short time but is much greater than any other force present. This simplification model

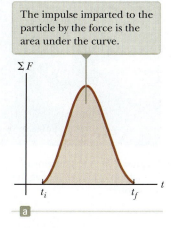

The impulse imparted to the particle by the force is the area under the curve.

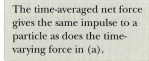

The time-averaged net force gives the same impulse to a particle as does the time-varying force in (a).

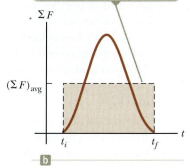

Figure 8.4 (a) A net force acting on a particle may vary in time. (b) The value of the constant force $(\Sigma F)_{\text{avg}}$ (horizontal dashed line) is chosen so that the area $(\Sigma F)_{\text{avg}}\Delta t$ of the rectangle is the same as the area under the curve in (a).

allows us to ignore the effects of other forces because these effects are small for the short time interval during which the large force acts. This approximation is especially useful in treating collisions in which the duration of the collision is very short. When this approximation is made, we refer to the force that is greater as an *impulsive force*. For example, when a baseball is struck with a bat, the duration of the collision is about 0.01 s and the average force the bat exerts on the ball during this time interval is typically several thousand newtons. This average force is much greater than the gravitational force, so we ignore any change in velocity related to the gravitational force during the collision. It is important to remember that $\vec{p}_i$ and $\vec{p}_f$ represent the momenta *immediately* before and after the collision, respectively. Therefore in the impulse approximation, very little motion of the particle takes place during the collision.

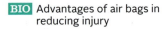
BIO Advantages of air bags in reducing injury

The concept of impulse helps us understand the value of air bags in stopping a passenger in an automobile accident (Fig. 8.5). The passenger experiences the same change in momentum and therefore the same impulse in a collision whether the car has air bags or not. The air bag allows the passenger to experience that change in momentum over a longer time interval, however, reducing the peak force on the passenger and increasing the chances of escaping without injury. Without the air bag, the passenger's head could move forward and be brought to rest in a short time interval by the steering wheel or the dashboard. In this case, the passenger undergoes the same change in momentum, but the short time interval results in a very large force that could cause severe head injury. Such injuries often result in spinal cord nerve damage where the nerves enter the base of the brain.

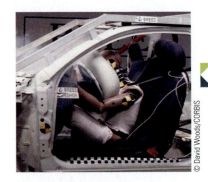
© David Woods/CORBIS

Figure 8.5 A test dummy is brought to rest by an air bag in an automobile.

QUICK QUIZ 8.3 Two objects are at rest on a frictionless surface. Object 1 has a greater mass than object 2. (**i**) When a constant force is applied to object 1, it accelerates through a distance d in a straight line. The force is removed from object 1 and is applied to object 2. At the moment when object 2 has accelerated through the same distance d, which statements are true? (**a**) $p_1 < p_2$ (**b**) $p_1 = p_2$ (**c**) $p_1 > p_2$ (**d**) $K_1 < K_2$ (**e**) $K_1 = K_2$ (**f**) $K_1 > K_2$ (**ii**) When a constant force is applied to object 1, it accelerates for a time interval Δt. The force is removed from object 1 and is applied to object 2. From the same list of choices, which statements are true after object 2 has accelerated for the same time interval Δt?

Example 8.4 | How Good Are the Bumpers?

In a particular crash test, a car of mass 1 500 kg collides with a wall as shown in Figure 8.6. The initial and final velocities of the car are $\vec{v}_i = -15.0\hat{i}$ m/s and $\vec{v}_f = 2.60\hat{i}$ m/s, respectively. If the collision lasts 0.150 s, find the impulse caused by the collision and the average net force exerted on the car.

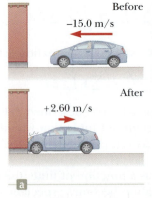

Before

−15.0 m/s

After

+2.60 m/s

a

Hyundai Motors/HO/Landov

b

Figure 8.6 (Example 8.4) (a) This car's momentum changes as a result of its collision with the wall. (b) In a crash test, much of the car's initial kinetic energy is transformed into energy associated with the damage to the car.

SOLUTION

Conceptualize The collision time is short, so we can imagine the car being brought to rest very rapidly and then moving back in the opposite direction with a reduced speed.

Categorize Let us assume the net force exerted on the car by the wall and friction from the ground is large compared with other forces on the car (such as air resistance). Furthermore, the gravitational force and the normal force exerted by the road on the car are perpendicular to the motion and therefore do not affect the horizontal momentum. Therefore, we categorize the problem as one in which we can apply the impulse approximation in the horizontal direction. We also see that the car's momentum changes due to an impulse from the environment. Therefore, we can apply the nonisolated system (momentum) model.

8.4 *cont.*

Analyze

Evaluate the initial and final momenta of the car:

$$\vec{\mathbf{p}}_i = m\vec{\mathbf{v}}_i = (1\,500 \text{ kg})(-15.0\hat{\mathbf{i}} \text{ m/s}) = -2.25 \times 10^4 \hat{\mathbf{i}} \text{ kg} \cdot \text{m/s}$$

$$\vec{\mathbf{p}}_f = m\vec{\mathbf{v}}_f = (1\,500 \text{ kg})(2.60\hat{\mathbf{i}} \text{ m/s}) = 0.39 \times 10^4 \hat{\mathbf{i}} \text{ kg} \cdot \text{m/s}$$

Use Equation 8.11 to find the impulse on the car:

$$\vec{\mathbf{I}} = \Delta\vec{\mathbf{p}} = \vec{\mathbf{p}}_f - \vec{\mathbf{p}}_i = 0.39 \times 10^4 \hat{\mathbf{i}} \text{ kg} \cdot \text{m/s} - (-2.25 \times 10^4 \hat{\mathbf{i}} \text{ kg} \cdot \text{m/s})$$

$$= \boxed{2.64 \times 10^4 \hat{\mathbf{i}} \text{ kg} \cdot \text{m/s}}$$

Use Equation 8.13 to evaluate the average net force exerted on the car:

$$\left(\sum \vec{\mathbf{F}}\right)_{\text{avg}} = \frac{\vec{\mathbf{I}}}{\Delta t} = \frac{2.64 \times 10^4 \hat{\mathbf{i}} \text{ kg} \cdot \text{m/s}}{0.150 \text{ s}} = \boxed{1.76 \times 10^5 \hat{\mathbf{i}} \text{ N}}$$

Finalize The net force found above is a combination of the normal force on the car from the wall and any friction force between the tires and the ground as the front of the car crumples. If the brakes are not operating while the crash occurs and the crumpling metal does not interfere with the free rotation of the tires, this friction force could be relatively small due to the freely rotating wheels. Notice that the signs of the velocities in this example indicate the reversal of directions. What would the mathematics be describing if both the initial and final velocities had the same sign?

What If? What if the car did not rebound from the wall? Suppose the final velocity of the car is zero and the time interval of the collision remains at 0.150 s. Would that represent a larger or a smaller net force on the car?

Answer In the original situation in which the car rebounds, the net force on the car does two things during the time interval: (1) it stops the car, and (2) it causes the car to move away from the wall at 2.60 m/s after the collision. If the car does not rebound, the net force is only doing the first of these steps—stopping the car—which requires a *smaller* force.

Mathematically, in the case of the car that does not rebound, the impulse is

$$\vec{\mathbf{I}} = \Delta\vec{\mathbf{p}} = \vec{\mathbf{p}}_f - \vec{\mathbf{p}}_i = 0 - (-2.25 \times 10^4 \hat{\mathbf{i}} \text{ kg} \cdot \text{m/s}) = 2.25 \times 10^4 \hat{\mathbf{i}} \text{ kg} \cdot \text{m/s}$$

The average net force exerted on the car is

$$\left(\sum \vec{\mathbf{F}}\right)_{\text{avg}} = \frac{\vec{\mathbf{I}}}{\Delta t} = \frac{2.25 \times 10^4 \hat{\mathbf{i}} \text{ kg} \cdot \text{m/s}}{0.150 \text{ s}} = 1.50 \times 10^5 \hat{\mathbf{i}} \text{ N}$$

which is indeed smaller than the previously calculated value, as was argued conceptually.

8.4 | Collisions in One Dimension

In this section, we use the law of conservation of momentum to describe what happens when two objects collide. The term **collision** represents an event during which two particles come close to each other and interact by means of forces. The forces due to the collision are assumed to be much larger than any external forces present, so we use the simplification model we call the impulse approximation. The general goal in collision problems is to relate the final conditions of the system to the initial conditions.

A collision may be the result of physical contact between two objects, as described in Figure 8.7a. This observation is common when two macroscopic objects collide, such as two billiard balls or a baseball and a bat.

The notion of what we mean by *collision* must be generalized because "contact" on a microscopic scale is ill defined. To understand the distinction between macroscopic and microscopic collisions, consider the collision of a proton with an alpha particle (the nucleus of the helium atom), illustrated in Figure 8.7b. Because the two particles are positively charged, they repel each other. A collision has occurred, but the colliding particles were never in "contact."

When two particles of masses m_1 and m_2 collide, the collision forces may vary in time in a complicated way, as seen in Figure 8.4. As a result, an analysis of the situation with Newton's second law could be very complicated. We find, however, that

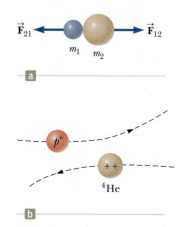

Figure 8.7 (a) A collision between two objects as the result of direct contact. (b) A "collision" between two charged particles that do not make contact.

the momentum concept is similar to the energy concept in Chapters 6 and 7 in that it provides us with a much easier method to solve problems involving isolated systems.

According to Equation 8.5, the momentum of an isolated system is conserved during some interaction event, such as a collision. The kinetic energy of the system, however, is generally *not* conserved in a collision. We define an **inelastic collision** as one in which the kinetic energy of the system is not conserved (even though momentum is conserved). The collision of a rubber ball with a hard surface is inelastic because some of the kinetic energy of the ball is transformed to internal energy when the ball is deformed while in contact with the surface.

A practical example of an inelastic collision is used to detect glaucoma, a disease in which the pressure inside the eye builds up and leads to blindness by damaging the cells of the retina. In this application, medical professionals use a device called a *tonometer* to measure the pressure inside the eye. This device releases a puff of air against the outer surface of the eye and measures the speed of the air after reflection from the eye. At normal pressure, the eye is slightly spongy and the pulse is reflected at low speed. As the pressure inside the eye increases, the outer surface becomes more rigid and the speed of the reflected pulse increases. Therefore, the speed of the reflected puff of air is used to measure the internal pressure of the eye.

Pitfall Prevention | 8.2
Inelastic Collisions
Generally, inelastic collisions are hard to analyze without additional information. Lack of this information appears in the mathematical representation as having more unknowns than equations.

When two objects collide and stick together after a collision, the maximum possible fraction of the initial kinetic energy is transformed or transferred away (by sound, for example); this collision is called a **perfectly inelastic collision.** For example, if two vehicles collide and become entangled, they move with some common velocity after the perfectly inelastic collision. If a meteorite collides with the Earth, it becomes buried in the ground and the collision is perfectly inelastic.

An **elastic collision** is defined as one in which the kinetic energy of the system is conserved (as well as momentum). Real collisions in the macroscopic world, such as those between billiard balls, are only approximately elastic because some transformation of kinetic energy takes place and some energy leaves the system by mechanical waves, sound. Imagine a billiard game with truly elastic collisions. The opening break would be completely silent! Truly elastic collisions do occur between atomic and subatomic particles. Elastic and perfectly inelastic collisions are *limiting* cases; a large number of collisions fall in the range between them.

In the remainder of this section, we treat collisions in one dimension and consider the two extreme cases: perfectly inelastic collisions and elastic collisions. The important distinction between these two types of collisions is that the momentum of the system is conserved in all cases, but the kinetic energy is conserved only in elastic collisions. When analyzing one-dimensional collisions, we can drop the vector notation and use positive and negative signs for velocities to denote directions, as we did in Chapter 2.

Before the collision, the particles move separately.

After the collision, the particles move together.

Active Figure 8.8 Schematic representation of a perfectly inelastic head-on collision between two particles.

Perfectly Inelastic Collisions

Consider two objects of masses m_1 and m_2 moving with initial velocities v_{1i} and v_{2i} along a straight line as in Active Figure 8.8. If the two objects collide head-on, stick together, and move with some common velocity v_f after the collision, the collision is perfectly inelastic. Because the total momentum of the two-object isolated system before the collision equals the total momentum of the combined-object system after the collision, we have

$$m_1 v_{1i} + m_2 v_{2i} = (m_1 + m_2) v_f \qquad \text{8.15} \blacktriangleleft$$

$$v_f = \frac{m_1 v_{1i} + m_2 v_{2i}}{m_1 + m_2} \qquad \text{8.16} \blacktriangleleft$$

Therefore, if we know the initial velocities of the two objects, we can use this single equation to determine the final common velocity.

Elastic Collisions

Now consider two objects that undergo an elastic head-on collision (Active Fig. 8.9) in one dimension. In this collision, both momentum and kinetic energy are conserved; therefore, we can write[2]

$$m_1 v_{1i} + m_2 v_{2i} = m_1 v_{1f} + m_2 v_{2f} \qquad \textbf{8.17} \blacktriangleleft$$

$$\tfrac{1}{2} m_1 v_{1i}^2 + \tfrac{1}{2} m_2 v_{2i}^2 = \tfrac{1}{2} m_1 v_{1f}^2 + \tfrac{1}{2} m_2 v_{2f}^2 \qquad \textbf{8.18} \blacktriangleleft$$

In a typical problem involving elastic collisions, two unknown quantities occur (such as v_{1f} and v_{2f}), and Equations 8.17 and 8.18 can be solved simultaneously to find them. An alternative approach, employing a little mathematical manipulation of Equation 8.18, often simplifies this process. Let us cancel the factor of $\tfrac{1}{2}$ in Equation 8.18 and rewrite the equation as

$$m_1(v_{1i}^2 - v_{1f}^2) = m_2(v_{2f}^2 - v_{2i}^2)$$

Here we have moved the terms containing m_1 to one side of the equation and those containing m_2 to the other. Next, let us factor both sides:

$$m_1(v_{1i} - v_{1f})(v_{1i} + v_{1f}) = m_2(v_{2f} - v_{2i})(v_{2f} + v_{2i}) \qquad \textbf{8.19} \blacktriangleleft$$

We now separate the terms containing m_1 and m_2 in the equation for conservation of momentum (Eq. 8.17) to obtain

$$m_1(v_{1i} - v_{1f}) = m_2(v_{2f} - v_{2i}) \qquad \textbf{8.20} \blacktriangleleft$$

To obtain our final result, we divide Equation 8.19 by Equation 8.20 and obtain

$$v_{1i} + v_{1f} = v_{2f} + v_{2i}$$

or, gathering initial and final values on opposite sides of the equation,

$$v_{1i} - v_{2i} = -(v_{1f} - v_{2f}) \qquad \textbf{8.21} \blacktriangleleft$$

This equation, in combination with the condition for conservation of momentum, Equation 8.17, can be used to solve problems dealing with one-dimensional elastic collisions between two objects. According to Equation 8.21, the relative speed[3] $v_{1i} - v_{2i}$ of the two objects before the collision equals the negative of their relative speed after the collision, $-(v_{1f} - v_{2f})$.

Suppose the masses and the initial velocities of both objects are known. Equations 8.17 and 8.21 can be solved for the final velocities in terms of the initial values because we have two equations and two unknowns:

$$v_{1f} = \left(\frac{m_1 - m_2}{m_1 + m_2}\right) v_{1i} + \left(\frac{2 m_2}{m_1 + m_2}\right) v_{2i} \qquad \textbf{8.22} \blacktriangleleft$$

$$v_{2f} = \left(\frac{2 m_1}{m_1 + m_2}\right) v_{1i} + \left(\frac{m_2 - m_1}{m_1 + m_2}\right) v_{2i} \qquad \textbf{8.23} \blacktriangleleft$$

It is important to remember that the appropriate signs for the numerical values of velocities v_{1i} and v_{2i} must be included in Equations 8.22 and 8.23. For example, if m_2 is moving to the left initially, as in Active Figure 8.9a, v_{2i} is negative.

---

[2]Notice that the kinetic energy of the system is the sum of the kinetic energies of the two particles. In our energy conservation examples in Chapter 7 involving a falling object and the Earth, we ignored the kinetic energy of the Earth because it is so small. Therefore, the kinetic energy of the *system* is just the kinetic energy of the falling *object*. That is a special case in which the mass of one of the objects (the Earth) is so immense that ignoring its kinetic energy introduces no measurable error. For problems such as those described here, however, and for the particle decay problems we will see in Chapters 30 and 31, we need to include the kinetic energies of *all* particles in the system.

[3]See Section 3.6 for a review of relative speed.

Pitfall Prevention | 8.3
Momentum and Kinetic Energy in Collisions
Linear momentum of an isolated system is conserved in *all* collisions. Kinetic energy of an isolated system is conserved *only* in elastic collisions. These statements are true because kinetic energy can be transformed into several types of energy or can be transferred out of the system (so that the system may *not* be isolated in terms of energy during the collision), but there is only one type of linear momentum.

Before the collision, the particles move separately.

After the collision, the particles continue to move separately with new velocities.

Active Figure 8.9 Schematic representation of an elastic head-on collision between two particles.

Pitfall Prevention | 8.4
Not a General Equation
We have spent some effort on deriving Equation 8.21, but remember that it can be used only in a very *specific* situation: a one-dimensional, elastic collision between two objects. The *general* concept is conservation of momentum (and conservation of kinetic energy if the collision is elastic) for an isolated system.

Let us consider some special cases. If $m_1 = m_2$, Equations 8.22 and 8.23 show us that $v_{1f} = v_{2i}$ and $v_{2f} = v_{1i}$. That is, the objects exchange speeds if they have equal masses. That is what one observes in head-on billiard ball collisions, assuming there is no spin on the ball: The initially moving ball stops and the initially stationary ball moves away with approximately the same speed.

If m_2 is initially at rest, $v_{2i} = 0$ and Equations 8.22 and 8.23 become

▶ Elastic collision in one dimension: particle 2 initially at rest

$$v_{1f} = \left(\frac{m_1 - m_2}{m_1 + m_2}\right)v_{1i} \qquad \textbf{8.24} \blacktriangleleft$$

$$v_{2f} = \left(\frac{2m_1}{m_1 + m_2}\right)v_{1i} \qquad \textbf{8.25} \blacktriangleleft$$

If m_1 is very large compared with m_2, we see from Equations 8.24 and 8.25 that $v_{1f} \approx v_{1i}$ and $v_{2f} \approx 2v_{1i}$. That is, when a very heavy object collides head-on with a very light one initially at rest, the heavy object continues its motion unaltered after the collision but the light object rebounds with a speed equal to about twice the initial speed of the heavy object. An example of such a collision is that of a moving heavy atom, such as uranium, with a light atom, such as hydrogen.

If m_2 is much larger than m_1 and if m_2 is initially at rest, we find from Equations 8.24 and 8.25 that $v_{1f} \approx -v_{1i}$ and $v_{2f} \approx 0$. That is, when a very light object collides head-on with a very heavy object initially at rest, the velocity of the light object is reversed and the heavy object remains approximately at rest. For example, imagine what happens when a table-tennis ball hits a stationary bowling ball.

> **QUICK QUIZ 8.4** A table-tennis ball is thrown at a stationary bowling ball. The table-tennis ball makes a one-dimensional elastic collision and bounces back along the same line. Compared with the bowling ball after the collision, does the table-tennis ball have (**a**) a larger magnitude of momentum and more kinetic energy, (**b**) a smaller magnitude of momentum and more kinetic energy, (**c**) a larger magnitude of momentum and less kinetic energy, (**d**) a smaller magnitude of momentum and less kinetic energy, or (**e**) the same magnitude of momentum and the same kinetic energy

> **PROBLEM-SOLVING STRATEGY: One-Dimensional Collisions**
>
> You should use the following approach when solving collision problems in one dimension:
>
> 1. **Conceptualize.** Imagine the collision occurring in your mind. Draw simple diagrams of the particles before and after the collision and include appropriate velocity vectors. At first, you may have to guess at the directions of the final velocity vectors.
>
> 2. **Categorize.** Is the system of particles isolated? If so, categorize the collision as elastic, inelastic, or perfectly inelastic.
>
> 3. **Analyze.** Set up the appropriate mathematical representation for the problem. If the collision is perfectly inelastic, use Equation 8.15. If the collision is elastic, use Equations 8.17 and 8.21. If the collision is inelastic, use Equation 8.17. To find the final velocities in this case, you will need some additional information.
>
> 4. **Finalize.** Once you have determined your result, check to see if your answers are consistent with the mental and pictorial representations and that your results are realistic.

Example 8.5 | **Kinetic Energy in a Perfectly Inelastic Collision**

We claimed that the maximum amount of kinetic energy was transformed to other forms in a perfectly inelastic collision. Prove this statement mathematically for a one-dimensional two-particle collision.

SOLUTION

Conceptualize We will assume that the maximum kinetic energy is transformed and prove that the collision must be perfectly inelastic.

Categorize We categorize the system of two particles as an isolated system. We also categorize the collision as one-dimensional.

Analyze Find an expression for the ratio of the final kinetic energy after the collision to the initial kinetic energy:

$$f = \frac{K_f}{K_i} = \frac{\frac{1}{2}m_1 v_{1f}^2 + \frac{1}{2}m_2 v_{2f}^2}{\frac{1}{2}m_1 v_{1i}^2 + \frac{1}{2}m_2 v_{2i}^2} = \frac{m_1 v_{1f}^2 + m_2 v_{2f}^2}{m_1 v_{1i}^2 + m_2 v_{2i}^2}$$

The *maximum* amount of energy transformed to other forms corresponds to the *minimum* value of f. For fixed initial conditions, imagine that the final velocities v_{1f} and v_{2f} are variables. Minimize the fraction f by taking the derivative of f with respect to v_{1f} and setting the result equal to zero:

$$\frac{df}{dv_{1f}} = \frac{d}{dv_{1f}}\left(\frac{m_1 v_{1f}^2 + m_2 v_{2f}^2}{m_1 v_{1i}^2 + m_2 v_{2i}^2}\right)$$

$$= \frac{2m_1 v_{1f} + 2m_2 v_{2f}\dfrac{dv_{2f}}{dv_{1f}}}{m_1 v_{1i}^2 + m_2 v_{2i}^2} = 0$$

$$\rightarrow \quad (1) \quad m_1 v_{1f} + m_2 v_{2f}\frac{dv_{2f}}{dv_{1f}} = 0$$

From the conservation of momentum condition, we can evaluate the derivative in (1). Differentiate Equation 8.17 with respect to v_{1f}:

$$\frac{d}{dv_{1f}}(m_1 v_{1i} + m_2 v_{2i}) = \frac{d}{dv_{1f}}(m_1 v_{1f} + m_2 v_{2f})$$

$$\rightarrow \quad 0 = m_1 + m_2\frac{dv_{2f}}{dv_{1f}} \quad \rightarrow \quad \frac{dv_{2f}}{dv_{1f}} = -\frac{m_1}{m_2}$$

Substitute this expression for the derivative into (1):

$$m_1 v_{1f} - m_2 v_{2f}\frac{m_1}{m_2} = 0 \quad \rightarrow \quad v_{1f} = v_{2f}$$

Finalize If the particles come out of the collision with the same velocities, they are joined together and it is a perfectly inelastic collision, which is what we set out to prove.

Example 8.6 | **Carry Collision Insurance!**

An 1 800-kg car stopped at a traffic light is struck from the rear by a 900-kg car. The two cars become entangled, moving along the same path as that of the originally moving car. If the smaller car were moving at 20.0 m/s before the collision, what is the velocity of the entangled cars after the collision?

SOLUTION

Conceptualize This kind of collision is easily visualized, and one can predict that after the collision both cars will be moving in the same direction as that of the initially moving car. Because the initially moving car has only half the mass of the stationary car, we expect the final velocity of the cars to be relatively small.

Categorize We identify the system of two cars as isolated in terms of momentum in the horizontal direction and apply the impulse approximation during the short time interval of the collision. The phrase "become entangled" tells us to categorize the collision as perfectly inelastic.

Analyze The magnitude of the total momentum of the system before the collision is equal to that of the smaller car because the larger car is initially at rest.

continued

8.6 *cont.*

Set the initial momentum of the system equal to the final momentum of the system:

$$p_i = p_f \quad \rightarrow \quad m_1 v_i = (m_1 + m_2) v_f$$

Solve for v_f and substitute numerical values:

$$v_f = \frac{m_1 v_i}{m_1 + m_2} = \frac{(900 \text{ kg})(20.0 \text{ m/s})}{900 \text{ kg} + 1\,800 \text{ kg}} = 6.67 \text{ m/s}$$

Finalize Because the final velocity is positive, the direction of the final velocity of the combination is the same as the velocity of the initially moving car as predicted. The speed of the combination is also much lower than the initial speed of the moving car.

What If? Suppose we reverse the masses of the cars. What if a stationary 900-kg car is struck by a moving 1 800-kg car? Is the final speed the same as before?

Answer Intuitively, we can guess that the final speed of the combination is higher than 6.67 m/s if the initially moving car is the more massive car. Mathematically, that should be the case because the system has a larger momentum if the initially moving car is the more massive one. Solving for the new final velocity, we find

$$v_f = \frac{m_1 v_i}{m_1 + m_2} = \frac{(1\,800 \text{ kg})(20.0 \text{ m/s})}{1\,800 \text{ kg} + 900 \text{ kg}} = 13.3 \text{ m/s}$$

which is two times greater than the previous final velocity.

Example 8.7 | Slowing Down Neutrons by Collisions

In a nuclear reactor, neutrons are produced when $^{235}_{92}\text{U}$ atoms split in a process called *fission*. These neutrons are moving at about 10^7 m/s and must be slowed down to about 10^3 m/s before they take part in another fission event. They are slowed down by being passed through a solid or liquid material called a *moderator*. The slowing-down process involves elastic collisions. Let us show that a neutron can lose most of its kinetic energy if it collides elastically with a moderator containing light nuclei, such as deuterium (in "heavy water," D_2O).

SOLUTION

Conceptualize Imagine a single neutron passing through the moderator material and repeatedly colliding with nuclei. The kinetic energy of the neutron will decrease in each collision and the neutron will eventually slow down to the desired 10^3 m/s.

Categorize We identify the neutron and a particular moderator nucleus as an isolated system and use the momentum version of the isolated system model. Let us assume that the moderator nucleus of mass m_m is at rest initially and that the neutron of mass m_n and initial speed v_{ni} collides head-on with it. Because the momentum and kinetic energy of this system are conserved in an elastic collision, Equations 8.24 and 8.25 can be applied to a one-dimensional collision of these two particles.

Analyze

Find an expression for the initial kinetic energy of the neutron:

$$K_{ni} = \tfrac{1}{2} m_n v_{ni}^2$$

Using Equation 8.24, find an expression for the final kinetic energy of the neutron:

$$K_{nf} = \tfrac{1}{2} m_n v_{nf}^2 = \tfrac{1}{2} m_n \left(\frac{m_n - m_m}{m_n + m_m} \right)^2 v_{ni}^2$$

Now find an expression for the fraction of the total kinetic energy possessed by the neutron after the collision:

$$(1) \quad f_n = \frac{K_{nf}}{K_{ni}} = \frac{\tfrac{1}{2} m_n \left(\dfrac{m_n - m_m}{m_n + m_m} \right)^2 v_{ni}^2}{\tfrac{1}{2} m_n v_{ni}^2} = \left(\frac{m_n - m_m}{m_n + m_m} \right)^2$$

Find an expression for the kinetic energy of the moderator nucleus after the collision using Equation 8.25:

$$(2) \quad K_{mf} = \tfrac{1}{2} m_m v_{mf}^2 = \frac{2 m_n^2 m_m}{(m_n + m_m)^2} v_{ni}^2$$

Use Equation (2) to find an expression for the fraction of the total kinetic energy transferred to the moderator nucleus:

$$(3) \quad f_{\text{trans}} = \frac{K_{mf}}{K_{ni}} = \frac{\dfrac{2 m_n^2 m_m}{(m_n + m_m)^2} v_{ni}^2}{\tfrac{1}{2} m_n v_{ni}^2} = \frac{4 m_n m_m}{(m_n + m_m)^2}$$

8.7 *cont.*

Finalize If $m_m \approx m_n$, we see that $f_{trans} \approx 1 = 100\%$. Because the system's kinetic energy is conserved, Equation (3) can also be obtained from Equation (1) with the condition that $f_n + f_m = 1$, so that $f_m = 1 - f_n$.

For collisions of the neutrons with deuterium nuclei in D_2O ($m_m = 2m_n$), $f_n = 1/9$ and $f_{trans} = 8/9$. That is, 89% of the neutron's kinetic energy is transferred to the deuterium nucleus. In practice, the moderator efficiency is reduced because head-on collisions are very unlikely to occur.

Example 8.8 | A Two-Body Collision with a Spring

A block of mass $m_1 = 1.60$ kg initially moving to the right with a speed of 4.00 m/s on a frictionless, horizontal track collides with a light spring attached to a second block of mass $m_2 = 2.10$ kg initially moving to the left with a speed of 2.50 m/s as shown in Figure 8.10a. The spring constant is 600 N/m.

(A) Find the velocities of the two blocks after the collision.

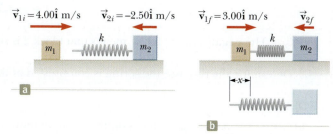

Figure 8.10 (Example 8.8) A moving block approaches a second moving block that is attached to a spring.

SOLUTION

Conceptualize With the help of Figure 8.10a, run an animation of the collision in your mind. Figure 8.10b shows an instant during the collision when the spring is compressed. Eventually, block 1 and the spring will again separate, so the system will look like Figure 8.10a again but with different velocity vectors for the two blocks.

Categorize Because the spring force is conservative, kinetic energy in the system of two blocks and the spring is not transformed to internal energy during the compression of the spring. Ignoring any sound made when the block hits the spring, we can categorize the collision as being elastic and the system as being isolated for both energy and momentum.

Analyze Because momentum of the system is conserved, apply Equation 8.17:	(1) $m_1 v_{1i} + m_2 v_{2i} = m_1 v_{1f} + m_2 v_{2f}$
Because the collision is elastic, apply Equation 8.21:	(2) $v_{1i} - v_{2i} = -(v_{1f} - v_{2f})$
Multiply Equation (2) by m_1:	(3) $m_1 v_{1i} - m_1 v_{2i} = -m_1 v_{1f} + m_1 v_{2f}$
Add Equations (1) and (3):	$2m_1 v_{1i} + (m_2 - m_1) v_{2i} = (m_1 + m_2) v_{2f}$
Solve for v_{2f}:	$v_{2f} = \dfrac{2m_1 v_{1i} + (m_2 - m_1) v_{2i}}{m_1 + m_2}$
Substitute numerical values:	$v_{2f} = \dfrac{2(1.60 \text{ kg})(4.00 \text{ m/s}) + (2.10 \text{ kg} - 1.60 \text{ kg})(-2.50 \text{ m/s})}{1.60 \text{ kg} + 2.10 \text{ kg}} = \boxed{3.12 \text{ m/s}}$
Solve Equation (2) for v_{1f} and substitute numerical values:	$v_{1f} = v_{2f} - v_{1i} + v_{2i} = 3.12 \text{ m/s} - 4.00 \text{ m/s} + (-2.50 \text{ m/s}) = \boxed{-3.38 \text{ m/s}}$

(B) Determine the velocity of block 2 during the collision, at the instant block 1 is moving to the right with a velocity of +3.00 m/s as in Figure 8.10b.

SOLUTION

Conceptualize Focus your attention now on Figure 8.10b, which represents the final configuration of the system for the time interval of interest.

continued

8.8 *cont.*

Categorize Because the momentum and mechanical energy of the system of two blocks and the spring are conserved *throughout* the collision, the collision can be categorized as elastic for *any* final instant of time. Let us now choose the final instant to be when block 1 is moving with a velocity of +3.00 m/s.

Analyze Apply Equation 8.17:

$$m_1 v_{1i} + m_2 v_{2i} = m_1 v_{1f} + m_2 v_{2f}$$

Solve for v_{2f}:

$$v_{2f} = \frac{m_1 v_{1i} + m_2 v_{2i} - m_1 v_{1f}}{m_2}$$

Substitute numerical values:

$$v_{2f} = \frac{(1.60 \text{ kg})(4.00 \text{ m/s}) + (2.10 \text{ kg})(-2.50 \text{ m/s}) - (1.60 \text{ kg})(3.00 \text{ m/s})}{2.10 \text{ kg}}$$

$$= -1.74 \text{ m/s}$$

Finalize The negative value for v_{2f} means that block 2 is still moving to the left at the instant we are considering.

(C) Determine the distance the spring is compressed at that instant.

SOLUTION

Conceptualize Once again, focus on the configuration of the system shown in Figure 8.10b.

Categorize For the system of the spring and two blocks, no friction or other nonconservative forces act within the system. Therefore, we categorize the system as isolated in terms of energy with no nonconservative forces acting. The system also remains isolated in terms of momentum.

Analyze We choose the initial configuration of the system to be that existing immediately before block 1 strikes the spring and the final configuration to be that when block 1 is moving to the right at 3.00 m/s.

Write a conservation of mechanical energy equation for the system:

$$K_i + U_i = K_f + U_f$$

Evaluate the energies, recognizing that two objects in the system have kinetic energy and that the potential energy is elastic:

$$\tfrac{1}{2} m_1 v_{1i}^2 + \tfrac{1}{2} m_2 v_{2i}^2 + 0 = \tfrac{1}{2} m_1 v_{1f}^2 + \tfrac{1}{2} m_2 v_{2f}^2 + \tfrac{1}{2} kx^2$$

Substitute the known values and the result of part (B):

$$\tfrac{1}{2}(1.60 \text{ kg})(4.00 \text{ m/s})^2 + \tfrac{1}{2}(2.10 \text{ kg})(2.50 \text{ m/s})^2 + 0$$

$$= \tfrac{1}{2}(1.60 \text{ kg})(3.00 \text{ m/s})^2 + \tfrac{1}{2}(2.10 \text{ kg})(1.74 \text{ m/s})^2 + \tfrac{1}{2}(600 \text{ N/m})x^2$$

Solve for x:

$$x = 0.173 \text{ m}$$

Finalize This answer is not the maximum compression of the spring because the two blocks are still moving toward each other at the instant shown in Figure 8.10b. Can you determine the maximum compression of the spring?

❮8.5❯ Collisions in Two Dimensions

In Section 8.1, we showed that the total momentum of a system is conserved when the system is isolated (i.e., when no external forces act on the system). For a general collision of two objects in three-dimensional space, the principle of conservation of momentum implies that the total momentum in each direction is conserved. An important subset of collisions takes place in a plane. The game of billiards is a familiar example involving multiple collisions of objects moving on a two-dimensional surface. Let us restrict our attention to a single two-dimensional collision between two objects that takes place in a plane. For such collisions, we obtain two component equations for the conservation of momentum:

$$m_1 v_{1ix} + m_2 v_{2ix} = m_1 v_{1fx} + m_2 v_{2fx}$$

$$m_1 v_{1iy} + m_2 v_{2iy} = m_1 v_{1fy} + m_2 v_{2fy}$$

where we use three subscripts in this general equation to represent, respectively, (1) the identification of the object, (2) initial and final values, and (3) the velocity component in the x or y direction.

Consider a two-dimensional problem in which an object of mass m_1 collides with an object of mass m_2 that is initially at rest as in Active Figure 8.11. After the collision, m_1 moves at an angle θ with respect to the horizontal and m_2 moves at an angle ϕ with respect to the horizontal. This collision is called a *glancing* collision. Applying the law of conservation of momentum in component form and noting that the initial y component of the momentum of the system is zero, we have

$$x \text{ component:} \quad m_1 v_{1i} + 0 = m_1 v_{1f} \cos\theta + m_2 v_{2f} \cos\phi \qquad \textbf{8.26} \blacktriangleleft$$

$$y \text{ component:} \quad 0 + 0 = m_1 v_{1f} \sin\theta - m_2 v_{2f} \sin\phi \qquad \textbf{8.27} \blacktriangleleft$$

If the collision is elastic, we can write a third equation for conservation of kinetic energy in the form

$$\tfrac{1}{2} m_1 v_{1i}^2 = \tfrac{1}{2} m_1 v_{1f}^2 + \tfrac{1}{2} m_2 v_{2f}^2 \qquad \textbf{8.28} \blacktriangleleft$$

If we know the initial velocity v_{1i} and the masses, we are left with four unknowns (v_{1f}, v_{2f}, θ, and ϕ). Because we have only three equations, one of the four remaining quantities must be given to determine the motion after the collision from conservation principles alone.

If the collision is inelastic, kinetic energy is *not* conserved and Equation 8.28 does *not* apply.

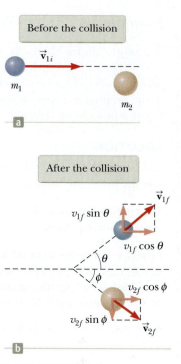

Active Figure 8.11 A glancing collision between two particles.

> ## PROBLEM-SOLVING STRATEGY: Two-Dimensional Collisions

The following procedure is recommended when dealing with problems involving collisions between two particles in two dimensions.

1. **Conceptualize.** Imagine the collisions occurring and predict the approximate directions in which the particles will move after the collision. Set up a coordinate system and define your velocities in terms of that system. It is convenient to have the x axis coincide with one of the initial velocities. Sketch the coordinate system, draw and label all velocity vectors, and include all the given information.

2. **Categorize.** Is the system of particles truly isolated? If so, categorize the collision as elastic, inelastic, or perfectly inelastic.

3. **Analyze.** Write expressions for the x and y components of the momentum of each object before and after the collision. Remember to include the appropriate signs for the components of the velocity vectors and pay careful attention to signs throughout the calculation.

   Write expressions for the *total* momentum in the x direction *before* and *after* the collision and equate the two. Repeat this procedure for the total momentum in the y direction.

   Proceed to solve the momentum equations for the unknown quantities. If the collision is inelastic, kinetic energy is *not* conserved and additional information is probably required. If the collision is perfectly inelastic, the final velocities of the two objects are equal.

   If the collision is elastic, kinetic energy is conserved and you can equate the total kinetic energy of the system before the collision to the total kinetic energy after the collision, providing an additional relationship between the velocity magnitudes.

4. **Finalize.** Once you have determined your result, check to see if your answers are consistent with the mental and pictorial representations and that your results are realistic.

Example 8.9 | Proton–Proton Collision

A proton collides elastically with another proton that is initially at rest. The incoming proton has an initial speed of 3.50×10^5 m/s and makes a glancing collision with the second proton as in Active Figure 8.11. (At close separations, the protons exert a repulsive electrostatic force on each other.) After the collision, one proton moves off at an angle of $37.0°$ to the original direction of motion and the second deflects at an angle of ϕ to the same axis. Find the final speeds of the two protons and the angle ϕ.

SOLUTION

Conceptualize This collision is like that shown in Active Figure 8.11, which will help you conceptualize the behavior of the system. We define the x axis to be along the direction of the velocity vector of the initially moving proton.

Categorize The pair of protons form an isolated system. Both momentum and kinetic energy of the system are conserved in this glancing elastic collision.

. . . .

Analyze Using the isolated system model for both momentum and energy for a two-dimensional elastic collision, set up the mathematical representation with Equations 8.26 through 8.28:

(1) $v_{1f} \cos \theta + v_{2f} \cos \phi = v_{1i}$

(2) $v_{1f} \sin \theta - v_{2f} \sin \phi = 0$

(3) $v_{1f}^2 + v_{2f}^2 = v_{1i}^2$

Rearrange Equations (1) and (2):

$v_{2f} \cos \phi = v_{1i} - v_{1f} \cos \theta$

$v_{2f} \sin \phi = v_{1f} \sin \theta$

Square these two equations and add them:

$v_{2f}^2 \cos^2 \phi + v_{2f}^2 \sin^2 \phi =$
$v_{1i}^2 - 2v_{1i}v_{1f} \cos \theta + v_{1f}^2 \cos^2 \theta + v_{1f}^2 \sin^2 \theta$

Incorporate that the sum of the squares of sine and cosine for *any* angle is equal to 1:

(4) $v_{2f}^2 = v_{1i}^2 - 2v_{1i}v_{1f} \cos \theta + v_{1f}^2$

Substitute Equation (4) into Equation (3):

$v_{1f}^2 + (v_{1i}^2 - 2v_{1i}v_{1f} \cos \theta + v_{1f}^2) = v_{1i}^2$

(5) $v_{1f}^2 - v_{1i}v_{1f} \cos \theta = 0$

One possible solution of Equation (5) is $v_{1f} = 0$, which corresponds to a head-on, one-dimensional collision in which the first proton stops and the second continues with the same speed in the same direction. That is not the solution we want.

Divide both sides of Equation (5) by v_{1f} and solve for the remaining factor of v_{1f}:

$v_{1f} = v_{1i} \cos \theta = (3.50 \times 10^5 \text{ m/s}) \cos 37.0° = \boxed{2.80 \times 10^5 \text{ m/s}}$

Use Equation (3) to find v_{2f}:

$v_{2f} = \sqrt{v_{1i}^2 - v_{1f}^2} = \sqrt{(3.50 \times 10^5 \text{ m/s})^2 - (2.80 \times 10^5 \text{ m/s})^2}$

$= \boxed{2.11 \times 10^5 \text{ m/s}}$

Use Equation (2) to find ϕ:

(2) $\phi = \sin^{-1}\left(\dfrac{v_{1f} \sin \theta}{v_{2f}}\right) = \sin^{-1}\left[\dfrac{(2.80 \times 10^5 \text{ m/s}) \sin 37.0°}{(2.11 \times 10^5 \text{ m/s})}\right]$

$= \boxed{53.0°}$

. . . .

Finalize It is interesting that $\theta + \phi = 90°$. This result is *not* accidental. Whenever two objects of equal mass collide elastically in a glancing collision and one of them is initially at rest, their final velocities are perpendicular to each other.

Example 8.10 | Collision at an Intersection

A 1 500-kg car traveling east with a speed of 25.0 m/s collides at an intersection with a 2 500-kg truck traveling north at a speed of 20.0 m/s as shown in Figure 8.12. Find the direction and magnitude of the velocity of the wreckage after the collision, assuming the vehicles stick together after the collision.

SOLUTION

Conceptualize Figure 8.12 should help you conceptualize the situation before and after the collision. Let us choose east to be along the positive x direction and north to be along the positive y direction.

8.10 *cont.*

Categorize Because we consider moments immediately before and immediately after the collision as defining our time interval, we ignore the small effect that friction would have on the wheels of the vehicles and model the system of two vehicles as isolated in terms of momentum. We also ignore the vehicles' sizes and model them as particles. The collision is perfectly inelastic because the car and the truck stick together after the collision.

....

Analyze Before the collision, the only object having momentum in the x direction is the car. Therefore, the magnitude of the total initial momentum of the system (car plus truck) in the x direction is that of only the car. Similarly, the total initial momentum of the system in the y direction is that of the truck. After the collision, let us assume the wreckage moves at an angle θ with respect to the x axis with speed v_f.

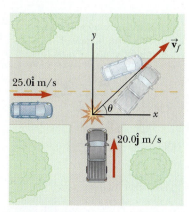

Figure 8.12 (Example 8.10) An eastbound car colliding with a northbound truck.

Equate the initial and final momenta of the system in the x direction:

$$\sum p_{xi} = \sum p_{xf} \quad \rightarrow \quad (1) \quad m_1 v_{1i} = (m_1 + m_2) v_f \cos \theta$$

Equate the initial and final momenta of the system in the y direction:

$$\sum p_{yi} = \sum p_{yf} \quad \rightarrow \quad (2) \quad m_2 v_{2i} = (m_1 + m_2) v_f \sin \theta$$

Divide Equation (2) by Equation (1):

$$\frac{m_2 v_{2i}}{m_1 v_{1i}} = \frac{\sin \theta}{\cos \theta} = \tan \theta$$

Solve for θ and substitute numerical values:

$$\theta = \tan^{-1}\left(\frac{m_2 v_{2i}}{m_1 v_{1i}}\right) = \tan^{-1}\left[\frac{(2\,500 \text{ kg})(20.0 \text{ m/s})}{(1\,500 \text{ kg})(25.0 \text{ m/s})}\right] = \boxed{53.1°}$$

Use Equation (2) to find the value of v_f and substitute numerical values:

$$v_f = \frac{m_2 v_{2i}}{(m_1 + m_2)\sin \theta} = \frac{(2\,500 \text{ kg})(20.0 \text{ m/s})}{(1\,500 \text{ kg} + 2\,500 \text{ kg})\sin 53.1°} = \boxed{15.6 \text{ m/s}}$$

....

Finalize Notice that the angle θ is qualitatively in agreement with Figure 8.12. Also notice that the final speed of the combination is less than the initial speeds of the two cars. This result is consistent with the kinetic energy of the system being reduced in an inelastic collision. It might help if you draw the momentum vectors of each vehicle before the collision and the two vehicles together after the collision.

8.6 | The Center of Mass

In this section, we describe the overall motion of a system of particles in terms of a very special point called the **center of mass** of the system. This notion gives us confidence in the particle model because we will see that the center of mass accelerates as if all the system's mass were concentrated at that point and all external forces act there.

Consider a system consisting of a pair of particles connected by a light, rigid rod (Active Fig. 8.13, page 254). The center of mass as indicated in the figure is located on the rod and is closer to the larger mass in the figure; we will see why soon. If a single force is applied at some point on the rod that is above the center of mass, the system rotates clockwise (Active Fig. 8.13a) as it translates through space. If the force is applied at a point on the rod below the center of mass, the system rotates counterclockwise (Active Fig. 8.13b). If the force is applied exactly at the center of mass, the system moves in the direction of $\vec{F}$ without rotating (Active Fig. 8.13c) as if the system is behaving as a particle. Therefore, in theory, the center of mass can be located with this experiment.

If we were to analyze the motion in Active Figure 8.13c, we would find that the system moves as if all its mass were concentrated at the center of mass. Furthermore, if the external net force on the system is $\sum \vec{F}$ and the total mass of the system is M, the

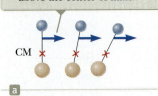

The system rotates clockwise when a force is applied above the center of mass.

CM

a

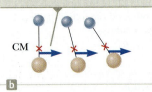

The system rotates counter-clockwise when a force is applied below the center of mass.

CM

b

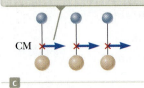

The system moves in the direction of the force without rotating when a force is applied at the center of mass.

CM

c

Active Figure 8.13 A force is applied to a system of two particles of unequal mass connected by a light, rigid rod.

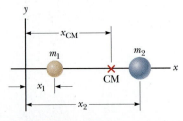

Active Figure 8.14 The center of mass of two particles having unequal mass is located on the x axis at x_{CM}, a point between the particles, closer to the one having the larger mass.

center of mass moves with an acceleration given by $\vec{a} = \Sigma\vec{F}/M$. That is, the system moves as if the resultant external force were applied to a single particle of mass M located at the center of mass, which justifies our particle model for extended objects. We have ignored all rotational effects for extended objects so far, implicitly assuming that forces were provided at just the right position so as to cause no rotation. We will study rotational motion in Chapter 10, where we will apply forces that do not pass through the center of mass.

The position of the center of mass of a system can be described as being the *average position* of the system's mass. For example, the center of mass of the pair of particles described in Active Figure 8.14 is located on the x axis, somewhere between the particles. The x coordinate of the center of mass in this case is

$$x_{CM} = \frac{m_1 x_1 + m_2 x_2}{m_1 + m_2} \qquad \textbf{8.29} \blacktriangleleft$$

For example, if $x_1 = 0$, $x_2 = d$, and $m_2 = 2m_1$, we find that $x_{CM} = \frac{2}{3}d$. That is, the center of mass lies closer to the more massive particle. If the two masses are equal, the center of mass lies midway between the particles.

We can extend the concept of center of mass to a system of many particles in three dimensions. The x coordinate of the center of mass of n particles is defined to be

$$x_{CM} \equiv \frac{m_1 x_1 + m_2 x_2 + m_3 x_3 + \cdots + m_n x_n}{m_1 + m_2 + m_3 + \cdots + m_n} = \frac{\sum\limits_i m_i x_i}{\sum\limits_i m_i} = \frac{\sum\limits_i m_i x_i}{M} \qquad \textbf{8.30} \blacktriangleleft$$

where x_i is the x coordinate of the ith particle and M is the *total mass* of the system. The y and z coordinates of the center of mass are similarly defined by the equations

$$y_{CM} \equiv \frac{\sum\limits_i m_i y_i}{M} \quad \text{and} \quad z_{CM} \equiv \frac{\sum\limits_i m_i z_i}{M} \qquad \textbf{8.31} \blacktriangleleft$$

The center of mass can also be located by its position vector, $\vec{r}_{CM}$. The rectangular coordinates of this vector are x_{CM}, y_{CM}, and z_{CM}, defined in Equations 8.30 and 8.31. Therefore,

$$\vec{r}_{CM} = x_{CM}\hat{\mathbf{i}} + y_{CM}\hat{\mathbf{j}} + z_{CM}\hat{\mathbf{k}} = \frac{\sum\limits_i m_i x_i \hat{\mathbf{i}} + \sum\limits_i m_i y_i \hat{\mathbf{j}} + \sum\limits_i m_i z_i \hat{\mathbf{k}}}{M}$$

$$\vec{r}_{CM} = \frac{\sum\limits_i m_i \vec{r}_i}{M} \qquad \textbf{8.32} \blacktriangleleft$$

where $\vec{r}_i$ is the position vector of the ith particle, defined by

$$\vec{r}_i \equiv x_i\hat{\mathbf{i}} + y_i\hat{\mathbf{j}} + z_i\hat{\mathbf{k}}$$

Equation 8.32 is useful for finding the center of mass of a relatively small number of discrete particles. What about an extended object, which has a *continuous* distribution of mass? Although locating the center of mass for an extended object is somewhat more cumbersome than locating the center of mass of a system of particles, this location is based on the same fundamental ideas. We can model the extended object as a system containing a large number of elements (Fig. 8.15). Each element is modeled as a particle of mass Δm_i, with coordinates x_i, y_i, z_i. The particle separation is very small, so this model is a good representation of the continuous mass distribution of the object. The x coordinate of the center of mass of the particles representing the object, and therefore of the approximate center of mass of the object, is

$$x_{CM} \approx \frac{\sum\limits_i x_i \Delta m_i}{M}$$

with similar expressions for y_{CM} and z_{CM}. If we let the number of elements approach infinity (and, as a consequence, the size and mass of each element approach zero), the model becomes indistinguishable from the continuous mass distribution and x_{CM} is given precisely. In this limit, we replace the sum by an integral and Δm_i by the differential element dm:

$$x_{CM} = \lim_{\Delta m_i \to 0} \frac{\sum_i x_i \Delta m_i}{M} = \frac{1}{M}\int x\, dm \qquad 8.33 ◄$$

where the integration is over the length of the object in the x direction. Likewise, for y_{CM} and z_{CM} we obtain

$$y_{CM} = \frac{1}{M}\int y\, dm \quad \text{and} \quad z_{CM} = \frac{1}{M}\int z\, dm \qquad 8.34 ◄$$

We can express the vector position of the center of mass of an extended object as

$$\vec{r}_{CM} = \frac{1}{M}\int \vec{r}\, dm \qquad 8.35 ◄$$

which is equivalent to the three expressions in Equations 8.33 and 8.34.

The center of mass of a homogeneous, symmetric object must lie on an axis of symmetry. For example, the center of mass of a homogeneous rod must lie midway between the ends of the rod. The center of mass of a homogeneous sphere or a homogeneous cube must lie at the geometric center of the object.

The center of mass of a system is often confused with the **center of gravity** of a system. Each portion of a system is acted on by the gravitational force. The net effect of all these forces is equivalent to the effect of a single force $M\vec{g}$ acting at a special point called the center of gravity. The center of gravity is the average position of the gravitational forces on all parts of the object. If $\vec{g}$ is uniform over the system, the center of gravity coincides with the center of mass. If the gravitational field over the system is not uniform, the center of gravity and the center of mass are different. In most cases, for objects or systems of reasonable size, the two points can be considered to be coincident.

One can experimentally determine the center of gravity of an irregularly shaped object, such as a wrench, by suspending the wrench from two different points (Fig. 8.16). An object of this size has virtually no variation in the gravitational field over its dimensions, so this method also locates the center of mass. The wrench is first hung from point A, and a vertical line AB is drawn (which can be established with a plumb bob) when the wrench is in equilibrium. The wrench is then hung from point C, and a second vertical line CD is drawn. The center of mass coincides with the intersection of these two lines. In fact, if the wrench is hung freely from any point, the vertical line through that point will pass through the center of mass.

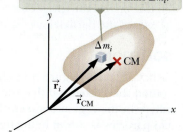

Figure 8.15 The center of mass of the object is located at the vector position $\vec{r}_{CM}$, which has coordinates x_{CM}, y_{CM}, and z_{CM}.

► Center of mass of a continuous mass distribution

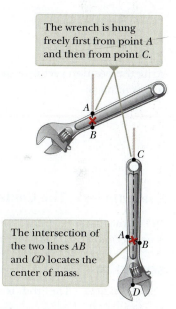

Figure 8.16 An experimental technique for determining the center of mass of a wrench.

QUICK QUIZ 8.5 A baseball bat of uniform denisty is cut at the location of its center of mass as shown in Figure 8.17. Which piece has the smaller mass? (a) the piece on the right (b) the piece on the left (c) both pieces have the same mass (d) impossible to determine

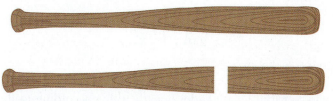

Figure 8.17 (Quick Quiz 8.5) A baseball bat cut at the location of its center of mass.

Example 8.11 | The Center of Mass of Three Particles

A system consists of three particles located as shown in Figure 8.18. Find the center of mass of the system. The masses of the particles are $m_1 = m_2 = 1.0$ kg and $m_3 = 2.0$ kg.

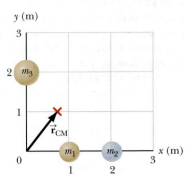

Figure 8.18
(Example 8.11) Two particles are located on the x axis, and a single particle is located on the y axis as shown. The vector indicates the location of the system's center of mass.

SOLUTION

Conceptualize Figure 8.18 shows the three masses. Your intuition should tell you that the center of mass is located somewhere in the region between the blue particle and the pair of tan particles as shown in the figure.

Categorize We categorize this example as a substitution problem because we will be using the equations for the center of mass developed in this section.

Use the defining equations for the coordinates of the center of mass and notice that $z_{CM} = 0$:

$$x_{CM} = \frac{1}{M}\sum_i m_i x_i = \frac{m_1 x_1 + m_2 x_2 + m_3 x_3}{m_1 + m_2 + m_3}$$

$$= \frac{(1.0 \text{ kg})(1.0 \text{ m}) + (1.0 \text{ kg})(2.0 \text{ m}) + (2.0 \text{ kg})(0)}{1.0 \text{ kg} + 1.0 \text{ kg} + 2.0 \text{ kg}} = \frac{3.0 \text{ kg} \cdot \text{m}}{4.0 \text{ kg}} = 0.75 \text{ m}$$

$$y_{CM} = \frac{1}{M}\sum_i m_i y_i = \frac{m_1 y_1 + m_2 y_2 + m_3 y_3}{m_1 + m_2 + m_3}$$

$$= \frac{(1.0 \text{ kg})(0) + (1.0 \text{ kg})(0) + (2.0 \text{ kg})(2.0 \text{ m})}{4.0 \text{ kg}} = \frac{4.0 \text{ kg} \cdot \text{m}}{4.0 \text{ kg}} = 1.0 \text{ m}$$

Write the position vector of the center of mass:

$$\vec{r}_{CM} \equiv x_{CM}\hat{i} + y_{CM}\hat{j} = (0.75\hat{i} + 1.0\hat{j}) \text{ m}$$

Example 8.12 | The Center of Mass of a Rod

(A) Show that the center of mass of a rod of mass M and length L lies midway between its ends, assuming the rod has a uniform mass per unit length.

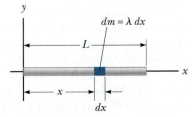

Figure 8.19 (Example 8.12) The geometry used to find the center of mass of a uniform rod.

SOLUTION

Conceptualize The rod is shown aligned along the x axis in Figure 8.19, so $y_{CM} = z_{CM} = 0$.

Categorize We categorize this example as an analysis problem because we need to divide the rod into small mass elements to perform the integration in Equation 8.33.

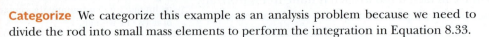

Analyze The mass per unit length (this quantity is called the *linear mass density*) can be written as $\lambda = M/L$ for the uniform rod. If the rod is divided into elements of length dx, the mass of each element is $dm = \lambda \, dx$.

Use Equation 8.33 to find an expression for x_{CM}:

$$x_{CM} = \frac{1}{M}\int x \, dm = \frac{1}{M}\int_0^L x\lambda \, dx = \frac{\lambda}{M}\frac{x^2}{2}\Big|_0^L = \frac{\lambda L^2}{2M}$$

Substitute $\lambda = M/L$:

$$x_{CM} = \frac{L^2}{2M}\left(\frac{M}{L}\right) = \tfrac{1}{2}L$$

One can also use symmetry arguments to obtain the same result.

(B) Suppose a rod is *nonuniform* such that its mass per unit length varies linearly with x according to the expression $\lambda = \alpha x$, where α is a constant. Find the x coordinate of the center of mass as a fraction of L.

8.12 *cont.*

SOLUTION

Conceptualize Because the mass per unit length is not constant in this case but is proportional to x, elements of the rod to the right are more massive than elements near the left end of the rod.

Categorize This problem is categorized similarly to part (A), with the added twist that the linear mass density is not constant.

. . . .

Analyze In this case, we replace dm in Equation 8.33 by $\lambda \, dx$, where $\lambda = \alpha x$.

Use Equation 8.33 to find an expression for x_{CM}:

$$x_{CM} = \frac{1}{M} \int x \, dm = \frac{1}{M} \int_0^L x\lambda \, dx = \frac{1}{M} \int_0^L x\alpha x \, dx$$

$$= \frac{\alpha}{M} \int_0^L x^2 \, dx = \frac{\alpha L^3}{3M}$$

Find the total mass of the rod:

$$M = \int dm = \int_0^L \lambda \, dx = \int_0^L \alpha x \, dx = \frac{\alpha L^2}{2}$$

Substitute M into the expression for x_{CM}:

$$x_{CM} = \frac{\alpha L^3}{3\alpha L^2/2} = \boxed{\tfrac{2}{3}L}$$

. . . .

Finalize Notice that the center of mass in part (B) is farther to the right than that in part (A). That result is reasonable because the elements of the rod become more massive as one moves to the right along the rod in part (B).

8.7 | Motion of a System of Particles

We can begin to understand the physical significance and utility of the center of mass concept by taking the time derivative of the position vector $\vec{\mathbf{r}}_{CM}$ of the center of mass, given by Equation 8.32. Assuming that M remains constant—that is, no particles enter or leave the system—we find the following expression for the **velocity of the center of mass** of the system:

$$\vec{\mathbf{v}}_{CM} = \frac{d\vec{\mathbf{r}}_{CM}}{dt} = \frac{1}{M}\sum_i m_i \frac{d\vec{\mathbf{r}}_i}{dt} = \frac{1}{M}\sum_i m_i\vec{\mathbf{v}}_i \qquad \textbf{8.36} \blacktriangleleft$$

▶ Velocity of the center of mass for a system of particles

where $\vec{\mathbf{v}}_i$ is the velocity of the ith particle. Rearranging Equation 8.36 gives

$$M\vec{\mathbf{v}}_{CM} = \sum_i m_i\vec{\mathbf{v}}_i = \sum_i \vec{\mathbf{p}}_i = \vec{\mathbf{p}}_{tot} \qquad \textbf{8.37} \blacktriangleleft$$

This result tells us that the total momentum of the system equals its total mass multiplied by the velocity of its center of mass. In other words, the total momentum of the system is equal to the momentum of a single particle of mass M moving with a velocity $\vec{\mathbf{v}}_{CM}$; this is the particle model.

If we now differentiate Equation 8.36 with respect to time, we find the **acceleration of the center of mass** of the system:

$$\vec{\mathbf{a}}_{CM} = \frac{d\vec{\mathbf{v}}_{CM}}{dt} = \frac{1}{M}\sum_i m_i \frac{d\vec{\mathbf{v}}_i}{dt} = \frac{1}{M}\sum_i m_i\vec{\mathbf{a}}_i \qquad \textbf{8.38} \blacktriangleleft$$

▶ Acceleration of the center of mass for a system of particles

Rearranging this expression and using Newton's second law, we have

$$M\vec{\mathbf{a}}_{CM} = \sum_i m_i\vec{\mathbf{a}}_i = \sum_i \vec{\mathbf{F}}_i \qquad \textbf{8.39} \blacktriangleleft$$

where $\vec{\mathbf{F}}_i$ is the force on particle i.

The forces on any particle in the system may include both external and internal forces. By Newton's third law, however, the force exerted by particle 1 on particle 2, for example, is equal in magnitude and opposite the force exerted by particle 2 on particle 1. When we sum over all internal forces in Equation 8.39, they cancel in

▶ Newton's second law for a system of particles

pairs. Therefore, the net force on the system is due *only* to external forces and we can write Equation 8.39 in the form

$$\sum \vec{\mathbf{F}}_{ext} = M \vec{\mathbf{a}}_{CM} = \frac{d\vec{\mathbf{p}}_{tot}}{dt}$$ **8.40** ◀

That is, the external net force on the system of particles equals the total mass of the system multiplied by the acceleration of the center of mass, or the time rate of change of the momentum of the system. Comparing Equation 8.40 with Newton's second law for a single particle, we see that the particle model we have used in several chapters can be described in terms of the center of mass:

The center of mass of a system of particles having combined mass M moves like an equivalent particle of mass M would move under the influence of the net external force on the system.

Let us integrate Equation 8.40 over a finite time interval:

$$\int \sum \vec{\mathbf{F}}_{ext}\, dt = \int M\vec{\mathbf{a}}_{CM}\, dt = \int M \frac{d\vec{\mathbf{v}}_{CM}}{dt}\, dt = M \int d\vec{\mathbf{v}}_{CM} = M\,\Delta\vec{\mathbf{v}}_{CM}$$

Notice that this equation can be written as

▶ Impulse–momentum theorem for a system of particles

$$\Delta\vec{\mathbf{p}}_{tot} = \vec{\mathbf{I}}$$ **8.41** ◀

where $\vec{\mathbf{I}}$ is the impulse imparted to the system by external forces and $\vec{\mathbf{p}}_{tot}$ is the momentum of the system. Equation 8.41 is the generalization of the impulse–momentum theorem for a particle (Eq. 8.11) to a system of many particles. It is also the mathematical representation of the nonisolated system (momentum) model for a system of many particles.

In the absence of external forces, the center of mass moves with uniform velocity as in the case of the translating and rotating wrench in Figure 8.20. If the net force acts along a line through the center of mass of an extended object such as the wrench, the object is accelerated without rotation. If the net force does not act through the center of mass, the object will undergo rotation in addition to translation. The linear acceleration of the center of mass is the same in either case, as given by Equation 8.40.

Finally, we see that if the external net force is zero, from Equation 8.40 it follows that

$$\frac{d\vec{\mathbf{p}}_{tot}}{dt} = M\vec{\mathbf{a}}_{CM} = 0$$

so that

$$\vec{\mathbf{p}}_{tot} = M\vec{\mathbf{v}}_{CM} = \text{constant} \quad (\text{when } \sum \vec{\mathbf{F}}_{ext} = 0)$$ **8.42** ◀

That is, the total linear momentum of a system of particles is constant if no external forces act on the system. It follows that, for an *isolated* system of particles, the total momentum is conserved. The law of conservation of momentum that was derived in Section 8.1 for a two-particle system is thus generalized to a many-particle system.

The center of mass of the wrench (marked with a white dot) moves in a straight line as the wrench rotates about this point.

Note the decreasing distance between the white dots.

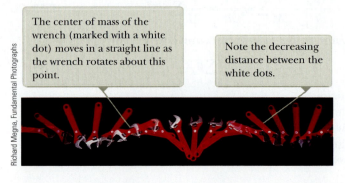

Richard Megna, Fundamental Photographs

Figure 8.20 Strobe photograph showing an overhead view of a wrench moving on a horizontal surface. The wrench moves from left to right in the photograph and is slowing down due to friction between the wrench and the supporting surface.

Figure 8.21 (Thinking Physics 8.1)
A boy takes a step in a boat. What happens to the boat?

> ### THINKING PHYSICS 8.1
>
> A boy stands at one end of a boat that is stationary relative to the dock (Fig. 8.21). He then walks to the opposite end of the boat, away from the dock. Does the boat move?
>
> **Reasoning** Yes, the boat moves toward the dock. Ignoring friction between the boat and water, no horizontal force acts on the system consisting of the boy and boat. The center of mass of the system therefore remains fixed relative to the dock (or any stationary point). As the boy moves away from the dock, the boat must move toward the dock such that the center of mass of the system remains fixed in position. ◀

QUICK QUIZ 8.6 A cruise ship is moving at constant speed through the water. The vacationers on the ship are eager to arrive at their next destination. They decide to try to speed up the cruise ship by gathering at the bow (the front) and running together toward the stern (the back) of the ship. (**i**) While they are running toward the stern, is the speed of the ship (a) higher than it was before, (b) unchanged, (c) lower than it was before, or (d) impossible to determine? (**ii**) The vacationers stop running when they reach the stern of the ship. After they have all stopped running, is the speed of the ship (a) higher than it was before they started running, (b) unchanged from what it was before they started running, (c) lower than it was before they started running, or (d) impossible to determine?

Example 8.13 | The Exploding Rocket

A rocket is fired vertically upward. At the instant it reaches an altitude of 1 000 m and a speed of $v_i = 300$ m/s, it explodes into three fragments having equal mass. One fragment moves upward with a speed of $v_1 = 450$ m/s following the explosion. The second fragment has a speed of $v_2 = 240$ m/s and is moving east right after the explosion. What is the velocity of the third fragment immediately after the explosion?

SOLUTION

Conceptualize Picture the explosion in your mind, with one piece going upward and a second piece moving horizontally toward the east. Do you have an intuitive feeling about the direction in which the third piece moves?

Categorize This example is a two-dimensional problem because we have two fragments moving in perpendicular directions after the explosion as well as a third fragment moving in an unknown direction in the plane defined by the velocity vectors of the other two fragments. We assume the time interval of the explosion is very short, so we use the impulse approximation in which we ignore the gravitational force and air resistance. Because the forces of the explosion are internal to the system (the rocket), the system is modeled as isolated in terms of momentum. Therefore, the total momentum $\vec{\mathbf{p}}_i$ of the rocket immediately before the explosion must equal the total momentum $\vec{\mathbf{p}}_f$ of the fragments immediately after the explosion.

Analyze Because the three fragments have equal mass, the mass of each fragment is $M/3$, where M is the total mass of the rocket. We will let $\vec{\mathbf{v}}_3$ represent the unknown velocity of the third fragment.

Using the isolated system (momentum) model, equate the initial and final momenta of the system and express the momenta in terms of masses and velocities:

$$\vec{\mathbf{p}}_i = \vec{\mathbf{p}}_f \quad \rightarrow \quad M\vec{\mathbf{v}}_i = \frac{M}{3}\vec{\mathbf{v}}_1 + \frac{M}{3}\vec{\mathbf{v}}_2 + \frac{M}{3}\vec{\mathbf{v}}_3$$

continued

8.13 *cont.*

Solve for $\vec{v}_3$: $\vec{v}_3 = 3\vec{v}_i - \vec{v}_1 - \vec{v}_2$

Substitute the numerical values: $\vec{v}_3 = 3(300\hat{j} \text{ m/s}) - (450\hat{j} \text{ m/s}) - (240\hat{i} \text{ m/s}) = \boxed{(-240\hat{i} + 450\hat{j}) \text{ m/s}}$

Finalize Notice that this event is the reverse of a perfectly inelastic collision. There is one object before the collision and three objects afterward. Imagine running a movie of the event backward: the three objects would come together and become a single object. In a perfectly inelastic collision, the kinetic energy of the system decreases. If you were to calculate the kinetic energy before and after the event in this example, you would find that the kinetic energy of the system increases. (Try it!) This increase in kinetic energy comes from the potential energy stored in whatever fuel exploded to cause the breakup of the rocket.

8.8 | Context Connection: Rocket Propulsion

On our trip to Mars, we will need to control our spacecraft by firing the rocket engines. When ordinary vehicles, such as the automobiles in Context 1, are propelled, the driving force for the motion is the friction force exerted by the road on the car. A rocket moving in space, however, has no road to "push" against. The source of the propulsion of a rocket must therefore be different. The operation of a rocket depends on the law of conservation of momentum as applied to a system, where the system is the rocket plus its ejected fuel.

The propulsion of a rocket can be understood by first considering the archer on ice in Example 8.2. As an arrow is fired from the bow, the arrow receives momentum $m\vec{v}$ in one direction and the archer receives a momentum of equal magnitude in the opposite direction. As additional arrows are fired, the archer moves faster, so a large velocity of the archer can be established by firing many arrows.

In a similar manner, as a rocket moves in free space (a vacuum), its momentum changes when some of its mass is released in the form of ejected gases. Because the ejected gases acquire some momentum, the rocket receives a compensating momentum in the opposite direction. The rocket therefore is accelerated as a result of the "push," or thrust, from the exhaust gases. Note that the rocket represents the *inverse* of an inelastic collision; that is, momentum is conserved, but the kinetic energy of the system is *increased* (at the expense of energy stored in the fuel of the rocket).

Suppose at some time t the magnitude of the momentum of the rocket plus the fuel is $(M + \Delta m)v$ (Fig. 8.22a). During a short time interval Δt, the rocket ejects fuel of mass Δm and the rocket's speed therefore increases to $v + \Delta v$ (Fig. 8.22b). If the fuel is ejected with velocity $\vec{v}_e$ *relative to the rocket*, the speed of the fuel relative to a stationary frame of reference is $v - v_e$ according to our discussion of relative velocity in Section 3.6. Therefore, if we equate the total initial momentum of the system with the total final momentum, we have

$$(M + \Delta m)v = M(v + \Delta v) + \Delta m(v - v_e)$$

Simplifying this expression gives

$$M \Delta v = \Delta m(v_e)$$

If we now take the limit as Δt goes to zero, $\Delta v \rightarrow dv$ and $\Delta m \rightarrow dm$. Furthermore, the increase dm in the exhaust mass corresponds to an equal decrease in the rocket mass, so $dm = -dM$. Note that the negative sign is introduced into the equation because dM represents a decrease in mass. Using this fact, we have

$$M \, dv = -v_e \, dM \qquad \qquad 8.43 \blacktriangleleft$$

Integrating this equation and taking the initial mass of the rocket plus fuel to be M_i and the final mass of the rocket plus its remaining fuel to be M_f, we have

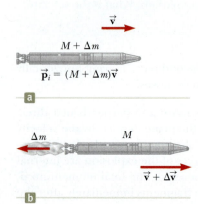

Figure 8.22 Rocket propulsion. (a) The initial mass of the rocket and all its fuel is $M + \Delta m$ at a time t, and its speed is v. (b) At a time $t + \Delta t$, the rocket's mass has been reduced to M, and an amount of fuel Δm has been ejected. The rocket's speed increases by an amount Δv.

$$\int_{v_i}^{v_f} dv = -v_e \int_{M_i}^{M_f} \frac{dM}{M}$$

$$v_f - v_i = v_e \ln\left(\frac{M_i}{M_f}\right)$$

8.44 ◀   ▶ Velocity change in rocket propulsion

which is the basic expression for rocket propulsion. It tells us that the increase in speed is proportional to the exhaust speed v_e. The exhaust speed should therefore be very high.

The **thrust** on the rocket is the force exerted on the rocket by the ejected exhaust gases. We can obtain an expression for the thrust from Equation 8.43:

$$\text{Thrust} = Ma = M\frac{dv}{dt} = \left|v_e \frac{dM}{dt}\right|$$

8.45 ◀   ▶ Rocket thrust

Here we see that the thrust increases as the exhaust speed increases and as the rate of change of mass (burn rate) increases.

We can now determine the amount of fuel needed to set us on our journey to Mars. The fuel requirements are well within the capabilities of current technology, as evidenced by the several missions to Mars that have already been accomplished. What if we wanted to visit another *star*, however, rather than another *planet*? This question raises many new technological challenges, including the requirement to consider the effects of relativity, which we investigate in the next chapter.

> ### THINKING PHYSICS 8.2

When Robert Goddard proposed the possibility of rocket-propelled vehicles, the *New York Times* agreed that such vehicles would be useful and successful within the Earth's atmosphere ("Topics of the Times," *New York Times*, January 13, 1920, p. 12). The *Times*, however, balked at the idea of using such a rocket in the vacuum of space, noting that "its flight would be neither accelerated nor maintained by the explosion of the charges it then might have left. To claim that it would be is to deny a fundamental law of dynamics, and only Dr. Einstein and his chosen dozen, so few and fit, are licensed to do that. . . . That Professor Goddard, with his 'chair' in Clark College and the countenancing of the Smithsonian Institution, does not know the relation of action to reaction, and of the need to have something better than a vacuum against which to react—to say that would be absurd. Of course, he only seems to lack the knowledge ladled out daily in high schools." What did the writer of this passage overlook?

Reasoning The writer of this passage was making a common mistake in believing that a rocket works by expelling gases that push on something, propelling the rocket forward. With this belief, it is impossible to see how a rocket fired in empty space would work.

Gases do not need to push on anything; it is the act itself of expelling the gases that pushes the rocket forward. This point can be argued from Newton's third law: The rocket pushes the gases backward, resulting in the gases pushing the rocket forward. It can also be argued from conservation of momentum: As the gases gain momentum in one direction, the rocket must gain momentum in the opposite direction to conserve the original momentum of the rocket–gas system.

The *New York Times* did publish a retraction 49 years later ("A Correction," *New York Times*, July 17, 1969, p. 43) while the *Apollo 11* astronauts were on their way to the Moon. It appeared on a page with two other articles entitled "Fundamentals of Space Travel" and "Spacecraft, Like Squid, Maneuver by 'Squirts'" and contained the following passages: "an editorial feature of the *New York Times* dismissed the notion that a rocket could function in a vacuum and commented on the ideas of Robert H. Goddard. . . . Further investigation and experimentation have confirmed the findings of Isaac Newton in the 17th century, and it is now definitely established that a rocket can function in a vacuum as well as in an atmosphere. The *Times* regrets the error." ◀

Example **8.14** | A Rocket in Space

A rocket moving in space, far from all other objects, has a speed of 3.0×10^3 m/s relative to the Earth. Its engines are turned on, and fuel is ejected in a direction opposite the rocket's motion at a speed of 5.0×10^3 m/s relative to the rocket.

(A) What is the speed of the rocket relative to the Earth once the rocket's mass is reduced to half its mass before ignition?

SOLUTION

Conceptualize Figure 8.22 shows the situation in this problem. From the discussion in this section and scenes from science fiction movies, we can easily imagine the rocket accelerating to a higher speed as the engine operates.

Categorize This problem is a substitution problem in which we use given values in the equations derived in this section.

Solve Equation 8.44 for the final velocity and substitute the known values:

$$v_f = v_i + v_e \ln\left(\frac{M_i}{M_f}\right)$$

$$= 3.0 \times 10^3 \text{ m/s} + (5.0 \times 10^3 \text{ m/s})\ln\left(\frac{M_i}{0.50 M_i}\right)$$

$$= \boxed{6.5 \times 10^3 \text{ m/s}}$$

(B) What is the thrust on the rocket if it burns fuel at the rate of 50 kg/s?

SOLUTION

Use Equation 8.45 and the result from part (A), noting that $dM/dt = 50$ kg/s:

$$\text{Thrust} = \left|v_e \frac{dM}{dt}\right| = (5.0 \times 10^3 \text{ m/s})(50 \text{ kg/s}) = \boxed{2.5 \times 10^5 \text{ N}}$$

> SUMMARY |

The linear momentum of any object of mass m moving with a velocity $\vec{\mathbf{v}}$ is

$$\vec{\mathbf{p}} \equiv m\vec{\mathbf{v}} \qquad \textbf{8.2}\blacktriangleleft$$

The **impulse** imparted to a particle by a net force $\Sigma\vec{\mathbf{F}}$ is equal to the time integral of the force:

$$\vec{\mathbf{I}} = \int_{t_i}^{t_f} \Sigma\vec{\mathbf{F}}\, dt \qquad \textbf{8.10}\blacktriangleleft$$

When two objects collide, the total momentum of the isolated system before the collision always equals the total momentum after the collision, regardless of the nature of the collision. An **inelastic collision** is one in which kinetic energy is not conserved. A **perfectly inelastic collision** is one in which the colliding objects stick together after the collision. An **elastic collision** is one in which both momentum and kinetic energy are conserved.

In a two- or three-dimensional collision, the components of momentum in each of the directions are conserved independently.

The vector position of the center of mass of a system of particles is defined as

$$\vec{\mathbf{r}}_{\text{CM}} = \frac{\sum_i m_i \vec{\mathbf{r}}_i}{M} \qquad \textbf{8.32}\blacktriangleleft$$

where M is the total mass of the system and $\vec{\mathbf{r}}_i$ is the position vector of the ith particle.

The **velocity of the center of mass for a system of particles** is

$$\vec{\mathbf{v}}_{\text{CM}} = \frac{1}{M}\sum_i m_i \vec{\mathbf{v}}_i \qquad \textbf{8.36}\blacktriangleleft$$

The total momentum of a system of particles equals the total mass multiplied by the velocity of the center of mass; that is, $\vec{\mathbf{p}}_{\text{tot}} = M\vec{\mathbf{v}}_{\text{CM}}$.

Newton's second law applied to a system of particles is

$$\Sigma\vec{\mathbf{F}}_{\text{ext}} = M\vec{\mathbf{a}}_{\text{CM}} = \frac{d\vec{\mathbf{p}}_{\text{tot}}}{dt} \qquad \textbf{8.40}\blacktriangleleft$$

where $\vec{\mathbf{a}}_{\text{CM}}$ is the acceleration of the center of mass and the sum is over all external forces. The center of mass therefore moves like an imaginary particle of mass M under the influence of the resultant external force on the system.

Analysis Models for Problem Solving

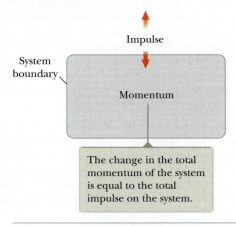

Nonisolated System (Momentum). If a system interacts with its environment in the sense that there is an external force on the system, the behavior of the system is described by the **impulse–momentum theorem:**

$$\Delta \vec{\mathbf{p}}_{tot} = \vec{\mathbf{I}} \qquad\qquad 8.11 \blacktriangleleft$$

The change in the total momentum of the system is equal to the total impulse on the system.

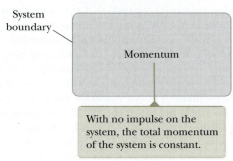

With no impulse on the system, the total momentum of the system is constant.

Isolated System (Momentum). The principle of **conservation of linear momentum** indicates that the total momentum of an isolated system (no external forces) is conserved regardless of the nature of the forces between the members of the system:

$$\vec{\mathbf{p}}_{tot} = M\vec{\mathbf{v}}_{CM} = \text{constant} \quad (\text{when } \sum \vec{\mathbf{F}}_{ext} = 0) \qquad 8.42 \blacktriangleleft$$

In the case of a two-particle system, this principle can be expressed as

$$\vec{\mathbf{p}}_{1i} + \vec{\mathbf{p}}_{2i} = \vec{\mathbf{p}}_{1f} + \vec{\mathbf{p}}_{2f} \qquad\qquad 8.6 \blacktriangleleft$$

The system may be isolated in terms of momentum but nonisolated in terms of energy, as in the case of inelastic collisions.

OBJECTIVE QUESTIONS

☐ denotes answer available in *Student Solutions Manual/Study Guide*

1. A 3-kg object moving to the right on a frictionless, horizontal surface with a speed of 2 m/s collides head-on and sticks to a 2-kg object that is initially moving to the left with a speed of 4 m/s. After the collision, which statement is true? (a) The kinetic energy of the system is 20 J. (b) The momentum of the system is 14 kg · m/s. (c) The kinetic energy of the system is greater than 5 J but less than 20 J. (d) The momentum of the system is −2 kg · m/s. (e) The momentum of the system is less than the momentum of the system before the collision.

2. A head-on, elastic collision occurs between two billiard balls of equal mass. If a red ball is traveling to the right with speed v and a blue ball is traveling to the left with speed $3v$ before the collision, what statement is true concerning their velocities subsequent to the collision? Neglect any effects of spin. (a) The red ball travels to the left with speed v, while the blue ball travels to the right with speed $3v$. (b) The red ball travels to the left with speed v, while the blue ball continues to move to the left with a speed $2v$. (c) The red ball travels to the left with speed $3v$, while the blue ball travels to the right with speed v. (d) Their final velocities cannot be determined because mo-

mentum is not conserved in the collision. (e) The velocities cannot be determined without knowing the mass of each ball.

3. A car of mass m traveling at speed v crashes into the rear of a truck of mass $2m$ that is at rest and in neutral at an intersection. If the collision is perfectly inelastic, what is the speed of the combined car and truck after the collision? (a) v (b) $v/2$ (c) $v/3$ (d) $2v$ (e) None of those answers is correct.

4. A 57.0-g tennis ball is traveling straight at a player at 21.0 m/s. The player volleys the ball straight back at 25.0 m/s. If the ball remains in contact with the racket for 0.060 s, what average force acts on the ball? (a) 22.6 N (b) 32.5 N (c) 43.7 N (d) 72.1 N (e) 102 N

5. A 5-kg cart moving to the right with a speed of 6 m/s collides with a concrete wall and rebounds with a speed of 2 m/s. What is the change in momentum of the cart? (a) 0 (b) 40 kg · m/s (c) −40 kg · m/s (d) −30 kg · m/s (e) −10 kg · m/s

6. A 2-kg object moving to the right with a speed of 4 m/s makes a head-on, elastic collision with a 1-kg object that is initially at rest. The velocity of the 1-kg object after the

collision is (a) greater than 4 m/s, (b) less than 4 m/s, (c) equal to 4 m/s, (d) zero, or (e) impossible to say based on the information provided.

7. The momentum of an object is increased by a factor of 4 in magnitude. By what factor is its kinetic energy changed? (a) 16 (b) 8 (c) 4 (d) 2 (e) 1

8. The kinetic energy of an object is increased by a factor of 4. By what factor is the magnitude of its momentum changed? (a) 16 (b) 8 (c) 4 (d) 2 (e) 1

9. A 10.0-g bullet is fired into a 200-g block of wood at rest on a horizontal surface. After impact, the block slides 8.00 m before coming to rest. If the coefficient of friction between the block and the surface is 0.400, what is the speed of the bullet before impact? (a) 106 m/s (b) 166 m/s (c) 226 m/s (d) 286 m/s (e) none of those answers is correct.

10. If two particles have equal kinetic energies, are their momenta equal? (a) yes, always (b) no, never (c) yes, as long as their masses are equal (d) yes, if both their masses and directions of motion are the same (e) yes, as long as they move along parallel lines

11. If two particles have equal momenta, are their kinetic energies equal? (a) yes, always (b) no, never (c) no, except when their speeds are the same (d) yes, as long as they move along parallel lines

12. Two particles of different mass start from rest. The same net force acts on both of them as they move over equal distances. How do their final kinetic energies compare? (a) The particle of larger mass has more kinetic energy. (b) The particle of smaller mass has more kinetic energy. (c) The particles have equal kinetic energies. (d) Either particle might have more kinetic energy.

13. Two particles of different mass start from rest. The same net force acts on both of them as they move over equal distances. How do the magnitudes of their final momenta compare? (a) The particle of larger mass has more momentum. (b) The particle of smaller mass has more momentum. (c) The particles have equal momenta. (d) Either particle might have more momentum.

14. A ball is suspended by a string that is tied to a fixed point above a wooden block standing on end. The ball is pulled back as shown in Figure OQ8.14 and released. In trial A, the ball rebounds elastically from the block. In trial B, two-sided tape causes the ball to stick to the block. In which case is the ball more likely to knock the

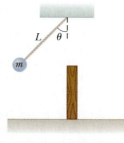

Figure OQ8.14

block over? (a) It is more likely in trial A. (b) It is more likely in trial B. (c) It makes no difference. (d) It could be either case, depending on other factors.

15. A massive tractor is rolling down a country road. In a perfectly inelastic collision, a small sports car runs into the machine from behind. (i) Which vehicle experiences a change in momentum of larger magnitude? (a) The car does. (b) The tractor does. (c) Their momentum changes are the same size. (d) It could be either vehicle. (ii) Which vehicle experiences a larger change in kinetic energy? (a) The car does. (b) The tractor does. (c) Their kinetic energy changes are the same size. (d) It could be either vehicle.

16. A basketball is tossed up into the air, falls freely, and bounces from the wooden floor. From the moment after the player releases it until the ball reaches the top of its bounce, what is the smallest system for which momentum is conserved? (a) the ball (b) the ball plus player (c) the ball plus floor (d) the ball plus the Earth (e) momentum is not conserved for any system.

17. You are standing on a saucer-shaped sled at rest in the middle of a frictionless ice rink. Your lab partner throws you a heavy Frisbee. You take different actions in successive experimental trials. Rank the following situations according to your final speed from largest to smallest. If your final speed is the same in two cases, give them equal rank. (a) You catch the Frisbee and hold onto it. (b) You catch the Frisbee and throw it back to your partner. (c) You bobble the catch, just touching the Frisbee so that it continues in its original direction more slowly. (d) You catch the Frisbee and throw it so that it moves vertically upward above your head. (e) You catch the Frisbee and set it down so that it remains at rest on the ice.

18. A boxcar at a rail yard is set into motion at the top of a hump. The car rolls down quietly and without friction onto a straight, level track where it couples with a flatcar of smaller mass, originally at rest, so that the two cars then roll together without friction. Consider the two cars as a system from the moment of release of the boxcar until both are rolling together. Answer the following questions yes or no. (a) Is mechanical energy of the system conserved? (b) Is momentum of the system conserved? Next, consider only the process of the boxcar gaining speed as it rolls down the hump. For the boxcar and the Earth as a system, (c) is mechanical energy conserved? (d) Is momentum conserved? Finally, consider the two cars as a system as the boxcar is slowing down in the coupling process. (e) Is mechanical energy of this system conserved? (f) Is momentum of this system conserved?

▶ CONCEPTUAL QUESTIONS

1. Does a larger net force exerted on an object always produce a larger change in the momentum of the object compared with a smaller net force? Explain.

2. While in motion, a pitched baseball carries kinetic energy and momentum. (a) Can we say that it carries a force that it can exert on any object it strikes? (b) Can the baseball

deliver more kinetic energy to the bat and batter than the ball carries initially? (c) Can the baseball deliver to the bat and batter more momentum than the ball carries initially? Explain each of your answers.

3. A bomb, initially at rest, explodes into several pieces. (a) Is linear momentum of the system (the bomb before the

explosion, the pieces after the explosion) conserved? Explain. (b) Is kinetic energy of the system conserved? Explain.

4. Does a larger net force always produce a larger change in kinetic energy than a smaller net force? Explain.

5. You are standing perfectly still and then take a step forward. Before the step, your momentum was zero, but afterward you have some momentum. Is the principle of conservation of momentum violated in this case? Explain your answer.

6. A juggler juggles three balls in a continuous cycle. Any one ball is in contact with one of his hands for one fifth of the time. (a) Describe the motion of the center of mass of the three balls. (b) What average force does the juggler exert on one ball while he is touching it?

7. Two students hold a large bed sheet vertically between them. A third student, who happens to be the star pitcher on the school baseball team, throws a raw egg at the center of the sheet. Explain why the egg does not break when it hits the sheet, regardless of its initial speed.

8. A sharpshooter fires a rifle while standing with the butt of the gun against her shoulder. If the forward momentum of a bullet is the same as the backward momentum of the gun, why isn't it as dangerous to be hit by the gun as by the bullet?

9. An airbag in an automobile inflates when a collision occurs, which protects the passenger from serious injury (see the photo on page 242). Why does the airbag soften the blow? Discuss the physics involved in this dramatic photograph.

10. On the subject of the following positions, state your own view and argue to support it. (a) The best theory of motion is that force causes acceleration. (b) The true measure of a force's effectiveness is the work it does, and the best theory of motion is that work done on an object changes its energy. (c) The true measure of a force's effect is impulse, and the best theory of motion is that impulse imparted to an object changes its momentum.

11. (a) Does the center of mass of a rocket in free space accelerate? Explain. (b) Can the speed of a rocket exceed the exhaust speed of the fuel? Explain.

12. In golf, novice players are often advised to be sure to "follow through" with their swing. Why does this advice make the ball travel a longer distance? If a shot is taken near the green, very little follow-through is required. Why?

13. An open box slides across a frictionless, icy surface of a frozen lake. What happens to the speed of the box as water from a rain shower falls vertically downward into the box? Explain.

PROBLEMS

Section 8.1 Linear Momentum
Section 8.2 Analysis Model: Isolated System (Momentum)

1. **BIO** In research in cardiology and exercise physiology, it is often important to know the mass of blood pumped by a person's heart in one stroke. This information can be obtained by means of a *ballistocardiograph*. The instrument works as follows. The subject lies on a horizontal pallet floating on a film of air. Friction on the pallet is negligible. Initially, the momentum of the system is zero. When the heart beats, it expels a mass m of blood into the aorta with speed v, and the body and platform move in the opposite direction with speed V. The blood velocity can be determined independently (e.g., by observing the Doppler shift of ultrasound). Assume that it is 50.0 cm/s in one typical trial. The mass of the subject plus the pallet is 54.0 kg. The pallet moves 6.00×10^{-5} m in 0.160 s after one heartbeat. Calculate the mass of blood that leaves the heart. Assume that the mass of blood is negligible compared with the total mass of the person. (This simplified example illustrates the principle of ballistocardiography, but in practice a more sophisticated model of heart function is used.)

2. **S** A particle of mass m moves with momentum of magnitude p. (a) Show that the kinetic energy of the particle is $K = p^2/2m$. (b) Express the magnitude of the particle's momentum in terms of its kinetic energy and mass.

3. A 3.00-kg particle has a velocity of $(3.00\hat{i} - 4.00\hat{j})$ m/s. (a) Find its x and y components of momentum. (b) Find the magnitude and direction of its momentum.

4. When you jump straight up as high as you can, what is the order of magnitude of the maximum recoil speed that you give to the Earth? Model the Earth as a perfectly solid object. In your solution, state the physical quantities you take as data and the values you measure or estimate for them.

5. **M** A 45.0-kg girl is standing on a 150-kg plank. Both are originally at rest on a frozen lake that constitutes a frictionless,

flat surface. The girl begins to walk along the plank at a constant velocity of $1.50\hat{\mathbf{i}}$ m/s relative to the plank. (a) What is the velocity of the plank relative to the ice surface? (b) What is the girl's velocity relative to the ice surface?

6. **S** A girl of mass m_g is standing on a plank of mass m_p. Both are originally at rest on a frozen lake that constitutes a frictionless, flat surface. The girl begins to walk along the plank at a constant velocity v_{gp} to the right relative to the plank. (The subscript gp denotes the girl relative to plank.) (a) What is the velocity v_{pi} of the plank relative to the surface of the ice? (b) What is the girl's velocity v_{gi} relative to the ice surface?

7. **Q|C** **W** Two blocks of masses m and $3m$ are placed on a frictionless, horizontal surface. A light spring is attached to the more massive block, and the blocks are pushed together with the spring between them (Fig. P8.7). A cord initially holding the blocks together is burned; after that happens, the block of mass $3m$ moves to the right with a speed of 2.00 m/s. (a) What is the velocity of the block of mass m? (b) Find the system's original elastic potential energy, taking $m = 0.350$ kg. (c) Is the original energy in the spring or in the cord? (d) Explain your answer to part (c). (e) Is the momentum of the system conserved in the bursting-apart process? Explain how that is possible considering (f) there are large forces acting and (g) there is no motion beforehand and plenty of motion afterward?

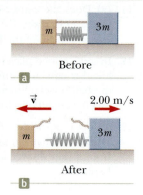

Figure P8.7

Section 8.3 Analysis Model: Nonisolated System (Momentum)

8. **Q|C** A man claims that he can hold onto a 12.0-kg child in a head-on collision as long as he has his seat belt on. Consider this man in a collision in which he is in one of two identical cars each traveling toward the other at 60.0 mi/h relative to the ground. The car in which he rides is brought to rest in 0.10 s. (a) Find the magnitude of the average force needed to hold onto the child. (b) Based on your result to part (a), is the man's claim valid? (c) What does the answer to this problem say about laws requiring the use of proper safety devices such as seat belts and special toddler seats?

9. **M** A 3.00-kg steel ball strikes a wall with a speed of 10.0 m/s at an angle of $\theta = 60.0°$ with the surface. It bounces off with the same speed and angle (Fig. P8.9). If the ball is in contact with the wall for 0.200 s, what is the average force exerted by the wall on the ball?

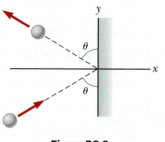

Figure P8.9

10. **W** A tennis player receives a shot with the ball (0.060 0 kg) traveling horizontally at 50.0 m/s and returns the shot with the ball traveling horizontally at 40.0 m/s in the opposite

direction. (a) What is the impulse delivered to the ball by the tennis racquet? (b) What work does the racquet do on the ball?

11. **W** An estimated force–time curve for a baseball struck by a bat is shown in Figure P8.11. From this curve, determine (a) the magnitude of the impulse delivered to the ball and (b) the average force exerted on the ball.

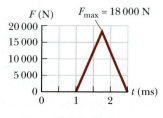

Figure P8.11

12. **Q|C** **S** A glider of mass m is free to slide along a horizontal air track. It is pushed against a launcher at one end of the track. Model the launcher as a light spring of force constant k compressed by a distance x. The glider is released from rest. (a) Show that the glider attains a speed of $v = x(k/m)^{1/2}$. (b) Show that the magnitude of the impulse imparted to the glider is given by the expression $I = x(km)^{1/2}$. (c) Is more work done on a cart with a large or a small mass?

13. A garden hose is held as shown in Figure P8.13. The hose is originally full of motionless water. What additional force is necessary to hold the nozzle stationary after the water flow is turned on if the discharge rate is 0.600 kg/s with a speed of 25.0 m/s?

Figure P8.13

14. In a slow-pitch softball game, a 0.200-kg softball crosses the plate at 15.0 m/s at an angle of 45.0° below the horizontal. The batter hits the ball toward center field, giving it a velocity of 40.0 m/s at 30.0° above the horizontal. (a) Determine the impulse delivered to the ball. (b) If the force on the ball increases linearly for 4.00 ms, holds constant for 20.0 ms, and then decreases linearly to zero in another 4.00 ms, what is the maximum force on the ball?

Section 8.4 Collisions in One Dimension

15. A railroad car of mass 2.50×10^4 kg is moving with a speed of 4.00 m/s. It collides and couples with three other coupled railroad cars, each of the same mass as the single car and moving in the same direction with an initial speed of 2.00 m/s. (a) What is the speed of the four cars after the collision? (b) How much mechanical energy is lost in the collision?

16. Four railroad cars, each of mass 2.50×10^4 kg, are coupled together and coasting along horizontal tracks at speed v_i toward the south. A very strong but foolish movie actor, riding on the second car, uncouples the front car and gives it a big push, increasing its speed to 4.00 m/s southward. The remaining three cars continue moving south, now at 2.00 m/s. (a) Find the initial speed of the four cars. (b) How much work did the actor do? (c) State the relationship between the process described here and the process in Problem 8.15.

17. **M** A 12.0-g wad of sticky clay is hurled horizontally at a 100-g wooden block initially at rest on a horizontal surface. The clay sticks to the block. After impact, the block slides

7.50 m before coming to rest. If the coefficient of friction between the block and the surface is 0.650, what was the speed of the clay immediately before impact?

18. **S** A wad of sticky clay of mass m is hurled horizontally at a wooden block of mass M initially at rest on a horizontal surface. The clay sticks to the block. After impact, the block slides a distance d before coming to rest. If the coefficient of friction between the block and the surface is μ, what was the speed of the clay immediately before impact?

19. **W** Two blocks are free to slide along the frictionless, wooden track shown in Figure P8.19. The block of mass $m_1 = 5.00$ kg is released from the position shown, at height $h = 5.00$ m above the flat part of the track. Protruding from its front end is the north pole of a strong magnet, which repels the north pole of an identical magnet embedded in the back end of the block of mass $m_2 = 10.0$ kg, initially at rest. The two blocks never touch. Calculate the maximum height to which m_1 rises after the elastic collision.

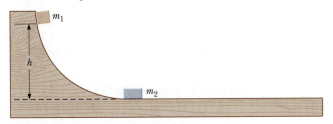

Figure P8.19

20. **S** As shown in Figure P8.20, a bullet of mass m and speed v passes completely through a pendulum bob of mass M. The bullet emerges with a speed of $v/2$. The pendulum bob is suspended by a stiff rod (*not* a string) of length ℓ and negligible mass. What is the minimum value of v such that the pendulum bob will barely swing through a complete vertical circle?

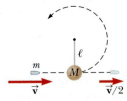

Figure P8.20

21. **M** A neutron in a nuclear reactor makes an elastic, head-on collision with the nucleus of a carbon atom initially at rest. (a) What fraction of the neutron's kinetic energy is transferred to the carbon nucleus? (b) The initial kinetic energy of the neutron is 1.60×10^{-13} J. Find its final kinetic energy and the kinetic energy of the carbon nucleus after the collision. (The mass of the carbon nucleus is nearly 12.0 times the mass of the neutron.)

22. **QC S** A tennis ball of mass m_t is held just above a basketball of mass m_b, as shown in Figure P8.22. With their centers vertically aligned, both are released from rest at the same moment so that the bottom of the basketball falls freely through a height h and strikes the floor. Assume an elastic collision with the ground instantaneously reverses the velocity of the basketball while the tennis ball is still moving down because the balls have separated a bit while falling. Next, the two balls meet in an elastic collision. (a) To what height does the tennis ball rebound? (b) How do you account for the height in (a) being larger than h? Does that seem like a violation of conservation of energy?

Figure P8.22

23. **QC** (a) Three carts of masses $m_1 = 4.00$ kg, $m_2 = 10.0$ kg, and $m_3 = 3.00$ kg move on a frictionless, horizontal track with speeds of $v_1 = 5.00$ m/s to the right, $v_2 = 3.00$ m/s to the right, and $v_3 = 4.00$ m/s to the left as shown in Figure P8.23. Velcro couplers make the carts stick together after colliding. Find the final velocity of the train of three carts. (b) **What If?** Does your answer in part (a) require that all the carts collide and stick together at the same moment? What if they collide in a different order?

Figure P8.23

24. A 7.00-g bullet, when fired from a gun into a 1.00-kg block of wood held in a vise, penetrates the block to a depth of 8.00 cm. This block of wood is next placed on a frictionless horizontal surface, and a second 7.00-g bullet is fired from the gun into the block. To what depth will the bullet penetrate the block in this case?

Section 8.5 Collisions in Two Dimensions

25. **W** An object of mass 3.00 kg, moving with an initial velocity of $5.00\hat{\mathbf{i}}$ m/s, collides with and sticks to an object of mass 2.00 kg with an initial velocity of $-3.00\hat{\mathbf{j}}$ m/s. Find the final velocity of the composite object.

26. Two automobiles of equal mass approach an intersection. One vehicle is traveling with speed 13.0 m/s toward the east, and the other is traveling north with speed v_{2i}. Neither driver sees the other. The vehicles collide in the intersection and stick together, leaving parallel skid marks at an angle of 55.0° north of east. The speed limit for both roads is 35 mi/h, and the driver of the northward-moving vehicle claims he was within the speed limit when the collision occurred. Is he telling the truth? Explain your reasoning.

27. **W** Two shuffleboard disks of equal mass, one orange and the other yellow, are involved in an elastic, glancing collision. The yellow disk is initially at rest and is struck by the orange disk moving with a speed of 5.00 m/s. After the collision, the orange disk moves along a direction that makes an angle of 37.0° with its initial direction of motion. The velocities of the two disks are perpendicular after the collision. Determine the final speed of each disk.

28. **S** Two shuffleboard disks of equal mass, one orange and the other yellow, are involved in an elastic, glancing collision. The yellow disk is initially at rest and is struck by the orange disk moving with a speed v_i. After the collision, the orange disk moves along a direction that makes an angle θ with its initial direction of motion. The velocities of the two disks are perpendicular after the collision. Determine the final speed of each disk.

29. **M** A billiard ball moving at 5.00 m/s strikes a stationary ball of the same mass. After the collision, the first ball moves at 4.33 m/s at an angle of 30.0° with respect to the original line of motion. Assuming an elastic collision (and ignoring friction and rotational motion), find the struck ball's velocity after the collision.

30. QC W A 90.0-kg fullback running east with a speed of 5.00 m/s is tackled by a 95.0-kg opponent running north with a speed of 3.00 m/s. (a) Explain why the successful tackle constitutes a perfectly inelastic collision. (b) Calculate the velocity of the players immediately after the tackle. (c) Determine the mechanical energy that disappears as a result of the collision. Account for the missing energy.

31. M An unstable atomic nucleus of mass 17.0×10^{-27} kg initially at rest disintegrates into three particles. One of the particles, of mass 5.00×10^{-27} kg, moves in the y direction with a speed of 6.00×10^6 m/s. Another particle, of mass 8.40×10^{-27} kg, moves in the x direction with a speed of 4.00×10^6 m/s. Find (a) the velocity of the third particle and (b) the total kinetic energy increase in the process.

32. S A proton, moving with a velocity of $v_i\hat{\mathbf{i}}$, collides elastically with another proton that is initially at rest. Assuming that the two protons have equal speeds after the collision, find (a) the speed of each proton after the collision in terms of v_i and (b) the direction of the velocity vectors after the collision.

33. A 0.300-kg puck, initially at rest on a horizontal, frictionless surface, is struck by a 0.200-kg puck moving initially along the x axis with a speed of 2.00 m/s. After the collision, the 0.200-kg puck has a speed of 1.00 m/s at an angle of $\theta = 53.0°$ to the positive x axis (see Active Fig. 8.11). (a) Determine the velocity of the 0.300-kg puck after the collision. (b) Find the fraction of kinetic energy lost in the collision.

Section 8.6 The Center of Mass

34. A uniform piece of sheet metal is shaped as shown in Figure P8.34. Compute the x and y coordinates of the center of mass of the piece.

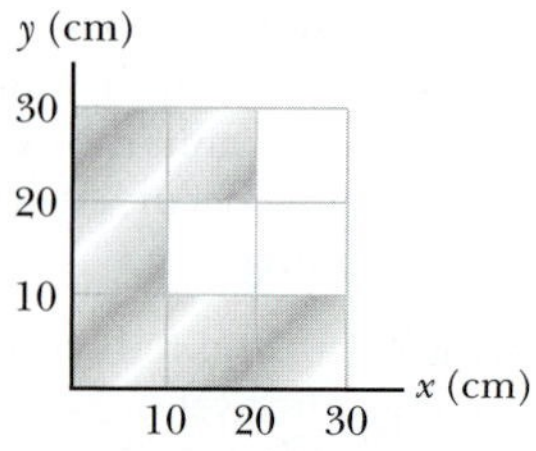

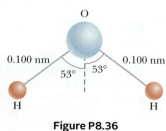

Figure P8.34

35. W Four objects are situated along the y axis as follows: a 2.00-kg object is at +3.00 m, a 3.00-kg object is at +2.50 m, a 2.50-kg object is at the origin, and a 4.00-kg object is at −0.500 m. Where is the center of mass of these objects?

36. A water molecule consists of an oxygen atom with two hydrogen atoms bound to it (Fig. P8.36). The angle between the two bonds is 106°. If the bonds are 0.100 nm long, where is the center of mass of the molecule?

Figure P8.36

37. Explorers in the jungle find an ancient monument in the shape of a large isosceles triangle as shown in Figure P8.37. The monument is made from tens of thousands of small stone blocks of density 3 800 kg/m³. The monument is 15.7 m high and 64.8 m wide at its base and is everywhere 3.60 m thick from front to back. Before the monument was built many years ago, all the stone blocks lay on the ground. How much work did laborers do on the blocks to put them in position while building the entire monument? *Note*: The gravitational potential energy of an object–Earth system is given by $U_g = Mgy_{CM}$, where M is the total mass of the object and y_{CM} is the elevation of its center of mass above the chosen reference level.

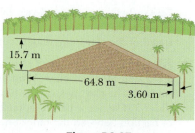

Figure P8.37

38. A rod of length 30.0 cm has linear density (mass per length) given by

$$\lambda = 50.0 + 20.0x$$

where x is the distance from one end, measured in meters, and λ is in grams/meter. (a) What is the mass of the rod? (b) How far from the $x = 0$ end is its center of mass?

Section 8.7 Motion of a System of Particles

39. A 2.00-kg particle has a velocity $(2.00\hat{\mathbf{i}} - 3.00\hat{\mathbf{j}})$ m/s, and a 3.00-kg particle has a velocity $(1.00\hat{\mathbf{i}} + 6.00\hat{\mathbf{j}})$ m/s. Find (a) the velocity of the center of mass and (b) the total momentum of the system.

40. A ball of mass 0.200 kg with a velocity of $1.50\hat{\mathbf{i}}$ m/s meets a ball of mass 0.300 kg with a velocity of $-0.400\hat{\mathbf{i}}$ m/s in a head-on, elastic collision. (a) Find their velocities after the collision. (b) Find the velocity of their center of mass before and after the collision.

41. M Romeo (77.0 kg) entertains Juliet (55.0 kg) by playing his guitar from the rear of their boat at rest in still water, 2.70 m away from Juliet, who is in the front of the boat. After the serenade, Juliet carefully moves to the rear of the boat (away from shore) to plant a kiss on Romeo's cheek. How far does the 80.0-kg boat move toward the shore it is facing?

42. Consider a system of two particles in the xy plane: $m_1 = 2.00$ kg is at the location $\vec{\mathbf{r}}_1 = (1.00\hat{\mathbf{i}} + 2.00\hat{\mathbf{j}})$ m and has a velocity of $(3.00\hat{\mathbf{i}} + 0.500\hat{\mathbf{j}})$ m/s; $m_2 = 3.00$ kg is at $\vec{\mathbf{r}}_2 = (-4.00\hat{\mathbf{i}} - 3.00\hat{\mathbf{j}})$ m and has velocity $(3.00\hat{\mathbf{i}} - 2.00\hat{\mathbf{j}})$ m/s. (a) Plot these particles on a grid or graph paper. Draw their position vectors and show their velocities. (b) Find the position of the center of mass of the system and mark it on the grid. (c) Determine the velocity of the center of mass and also show it on the diagram. (d) What is the total linear momentum of the system?

Section 8.8 Context Connection: Rocket Propulsion

43. An orbiting spacecraft is described not as a "zero-g" but rather as a "microgravity" environment for its occupants and for onboard experiments. Astronauts experience slight lurches due to the motions of equipment and other astronauts and as a result of venting of materials from the craft. Assume that a 3 500-kg spacecraft undergoes an acceleration of $2.50 \ \mu g = 2.45 \times 10^{-5}$ m/s² due to a leak from one of its hydraulic control systems. The fluid is known to escape with a speed of 70.0 m/s into the vacuum of space. How much fluid will be lost in 1.00 h if the leak is not stopped?

44. **Q|C** **M** A rocket for use in deep space is to be capable of boosting a total load (payload plus rocket frame and engine) of 3.00 metric tons to a speed of 10 000 m/s. (a) It has an engine and fuel designed to produce an exhaust speed of 2 000 m/s. How much fuel plus oxidizer is required? (b) If a different fuel and engine design could give an exhaust speed of 5 000 m/s, what amount of fuel and oxidizer would be required for the same task? (c) Noting that the exhaust speed in part (b) is 2.50 times higher than that in part (a), explain why the required fuel mass is not simply smaller by a factor of 2.50.

45. **Review.** The first stage of a Saturn V space vehicle consumed fuel and oxidizer at the rate of 1.50×10^4 kg/s with an exhaust speed of 2.60×10^3 m/s. (a) Calculate the thrust produced by this engine. (b) Find the acceleration the vehicle had just as it lifted off the launch pad on the Earth, taking the vehicle's initial mass as 3.00×10^6 kg.

46. A rocket has total mass $M_i = 360$ kg, including $M_f = 330$ kg of fuel and oxidizer. In interstellar space, it starts from rest at the position $x = 0$, turns on its engine at time $t = 0$, and puts out exhaust with relative speed $v_e = 1\ 500$ m/s at the constant rate $k = 2.50$ kg/s. The fuel will last for a burn time of $T_b = M_f/k = 330$ kg/(2.5 kg/s) = 132 s. (a) Show that during the burn the velocity of the rocket as a function of time is given by

$$v(t) = -v_e \ln\left(1 - \frac{kt}{M_i}\right)$$

(b) Make a graph of the velocity of the rocket as a function of time for times running from 0 to 132 s. (c) Show that the acceleration of the rocket is

$$a(t) = \frac{kv_e}{M_i - kt}$$

(d) Graph the acceleration as a function of time. (e) Show that the position of the rocket is

$$x(t) = v_e\left(\frac{M_i}{k} - t\right)\ln\left(1 - \frac{kt}{M_i}\right) + v_e t$$

(f) Graph the position during the burn as a function of time.

47. A model rocket engine has an average thrust of 5.26 N. It has an initial mass of 25.5 g, which includes fuel mass of 12.7 g. The duration of its burn is 1.90 s. (a) What is the average exhaust speed of the engine? (b) This engine is placed in a rocket body of mass 53.5 g. What is the final velocity of the rocket if it were to be fired from rest in outer space by an astronaut on a spacewalk? Assume the fuel burns at a constant rate.

Additional Problems

48. **S** Two gliders are set in motion on a horizontal air track. A spring of force constant k is attached to the back end of the second glider. As shown in Figure P8.48, the first glider, of mass m_1, moves to the right with speed v_1, and the second glider, of mass m_2, moves more slowly to the right with speed

v_2. When m_1 collides with the spring attached to m_2, the spring compresses by a distance x_{max}, and the gliders then move apart again. In terms of v_1, v_2, m_1, m_2, and k, find (a) the speed v at maximum compression, (b) the maximum compression x_{max}, and (c) the velocity of each glider after m_1 has lost contact with the spring.

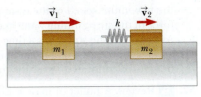

Figure P8.48

49. **Review.** A 60.0-kg person running at an initial speed of 4.00 m/s jumps onto a 120-kg cart initially at rest (Fig. P8.49). The person slides on the cart's top surface and finally comes to rest relative to the cart. The coefficient of kinetic friction between the person and the cart is 0.400. Friction between the cart and ground can be ignored. (a) Find the final velocity of the person and cart relative to the ground. (b) Find the friction force acting on the person while he is sliding across the top surface of the cart. (c) How long does the friction force act on the person? (d) Find the change in momentum of the person and the change in momentum of the cart. (e) Determine the displacement of the person relative to the ground while he is sliding on the cart. (f) Determine the displacement of the cart relative to the ground while the person is sliding. (g) Find the change in kinetic energy of the person. (h) Find the change in kinetic energy of the cart. (i) Explain why the answers to (g) and (h) differ. (What kind of collision is this one, and what accounts for the loss of mechanical energy?)

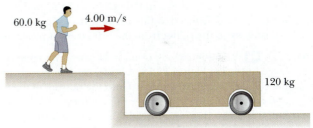

Figure P8.49

50. *Why is the following situation impossible?* An astronaut, together with the equipment he carries, has a mass of 150 kg. He is taking a space walk outside his spacecraft, which is drifting through space with a constant velocity. The astronaut accidentally pushes against the spacecraft and begins moving away at 20.0 m/s, relative to the spacecraft, without a tether. To return, he takes equipment off his space suit and throws it in the direction away from the spacecraft. Because of his bulky space suit, he can throw equipment at a maximum speed of 5.00 m/s relative to himself. After throwing enough equipment, he starts moving back to the spacecraft and is able to grab onto it and climb inside.

51. **BIO** When it is threatened, a squid can escape by expelling a jet of water, sometimes colored with camouflaging ink. Consider a squid originally at rest in ocean water of constant density 1 030 kg/m³. Its original mass is 90.0 kg, of which a significant fraction is water inside its mantle. It expels this water through its siphon, a circular opening of diameter 3.00 cm, at a speed of 16.0 m/s. (a) As the squid is just starting to move, the surrounding water exerts no drag force on it. Find the squid's initial acceleration. (b) To estimate

the maximum speed of the escaping squid, model the drag force of the surrounding water as described by Equation 5.7. Assume that the squid has a drag coefficient of 0.300 and a cross-sectional area of 800 cm². Find the speed at which the drag force counterbalances the thrust of its jet.

52. **S** Review. A bullet of mass m is fired into a block of mass M initially at rest at the edge of a frictionless table of height h (Fig. P8.52). The bullet remains in the block, and after impact the block lands a distance d from the bottom of the table. Determine the initial speed of the bullet.

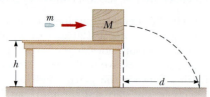

Figure P8.52

53. **Q|C** A 1.25-kg wooden block rests on a table over a large hole as in Figure P8.53. A 5.00-g bullet with an initial velocity v_i is fired upward into the bottom of the block and remains in the block after the collision. The block and bullet rise to a maximum height of 22.0 cm. (a) Describe how you would find the initial velocity of the bullet using ideas you have learned in this chapter. (b) Calculate the initial velocity of the bullet from the information provided.

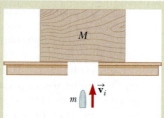

Figure P8.53 Problems 53 and 54.

54. **Q|C** **S** A wooden block of mass M rests on a table over a large hole as in Figure P8.53. A bullet of mass m with an initial velocity of v_i is fired upward into the bottom of the block and remains in the block after the collision. The block and bullet rise to a maximum height of h. (a) Describe how you would find the initial velocity of the bullet using ideas you have learned in this chapter. (b) Find an expression for the initial velocity of the bullet.

55. **W** A small block of mass $m_1 = 0.500$ kg is released from rest at the top of a frictionless, curve-shaped wedge of mass $m_2 = 3.00$ kg, which sits on a frictionless, horizontal surface as shown in Figure P8.55a. When the block leaves the wedge, its velocity is measured to be 4.00 m/s to the right as shown in Figure P8.55b. (a) What is the velocity of the wedge after the block reaches the horizontal surface? (b) What is the height h of the wedge?

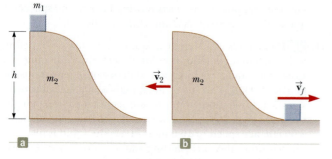

Figure P8.55

56. A jet aircraft is traveling at 500 mi/h (223 m/s) in horizontal flight. The engine takes in air at a rate of 80.0 kg/s and burns fuel at a rate of 3.00 kg/s. The exhaust gases are ejected at 600 m/s relative to the aircraft. Find the thrust of the jet engine and the delivered power.

57. **M** A 5.00-g bullet moving with an initial speed of $v = 400$ m/s is fired into and passes through a 1.00-kg block as shown in Figure P8.57. The block, initially at rest on a frictionless, horizontal surface, is connected to a spring with force constant 900 N/m. The block moves $d = 5.00$ cm to the right after impact before being brought to rest by the spring. Find (a) the speed at which the bullet emerges from the block and (b) the amount of initial kinetic energy of the bullet that is converted into internal energy in the bullet–block system during the collision.

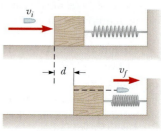

Figure P8.57

58. **GP** **Q|C** Review. There are (one can say) three coequal theories of motion for a single particle: Newton's second law, stating that the total force on the particle causes its acceleration; the work–kinetic energy theorem, stating that the total work on the particle causes its change in kinetic energy; and the impulse–momentum theorem, stating that the total impulse on the particle causes its change in momentum. In this problem, you compare predictions of the three theories in one particular case. A 3.00-kg object has velocity $7.00\hat{j}$ m/s. Then, a constant net force $12.0\hat{i}$ N acts on the object for 5.00 s. (a) Calculate the object's final velocity, using the impulse–momentum theorem. (b) Calculate its acceleration from $\vec{a} = (\vec{v}_f - \vec{v}_i)/\Delta t$. (c) Calculate its acceleration from $\vec{a} = \Sigma \vec{F}/m$. (d) Find the object's vector displacement from $\Delta r = \vec{v}_i t + \frac{1}{2}\vec{a}t^2$. (e) Find the work done on the object from $W = \vec{F} \cdot \Delta \vec{r}$. (f) Find the final kinetic energy from $\frac{1}{2}mv_f^2 = \frac{1}{2}m\vec{v}_f \cdot \vec{v}_f$. (g) Find the final kinetic energy from $\frac{1}{2}mv_i^2 + W$. (h) State the result of comparing the answers to parts (b) and (c), and the answers to parts (f) and (g).

59. Review. A light spring of force constant 3.85 N/m is compressed by 8.00 cm and held between a 0.250-kg block on the left and a 0.500-kg block on the right. Both blocks are at rest on a horizontal surface. The blocks are released simultaneously so that the spring tends to push them apart. Find the maximum velocity each block attains if the coefficient of kinetic friction between each block and the surface is (a) 0, (b) 0.100, and (c) 0.462. Assume the coefficient of static friction is greater than the coefficient of kinetic friction in every case.

60. A cannon is rigidly attached to a carriage, which can move along horizontal rails but is connected to a post by a large spring, initially unstretched and with force constant $k = 2.00 \times 10^4$ N/m, as shown in Figure P8.60. The cannon fires a 200-kg projectile at a velocity of 125 m/s directed 45.0° above the horizontal. (a) Assuming that the mass of the cannon and its carriage is 5 000 kg, find the recoil speed of the cannon. (b) Determine the maximum extension of the spring. (c) Find the maximum force the spring exerts on the carriage. (d) Consider the system consisting of the cannon, carriage, and projectile. Is the momentum of this system conserved during the firing? Why or why not?

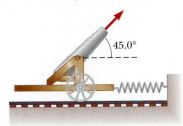

Figure P8.60

61. Two particles with masses m and $3m$ are moving toward each other along the x axis with the same initial speeds v_i. The particle with mass m is traveling to the left, and particle $3m$ is traveling to the right. They undergo a head-on elastic collision and each rebounds along the same line as it approached. Find the final speeds of the particles.

62. **S** Two particles with masses m and $3m$ are moving toward each other along the x axis with the same initial speeds v_i. Particle m is traveling to the left, and particle $3m$ is traveling to the right. They undergo an elastic glancing collision such that particle m is moving in the negative y direction after the collision at a right angle from its initial direction. (a) Find the final speeds of the two particles in terms of v_i. (b) What is the angle θ at which the particle $3m$ is scattered?

63. George of the Jungle, with mass m, swings on a light vine hanging from a stationary tree branch. A second vine of equal length hangs from the same point, and a gorilla of larger mass M swings in the opposite direction on it. Both vines are horizontal when the primates start from rest at the same moment. George and the gorilla meet at the lowest point of their swings. Each is afraid that one vine will break, so they grab each other and hang on. They swing upward together, reaching a point where the vines make an angle of 35.0° with the vertical. Find the value of the ratio m/M.

64. **Q C** Sand from a stationary hopper falls onto a moving conveyor belt at the rate of 5.00 kg/s as shown in Figure P8.64. The conveyor belt is supported by frictionless rollers and moves at a constant speed of $v = 0.750$ m/s under the action of a constant horizontal external force $\vec{F}_{ext}$ supplied by the motor that drives the belt. Find (a) the sand's rate of change of momentum in the horizontal direction, (b) the force of friction exerted by the belt on the sand, (c) the external force $\vec{F}_{ext}$, (d) the work done by $\vec{F}_{ext}$ in 1 s, and (e) the kinetic energy acquired by the falling sand each second due to the change in its horizontal motion. (f) Why are the answers to parts (d) and (e) different?

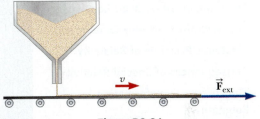

Figure P8.64

65. **S** **Review.** A chain of length L and total mass M is released from rest with its lower end just touching the top of a table as shown in Figure P8.65a. Find the force exerted by the table on the chain after the chain has fallen through a distance x as shown in Figure P8.65b. (Assume each link comes to rest the instant it reaches the table.)

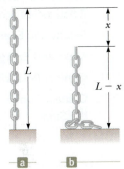

Figure P8.65

Relativity

Chapter Outline

Emily Serway

Standing on the shoulders of a giant. David Serway, son of one of the authors, watches over his children, Nathan and Kaitlyn, as they frolic in the arms of Albert Einstein at the Einstein memorial in Washington, D.C. It is well known that Einstein, the principal architect of relativity, was very fond of children.

Our everyday experiences and observations are associated with objects that move at speeds much less than that of light in a vacuum, $c = 3.00 \times 10^8$ m/s. Analysis models and definitions of quantities based on Newtonian mechanics and early concepts of space and time were formulated to describe the motion of such objects. This formalism is very successful in describing a wide range of phenomena that occur at low speeds, as we have seen in previous chapters. It fails, however, when applied to objects whose speeds approach that of light. Experimentally, the predictions of Newtonian theory can be tested by accelerating electrons or other particles to very high speeds. For example, it is possible to accelerate an electron to a speed of 0.99c. According to the Newtonian definition of kinetic energy, if the energy transferred to such an electron were increased by a factor of 4, the electron speed should double to 1.98c. Relativistic calculations, however, show that the speed of the electron—as well as the speeds of all other objects in the Universe—remains less than the speed of light. Because it places no upper limit on speed, Newtonian mechanics is contrary to modern theoretical predictions and experimental results, and the Newtonian models that we have developed are limited to objects moving much slower than the speed of light. Because Newtonian mechanics does not correctly predict the results of experiments carried out on objects moving at high speeds, we need a new formalism that is valid for these objects.

In 1905, at the age of only 26, Albert Einstein published his *special theory of relativity*, which is the subject of most of this chapter. Regarding the theory, Einstein wrote:

> The relativity theory arose from necessity, from serious and deep contradictions in the old theory from which there seemed no escape. The strength of the new theory lies in the consistency and simplicity with which it solves all these difficulties, using only a few very convincing assumptions.[1]

Although Einstein made many important contributions to science, special relativity alone represents one of the greatest intellectual achievements of the 20th century. With special relativity, experimental observations can be correctly predicted for objects over the range of all possible speeds, from rest to speeds approaching the speed of light. This chapter gives an introduction to special relativity, with emphasis on some of its consequences.

9.1 | The Principle of Galilean Relativity

We will begin by considering the notion of relativity at low speeds. This discussion was actually begun in Section 3.6 when we discussed relative velocity. At that time, we discussed the importance of the observer and the significance of his or her motion with respect to what is being observed. In a similar way here, we will generate equations that allow us to express one observer's measurements in terms of the other's. This process will lead to some rather unexpected and startling results about our understanding of space and time.

As we have mentioned previously, it is necessary to establish a frame of reference when describing a physical event. You should recall from Chapter 4 that an inertial frame is one in which an object is measured to have no acceleration if no forces act on it. Furthermore, any frame moving with constant velocity with respect to an inertial frame must also be an inertial frame. The laws predicting the results of an experiment performed in a vehicle moving with uniform velocity will be identical for the driver of the vehicle and a hitchhiker on the side of the road. The formal statement of this result is called the **principle of Galilean relativity:**

The laws of mechanics must be the same in all inertial frames of reference.

▶ Principle of Galilean relativity

The following observation illustrates the equivalence of the laws of mechanics in different inertial frames. Consider a pickup truck moving with a constant velocity as in Figure 9.1a. If a passenger in the truck throws a ball straight up in the air, the

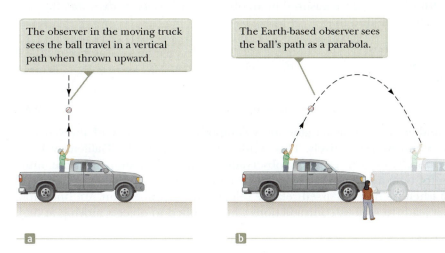

The observer in the moving truck sees the ball travel in a vertical path when thrown upward.

The Earth-based observer sees the ball's path as a parabola.

a

b

Figure 9.1 Two observers watch the path of a thrown ball and obtain different results.

---

[1]A. Einstein and L. Infeld, *The Evolution of Physics* (New York, Simon and Schuster, 1966), p. 192.

passenger observes that the ball moves in a vertical path (ignoring air resistance). The motion of the ball appears to be precisely the same as if the ball were thrown by a person at rest on the Earth and observed by that person. The kinematic equations of Chapter 2 describe the results correctly whether the truck is at rest or in uniform motion. Now consider the ball thrown in the truck as viewed by an observer at rest on the Earth. This observer sees the path of the ball as a parabola as in Figure 9.1b. Furthermore, according to this observer, the ball has a horizontal component of velocity equal to the speed of the truck. Although the two observers measure different velocities and see different paths of the ball, they see the same forces on the ball and agree on the validity of Newton's laws as well as classical principles such as conservation of energy and conservation of momentum. Their measurements differ, but the measurements they make satisfy the same laws. All differences between the two views stem from the relative motion of one frame with respect to the other.

Suppose some physical phenomenon, which we call an **event**, occurs. The event's location in space and time of occurrence can be specified by an observer with the coordinates (x, y, z, t). We would like to be able to transform these coordinates from one inertial frame to another moving with uniform relative velocity, which will allow us to express one observer's measurements in terms of the other's.

Consider two inertial frames S and S′ (Fig. 9.2). The frame S′ moves with a constant velocity $\vec{v}$ along the common x and x' axes, where $\vec{v}$ is measured relative to S. We assume that the origins of S and S′ coincide at $t = 0$. Therefore, at time t, the origin of frame S′ is to the right of the origin of S by a distance vt. An event occurs at point P and time t. An observer in S describes the event with space–time coordinates (x, y, z, t), and an observer in S′ describes the same event with coordinates (x', y', z', t'). As we can see from Figure 9.2, a simple geometric argument shows that the space coordinates are related by the equations

$$x' = x - vt \qquad y' = y \qquad z' = z \qquad \textbf{9.1}◀$$

Time is assumed to be the same in both inertial frames. That is, within the framework of classical mechanics, all clocks run at the same rate, regardless of their velocity, so that the time at which an event occurs for an observer in S is the same as the time for the same event in S′:

$$t' = t \qquad \textbf{9.2}◀$$

Equations 9.1 and 9.2 constitute what is known as the **Galilean transformation of coordinates.**

Now suppose a particle moves through a displacement dx in a time interval dt as measured by an observer in S. It follows from the first of Equations 9.1 that the corresponding displacement dx' measured by an observer in S′ is $dx' = dx - v\,dt$. Because $dt = dt'$ (Eq. 9.2), we find that

$$\frac{dx'}{dt'} = \frac{dx}{dt} - v$$

or

$$u_x' = u_x - v \qquad \textbf{9.3}◀$$

where u_x and u_x' are the instantaneous x components of velocity of the particle[2] relative to S and S′, respectively. This result, which is called the **Galilean velocity transformation,** is used in everyday observations and is consistent with our intuitive notion of time and space. It is the same equation we generated in Section 3.6 (Eq. 3.22) when we first discussed relative velocity in one dimension. We find, however, that it leads to serious contradictions when applied to objects moving at high speeds.

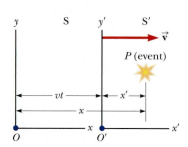

Figure 9.2 An event occurs at a point P and time t. The event is seen by two observers O and O' in inertial frames S and S′, where S′ moves with a velocity $\vec{v}$ relative to S.

Pitfall Prevention | 9.1
The Relationship Between the S and S′ Frames
Many of the mathematical representations in this chapter are true only for the specified relationship between the S and S′ frames. The x and x' axes coincide, except their origins are different. The y and y' axes (and the z and z' axes) are parallel, but they only coincide at one instant due to the time-varying displacement of the origin of S′ with respect to that of S. We choose the time $t = 0$ to be the instant at which the origins of the two coordinate systems coincide. If the S′ frame is moving in the positive x direction relative to S, then v is positive; otherwise, it is negative.

---

[2]We have used v for the speed of the S′ frame relative to the S frame. To avoid confusion, we will use u for the speed of an object or particle.

9.2 | The Michelson–Morley Experiment

Many experiments similar to throwing the ball in the pickup truck, described in the preceding section, show us that the laws of classical mechanics are the same in all inertial frames of reference. When similar inquiries are made into the laws of other branches of physics, however, the results are contradictory. In particular, the laws of electricity and magnetism are found to depend on the frame of reference used. It might be argued that these laws are wrong, but that is difficult to accept because the laws are in total agreement with known experimental results. The Michelson–Morley experiment was one of many attempts to investigate this dilemma.

The experiment stemmed from a misconception early physicists had concerning the manner in which light propagates. The properties of mechanical waves, such as water and sound waves, were well known, and all these waves require a *medium* to support the propagation of the disturbance, as we shall discuss in Chapter 13. For sound from your stereo system, the medium is the air, and for ocean waves, the medium is the water surface. In the 19th century, physicists subscribed to a model for light in which electromagnetic waves also require a medium through which to propagate. They proposed that such a medium exists, filling all space, and they named it the **luminiferous ether.** The ether would define an **absolute frame of reference** in which the speed of light is c.

The most famous experiment designed to show the presence of the ether was performed in 1887 by A. A. Michelson (1852–1931) and E. W. Morley (1838–1923). The objective was to determine the speed of the Earth through space with respect to the ether, and the experimental tool used was a device called the *interferometer,* shown schematically in Active Figure 9.3.

Light from the source at the left encounters a beam splitter M_0, which is a partially silvered mirror. Part of the light passes through toward mirror M_2, and the other part is reflected upward toward mirror M_1. Both mirrors are the same distance from the beam splitter. After reflecting from these mirrors, the light returns to the beam splitter, and part of each light beam propagates toward the observer at the bottom.

Suppose one arm of the interferometer (Arm 2, in Active Fig. 9.3) is aligned along the direction of the velocity $\vec{v}$ of the Earth through space and therefore through the ether. The "ether wind" blowing in the direction opposite the Earth's motion should cause the speed of light, as measured in the Earth's frame of reference, to be $c - v$ as the light approaches mirror M_2 in Active Figure 9.3 and $c + v$ after reflection.

The other arm (Arm 1) is perpendicular to the ether wind. For light to travel in this direction, the vector $\vec{c}$ must be aimed "upstream" so that the vector addition of $\vec{c}$ and $\vec{v}$ gives the speed of the light perpendicular to the ether wind as $\sqrt{c^2 - v^2}$. This situation is similar to Example 3.6, in which a boat crosses a river with a current. The boat is a model for the light beam in the Michelson–Morley experiment, and the river current is a model for the ether wind.

Because they travel in perpendicular directions with different speeds, light beams leaving the beam splitter simultaneously will arrive back at the beam splitter at different times. The interferometer is designed to detect this time difference. Measurements failed, however, to show any time difference! The Michelson–Morley experiment was repeated by other researchers under varying conditions and at different locations, but the results were always the same: *No time difference of the magnitude required was ever observed.*[3]

The negative result of the Michelson–Morley experiment not only contradicted the ether hypothesis, but it also meant that it was impossible to measure the absolute speed of the Earth with respect to the ether frame. From a theoretical

---

[3]From an Earth observer's point of view, changes in the Earth's speed and direction of motion in the course of a year are viewed as ether wind shifts. Even if the speed of the Earth with respect to the ether were zero at some time, six months later the Earth is moving in the opposite direction, the speed of the Earth with respect to the ether would be nonzero, and a clear time difference should be detected. None has ever been observed, however.

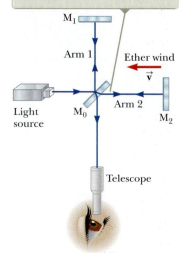

According to the ether wind theory, the speed of light should be $c - v$ as the beam approaches mirror M_2 and $c + v$ after reflection.

Active Figure 9.3 In the Michelson interferometer, the ether theory claims that the time interval for a light beam to travel from the beam splitter to mirror M_1 and back will be different from that for a light beam to travel from the beam splitter to mirror M_2 and back. The interferometer is sufficiently sensitive to detect this difference.

Albert A. Michelson
(1852–1931)
Michelson was born in Prussia in a town that later became part of Poland. He moved to the United States as a small child and spent much of his adult life making accurate measurements of the speed of light. In 1907, he was the first American to be awarded the Nobel Prize in Physics, which he received for his work in optics. His most famous experiment, conducted with Edward Morley in 1887, indicated that it was impossible to measure the absolute velocity of the Earth with respect to the ether.

viewpoint, it was impossible to find the absolute frame. As we shall see in the next section, however, Einstein offered a postulate that places a different interpretation on the negative result. In later years, when more was known about the nature of light, the idea of an ether that permeates all space was abandoned. Light is now understood to be an electromagnetic wave that requires no medium for its propagation. As a result, an ether through which light travels is an unnecessary construct.

Modern versions of the Michelson–Morley experiment have placed an upper limit of about 5 cm/s = 0.05 m/s on ether wind velocity. We can show that the speed of the Earth in its orbit around the Sun is 2.97×10^4 m/s, six orders of magnitude larger than the upper limit of ether wind velocity! These results have shown quite conclusively that the motion of the Earth has no effect on the measured speed of light.

9.3 | Einstein's Principle of Relativity

In the preceding section, we noted the failure of experiments to measure the relative speed between the ether and the Earth. Einstein proposed a theory that boldly removed these difficulties and at the same time completely altered our notion of space and time.[4] He based his relativity theory on two postulates:

1. **The principle of relativity:** All the laws of physics are the same in all inertial reference frames.
2. **The constancy of the speed of light:** The speed of light in vacuum has the same value in all inertial frames, regardless of the velocity of the observer or the velocity of the source emitting the light.

These postulates form the basis of **special relativity,** which is the relativity theory applied to observers moving with constant velocity. The first postulate asserts that *all* the laws of physics—those dealing with mechanics, electricity and magnetism, optics, thermodynamics, and so on—are the same in all reference frames moving with constant velocity relative to each other. This postulate is a sweeping generalization of the principle of Galilean relativity, which only refers to the laws of mechanics. From an experimental point of view, Einstein's principle of relativity means that any kind of experiment performed in a laboratory at rest must agree with the same laws of physics as when performed in a laboratory moving at constant velocity relative to the first one. Hence, no preferred inertial reference frame exists and it is impossible to detect absolute motion.

Note that postulate 2, the principle of the constancy of the speed of light, is required by postulate 1: If the speed of light were not the same in all inertial frames, it would be possible to experimentally distinguish between inertial frames and a preferred, absolute frame in which the speed of light is c, in contradiction to postulate 1. Postulate 2 also eliminates the problem of measuring the speed of the ether by denying the existence of the ether and boldly asserting that light always moves with speed c relative to all inertial observers.

9.4 | Consequences of Special Relativity

If we accept the postulates of special relativity, we must conclude that relative motion is unimportant when measuring the speed of light, which is the lesson of the Michelson–Morley experiment. At the same time, we must alter our commonsense notion of space and time and be prepared for some very unexpected consequences, as we shall see now.

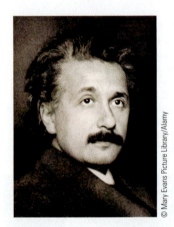

Albert Einstein
**German-American Physicist
(1879–1955)**
Einstein, one of the greatest physicists of all time, was born in Ulm, Germany. In 1905, at age 26, he published four scientific papers that revolutionized physics. Two of these papers were concerned with what is now considered his most important contribution: the special theory of relativity.

In 1916, Einstein published his work on the general theory of relativity. The most dramatic prediction of this theory is the degree to which light is deflected by a gravitational field. Measurements made by astronomers on bright stars in the vicinity of the eclipsed Sun in 1919 confirmed Einstein's prediction, and Einstein became a world celebrity as a result. Einstein was deeply disturbed by the development of quantum mechanics in the 1920s despite his own role as a scientific revolutionary. In particular, he could never accept the probabilistic view of events in nature that is a central feature of quantum theory. The last few decades of his life were devoted to an unsuccessful search for a unified theory that would combine gravitation and electromagnetism.

---

[4]A. Einstein, "On the Electrodynamics of Moving Bodies," *Ann. Physik* 17:891, 1905. For an English translation of this article and other publications by Einstein, see the book by H. Lorentz, A. Einstein, H. Minkowski, and H. Weyl, *The Principle of Relativity* (New York: Dover, 1958).

Simultaneity and the Relativity of Time

A basic premise of Newtonian mechanics is that a universal time scale exists that is the same for all observers. In fact, Newton wrote, "Absolute, true, and mathematical time, of itself, and from its own nature, flows equably without relation to anything external." Therefore, Newton and his followers simply took simultaneity for granted. In his development of special relativity, Einstein abandoned the notion that two events that appear simultaneous to one observer appear simultaneous to all observers. According to Einstein, a time measurement depends on the reference frame in which the measurement is made.

Einstein devised the following thought experiment to illustrate this point. A boxcar moves with uniform velocity and two lightning bolts strike its ends, as in Figure 9.4a, leaving marks on the boxcar and on the ground. The marks on the boxcar are labeled A' and B', and those on the ground are labeled A and B. An observer at O' moving with the boxcar is midway between A' and B', and a ground observer at O is midway between A and B. The events recorded by the observers are the arrivals of light signals from the lightning bolts.

The two light signals reach observer O at the same time as indicated in Figure 9.4b. As a result, O concludes that the events at A and B occurred simultaneously. Now consider the same events as viewed by the observer on the boxcar at O'. From our frame of reference, at rest with respect to the tracks in Figure 9.4, we see the lightning strikes occur as A' passes A, O' passes O, and B' passes B. By the time the light has reached observer O, observer O' has moved as indicated in Figure 9.4b. Therefore, the light signal from B' has already swept past O' because it had less distance to travel, but the light from A' has not yet reached O'. According to Einstein, observer O' must find that light travels at the same speed as that measured by observer O. Observer O' therefore concludes that the lightning struck the front of the boxcar before it struck the back. This thought experiment clearly demonstrates that the two events, which appear to be simultaneous to observer O, do not appear to be simultaneous to observer O'. In general, two events separated in space and observed to be simultaneous in one reference frame are not observed to be simultaneous in a second frame moving relative to the first. That is, simultaneity is not an absolute concept but one that depends on the state of motion of the observer.

Einstein's thought experiment demonstrates that two observers can disagree on the simultaneity of two events. This disagreement, however, depends on the transit time of light to the observers and therefore does *not* demonstrate the deeper meaning of relativity. In relativistic analyses of high-speed situations, relativity shows that simultaneity is relative *even when the transit time is subtracted out*. In fact, all the relativistic effects that we will discuss from here on will assume that we are ignoring differences caused by the transit time of light to the observers.

The events appear to be simultaneous to the stationary observer O who is standing midway between A and B.

The events do not appear to be simultaneous to observer O', who claims that the front of the car is struck before the rear.

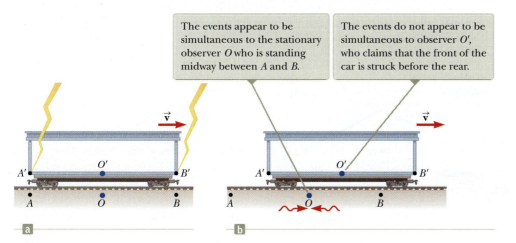

a

b

Figure 9.4 (a) Two lightning bolts strike the ends of a moving boxcar. (b) Note that the leftward-traveling light signal from B' has already passed observer O', but the rightward-traveling light signal from A' has not yet reached O'.

Time Dilation

According to the preceding paragraph, observers in different inertial frames measure different time intervals between a pair of events, independent of the transit time of the light. This situation can be illustrated by considering a vehicle moving to the right with a speed v as in the pictorial representation in Active Figure 9.5a. A mirror is fixed to the ceiling of the vehicle, and observer O', at rest in a frame attached to the vehicle, holds a flashlight a distance d below the mirror. At some instant, the flashlight is turned on momentarily and emits a pulse of light (event 1) directed toward the mirror. At some later time after reflecting from the mirror, the pulse arrives back at the flashlight (event 2). Observer O' carries a clock that she uses to measure the time interval Δt_p between these two events. (The subscript p stands for "proper," as will be discussed shortly.) Because the light pulse has a constant speed c, the time interval required for the pulse to travel from O' to the mirror and back to O' (a distance of $2d$) can be found by modeling the light pulse as a particle under constant speed as discussed in Chapter 2:

$$\Delta t_p = \frac{2d}{c}$$

9.4◀

This time interval Δt_p is measured by O', for whom the two events occur at the same spatial position.

Now consider the same pair of events as viewed by observer O at rest with respect to a second frame attached to the ground as in Active Figure 9.5b. According to this observer, the mirror and flashlight are moving to the right with a speed v. The geometry appears to be entirely different as viewed by this observer. By the time the light from the flashlight reaches the mirror, the mirror has moved horizontally a distance $v\Delta t/2$, where Δt is the time interval required for the light to travel from the flashlight to the mirror and back to the flashlight as measured by observer O. In other words, the second observer concludes that because of the motion of the vehicle, if the light is to hit the mirror, it must leave the flashlight at an angle with respect to the vertical direction. Comparing Active Figures 9.5a and 9.5b, we see that the light must travel farther to arrive back at the mirror when observed in the second frame than in the first frame.

According to the second postulate of special relativity, both observers must measure c for the speed of light. Because the light travels farther in the second frame but at the same speed, it follows that the time interval Δt measured by the observer in the second frame is longer than the time interval Δt_p measured by the observer in

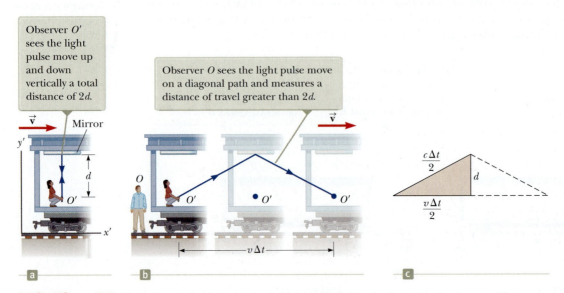

Observer O' sees the light pulse move up and down vertically a total distance of $2d$.

Observer O sees the light pulse move on a diagonal path and measures a distance of travel greater than $2d$.

Active Figure 9.5 (a) A mirror is fixed to a moving vehicle, and a light pulse is sent out by observer O' at rest in the vehicle. (b) Relative to a stationary observer O standing alongside the vehicle, the mirror and O' move with a speed v. (c) The right triangle for calculating the relationship between Δt and Δt_p.

the first frame. To obtain a relationship between these two time intervals, it is convenient to use the right triangle geometric model shown in Active Figure 9.5c. The Pythagorean theorem applied to the triangle gives

$$\left(\frac{c\,\Delta t}{2}\right)^2 = \left(\frac{v\,\Delta t}{2}\right)^2 + d^2$$

Solving for Δt gives

$$\Delta t = \frac{2d}{\sqrt{c^2 - v^2}} = \frac{2d}{c\sqrt{1 - \dfrac{v^2}{c^2}}} \qquad \text{9.5} \blacktriangleleft$$

Because $\Delta t_p = 2d/c$, we can express Equation 9.5 as

$$\Delta t = \frac{\Delta t_p}{\sqrt{1 - \dfrac{v^2}{c^2}}} = \gamma\,\Delta t_p \qquad \text{9.6} \blacktriangleleft$$

where $\gamma = (1 - v^2/c^2)^{-1/2}$. This result says that the time interval Δt measured by O is longer than the time interval Δt_p measured by O' because γ is always greater than one. That is, $\Delta t > \Delta t_p$. This effect is known as **time dilation.**

We can see that time dilation is not observed in our everyday lives by considering the factor γ. This factor deviates significantly from a value of 1 only for very high speeds, as shown in Table 9.1. For example, for a speed of $0.1c$, the value of γ is 1.005. Therefore, a time dilation of only 0.5% occurs at one-tenth the speed of light. Speeds we encounter on an everyday basis are far slower than that, so we do not see time dilation in normal situations.

The time interval Δt_p in Equation 9.6 is called the **proper time interval.** In general, the proper time interval is defined as **the time interval between two events as measured by an observer for whom the events occur at the same point in space.** In our case, observer O' measures the proper time interval. For us to be able to use Equation 9.6, the events must occur at the same spatial position in *some* inertial frame. Therefore, for instance, this equation cannot be used to relate the measurements made by the two observers in the lightning example described at the beginning of this section because the lightning strikes occur at different positions for both observers.

If a clock is moving with respect to you, the time interval between ticks of the moving clock is measured to be longer than the time interval between ticks of an identical clock in your reference frame. Therefore, it is often said that a moving clock is measured to run more slowly than a clock in your reference frame by a factor γ. That is true for mechanical clocks as well as for the light clock just described. We can generalize this result by stating that all physical processes, including chemical and biological ones, slow down relative to a stationary clock when those processes occur in a frame moving with respect to the clock. For example, the heartbeat of an astronaut moving through space would keep time with a clock inside the spaceship. Both the astronaut's clock and heartbeat would be measured to be slowed down according to an observer on the Earth comparing time intervals with his own clock at rest with respect to him (although the astronaut would have no sensation of life slowing down in the spaceship).

Time dilation is a verifiable phenomenon; let us look at one situation in which the effects of time dilation can be observed and that served as an important historical confirmation of the predictions of relativity. Muons are unstable elementary particles that have a charge equal to that of the electron and a mass 207 times that of the electron. They decay into electrons and neutrinos, which we will study in Chapters 30 and 31. Muons can be produced as a result of collisions of cosmic radiation with atoms high in the atmosphere. Slow-moving muons in the laboratory have a lifetime measured to be the proper time interval $\Delta t_p = 2.2\ \mu s$. If we assume that the speed of atmospheric muons is close to the speed of

TABLE 9.1 |

Approximate Values for γ at Various Speeds

v/c	γ
0	1
0.001 0	1.000 000 5
0.010	1.000 05
0.10	1.005
0.20	1.021
0.30	1.048
0.40	1.091
0.50	1.155
0.60	1.250
0.70	1.400
0.80	1.667
0.90	2.294
0.92	2.552
0.94	2.931
0.96	3.571
0.98	5.025
0.99	7.089
0.995	10.01
0.999	22.37

Figure 9.6 Travel of muons according to an Earth-based observer.

Without relativistic considerations, according to an observer on the Earth, muons created in the atmosphere and traveling downward with a speed close to c travel only about 6.6×10^2 m before decaying with an average lifetime of 2.2 μs. Therefore, very few muons would reach the surface of the Earth.

With relativistic considerations, the muon's lifetime is dilated according to an observer on the Earth. Hence, according to this observer, the muon can travel about 4.8×10^3 m before decaying. The result is many of them arriving at the surface.

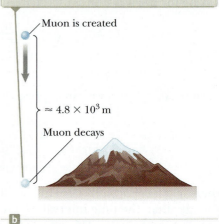

Muon is created

≈ 6.6×10^2 m

Muon decays

Muon is created

≈ 4.8×10^3 m

Muon decays

a   b

light, we find that these particles can travel a distance during their lifetime of approximately $(3.0 \times 10^8 \text{ m/s})(2.2 \times 10^{-6} \text{ s}) \approx 6.6 \times 10^2$ m before they decay (Fig. 9.6a). Hence, they are unlikely to reach the surface of the Earth from high in the atmosphere where they are produced; nonetheless, experiments show that a large number of muons *do* reach the surface. The phenomenon of time dilation explains this effect. As measured by an observer on the Earth, the muons have a dilated lifetime equal to $\gamma \Delta t_p$. For example, for $v = 0.99c$, $\gamma \approx 7.1$ and $\gamma \Delta t_p \approx 16$ μs. Hence, the average distance traveled by the muons in this time interval as measured by an observer on the Earth is approximately $(3.0 \times 10^8 \text{ m/s})$ $(16 \times 10^{-6} \text{ s}) \approx 4.8 \times 10^3$ m, as shown in Figure 9.6b.

The results of an experiment reported by J. C. Hafele and R. E. Keating provided direct evidence of time dilation.[5] The experiment involved the use of very stable atomic clocks. Time intervals measured with four such clocks in jet flight were compared with time intervals measured by reference clocks located at the U.S. Naval Observatory. Their results were in good agreement with the predictions of special relativity and can be explained in terms of the relative motion between the Earth's rotation and the jet aircraft. In their paper, Hafele and Keating report the following: "Relative to the atomic time scale of the U.S. Naval Observatory, the flying clocks lost 59 ± 10 ns during the eastward trip and gained 273 ± 7 ns during the westward trip."

In a more recent experiment, Chou, Hume, Rosenband, and Wineland[6] demonstrated time dilation with speeds as low as 10 m/s. Their experimental design included laser cooling of trapped ions, which we will discuss in Chapter 24.

QUICK QUIZ 9.1 Suppose the observer O' on the train in Active Figure 9.5 aims her flashlight at the far wall of the boxcar and turns it on and off, sending a pulse of light toward the far wall. Both O' and O measure the time interval between when the pulse leaves the flashlight and when it hits the far wall. Which observer measures the proper time interval between these two events? **(a)** O' **(b)** O **(c)** both observers **(d)** neither observer

[5] J. C. Hafele and R. E. Keating, "Around the World Atomic Clocks: Relativistic Time Gains Observed," *Science*, July 14, 1972, p. 168.

[6] C. Chou, D. Hume, T. Rosenband, and D. Wineland, "Optical Clocks and Relativity," *Science*, September 24, 2010, p. 1630

A crew on a spacecraft watches a movie that is two hours long. The spacecraft is moving at high speed through space. Does an Earth-based observer watching the movie screen on the spacecraft through a powerful telescope measure the duration of the movie to be (**a**) longer than, (**b**) shorter than, or (**c**) equal to two hours?

The Twin Paradox

An intriguing consequence of time dilation is the so-called twin paradox (Fig. 9.7). Consider an experiment involving a set of twins named Speedo and Goslo. At age 20, Speedo, the more adventuresome of the two, sets out on an epic journey to Planet X, located 20 light-years (ly) from the Earth. (Note that 1 ly is the distance light travels through free space in 1 year. It is equal to 9.46×10^{15} m.) Furthermore, his spaceship is capable of reaching a speed of $0.95c$ relative to the inertial frame of his twin brother back home. After reaching Planet X, Speedo becomes homesick and immediately returns to the Earth at the same speed $0.95c$. Upon his return, Speedo is shocked to discover that Goslo has aged 42 yr and is now 62 yr old. Speedo, on the other hand, has aged only 13 yr.

At this point, it is fair to raise the following question: Which twin is the traveler and which is really younger as a result of this experiment? From Goslo's frame of reference, he was at rest while his brother traveled at a high speed away from him and then came back. According to Speedo, however, he himself remained stationary while Goslo and the Earth raced away from him and then headed back. There is an apparent contradiction due to the apparent symmetry of the observations. Which twin has developed signs of excess aging?

The situation in this problem is actually not symmetrical. To resolve this apparent paradox, recall that the special theory of relativity describes observations made in inertial frames of reference moving relative to each other. Speedo, the space traveler, must experience a series of accelerations during his journey because he must fire his rocket engines to slow down and start moving back toward the Earth. As a result, his speed is not always uniform, and consequently he is not always in a single inertial frame. Therefore, there is no paradox because only Goslo, who is always in a single inertial frame, can make correct predictions based on special relativity. During each passing year noted by Goslo, slightly less than 4 months elapses for Speedo.

Only Goslo, who is in a single inertial frame, can apply the simple time dilation equation to Speedo's trip. Therefore, Goslo finds that instead of aging 42 yr, Speedo ages only $(1 - v^2/c^2)^{1/2}(42 \text{ yr}) = 13$ yr. According to both twins, Speedo spends 6.5 yr traveling to Planet X and 6.5 yr returning, for a total travel time of 13 yr.

As Speedo (on the left) leaves his brother on Earth, both twins are the same age.

When Speedo returns from his journey, Goslo (on the right) is much older than Speedo.

a

b

Figure 9.7 The twin paradox. Speedo takes a journey to a star 20 light-years away and returns to the Earth.

QUICK QUIZ 9.3 Suppose astronauts are paid according to the amount of time they spend traveling in space. After a long voyage traveling at a speed approaching c, would a crew rather be paid according to (a) an Earth-based clock, (b) their spacecraft's clock, or (c) either clock?

THINKING PHYSICS 9.1

Suppose a student explains time dilation with the following argument: If I start running away from a clock at 12:00 at a speed very close to the speed of light, I would not see the time change, because the light from the clock representing 12:01 would never reach me. What is the flaw in this argument?

Reasoning The implication in this argument is that the velocity of light relative to the runner is approximately *zero* because "the light . . . would never reach me." In this Galilean point of view, the relative velocity is a simple subtraction of running velocity from the light velocity. From the point of view of special relativity, one of the fundamental postulates is that the speed of light is the same for all observers, *including one running away from the light source at the speed of light*. Therefore, the light from 12:01 will move toward the runner at the speed of light, as measured by all observers, including the runner. ◄

Example 9.1 | What Is the Period of the Pendulum?

The period of a pendulum is measured to be 3.00 s in the reference frame of the pendulum. What is the period when measured by an observer moving at a speed of $0.960c$ relative to the pendulum?

SOLUTION

Conceptualize Let's change frames of reference. Instead of the observer moving at $0.960c$, we can take the equivalent point of view that the observer is at rest and the pendulum is moving at $0.960c$ past the stationary observer. Hence, the pendulum is an example of a clock moving at high speed with respect to an observer.

Categorize Based on the Conceptualize step, we can categorize this problem as one involving time dilation.

Analyze The proper time interval, measured in the rest frame of the pendulum, is $\Delta t_p = 3.00$ s.

Use Equation 9.6 to find the dilated time interval:

$$\Delta t = \gamma \Delta t_p = \frac{1}{\sqrt{1 - \dfrac{(0.960c)^2}{c^2}}} \Delta t_p = \frac{1}{\sqrt{1 - 0.921\,6}} \Delta t_p$$

$$= 3.57(3.00 \text{ s}) = \boxed{10.7 \text{ s}}$$

Finalize This result shows that a moving pendulum is indeed measured to take longer to complete a period than a pendulum at rest does. The period increases by a factor of $\gamma = 3.57$.

What If? What if the speed of the observer increases by 4.00%? Does the dilated time interval increase by 4.00%?

Answer Based on the highly nonlinear behavior of γ as a function of v exhibited in Table 9.1, we would guess that the increase in Δt would be different from 4.00%.

Find the new speed if it increases by 4.00%:

$$v_{\text{new}} = (1.040\,0)(0.960c) = 0.998\,4c$$

Perform the time dilation calculation again:

$$\Delta t = \gamma \Delta t_p = \frac{1}{\sqrt{1 - \dfrac{(0.998\,4c)^2}{c^2}}} \Delta t_p = \frac{1}{\sqrt{1 - 0.996\,8}} \Delta t_p$$

$$= 17.68(3.00 \text{ s}) = 53.1 \text{ s}$$

Therefore, the 4.00% increase in speed results in almost a 400% increase in the dilated time!

Length Contraction

The measured distance between two points also depends on the frame of reference. The **proper length** of an object is defined as **the distance in space between the end points of the object measured by someone who is at rest relative to the object.** An observer in a reference frame that is moving with respect to the object will measure a length along the direction of the velocity that is always less than the proper length. This effect is known as **length contraction.** Although we have introduced this effect through the mental representation of an object, the object is not necessary. The distance between *any* two points in space is measured by an observer to be contracted along the direction of the velocity of the observer relative to the points.

Consider a spacecraft traveling with a speed v from one star to another. We will consider the time interval between two events: (1) the leaving of the spacecraft from the first star and (2) the arrival of the spacecraft at the second star. There are two observers: one on the Earth and the other in the spacecraft. The observer at rest on the Earth (and also at rest with respect to the two stars) measures the distance between the stars to be L_p, the proper length. Using the particle under constant velocity model, according to this observer, the time interval required for the spacecraft to complete the voyage is $\Delta t = L_p/v$. What does an observer in the moving spacecraft measure for the distance between the stars? This observer measures the proper time interval because the passage of each of the two stars by his spacecraft occurs at the same position in his reference frame, at his spacecraft. Therefore, because of time dilation, the time interval required to travel between the stars as measured by the space traveler will be smaller than that for the Earth-bound observer, who is in motion with respect to the space traveler. Using the time dilation expression, the proper time interval between events is $\Delta t_p = \Delta t/\gamma$. The space traveler claims to be at rest and sees the destination star moving toward the spacecraft with speed v. Because the space traveler reaches the star in the time interval $\Delta t_p < \Delta t$, he concludes that the distance L between the stars is shorter than L_p. This distance measured by the space traveler is

$$L = v\,\Delta t_p = v\frac{\Delta t}{\gamma}$$

Because $L_p = v\Delta t$, we see that

$$L = \frac{L_p}{\gamma} = L_p\sqrt{1 - \frac{v^2}{c^2}} \qquad \textbf{9.7} \blacktriangleleft$$

Because $(1 - v^2/c^2)^{1/2}$ is less than 1, the space traveler measures a length that is shorter than the proper length. Therefore, an observer in motion with respect to two points in space measures the length L between the points (along the direction of motion) to be shorter than the length L_p measured by an observer at rest with respect to the points (the proper length).

Note that length contraction takes place only along the direction of motion. For example, suppose a meterstick moves past an Earth observer with speed v as in Active Figure 9.8. The length of the meterstick as measured by an observer in a frame attached to the stick is the proper length L_p as in Active Figure 9.8a. The length L of the stick measured by the Earth observer is shorter than L_p by the factor $(1 - v^2/c^2)^{1/2}$, but the width is the same. Furthermore, length contraction is a symmetric effect. If the stick is at rest on the Earth, an observer in the moving frame would also measure its length to be shorter by the same factor $(1 - v^2/c^2)^{1/2}$.

It is important to emphasize that the proper length and proper time interval are defined differently. The proper length is measured by an observer at rest with respect to the end points of the length. The proper time interval between two events is measured by someone for whom the events occur at the same position. Often, the proper time interval and the proper length are not measured by the same observer. As an example, let us return to the decaying muons moving at speeds close to the speed of light. An observer in the muon's reference frame would measure the proper lifetime, and an Earth-based observer would measure the proper length

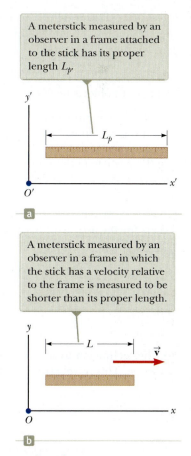

A meterstick measured by an observer in a frame attached to the stick has its proper length L_p.

a

A meterstick measured by an observer in a frame in which the stick has a velocity relative to the frame is measured to be shorter than its proper length.

b

Active Figure 9.8 The length of a meterstick is measured by two observers.

(the distance from creation to decay in Fig. 9.6). In the muon's reference frame, no time dilation occurs, but the distance of travel is observed to be shorter when measured in this frame. Likewise, in the Earth observer's reference frame, a time dilation does occur, but the distance of travel is measured to be the proper length. Therefore, when calculations on the muon are performed in both frames, the outcome of the experiment in one frame is the same as the outcome in the other frame: More muons reach the surface than would be predicted without relativistic calculations.

> **QUICK QUIZ 9.4** You are packing for a trip to another star. During the journey, you will be traveling at $0.99c$. You are trying to decide whether you should buy smaller sizes of your clothing because you will be thinner on your trip due to length contraction. You also plan to save money by reserving a smaller cabin to sleep in because you will be shorter when you lie down. Should you (a) buy smaller sizes of clothing, (b) reserve a smaller cabin, (c) do neither of these things, or (d) do both of these things?

Example 9.2 | A Voyage to Sirius

An astronaut takes a trip to Sirius, which is located a distance of 8 light-years from the Earth. The astronaut measures the time of the one-way journey to be 6 years. If the spaceship moves at a constant speed of $0.8c$, how can the 8-ly distance be reconciled with the 6-year trip time measured by the astronaut?

SOLUTION

Conceptualize An observer on the Earth measures light to require 8 years to travel between Sirius and the Earth. The astronaut measures a time interval for his travel of only 6 years. Is the astronaut traveling faster than light?

Categorize Because the astronaut is measuring a length of space between the Earth and Sirius that is in motion with respect to her, we categorize this example as a length contraction problem. We also model the astronaut as a particle moving with constant velocity.

Analyze The distance of 8 ly represents the proper length from the Earth to Sirius measured by an observer on the Earth seeing both objects nearly at rest.

Calculate the contracted length measured by the astronaut using Equation 9.7:

$$L = \frac{8 \text{ ly}}{\gamma} = (8 \text{ ly})\sqrt{1 - \frac{v^2}{c^2}} = (8 \text{ ly})\sqrt{1 - \frac{(0.8c)^2}{c^2}} = 5 \text{ ly}$$

Use the particle under constant velocity model to find the travel time measured on the astronaut's clock:

$$\Delta t = \frac{L}{v} = \frac{5 \text{ ly}}{0.8c} = \frac{5 \text{ ly}}{0.8(1 \text{ ly/yr})} = 6 \text{ yr}$$

Finalize Notice that we have used the value for the speed of light as $c = 1$ ly/yr. The trip takes a time interval shorter than 8 years for the astronaut because, to her, the distance between the Earth and Sirius is measured to be shorter.

What If? What if this trip is observed with a very powerful telescope by a technician in Mission Control on the Earth? At what time will this technician *see* that the astronaut has arrived at Sirius?

Answer The time interval the technician measures for the astronaut to arrive is

$$\Delta t = \frac{L_p}{v} = \frac{8 \text{ ly}}{0.8c} = 10 \text{ yr}$$

For the technician to *see* the arrival, the light from the scene of the arrival must travel back to the Earth and enter the telescope. This travel requires a time interval of

$$\Delta t = \frac{L_p}{v} = \frac{8 \text{ ly}}{c} = 8 \text{ yr}$$

Therefore, the technician sees the arrival after 10 yr + 8 yr = 18 yr. If the astronaut immediately turns around and comes back home, she arrives, according to the technician, 20 years after leaving, only 2 years *after the technician saw her arrive!* In addition, the astronaut would have aged by only 12 years.

Example 9.3 | Speedy Plunge

An observer on Earth sees a spaceship at an altitude of 4 350 km moving downward toward the Earth with a speed of 0.970c.

(A) What is the distance from the spaceship to the Earth as measured by the spaceship's captain?

SOLUTION

Conceptualize Imagine you are the captain, at rest in a reference frame attached to the spaceship: the Earth is rushing toward you at 0.970c; hence, the distance between the spaceship and the Earth is contracted.

Cagegorize We have an observer (the captain) and a moving length in space (the Earth–spaceship distance), so we categorize this example as a length contraction problem. The proper length is 4 350 km as measured by the Earth observer.

Analyze Use Equation 9.7 to find the contracted length, which represents the altitude of the spaceship above the surface of the Earth as measured by the captain:

$$L = L_p\sqrt{1 - v^2/c^2} = (4\ 350\ \text{km})\sqrt{1 - (0.970c)^2/c^2}$$

$$= 1.06 \times 10^3\ \text{km}$$

(B) After firing his engines for a time interval to slow down, the captain measures the ship's altitude as 267 km, whereas the observer on Earth measures it to be 625 km. What is the speed of the spaceship at this instant?

SOLUTION

Analyze Write the length-contraction equation (Eq.9.7):

$$L = L_p\sqrt{1 - v^2/c^2}$$

Square both sides of this equation and solve for v:

$$L^2 = L_p^2(1 - v^2/c^2) \quad \rightarrow \quad 1 - v^2/c^2 = \left(\frac{L}{L_p}\right)^2$$

$$v = c\sqrt{1 - (L/L_p)^2} = c\sqrt{1 - (267\ \text{km}/625\ \text{km})^2}$$

$$v = 0.904c$$

Finalize The answers are consistent with our expectations. The length in part (A) is shorter than the proper length, as we would expect from the length contraction phenomenon. In part (B), the calculated speed is indeed lower than the original speed, consistent with the fact that the captain fired the rocket engines to slow down.

9.5 | The Lorentz Transformation Equations

Suppose an event that occurs at some location and time is reported by two observers: one at rest in a frame S and another in a frame S′ that is moving to the right with speed v as in Figure 9.9. The observer in S reports the event with space–time coordinates (x, y, z, t), and the observer in S′ reports the same event using the coordinates (x', y', z', t'). If two events occur at P and Q in Figure 9.9, Equation 9.1 predicts that $\Delta x = \Delta x'$; that is, the distance between the two points in space at which the events occur does not depend on the motion of the observer. Because this notion is contradictory to that of length contraction, the Galilean transformation is not valid when v approaches the speed of light. In this section, we state the correct transformation equations that apply for all speeds in the range $0 \leq v < c$.

The equations that relate these measurements and enable us to transform coordinates from S to S′ are the **Lorentz transformation equations:**

$$x' = \gamma(x - vt) \qquad y' = y \qquad z' = z \qquad t' = \gamma\left(t - \frac{v}{c^2}x\right) \qquad \text{9.8} \blacktriangleleft$$

These transformation equations were developed by Hendrik A. Lorentz (1853–1928) in 1890 in connection with electromagnetism. Einstein, however, recognized their physical significance and took the bold step of interpreting them within the framework of special relativity.

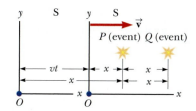

Figure 9.9 Events occur at points P and Q and are observed by an observer at rest in the S frame and another in the S′ frame, which is moving to the right with a speed v.

▶ Lorentz transformation for S → S′

We see that the value for t' assigned to an event by observer O' depends both on the time t and on the coordinate x as measured by observer O. Therefore, in relativity, space and time are not separate concepts but rather are closely interwoven with each other in what we call **space–time.** This case is unlike that of the Galilean transformation in which $t = t'$.

If we wish to transform coordinates in the S′ frame to coordinates in the S frame, we simply replace v by $-v$ and interchange the primed and unprimed coordinates in Equation 9.8:

► Inverse Lorentz transformation for S′ → S

$$x = \gamma(x' + vt') \qquad y = y' \qquad z = z' \qquad t = \gamma\left(t' + \frac{v}{c^2}x'\right) \qquad \textbf{9.9}◄$$

When $v \ll c$, the Lorentz transformation reduces to the Galilean transformation. To check, note that if $v \ll c$, $v^2/c^2 \ll 1$, so γ approaches 1 and Equation 9.8 reduces in this limit to Equations 9.1 and 9.2:

$$x' = x - vt \qquad y' = y \qquad z' = z \qquad t' = t$$

Lorentz Velocity Transformation

Let us now derive the **Lorentz velocity transformation,** which is the relativistic counterpart of the Galilean velocity transformation, Equation 9.3. Once again S′ is a frame of reference that moves at a speed v relative to another frame S along the common x and x' axes. Suppose an object is measured in S′ to have an instantaneous velocity component u'_x given by

$$u'_x = \frac{dx'}{dt'} \qquad \textbf{9.10}◄$$

Using Equations 9.8, we have

$$dx' = \gamma(dx - v\,dt) \qquad \text{and} \qquad dt' = \gamma\left(dt - \frac{v}{c^2}dx\right)$$

Substituting these values into Equation 9.10 gives

$$u'_x = \frac{dx'}{dt'} = \frac{dx - v\,dt}{dt - \frac{v}{c^2}dx} = \frac{\frac{dx}{dt} - v}{1 - \frac{v}{c^2}\frac{dx}{dt}}$$

Note, though, that dx/dt is the velocity component u_x of the object measured in S, so this expression becomes

► Lorentz velocity transformation for S → S′

$$u'_x = \frac{u_x - v}{1 - \frac{u_x v}{c^2}} \qquad \textbf{9.11}◄$$

Similarly, if the object has velocity components along y and z, the components in S′ are

$$u'_y = \frac{u_y}{\gamma\left(1 - \frac{u_x v}{c^2}\right)} \qquad \text{and} \qquad u'_z = \frac{u_z}{\gamma\left(1 - \frac{u_x v}{c^2}\right)} \qquad \textbf{9.12}◄$$

When u_x or v is much smaller than c (the nonrelativistic case), the denominator of Equation 9.11 approaches unity and so $u'_x \approx u_x - v$. This result corresponds to the Galilean velocity transformations. In the other extreme, when $u_x = c$, Equation 9.11 becomes

$$u'_x = \frac{c - v}{1 - \frac{cv}{c^2}} = \frac{c\left(1 - \frac{v}{c}\right)}{1 - \frac{v}{c}} = c$$

Pitfall Prevention | 9.3
What Can the Observers Agree On?
We have seen several measurements that the two observers O and O' do *not* agree on: (1) the time interval between events that take place in the same position in one of their frames, (2) the distance between two points that remain fixed in one of their frames, (3) the velocity components of a moving particle, and (4) whether two events occurring at different locations in both frames are simultaneous or not. The two observers *can agree* on: (1) their relative speed of motion v with respect to each other, (2) the speed c of any ray of light, and (3) the simultaneity of two events taking place at the same position *and* time in some frame.

From this result, we see that an object whose speed approaches c relative to an observer in S also has a speed approaching c relative to an observer in S′, independent of the relative motion of S and S′. Note that this conclusion is consistent with Einstein's second postulate, namely, that the speed of light must be c relative to all inertial frames of reference.

To obtain u_x in terms of u_x', we replace v by $-v$ in Equation 9.11 and interchange the roles of primed and unprimed variables:

$$u_x = \frac{u_x' + v}{1 + \dfrac{u_x' v}{c^2}}$$

9.13 ◄   ▶ Inverse Lorentz velocity transformation for S′ → S

QUICK QUIZ 9.5 You are driving on a freeway at a relativistic speed. Straight ahead of you, a technician standing on the ground turns on a searchlight and a beam of light moves exactly vertically upward, as seen by the technician. As you observe the beam of light, you measure the magnitude of the vertical component of its velocity as (a) equal to c, (b) greater than c, or (c) less than c. If the technician aims the searchlight directly at you instead of upward, you measure the magnitude of the horizontal component of its velocity as (d) equal to c, (e) greater than c, or (f) less than c.

Example 9.4 | Relative Velocity of Two Spacecraft

Two spacecraft A and B are moving in opposite directions as shown in Figure 9.10. An observer on the Earth measures the speed of spacecraft A to be $0.750c$ and the speed of spacecraft B to be $0.850c$. Find the velocity of spacecraft B as observed by the crew on spacecraft A.

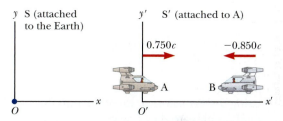

Figure 9.10 (Example 9.4) Two spacecraft A and B move in opposite directions. The speed of spacecraft B relative to spacecraft A is *less* than c and is obtained from the relativistic velocity transformation equation.

SOLUTION

Conceptualize There are two observers, one (O) on the Earth and one (O') on spacecraft A. The event is the motion of spacecraft B.

Categorize Because the problem asks to find an observed velocity, we categorize this example as one requiring the Lorentz velocity transformation.

Analyze The Earth-based observer at rest in the S frame makes two measurements, one of each spacecraft. We want to find the velocity of spacecraft B as measured by the crew on spacecraft A. Therefore, $u_x = -0.850c$. The velocity of spacecraft A is also the velocity of the observer at rest in spacecraft A (the S′ frame) relative to the observer at rest on the Earth. Therefore, $v = 0.750c$.

Obtain the velocity u_x' of spacecraft B relative to spacecraft A using Equation 9.11:

$$u_x' = \frac{u_x - v}{1 - \dfrac{u_x v}{c^2}} = \frac{-0.850c - 0.750c}{1 - \dfrac{(-0.850c)(0.750c)}{c^2}} = \boxed{-0.977c}$$

Finalize The negative sign indicates that spacecraft B is moving in the negative x direction as observed by the crew on spacecraft A. Is that consistent with your expectation from Figure 9.10? Notice that the speed is less than c. That is, an object whose speed is less than c in one frame of reference must have a speed less than c in any other frame. (Had you used the Galilean velocity transformation equation in this example, you would have found that $u_x' = u_x - v = -0.850c - 0.750c = -1.60c$, which is impossible. The Galilean transformation equation does not work in relativistic situations.)

What If? What if the two spacecraft pass each other? What is their relative speed now?

Answer The calculation using Equation 9.11 involves only the velocities of the two spacecraft and does not depend on their locations. After they pass each other, they have the same velocities, so the velocity of spacecraft B as observed by the crew on spacecraft A is the same, $-0.977c$. The only difference after they pass is that spacecraft B is receding from spacecraft A, whereas it was approaching spacecraft A before it passed.

Example 9.5 | Relativistic Leaders of the Pack

Two motorcycle pack leaders named David and Emily are racing at relativistic speeds along perpendicular paths as shown in Figure 9.11. How fast does Emily recede as seen by David over his right shoulder?

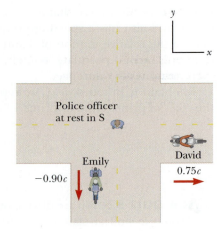

SOLUTION

Conceptualize The two observers are David and the police officer in Figure 9.11. The event is the motion of Emily. Figure 9.11 represents the situation as seen by the police officer at rest in frame S. Frame S′ moves along with David.

Figure 9.11 (Example 9.5) David moves east with a speed 0.75c relative to the police officer, and Emily travels south at a speed 0.90c relative to the officer.

Categorize Because the problem asks to find an observed velocity, we categorize this problem as one requiring the Lorentz velocity transformation. The motion takes place in two dimensions.

Analyze Identify the velocity components for David and Emily according to the police officer:

David: $v_x = v = 0.75c \quad v_y = 0$

Emily: $u_x = 0 \quad u_y = -0.90c$

Using Equations 9.11 and 9.12, calculate u'_x and u'_y for Emily as measured by David:

$$u'_x = \frac{u_x - v}{1 - \dfrac{u_x v}{c^2}} = \frac{0 - 0.75c}{1 - \dfrac{(0)(0.75c)}{c^2}} = -0.75c$$

$$u'_y = \frac{u_y}{\gamma\left(1 - \dfrac{u_x v}{c^2}\right)} = \frac{\sqrt{1 - \dfrac{(0.75c)^2}{c^2}}\,(-0.90c)}{1 - \dfrac{(0)(0.75c)}{c^2}} = -0.60c$$

Using the Pythagorean theorem, find the speed of Emily as measured by David:

$$u' = \sqrt{(u'_x)^2 + (u'_y)^2} = \sqrt{(-0.75c)^2 + (-0.60c)^2} = \boxed{0.96c}$$

Finalize This speed is less than c, as required by the special theory of relativity.

Pitfall Prevention | 9.4
Watch Out for "Relativistic Mass"
Some older treatments of relativity maintained the conservation of momentum principle at high speeds by using a model in which a particle's mass increases with speed. You might still encounter this notion of "relativistic mass" in your outside reading, especially in older books. Be aware that this notion is no longer widely accepted; today, mass is considered as *invariant*, independent of speed. The mass of an object in all frames is considered to be the mass as measured by an observer at rest with respect to the object.

▶9.6 | Relativistic Momentum and the Relativistic Form of Newton's Laws

We have seen that to describe the motion of particles within the framework of special relativity properly, the Galilean transformation must be replaced by the Lorentz transformation. Because the laws of physics must remain unchanged under the Lorentz transformation, we must generalize Newton's laws and the definitions of momentum and energy to conform to the Lorentz transformation and the principle of relativity. These generalized definitions should reduce to the classical (nonrelativistic) definitions for $v \ll c$ or $u \ll c$. (As we have done previously, we will use v for the speed of one reference frame relative to another and u for the speed of a particle.)

First, recall one of our isolated system models: the total momentum of an isolated system of particles is conserved. Suppose a collision between two particles is described in a reference frame S in which the momentum of the system is measured to be conserved. If the velocities in a second reference frame S′ are calculated using the Lorentz transformation and the Newtonian definition of momentum, $\vec{\mathbf{p}} = m\vec{\mathbf{u}}$, is used, it is found that the momentum of the system is *not* measured to be conserved in the second reference frame. This finding violates one of Einstein's postulates: The laws of physics are the same in all inertial frames. Therefore, assuming the Lorentz transformation to be correct, we must modify the definition of momentum.

The relativistic equation for the momentum of a particle of mass m that maintains the principle of conservation of momentum is

$$\vec{\mathbf{p}} \equiv \frac{m\vec{\mathbf{u}}}{\sqrt{1 - \dfrac{u^2}{c^2}}}$$

9.14 ◀   ▶ Definition of relativistic momentum

where $\vec{\mathbf{u}}$ is the velocity of the particle. When u is much less than c, the denominator of Equation 9.14 approaches unity, so $\vec{\mathbf{p}}$ approaches $m\vec{\mathbf{u}}$. Therefore, the relativistic equation for $\vec{\mathbf{p}}$ reduces to the classical expression when u is small compared with c. Equation 9.14 is often written in simpler form as

$$\vec{\mathbf{p}} = \gamma m\vec{\mathbf{u}}$$

9.15 ◀

using our previously defined expression[7] for γ.

The relativistic force $\vec{\mathbf{F}}$ on a particle whose momentum is $\vec{\mathbf{p}}$ is defined as

$$\vec{\mathbf{F}} \equiv \frac{d\vec{\mathbf{p}}}{dt}$$

9.16 ◀

where $\vec{\mathbf{p}}$ is given by Equation 9.14. This expression preserves both classical mechanics in the limit of low velocities and conservation of momentum for an isolated system ($\Sigma \vec{\mathbf{F}}_{\text{ext}} = 0$) both relativistically and classically.

We leave it to Problem 56 at the end of the chapter to show that the acceleration $\vec{\mathbf{a}}$ of a particle decreases under the action of a constant force, in which case $a \propto (1 - u^2/c^2)^{3/2}$. From this proportionality, note that as the particle's speed approaches c, the acceleration caused by any finite force approaches zero. It is therefore impossible to accelerate a particle from rest to a speed $u \geq c$.

Hence, c is an upper limit for the speed of any particle. In fact, it is possible to show that no *matter, energy,* or *information* can travel through space faster than c. Note that the relative speeds of the two spacecraft in Example 9.4 and the two motorcyclists in Example 9.5 were both less than c. If we had attempted to solve these examples with Galilean transformations, we would have obtained relative speeds larger than c in both cases.

Example 9.6 | Linear Momentum of an Electron

An electron, which has a mass of 9.11×10^{-31} kg, moves with a speed of $0.750c$. Find the magnitude of its relativistic momentum and compare this value with the momentum calculated from the classical expression.

SOLUTION

Conceptualize Imagine an electron moving with high speed. The electron carries momentum, but the magnitude of its momentum is not given by $p = mu$ because the speed is relativistic.

Categorize We categorize this example as a substitution problem involving a relativistic equation.

Use Equation 9.14 with $u = 0.750c$ to find the magnitude of the momentum:

$$p = \frac{m_e u}{\sqrt{1 - \dfrac{u^2}{c^2}}}$$

$$p = \frac{(9.11 \times 10^{-31}\ \text{kg})(0.750)(3.00 \times 10^8\ \text{m/s})}{\sqrt{1 - \dfrac{(0.750c)^2}{c^2}}}$$

$$= 3.10 \times 10^{-22}\ \text{kg} \cdot \text{m/s}$$

The classical expression (used incorrectly here) gives $p_{\text{classical}} = m_e u = 2.05 \times 10^{-22}$ kg · m/s. Hence, the correct relativistic result is 50% greater than the classical result!

---

[7] We defined γ previously in terms of the speed v of one frame relative to another frame. The same symbol is also used for $(1 - u^2/c^2)^{-1/2}$, where u is the speed of a particle.

9.7 | Relativistic Energy

We have seen that the definition of momentum requires generalization to make it compatible with the principle of relativity. We find that the definition of kinetic energy must also be modified.

To derive the relativistic form of the work–kinetic energy theorem, let us start with the definition of the work done by a force of magnitude F on a particle initially at rest. Recall from Chapter 6 that the work–kinetic energy theorem states that in the appropriate simple situation, the work done by a net force acting on a particle equals the change in kinetic energy of the particle. Because the initial kinetic energy is zero, we conclude that the work W done in accelerating a particle from rest is equivalent to the relativistic kinetic energy K of the particle:

$$W = \Delta K = K - 0 = K = \int_{x_1}^{x_2} F\, dx = \int_{x_1}^{x_2} \frac{dp}{dt}\, dx \qquad \text{9.17} \blacktriangleleft$$

where we are considering the special case of force and displacement vectors along the x axis for simplicity. To perform this integration and find the relativistic kinetic energy as a function of u, we first evaluate dp/dt, using Equation 9.14:

$$\frac{dp}{dt} = \frac{d}{dt} \frac{mu}{\sqrt{1 - \frac{u^2}{c^2}}} = \frac{m(du/dt)}{\left(1 - \frac{u^2}{c^2}\right)^{3/2}}$$

Substituting this expression for dp/dt and $dx = u\, dt$ into Equation 9.17 gives

$$K = \int_0^t \frac{m(du/dt)\, u\, dt}{\left(1 - \frac{u^2}{c^2}\right)^{3/2}} = m \int_0^u \frac{u}{\left(1 - \frac{u^2}{c^2}\right)^{3/2}}\, du$$

Evaluating the integral, we find that

$$K = \frac{mc^2}{\sqrt{1 - \frac{u^2}{c^2}}} - mc^2 = \gamma mc^2 - mc^2 = (\gamma - 1)mc^2 \qquad \text{9.18} \blacktriangleleft$$

▶ Relativistic kinetic energy

At low speeds, where $u/c \ll 1$, Equation 9.18 should reduce to the classical expression $K = \frac{1}{2}mu^2$. We can show this reduction by using the binomial expansion $(1 - x^2)^{-1/2} \approx 1 + \frac{1}{2}x^2 + \cdots$ for $x \ll 1$, where the higher-order powers of x are ignored in the expansion because they are so small. In our case, $x = u/c$, so

$$\gamma = \frac{1}{\sqrt{1 - \frac{u^2}{c^2}}} = \left(1 - \frac{u^2}{c^2}\right)^{-1/2} \approx 1 + \frac{1}{2}\frac{u^2}{c^2} + \cdots$$

Substituting into Equation 9.18 gives

$$K \approx \left(1 + \frac{1}{2}\frac{u^2}{c^2} + \cdots\right)mc^2 - mc^2 = \frac{1}{2}mu^2$$

which agrees with the classical result. Figure 9.12 shows a comparison of the speed–kinetic energy relationships for a particle using the nonrelativistic expression for K (the blue curve) and the relativistic expression for K (the brown curve). The curves are in good agreement at low speeds, but deviate at higher speeds. The nonrelativistic expression indicates a violation of special relativity because it suggests that sufficient energy can be added to the particle to accelerate it to a speed larger than c. In the relativistic case, the particle speed never exceeds c, regardless of the kinetic energy, which is consistent with experimental results. When an object's speed is less than one-tenth the speed of light, the classical kinetic energy equation differs by

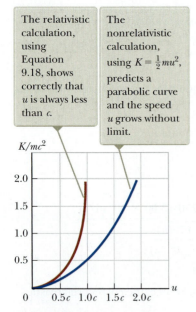

The relativistic calculation, using Equation 9.18, shows correctly that u is always less than c.

The nonrelativistic calculation, using $K = \frac{1}{2}mu^2$, predicts a parabolic curve and the speed u grows without limit.

Figure 9.12 A graph comparing relativistic and nonrelativistic kinetic energy of a moving particle. The energies are plotted as a function of speed u.

less than 1% from the relativistic equation (which is experimentally verified at all speeds). Therefore, for practical calculations it is valid to use the classical equation when the object's speed is less than 0.1c.

The constant term mc^2 in Equation 9.18, which is independent of the speed, is called the **rest energy** E_R of the particle:

$$E_R = mc^2$$

9.19 ◀ ▶ Rest energy

The term γmc^2 in Equation 9.18 depends on the particle speed and is the sum of the kinetic and rest energies. We define γmc^2 to be the **total energy** E; that is, total energy = kinetic energy + rest energy:

$$E = \gamma mc^2 = K + mc^2 = K + E_R$$

9.20 ◀

or, when γ is replaced by its equivalent,

$$E = \frac{mc^2}{\sqrt{1 - \frac{u^2}{c^2}}}$$

9.21 ◀ ▶ Total energy of a relativistic particle

The relation $E_R = mc^2$ shows that **mass is a manifestation of energy.** It also shows that a small mass corresponds to an enormous amount of energy. This concept is fundamental to much of the field of nuclear physics.

In many situations, the momentum or energy of a particle is measured rather than its speed. It is therefore useful to have an expression relating the total energy E to the relativistic momentum p, which is accomplished by using the expressions $E = \gamma mc^2$ and $p = \gamma mu$. By squaring these equations and subtracting, we can eliminate u (see Problem 9.36). The result, after some algebra, is

$$E^2 = p^2c^2 + (mc^2)^2$$

9.22 ◀ ▶ Energy–momentum relationship for a relativistic particle

When the particle is at rest, $p = 0$, and so $E = E_R = mc^2$. That is, the total energy equals the rest energy.

For the case of particles that have zero mass, such as photons (massless, chargeless particles of light to be discussed further in Chapter 28), we set $m = 0$ in Equation 9.22 and see that

$$E = pc$$

9.23 ◀

This equation is an exact expression relating energy and momentum for photons, which always travel at the speed of light.

When dealing with subatomic particles, it is convenient to express their energy in a unit called an *electron volt* (eV). The equality between electron volts and our standard energy unit is

$$1 \text{ eV} = 1.602 \times 10^{-19} \text{ J}$$

For example, the mass of an electron is 9.11×10^{-31} kg. Hence, the rest energy of the electron is

$$E_R = m_ec^2 = (9.11 \times 10^{-31} \text{ kg})(3.00 \times 10^8 \text{ m/s})^2 = 8.20 \times 10^{-14} \text{ J}$$

Converting to eV, we have

$$E_R = m_ec^2 = (8.20 \times 10^{-14} \text{ J})\left(\frac{1 \text{ eV}}{1.602 \times 10^{-19} \text{ J}}\right) = 0.511 \text{ MeV}$$

QUICK QUIZ 9.6 The following *pairs* of energies—particle 1: E, $2E$; particle 2: E, $3E$; particle 3: $2E$, $4E$—represent the rest energy and total energy of three different particles. Rank the particles from greatest to least according to their (a) mass, (b) kinetic energy, and (c) speed.

Example 9.7 | **The Energy of a Speedy Proton**

(A) Find the rest energy of a proton in units of electron volts.

SOLUTION

Conceptualize Even if the proton is not moving, it has energy associated with its mass. If it moves, the proton possesses more energy, with the total energy being the sum of its rest energy and its kinetic energy.

Categorize The phrase "rest energy" suggests we must take a relativistic rather than a classical approach to this problem.

Analyze Use Equation 9.19 to find the rest energy:
$$E_R = m_p c^2 = (1.673 \times 10^{-27}\text{ kg})(2.998 \times 10^8\text{ m/s})^2$$
$$= (1.504 \times 10^{-10}\text{ J})\left(\frac{1.00\text{ eV}}{1.602 \times 10^{-19}\text{ J}}\right) = 938\text{ MeV}$$

(B) If the total energy of a proton is three times its rest energy, what is the speed of the proton?

SOLUTION

Use Equation 9.21 to relate the total energy of the proton to the rest energy:
$$E = 3m_p c^2 = \frac{m_p c^2}{\sqrt{1 - \dfrac{u^2}{c^2}}} \rightarrow 3 = \frac{1}{\sqrt{1 - \dfrac{u^2}{c^2}}}$$

Solve for u:
$$1 - \frac{u^2}{c^2} = \frac{1}{9} \rightarrow \frac{u^2}{c^2} = \frac{8}{9}$$
$$u = \frac{\sqrt{8}}{3}c = 0.943c = 2.83 \times 10^8\text{ m/s}$$

(C) Determine the kinetic energy of the proton in units of electron volts.

SOLUTION

Use Equation 9.20 to find the kinetic energy of the proton:
$$K = E - m_p c^2 = 3m_p c^2 - m_p c^2 = 2m_p c^2$$
$$= 2(938\text{ MeV}) = 1.88 \times 10^3\text{ MeV}$$

(D) What is the proton's momentum?

SOLUTION

Use Equation 9.22 to calculate the momentum:
$$E^2 = p^2 c^2 + (m_p c^2)^2 = (3m_p c^2)^2$$
$$p^2 c^2 = 9(m_p c^2)^2 - (m_p c^2)^2 = 8(m_p c^2)^2$$
$$p = \sqrt{8}\frac{m_p c^2}{c} = \sqrt{8}\frac{938\text{ MeV}}{c} = 2.65 \times 10^3\text{ MeV}/c$$

Finalize The unit of momentum in part (D) is written MeV/c, which is a common unit in particle physics. For comparison, you might want to solve this example using classical equations.

9.8 | Mass and Energy

Equation 9.20, $E = \gamma mc^2$, which represents the total energy of a particle, suggests that even when a particle is at rest ($\gamma = 1$) it still possesses enormous energy through its mass. The clearest experimental proof of the equivalence of mass and energy occurs in nuclear and elementary particle interactions in which the conversion of mass into kinetic energy takes place. Hence, we cannot use the principle of conservation of energy in relativistic situations exactly as it is outlined in Chapter 7. We must include rest energy as another form of energy storage.

This concept is important in atomic and nuclear processes, in which the change in mass during the process is on the order of the initial mass. For example, in a conventional nuclear reactor, the uranium nucleus undergoes *fission*, a reaction that results in several lighter fragments having considerable kinetic energy. In the case of a ^{235}U atom, which is used as fuel in nuclear power plants, the fragments are two lighter nuclei and a few neutrons. The total mass of the fragments is less than that of the ^{235}U by an amount Δm. The corresponding energy Δmc^2 associated with this mass difference is exactly equal to the total kinetic energy of the fragments. The kinetic energy is transferred by collisions with water molecules as the fragments move through water, raising the internal energy of the water. This internal energy is used to produce steam for the generation of electric power.

Next, consider a basic *fusion* reaction in which two deuterium atoms combine to form one helium atom. The decrease in mass that results from the creation of one helium atom from two deuterium atoms is $\Delta m = 4.25 \times 10^{-29}$ kg. Hence, the corresponding energy that results from one fusion reaction is calculated to be $\Delta mc^2 = 3.83 \times 10^{-12}$ J $= 23.9$ MeV. To appreciate the magnitude of this result, consider that if 1 g of deuterium is converted to helium, the energy released is on the order of 10^{12} J! At the year 2012 cost of energy transferred from power plants in the United States by electrical transmission, this energy would be worth about $32 000. We will see more details of these nuclear processes in Chapter 30.

Example 9.8 | Mass Change in a Radioactive Decay

The ^{216}Po nucleus is unstable and exhibits radioactivity (Chapter 30). It decays to ^{212}Pb by emitting an alpha particle, which is a helium nucleus, ^4He. The relevant masses are $m_i = m(^{216}Po) = 216.001\ 915$ u and $m_f = m(^{212}Pb) + m(^4He) = 211.991\ 898$ u $+ 4.002\ 603$ u. The unit u is an *atomic mass unit*, where 1 u $= 1.660 \times 10^{-27}$ kg.

(A) Find the mass change of the system in this decay.

SOLUTION

Conceptualize The initial system is the ^{216}Po nucleus. Imagine the mass of the system decreasing during the decay and transforming to kinetic energy of the alpha particle and the ^{212}Pb nucleus after the decay.

Categorize We use concepts discussed in this section, so we categorize this example as a substitution problem.

Calculate the change in mass using the mass values given in the problem statement:

$$\Delta m = 216.001\ 915\ \text{u} - (211.991\ 898\ \text{u} + 4.002\ 603\ \text{u})$$

$$= 0.007\ 414\ \text{u} = \boxed{1.23 \times 10^{-29}\ \text{kg}}$$

(B) Find the energy this mass change represents.

SOLUTION

Use Equation 9.19 to find the energy associated with this mass change:

$$E = \Delta mc^2 = (1.23 \times 10^{-29}\ \text{kg})(3.00 \times 10^8\ \text{m/s})^2$$

$$= 1.11 \times 10^{-12}\ \text{J} = \boxed{6.92\ \text{MeV}}$$

9.9 | General Relativity

Up to this point, we have sidestepped a curious puzzle. Mass has two seemingly different properties: It determines a force of mutual gravitational attraction between two objects (Newton's law of universal gravitation), and it also represents the resistance of a single object to being accelerated (Newton's second law), regardless of the type of force producing the acceleration. How can one quantity have two such different properties? An answer to this question, which puzzled

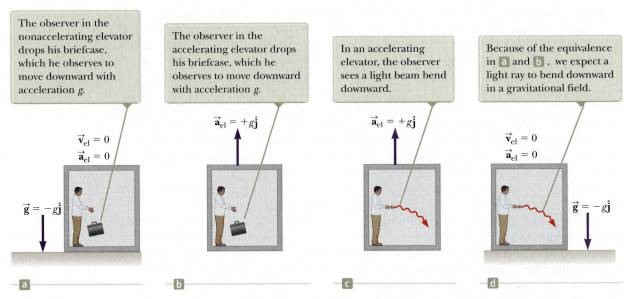

The observer in the nonaccelerating elevator drops his briefcase, which he observes to move downward with acceleration *g*.

The observer in the accelerating elevator drops his briefcase, which he observes to move downward with acceleration *g*.

In an accelerating elevator, the observer sees a light beam bend downward.

Because of the equivalence in **a** and **b**, we expect a light ray to bend downward in a gravitational field.

$\vec{\mathbf{v}}_{el} = 0$
$\vec{\mathbf{a}}_{el} = 0$

$\vec{\mathbf{a}}_{el} = +g\hat{\mathbf{j}}$

$\vec{\mathbf{a}}_{el} = +g\hat{\mathbf{j}}$

$\vec{\mathbf{v}}_{el} = 0$
$\vec{\mathbf{a}}_{el} = 0$

$\vec{\mathbf{g}} = -g\hat{\mathbf{j}}$

$\vec{\mathbf{g}} = -g\hat{\mathbf{j}}$

a   b   c   d

Figure 9.13 (a) The observer is at rest in an elevator in a uniform gravitational field $\vec{\mathbf{g}} = -g\hat{\mathbf{j}}$, directed downward. (b) The observer is in a region where gravity is negligible, but the elevator moves upward with an acceleration $\vec{\mathbf{a}}_{el} = +g\hat{\mathbf{j}}$. According to Einstein, the frames of reference in parts (a) and (b) are equivalent in every way. No local experiment can distinguish any difference between the two frames. (c) An observer watches a beam of light in an accelerating elevator. (d) Einstein's prediction of the behavior of a beam of light in a gravitational field.

Newton and many other physicists over the years, was provided when Einstein published his theory of gravitation, known as *general relativity*, in 1916. Because it is a mathematically complex theory, we offer merely a hint of its elegance and insight.

In Einstein's view, the dual behavior of mass was evidence for a very intimate and basic connection between the two behaviors. He pointed out that no mechanical experiment (e.g., dropping an object) could distinguish between the two situations illustrated in Figures 9.13a and 9.13b. In Figure 9.13a, a person is standing in an elevator on the surface of a planet and feels pressed into the floor due to the gravitational force. If he releases his briefcase, he observes it moving toward the floor with acceleration $\vec{\mathbf{g}} = -g\hat{\mathbf{j}}$. In Figure 9.13b, the person is in an elevator in empty space accelerating upward with $\vec{\mathbf{a}}_{el} = +g\hat{\mathbf{j}}$. The person feels pressed into the floor with the same force as in Figure 9.13a. If he releases his briefcase, he observes it moving toward the floor with acceleration *g*, just as in the previous situation. In each case, an object released by the observer undergoes a downward acceleration of magnitude *g* relative to the floor. In Figure 9.13a, the person is at rest in an inertial frame in a gravitational field due to the planet. (A gravitational field exists around any object with mass, such as a planet. We will define the gravitational field formally in Chapter 11.) In Figure 9.13b, the person is in a noninertial frame accelerating in gravity-free space. Einstein's claim is that these two situations are completely equivalent.

Einstein carried this idea further and proposed that *no* experiment, mechanical or otherwise, could distinguish between the two cases. This extension to include all phenomena (not just mechanical ones) has interesting consequences. For example, suppose a light pulse is sent horizontally across the elevator as in Figure 9.13c, in which the elevator is accelerating upward in empty space. From the point of view of an observer in an inertial frame outside the elevator, the light travels in a straight line while the floor of the elevator accelerates upward. According to the observer on the elevator, however, the trajectory of the light pulse bends downward as the floor of the elevator (and the observer) accelerates upward. Therefore, based on the equality of parts (a) and (b) of the figure for all phenomena, Einstein proposed that a beam of light should also be bent downward by a gravitational field, as in Figure 9.13d.

The two postulates of Einstein's **general theory of relativity** are as follows:

- All the laws of nature have the same form for observers in any frame of reference, whether accelerated or not.
- In the vicinity of any given point, a gravitational field is equivalent to an accelerated frame of reference in the absence of gravitational effects. (This postulate is known as the **principle of equivalence.**)

▶ Postulates of general relativity

One interesting effect predicted by general relativity is that the passage of time is altered by gravity. A clock in the presence of gravity runs more slowly than one for which gravity is negligible. Consequently, the frequencies of radiation emitted by atoms in the presence of a strong gravitational field are shifted to lower values compared with the same emissions in a weak field. This gravitational shift has been detected in light emitted by atoms in massive stars. It has also been verified on the Earth by comparing the frequencies of radiation emitted from laser-cooled ions separated vertically by less than 1 m.

The second postulate suggests that a gravitational field may be "transformed away" at any point if we choose an appropriate accelerated frame of reference, a freely falling one. Einstein developed an ingenious method of describing the acceleration necessary to make the gravitational field "disappear." He specified a certain quantity, the *curvature of space–time,* that describes the gravitational effect of a mass. In fact, the curvature of space–time completely replaces Newton's gravitational theory. According to Einstein, there is no such thing as a gravitational force. Rather, the presence of a mass causes a curvature of space–time in the vicinity of the mass, and this curvature dictates the space–time path that all freely moving objects must follow.

One important test of general relativity is the prediction that a light ray passing near the Sun should be deflected by some angle. This prediction was confirmed by astronomers as the bending of starlight during a total solar eclipse shortly following World War I (Fig. 9.14).

As an example of the effects of curved space–time, imagine two travelers moving on parallel paths a few meters apart on the surface of the Earth and maintaining an exact northward heading along two longitude lines. As they observe each other near the equator, they will claim that their paths are exactly parallel. As they approach the North Pole, however, they will notice that they are moving closer together and that they will actually meet at the North Pole. Therefore, they will claim that they moved along parallel paths, but moved toward each other, *as if there were an attractive force between them.* They will make this conclusion based on their everyday experience of moving on flat surfaces. From our mental representation, however, we realize that they are walking on a curved surface, and the geometry of the curved surface, rather than an attractive force, causes them to converge. In a similar way, general relativity replaces the notion of forces with the movement of objects through curved space–time.

If a concentration of mass in space becomes very great, as is believed to occur when a large star exhausts its nuclear fuel and collapses to a very small volume, a **black hole** may form. Here the curvature of space–time is so extreme that, within a certain distance from the center of the black hole, all matter and light become trapped. We will say more about black holes in Chapter 11.

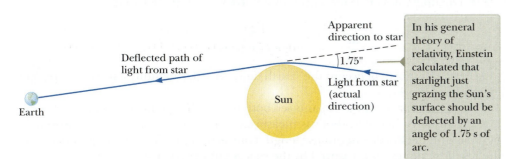

Apparent direction to star

Deflected path of light from star

1.75"

Sun

Light from star (actual direction)

Earth

In his general theory of relativity, Einstein calculated that starlight just grazing the Sun's surface should be deflected by an angle of 1.75 s of arc.

Figure 9.14 Deflection of starlight passing near the Sun. Because of this effect, the Sun or other remote objects can act as a *gravitational lens.*

> **THINKING PHYSICS 9.2**
>
> Atomic clocks are extremely accurate; in fact, an error of 1 s in 3 million years is typical. This error can be described as about 1 part in 10^{14}. On the other hand, the atomic clock in Boulder, Colorado, near Denver, is often 15 ns faster than the one in Washington, D.C., after only one day. This error is one of about 1 part in 6×10^{12}, which is about 17 times larger than the previously expressed error. If atomic clocks are so accurate, why does a clock in Boulder not remain in synchronization with one in Washington, D.C.? (*Hint:* Denver is known as the Mile High City.)
>
> **Reasoning** According to the general theory of relativity, the passage of time depends on gravity. Time is measured to run more slowly in strong gravitational fields. Washington, D.C., is at an elevation very close to sea level, but Boulder is about a mile higher in altitude. This difference results in a weaker gravitational field at Boulder than at Washington, D.C. As a result, time is measured to run more rapidly in Boulder than in Washington, D.C. ◄

9.10 | Context Connection: From Mars to the Stars

In this chapter, we have discussed the strange effects of traveling at high speeds. Do we need to consider these effects in our planned mission to Mars?

To answer this question, let us consider a typical spacecraft speed necessary to travel from the Earth to Mars. This speed is on the order of 10^4 m/s. Let us evaluate γ for this speed:

$$\gamma = \frac{1}{\sqrt{1 - \dfrac{u^2}{c^2}}} = \frac{1}{\sqrt{1 - \dfrac{(10^4 \text{ m/s})^2}{(3.00 \times 10^8 \text{ m/s})^2}}} = 1.000\ 000\ 000\ 6$$

where we have completely ignored the rules of significant figures so that we could find the first nonzero digit to the right of the decimal place!

It is clear from this result that relativistic considerations are not critically important for our trip to Mars. Yet what about deeper travels into space? Suppose we wish to travel to another star. This distance is several orders of magnitude larger. The nearest star is about 4.2 ly from the Earth. In comparison, Mars is 4.0×10^{-5} ly at its farthest from the Earth. Therefore, we are talking about a distance to the nearest star that is five orders of magnitude larger than the distance to Mars. Very long time intervals will be needed to reach even the nearest star. At the escape speed from the Sun, for example, assuming that this speed is maintained during the entire trip, the time interval is 30 000 years for travel to the nearest star. This time interval is clearly prohibitive, especially if we would like the people who leave the Earth to be the same people who arrive at the star!

We can use the principles of relativity to reduce this time interval significantly by traveling at very high speeds. Suppose our spacecraft travels at a constant speed of $0.99c$. The time interval as measured by an observer on the Earth then is

$$\Delta t = \frac{L_p}{u} = \frac{4.2 \text{ ly}}{0.99(1.0 \text{ ly/yr})} = 4.2 \text{ yr}$$

where the distance between the Earth and the destination star is the proper length L_p.

Because the spacecraft occupants see both the Earth and the destination star moving, the distance between them is measured to be shorter than that measured by observers on the Earth. We can use length contraction to calculate the distance from the Earth to the star as measured by the spacecraft occupants:

$$L = \frac{L_p}{\gamma} = L_p\sqrt{1 - \frac{u^2}{c^2}} = (4.2 \text{ ly})\sqrt{1 - \frac{(0.99c)^2}{c^2}} = 0.59 \text{ ly}$$

The time interval required to reach the star is now

$$\Delta t = \frac{L}{u} = \frac{0.59 \text{ ly}}{0.99(1.0 \text{ ly/yr})} = 0.60 \text{ yr}$$

which is clearly a reduction from the low-speed trip!

There are three major technical problems with this scenario, however. The first is the technological challenge of designing and building a spacecraft and rocket engine assembly that can attain a speed of 0.99c. Second is the design of a safety system that will provide early warnings about running into asteroids, meteoroids, or other bits of matter while traveling at almost light speed through space. Even a small piece of rock could be disastrous if struck at 0.99c. The third problem is related to the twin paradox discussed earlier in this chapter. During the trip to the star, 4.2 yr will pass on the Earth. If the travelers return to the Earth, another 4.2 yr will pass. Therefore, the travelers will have aged by only 2(0.6 yr) = 1.2 yr, but 8.4 yr will have passed on the Earth. For stars farther away than the nearest star, these effects could result in the personnel assisting with the liftoff from the Earth no longer being alive when the travelers return. We see that travel to the stars will be an enormous challenge!

There is also a biological consideration associated with the prospect of traveling to a star at 0.99c. Attaining this speed will require a large acceleration in order that the time interval to reach 0.99c is short compared to the time interval traveling at this speed. The human body has certain limits, however, to being accelerated. As found in Problem 41 in Chapter 2, an Air Force officer experienced accelerations of magnitude 20g for very short time intervals as his rocket sled was brought to rest. In other experiments, he survived brief accelerations up to 46g. In our proposed space travel, the inhabitants of the spacecraft will have to experience *sustained* large accelerations.

BIO Human limits on acceleration

If the spacecraft inhabitants have the tops of their heads pointing in the direction of the acceleration, blood will be driven downward toward their feet, just as if they were in an increased gravitational field. This can cause loss of consciousness at accelerations as low as 5g. If the spacecraft inhabitants are traveling feetfirst, the limits are even lower. Accelerations in the range of 2g to 3g can drive blood to the head and cause capillaries in the eyes to burst.

Sustained accelerations can cause serious symptoms, including death, if the acceleration is higher than about 10g. Pilots have used specially designed suits and have learned muscle techniques that allow tolerance of accelerations of up to 9g. Further technological advances will need to be made, however, in order for space travelers to be protected from injury during accelerations up to 0.99c.

SUMMARY

The two basic postulates of **special relativity** are:

- All the laws of physics are the same in all inertial reference frames.
- The speed of light in vacuum has the same value, in all inertial frames, regardless of the velocity of the observer or the velocity of the source emitting the light.

Three consequences of special relativity are:

- Events that are simultaneous for one observer may not be simultaneous for another observer who is in motion relative to the first.
- Clocks in motion relative to an observer are measured to be slowed down by a factor γ. This phenomenon is known as **time dilation.**

- Lengths of objects in motion are measured to be shorter in the direction of motion. This phenomenon is known as **length contraction.**

To satisfy the postulates of special relativity, the Galilean transformations must be replaced by the **Lorentz transformation equations:**

$$x' = \gamma(x - vt)$$
$$y' = y$$
$$z' = z$$
$$t' = \gamma\left(t - \frac{v}{c^2}x\right)$$

9.8◀

where $\gamma = (1 - v^2/c^2)^{-1/2}$.

The relativistic form of the **Lorentz velocity transformation** is

$$u'_x = \frac{u_x - v}{1 - \frac{u_x v}{c^2}} \qquad 9.11 ◀$$

where u_x is the speed of an object as measured in the S frame and u'_x is its speed measured in the S′ frame.

The relativistic expression for the momentum of a particle moving with a velocity $\vec{u}$ is

$$\vec{p} \equiv \frac{m\vec{u}}{\sqrt{1 - \frac{u^2}{c^2}}} = \gamma m\vec{u} \qquad 9.14, 9.15 ◀$$

The relativistic expression for the kinetic energy of a particle is

$$K = \gamma mc^2 - mc^2 = (\gamma - 1)mc^2 \qquad 9.18 ◀$$

where $E_R = mc^2$ is the **rest energy** of the particle.

The **total energy** E of a particle is given by the expression

$$E = \frac{mc^2}{\sqrt{1 - \frac{u^2}{c^2}}} \qquad 9.21 ◀$$

The total energy of a particle is the sum of its rest energy and its kinetic energy: $E = E_R + K$.

The relativistic momentum of a particle is related to its total energy through the equation

$$E^2 = p^2 c^2 + (mc^2)^2 \qquad 9.22 ◀$$

The **general theory of relativity** claims that no experiment can distinguish between a gravitational field and an accelerating reference frame. It correctly predicts that the path of light is affected by a gravitational field.

▶ OBJECTIVE QUESTIONS

☐ denotes answer available in *Student Solutions Manual/Study Guide*

1. An astronaut is traveling in a spacecraft in outer space in a straight line at a constant speed of $0.500c$. Which of the following effects would she experience? (a) She would feel heavier. (b) She would find it harder to breathe. (c) Her heart rate would change. (d) Some of the dimensions of her spacecraft would be shorter. (e) None of those answers is correct.

2. A distant astronomical object (a quasar) is moving away from us at half the speed of light. What is the speed of the light we receive from this quasar? (a) greater than c (b) c (c) between $c/2$ and c (d) $c/2$ (e) between 0 and $c/2$

3. As a car heads down a highway traveling at a speed v away from a ground observer, which of the following statements are true about the measured speed of the light beam from the car's headlights? More than one statement may be correct. (a) The ground observer measures the light speed to be $c + v$. (b) The driver measures the light speed to be c. (c) The ground observer measures the light speed to be c. (d) The driver measures the light speed to be $c - v$. (e) The ground observer measures the light speed to be $c - v$.

4. A spacecraft zooms past the Earth with a constant velocity. An observer on the Earth measures that an undamaged clock on the spacecraft is ticking at one-third the rate of an identical clock on the Earth. What does an observer on the spacecraft measure about the Earth-based clock's ticking rate? (a) It runs more than three times faster than his own clock. (b) It runs three times faster than his own. (c) It runs at the same rate as his own. (d) It runs at one-third the rate of his own. (e) It runs at less than one-third the rate of his own.

5. Which of the following statements are fundamental postulates of the special theory of relativity? More than one statement may be correct. (a) Light moves through a substance called the ether. (b) The speed of light depends on the inertial reference frame in which it is measured. (c) The laws of physics depend on the inertial reference frame in which they are used. (d) The laws of physics are the same in all

inertial reference frames. (e) The speed of light is independent of the inertial reference frame in which it is measured.

6. A spacecraft built in the shape of a sphere moves past an observer on the Earth with a speed of $0.500c$. What shape does the observer measure for the spacecraft as it goes by? (a) a sphere (b) a cigar shape, elongated along the direction of motion (c) a round pillow shape, flattened along the direction of motion (d) a conical shape, pointing in the direction of motion

7. (i) Does the speed of an electron have an upper limit? (a) yes, the speed of light c (b) yes, with another value (c) no (ii) Does the magnitude of an electron's momentum have an upper limit? (a) yes, $m_e c$ (b) yes, with another value (c) no (iii) Does the electron's kinetic energy have an upper limit? (a) yes, $m_e c^2$ (b) yes, $\frac{1}{2} m_e c^2$ (c) yes, with another value (d) no

8. The following three particles all have the same total energy E: (a) a photon, (b) a proton, and (c) an electron. Rank the magnitudes of the particles' momenta from greatest to smallest.

9. Two identical clocks are set side by side and synchronized. One remains on the Earth. The other is put into orbit around the Earth moving rapidly toward the east. (i) As measured by an observer on the Earth, does the orbiting clock (a) run faster than the Earth-based clock, (b) run at the same rate, or (c) run slower? (ii) The orbiting clock is returned to its original location and brought to rest relative to the Earth-based clock. Thereafter, what happens? (a) Its reading lags farther and farther behind the Earth-based clock. (b) It lags behind the Earth-based clock by a constant amount. (c) It is synchronous with the Earth-based clock. (d) It is ahead of the Earth-based clock by a constant amount. (e) It gets farther and farther ahead of the Earth-based clock.

10. You measure the volume of a cube at rest to be V_0. You then measure the volume of the same cube as it passes you in a direction parallel to one side of the cube. The speed of the cube is $0.980c$, so $\gamma \approx 5$. Is the volume you measure close to (a) $V_0/25$, (b) $V_0/5$, (c) V_0, (d) $5V_0$, or (e) $25V_0$?

CONCEPTUAL QUESTIONS

1. A train is approaching you at very high speed as you stand next to the tracks. Just as an observer on the train passes you, you both begin to play the same recorded version of a Beethoven symphony on identical MP3 players. (a) According to you, whose MP3 player finishes the symphony first? (b) **What If?** According to the observer on the train, whose MP3 player finishes the symphony first? (c) Whose MP3 player actually finishes the symphony first?

2. Explain why, when defining the length of a rod, it is necessary to specify that the positions of the ends of the rod are to be measured simultaneously.

3. The speed of light in water is 230 Mm/s. Suppose an electron is moving through water at 250 Mm/s. Does that violate the principle of relativity? Explain.

4. A particle is moving at a speed less than $c/2$. If the speed of the particle is doubled, what happens to its momentum?

5. Two identical clocks are in the same house, one upstairs in a bedroom and the other downstairs in the kitchen. Which clock runs slower? Explain.

6. List three ways our day-to-day lives would change if the speed of light were only 50 m/s.

7. It is said that Einstein, in his teenage years, asked the question, "What would I see in a mirror if I carried it in my hands and ran at a speed near that of light?" How would you answer this question?

8. (a) "Newtonian mechanics correctly describes objects moving at ordinary speeds, and relativistic mechanics correctly describes objects moving very fast." (b) "Relativistic mechanics must make a smooth transition as it reduces to Newtonian mechanics in a case in which the speed of an object becomes small compared with the speed of light." Argue for or against statements (a) and (b).

9. Give a physical argument that shows it is impossible to accelerate an object of mass m to the speed of light, even with a continuous force acting on it.

10. **(i)** An object is placed at a position $p > f$ from a concave mirror as shown in Figure CQ9.10a, where f is the focal length of the mirror. In such a situation, an image is formed at a distance q from the mirror, as we discuss in Chapter 26. The distances are related by the mirror equation:

$$\frac{1}{p} + \frac{1}{q} = \frac{1}{f}$$

In a finite time interval, the object is moved to the right to a position at the focal point F of the mirror. Show that the image of the object moves at a speed greater than the speed of light. **(ii)** A laser pointer is suspended in a horizontal plane and set into rapid rotation as shown in Figure CQ9.10b. Show that the spot of light it produces on a distant screen can move across the screen at a speed greater than the speed of light. (If you carry out this experiment, make sure the direct laser light cannot enter a person's eyes.) **(iii)** Argue that the experiments in parts (i) and (ii) do not invalidate the principle that no material, no energy, and no information can move faster than light moves in a vacuum.

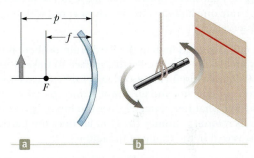

Figure CQ9.10

11. With regard to reference frames, how does general relativity differ from special relativity?

PROBLEMS

Section 9.1 The Principle of Galilean Relativity

1. In a laboratory frame of reference, an observer notes that Newton's second law is valid. Assume forces and masses are measured to be the same in any reference frame for speeds small compared with the speed of light. (a) Show that Newton's second law is also valid for an observer moving at a constant speed, small compared with the speed of light, relative to the laboratory frame. (b) Show that Newton's second law is *not* valid in a reference frame moving past the laboratory frame with a constant acceleration.

2. A car of mass 2 000 kg moving with a speed of 20.0 m/s collides and locks together with a 1 500-kg car at rest at a stop sign. Show that momentum is conserved in a reference frame moving at 10.0 m/s in the direction of the moving car.

Section 9.2 The Michelson–Morley Experiment

Section 9.3 Einstein's Principle of Relativity

Section 9.4 Consequences of Special Relativity

Note: Problem 40 in Chapter 3 can be assigned with this section.

3. **W** How fast must a meterstick be moving if its length is measured to shrink to 0.500 m?

4. **BIO** An astronaut is traveling in a space vehicle moving at $0.500c$ relative to the Earth. The astronaut measures her pulse rate at 75.0 beats per minute. Signals generated by the astronaut's pulse are radioed to the Earth when the vehicle is moving in a direction perpendicular to the line that connects the vehicle with an observer on the Earth. (a) What pulse rate does the Earth-based observer measure? (b) **What If?** What would be the pulse rate if the speed of the space vehicle were increased to $0.990c$?

5. **W** At what speed does a clock move if it is measured to run at a rate one-half the rate of a clock at rest with respect to an observer?

6. A muon formed high in the Earth's atmosphere is measured by an observer on the Earth's surface to travel at speed $v = 0.990c$ for a distance of 4.60 km before it decays into an electron, a neutrino, and an antineutrino ($\mu^- \rightarrow e^- + \nu + \bar{\nu}$). (a) For what time interval does the muon live as measured in its reference frame? (b) How far does the Earth travel as measured in the frame of the muon?

7. An astronomer on the Earth observes a meteoroid in the southern sky approaching the Earth at a speed of $0.800c$. At the time of its discovery the meteoroid is 20.0 ly from the Earth. Calculate (a) the time interval required for the meteoroid to reach the Earth as measured by the Earthbound astronomer, (b) this time interval as measured by a tourist on the meteoroid, and (c) the distance to the Earth as measured by the tourist.

8. For what value of v does $\gamma = 1.010\ 0$? Observe that for speeds lower than this value, time dilation and length contraction are effects amounting to less than 1%.

9. An atomic clock moves at 1 000 km/h for 1.00 h as measured by an identical clock on the Earth. At the end of the 1.00-h interval, how many nanoseconds slow will the moving clock be compared with the Earth-based clock? .

10. The identical twins Speedo and Goslo join a migration from the Earth to Planet X, 20.0 ly away in a reference frame in which both planets are at rest. The twins, of the same age, depart at the same moment on different spacecraft. Speedo's spacecraft travels steadily at $0.950c$ and Goslo's at $0.750c$. (a) Calculate the age difference between the twins after Goslo's spacecraft lands on Planet X. (b) Which twin is older?

11. **M** A spacecraft with a proper length of 300 m passes by an observer on the Earth. According to this observer, it takes 0.750 µs for the spacecraft to pass a fixed point. Determine the speed of the spacecraft as measured by the Earth-based observer.

12. **S** A spacecraft with a proper length of L_p passes by an observer on the Earth. According to this observer, it takes a time interval Δt for the spacecraft to pass a fixed point. Determine the speed of the object as measured by the Earth-based observer.

13. A friend passes by you in a spacecraft traveling at a high speed. He tells you that his craft is 20.0 m long and that the identically constructed craft you are sitting in is 19.0 m long. According to your observations, (a) how long is your spacecraft, (b) how long is your friend's craft, and (c) what is the speed of your friend's craft?

14. An interstellar space probe is launched from the Earth. After a brief period of acceleration it moves with a constant velocity, with a magnitude of 70.0% of the speed of light. Its nuclear-powered batteries supply the energy to keep its data transmitter active continuously. The batteries have a lifetime of 15.0 yr as measured in a rest frame. (a) How long do the batteries on the space probe last as measured by Mission Control on the Earth? (b) How far is the probe from the Earth when its batteries fail as measured by Mission Control? (c) How far is the probe from the Earth when its batteries fail as measured by its built-in trip odometer? (d) For what total time interval after launch are data received from the probe by Mission Control? Note that radio waves travel at the speed of light and fill the space between the probe and the Earth at the time of battery failure.

15. A moving rod is observed to have a length of $\ell = 2.00$ m and to be oriented at an angle of $\theta = 30.0°$ with respect to the direction of motion as shown in Figure P9.15. The rod has a speed of $0.995c$. (a) What is the proper length of the rod? (b) What is the orientation angle in the proper frame?

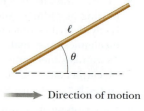

Direction of motion

Figure P9.15

Section 9.5 The Lorentz Transformation Equations

16. Shannon observes two light pulses to be emitted from the same location, but separated in time by 3.00 µs. Kimmie observes the emission of the same two pulses to be separated in time by 9.00 µs. (a) How fast is Kimmie moving relative to Shannon? (b) According to Kimmie, what is the separation in space of the two pulses?

17. **W** A red light flashes at position $x_R = 3.00$ m and time $t_R = 1.00 \times 10^{-9}$ s, and a blue light flashes at $x_B = 5.00$ m and $t_B = 9.00 \times 10^{-9}$ s, all measured in the S reference frame. Reference frame S′ moves uniformly to the right and has its origin at the same point as S at $t = t' = 0$. Both flashes are observed to occur at the same place in S′. (a) Find the relative speed between S and S′. (b) Find the location of the two flashes in frame S′. (c) At what time does the red flash occur in the S′ frame?

18. Keilah, in reference frame S, measures two events to be simultaneous. Event A occurs at the point (50.0 m, 0, 0) at the instant 9:00:00 Universal time on January 15, 2010.

Event B occurs at the point (150 m, 0, 0) at the same moment. Torrey, moving past with a velocity of $0.800c\,\hat{\mathbf{i}}$, also observes the two events. In her reference frame S′, which event occurred first and what time interval elapsed between the events?

19. An enemy spacecraft moves away from the Earth at a speed of $v = 0.800c$ (Fig. P9.19). A galactic patrol spacecraft pursues at a speed of $u = 0.900c$ relative to the Earth. Observers on the Earth measure the patrol craft to be overtaking the enemy craft at a relative speed of $0.100c$. With what speed is the patrol craft overtaking the enemy craft as measured by the patrol craft's crew?

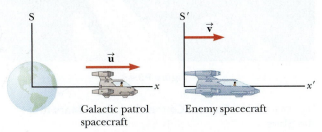

Galactic patrol spacecraft   Enemy spacecraft

Figure P9.19

20. A spacecraft is launched from the surface of the Earth with a velocity of $0.600c$ at an angle of $50.0°$ above the horizontal positive x axis. Another spacecraft is moving past with a velocity of $0.700c$ in the negative x direction. Determine the magnitude and direction of the velocity of the first spacecraft as measured by the pilot of the second spacecraft.

21. **M** Figure P9.21 shows a jet of material (at the upper right) being ejected by galaxy M87 (at the lower left). Such jets are believed to be evidence of supermassive black holes at the center of a galaxy. Suppose two jets of material from the center of a galaxy are ejected in opposite directions. Both jets move at $0.750c$ relative to the galaxy center. Determine the speed of one jet relative to the other.

NASA/STSci

Figure P9.21

Section 9.6 Relativistic Momentum and the Relativistic Form of Newton's Laws

22. The nonrelativistic expression for the momentum of a particle, $p = mu$, agrees with experiment if $u \ll c$. For what speed does the use of this equation give an error in the measured momentum of (a) 1.00% and (b) 10.0%?

23. **W** Calculate the momentum of an electron moving with a speed of (a) $0.010\ 0c$, (b) $0.500c$, and (c) $0.900c$.

24. **S** Show that the speed of an object having momentum of magnitude p and mass m is

$$u = \frac{c}{\sqrt{1 + (mc/p)^2}}$$

25. A golf ball travels with a speed of 90.0 m/s. By what fraction does its relativistic momentum magnitude p differ from its classical value mu? That is, find the ratio $(p - mu)/mu$.

26. The speed limit on a certain roadway is 90.0 km/h. Suppose speeding fines are made proportional to the amount by which a vehicle's momentum exceeds the momentum it would have when traveling at the speed limit. The fine for driving at 190 km/h (that is, 100 km/h over the speed limit) is \$80.0. What, then, is the fine for traveling (a) at 1 090 km/h? (b) At 1 000 000 090 km/h?

27. **M** An unstable particle at rest spontaneously breaks into two fragments of unequal mass. The mass of the first fragment is 2.50×10^{-28} kg, and that of the other is 1.67×10^{-27} kg. If the lighter fragment has a speed of $0.893c$ after the breakup, what is the speed of the heavier fragment?

Section 9.7 Relativistic Energy

28. Show that for any object moving at less than one-tenth the speed of light, the relativistic kinetic energy agrees with the result of the classical equation $K = \frac{1}{2}mu^2$ to within less than 1%. Therefore, for most purposes, the classical equation is sufficient to describe these objects.

29. Determine the energy required to accelerate an electron from (a) $0.500c$ to $0.900c$ and (b) $0.900c$ to $0.990c$.

30. **Q|C** (a) Find the kinetic energy of a 78.0-kg spacecraft launched out of the solar system with speed 106 km/s by using the classical equation $K = \frac{1}{2}mu^2$. (b) **What If?** Calculate its kinetic energy using the relativistic equation. (c) Explain the result of comparing the answers of parts (a) and (b).

31. An electron has a kinetic energy five times greater than its rest energy. Find (a) its total energy and (b) its speed.

32. A cube of steel has a volume of 1.00 cm^3 and a mass of 8.00 g when at rest on the Earth. If this cube is now given a speed $u = 0.900c$, what is its density as measured by a stationary observer? Note that relativistic density is defined as E_R/c^2V.

33. The rest energy of an electron is 0.511 MeV. The rest energy of a proton is 938 MeV. Assume both particles have kinetic energies of 2.00 MeV. Find the speed of (a) the electron and (b) the proton. (c) By what factor does the speed of the electron exceed that of the proton? (d) Repeat the calculations in parts (a) through (c) assuming both particles have kinetic energies of 2 000 MeV.

34. **GP** An unstable particle with mass $m = 3.34 \times 10^{-27}$ kg is initially at rest. The particle decays into two fragments that fly off along the x axis with velocity components $u_1 = 0.987c$ and $u_2 = -0.868c$. From this information, we wish to determine the masses of fragments 1 and 2. (a) Is the initial system of the unstable particle, which becomes the system of the two fragments, isolated or nonisolated? (b) Based on your answer to part (a), what two analysis models are appropriate for this situation? (c) Find the values of γ for the two fragments after the decay. (d) Using one of the analysis models in part (b), find a relationship between the masses

m_1 and m_2 of the fragments. (e) Using the second analysis model in part (b), find a second relationship between the masses m_1 and m_2. (f) Solve the relationships in parts (d) and (e) simultaneously for the masses m_1 and m_2.

35. A proton moves at $0.950c$. Calculate its (a) rest energy, (b) total energy, and (c) kinetic energy.

36. **S** Show that the energy–momentum relationship in Equation 9.22, $E^2 = p^2c^2 + (mc^2)^2$, follows from the expressions $E = \gamma mc^2$ and $p = \gamma mu$.

37. **M** A proton in a high-energy accelerator moves with a speed of $c/2$. Use the work–kinetic energy theorem to find the work required to increase its speed to (a) $0.750c$ and (b) $0.995c$.

38. An object having mass 900 kg and traveling at speed $0.850c$ collides with a stationary object having mass 1 400 kg. The two objects stick together. Find (a) the speed and (b) the mass of the composite object.

39. **M** A pion at rest ($m_\pi = 273m_e$) decays to a muon ($m_\mu = 207m_e$) and an antineutrino ($m_{\bar{\nu}} \approx 0$). The reaction is written $\pi^- \rightarrow \mu^- + \bar{\nu}$. Find (a) the kinetic energy of the muon and (b) the energy of the antineutrino in electron volts.

Section 9.8 Mass and Energy

40. **W** In a nuclear power plant, the fuel rods last 3 yr before they are replaced. The plant can transform energy at a maximum possible rate of 1.00 GW. Supposing it operates at 80.0% capacity for 3.00 yr, what is the loss of mass of the fuel?

41. The power output of the Sun is 3.85×10^{26} W. By how much does the mass of the Sun decrease each second?

42. **Q|C** **W** When 1.00 g of hydrogen combines with 8.00 g of oxygen, 9.00 g of water is formed. During this chemical reaction, 2.86×10^5 J of energy is released. (a) Is the mass of the water larger or smaller than the mass of the reactants? (b) What is the difference in mass? (c) Explain whether the change in mass is likely to be detectable.

Section 9.9 General Relativity

43. **Review.** A global positioning system (GPS) satellite moves in a circular orbit with period 11 h 58 min. (a) Determine the radius of its orbit. (b) Determine its speed. (c) The nonmilitary GPS signal is broadcast at a frequency of 1 575.42 MHz in the reference frame of the satellite. When it is received on the Earth's surface by a GPS receiver (Fig. P9.43), what is the fractional change in this frequency due to time dilation as described by special relativity? (d) The gravitational "blueshift" of the frequency according to general relativity is a separate effect. It is called a blueshift to indicate a change to a higher frequency. The magnitude of that fractional change is given by

$$\frac{\Delta f}{f} = \frac{\Delta U_g}{mc^2}$$

where U_g is the change in gravitational potential energy of an object–Earth system when the object of mass m is moved between the two points where the signal is observed. Calculate this fractional change in frequency due to the change in position of the satellite from the Earth's surface to its orbital position. (e) What is the overall fractional change

in frequency due to both time dilation and gravitational blueshift?

Figure P9.43

Section 9.10 Context Connection: From Mars to the Stars

44. **Q|C** **Review.** In 1963, astronaut Gordon Cooper orbited the Earth 22 times. The press stated that for each orbit, he aged two-millionths of a second less than he would have had he remained on the Earth. (a) Assuming Cooper was 160 km above the Earth in a circular orbit, determine the difference in elapsed time between someone on the Earth and the orbiting astronaut for the 22 orbits. You may use the approximation

$$\frac{1}{\sqrt{1-x}} \approx 1 + \frac{x}{2}$$

for small x. (b) Did the press report accurate information? Explain.

45. An astronaut wishes to visit the Andromeda galaxy, making a one-way trip that will take 30.0 yr in the spacecraft's frame of reference. Assume the galaxy is 2.00×10^6 ly away and the astronaut's speed is constant. (a) How fast must he travel relative to the Earth? (b) What will be the kinetic energy of his 1 000-metric-ton spacecraft? (c) What is the cost of this energy if it is purchased at a typical consumer price for electric energy of $0.110/kWh?

Additional Problems

46. An object disintegrates into two fragments. One fragment has mass 1.00 MeV/c^2 and momentum 1.75 MeV/c in the positive x direction, and the other has mass 1.50 MeV/c^2 and momentum 2.00 MeV/c in the positive y direction. Find (a) the mass and (b) the speed of the original object.

47. **M** The net nuclear fusion reaction inside the Sun can be written as $4^1\text{H} \rightarrow {}^4\text{He} + E$. The rest energy of each hydrogen atom is 938.78 MeV, and the rest energy of the helium-4 atom is 3 728.4 MeV. Calculate the percentage of the starting mass that is transformed to other forms of energy.

48. *Why is the following situation impossible?* On their 40th birthday, twins Speedo and Goslo say good-bye as Speedo takes off for a planet that is 50 ly away. He travels at a constant speed of $0.85c$ and immediately turns around and comes back to the Earth after arriving at the planet. Upon arriving back at the Earth, Speedo has a joyous reunion with Goslo.

49. Massive stars ending their lives in supernova explosions produce the nuclei of all the atoms in the bottom half of the periodic table by fusion of smaller nuclei. This problem roughly models that process. A particle of mass $m = 1.99 \times 10^{-26}$ kg moving with a velocity $\vec{u} = 0.500c\,\hat{i}$ collides head-on and sticks to a particle of mass $m' = m/3$ moving with the velocity $\vec{u} = -0.500c\,\hat{i}$. What is the mass of the resulting particle?

50. **QC** **S** Massive stars ending their lives in supernova explosions produce the nuclei of all the atoms in the bottom half of the periodic table by fusion of smaller nuclei. This problem roughly models that process. A particle of mass m moving along the x axis with a velocity component $+u$ collides head-on and sticks to a particle of mass $m/3$ moving along the x axis with the velocity component $-u$. (a) What is the mass M of the resulting particle? (b) Evaluate the expression from part (a) in the limit $u \to 0$. (c) Explain whether the result agrees with what you should expect from nonrelativistic physics.

51. **M** The cosmic rays of highest energy are protons that have kinetic energy on the order of 10^{13} MeV. (a) As measured in the proton's frame, what time interval would a proton of this energy require to travel across the Milky Way galaxy, which has a proper diameter $\sim 10^5$ ly? (b) From the point of view of the proton, how many kilometers across is the galaxy?

52. An electron has a speed of $0.750c$. (a) Find the speed of a proton that has the same kinetic energy as the electron. (b) **What If?** Find the speed of a proton that has the same momentum as the electron.

53. An alien spaceship traveling at $0.600c$ toward the Earth launches a landing craft. The landing craft travels in the same direction with a speed of $0.800c$ relative to the mother ship. As measured on the Earth, the spaceship is 0.200 ly from the Earth when the landing craft is launched. (a) What speed do the Earth-based observers measure for the approaching landing craft? (b) What is the distance to the Earth at the moment of the landing craft's launch as measured by the aliens? (c) What travel time is required for the landing craft to reach the Earth as measured by the aliens on the mother ship? (d) If the landing craft has a mass of 4.00×10^5 kg, what is its kinetic energy as measured in the Earth reference frame?

54. (a) Prepare a graph of the relativistic kinetic energy and the classical kinetic energy, both as a function of speed, for an object with a mass of your choice. (b) At what speed does the classical kinetic energy underestimate the experimental value by 1%? (c) By 5%? (d) By 50%?

55. A supertrain with a proper length of 100 m travels at a speed of $0.950c$ as it passes through a tunnel having a proper length of 50.0 m. As seen by a trackside observer, is the train ever completely within the tunnel? If so, by how much do the train's ends clear the ends of the tunnel?

56. **QC** **S** A particle with electric charge q moves along a straight line in a uniform electric field $\vec{E}$ with speed u. The electric force exerted on the charge is $q\vec{E}$. The velocity of the particle and the electric field are both in the x direction. (a) Show that the acceleration of the particle in the x direction is given by

$$a = \frac{du}{dt} = \frac{qE}{m}\left(1 - \frac{u^2}{c^2}\right)^{3/2}$$

(b) Discuss the significance of the dependence of the acceleration on the speed. (c) **What If?** If the particle starts from rest at $x = 0$ at $t = 0$, how would you proceed to find the speed of the particle and its position at time t?

57. An observer in a coasting spacecraft moves toward a mirror at speed $v = 0.650c$ relative to the reference frame labeled S in Figure P9.57. The mirror is stationary with respect to S. A light pulse emitted by the spacecraft travels toward the mirror and is reflected back to the spacecraft. The spacecraft is a distance $d = 5.66 \times 10^{10}$ m from the mirror (as measured by observers in S) at the moment the light pulse leaves the spacecraft. What is the total travel time of the pulse as measured by observers in (a) the S frame and (b) the spacecraft?

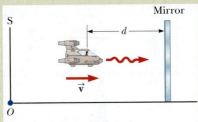

Figure P9.57 Problems 57 and 58.

58. **S** An observer in a coasting spacecraft moves toward a mirror at speed v relative to the reference frame labeled S in Figure P9.57. The mirror is stationary with respect to S. A light pulse emitted by the spacecraft travels toward the mirror and is reflected back to the spacecraft. The spacecraft is a distance d from the mirror (as measured by observers in S) at the moment the light pulse leaves the spacecraft. What is the total travel time of the pulse as measured by observers in (a) the S frame and (b) the spacecraft?

59. **M** Spacecraft I, containing students taking a physics exam, approaches the Earth with a speed of $0.600c$ (relative to the Earth), while spacecraft II, containing professors proctoring the exam, moves at $0.280c$ (relative to the Earth) directly toward the students. If the professors stop the exam after 50.0 min have passed on their clock, for what time interval does the exam last as measured by (a) the students and (b) an observer on the Earth?

60. **S** A physics professor on the Earth gives an exam to her students, who are in a spacecraft traveling at speed v relative to the Earth. The moment the craft passes the professor, she signals the start of the exam. She wishes her students to have a time interval T_0 (spacecraft time) to complete the exam. Show that she should wait a time interval (Earth time) of

$$T = T_0\sqrt{\frac{1 - v/c}{1 + v/c}}$$

before sending a light signal telling them to stop. (*Suggestion:* Remember that it takes some time for the second light signal to travel from the professor to the students.)

61. A gamma ray (a high-energy photon) can produce an electron (e^-) and a positron (e^+) of equal mass when it enters the electric field of a heavy nucleus: $\gamma \to e^+ + e^-$. What minimum gamma-ray energy is required to accomplish this task?

62. Imagine that the entire Sun, of mass M_S, collapses to a sphere of radius R_g such that the work required to remove a small mass m from the surface would be equal to its rest energy mc^2. This radius is called the *gravitational radius* for

the Sun. (a) Use this approach to show that $R_g = GM_S/c^2$. (b) Find a numerical value for R_g.

63. Owen and Dina are at rest in frame S′, which is moving at $0.600c$ with respect to frame S. They play a game of catch while Ed, at rest in frame S, watches the action (Fig. P9.63). Owen throws the ball to Dina at $0.800c$ (according to Owen), and their separation (measured in S′) is equal to 1.80×10^{12} m. (a) According to Dina, how fast is the ball moving? (b) According to Dina, what time interval is required for the ball to reach her? According to Ed, (c) how far apart are Owen and Dina, (d) how fast is the ball moving, and (e) what time interval is required for the ball to reach Dina?

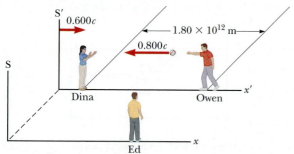

Figure P9.63

64. **S** A rod of length L_0 moving with a speed v along the horizontal direction makes an angle θ_0 with respect to the x' axis. (a) Show that the length of the rod as measured by a stationary observer is $L = L_0[1 - (v^2/c^2)\cos^2\theta_0]^{1/2}$. (b) Show that the angle that the rod makes with the x axis is given by $\tan\theta = \gamma\tan\theta_0$. These results show that the rod is both contracted and rotated. (Take the lower end of the rod to be at the origin of the primed coordinate system.)

65. **Q C** Suppose our Sun is about to explode. In an effort to escape, we depart in a spacecraft at $v = 0.800c$ and head toward the star Tau Ceti, 12.0 ly away. When we reach the midpoint of our journey from the Earth, we see our Sun explode, and, unfortunately, at the same instant, we see Tau Ceti explode as well. (a) In the spacecraft's frame of reference, should we conclude that the two explosions occurred simultaneously? If not, which occurred first? (b) **What If?** In a frame of reference in which the Sun and Tau Ceti are at rest, did they explode simultaneously? If not, which exploded first?

Rotational Motion

Chapter Outline

Courtesy Tourism Malaysia

The Malaysian pastime of *gasing* involves the spinning of tops that can have masses up to 5 kg. Professional spinners can spin their tops so that they might rotate for 1 to 2 h before stopping. We will study the rotational motion of objects such as these tops in this chapter.

When an extended object, such as a wheel, rotates about its axis, the motion cannot be analyzed by modeling the object as a particle because at any given time different parts of the object are moving with different speeds and in different directions. We can, however, analyze the motion by considering an extended object to be composed of a *collection* of moving particles.

In dealing with a rotating object, analysis is greatly simplified by assuming that the object is rigid. A **rigid object** is one that is nondeformable; that is, it is an object in which the relative locations of all particles of which the object is composed remain constant. All real objects are deformable to some extent; our rigid-object model, however, is useful in many situations in which deformation is negligible.

⟨10.1 | Angular Position, Speed, and Acceleration

▶ The radian

We began our study of translational motion in Chapter 2 by defining the terms *position*, *velocity*, and *acceleration*. For example, we locate a particle in one-dimensional space with the position variable x. In this chapter, we will insert the word *translational* before our previously studied kinematic variables to distinguish them from the analogous *rotational* variables that we will develop.

Figure 10.1 illustrates an overhead view of a rotating compact disc. The disc rotates about a fixed axis that is perpendicular to the page and passes through the center of the disc at O. A particle at P is at a fixed distance r from the origin and rotates about it in a circle of radius r. (In fact, *every* particle on the disc undergoes circular motion about O.) It is convenient to represent the position of P with its polar coordinates (r, θ), where r is the distance from the origin to P and θ is measured *counterclockwise* from some reference line shown in Figure 10.1. In this representation, the only coordinate for the particle that changes in time is the angle θ, while r remains constant. As the particle moves along the circular path from the reference line, which is at $\theta = 0$, it moves through an arc of length s as in Figure 10.1b. The arc length s is related to the angle θ through the relationship

$$s = r\theta \qquad \textbf{10.1a} \blacktriangleleft$$

$$\theta = \frac{s}{r} \qquad \textbf{10.1b} \blacktriangleleft$$

Because θ is the ratio of an arc length and the radius of the circle, it is a pure number. Usually, we give θ the artificial unit **radian** (rad), where one radian is the angle subtended by an arc length equal to the radius of the arc. Because the circumference of a circle is $2\pi r$, it follows from Equation 10.1b that 360° corresponds to an angle of $(2\pi r/r)$ rad $= 2\pi$ rad. (Also note that 2π rad corresponds to one complete revolution.) Hence, 1 rad $= 360°/2\pi \approx 57.3°$. To convert an angle in degrees to an angle in radians, we use π rad $= 180°$, so

$$\theta(\text{rad}) = \frac{\pi}{180°}\theta(\text{deg})$$

For example, 60° equals $\pi/3$ rad, and 45° equals $\pi/4$ rad.

We will focus much of our attention in this chapter on rigid objects. Approximating a real object as a rigid object is a simplification model, the **rigid object model.** We will base several analysis models on this simplification model, as we did for the particle model.

Because the disc in Figure 10.1 is a rigid object, as the particle at P moves along the circular path from the reference line every other particle on the object rotates through the same angle θ. Therefore, we can associate the angle θ with the entire rigid object as well as with an individual particle, which allows us to define the angular position of a rigid object in its rotational motion. We choose a radial line on the object, such as a line connecting O and a chosen particle on the object. The **angular position** of the rigid object is the angle θ between this radial line on the object and the fixed reference line in space, which is often chosen as the x axis. This process is similar to the way we identify the position of an object in translational motion as the distance x between the object and the reference position, which is the origin, $x = 0$. Therefore, the angle θ plays the same role in rotational motion that the position x does in translational motion.

As a particle on a rigid object travels from position Ⓐ to position Ⓑ in a time interval Δt as in Figure 10.2, the reference line fixed to the object sweeps out an angle $\Delta\theta = \theta_f - \theta_i$. This quantity $\Delta\theta$ is defined as the **angular displacement** of the rigid object:

$$\Delta\theta \equiv \theta_f - \theta_i$$

To define angular position for the disc, a fixed reference line is chosen. A particle at P is located at a distance r from the rotation axis through O.

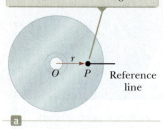

a

As the disc rotates, a particle at P moves through an arc length s on a circular path of radius r. The angular position of P is θ.

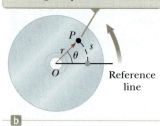

b

Figure 10.1 A compact disc rotating about a fixed axis through O perpendicular to the plane of the figure.

The rate at which this angular displacement occurs can vary. If the rigid object spins rapidly, this displacement can occur in a short time interval. If it rotates slowly, this displacement occurs in a longer time interval. These different rotation rates can be quantified by defining the **average angular speed** ω_{avg} (Greek letter omega) as the ratio of the angular displacement of a rigid object to the time interval Δt during which the displacement occurs:

$$\omega_{avg} \equiv \frac{\theta_f - \theta_i}{t_f - t_i} = \frac{\Delta \theta}{\Delta t}$$

10.2 ◀ ▶ Average angular speed

In analogy to instantaneous translational speed, the **instantaneous angular speed** ω is defined as the limit of the average angular speed as Δt approaches zero:

$$\omega \equiv \lim_{\Delta t \to 0} \frac{\Delta \theta}{\Delta t} = \frac{d\theta}{dt}$$

10.3 ◀ ▶ Instantaneous angular speed

Angular speed has units of rad/s, which can be written as s^{-1} because radians are not dimensional. We take ω to be positive when θ is increasing (counterclockwise motion in Fig. 10.2) and negative when θ is decreasing (clockwise motion in Fig. 10.2).

If the instantaneous angular speed of an object changes from ω_i to ω_f in the time interval Δt, the object has an angular acceleration. The **average angular acceleration** α_{avg} (Greek letter alpha) of a rotating rigid object is defined as the ratio of the change in the angular speed to the time interval Δt during which the change in angular speed occurs:

$$\alpha_{avg} \equiv \frac{\omega_f - \omega_i}{t_f - t_i} = \frac{\Delta \omega}{\Delta t}$$

10.4 ◀ ▶ Average angular acceleration

In analogy to instantaneous translational acceleration, the **instantaneous angular acceleration** is defined as the limit of the average angular acceleration as Δt approaches zero:

$$\alpha \equiv \lim_{\Delta t \to 0} \frac{\Delta \omega}{\Delta t} = \frac{d\omega}{dt}$$

10.5 ◀ ▶ Instantaneous angular acceleration

Angular acceleration has units of radians per second squared (rad/s^2), or simply s^{-2}. Notice that α is positive when a rigid object rotating counterclockwise is speeding up or when a rigid object rotating clockwise is slowing down during some time interval.

When a rigid object is rotating about a *fixed* axis, every particle on the object rotates about that axis through the same angle in a given time interval and has the same angular speed and the same angular acceleration. That is, the quantities θ, ω, and α characterize the rotational motion of the entire rigid object as well as individual particles in the object.

The angular position (θ), angular speed (ω), and angular acceleration (α) of a rigid object are analogous to translational position (x), translational speed (v), and translational acceleration (a), respectively, for the corresponding one-dimensional motion of a particle discussed in Chapter 2. The variables θ, ω, and α differ dimensionally from the variables x, v, and a only by a factor having the unit of length. (See Section 10.3.)

We have not associated any direction with the angular speed and angular acceleration.[1] Strictly speaking, ω and α are the magnitudes of the angular velocity and angular acceleration vectors $\vec{\omega}$ and $\vec{\alpha}$ and they should always be positive. Because we are considering rotation about a fixed axis, we can indicate the directions of these vectors by assigning a positive or negative sign to ω and α, as discussed for ω after Equation 10.3. For rotation about a fixed axis, the only direction in space that uniquely specifies the rotational motion is the direction along the axis of rotation. Therefore, the direction of $\vec{\omega}$ is along this axis. If a particle rotates in the xy plane

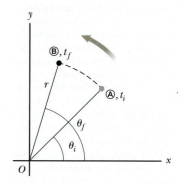

Figure 10.2 A particle on a rotating rigid object moves from Ⓐ to Ⓑ along the arc of a circle. In the time interval $\Delta t = t_f - t_i$, the radial line of length r sweeps out an angle $\Delta \theta = \theta_f - \theta_i$.

---

[1]Although we do not verify it here, the instantaneous angular velocity and instantaneous angular acceleration are vector quantities, but the corresponding average values are not because angular displacements do not add as vector quantities for finite rotations.

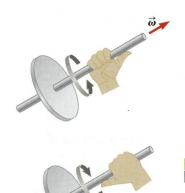

Figure 10.3 The right-hand rule for determining the direction of the angular velocity vector.

as in Figure 10.2, the direction of $\vec{\omega}$ is *out of* the plane of the diagram when the rotation is counterclockwise and *into* the plane of the diagram when the rotation is clockwise. To illustrate this convention, it is convenient to use the **right-hand rule** demonstrated in Figure 10.3. When the four fingers of the right hand are wrapped in the direction of the rotation, the extended right thumb points in the direction of $\vec{\omega}$. The direction of $\vec{\alpha}$ follows from its definition $d\vec{\omega}/dt$. The direction of $\vec{\alpha}$ is the same as $\vec{\omega}$ if the angular speed is increasing in time and is antiparallel to $\vec{\omega}$ if the angular speed is decreasing in time.

> **QUICK QUIZ 10.1** A rigid object is rotating in a counterclockwise sense around a fixed axis. Each of the following pairs of quantities represents an initial angular position and a final angular position of the rigid object. (**i**) Which of the sets can *only* occur if the rigid object rotates through more than 180°?   (**a**) 3 rad, 6 rad   (**b**) −1 rad, 1 rad   (**c**) 1 rad, 5 rad   (**ii**) Suppose the change in angular position for each of these pairs of values occurs in 1 s. Which choice represents the lowest average angular speed?

10.2 | Analysis Model: Rigid Object Under Constant Angular Acceleration

Imagine that a rigid object rotates about a fixed axis and that it has a constant angular acceleration. In this case, we generate a new analysis model for rotational motion called the **rigid object under constant angular acceleration.** This model is the rotational analog to the particle under constant acceleration model. We develop kinematic relationships for this model in this section. Writing Equation 10.5 in the form $d\omega = \alpha\,dt$ and integrating from $t_i = 0$ to $t_f = t$ gives

$$\omega_f = \omega_i + \alpha t \quad \text{(for constant } \alpha\text{)} \qquad \textbf{10.6} \blacktriangleleft$$

where ω_i is the angular speed of the rigid object at time $t = 0$. Equation 10.6 allows us to find the angular speed ω_f of the object at any later time t. Substituting Equation 10.6 into Equation 10.3 and integrating once more, we obtain

$$\theta_f = \theta_i + \omega_i t + \tfrac{1}{2}\alpha t^2 \quad \text{(for constant } \alpha\text{)} \qquad \textbf{10.7} \blacktriangleleft$$

where θ_i is the angular position of the rigid object at time $t = 0$. Equation 10.7 allows us to find the angular position θ_f of the object at any later time t. Eliminating t from Equations 10.6 and 10.7 gives

$$\omega_f^2 = \omega_i^2 + 2\alpha(\theta_f - \theta_i) \quad \text{(for constant } \alpha\text{)} \qquad \textbf{10.8} \blacktriangleleft$$

This equation allows us to find the angular speed ω_f of the rigid object for any value of its angular position θ_f. If we eliminate α between Equations 10.6 and 10.7, we obtain

$$\theta_f = \theta_i + \tfrac{1}{2}(\omega_i + \omega_f)t \quad \text{(for constant } \alpha\text{)} \qquad \textbf{10.9} \blacktriangleleft$$

Notice that these kinematic expressions for the rigid object under constant angular acceleration are of the same mathematical form as those for a particle under constant acceleration (Chapter 2). They can be generated from the equations for translational motion by making the substitutions $x \rightarrow \theta$, $v \rightarrow \omega$, and $a \rightarrow \alpha$. Table 10.1 compares the kinematic equations for rotational and translational motion.

> **Pitfall Prevention | 10.3**
> **Just Like Translation?**
> Equations 10.6 to 10.9 and Table 10.1 might suggest that rotational kinematics is just like translational kinematics. That is almost true, with two key differences. (1) In rotational kinematics, you must specify a rotation axis (per Pitfall Prevention 10.1). (2) In rotational motion, the object keeps returning to its original orientation; therefore, you may be asked for the number of revolutions made by a rigid object. This concept has no analog in translational motion.

TABLE 10.1 | Kinematic Equations for Rotational and Translational Motion

Rigid Body Under Constant Angular Acceleration	Particle Under Constant Acceleration
$\omega_f = \omega_i + \alpha t$	$v_f = v_i + at$
$\theta_f = \theta_i + \omega_i t + \tfrac{1}{2}\alpha t^2$	$x_f = x_i + v_i t + \tfrac{1}{2}at^2$
$\omega_f^2 = \omega_i^2 + 2\alpha(\theta_f - \theta_i)$	$v_f^2 = v_i^2 + 2a(x_f - x_i)$
$\theta_f = \theta_i + \tfrac{1}{2}(\omega_i + \omega_f)t$	$x_f = x_i + \tfrac{1}{2}(v_i + v_f)t$

> **QUICK QUIZ 10.2** consider again the pairs of angular positions for the rigid object in Quick Quiz 10.1. If the object starts from rest at the initial angular position, moves counterclockwise with constant angular acceleration, and arrives at the final angular position with the same angular speed in all three cases, for which choice is the angular acceleration the highest?

Example 10.1 | Rotating Wheel

A wheel rotates with a constant angular acceleration of 3.50 rad/s².

(A) If the angular speed of the wheel is 2.00 rad/s at $t = 0$, through what angular displacement does the wheel rotate in 2.00 s?

SOLUTION

Conceptualize Look again at Figure 10.1. Imagine that the compact disc rotates with its angular speed increasing at a constant rate. You start your stopwatch when the disc is rotating at 2.00 rad/s. This mental image is a model for the motion of the wheel in this example.

Categorize The phrase "with a constant angular acceleration" tells us to use the rigid object under constant angular acceleration model.

Analyze
Arrange Equation 10.7 so that it expresses the angular displacement of the object:

$$\Delta\theta = \theta_f - \theta_i = \omega_i t + \tfrac{1}{2}\alpha t^2$$

Substitute the known values to find the angular displacement at $t = 2.00$ s:

$$\Delta\theta = (2.00 \text{ rad/s})(2.00 \text{ s}) + \tfrac{1}{2}(3.50 \text{ rad/s}^2)(2.00 \text{ s})^2$$

$$= 11.0 \text{ rad} = (11.0 \text{ rad})(180°/\pi \text{ rad}) = 630°$$

(B) Through how many revolutions has the wheel turned during this time interval?

SOLUTION

Multiply the angular displacement found in part (A) by a conversion factor to find the number of revolutions:

$$\Delta\theta = 630°\left(\frac{1 \text{ rev}}{360°}\right) = 1.75 \text{ rev}$$

(C) What is the angular speed of the wheel at $t = 2.00$ s?

SOLUTION

Use Equation 10.6 to find the angular speed at $t = 2.00$ s:

$$\omega_f = \omega_i + \alpha t = 2.00 \text{ rad/s} + (3.50 \text{ rad/s}^2)(2.00 \text{ s})$$

$$= 9.00 \text{ rad/s}$$

Finalize We could also obtain this result using Equation 10.8 and the results of part (A). (Try it!)

What If? Suppose a particle moves along a straight line with a constant acceleration of 3.50 m/s². If the velocity of the particle is 2.00 m/s at $t = 0$, through what displacement does the particle move in 2.00 s? What is the velocity of the particle at $t = 2.00$ s?

Answer Notice that these questions are translational analogs to parts (A) and (C) of the original problem. The mathematical solution follows exactly the same form. For the displacement,

$$\Delta x = x_f - x_i = v_i t + \tfrac{1}{2}at^2$$

$$= (2.00 \text{ m/s})(2.00 \text{ s}) + \tfrac{1}{2}(3.50 \text{ m/s}^2)(2.00 \text{ s})^2 = 11.0 \text{ m}$$

and for the velocity,

$$v_f = v_i + at = 2.00 \text{ m/s} + (3.50 \text{ m/s}^2)(2.00 \text{ s}) = 9.00 \text{ m/s}$$

There is no translational analog to part (B) because translational motion under constant acceleration is not repetitive.

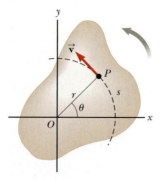

Active Figure 10.4 As a rigid object rotates about the fixed axis through O, the point P has a tangential velocity $\vec{v}$ that is always tangent to the circular path of radius r.

10.3 | Relations Between Rotational and Translational Quantities

In this section, we derive some useful relations between the angular speed and angular acceleration of a rotating rigid object and the translational speed and translational acceleration of a point in the object. To do so, we must keep in mind that when a rigid object rotates about a fixed axis, as in Active Figure 10.4, *every* particle of the object moves in a circle whose center is on the axis of rotation.

Because the point P in Active Figure 10.4 moves in a circle of radius r, its translational velocity vector $\vec{v}$ is always tangent to the path and hence is called **tangential velocity.** The magnitude of the tangential velocity of the particle is, by definition, the **tangential speed,** $v = ds/dt$, where s is the distance traveled by the particle along the circular path. Recalling that $s = r\theta$ (Eq. 10.1a) and noting that r is a constant, we obtain

$$v = \frac{ds}{dt} = r\frac{d\theta}{dt}$$

$$v = r\omega \qquad \text{10.10} \blacktriangleleft$$

That is, the tangential speed of a point on a rotating rigid object equals the perpendicular distance of that point from the axis of rotation multiplied by the angular speed. Therefore, although every point on the rigid object has the same *angular* speed, not every point has the same *tangential* speed because r is not the same for all points on the object. Equation 10.10 shows that the tangential speed of a point on the rotating object increases as one moves outward from the center of rotation, as we would intuitively expect. For example, the outer end of a swinging golf club moves much faster than the handle.

We can relate the angular acceleration of the rotating rigid object to the tangential acceleration of the point P by taking the time derivative of v:

$$a_t = \frac{dv}{dt} = r\frac{d\omega}{dt}$$

$$a_t = r\alpha \qquad \text{10.11} \blacktriangleleft$$

That is, the tangential component of the translational acceleration of a point on a rotating object equals the point's perpendicular distance from the axis of rotation multiplied by the angular acceleration.

In Chapter 3, we found that a particle rotating in a circular path undergoes a centripetal, or radial, acceleration of magnitude v^2/r directed toward the center of rotation (Fig. 10.5). Because $v = r\omega$, we can express the centripetal acceleration of that point in terms of the angular speed as

$$a_c = \frac{v^2}{r} = r\omega^2 \qquad \text{10.12} \blacktriangleleft$$

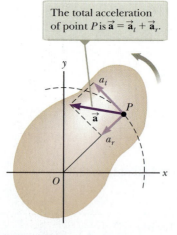

The total acceleration of point P is $\vec{a} = \vec{a}_t + \vec{a}_r$.

Figure 10.5 As a rigid object rotates about a fixed axis through O, the point P experiences a tangential component a_t and a radial component a_r of translational acceleration.

The total acceleration vector at the point is $\vec{a} = \vec{a}_t + \vec{a}_r$, where the magnitude of $\vec{a}_r$ is the centripetal acceleration a_c. Because $\vec{a}$ is a vector having a radial and a tangential component, the magnitude of $\vec{a}$ at the point P on the rotating rigid object is

$$a = \sqrt{a_t^2 + a_r^2} = \sqrt{r^2\alpha^2 + r^2\omega^4} = r\sqrt{\alpha^2 + \omega^4} \qquad \text{10.13} \blacktriangleleft$$

QUICK QUIZ 10.3 Ethan and Joseph are riding on a merry-go-round. Ethan rides on a horse at the outer rim of the circular platform, twice as far from the center of the circular platform as Joseph, who rides on an inner horse. **(i)** When the merry-go-round is rotating at a constant angular speed, what is Ethan's angular speed? **(a)** twice Joseph's **(b)** the same as Joseph's **(c)** half of Joseph's **(d)** impossible to determine **(ii)** When the merry-go-round is rotating at a constant angular speed, describe Ethan's tangential speed from the same list of choices.

THINKING PHYSICS 10.1

A phonograph record (LP, for *long-playing*) rotates at a constant *angular* speed. A compact disc (CD) rotates so that the surface sweeps past the laser at a constant *tangential* speed. Consider two circular grooves of information on an LP, one near the outer edge and one near the inner edge. Suppose the outer groove "contains" 1.8 s of music. Does the inner groove also contain 1.8 s of music? And for the CD, do the inner and outer "grooves" contain the same time interval of music?

Reasoning On the LP the inner and outer grooves must both rotate once in the same time interval. Therefore, each groove, regardless of where it is on the record, contains the same time interval of information. Of course, on the inner grooves, this same information must be compressed into a smaller circumference. On a CD, the constant tangential speed requires that no such compression occur; the digital pits representing the information are spaced uniformly everywhere on the surface. Therefore, there is more information in an outer "groove," because of its larger circumference and, as a result, a longer time interval of music than in the inner "groove." ◄

THINKING PHYSICS 10.2

The launch area for the European Space Agency is not in Europe, but rather in South America. Why?

Reasoning Placing a satellite in Earth orbit requires providing a large tangential speed to the satellite, which is the task of the rocket propulsion system. Anything that reduces the requirements on the propulsion system is a welcome contribution. The surface of the Earth is already traveling toward the east at a high speed due to the rotation of the Earth. Therefore, if rockets are launched toward the east, the rotation of the Earth provides some initial tangential speed, reducing somewhat the requirements on the propulsion system. If rockets were launched from Europe, which is at a relatively large latitude, the contribution of the Earth's rotation is relatively small because the distance between Europe and the rotation axis of the Earth is relatively small. The ideal place for launching is at the equator, which is as far as one can be from the rotation axis of the Earth and still be on the surface of the Earth. This location results in the largest possible tangential speed due to the Earth's rotation. The European Space Agency exploits this advantage by launching from French Guiana, which is only a few degrees north of the equator.

A second advantage of this location is that launching toward the east takes the spacecraft over water. In the event of an accident or a failure, the wreckage will fall into the ocean rather than into populated areas as it would if launched to the east from Europe. Similarly, the United States launches spacecraft from Florida rather than California, despite the more favorable weather conditions in California. ◄

10.4 | Rotational Kinetic Energy

Let us consider an object as a system of particles and assume it rotates about a fixed z axis with an angular speed ω. Figure 10.6 shows the rotating object and identifies one particle on the object located at a distance r_i from the rotation axis. If the mass of the ith particle is m_i and its tangential speed is v_i, its kinetic energy is

$$K_i = \tfrac{1}{2}m_i v_i^2$$

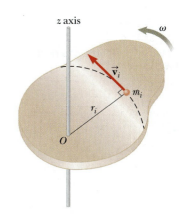

Figure 10.6 A rigid object rotating about the z axis with angular speed ω. The kinetic energy of the particle of mass m_i is $\tfrac{1}{2}m_i v_i^2$. The total kinetic energy of the rigid object is called its rotational kinetic energy.

To proceed further, recall that although every particle in the rigid object has the same angular speed ω, the individual tangential speeds depend on the distance r_i from the axis of rotation according to Equation 10.10. The *total* kinetic energy of the rotating rigid object is the sum of the kinetic energies of the individual particles:

$$K_R = \sum_i K_i = \sum_i \tfrac{1}{2} m_i v_i^2 = \tfrac{1}{2} \sum_i m_i r_i^2 \omega^2$$

We can write this expression in the form

$$K_R = \tfrac{1}{2} \left(\sum_i m_i r_i^2 \right) \omega^2 \qquad \textbf{10.14} \blacktriangleleft$$

where we have factored ω^2 from the sum because it is common to every particle. We simplify this expression by defining the quantity in parentheses as the **moment of inertia** I of the rigid object:

▶ Moment of inertia

$$I \equiv \sum_i m_i r_i^2 \qquad \textbf{10.15} \blacktriangleleft$$

From the definition of moment of inertia,[2] we see that it has dimensions of ML^2 ($kg \cdot m^2$ in SI units). With this notation, Equation 10.14 becomes

▶ Rotational kinetic energy

$$K_R = \tfrac{1}{2} I \omega^2 \qquad \textbf{10.16} \blacktriangleleft$$

The moment of inertia is a measure of an object's *resistance to change in its angular speed*. Therefore, it plays a role in rotational motion identical to the role mass plays in translational motion. Notice that moment of inertia depends not only on the mass of the rigid object but also on *how the mass is distributed around the rotation axis*.

Although we shall commonly refer to the quantity $\tfrac{1}{2} I \omega^2$ in Equation 10.16 as the **rotational kinetic energy**, it is not a new form of energy. It is ordinary kinetic energy because it was derived from a sum over individual kinetic energies of the particles contained in the rigid object. It is a new role for kinetic energy for us, however, because we have only considered kinetic energy associated with translation through space so far. On the storage side of the conservation of energy equation (see Eq. 7.2), we should now consider that the kinetic energy term should be the sum of the changes in both translational and rotational kinetic energy. Therefore, in energy versions of system models, we should keep in mind the possibility of rotational kinetic energy.

The moment of inertia of a system of discrete particles can be calculated in a straightforward way with Equation 10.15. We can evaluate the moment of inertia of a continuous rigid object by imagining the object to be divided into many small elements, each of which has mass Δm_i. We use the definition $I = \sum_i r_i^2 \Delta m_i$ and take the limit of this sum as $\Delta m_i \rightarrow 0$. In this limit, the sum becomes an integral over the volume of the object:

▶ Moment of inertia of a rigid object

$$I = \lim_{\Delta m_i \to 0} \sum_i r_i^2 \Delta m_i = \int r^2 \, dm \qquad \textbf{10.17} \blacktriangleleft$$

It is usually easier to calculate moments of inertia in terms of the volume of the elements rather than their mass, and we can easily make this change by using Equation 1.1, $\rho = m/V$, where ρ is the density of the object and V is its volume. We can express the mass of an element by writing Equation 1.1 in differential form, $dm = \rho \, dV$. Substituting this result into Equation 10.17 gives

$$I = \int \rho r^2 \, dV \qquad \textbf{10.18} \blacktriangleleft$$

If the object is homogenous, ρ is uniform over the volume of the object and the integral can be evaluated for a known geometry. If ρ is not uniform over the volume of the object, its variation with position must be known in order to perform the integration.

Pitfall Prevention | 10.4
No Single Moment of Inertia
We have pointed out that moment of inertia is analogous to mass, but there is one major difference. Mass is an inherent property of an object and has a single value. The moment of inertia of an object depends on your choice of rotation axis; therefore, an object has no single value of the moment of inertia. An object does have a *minimum* value of the moment of inertia, which is that calculated around an axis passing through the center of mass of the object.

---

[2] Civil engineers use moment of inertia to characterize the elastic properties (rigidity) of such structures as loaded beams. Hence, it is often useful even in a nonrotational context.

Moments of Inertia of Homogeneous Rigid Objects with Different Geometries

Hoop or thin cylindrical shell
$I_{CM} = MR^2$

Hollow cylinder
$I_{CM} = \frac{1}{2}M(R_1^2 + R_2^2)$

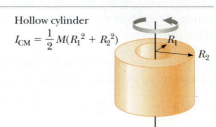

Solid cylinder or disk
$I_{CM} = \frac{1}{2}MR^2$

Rectangular plate
$I_{CM} = \frac{1}{12}M(a^2 + b^2)$

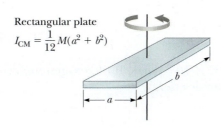

Long, thin rod with rotation axis through center
$I_{CM} = \frac{1}{12}ML^2$

Long, thin rod with rotation axis through end
$I = \frac{1}{3}ML^2$

Solid sphere
$I_{CM} = \frac{2}{5}MR^2$

Thin spherical shell
$I_{CM} = \frac{2}{3}MR^2$

For symmetric objects, the moment of inertia can be expressed in terms of the total mass of the object and one or more dimensions of the object. Table 10.2 shows the moments of inertia of various common symmetric objects.

QUICK QUIZ 10.4 A section of hollow pipe and a solid cylinder have the same radius, mass, and length. They both rotate about their long central axes with the same angular speed. Which object has the higher rotational kinetic energy? **(a)** The hollow pipe does. **(b)** The solid cylinder does. **(c)** They have the same rotational kinetic energy. **(d)** It is impossible to determine.

Example 10.2 | The Oxygen Molecule

Consider the diatomic oxygen molecule O_2, which is rotating in the *xy* plane about the *z* axis passing through its center, perpendicular to its length. The mass of each oxygen atom is 2.66×10^{-26} kg, and at room temperature, the average separation between the two oxygen atoms is $d = 1.21 \times 10^{-10}$ m.

(A) Calculate the moment of inertia of the molecule about the *z* axis.

continued

10.2 *cont.*

SOLUTION

Conceptualize Imagine the thin rod rotating about its center on the left side of Table 10.2. Now imagine placing two identical small spheres on each end of the rod and letting the mass of the rod become infinitesimally small. The result of this imaginary process is a macroscopic mental model for the oxygen molecule.

Categorize We model the molecule as a rigid object, consisting of two particles (the two oxygen atoms), in rotation. We will evaluate results from definitions developed in this section, so we categorize this example as a substitution problem.

Note that the distance of each particle from the z axis is $d/2$. Find the moment of inertia about the z axis:

$$I = \sum_i m_i r_i^2 = m\left(\frac{d}{2}\right)^2 + m\left(\frac{d}{2}\right)^2 = \frac{md^2}{2}$$

$$= \frac{(2.66 \times 10^{-26} \text{ kg})(1.21 \times 10^{-10} \text{ m})^2}{2}$$

$$= \boxed{1.95 \times 10^{-46} \text{ kg} \cdot \text{m}^2}$$

(B) A typical angular speed of a molecule is 4.60×10^{12} rad/s. If the oxygen molecule is rotating with this angular speed about the z axis, what is its rotational kinetic energy?

SOLUTION

Use Equation 10.16 to find the rotational kinetic energy:

$$K_R = \tfrac{1}{2}I\omega^2$$

$$= \tfrac{1}{2}(1.95 \times 10^{-46} \text{ kg} \cdot \text{m}^2)(4.60 \times 10^{12} \text{ rad/s})^2$$

$$= \boxed{2.06 \times 10^{-21} \text{ J}}$$

Example 10.3 | An Unusual Baton

Four tiny spheres are fastened to the ends of two rods of negligible mass lying in the xy plane to form an unusual baton (Fig. 10.7). We shall assume the radii of the spheres are small compared with the dimensions of the rods.

(A) If the system rotates about the y axis (Fig. 10.7a) with an angular speed ω, find the moment of inertia and the rotational kinetic energy of the system about this axis.

SOLUTION

Conceptualize Figure 10.7 is a pictorial representation that helps conceptualize the system of spheres and how it spins.

Categorize This example is a substitution problem because it is a straightforward application of the definitions discussed in this section.

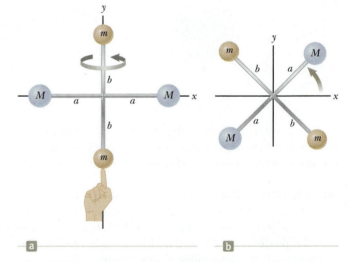

Figure 10.7 (Example 10.3) Four spheres form an unusual baton. (a) The baton is rotated about the y axis. (b) The baton is rotated about the z axis.

Apply Equation 10.15 to the system:

$$I_y = \sum_i m_i r_i^2 = Ma^2 + Ma^2 = \boxed{2Ma^2}$$

Evaluate the rotational kinetic energy using Equation 10.16:

$$K_R = \tfrac{1}{2}I_y\omega^2 = \tfrac{1}{2}(2Ma^2)\omega^2 = \boxed{Ma^2\omega^2}$$

That the two spheres of mass m do not enter into this result makes sense because they have no motion about the axis of rotation; hence, they have no rotational kinetic energy. By similar logic, we expect the moment of inertia about the x axis to be $I_x = 2mb^2$ with a rotational kinetic energy about that axis of $K_R = mb^2\omega^2$.

10.3 *cont.*

(B) Suppose the system rotates in the *xy* plane about an axis (the *z* axis) through the center of the baton (Fig. 10.7b). Calculate the moment of inertia and rotational kinetic energy about this axis.

SOLUTION

Apply Equation 10.15 for this new rotation axis:

$$I_z = \sum_i m_i r_i^2 = Ma^2 + Ma^2 + mb^2 + mb^2 = \boxed{2Ma^2 + 2mb^2}$$

Evaluate the rotational kinetic energy using Equation 10.16:

$$K_R = \tfrac{1}{2}I_z\omega^2 = \tfrac{1}{2}(2Ma^2 + 2mb^2)\omega^2 = \boxed{(Ma^2 + mb^2)\omega^2}$$

Comparing the results for parts (A) and (B), we conclude that the moment of inertia and therefore the rotational kinetic energy associated with a given angular speed depend on the axis of rotation. In part (B), we expect the result to include all four spheres and distances because all four spheres are rotating in the *xy* plane. Based on the work–kinetic energy theorem, the smaller rotational kinetic energy in part (A) than in part (B) indicates it would require less work to set the system into rotation about the *y* axis than about the *z* axis.

What If? What if the mass *M* is much larger than *m*? How do the answers to parts (A) and (B) compare?

Answer If $M \gg m$, then *m* can be neglected and the moment of inertia and the rotational kinetic energy in part (B) become

$$I_z = 2Ma^2 \quad \text{and} \quad K_R = Ma^2\omega^2$$

which are the same as the answers in part (A). If the masses *m* of the two tan spheres in Figure 10.7 are negligible, these spheres can be removed from the figure and rotations about the *y* and *z* axes are equivalent.

Example **10.4** | **Uniform Rigid Rod**

Calculate the moment of inertia of a uniform rigid rod of length *L* and mass *M* (Fig. 10.8) about an axis perpendicular to the rod (the *y′* axis) and passing through its center of mass.

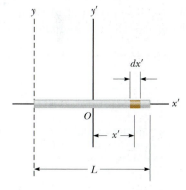

Figure 10.8 (Example 10.4) A uniform rigid rod of length *L*. The moment of inertia about the *y′* axis is less than that about the *y* axis.

SOLUTION

Conceptualize Imagine twirling the rod in Figure 10.8 with your fingers around its midpoint. If you have a meterstick handy, use it to simulate the spinning of a thin rod and feel the resistance it offers to being spun.

Categorize This example is a substitution problem, using the definition of moment of inertia in Equation 10.17. As with any calculus problem, the solution involves reducing the integrand to a single variable.

The shaded length element dx' in Figure 10.8 has a mass dm equal to the mass per unit length λ multiplied by dx'.

Express dm in terms of dx':

$$dm = \lambda\, dx' = \frac{M}{L}\, dx'$$

Substitute this expression into Equation 10.17, with $r^2 = (x')^2$:

$$I_y = \int r^2\, dm = \int_{-L/2}^{L/2} (x')^2 \frac{M}{L}\, dx' = \frac{M}{L} \int_{-L/2}^{L/2} (x')^2\, dx'$$

$$= \frac{M}{L} \left[\frac{(x')^3}{3} \right]_{-L/2}^{L/2} = \boxed{\tfrac{1}{12}ML^2}$$

Check this result in Table 10.2. For practice, calculate the moment of inertia about an axis *y* passing through the end of the rod in Figure 10.8.

Example 10.5 | Uniform Solid Cylinder

A uniform solid cylinder has a radius R, mass M, and length L. Calculate its moment of inertia about its central axis (the z axis in Fig. 10.9).

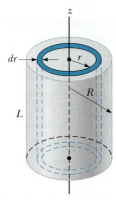

SOLUTION

Conceptualize To simulate this situation, imagine twirling a can of frozen juice around its central axis. Don't twirl a can of vegetable soup; it is not a rigid object! The liquid is able to move relative to the metal can.

Categorize This example is a substitution problem, using the definition of moment of inertia. As with Example 10.4, we must reduce the integrand to a single variable.

It is convenient to divide the cylinder into many cylindrical shells, each having radius r, thickness dr, and length L as shown in Figure 10.9. The density of the cylinder is ρ. The volume dV of each shell is its cross-sectional area multiplied by its length: $dV = L\, dA = L(2\pi r)\, dr$.

Figure 10.9 (Example 10.5) Calculating I about the z axis for a uniform solid cylinder.

Express dm in terms of dr:

$$dm = \rho\, dV = \rho L(2\pi r)\, dr$$

Substitute this expression into Equation 10.17:

$$I_z = \int r^2\, dm = \int r^2 [\rho L(2\pi r)\, dr] = 2\pi \rho L \int_0^R r^3\, dr = \tfrac{1}{2}\pi \rho L R^4$$

Use the total volume $\pi R^2 L$ of the cylinder to express its density:

$$\rho = \frac{M}{V} = \frac{M}{\pi R^2 L}$$

Substitute this value into the expression for I_z:

$$I_z = \tfrac{1}{2}\pi \left(\frac{M}{\pi R^2 L}\right) L R^4 = \tfrac{1}{2}MR^2$$

Check this result in Table 10.2.

What If? What if the length of the cylinder in Figure 10.9 is increased to $2L$, while the mass M and radius R are held fixed? How does that change the moment of inertia of the cylinder?

Answer Notice that the result for the moment of inertia of a cylinder does not depend on L, the length of the cylinder. It applies equally well to a long cylinder and a flat disk having the same mass M and radius R. Therefore, the moment of inertia of the cylinder would not be affected by changing its length.

▶10.5 | Torque and the Vector Product

The component $F \sin \phi$ tends to rotate the wrench about an axis through O.

When a net force is exerted on a rigid object pivoted about some axis and the line of action[3] of the force does not pass through the pivot, the object tends to rotate about that axis. For example, when you push on a door, the door rotates about an axis through the hinges. The tendency of a force to rotate an object about some axis is measured by a vector quantity called **torque**. Torque is the cause of changes in rotational motion and is analogous to force, which causes changes in translational motion. Consider the wrench pivoted about the axis through O in Figure 10.10. The applied force $\vec{F}$ generally can act at an angle ϕ with respect to the position vector $\vec{r}$ locating the point of application of the force. We define the torque τ resulting from the force $\vec{F}$ with the expression[4]

$$\tau \equiv rF \sin \phi \qquad\qquad \text{10.19}◀$$

Torque has units of newton · meters (N · m) in the SI system.[5]

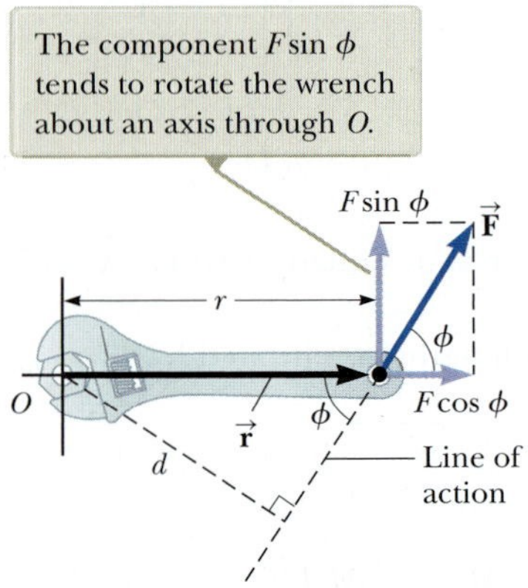

Figure 10.10 A force $\vec{F}$ is applied to a wrench in an effort to loosen a bolt. The force has a greater rotating tendency about an axis through O as F increases and as the moment arm d increases.

---

[3]The line of action of a force is an imaginary line colinear with the force vector and extending to infinity in both directions.

[4]In general, torque is a vector. For rotation about a fixed axis, however, we will use italic, nonbold notation and specify the direction with a positive or a negative sign as we did for angular speed and acceleration in Section 10.1. We will briefly treat the vector nature of torque in a short while.

[5]In Chapter 6, we saw the product of newtons and meters when we defined work and called this product a *joule*. We do not use this term here because the joule is only to be used when discussing energy. For torque, the unit is simply the newton · meter, or N · m.

It is very important to recognize that torque is defined only when a reference axis is specified, from which the distance r is determined. We can interpret Equation 10.19 in two different ways. Looking at the force components in Figure 10.10, we see that the component $F \cos \phi$ parallel to $\vec{r}$ will not cause a rotation of the wrench around the pivot point because its line of action passes right through the pivot point. Similarly, you cannot open a door by pushing on the hinges! Therefore, only the perpendicular component $F \sin \phi$ causes a rotation of the wrench about the pivot. In this case, we can write Equation 10.19 as

$$\tau = r(F \sin \phi)$$

so that the torque is the product of the distance to the point of application of the force and the perpendicular component of the force. In some problems, this method is the easiest way to interpret the calculation of the torque.

The second way to interpret Equation 10.19 is to associate the sine function with the distance r so that we can write

$$\tau = F(r \sin \phi) = Fd$$

The quantity $d = r \sin \phi$, called the **moment arm** (or *lever arm*) of the force $\vec{F}$, represents the perpendicular distance from the rotation axis to the line of action of $\vec{F}$. In some problems, this approach to the calculation of the torque is easier than that of resolving the force into components.

If two or more forces are acting on a rigid object, as in Active Figure 10.11, each has a tendency to produce a rotation about the pivot at O. For example, if the object is initially at rest, $\vec{F}_2$ tends to rotate the object clockwise and $\vec{F}_1$ tends to rotate the object counterclockwise. We shall use the convention that the sign of the torque resulting from a force is positive if its turning tendency is counterclockwise around the rotation axis and negative if its turning tendency is clockwise. For example, in Active Figure 10.11, the torque resulting from $\vec{F}_1$, which has a moment arm of d_1, is *positive* and equal to $+F_1 d_1$; the torque from $\vec{F}_2$ is *negative* and equal to $-F_2 d_2$. Hence, the *net* torque acting on the rigid object about an axis through O is

$$\tau_{net} = \tau_1 + \tau_2 = F_1 d_1 - F_2 d_2$$

From the definition of torque, we see that the rotating tendency increases as F increases and as d increases. For example, we cause more rotation of a door if (a) we push harder or (b) we push at the doorknob rather than at a point close to the hinges. Torque should *not* be confused with force. Torque *depends* on force, but it also depends on *where the force is applied*.

So far, we have not discussed the vector nature of torque aside from assigning a positive or negative value to τ. Consider a force $\vec{F}$ acting on a particle located at the vector position $\vec{r}$ (Active Fig. 10.12). The *magnitude* of the torque due to this force relative to an axis through the origin is $|rF \sin \phi|$, where ϕ is the angle between $\vec{r}$ and $\vec{F}$. The axis about which $\vec{F}$ would tend to produce rotation is perpendicular to the plane formed by $\vec{r}$ and $\vec{F}$. If the force lies in the xy plane, as in Active Figure 10.12, the torque is represented by a vector parallel to the z axis. The force in Active Figure 10.12 creates a torque that tends to rotate the particle counterclockwise when we are looking down the z axis. We define the direction of torque such that the vector $\vec{\tau}$ is in the positive z direction (i.e., coming toward your eyes). If we reverse the direction of $\vec{F}$ in Active Figure 10.12, $\vec{\tau}$ is in the negative z direction. With this choice, the torque vector can be defined to be equal to the **vector product**, or **cross product**, of $\vec{r}$ and $\vec{F}$:

$$\vec{\tau} \equiv \vec{r} \times \vec{F} \qquad \text{10.20} \blacktriangleleft$$

We now give a formal definition of the vector product, first introduced in Section 1.8. Given any two vectors $\vec{A}$ and $\vec{B}$, the vector product $\vec{A} \times \vec{B}$ is defined as a

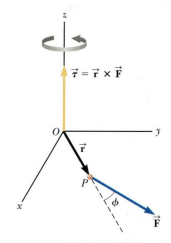

Active Figure 10.11 The force $\vec{F}_1$ tends to rotate the object counterclockwise about an axis through O, and $\vec{F}_2$ tends to rotate the object clockwise.

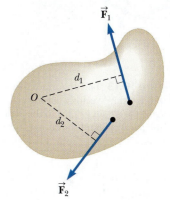

Active Figure 10.12 The torque vector $\vec{\tau}$ lies in a direction perpendicular to the plane formed by the position vector $\vec{r}$ and the applied force vector $\vec{F}$.

▶ Definition of torque using the cross product

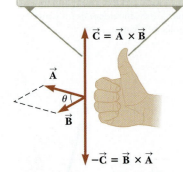

The direction of $\vec{\mathbf{C}}$ is perpendicular to the plane formed by $\vec{\mathbf{A}}$ and $\vec{\mathbf{B}}$, and its direction is determined by the right-hand rule.

$\vec{\mathbf{C}} = \vec{\mathbf{A}} \times \vec{\mathbf{B}}$

$-\vec{\mathbf{C}} = \vec{\mathbf{B}} \times \vec{\mathbf{A}}$

Figure 10.13 The vector product $\vec{\mathbf{A}} \times \vec{\mathbf{B}}$ is a third vector $\vec{\mathbf{C}}$ having a magnitude $AB \sin \theta$ equal to the area of the parallelogram shown.

third vector $\vec{\mathbf{C}}$, the *magnitude* of which is $AB \sin \theta$, where θ is the angle between $\vec{\mathbf{A}}$ and $\vec{\mathbf{B}}$:

$$\vec{\mathbf{C}} = \vec{\mathbf{A}} \times \vec{\mathbf{B}} \qquad \text{10.21} \blacktriangleleft$$

$$C = |\vec{\mathbf{C}}| \equiv AB \sin \theta \qquad \text{10.22} \blacktriangleleft$$

Note that the quantity $AB \sin \theta$ is equal to the area of the parallelogram formed by $\vec{\mathbf{A}}$ and $\vec{\mathbf{B}}$, as shown in Figure 10.13. The *direction* of $\vec{\mathbf{A}} \times \vec{\mathbf{B}}$ is perpendicular to the plane formed by $\vec{\mathbf{A}}$ and $\vec{\mathbf{B}}$ and is determined by the right-hand rule illustrated in Figure 10.13. The four fingers of the right hand are pointed along $\vec{\mathbf{A}}$ and then "wrapped" into $\vec{\mathbf{B}}$ through the angle θ. The direction of the upright thumb is the direction of $\vec{\mathbf{A}} \times \vec{\mathbf{B}}$. Because of the notation, $\vec{\mathbf{A}} \times \vec{\mathbf{B}}$ is often read "$\vec{\mathbf{A}}$ cross $\vec{\mathbf{B}}$," hence the term *cross product*.

Some properties of the vector product follow from its definition:

- Unlike the case of the scalar product, the vector product is not commutative; in fact,

$$\vec{\mathbf{A}} \times \vec{\mathbf{B}} = -\vec{\mathbf{B}} \times \vec{\mathbf{A}} \qquad \text{10.23} \blacktriangleleft$$

Therefore, if you change the order of the vector product, you must change the sign. You can easily verify this relation with the right-hand rule (see Fig. 10.13).
- If $\vec{\mathbf{A}}$ is parallel to $\vec{\mathbf{B}}$ ($\theta = 0°$ or $180°$), then $\vec{\mathbf{A}} \times \vec{\mathbf{B}} = 0$; therefore, it follows that $\vec{\mathbf{A}} \times \vec{\mathbf{A}} = 0$.
- If $\vec{\mathbf{A}}$ is perpendicular to $\vec{\mathbf{B}}$, then $|\vec{\mathbf{A}} \times \vec{\mathbf{B}}| = AB$.
- The vector product obeys the distributive law:

$$\vec{\mathbf{A}} \times (\vec{\mathbf{B}} + \vec{\mathbf{C}}) = \vec{\mathbf{A}} \times \vec{\mathbf{B}} + \vec{\mathbf{A}} \times \vec{\mathbf{C}} \qquad \text{10.24} \blacktriangleleft$$

- The derivative of the vector product with respect to some variable such as t is

$$\frac{d}{dt}(\vec{\mathbf{A}} \times \vec{\mathbf{B}}) = \frac{d\vec{\mathbf{A}}}{dt} \times \vec{\mathbf{B}} + \vec{\mathbf{A}} \times \frac{d\vec{\mathbf{B}}}{dt} \qquad \text{10.25} \blacktriangleleft$$

where it is important to preserve the multiplicative order of the terms on the right side in view of Equation 10.23.

It is left to Problem 10.26 to show, from Equations 10.21 and 10.22 and the definition of unit vectors, that the vector products of the unit vectors $\hat{\mathbf{i}}$, $\hat{\mathbf{j}}$, and $\hat{\mathbf{k}}$ obey the following expressions:

$$\hat{\mathbf{i}} \times \hat{\mathbf{i}} = \hat{\mathbf{j}} \times \hat{\mathbf{j}} = \hat{\mathbf{k}} \times \hat{\mathbf{k}} = 0$$

$$\hat{\mathbf{i}} \times \hat{\mathbf{j}} = -\hat{\mathbf{j}} \times \hat{\mathbf{i}} = \hat{\mathbf{k}} \qquad \text{10.26} \blacktriangleleft$$

$$\hat{\mathbf{j}} \times \hat{\mathbf{k}} = -\hat{\mathbf{k}} \times \hat{\mathbf{j}} = \hat{\mathbf{i}}$$

$$\hat{\mathbf{k}} \times \hat{\mathbf{i}} = -\hat{\mathbf{i}} \times \hat{\mathbf{k}} = \hat{\mathbf{j}}$$

Signs are interchangeable. For example, $\hat{\mathbf{i}} \times (-\hat{\mathbf{j}}) = -\hat{\mathbf{i}} \times \hat{\mathbf{j}} = -\hat{\mathbf{k}}$.

QUICK QUIZ 10.5 (i) If you are trying to loosen a stubborn screw from a piece of wood with a screwdriver and fail, should you find a screwdriver for which the handle is (a) longer or (b) fatter? (ii) If you are trying to loosen a stubborn bolt from a piece of metal with a wrench and fail, should you find a wrench for which the handle is (a) longer or (b) fatter?

Example 10.6 | The Net Torque on a Cylinder

A one-piece cylinder is shaped as shown in Figure 10.14, with a core section protruding from the larger drum. The cylinder is free to rotate about the central z axis shown in the drawing. A rope wrapped around the drum, which has radius R_1, exerts a force $\vec{\mathbf{T}}_1$ to the right on the cylinder. A rope wrapped around the core, which has radius R_2, exerts a force $\vec{\mathbf{T}}_2$ downward on the cylinder.

(A) What is the net torque acting on the cylinder about the rotation axis (which is the z axis in Fig. 10.14)?

SOLUTION

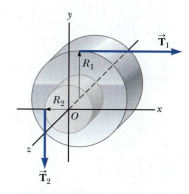

Figure 10.14 (Example 10.6) A solid cylinder pivoted about the z axis through O. The moment arm of $\vec{\mathbf{T}}_1$ is R_1, and the moment arm of $\vec{\mathbf{T}}_2$ is R_2.

Conceptualize Imagine that the cylinder in Figure 10.14 is a shaft in a machine. The force $\vec{\mathbf{T}}_1$ could be applied by a drive belt wrapped around the drum. The force $\vec{\mathbf{T}}_2$ could be applied by a friction brake at the surface of the core.

Categorize This example is a substitution problem in which we evaluate the net torque using Equation 10.19.

The torque due to $\vec{\mathbf{T}}_1$ about the rotation axis is $-R_1 T_1$. (The sign is negative because the torque tends to produce clockwise rotation.) The torque due to $\vec{\mathbf{T}}_2$ is $+R_2 T_2$. (The sign is positive because the torque tends to produce counterclockwise rotation of the cylinder.)

Evaluate the net torque about the rotation axis:

$$\sum \tau = \tau_1 + \tau_2 = \boxed{R_2 T_2 - R_1 T_1}$$

As a quick check, notice that if the two forces are of equal magnitude, the net torque is negative because $R_1 > R_2$. Starting from rest with both forces of equal magnitude acting on it, the cylinder would rotate clockwise because $\vec{\mathbf{T}}_1$ would be more effective at turning it than would $\vec{\mathbf{T}}_2$.

(B) Suppose $T_1 = 5.0$ N, $R_1 = 1.0$ m, $T_2 = 15$ N, and $R_2 = 0.50$ m. What is the net torque about the rotation axis, and which way does the cylinder rotate starting from rest?

SOLUTION

Substitute the given values:

$$\sum \tau = (0.50 \text{ m})(15 \text{ N}) - (1.0 \text{ m})(5.0 \text{ N}) = \boxed{2.5 \text{ N} \cdot \text{m}}$$

Because this net torque is positive, the cylinder begins to rotate in the counterclockwise direction.

Example 10.7 | The Vector Product

Two vectors lying in the xy plane are given by the equations $\vec{\mathbf{A}} = 2\hat{\mathbf{i}} + 3\hat{\mathbf{j}}$ and $\vec{\mathbf{B}} = -\hat{\mathbf{i}} + 2\hat{\mathbf{j}}$. Find $\vec{\mathbf{A}} \times \vec{\mathbf{B}}$ and verify that $\vec{\mathbf{A}} \times \vec{\mathbf{B}} = -\vec{\mathbf{B}} \times \vec{\mathbf{A}}$.

SOLUTION

Conceptualize Given the unit-vector notations of the vectors, think about the directions the vectors point in space. Imagine the parallelogram shown in Figure 10.13 for these vectors.

Categorize Because we use the definition of the cross product discussed in this section, we categorize this example as a substitution problem.

Write the cross product of the two vectors:

$$\vec{\mathbf{A}} \times \vec{\mathbf{B}} = (2\hat{\mathbf{i}} + 3\hat{\mathbf{j}}) \times (-\hat{\mathbf{i}} + 2\hat{\mathbf{j}})$$

Perform the multiplication:

$$\vec{\mathbf{A}} \times \vec{\mathbf{B}} = 2\hat{\mathbf{i}} \times (-\hat{\mathbf{i}}) + 2\hat{\mathbf{i}} \times 2\hat{\mathbf{j}} + 3\hat{\mathbf{j}} \times (-\hat{\mathbf{i}}) + 3\hat{\mathbf{j}} \times 2\hat{\mathbf{j}}$$

Use Equation 10.26 to evaluate the various terms:

$$\vec{\mathbf{A}} \times \vec{\mathbf{B}} = 0 + 4\hat{\mathbf{k}} + 3\hat{\mathbf{k}} + 0 = \boxed{7\hat{\mathbf{k}}}$$

To verify that $\vec{\mathbf{A}} \times \vec{\mathbf{B}} = -\vec{\mathbf{B}} \times \vec{\mathbf{A}}$, evaluate $\vec{\mathbf{B}} \times \vec{\mathbf{A}}$:

$$\vec{\mathbf{B}} \times \vec{\mathbf{A}} = (-\hat{\mathbf{i}} + 2\hat{\mathbf{j}}) \times (2\hat{\mathbf{i}} + 3\hat{\mathbf{j}})$$

Perform the multiplication:

$$\vec{\mathbf{B}} \times \vec{\mathbf{A}} = (-\hat{\mathbf{i}}) \times 2\hat{\mathbf{i}} + (-\hat{\mathbf{i}}) \times 3\hat{\mathbf{j}} + 2\hat{\mathbf{j}} \times 2\hat{\mathbf{i}} + 2\hat{\mathbf{j}} \times 3\hat{\mathbf{j}}$$

Use Equation 10.26 to evaluate the various terms:

$$\vec{\mathbf{B}} \times \vec{\mathbf{A}} = 0 - 3\hat{\mathbf{k}} - 4\hat{\mathbf{k}} + 0 = \boxed{-7\hat{\mathbf{k}}}$$

Therefore, $\vec{\mathbf{A}} \times \vec{\mathbf{B}} = -\vec{\mathbf{B}} \times \vec{\mathbf{A}}$.

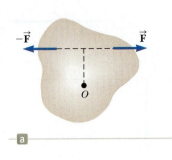

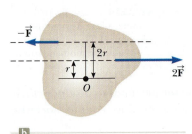

Figure 10.15 (a) The two forces acting on the object are equal in magnitude and opposite in direction. Because they also act along the same line of action, the net torque is zero and the object is in equilibrium. (b) Another situation in which two forces act on an object to produce zero net torque about O (but *not* zero net force).

⟨10.6 | Analysis Model: Rigid Object in Equilibrium

We have defined a rigid object and have discussed torque as the cause of changes in rotational motion of a rigid object. We can now establish analysis models for a rigid object subject to torques that are analogous to those for a particle subject to forces. We begin by imagining a rigid object with balanced torques, which will give us an analysis model that we call the **rigid object in equilibrium.**

Consider two forces of equal magnitude and opposite direction applied to an object as shown in Figure 10.15a. The force directed to the right tends to rotate the object clockwise about an axis perpendicular to the diagram through O, whereas the force directed to the left tends to rotate it counterclockwise about that axis. Because the forces are of equal magnitude and act at the same perpendicular distance from O, their torques are equal in magnitude. Therefore, the net torque on the rigid object is zero. The situation shown in Figure 10.15b is another case in which the net torque about O is zero (although the net *force* on the object is not zero), and we can devise many more cases.

With no net torque, no change occurs in rotational motion and the rotational motion of the rigid object remains in its original state. This state is an equilibrium situation, analogous to translational equilibrium, discussed in Chapter 4.

We now have two conditions for complete equilibrium of an object, which can be stated as follows:

- The net external force must equal zero:

$$\sum \vec{F}_{\text{ext}} = 0 \qquad \text{10.27} \blacktriangleleft$$

- The net external torque must be zero about *any* axis:

$$\sum \vec{\tau}_{\text{ext}} = 0 \qquad \text{10.28} \blacktriangleleft$$

The first condition is a statement of translational equilibrium. The second condition is a statement of rotational equilibrium. In the special case of **static equilibrium**, the object is at rest, so it has no translational or angular speed (i.e., $v_{\text{CM}} = 0$ and $\omega = 0$).

The two vector expressions given by Equations 10.27 and 10.28 are equivalent, in general, to six scalar equations: three from the first condition of equilibrium and three from the second (corresponding to x, y, and z components). Hence, in a complex system involving several forces acting in various directions, you would be faced with solving a set of equations with many unknowns. Here, we restrict our discussion to situations in which all the forces on an object lie in the xy plane. (Forces whose vector representations are in the same plane are said to be *coplanar.*) With this restriction, we need to deal with only three scalar equations. Two of them come from balancing the forces on the object in the x and y directions. The third comes from the torque equation, namely, that the net torque on the object about an axis through *any* point in the xy plane must be zero. Hence, the two conditions of equilibrium provide the equations

$$\sum F_x = 0 \qquad \sum F_y = 0 \qquad \sum \tau_z = 0 \qquad \text{10.29} \blacktriangleleft$$

where the axis of the torque equation is arbitrary. Use these equations when you determine that the rigid object in equilibrium model is appropriate and the forces on the rigid object are in the xy plane.

In working static equilibrium problems, it is important to recognize all external forces acting on the object. Failure to do so will result in an incorrect analysis. The following procedure is recommended when analyzing an object in equilibrium under the action of several external forces:

> **PROBLEM-SOLVING STRATEGY:** Rigid Object in Equilibrium

When analyzing a rigid object in equilibrium under the action of several external forces, use the following procedure.

1. **Conceptualize.** Think about the object that is in equilibrium and identify all the forces on it. Imagine what effect each force would have on the rotation of the object if it were the only force acting.

2. **Categorize.** Confirm that the object under consideration is indeed a rigid object in equilibrium. The object must have zero translational acceleration and zero angular acceleration.

3. **Analyze.** Draw a diagram and label all external forces acting on the object. Try to guess the correct direction for any forces that are not specified. When using the particle under a net force model, the object on which forces act can be represented in a free-body diagram with a dot because it does not matter where on the object the forces are applied. When using the rigid object in equilibrium model, however, we cannot use a dot to represent the object because the location where forces act is important in the calculation. Therefore, in a diagram showing the forces on an object, we must show the actual object or a simplified version of it.

   Resolve all forces into rectangular components, choosing a convenient coordinate system. Then apply the first condition for equilibrium, Equation 10.27. Remember to keep track of the signs of the various force components.

   Choose a convenient axis for calculating the net torque on the rigid object. Remember that the choice of the axis for the torque equation is arbitrary; therefore, choose an axis that simplifies your calculation as much as possible. Usually, the most convenient axis for calculating torques is one through a point at which several forces act, so their torques around this axis are zero. If you don't know a force or don't need to know a force, it is often beneficial to choose an axis through the point at which this force acts. Apply the second condition for equilibrium, Equation 10.28.

   Solve the simultaneous equations for the unknowns in terms of the known quantities.

4. **Finalize.** Make sure your results are consistent with your diagram. If you selected a direction that leads to a negative sign in your solution for a force, do not be alarmed; it merely means that the direction of the force is the opposite of what you guessed. Add up the vertical and horizontal forces on the object and confirm that each set of components adds to zero. Add up the torques on the object and confirm that the sum equals zero.

Example 10.8 | Standing on a Horizontal Beam

A uniform horizontal beam with a length of $\ell = 8.00$ m and a weight of $W_b = 200$ N is attached to a wall by a pin connection. Its far end is supported by a cable that makes an angle of $\phi = 53.0°$ with the beam (Fig. 10.16a, page 322). A person of weight $W_p = 600$ N stands a distance $d = 2.00$ m from the wall. Find the tension in the cable as well as the magnitude and direction of the force exerted by the wall on the beam.

SOLUTION

Conceptualize Imagine that the person in Figure 10.16a moves outward on the beam. It seems reasonable that the farther he moves outward, the larger the torque he applies about the pivot and the larger the tension in the cable must be to balance this torque.

Categorize Because the system is at rest, we categorize the beam as a rigid object in equilibrium.

continued

10.8 *cont.*

Analyze We identify all the external forces acting on the beam: the 200-N gravitational force, the force $\vec{T}$ exerted by the cable, the force $\vec{R}$ exerted by the wall at the pivot, and the 600-N force that the person exerts on the beam. These forces are all indicated in the force diagram for the beam shown in Figure 10.16b. When we assign directions for forces, it is sometimes helpful to imagine what would happen if a force were suddenly removed. For example, if the wall were to vanish suddenly, the left end of the beam would move to the left as it begins to fall. This scenario tells us that the wall is not only holding the beam up but is also pressing outward against it. Therefore, we draw the vector $\vec{R}$ in the direction shown in Figure 10.16b. Figure 10.16c shows the horizontal and vertical components of $\vec{T}$ and $\vec{R}$.

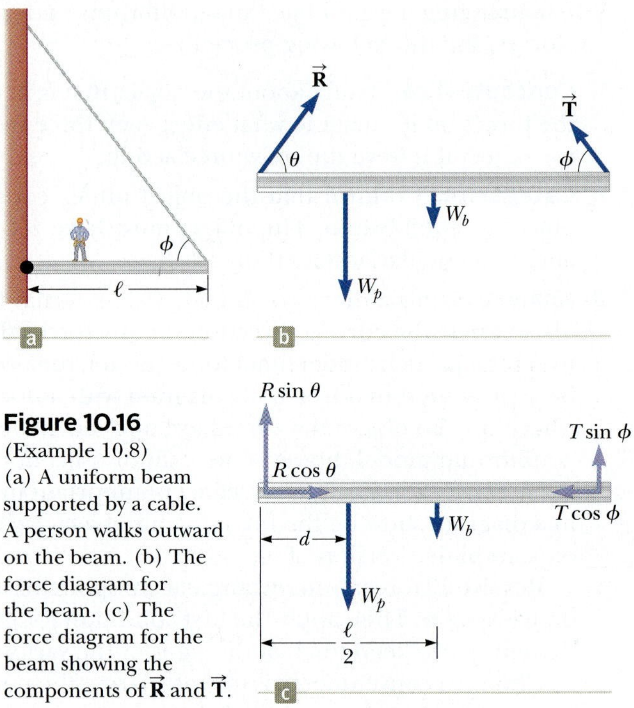

Substitute expressions for the forces on the beam into Equation 10.27:

(1) $\sum F_x = R\cos\theta - T\cos\phi = 0$

(2) $\sum F_y = R\sin\theta + T\sin\phi - W_p - W_b = 0$

where we have chosen rightward and upward as our positive directions. Because R, T, and θ are all unknown, we cannot obtain a solution from these expressions alone. (To solve for the unknowns, the number of simultaneous equations must generally equal the number of unknowns.)

Now let's invoke the condition for rotational equilibrium. A convenient axis to choose for our torque equation is the one that passes through the pin connection. The feature that makes this axis so convenient is that the force $\vec{R}$ and the horizontal component of $\vec{T}$ both have a moment arm of zero; hence, these forces produce no torque about this axis.

Figure 10.16
(Example 10.8)
(a) A uniform beam supported by a cable. A person walks outward on the beam. (b) The force diagram for the beam. (c) The force diagram for the beam showing the components of $\vec{R}$ and $\vec{T}$.

Substitute expressions for the torques on the beam into Equation 10.28:

$$\sum\tau_z = (T\sin\phi)(\ell) - W_pd - W_b\left(\frac{\ell}{2}\right) = 0$$

This equation contains only T as an unknown because of our choice of rotation axis. Solve for T and substitute numerical values:

$$T = \frac{W_pd + W_b(\ell/2)}{\ell\sin\phi} = \frac{(600\text{ N})(2.00\text{ m}) + (200\text{ N})(4.00\text{ m})}{(8.00\text{ m})\sin 53.0°} = \boxed{313\text{ N}}$$

Rearrange Equations (1) and (2) and then divide:

$$\frac{R\sin\theta}{R\cos\theta} = \tan\theta = \frac{W_p + W_b - T\sin\phi}{T\cos\phi}$$

Solve for θ and substitute numerical values:

$$\theta = \tan^{-1}\left(\frac{W_p + W_b - T\sin\phi}{T\cos\phi}\right)$$

$$= \tan^{-1}\left[\frac{600\text{ N} + 200\text{ N} - (313\text{ N})\sin 53.0°}{(313\text{ N})\cos 53.0°}\right] = \boxed{71.1°}$$

Solve Equation (1) for R and substitute numerical values:

$$R = \frac{T\cos\phi}{\cos\theta} = \frac{(313\text{ N})\cos 53.0°}{\cos 71.1°} = \boxed{581\text{ N}}$$

Finalize The positive value for the angle θ indicates that our estimate of the direction of $\vec{R}$ was accurate.

Had we selected some other axis for the torque equation, the solution might differ in the details but the answers would be the same. For example, had we chosen an axis through the center of gravity of the beam, the torque equation would involve both T and R. This equation, coupled with Equations (1) and (2), however, could still be solved for the unknowns. Try it!

Example **10.9** | **The Leaning Ladder**

A uniform ladder of length ℓ rests against a smooth, vertical wall (Fig. 10.17a). The mass of the ladder is m, and the coefficient of static friction between the ladder and the ground is $\mu_s = 0.40$. Find the minimum angle θ_{min} at which the ladder does not slip.

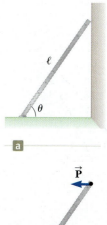

SOLUTION

Conceptualize Think about any ladders you have climbed. Do you want a large friction force between the bottom of the ladder and the surface or a small one? If the friction force is zero, will the ladder stay up? Simulate a ladder with a ruler leaning against a vertical surface. Does the ruler slip at some angles and stay up at others?

Categorize We do not wish the ladder to slip, so we model it as a rigid object in equilibrium.

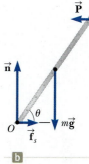

Analyze A diagram showing all the external forces acting on the ladder is illustrated in Figure 10.17b. The force exerted by the ground on the ladder is the vector sum of a normal force $\vec{n}$ and the force of static friction $\vec{f}_s$. The force $\vec{P}$ exerted by the wall on the ladder is horizontal because the wall is frictionless.

Figure 10.17 (Example 10.9) (a) A uniform ladder at rest, leaning against a smooth wall. The ground is rough. (b) The forces on the ladder.

Apply the first condition for equilibrium to the ladder:

(1) $\sum F_x = f_s - P = 0$

(2) $\sum F_y = n - mg = 0$

Solve Equation (1) for P:

(3) $P = f_s$

Solve Equation (2) for n:

(4) $n = mg$

When the ladder is on the verge of slipping, the force of static friction must have its maximum value, which is given by $f_{s,max} = \mu_s n$. Combine this equation with Equations (3) and (4):

(5) $P = f_{s,max} = \mu_s n = \mu_s mg$

Apply the second condition for equilibrium to the ladder, taking torques about an axis through O:

$$\sum \tau_O = P\ell \sin \theta_{min} - mg\frac{\ell}{2} \cos \theta_{min} = 0$$

Solve for $\tan \theta_{min}$ and substitute for P from Equation (5):

$$\frac{\sin \theta_{min}}{\cos \theta_{min}} = \tan \theta_{min} = \frac{mg}{2P} = \frac{mg}{2\mu_s mg} = \frac{1}{2\mu_s}$$

Solve for the angle θ_{min}:

$$\theta_{min} = \tan^{-1}\left(\frac{1}{2\mu_s}\right) = \tan^{-1}\left[\frac{1}{2(0.40)}\right] = \boxed{51°}$$

Finalize Notice that the angle depends only on the coefficient of friction, not on the mass or length of the ladder.

10.7 | Analysis Model: Rigid Object Under a Net Torque

In the preceding section, we investigated the equilibrium situation in which the net torque on a rigid object is zero. What if the net torque on a rigid object is not zero? In analogy with Newton's second law for translational motion, we should expect the angular speed of the rigid object to change. The net torque will cause angular acceleration of the rigid object. We describe this situation with a new analysis model, the **rigid object under a net torque,** and investigate this model in this section.

Let us imagine a rotating rigid object again as a collection of particles. The rigid object will be subject to a number of forces applied at various locations on the rigid object at which individual particles will be located. Therefore, we can imagine that the forces on the rigid object are exerted on individual particles of the rigid object. We will calculate the net torque on the object due to the torques resulting from these forces around the rotation axis of the rotating object. Any applied force can be represented by its radial component and its tangential component. The radial component of an applied force provides no torque because its line of action goes through the rotation axis. Therefore, only the tangential component of an applied force contributes to the torque.

On any given particle, described by index variable i, within the rigid object, we can use Newton's second law to describe the tangential acceleration of the particle:

$$F_{ti} = m_i a_{ti}$$

where the t subscript refers to tangential components. Let us multiply both sides of this expression by r_i, the distance of the particle from the rotation axis:

$$r_i F_{ti} = r_i m_i a_{ti}$$

Using Equation 10.11 and recognizing the definition of torque ($\tau = rF\sin\phi = rF_t$ in this case), we can rewrite this expression as

$$\tau_i = m_i r_i^2 \alpha_i$$

Now, let us add up the torques on all particles of the rigid object:

$$\sum_i \tau_i = \sum_i m_i r_i^2 \alpha_i$$

The left side is the net torque on all particles of the rigid object. The net torque associated with *internal* forces is zero, however. To understand why, recall that Newton's third law tells us that the internal forces occur in equal and opposite pairs that lie along the line of separation of each pair of particles. The torque due to each action–reaction force pair is therefore zero. On summation of all torques, we see that the *net internal torque vanishes*. The term on the left, then, reduces to the net *external* torque.

On the right, we adopt the rigid object model by demanding that all particles have the same angular acceleration α. Therefore, this equation becomes

$$\sum \tau_{\text{ext}} = \left(\sum_i m_i r_i^2\right)\alpha$$

where the torque and angular acceleration no longer have subscripts because they refer to quantities associated with the rigid object as a whole rather than to individual particles. We recognize the quantity in parentheses as the moment of inertia I of the rigid object. Therefore,

▶ Rotational analog to Newton's
second law

$$\boxed{\sum \tau_{\text{ext}} = I\alpha} \qquad \qquad \textbf{10.30} ◀$$

That is, the net torque acting on the rigid object is proportional to its angular acceleration, and the proportionality constant is the moment of inertia. It is important to note that $\sum \tau_{\text{ext}} = I\alpha$ is the rotational analog of Newton's second law of motion for a system of particles (Eq. 8.40), $\sum F_{\text{ext}} = Ma_{\text{CM}}$.

QUICK QUIZ 10.6 You turn off your electric drill and find that the time interval for the rotating bit to come to rest due to frictional torque in the drill is Δt. You replace the bit with a larger one that results in a doubling of the moment of inertia of the drill's entire rotating mechanism. When this larger bit is rotated at the same angular speed as the first, and the drill is turned off, the frictional torque remains the same as that for the previous situation. What is the time interval for this second bit to come to rest?   (a) $4\Delta t$ (b) $2\Delta t$ (c) Δt (d) $0.5\Delta t$ (e) $0.25\Delta t$ (f) impossible to determine

Example 10.10 | Angular Acceleration of a Wheel

A wheel of radius R, mass M, and moment of inertia I is mounted on a frictionless, horizontal axle as in Figure 10.18. A light cord wrapped around the wheel supports an object of mass m. When the wheel is released, the object accelerates downward, the cord unwraps off the wheel, and the wheel rotates with an angular acceleration. Find expressions for the angular acceleration of the wheel, the translational acceleration of the object, and the tension in the cord.

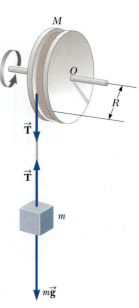

SOLUTION

Conceptualize Imagine that the object is a bucket in an old-fashioned wishing well. It is tied to a cord that passes around a cylinder equipped with a crank for raising the bucket. After the bucket has been raised, the system is released and the bucket accelerates downward while the cord unwinds off the cylinder.

Categorize We apply two analysis models here. The object is modeled as a particle under a net force. The wheel is modeled as a rigid object under a net torque.

Analyze The magnitude of the torque acting on the wheel about its axis of rotation is $\tau = TR$, where T is the force exerted by the cord on the rim of the wheel. (The gravitational force exerted by the Earth on the wheel and the normal force exerted by the axle on the wheel both pass through the axis of rotation and therefore produce no torque.)

Figure 10.18 (Example 10.10) An object hangs from a cord wrapped around a wheel.

Write Equation 10.30:

$$\sum \tau_{ext} = I\alpha$$

Solve for α and substitute the net torque:

$$(1) \ \alpha = \frac{\sum \tau_{ext}}{I} = \frac{TR}{I}$$

Apply Newton's second law to the motion of the object, taking the downward direction to be positive:

$$\sum F_y = mg - T = ma$$

Solve for the acceleration a:

$$(2) \ a = \frac{mg - T}{m}$$

Equations (1) and (2) have three unknowns: α, a, and T. Because the object and wheel are connected by a cord that does not slip, the translational acceleration of the suspended object is equal to the tangential acceleration of a point on the wheel's rim. Therefore, the angular acceleration α of the wheel and the translational acceleration of the object are related by $a = R\alpha$.

Use this fact together with Equations (1) and (2):

$$(3) \ a = R\alpha = \frac{TR^2}{I} = \frac{mg - T}{m}$$

Solve for the tension T:

$$(4) \ T = \frac{mg}{1 + (mR^2/I)}$$

Substitute Equation (4) into Equation (2) and solve for a:

$$(5) \ a = \frac{g}{1 + (I/mR^2)}$$

Use $a = R\alpha$ and Equation (5) to solve for α:

$$\alpha = \frac{a}{R} = \frac{g}{R + (I/mR)}$$

Finalize We finalize this problem by imagining the behavior of the system in some extreme limits.

What If? What if the wheel were to become very massive so that I becomes very large? What happens to the acceleration a of the object and the tension T?

Answer If the wheel becomes infinitely massive, we can imagine that the object of mass m will simply hang from the cord without causing the wheel to rotate.

We can show that mathematically by taking the limit $I \rightarrow \infty$. Equation (5) then becomes

$$a = \frac{g}{1 + (I/mR^2)} \ \rightarrow \ 0$$

which agrees with our conceptual conclusion that the object will hang at rest. Also, Equation (4) becomes

$$T = \frac{mg}{1 + (mR^2/I)} \ \rightarrow \ mg$$

which is consistent because the object simply hangs at rest in equilibrium between the gravitational force and the tension in the string.

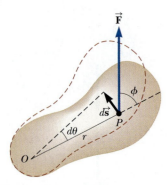

Figure 10.19 A rigid object rotates about an axis through O under the action of an external force $\vec{F}$ applied at P.

⟨10.8 | Energy Considerations in Rotational Motion

In translational motion, we found energy concepts, and in particular the reduction of the conservation of energy equation called the work–kinetic energy theorem, to be extremely useful in describing the motion of a system. Energy concepts can be equally useful in simplifying the analysis of rotational motion. From the conservation of energy equation, we expect that for rotation of an object about a fixed axis, the work done by external forces on the object will equal the change in the rotational kinetic energy as long as energy is not stored by any other means. To show that this case is in fact true, we begin by finding an expression for the work done by a torque.

Consider a rigid object pivoted at the point O in Figure 10.19. Suppose a single external force $\vec{F}$ is applied at the point P and $d\vec{s}$ is the displacement of the point of application of the force. The small amount of work dW done on the object by $\vec{F}$ as the point of application rotates through an infinitesimal distance $ds = r\,d\theta$ in a time interval dt is

$$dW = \vec{F} \cdot d\vec{s} = (F\sin\phi)\,r\,d\theta$$

where $F\sin\phi$ is the tangential component of $\vec{F}$, or the component of the force along the displacement. Note from Figure 10.19 that the radial component of $\vec{F}$ does no work because it is perpendicular to the displacement of the point of application of the force.

Because the magnitude of the torque due to $\vec{F}$ about the origin is defined as $rF\sin\phi$, we can write the work done on the object for the infinitesimal rotation in the form

$$dW = \tau\,d\theta \qquad \qquad \textbf{10.31}◀$$

Notice that this expression is the product of torque and angular displacement, making it analogous to the work done on the object in translational motion, which is the product of force and translational displacement.

Now, we will combine this result with the rotational form of Newton's second law, $\tau = I\alpha$. Using the chain rule from calculus, we can express the torque as

$$\tau = I\alpha = I\frac{d\omega}{dt} = I\frac{d\omega}{d\theta}\frac{d\theta}{dt} = I\frac{d\omega}{d\theta}\omega$$

Rearranging this expression and noting that $\tau\,d\theta = dW$ from Equation 10.31, we have

$$\tau\,d\theta = dW = I\omega\,d\omega$$

Integrating this expression, we find the total work done by the torque:

$$W = \int_{\theta_i}^{\theta_f} \tau\,d\theta = \int_{\omega_i}^{\omega_f} I\omega\,d\omega$$

$$\boxed{W = \tfrac{1}{2}I\omega_f^{\,2} - \tfrac{1}{2}I\omega_i^{\,2} = \Delta K_R} \qquad \textbf{10.32}◀$$

▶ Work–kinetic energy theorem for pure rotation

Notice that this equation has exactly the same mathematical form as the work–kinetic energy theorem for translation. Equation 10.32 is a form of the nonisolated system (energy) model discussed in Chapter 7. Work is done on the system of the rigid object, which represents a transfer of energy across the boundary of the system that appears as an increase in the object's rotational kinetic energy.

In general, we can combine this theorem with the translational form of the work–kinetic energy theorem from Chapter 6. Therefore, the net work done by external forces on an object is the change in its *total* kinetic energy, which is the sum of the translational and rotational kinetic energies. For example, when a pitcher throws a baseball, the work done by the pitcher's hands appears as kinetic energy associated with the ball moving through space as well as rotational kinetic energy associated with the spinning of the ball.

In addition to the work–kinetic energy theorem, other energy principles can also be applied to rotational situations. For example, if a system involving rotating objects is isolated and no nonconservative forces act within the system, the isolated system model and the principle of conservation of mechanical energy can be used to analyze the system as in Example 10.11 below.

We finish this discussion of energy concepts for rotation by investigating the *rate* at which work is being done by $\vec{F}$ on an object rotating about a fixed axis. This rate is obtained by dividing the left and right sides of Equation 10.31 by dt:

$$\frac{dW}{dt} = \tau \frac{d\theta}{dt}$$

10.33 ◀

The quantity dW/dt is, by definition, the instantaneous power P delivered by the force. Furthermore, because $d\theta/dt = \omega$, Equation 10.33 reduces to

$$P = \tau\omega$$

10.34 ◀   ▶ Power delivered to a rotating object

This expression is analogous to $P = Fv$ in the case of translational motion.

Example 10.11 | Rotating Rod

A uniform rod of length L and mass M is free to rotate on a frictionless pin passing through one end (Fig. 10.20). The rod is released from rest in the horizontal position.

(A) What is its angular speed when the rod reaches its lowest position?

SOLUTION

Conceptualize Consider Figure 10.20 and imagine the rod rotating downward through a quarter turn about the pivot at the left end.

Categorize The angular acceleration of the rod is not constant. Therefore, the kinematic equations for rotation (Section 10.2) cannot be used to solve this example. We categorize the system of the rod and the Earth as an isolated system in terms of energy with no nonconservative forces acting and use the principle of conservation of mechanical energy.

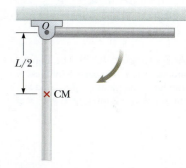

Figure 10.20 (Example 10.11) A uniform rigid rod pivoted at O rotates in a vertical plane under the action of the gravitational force.

Analyze We choose the configuration in which the rod is hanging straight down as the reference configuration for gravitational potential energy and assign a value of zero for this configuration. When the rod is in the horizontal position, it has no rotational kinetic energy. The potential energy of the system in this configuration relative to the reference configuration is $MgL/2$ because the center of mass of the rod is at a height $L/2$ higher than its position in the reference configuration. When the rod reaches its lowest position, the energy of the system is entirely rotational energy $\frac{1}{2}I\omega^2$, where I is the moment of inertia of the rod about an axis passing through the pivot.

Using the isolated system (energy) model, write a conservation of mechanical energy equation for the system:

$$K_f + U_f = K_i + U_i$$

Substitute for each of the energies:

$$\tfrac{1}{2}I\omega^2 + 0 = 0 + \tfrac{1}{2}MgL$$

Solve for ω and use $I = \frac{1}{3}ML^2$ (see Table 10.2) for the rod:

$$\omega = \sqrt{\frac{MgL}{I}} = \sqrt{\frac{MgL}{\frac{1}{3}ML^2}} = \sqrt{\frac{3g}{L}}$$

(B) Determine the tangential speed of the center of mass and the tangential speed of the lowest point on the rod when it is in the vertical position.

SOLUTION

Use Equation 10.10 and the result from part (A):

$$v_{CM} = r\omega = \frac{L}{2}\omega = \tfrac{1}{2}\sqrt{3gL}$$

Because r for the lowest point on the rod is twice what it is for the center of mass, the lowest point has a tangential speed twice that of the center of mass:

$$v = 2v_{CM} = \sqrt{3gL}$$

Finalize Applying an energy approach allows us to find the angular speed of the rod at the lowest point. Convince yourself that you could find the angular speed of the rod at any angular position by knowing the location of the center of mass at this position.

10.9 | Analysis Model: Nonisolated System (Angular Momentum)

> The angular momentum $\vec{\mathbf{L}}$ depends on the origin about which it is measured and is a vector perpendicular to both $\vec{\mathbf{r}}$ and $\vec{\mathbf{p}}$.

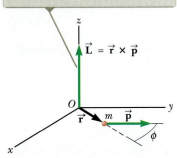

Active Figure 10.21 The angular momentum $\vec{\mathbf{L}}$ of a particle of mass m and linear momentum $\vec{\mathbf{p}}$ located at the position $\vec{\mathbf{r}}$ is given by $\vec{\mathbf{L}} = \vec{\mathbf{r}} \times \vec{\mathbf{p}}$.

> **Pitfall Prevention | 10.6**
> **Is Rotation Necessary for Angular Momentum?**
> We can define angular momentum even if the particle is not moving in a circular path. Even a particle moving in a straight line has angular momentum about any axis displaced from the path of the particle.

Imagine an object rotating in space with no motion of its center of mass. Each particle in the object is moving in a circular path, so momentum is associated with the motion of each particle. Although the object has no linear momentum (its center of mass is not moving through space), particles in the object are in motion, so a "quantity of motion" is associated with its rotation. We will investigate the **angular momentum** that the object has in this section.

Consider a particle of mass m, situated at the vector position $\vec{\mathbf{r}}$ and moving with a momentum $\vec{\mathbf{p}}$, as shown in Active Figure 10.21. For now, we don't consider it as a particle on a rigid object; it is any particle moving with momentum $\vec{\mathbf{p}}$. We will apply the result to a rotating rigid object shortly. The **instantaneous angular momentum $\vec{\mathbf{L}}$** of the particle relative to the origin O is defined by the vector product of its instantaneous position vector $\vec{\mathbf{r}}$ and the instantaneous linear momentum $\vec{\mathbf{p}}$:

$$\vec{\mathbf{L}} \equiv \vec{\mathbf{r}} \times \vec{\mathbf{p}} \qquad \text{10.35} \blacktriangleleft$$

The SI units of angular momentum are $\text{kg} \cdot \text{m}^2/\text{s}$. Note that both the magnitude and the direction of $\vec{\mathbf{L}}$ depend on the choice of origin. The direction of $\vec{\mathbf{L}}$ is perpendicular to the plane formed by $\vec{\mathbf{r}}$ and $\vec{\mathbf{p}}$, and the sense of $\vec{\mathbf{L}}$ is governed by the right-hand rule. For example, in Active Figure 10.21, $\vec{\mathbf{r}}$ and $\vec{\mathbf{p}}$ are assumed to be in the xy plane and $\vec{\mathbf{L}}$ points in the z direction. Because $\vec{\mathbf{p}} = m\vec{\mathbf{v}}$, the magnitude of $\vec{\mathbf{L}}$ is

$$L = mvr\sin\phi \qquad \text{10.36} \blacktriangleleft$$

where ϕ is the angle between $\vec{\mathbf{r}}$ and $\vec{\mathbf{p}}$. It follows that $\vec{\mathbf{L}}$ is zero when $\vec{\mathbf{r}}$ is parallel to $\vec{\mathbf{p}}$ ($\phi = 0°$ or $180°$). In other words, when the particle moves along a line that passes through the origin, it has zero angular momentum with respect to the origin, which is equivalent to stating that the momentum vector is not tangent to *any* circle drawn about the origin. On the other hand, if $\vec{\mathbf{r}}$ is perpendicular to $\vec{\mathbf{p}}$ ($\phi = 90°$), L is a maximum and equal to mvr. In fact, at that instant the particle moves exactly as though it were on the rim of a wheel of radius r rotating at angular speed $\omega = v/r$ about an axis through the origin in a plane defined by $\vec{\mathbf{r}}$ and $\vec{\mathbf{p}}$. A particle has nonzero angular momentum about some point if the position vector of the particle measured from that point rotates about the point as the particle moves.

For translational motion, we found that the net force on a particle equals the time rate of change of the particle's linear momentum (Eq. 8.4). We shall now show that Newton's second law implies an analogous situation for rotation: that the net torque acting on a particle equals the time rate of change of the particle's angular momentum. Let us start by writing the torque on the particle in the form

$$\vec{\boldsymbol{\tau}} = \vec{\mathbf{r}} \times \vec{\mathbf{F}} = \vec{\mathbf{r}} \times \frac{d\vec{\mathbf{p}}}{dt} \qquad \text{10.37} \blacktriangleleft$$

where we have used $\vec{\mathbf{F}} = d\vec{\mathbf{p}}/dt$ (Eq. 8.4). Now let us differentiate Equation 10.35 with respect to time, using the product rule for differentiation (Eq. 10.25):

$$\frac{d\vec{\mathbf{L}}}{dt} = \frac{d}{dt}(\vec{\mathbf{r}} \times \vec{\mathbf{p}}) = \frac{d\vec{\mathbf{r}}}{dt} \times \vec{\mathbf{p}} + \vec{\mathbf{r}} \times \frac{d\vec{\mathbf{p}}}{dt}$$

It is important to adhere to the order of factors in the vector product because the vector product is not commutative, as we saw in Section 10.5.

The first term on the right in the preceding equation is zero because $\vec{\mathbf{v}} = d\vec{\mathbf{r}}/dt$ is parallel to $\vec{\mathbf{p}}$. Therefore,

$$\frac{d\vec{\mathbf{L}}}{dt} = \vec{\mathbf{r}} \times \frac{d\vec{\mathbf{p}}}{dt} \qquad \text{10.38} \blacktriangleleft$$

Comparing Equations 10.37 and 10.38, we see that

> ▶ Torque on a particle equals time rate of change of angular momentum of the particle

$$\vec{\boldsymbol{\tau}} = \frac{d\vec{\mathbf{L}}}{dt} \qquad \text{10.39} \blacktriangleleft$$

This result is the rotational analog of Newton's second law, $\vec{F} = d\vec{p}/dt$. Equation 10.39 says that the torque acting on a particle is equal to the time rate of change of the particle's angular momentum. Note that Equation 10.39 is valid only if the axes used to define $\vec{\tau}$ and $\vec{L}$ are the *same*. Equation 10.39 is also valid when several forces are acting on the particle, in which case $\vec{\tau}$ is the *net* torque on the particle. Of course, the same origin must be used in calculating all torques as well as the angular momentum.

Now, let us apply these ideas to a system of particles. The total angular momentum $\vec{L}$ of the system of particles about some point is defined as the vector sum of the angular momenta of the individual particles:

$$\vec{L} = \vec{L}_1 + \vec{L}_2 + \cdots + \vec{L}_n = \sum_i \vec{L}_i$$

where the vector sum is over all the n particles in the system.

Because the individual angular momenta of the particles may change in time, the total angular momentum may also vary in time. In fact, the time rate of change of the total angular momentum of the system equals the vector sum of *all* torques, including those associated with internal forces between particles and those associated with external forces.

As we found in our discussion of the rigid object under a net torque, however, the sum of the internal torques is zero. Therefore, we conclude that the total angular momentum can vary with time *only* if there is a net *external* torque on the system, so that we have

$$\sum \vec{\tau}_{ext} = \sum_i \frac{d\vec{L}_i}{dt} = \frac{d}{dt} \sum_i \vec{L}_i$$

$$\sum \vec{\tau}_{ext} = \frac{d\vec{L}_{tot}}{dt} \qquad \textbf{10.40} \blacktriangleleft$$

▶ Net external torque on a system equals time rate of change of angular momentum of the system

That is, the time rate of change of the total angular momentum of the system about some origin in an inertial frame equals the net external torque acting on the system about that origin. Note that Equation 10.40 is the rotational analog of $\sum \vec{F}_{ext} = d\vec{p}_{tot}/dt$ (Eq. 8.40) for a system of particles.

This result is valid for a system of particles that change their positions with respect to one another, that is, a nonrigid object. In this discussion of angular momentum of a system of particles, notice that we never imposed the rigid-object condition.

Equation 10.40 is the primary equation in the **angular momentum version of the nonisolated system model.** The system's angular momentum changes in response to an interaction with the environment, described by means of the net torque on the system.

One final result can be obtained for angular momentum, which will serve as an analog to the definition of linear momentum. Let us imagine a rigid object rotating about an axis. Each particle of mass m_i in the rigid object moves in a circular path of radius r_i, with a tangential speed v_i. Therefore, the total angular momentum of the rigid object is

$$L = \sum_i m_i v_i r_i$$

Let us now replace the tangential speed with the product of the radial distance and the angular speed (Eq. 10.10):

$$L = \sum_i m_i v_i r_i = \sum_i m_i (r_i \omega) r_i = \left(\sum_i m_i r_i^2 \right) \omega$$

We recognize the combination in the parentheses as the moment of inertia, so we can write the angular momentum of the rigid object as

$$L = I\omega \qquad \textbf{10.41} \blacktriangleleft$$

▶ Angular momentum of an object with moment of inertia I

> **TABLE 10.3** | A Comparison of Equations for Rotational and Translational Motion: Dynamic Equations

	Rotational Motion About a Fixed Axis	Translational Motion
Kinetic energy	$K_R = \frac{1}{2}I\omega^2$	$K = \frac{1}{2}mv^2$
Equilibrium	$\sum \vec{\tau}_{\text{ext}} = 0$	$\sum \vec{F}_{\text{ext}} = 0$
Newton's second law	$\sum \tau_{\text{ext}} = I\alpha$	$\sum \vec{F}_{\text{ext}} = m\vec{a}$
Nonisolated system	$\vec{\tau}_{\text{ext}} = \dfrac{d\vec{L}_{\text{tot}}}{dt}$	$\vec{F}_{\text{ext}} = \dfrac{d\vec{p}_{\text{tot}}}{dt}$
Momentum	$L = I\omega$	$\vec{p} = m\vec{v}$
Isolated system	$\vec{L}_i = \vec{L}_f$	$\vec{p}_i = \vec{p}_f$
Power	$P = \tau\omega$	$P = Fv$

Note: Equations in translation motion expressed in terms of vectors have rotational analogs in terms of vectors. Because the full vector treatment of rotation is beyond the scope of this book, however, some rotational equations are given in nonvector form.

which is the rotational analog to $p = mv$. Table 10.3 is a continuation of Table 10.1, with additional translational and rotational analogs that we have developed in the past few sections and one that we will develop in the next section.

> **QUICK QUIZ 10.7** A solid sphere and a hollow sphere have the same mass and radius. They are rotating with the same angular speed. Which is the one with the higher angular momentum? (a) the solid sphere (b) the hollow sphere (c) both have the same angular momentum (d) impossible to determine

Example 10.12 | A System of Objects

A sphere of mass m_1 and a block of mass m_2 are connected by a light cord that passes over a pulley as shown in Figure 10.22. The radius of the pulley is R, and the mass of the thin rim is M. The spokes of the pulley have negligible mass. The block slides on a frictionless, horizontal surface. Find an expression for the linear acceleration of the two objects, using the concepts of angular momentum and torque.

SOLUTION

Conceptualize When the system is released, the block slides to the left, the sphere drops downward, and the pulley rotates counterclockwise. This situation is similar to problems we have solved earlier except that now we want to use an angular momentum approach.

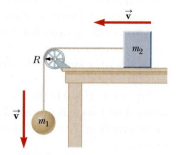

Figure 10.22 (Example 10.12) When the system is released, the sphere moves downward and the block moves to the left.

Categorize We identify the block, pulley, and sphere as a nonisolated system, subject to the external torque due to the gravitational force on the sphere. We shall calculate the angular momentum about an axis that coincides with the axle of the pulley. The angular momentum of the system includes that of two objects moving translationally (the sphere and the block) and one object undergoing pure rotation (the pulley).

Analyze At any instant of time, the sphere and the block have a common speed v, so the angular momentum of the sphere is m_1vR and that of the block is m_2vR. At the same instant, all points on the rim of the pulley also move with speed v, so the angular momentum of the pulley is MvR.

Now let's address the total external torque acting on the system about the pulley axle. Because it has a moment arm of zero, the force exerted by the axle on the pulley does not contribute to the torque. Furthermore, the normal force acting on the block is balanced by the gravitational force $m_2\vec{g}$, so these forces do not contribute to the torque. The gravitational force $m_1\vec{g}$ acting on the sphere produces a torque about the axle equal in magnitude to m_1gR, where R is the moment arm of the force about the axle. This result is the total external torque about the pulley axle; that is, $\sum \tau_{\text{ext}} = m_1gR$.

10.12 *cont.*

Write an expression for the total angular momentum of the system:

$$(1) \quad L_{\text{tot}} = m_1 vR + m_2 vR + MvR = (m_1 + m_2 + M)vR$$

Substitute this expression and the total external torque into Equation 10.40:

$$\sum \tau_{\text{ext}} = \frac{dL_{\text{tot}}}{dt}$$

$$m_1 gR = \frac{d}{dt}[(m_1 + m_2 + M)vR]$$

$$(2) \quad m_1 gR = (m_1 + m_2 + M)R\frac{dv}{dt}$$

Recognizing that $dv/dt = a$, solve Equation (2) for a:

$$(3) \quad a = \frac{m_1 g}{m_1 + m_2 + M}$$

Finalize When we evaluated the net torque about the axle, we did not include the forces that the cord exerts on the objects because these forces are internal to the system under consideration. Instead, we analyzed the system as a whole. Only *external* torques contribute to the change in the system's angular momentum.

10.10 | Analysis Model: Isolated System (Angular Momentum)

In Chapter 8, we found that the total linear momentum of a system of particles remains constant if the system is isolated, that is, if the net external force acting on the system is zero. We have an analogous conservation law in rotational motion:

> The total angular momentum of a system is constant in both magnitude and direction if the net external torque acting on the system is zero, that is, if the system is isolated.

▶ Conservation of angular momentum

This statement is often called[6] the principle of **conservation of angular momentum** and is the basis of the **angular momentum version of the isolated system model.** This principle follows directly from Equation 10.40, which indicates that if

$$\sum \vec{\tau}_{\text{ext}} = \frac{d\vec{L}_{\text{tot}}}{dt} = 0 \qquad \textbf{10.42} \blacktriangleleft$$

then

$$\vec{L}_{\text{tot}} = \text{constant} \quad \text{or} \quad \vec{L}_i = \vec{L}_f \qquad \textbf{10.43} \blacktriangleleft$$

For an isolated system consisting of a number of particles, we write this conservation law as $\vec{L}_{\text{tot}} = \sum \vec{L}_n = \text{constant}$, where the index n denotes the nth particle in the system.

If an isolated rotating system is deformable so that its mass undergoes redistribution in some way, the system's moment of inertia changes. Because the magnitude of the angular momentum of the system is $L = I\omega$ (Eq. 10.41), conservation of angular momentum requires that the product of I and ω must remain constant. Therefore, a change in I for an isolated system requires a change in ω. In this case, we can express the principle of conservation of angular momentum as

$$I_i \omega_i = I_f \omega_f = \text{constant} \qquad \textbf{10.44} \blacktriangleleft$$

This expression is valid both for rotation about a fixed axis and for rotation about an axis through the center of mass of a moving system as long as that axis remains fixed in direction. We require only that the net external torque be zero.

[6]The most general conservation of angular momentum equation is Equation 10.40, which describes how the system interacts with its environment.

When his arms and legs are close to his body, the skater's moment of inertia is small and his angular speed is large.

To slow down for the finish of his spin, the skater moves his arms and legs outward, increasing his moment of inertia.

Figure 10.23 Angular momentum is conserved as Russian gold medalist Evgeni Plushenko performs during the Turin 2006 Winter Olympic Games.

Many examples demonstrate conservation of angular momentum for a deformable system. You may have observed a figure skater spinning in the finale of a program (Fig. 10.23). The angular speed of the skater is large when his hands and feet are close to the trunk of his body. (Notice the skater's hair!) Ignoring friction between skater and ice, there are no external torques on the skater. The moment of inertia of his body increases as his hands and feet are moved away from his body at the finish of the spin. According to the principle of conservation of angular momentum, his angular speed must decrease. In a similar way, when divers or acrobats wish to make several somersaults, they pull their hands and feet close to their bodies to rotate at a higher rate. In these cases, the external force due to gravity acts through the center of mass and hence exerts no torque about an axis through this point. Therefore, the angular momentum about the center of mass must be conserved; that is, $I_i\omega_i = I_f\omega_f$. For example, when divers wish to double their angular speed, they must reduce their moment of inertia to half its initial value.

In Equation 10.43, we have a third version of the isolated system model. We can now state that the energy, linear momentum, and angular momentum of an isolated system are all constant:

$$E_i = E_f \quad \text{(if there are no energy transfers across the system boundary)}$$
$$\vec{p}_i = \vec{p}_f \quad \text{(if the net external force on the system is zero)}$$
$$\vec{L}_i = \vec{L}_f \quad \text{(if the net external torque on the system is zero)}$$

A system may be isolated in terms of one of these quantities but not in terms of another. If a system is nonisolated in terms of momentum or angular momentum, it will often be nonisolated also in terms of energy because the system has a net force or torque on it and the net force or torque will do work on the system. We can, however, identify systems that are nonisolated in terms of energy but isolated in terms of momentum. For example, imagine pushing inward on a balloon (the system) between your hands. Work is done in compressing the balloon, so the system is nonisolated in terms of energy, but there is zero net force on the system, so the system is isolated in terms of momentum. A similar statement could be made about twisting the ends of a long, springy piece of metal with both hands. Work is done on the metal (the system), so energy is stored in the nonisolated system as elastic potential energy, but the net torque on the system is zero. Therefore, the system is isolated in terms of angular momentum. Other examples are collisions of macroscopic objects, which represent isolated systems in terms of momentum but nonisolated systems in terms of energy because of the output of energy from the system by mechanical waves (sound).

An interesting astrophysical example of conservation of angular momentum occurs when, at the end of its lifetime, a massive star uses up all its fuel and collapses under the influence of gravitational forces, causing a gigantic outburst of energy called a supernova explosion. The best-studied example of a remnant of a supernova explosion is the Crab Nebula, a chaotic, expanding mass of gas (Fig. 10.24). In a supernova, part of the star's mass is released into space, where it eventually condenses into new stars and planets. Most of what is left behind typically collapses into a **neutron star,** an extremely dense sphere of matter with a diameter of about 10 km in comparison with the 10^6-km diameter of the original star and containing a large fraction of the star's original mass. As the moment of inertia of the system decreases during the collapse, the star's rotational speed increases, similar to the change in speed of the skater in Figure 10.23. About 2 000 rapidly rotating neutron stars have been identified since the first discovery of such astronomical bodies in 1967, with periods of rotation ranging from a millisecond to several seconds. The neutron star—an object with a mass greater than the Sun, rotating about its axis many times each second—is a most dramatic system!

We can also detect the effects of conservation of momentum on the rotation of the Earth when an earthquake occurs. An earthquake causes the mass distribution of the Earth to change, and the result is a change in the moment of inertia of the Earth. Just as with the spinning skater, this change will cause the angular speed of the Earth

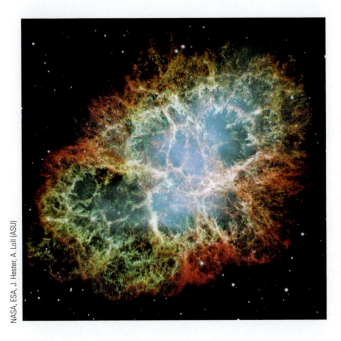

Figure 10.24 The Crab Nebula, in the constellation Taurus. This nebula is the remnant of a supernova explosion, which was seen on Earth in the year 1054. It is located some 6 300 light-years away and is approximately 6 light-years in diameter, still expanding outward.

to change. An earthquake of magnitude 8.8 in Chile in February 2010 caused the Earth's period to decrease by 1.3 μs. Similarly, the magnitude-9.0 earthquake off the coast of Japan in March 2011 caused a further decrease by 1.8 μs.

QUICK QUIZ 10.8 A competitive diver leaves the diving board and falls toward the water with her body straight and rotating slowly. She pulls her arms and legs into a tight tuck position. What happens to her rotational kinetic energy? **(a)** It increases. **(b)** It decreases. **(c)** It stays the same. **(d)** It is impossible to determine.

Example **10.13** | **A Revolving Puck on a Horizontal, Frictionless Surface**

A puck of mass m on a horizontal, frictionless table is connected to a string that passes through a small hole in the table. The puck is set into circular motion of radius R, at which time its speed is v_i (Fig. 10.25).

(A) If the string is pulled from the bottom so that the radius of the circular path is decreased to r, find an expression for the final speed v_f of the puck.

SOLUTION

Conceptualize Imagine the puck in Figure 10.25 moving in its circular path. Now imagine pulling downward on the string so that the puck moves into a circular path with a smaller radius. Do you expect it to move faster or slower? What happens to a spinning skater when he brings his arms in close to his body?

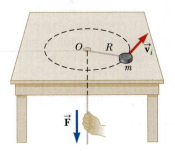

Figure 10.25 (Example 10.13) When the string is pulled downward, the speed of the puck changes.

Categorize We identify the puck as the system. Is the system isolated or nonisolated? The gravitational force acting on the puck is balanced by the upward normal force, so these forces cancel, resulting in zero net torque from these forces. The force $\vec{\mathbf{F}}$ of the string on the puck acts toward the center of rotation, and the vector position $\vec{\mathbf{r}}$ is directed away from O. Therefore, we see that the torque about the center of rotation due to this force is $\vec{\boldsymbol{\tau}} = \vec{\mathbf{r}} \times \vec{\mathbf{F}} = 0$. Although three forces act on the puck, the net torque exerted on it is zero. Therefore, the puck is an isolated system in terms of angular momentum!

Analyze

From the isolated system model, set the initial angular momentum equal to the final angular momentum:

$$L = mv_i R = mv_f r$$

continued

10.13 *cont.*

Solve for the final speed:

$$v_f = \frac{v_i R}{r}$$

From this result we see that as r decreases, the speed v increases.

(B) Show that the kinetic energy of the puck is not conserved in this process.

SOLUTION

Find an expression for the ratio of the final kinetic energy to the initial kinetic energy:

$$\frac{K_f}{K_i} = \frac{\frac{1}{2}mv_f^2}{\frac{1}{2}mv_i^2} = \frac{1}{v_i^2}\left(\frac{v_i R}{r}\right)^2 = \frac{R^2}{r^2}$$

Because this ratio is not equal to 1, kinetic energy is not conserved.

Finalize Because $R > r$, the kinetic energy of the puck has increased. This increase corresponds to energy entering the system of the puck by means of the work done by the person pulling the string. Even though the system is isolated in terms of angular momentum, it is nonisolated in terms of energy!

Example 10.14 | **Formation of a Neutron Star**

A star rotates with a period of 30 days about an axis through its center. The period is the time interval required for a point on the star's equator to make one complete revolution around the axis of rotation. After the star undergoes a supernova explosion, the stellar core, which had a radius of 1.0×10^4 km, collapses into a neutron star of radius 3.0 km. Determine the period of rotation of the neutron star.

SOLUTION

Conceptualize The change in the neutron star's motion is similar to that of the skater described earlier, but in the reverse direction. As the mass of the star moves closer to the rotation axis, we expect the star to spin faster.

Categorize Let us assume that during the collapse of the stellar core, (1) no external torque acts on it, (2) it remains spherical with the same relative mass distribution, and (3) its mass remains constant. We categorize the star as an isolated system in terms of angular momentum. We do not know the mass distribution of the star, but we have assumed the distribution is symmetric, so the moment of inertia can be expressed as kMR^2, where k is some numerical constant. (From Table 10.2, for example, we see that $k = \frac{2}{5}$ for a solid sphere and $k = \frac{2}{3}$ for a spherical shell.)

Analyze Let's use the symbol T for the period, with T_i being the initial period of the star and T_f being the period of the neutron star. The star's angular speed is given by $\omega = 2\pi/T$.

Write Equation 10.44:

$$I_i \omega_i = I_f \omega_f$$

Use $\omega = 2\pi/T$ to rewrite this equation in terms of the initial and final periods:

$$I_i\left(\frac{2\pi}{T_i}\right) = I_f\left(\frac{2\pi}{T_f}\right)$$

Substitute the moments of inertia in the preceding equation:

$$kMR_i^2\left(\frac{2\pi}{T_i}\right) = kMR_f^2\left(\frac{2\pi}{T_f}\right)$$

Solve for the final period of the star:

$$T_f = \left(\frac{R_f}{R_i}\right)^2 T_i$$

Substitute numerical values:

$$T_f = \left(\frac{3.0 \text{ km}}{1.0 \times 10^4 \text{ km}}\right)^2 (30 \text{ days}) = 2.7 \times 10^{-6} \text{ days} = \boxed{0.23 \text{ s}}$$

Finalize The neutron star does indeed rotate faster after it collapses, as predicted. It moves very fast, in fact, rotating about four times each second.

10.11 | Precessional Motion of Gyroscopes

An unusual and fascinating type of motion you have probably observed is that of a top spinning about its axis of symmetry as shown in Figure 10.26a. If the top spins rapidly, the symmetry axis rotates about the z axis, sweeping out a cone (see Fig. 10.26b). The motion of the symmetry axis about the vertical—known as **precessional motion**—is usually slow relative to the spinning motion of the top.

It is quite natural to wonder why the top does not fall over. Because the center of mass is not directly above the pivot point O, a net torque is acting on the top about an axis passing through O, a torque resulting from the gravitational force $M\vec{g}$. The top would certainly fall over if it were not spinning. Because it is spinning, however, it has an angular momentum $\vec{L}$ directed along its symmetry axis. We shall show that this symmetry axis moves about the z axis (precessional motion occurs) because the torque produces a change in the *direction* of the symmetry axis. This illustration is an excellent example of the importance of the vector nature of angular momentum.

The essential features of precessional motion can be illustrated by considering the simple gyroscope shown in Figure 10.27a. The two forces acting on the gyroscope are shown in Figure 10.27b: the downward gravitational force $M\vec{g}$ and the normal force $\vec{n}$ acting upward at the pivot point O. The normal force produces no torque about an axis passing through the pivot because its moment arm through that point is zero. The gravitational force, however, produces a torque $\vec{\tau} = \vec{r} \times M\vec{g}$ about an axis passing through O, where the direction of $\vec{\tau}$ is perpendicular to the plane formed by $\vec{r}$ and $M\vec{g}$. By necessity, the vector $\vec{\tau}$ lies in a horizontal xy plane perpendicular to the angular momentum vector. The net torque and angular momentum of the gyroscope are related through Equation 10.40:

$$\sum \vec{\tau}_{ext} = \frac{d\vec{L}}{dt}$$

This expression shows that in the infinitesimal time interval dt, the nonzero torque produces a change in angular momentum $d\vec{L}$, a change that is in the same direction as $\vec{\tau}$. Therefore, like the torque vector, $d\vec{L}$ must also be perpendicular to $\vec{L}$. Figure 10.27c illustrates the resulting precessional motion of the symmetry axis of the gyroscope. In a time interval dt, the change in angular momentum is $d\vec{L} = \vec{L}_f - \vec{L}_i = \vec{\tau}\, dt$. Because $d\vec{L}$ is perpendicular to $\vec{L}$, the magnitude of $\vec{L}$ does not change ($|\vec{L}_i| = |\vec{L}_f|$). Rather, what is changing is the *direction* of $\vec{L}$. Because the change in

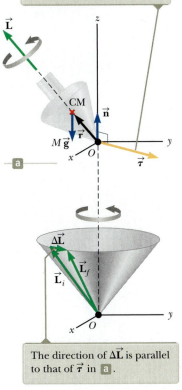

The right-hand rule indicates that $\vec{\tau} = \vec{r} \times \vec{F} = \vec{r} \times M\vec{g}$ is in the xy plane.

The direction of $\Delta\vec{L}$ is parallel to that of $\vec{\tau}$ in a.

Figure 10.26 Precessional motion of a top spinning about its symmetry axis. (a) The only external forces acting on the top are the normal force $\vec{n}$ and the gravitational force $M\vec{g}$. The direction of the angular momentum $\vec{L}$ is along the axis of symmetry. (b) Because $\vec{L}_f = \Delta\vec{L} + \vec{L}_i$, the top precesses about the z axis.

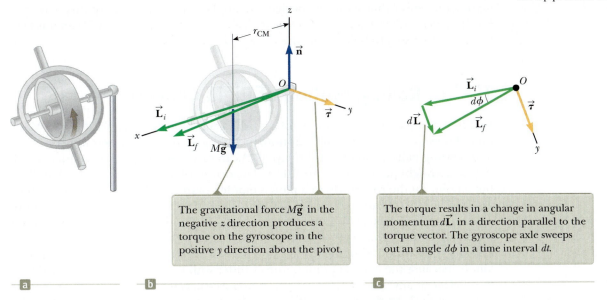

The gravitational force $M\vec{g}$ in the negative z direction produces a torque on the gyroscope in the positive y direction about the pivot.

The torque results in a change in angular momentum $d\vec{L}$ in a direction parallel to the torque vector. The gyroscope axle sweeps out an angle $d\phi$ in a time interval dt.

Figure 10.27 (a) A spinning gyroscope is placed on a pivot at the right end. (b) Diagram for the spinning gyroscope showing forces, torque, and angular momentum. (c) Overhead view (looking down the z axis) of the gyroscope's initial and final angular momentum vectors for an infinitesimal time interval dt.

angular momentum $d\vec{\mathbf{L}}$ is in the direction of $\vec{\tau}$, which lies in the xy plane, the gyroscope undergoes precessional motion.

To simplify the description of the system, we assume the total angular momentum of the precessing wheel is the sum of the angular momentum $I\vec{\boldsymbol{\omega}}$ due to the spinning and the angular momentum due to the motion of the center of mass about the pivot. In our treatment, we shall neglect the contribution from the center-of-mass motion and take the total angular momentum to be simply $I\vec{\boldsymbol{\omega}}$. In practice, this approximation is good if $\vec{\boldsymbol{\omega}}$ is made very large.

The vector diagram in Figure 10.27c shows that in the time interval dt, the angular momentum vector rotates through an angle $d\phi$, which is also the angle through which the gyroscope axle rotates. From the vector triangle formed by the vectors $\vec{\mathbf{L}}_i$, $\vec{\mathbf{L}}_f$, and $d\vec{\mathbf{L}}$, we see that

$$d\phi = \frac{dL}{L} = \frac{\sum \tau_{ext} \, dt}{L} = \frac{(Mgr_{CM}) \, dt}{L}$$

Dividing through by dt and using the relationship $L = I\omega$, we find that the rate at which the axle rotates about the vertical axis is

$$\omega_p = \frac{d\phi}{dt} = \frac{Mgr_{CM}}{I\omega}$$

10.45 ◀

The angular speed ω_p is called the **precessional frequency.** This result is valid only when $\omega_p \ll \omega$. Otherwise, a much more complicated motion is involved. As you can see from Equation 10.45, the condition $\omega_p \ll \omega$ is met when ω is large, that is, when the wheel spins rapidly. Furthermore, notice that the precessional frequency decreases as ω increases, that is, as the wheel spins faster about its axis of symmetry.

With careful manufacturing tolerances, precession due to gravitational torque can be made very small and gyroscopes can be used for guidance systems in vehicles, whereby a change in the direction of the velocity of a vehicle is detected as a change between the direction of the angular momentum of the gyroscope and a reference direction attached to the vehicle. With proper electronic feedback, the deviation from the desired direction of motion can be removed, bringing the angular momentum back in line with the reference direction. Precession rates for highly specialized military gyroscopes are as low as 0.02° per day.

Gyroscopes are becoming increasingly involved in everyday life applications. Anyone who has ridden a Segway electric vehicle has been kept upright by a system of five gyroscopes in the control system of the device. The Apple iPhone 4 includes a gyroscopic sensor that assists the device with applications involving advanced motion sensing. As another example, image stabilization technology in digital cameras uses gyroscopic sensors to help clarify the images taken by the camera.

❮10.12 | Rolling Motion of Rigid Objects

In this section, we treat the motion of a rigid object rolling along a flat surface. Many everyday examples exist for such motion, including automobile tires rolling on roads and bowling balls rolling toward the pins. As an example, suppose a cylinder is rolling on a straight path as in Figure 10.28. The center of mass moves in a straight line, but a point on the rim moves in a more complex path called a *cycloid.* Let us further assume that the cylinder of radius R is uniform and rolls on a surface with friction. The surfaces must exert friction forces on each other; otherwise, the cylinder would simply slide rather than roll. If the friction force on the cylinder is large enough, the cylinder rolls without slipping. In this situation, the friction force is static rather than kinetic because the contact point of the cylinder with the surface is at rest relative to the surface at any instant. The static friction force acts through no displacement, so it does no work on the cylinder and causes no decrease in mechanical energy of the cylinder. In real rolling objects, deformations of the surfaces result in some rolling resistance. If both surfaces are hard, however, they will deform very little, and rolling

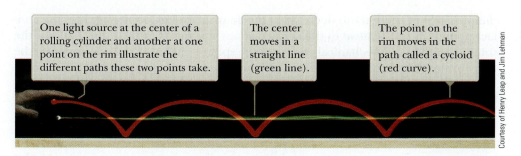

One light source at the center of a rolling cylinder and another at one point on the rim illustrate the different paths these two points take.

The center moves in a straight line (green line).

The point on the rim moves in the path called a cycloid (red curve).

Figure 10.28 Two points on a rolling object take different paths through space.

resistance can be negligibly small. Therefore, we can model the rolling motion as maintaining constant mechanical energy. The wheel was a great invention!

As the cylinder rotates through an angle θ, its center of mass moves a distance of $s = r\theta$. Therefore, the speed and acceleration of the center of mass for pure rolling motion are

$$v_{CM} = \frac{ds}{dt} = R\frac{d\theta}{dt} = R\omega \qquad \textbf{10.46} \blacktriangleleft$$

$$a_{CM} = \frac{dv_{CM}}{dt} = R\frac{d\omega}{dt} = R\alpha \qquad \textbf{10.47} \blacktriangleleft$$

▶ Relations between translational and rotational variables for a rolling object

The translational velocities of various points on the rolling cylinder are illustrated in Figure 10.29. Note that the translational velocity of any point is in a direction perpendicular to the line from that point to the contact point. At any instant, the point P is at rest relative to the surface because sliding does not occur.

We can express the **total kinetic energy** of a rolling object of mass M and moment of inertia I as the combination of the rotational kinetic energy around the center of mass plus the translational kinetic energy of the center of mass:

$$K = \tfrac{1}{2}I_{CM}\omega^2 + \tfrac{1}{2}Mv_{CM}^2 \qquad \textbf{10.48} \blacktriangleleft$$

▶ Total kinetic energy of a rolling object

A useful theorem called the **parallel axis theorem** enables us to express this energy in terms of the moment of inertia I_p through any axis parallel to the axis through the center of mass of an object. This theorem states that

$$I_p = I_{CM} + MD^2 \qquad \textbf{10.49} \blacktriangleleft$$

where D is the distance from the center-of-mass axis to the parallel axis and M is the total mass of the object. Let us use this theorem to express the moment of inertia around an axis passing through the contact point P between the rolling object and the surface. The distance from this point to the center of mass of the symmetric object is its radius, so

$$I_P = I_{CM} + MR^2$$

If we write the translational speed of the center of mass of the object in Equation 10.48 in terms of the angular speed, we have

$$K = \tfrac{1}{2}I_{CM}\omega^2 + \tfrac{1}{2}MR^2\omega^2 = \tfrac{1}{2}(I_{CM} + MR^2)\omega^2 = \tfrac{1}{2}I_P\omega^2 \qquad \textbf{10.50} \blacktriangleleft$$

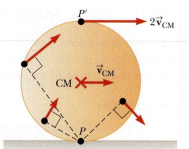

Figure 10.29 All points on a rolling object move in a direction perpendicular to a line through the instantaneous point of contact P. The center of the object moves with a velocity $\vec{v}_{CM}$, whereas the point P' moves with a velocity $2\vec{v}_{CM}$.

Therefore, the kinetic energy of the rolling object can be considered as equivalent to a purely rotational kinetic energy of the object rotating around its contact point.

We can use the energy version of the isolated system model to treat a class of problems concerning the rolling motion of a rigid object down a rough incline. In these types of problems, gravitational potential energy of the object–Earth system decreases as the rotational and translational kinetic energies of the object increase. For example, consider a sphere rolling without slipping after being released from rest at the top of an incline and descending through a vertical height h. Note that accelerated rolling motion is possible only if a friction force is present between the sphere and the incline to produce a net torque about the center of mass. Despite the presence of friction, no decrease in mechanical energy occurs because the contact point is at rest relative to the surface at any instant. (On the other hand, if the sphere were to slip, mechanical energy of the sphere–incline–Earth system would be transformed to internal energy due to the nonconservative force of kinetic friction.)

Using $v_{CM} = R\omega$ for pure rolling motion, we can express Equation 10.48 as

$$K = \tfrac{1}{2}I_{CM}\left(\frac{v_{CM}}{R}\right)^2 + \tfrac{1}{2}Mv_{CM}{}^2$$

$$K = \tfrac{1}{2}\left(\frac{I_{CM}}{R^2} + M\right)v_{CM}{}^2 \qquad \textbf{10.51}\blacktriangleleft$$

For the system of the sphere and the Earth, we define the zero configuration of gravitational potential energy to be when the sphere is at the bottom of the incline. Therefore, conservation of mechanical energy gives us

$$K_f + U_f = K_i + U_i$$

$$\tfrac{1}{2}\left(\frac{I_{CM}}{R^2} + M\right)v_{CM}{}^2 + 0 = 0 + Mgh$$

$$v_{CM} = \left(\frac{2gh}{1 + I_{CM}/MR^2}\right)^{1/2} \qquad \textbf{10.52}\blacktriangleleft$$

> **QUICK QUIZ 10.9** Two items A and B are placed at the top of an incline and released from rest. For *each* of the three pairs of items in (i), (ii), and (iii), which item arrives at the bottom of the incline first? (**i**) a ball A rolling without slipping and a box B sliding on a frictionless portion of the incline (**ii**) a sphere A that has twice the mass and twice the radius of a sphere B, where both roll without slipping (**iii**) a sphere A that has the same mass and radius as a sphere B, but sphere A is solid while sphere B is hollow and both roll without slipping. Choose from the following list for each of the three pairs of items. (**a**) item A (**b**) item B (**c**) items A and B arrive at the same time (**d**) impossible to determine

Example 10.15 | Sphere Rolling Down an Incline

For the solid sphere shown in Active Figure 10.30, calculate the translational speed of the center of mass at the bottom of the incline and the magnitude of the translational acceleration of the center of mass.

SOLUTION

Conceptualize Imagine rolling the sphere down the incline. Compare it in your mind to a book sliding down a frictionless incline. You probably have experience with objects rolling down inclines and may be tempted to think that the sphere would move down the incline faster than the book. You do *not*, however, have experience with objects sliding down *frictionless* inclines! So, which object will reach the bottom first?

Categorize We model the sphere and the Earth as an isolated system in terms of energy with no nonconservative forces acting. This model is the one that led to Equation 10.52, so we can use that result.

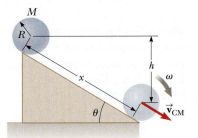

Figure 10.30 (Example 10.15) A sphere rolling down an incline. Mechanical energy of the sphere–Earth system is conserved if no slipping occurs.

Analyze Evaluate the speed of the center of mass of the sphere from Equation 10.52:

$$(1) \quad v_{CM} = \left[\frac{2gh}{1 + \left(\tfrac{2}{5}MR^2/MR^2\right)}\right]^{1/2} = \left(\tfrac{10}{7}gh\right)^{1/2}$$

This result is less than $\sqrt{2gh}$, which is the speed an object would have if it simply slid down the incline without rotating. (Eliminate the rotation by setting $I_{CM} = 0$ in Eq. 10.52.)

To calculate the translational acceleration of the center of mass, notice that the vertical displacement of the sphere is related to the distance x it moves along the incline through the relationship $h = x \sin\theta$.

10.15 *cont.*

Use this relationship to rewrite Equation (1): $\quad\quad\quad\quad\quad\quad v_{CM}^2 = \frac{10}{7}gx\sin\theta$

Write Equation 2.14 for an object starting from rest and $\quad v_{CM}^2 = 2a_{CM}x$
moving through a distance x under constant acceleration:

Equate the preceding two expressions to find a_{CM}: $\quad\quad\quad a_{CM} = \frac{5}{7}g\sin\theta$

Finalize Both the speed and the acceleration of the center of mass are *independent* of the mass and the radius of the sphere. That is, all homogeneous solid spheres experience the same speed and acceleration on a given incline. Try to verify this statement experimentally with balls of different sizes, such as a marble and a croquet ball.

If we were to repeat the acceleration calculation for a hollow sphere, a solid cylinder, or a hoop, we would obtain similar results in which only the factor in front of $g\sin\theta$ would differ. The constant factors that appear in the expressions for v_{CM} and a_{CM} depend only on the moment of inertia about the center of mass for the specific object. In all cases, the acceleration of the center of mass is *less* than $g\sin\theta$, the value the acceleration would have if the incline were frictionless and no rolling occurred.

10.13 | Context Connection: Turning the Spacecraft

In the Context Connection of Chapter 8, we discussed how to make a spacecraft move in empty space by firing its rocket engines. Let us now consider how to make the spacecraft turn in empty space.

One way to change the orientation of a spacecraft is to have small rocket engines that fire perpendicularly out the side of the spacecraft, providing a torque around its center of mass. This torque causes an angular acceleration around the center of mass of the spacecraft and therefore an angular speed. This rotation can be stopped to give the spacecraft the desired final orientation by firing the sideward-mounted rocket engines in the opposite direction. This option is desirable, and many spacecraft have such sideward-mounted rocket engines. An undesirable feature of this technique is that it consumes nonrenewable fuel on the spacecraft, both to initiate and to stop the rotation.

The torque exerted on the spacecraft in this situation is not an external torque, so this is not an example of the rigid object under a net torque model. The torque on the spacecraft arises from internal forces between components of the system. The spacecraft exerts forces on the exhaust gases to expel them from the spacecraft, and, by Newton's third law, the gases exert a force back on the spacecraft. Therefore, this is an application of the isolated system model for angular momentum. The gases are given an angular momentum in one direction and the spacecraft turns in the other direction. It is a rotational analog to the archer discussed in Example 8.2 or the rocket propulsion discussed in Section 8.8.

Let us consider another possibility related to the angular momentum version of the isolated system model that does not involve expelling gases. Suppose the spacecraft carries a gyroscope that is not rotating, as in Figure 10.31a (page 340). In this case, the angular momentum of the spacecraft about its center of mass is zero. Suppose the gyroscope is set into rotation. Now, it would appear that the spacecraft system has a nonzero angular momentum because of the rotation of the gyroscope. Yet there is no external torque on the system, so the angular momentum of the isolated system must remain zero according to the principle of conservation of angular momentum. This principle can be satisfied by realizing that the spacecraft will turn in the direction opposite to that of the gyroscope so that the angular momentum vectors of the gyroscope and the spacecraft cancel, resulting in no angular momentum of the system. The result of rotating the gyroscope, as in Figure 10.31b, is that the spacecraft turns! By including three gyroscopes with mutually perpendicular axles, any desired rotation in space can be achieved. Once the desired orientation is achieved, the rotation of the gyroscope is halted.

This effect occurred in an undesirable situation with the *Voyager 2* spacecraft during its flight. The spacecraft carried a tape recorder whose reels rotated at high speeds.

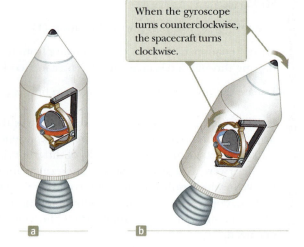

When the gyroscope turns counterclockwise, the spacecraft turns clockwise.

Figure 10.31 (a) A spacecraft carries a gyroscope that is not spinning. (b) When the gyroscope is set into rotation, the spacecraft turns the other way so that the angular momentum of the system is conserved.

Each time the tape recorder was turned on, the reels acted as gyroscopes and the spacecraft started an undesirable rotation in the opposite direction. This rotation had to be counteracted by Mission Control by using the sideward-firing jets to stop the rotation!

SUMMARY

The **instantaneous angular speed** of a particle rotating in a circle or of a rigid object rotating about a fixed axis is

$$\omega \equiv \frac{d\theta}{dt} \qquad \text{10.3}$$

where ω is in rad/s or s^{-1}.

The **instantaneous angular acceleration** of a particle rotating in a circle or of a rigid object rotating about a fixed axis is

$$\alpha \equiv \frac{d\omega}{dt} \qquad \text{10.5}$$

and has units of rad/s² or s⁻².

When a rigid object rotates about a fixed axis, every part of the object has the same angular speed and the same angular acceleration. Different parts of the object, in general, have different translational speeds and different translational accelerations, however.

When a particle rotates about a fixed axis, the angular position, the angular speed, and the angular acceleration are related to the tangential position, the tangential speed, and the tangential acceleration through the relationships

$$s = r\theta \qquad \text{10.1a}$$

$$v = r\omega \qquad \text{10.10}$$

$$a_t = r\alpha \qquad \text{10.11}$$

The **moment of inertia** of a system of particles is

$$I = \sum_i m_i r_i^2 \qquad \text{10.15}$$

If a rigid object rotates about a fixed axis with angular speed ω, its **rotational kinetic energy** can be written

$$K_R = \tfrac{1}{2}I\omega^2 \qquad \text{10.16}$$

where I is the moment of inertia about the axis of rotation.

The moment of inertia of a continuous object of density ρ is

$$I = \int \rho r^2 dV \qquad \text{10.18}$$

The **torque** $\vec{\tau}$ due to a force $\vec{F}$ about an origin in an inertial frame is defined to be

$$\vec{\tau} \equiv \vec{r} \times \vec{F} \qquad \text{10.20}$$

where $\vec{r}$ is the position vector of the point of application of the force.

Given two vectors $\vec{A}$ and $\vec{B}$, their **vector product** or **cross product** $\vec{A} \times \vec{B}$ is a vector $\vec{C}$ having the magnitude

$$C \equiv AB \sin\theta \qquad \text{10.22}$$

where θ is the angle between $\vec{A}$ and $\vec{B}$. The direction of $\vec{C}$ is perpendicular to the plane formed by $\vec{A}$ and $\vec{B}$, and is determined by the right-hand rule.

The **angular momentum** $\vec{L}$ of a particle with linear momentum $\vec{p} = m\vec{v}$ is

$$\vec{L} \equiv \vec{r} \times \vec{p} \qquad \text{10.35}$$

where $\vec{r}$ is the vector position of the particle relative to the origin. If ϕ is the angle between $\vec{r}$ and $\vec{p}$, the magnitude of $\vec{L}$ is

$$L = mvr \sin\phi \qquad \text{10.36}$$

The **total kinetic energy** of a rigid object, such as a cylinder, that is rolling on a rough surface without slipping equals the rotational kinetic energy $\tfrac{1}{2}I_{CM}\omega^2$ about the object's center of mass plus the translational kinetic energy $\tfrac{1}{2}Mv_{CM}^2$ of the center of mass:

$$K = \tfrac{1}{2}I_{CM}\omega^2 + \tfrac{1}{2}Mv_{CM}^2 \qquad \text{10.48}$$

In this expression, v_{CM} is the speed of the center of mass and $v_{CM} = R\omega$ for pure rolling motion.

❯Analysis Models for Problem Solving

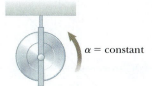
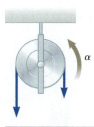

Rigid Object Under Constant Angular Acceleration. If a rigid object rotates about a fixed axis under constant angular acceleration, one can apply equations of kinematics that are analogous to those for translational motion of a particle under constant acceleration:

$$\omega_f = \omega_i + \alpha t \qquad \text{10.6}◀$$

$$\theta_f = \theta_i + \omega_i t + \tfrac{1}{2}\alpha t^2 \qquad \text{10.7}◀$$

$$\omega_f{}^2 = \omega_i{}^2 + 2\alpha(\theta_f - \theta_i) \qquad \text{10.8}◀$$

$$\theta_f = \theta_i + \tfrac{1}{2}(\omega_i + \omega_f)t \qquad \text{10.9}◀$$

$\alpha = \text{constant}$

Rigid Object Under a Net Torque. If a rigid object free to rotate about a fixed axis has a net external torque acting on it, the object undergoes an angular acceleration α, where

$$\sum \tau_{\text{ext}} = I\alpha \qquad \text{10.30}◀$$

This equation is the rotational analog to Newton's second law in the particle under a net force model.

α

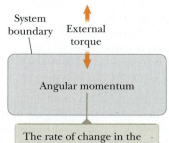

Nonisolated System (Angular Momentum). If a system interacts with its environment in the sense that there is an external torque on the system, the net external torque acting on the system is equal to the time rate of change of its angular momentum:

$$\sum \vec{\tau}_{\text{ext}} = \frac{d\vec{L}_{\text{tot}}}{dt} \qquad \text{10.40}◀$$

System boundary — External torque

Angular momentum

The rate of change in the angular momentum of the nonisolated system is equal to the net external torque on the system.

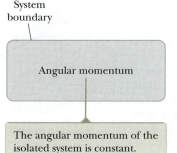

Isolated System (Angular Momentum). If a system experiences no external torque from the environment, the total angular momentum of the system is conserved:

$$\vec{L}_i = \vec{L}_f \qquad \text{10.43}◀$$

Applying this law of conservation of angular momentum to a system whose moment of inertia changes gives

$$I_i\omega_i = I_f\omega_f = \text{constant} \qquad \text{10.44}◀$$

System boundary

Angular momentum

The angular momentum of the isolated system is constant.

$a = 0 \qquad \alpha = 0$
$\Sigma F_x = 0 \qquad \Sigma \tau_z = 0$
$\Sigma F_y = 0$

Rigid Object in Equilibrium A rigid object in equilibrium exhibits no translational or angular acceleration. The net external force acting on it is zero, and the net external torque on it is zero about any axis:

$$\sum \vec{F}_{\text{ext}} = 0 \qquad \text{10.27}◀$$

$$\sum \vec{\tau}_{\text{ext}} = 0 \qquad \text{10.28}◀$$

The first condition is the condition for translational equilibrium, and the second is the condition for rotational equilibrium.

▶ OBJECTIVE QUESTIONS |

☐ denotes answer available in *Student Solutions Manual/Study Guide*

1. A cyclist rides a bicycle with a wheel radius of 0.500 m across campus. A piece of plastic on the front rim makes a clicking sound every time it passes through the fork. If the cyclist counts 320 clicks between her apartment and the cafeteria, how far has she traveled? (a) 0.50 km (b) 0.80 km (c) 1.0 km (d) 1.5 km (e) 1.8 km

2. A grindstone increases in angular speed from 4.00 rad/s to 12.00 rad/s in 4.00 s. Through what angle does it turn during that time interval if the angular acceleration is constant? (a) 8.00 rad (b) 12.0 rad (c) 16.0 rad (d) 32.0 rad (e) 64.0 rad

3. A wheel is rotating about a fixed axis with constant angular acceleration 3 rad/s². At different moments, its angular speed is −2 rad/s, 0, and +2 rad/s. For a point on the rim of the wheel, consider at these moments the magnitude of the tangential component of acceleration and the magnitude of the radial component of acceleration. Rank the following five items from largest to smallest: (a) $|a_t|$ when $\omega = -2$ rad/s, (b) $|a_r|$ when $\omega = -2$ rad/s, (c) $|a_r|$ when $\omega = 0$, (d) $|a_t|$ when $\omega = 2$ rad/s, and (e) $|a_r|$ when $\omega = 2$ rad/s. If two items are equal, show them as equal in your ranking. If a quantity is equal to zero, show that fact in your ranking.

4. Vector $\vec{A}$ is in the negative y direction, and vector $\vec{B}$ is in the negative x direction. (i) What is the direction of $\vec{A} \times \vec{B}$? (a) no direction because it is a scalar (b) x (c) $-y$ (d) z (e) $-z$ (ii) What is the direction of $\vec{B} \times \vec{A}$? Choose from the same possibilities (a) through (e).

5. Assume a single 300-N force is exerted on a bicycle frame as shown in Figure OQ10.5. Consider the torque produced by this force about axes perpendicular to the plane of the paper and through each of the points A through E, where E is the center of mass of the frame. Rank the torques τ_A, τ_B, τ_C, τ_D, and τ_E from largest to smallest, noting that zero is greater than a negative quantity. If two torques are equal, note their equality in your ranking.

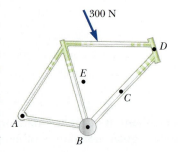

Figure OQ10.5

6. Consider an object on a rotating disk a distance r from its center, held in place on the disk by static friction. Which of the following statements is *not* true concerning this object? (a) If the angular speed is constant, the object must have constant tangential speed. (b) If the angular speed is constant, the object is not accelerated. (c) The object has a tangential acceleration only if the disk has an angular acceleration. (d) If the disk has an angular acceleration, the object has both a centripetal acceleration and a tangential acceleration. (e) The object always has a centripetal acceleration except when the angular speed is zero.

7. Answer yes or no to the following questions. (a) Is it possible to calculate the torque acting on a rigid object without specifying an axis of rotation? (b) Is the torque independent of the location of the axis of rotation?

8. Figure OQ10.8 shows a system of four particles joined by light, rigid rods. Assume $a = b$ and M is larger than m. About which of the coordinate axes does the system have (i) the smallest and (ii) the largest moment of inertia? (a) the x axis (b) the y axis (c) the z axis. (d) The moment of inertia has the same small value for two axes. (e) The moment of inertia is the same for all three axes.

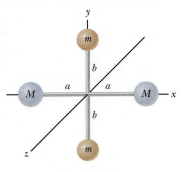

Figure OQ10.8

9. As shown in Figure OQ10.9, a cord is wrapped onto a cylindrical reel mounted on a fixed, frictionless, horizontal axle. When does the reel have a greater magnitude of angular acceleration? (a) When the cord is pulled down with a constant force of 50 N. (b) When an object of weight 50 N is hung from the cord and released. (c) The angular accelerations in parts (a) and (b) are equal. (d) It is impossible to determine.

Figure OQ10.9

10. Two forces are acting on an object. Which of the following statements is correct? (a) The object is in equilibrium if the forces are equal in magnitude and opposite in direction. (b) The object is in equilibrium if the net torque on the object is zero. (c) The object is in equilibrium if the forces act at the same point on the object. (d) The object is in equilibrium if the net force and the net torque on the object are both zero. (e) The object cannot be in equilibrium because more than one force acts on it.

11. Consider the object in Figure OQ10.11. A single force is exerted on the object. The line of action of the force does not pass through the object's center of mass. The acceleration of the object's center of mass due to this force (a) is the same as if the force were applied at the center of mass, (b) is larger than the acceleration would be if the force were applied at the center of mass, (c) is smaller than the acceleration would be if the force were applied at the center of mass, or (d) is zero because the force causes only angular acceleration about the center of mass.

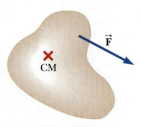

Figure OQ10.11

12. A constant net torque is exerted on an object. Which of the following quantities for the object cannot be constant? Choose all that apply. (a) angular position (b) angular velocity (c) angular acceleration (d) moment of inertia (e) kinetic energy

13. Let us name three perpendicular directions as right, up, and toward you as you might name them when you are facing a

television screen that lies in a vertical plane. Unit vectors for these directions are $\hat{\mathbf{r}}$, $\hat{\mathbf{u}}$, and $\hat{\mathbf{t}}$, respectively. Consider the quantity $(-3\hat{\mathbf{u}} \times 2\hat{\mathbf{t}})$. **(i)** Is the magnitude of this vector (a) 6, (b) 3, (c) 2, or (d) 0? **(ii)** Is the direction of this vector (a) down, (b) toward you, (c) up, (d) away from you, or (e) left?

14. A rod 7.0 m long is pivoted at a point 2.0 m from the left end. A downward force of 50 N acts at the left end, and a downward force of 200 N acts at the right end. At what distance to the right of the pivot can a third force of 300 N acting upward be placed to produce rotational equilibrium? *Note:* Neglect the weight of the rod. (a) 1.0 m (b) 2.0 m (c) 3.0 m (d) 4.0 m (e) 3.5 m

CONCEPTUAL QUESTIONS

1. (a) What is the angular speed of the second hand of a clock? (b) What is the direction of $\vec{\omega}$ as you view a clock hanging on a vertical wall? (c) What is the magnitude of the angular acceleration vector $\vec{\alpha}$ of the second hand?

2. Explain why changing the axis of rotation of an object changes its moment of inertia.

3. Why does a long pole help a tightrope walker stay balanced?

4. Which of the entries in Table 10.2 applies to finding the moment of inertia (a) of a long, straight sewer pipe rotating about its axis of symmetry? (b) Of an embroidery hoop rotating about an axis through its center and perpendicular to its plane? (c) Of a uniform door turning on its hinges? (d) Of a coin turning about an axis through its center and perpendicular to its faces?

5. (a) Give an example in which the net force acting on an object is zero and yet the net torque is nonzero. (b) Give an example in which the net torque acting on an object is zero and yet the net force is nonzero.

6. Suppose just two external forces act on a stationary, rigid object and the two forces are equal in magnitude and opposite in direction. Under what condition does the object start to rotate?

7. If you see an object rotating, is there necessarily a net torque acting on it?

8. A scientist arriving at a hotel asks a bellhop to carry a heavy suitcase. When the bellhop rounds a corner, the suitcase suddenly swings away from him for some unknown reason. The alarmed bellhop drops the suitcase and runs away. What might be in the suitcase?

9. Three objects of uniform density—a solid sphere, a solid cylinder, and a hollow cylinder—are placed at the top of an incline (Fig. CQ10.9). They are all released from rest at the same elevation and roll without slipping. (a) Which object reaches the bottom first? (b) Which reaches it last? *Note:* The result is independent of the masses and the radii of the objects. (Try this activity at home!)

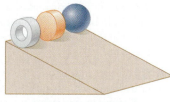
Figure CQ10.9

15. An ice skater starts a spin with her arms stretched out to the sides. She balances on the tip of one skate to turn without friction. She then pulls her arms in so that her moment of inertia decreases by a factor of 2. In the process of her doing so, what happens to her kinetic energy? (a) It increases by a factor of 4. (b) It increases by a factor of 2. (c) It remains constant. (d) It decreases by a factor of 2. (e) It decreases by a factor of 4.

16. A 20.0-kg horizontal plank 4.00 m long rests on two supports, one at the left end and a second 1.00 m from the right end. What is the magnitude of the force exerted on the plank by the support near the right end? (a) 32.0 N (b) 45.2 N (c) 112 N (d) 131 N (e) 98.2 N

denotes answer available in *Student Solutions Manual/Study Guide*

10. A person balances a meterstick in a horizontal position on the extended index fingers of her right and left hands. She slowly brings the two fingers together. The stick remains balanced, and the two fingers always meet at the 50-cm mark regardless of their original positions. (Try it!) Explain why that occurs.

11. If the torque acting on a particle about an axis through a certain origin is zero, what can you say about its angular momentum about that axis?

12. A cat usually lands on its feet regardless of the position from which it is dropped. A slow-motion film of a cat falling shows that the upper half of its body twists in one direction while the lower half twists in the opposite direction. (See Fig. CQ10.12.) Why does this type of rotation occur?

© Biosphoto/Labat J.-M. & Roquette F./Peter Arnold/Photo Library

Figure CQ10.12

13. Stars originate as large bodies of slowly rotating gas. Because of gravity, these clumps of gas slowly decrease in size. What happens to the angular speed of a star as it shrinks? Explain.

14. A girl has a large, docile dog she wishes to weigh on a small bathroom scale. She reasons that she can determine her dog's weight with the following method. First she puts the dog's two front feet on the scale and records the scale reading. Then she places only the dog's two back feet on the scale and records the reading. She thinks that the sum of the readings will be the dog's weight. Is she correct? Explain your answer.

15. If global warming continues over the next one hundred years, it is likely that some polar ice will melt and the water will be distributed closer to the equator. (a) How would that change the moment of inertia of the Earth? (b) Would the duration of the day (one revolution) increase or decrease?

16. Can an object be in equilibrium if it is in motion? Explain.

17. A ladder stands on the ground, leaning against a wall. Would you feel safer climbing up the ladder if you were told that the ground is frictionless but the wall is rough or if you were told that the wall is frictionless but the ground is rough? Explain your answer.

PROBLEMS

Section 10.1 Angular Position, Speed, and Acceleration

1. **W** During a certain time interval, the angular position of a swinging door is described by $\theta = 5.00 + 10.0t + 2.00t^2$, where θ is in radians and t is in seconds. Determine the angular position, angular speed, and angular acceleration of the door (a) at $t = 0$ and (b) at $t = 3.00$ s.

2. A bar on a hinge starts from rest and rotates with an angular acceleration $\alpha = (10 + 6t)$, where α is in rad/s² and t is in seconds. Determine the angle in radians through which the bar turns in the first 4.00 s.

3. A potter's wheel moves uniformly from rest to an angular speed of 1.00 rev/s in 30.0 s. (a) Find its average angular acceleration in radians per second per second. (b) Would doubling the angular acceleration during the given period have doubled the final angular speed?

Section 10.2 Analysis Model: Rigid Object Under Constant Angular Acceleration

4. A dentist's drill starts from rest. After 3.20 s of constant angular acceleration, it turns at a rate of 2.51×10^4 rev/min. (a) Find the drill's angular acceleration. (b) Determine the angle (in radians) through which the drill rotates during this period.

5. The tub of a washer goes into its spin cycle, starting from rest and gaining angular speed steadily for 8.00 s, at which time it is turning at 5.00 rev/s. At this point, the person doing the laundry opens the lid, and a safety switch turns off the washer. The tub smoothly slows to rest in 12.0 s. Through how many revolutions does the tub turn while it is in motion?

6. *Why is the following situation impossible?* Starting from rest, a disk rotates around a fixed axis through an angle of 50.0 rad in a time interval of 10.0 s. The angular acceleration of the disk is constant during the entire motion, and its final angular speed is 8.00 rad/s.

7. **M** An electric motor rotating a workshop grinding wheel at 1.00×10^2 rev/min is switched off. Assume the wheel has a constant negative angular acceleration of magnitude 2.00 rad/s². (a) How long does it take the grinding wheel to stop? (b) Through how many radians has the wheel turned during the time interval found in part (a)?

8. A centrifuge in a medical laboratory rotates at an angular speed of 3 600 rev/min. When switched off, it rotates through 50.0 revolutions before coming to rest. Find the constant angular acceleration of the centrifuge.

9. **W** A rotating wheel requires 3.00 s to rotate through 37.0 revolutions. Its angular speed at the end of the 3.00-s interval is 98.0 rad/s. What is the constant angular acceleration of the wheel?

Section 10.3 Relations Between Rotational and Translational Quantities

10. **M** A wheel 2.00 m in diameter lies in a vertical plane and rotates about its central axis with a constant angular acceleration of 4.00 rad/s². The wheel starts at rest at $t = 0$, and the radius vector of a certain point P on the rim makes an angle of 57.3° with the horizontal at this time. At $t = 2.00$ s, find (a) the angular speed of the wheel and, for point P, (b) the tangential speed, (c) the total acceleration, and (d) the angular position.

11. **M** A disk 8.00 cm in radius rotates at a constant rate of 1 200 rev/min about its central axis. Determine (a) its angular speed in radians per second, (b) the tangential speed at a point 3.00 cm from its center, (c) the radial acceleration of a point on the rim, and (d) the total distance a point on the rim moves in 2.00 s.

12. Make an order-of-magnitude estimate of the number of revolutions through which a typical automobile tire turns in one year. State the quantities you measure or estimate and their values.

13. **W** A car traveling on a flat (unbanked), circular track accelerates uniformly from rest with a tangential acceleration of 1.70 m/s². The car makes it one-quarter of the way around the circle before it skids off the track. From these data, determine the coefficient of static friction between the car and the track.

14. **S** A car traveling on a flat (unbanked), circular track accelerates uniformly from rest with a tangential acceleration of a. The car makes it one-quarter of the way around the circle before it skids off the track. From these data, determine the coefficient of static friction between the car and the track.

15. A digital audio compact disc carries data, each bit of which occupies 0.6 μm along a continuous spiral track from the

inner circumference of the disc to the outside edge. A CD player turns the disc to carry the track counterclockwise above a lens at a constant speed of 1.30 m/s. Find the required angular speed (a) at the beginning of the recording, where the spiral has a radius of 2.30 cm, and (b) at the end of the recording, where the spiral has a radius of 5.80 cm. (c) A full-length recording lasts for 74 min 33 s. Find the average angular acceleration of the disc. (d) Assuming that the acceleration is constant, find the total angular displacement of the disc as it plays. (e) Find the total length of the track.

16. **Q|C** **W** Figure P10.16 shows the drive train of a bicycle that has wheels 67.3 cm in diameter and pedal cranks 17.5 cm long. The cyclist pedals at a steady cadence of 76.0 rev/min. The chain engages with a front sprocket 15.2 cm in diameter and a rear sprocket 7.00 cm in diameter. Calculate (a) the speed of a link of the chain relative to the bicycle frame, (b) the angular speed of the bicycle wheels, and (c) the speed of the bicycle relative to the road. (d) What pieces of data, if any, are not necessary for the calculations?

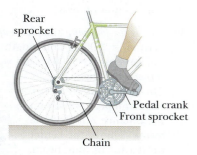

Figure P10.16

Section 10.4 Rotational Kinetic Energy

17. Big Ben, the Parliament tower clock in London, has an hour hand 2.70 m long with a mass of 60.0 kg and a minute hand 4.50 m long with a mass of 100 kg (Fig. P10.17). Calculate the total rotational kinetic energy of the two hands about the axis of rotation. (You may model the hands as long, thin rods rotated about one end. Assume the hour and minute hands are rotating at a constant rate of one revolution per 12 hours and 60 minutes, respectively.)

Figure P10.17 Problems 17, 49, and 66.

18. **Q|C** **W** Rigid rods of negligible mass lying along the y axis connect three particles (Fig. P10.18). The system rotates about the x axis with an angular speed of 2.00 rad/s. Find (a) the moment of inertia about the x axis, (b) the total rotational kinetic energy evaluated from $\frac{1}{2}I\omega^2$, (c) the tangential speed of each particle, and (d) the total kinetic energy evaluated from $\sum \frac{1}{2}m_i v_i^2$. (e) Compare the answers for kinetic energy in parts (b) and (d).

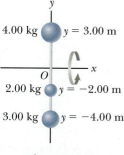

4.00 kg $y = 3.00$ m
2.00 kg $y = -2.00$ m
3.00 kg $y = -4.00$ m

Figure P10.18

19. **Q|C** A *war-wolf*, or *trebuchet*, is a device used during the Middle Ages to throw rocks at castles and now sometimes used to fling large vegetables and pianos as a sport. A simple trebuchet is shown in Figure P10.19. Model it as a stiff rod of negligible mass, 3.00 m long, joining particles of mass $m_1 = 0.120$ kg and $m_2 = 60.0$ kg at its ends. It can turn on a frictionless, horizontal axle perpendicular to the rod and 14.0 cm from the large-mass particle. The operator releases the trebuchet from rest in a horizontal orientation. (a) Find the maximum speed that the small-mass object attains. (b) While the small-mass object is gaining speed, does it move with constant acceleration? (c) Does it move with constant tangential acceleration? (d) Does the trebuchet move with constant angular acceleration? (e) Does it have constant momentum? (f) Does the trebuchet–Earth system have constant mechanical energy?

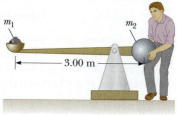

Figure P10.19

20. As a gasoline engine operates, a flywheel turning with the crankshaft stores energy after each fuel explosion, providing the energy required to compress the next charge of fuel and air. For the engine of a certain lawn tractor, suppose a flywheel must be no more than 18.0 cm in diameter. Its thickness, measured along its axis of rotation, must be no larger than 8.00 cm. The flywheel must release energy 60.0 J when its angular speed drops from 800 rev/min to 600 rev/min. Design a sturdy steel (density 7.85×10^3 kg/m³) flywheel to meet these requirements with the smallest mass you can reasonably attain. Specify the shape and mass of the flywheel.

21. **Q|C** **Review.** Consider the system shown in Figure P10.21 with $m_1 = 20.0$ kg, $m_2 = 12.5$ kg, $R = 0.200$ m, and the mass of the pulley $M = 5.00$ kg. Object m_2 is resting on the floor, and object m_1 is 4.00 m above the floor when it is released from rest. The pulley axis is frictionless. The cord is light, does not stretch, and does not slip on the pulley. (a) Calculate the time interval required for m_1 to hit the floor. (b) How would your answer change if the pulley were massless?

Figure P10.21

Section 10.5 Torque and the Vector Product

22. **W** The fishing pole in Figure P10.22 makes an angle of 20.0° with the horizontal. What is the torque exerted by the fish about

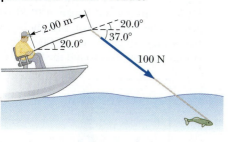

Figure P10.22

an axis perpendicular to the page and passing through the angler's hand if the fish pulls on the fishing line with a force $\vec{F} = 100$ N at an angle 37.0° below the horizontal? The force is applied at a point 2.00 m from the angler's hands.

23. **M** Find the net torque on the wheel in Figure P10.23 about the axle through O, taking $a = 10.0$ cm and $b = 25.0$ cm.

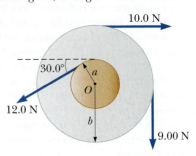

Figure P10.23

24. Two vectors are given by $\vec{A} = -3\hat{i} + 7\hat{j} - 4\hat{k}$ and $\vec{B} = 6\hat{i} - 10\hat{j} + 9\hat{k}$. Evaluate the following quantities (a) $\cos^{-1}[\vec{A} \cdot \vec{B}/AB]$ and (b) $\sin^{-1}[|\vec{A} \times \vec{B}|/AB]$. (c) Which give(s) the angle between the vectors?

25. **W** Given $\vec{M} = 2\hat{i} - 3\hat{j} + \hat{k}$ and $\vec{N} = 4\hat{i} + 5\hat{j} - 2\hat{k}$, calculate the vector product $\vec{M} \times \vec{N}$.

26. **S** Use the definition of the vector product and the definitions of the unit vectors $\hat{i}$, $\hat{j}$, and $\hat{k}$ to prove Equations 10.26 You may assume the x axis points to the right, the y axis up, and the z axis horizontally toward you (not away from you). This choice is said to make the coordinate system a *right-handed system*.

27. A force of $\vec{F} = (2.00\hat{i} + 3.00\hat{j})$ N is applied to an object that is pivoted about a fixed axle aligned along the z coordinate axis. The force is applied at the point $\vec{r} = (4.00\hat{i} + 5.00\hat{j})$ m. Find (a) the magnitude of the net torque about the z axis and (b) the direction of the torque vector $\vec{\tau}$.

Section 10.6 Analysis Model: Rigid Object in Equilibrium

28. **GP** A uniform beam resting on two pivots has a length $L = 6.00$ m and mass $M = 90.0$ kg. The pivot under the left end exerts a normal force n_1 on the beam, and the second pivot located a distance $\ell = 4.00$ m from the left end exerts a normal force n_2. A woman of mass $m = 55.0$ kg steps onto the left end of the beam and begins walking to the right as in Figure P10.28. The goal is to find the woman's position when the beam begins to tip. (a) What is the appropriate analysis model for the beam before it begins to tip? (b) Sketch a force diagram for the beam, labeling the gravitational and normal forces acting on the beam and placing the woman a distance x to the right of the first pivot, which is the origin. (c) Where is the woman when the normal force n_1 is the greatest? (d) What is n_1 when the beam is about to tip? (e) Use Equation 10.27 to find the value of n_2 when the beam is about to tip. (f) Using the result of part (d) and Equation 10.28, with torques computed around the second pivot, find the woman's position x when the beam is about to tip. (g) Check the answer to part (e) by computing torques around the first pivot point.

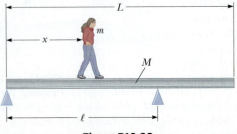

Figure P10.28

29. **BIO** **W** In exercise physiology studies, it is sometimes important to determine the location of a person's center of mass. This determination can be done with the arrangement shown in Figure P10.29. A light plank rests on two scales, which read $F_{g1} = 380$ N and $F_{g2} = 320$ N. A distance of 1.65 m separates the scales. How far from the woman's feet is her center of mass?

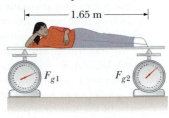

Figure P10.29

30. *Why is the following situation impossible?* A uniform beam of mass $m_b = 3.00$ kg and length $\ell = 1.00$ m supports blocks with masses $m_1 = 5.00$ kg and $m_2 = 15.0$ kg at two positions as shown in Figure P10.30. The beam rests on two triangular blocks, with point P a distance $d = 0.300$ m to the right of the center of gravity of the beam. The position of the object of mass m_2 is adjusted along the length of the beam until the normal force on the beam at O is zero.

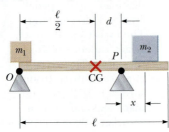

Figure P10.30

31. **W** Figure P10.31 shows a claw hammer being used to pull a nail out of a horizontal board. The mass of the hammer is 1.00 kg. A force of 150 N is exerted horizontally as shown, and the nail does not yet move relative to the board. Find (a) the force exerted by the hammer claws on the nail and (b) the force exerted by the surface on the point of contact with the hammer head. Assume the force the hammer exerts on the nail is parallel to the nail.

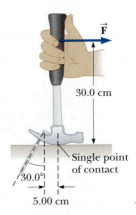

Figure P10.31

32. **S** A uniform sign of weight F_g and width $2L$ hangs from a light, horizontal beam hinged at the wall and supported by a cable (Fig. P10.32). Determine (a) the tension in the cable and (b) the components of the reaction force exerted by the wall on the beam in terms of F_g, d, L, and θ.

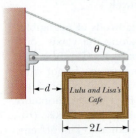

Figure P10.32

33. **M** A 15.0-m uniform ladder weighing 500 N rests against a frictionless wall. The ladder makes a 60.0° angle with the horizontal. (a) Find the horizontal and vertical forces the ground exerts on the base of the ladder when an 800-N firefighter has climbed 4.00 m along the ladder from the bottom. (b) If the ladder is just on the verge of slipping when the firefighter is 9.00 m from the bottom, what is the coefficient of static friction between ladder and ground?

34. **S** A uniform ladder of length L and mass m_1 rests against a frictionless wall. The ladder makes an angle θ with the horizontal. (a) Find the horizontal and vertical forces the ground exerts on the base of the ladder when a firefighter of mass m_2 has climbed a distance x along the ladder from the bottom. (b) If the ladder is just on the verge of slipping when the firefighter is a distance d along the ladder from the bottom, what is the coefficient of static friction between ladder and ground?

35. **BIO** The arm in Figure P10.35 weighs 41.5 N. The gravitational force on the arm acts through point A. Determine the magnitudes of the tension force $\vec{F}_t$ in the deltoid muscle and the force $\vec{F}_s$ exerted by the shoulder on the humerus (upper-arm bone) to hold the arm in the position shown.

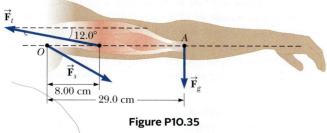

Figure P10.35

36. A crane of mass $m_1 = 3\,000$ kg supports a load of mass $m_2 = 10\,000$ kg as shown in Figure P10.36. The crane is pivoted with a frictionless pin at A and rests against a smooth support at B. Find the reaction forces at (a) point A and (b) point B.

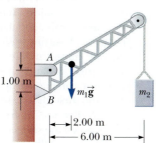

Figure P10.36

Section 10.7 Analysis Model: Rigid Object Under a Net Torque

Section 10.8 Energy Considerations in Rotational Motion

37. An electric motor turns a flywheel through a drive belt that joins a pulley on the motor and a pulley that is rigidly attached to the flywheel as shown in Figure P10.37. The flywheel is a solid disk with a mass of 80.0 kg and a radius $R = 0.625$ m. It turns on a frictionless axle. Its pulley has much smaller mass and a radius of $r = 0.230$ m. The tension T_u in the upper (taut) segment of the belt is 135 N, and the flywheel has a clockwise angular acceleration of 1.67 rad/s². Find the tension in the lower (slack) segment of the belt.

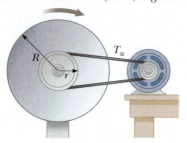

Figure P10.37

38. **S** This problem describes one experimental method for determining the moment of inertia of an irregularly shaped object such as the payload for a satellite. Figure P10.38 shows a counterweight of mass m suspended by a cord wound around a spool of radius r, forming part of a turntable supporting the object. The turntable can rotate without friction. When the counterweight is released from rest, it descends through a distance h, acquiring a speed v. Show that the moment of inertia I of the rotating apparatus (including the turntable) is $mr^2(2gh/v^2 - 1)$.

Figure P10.38

39. The combination of an applied force and a friction force produces a constant total torque of 36.0 N · m on a wheel rotating about a fixed axis. The applied force acts for 6.00 s. During this time, the angular speed of the wheel increases from 0 to 10.0 rad/s. The applied force is then removed, and the wheel comes to rest in 60.0 s. Find (a) the moment of inertia of the wheel, (b) the magnitude of the torque due to friction, and (c) the total number of revolutions of the wheel during the entire interval of 66.0 s.

40. In Figure P10.40, the hanging object has a mass of $m_1 = 0.420$ kg; the sliding block has a mass of $m_2 = 0.850$ kg; and the pulley is a hollow cylinder with a mass of $M = 0.350$ kg, an inner radius of $R_1 = 0.020\,0$ m, and an outer radius of $R_2 = 0.030\,0$ m. Assume the mass of the spokes is negligible. The coefficient of kinetic friction between the block and the horizontal surface is $\mu_k = 0.250$. The pulley turns without friction on its axle. The light cord does not stretch and does not slip on the pulley. The block has a velocity of $v_i = 0.820$ m/s toward the pulley when it passes a reference point on the table. (a) Use energy methods to predict its speed after it has moved to a second point, 0.700 m away. (b) Find the angular speed of the pulley at the same moment.

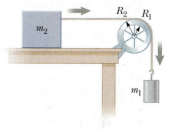

Figure P10.40

41. **W** A potter's wheel—a thick stone disk of radius 0.500 m and mass 100 kg—is freely rotating at 50.0 rev/min. The potter can stop the wheel in 6.00 s by pressing a wet rag against the rim and exerting a radially inward force of 70.0 N. Find the effective coefficient of kinetic friction between wheel and rag.

42. **M** A model airplane with mass 0.750 kg is tethered to the ground by a wire so that it flies in a horizontal circle 30.0 m in radius. The airplane engine provides a net thrust of 0.800 N perpendicular to the tethering wire. (a) Find the torque the net thrust produces about the center of the circle. (b) Find the angular acceleration of the airplane. (c) Find the translational acceleration of the airplane tangent to its flight path.

43. **GP** Review. As shown in Figure P10.43 (page 348), two blocks are connected by a string of negligible mass passing over a pulley of radius $r = 0.250$ m and moment of inertia I. The block on the frictionless incline is moving with a constant acceleration of magnitude $a = 2.00$ m/s^2. From this information, we wish to find the moment of inertia of the pulley. (a) What analysis model is appropriate for the blocks? (b) What analysis model is appropriate for the pulley? (c) From the analysis model in part (a), find the tension T_1. (d) Similarly, find the tension T_2. (e) From the analysis model in part (b), find a symbolic expression for the moment of inertia of the pulley in terms of the tensions T_1 and T_2, the pulley radius r, and the acceleration a. (f) Find the numerical value of the moment of inertia of the pulley.

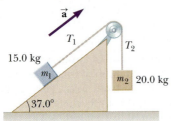

Figure P10.43

44. Consider two objects with $m_1 > m_2$ connected by a light string that passes over a pulley having a moment of inertia of I about its axis of rotation as shown in Figure P10.44. The string does not slip on the pulley or stretch. The pulley turns without friction. The two objects are released from rest separated by a vertical distance $2h$. (a) Use the principle of conservation of energy to find the translational speeds of the objects as they pass each other. (b) Find the angular speed of the pulley at this time.

Figure P10.44

45. **M** **Review.** An object with a mass of $m = 5.10$ kg is attached to the free end of a light string wrapped around a reel of radius $R = 0.250$ m and mass $M = 3.00$ kg. The reel is a solid disk, free to rotate in a vertical plane about the horizontal axis passing through its center as shown in Figure P10.45. The suspended object is released from rest 6.00 m above the floor. Determine (a) the tension in the string, (b) the acceleration of the object, and (c) the speed with which the object hits the floor. (d) Verify your answer to part (c) by using the isolated system (energy) model.

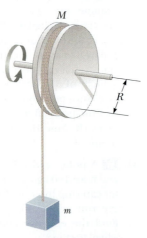

Figure P10.45

Section 10.9 Analysis Model: Nonisolated System (Angular Momentum)

Section 10.10 Analysis Model: Isolated System (Angular Momentum)

46. **W** A playground merry-go-round of radius $R = 2.00$ m has a moment of inertia $I = 250$ kg · m^2 and is rotating at 10.0 rev/min about a frictionless, vertical axle. Facing the axle, a 25.0-kg child hops onto the merry-go-round and manages to sit down on the edge. What is the new angular speed of the merry-go-round?

47. **M** The position vector of a particle of mass 2.00 kg as a function of time is given by $\vec{r} = (6.00\hat{i} + 5.00t\hat{j})$, where $\vec{r}$ is in meters and t is in seconds. Determine the angular momentum of the particle about the origin as a function of time.

48. Heading straight toward the summit of Pike's Peak, an airplane of mass 12 000 kg flies over the plains of Kansas at nearly constant altitude 4.30 km with constant velocity 175 m/s west. (a) What is the airplane's vector angular momentum relative to a wheat farmer on the ground directly below the airplane? (b) Does this value change as the airplane continues its motion along a straight line? (c) **What If?** What is its angular momentum relative to the summit of Pike's Peak?

49. Big Ben (Fig. P10.17), the Parliament tower clock in London, has hour and minute hands with lengths of 2.70 m and 4.50 m and masses of 60.0 kg and 100 kg, respectively. Calculate the total angular momentum of these hands about the center point. (You may model the hands as long, thin rods rotating about one end. Assume the hour and minute hands are rotating at a constant rate of one revolution per 12 hours and 60 minutes, respectively.)

50. **S** **W** A disk with moment of inertia I_1 rotates about a frictionless, vertical axle with angular speed ω_i. A second disk, this one having moment of inertia I_2 and initially not rotating, drops onto the first disk (Fig. P10.50). Because of friction between the surfaces, the two eventually reach the same angular speed ω_f. (a) Calculate ω_f. (b) Calculate the ratio of the final to the initial rotational energy.

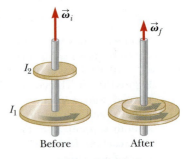

Figure P10.50

51. **M** A particle of mass 0.400 kg is attached to the 100-cm mark of a meterstick of mass 0.100 kg. The meterstick rotates on the surface of a frictionless, horizontal table with an angular speed of 4.00 rad/s. Calculate the angular momentum of the system when the stick is pivoted about an axis (a) perpendicular to the table through the 50.0-cm mark and (b) perpendicular to the table through the 0-cm mark.

52. A space station is constructed in the shape of a hollow ring of mass 5.00 × 10^4 kg. Members of the crew walk on a deck formed by the inner surface of the outer cylindrical wall of the ring, with radius $r = 100$ m. At rest when constructed, the ring is set rotating about its axis

Figure P10.52 Problems 52 and 54.

so that the people inside experience an effective free-fall acceleration equal to g. (See Fig. P10.52.) The rotation is achieved by firing two small rockets attached tangentially to opposite points on the rim of the ring. (a) What angular momentum does the space station acquire? (b) For what time interval must the rockets be fired if each exerts a thrust of 125 N?

53. A puck of mass m_1 = 80.0 g and radius r_1 = 4.00 cm glides across an air table at a speed v = 1.50 m/s as shown in Figure P10.53a. It makes a glancing collision with a second puck of radius r_2 = 6.00 cm and mass m_2 = 120 g (initially at rest) such that their rims just touch. Because their rims are coated with instant-acting glue, the pucks stick together and rotate after the collision (Fig. P10.53b). (a) What is the angular momentum of the system relative to the center of mass? (b) What is the angular speed about the center of mass?

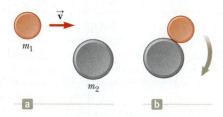

m_1

$\vec{v}$

m_2

a   b

Figure P10.53

54. *Why is the following situation impossible?* A space station shaped like a giant wheel has a radius of r = 100 m and a moment of inertia of 5.00×10^8 kg · m². A crew of 150 people of average mass 65.0 kg is living on the rim, and the station's rotation causes the crew to experience an apparent free-fall acceleration of g (Fig. P10.52). A research technician is assigned to perform an experiment in which a ball is dropped at the rim of the station every 15 minutes and the time interval for the ball to drop a given distance is measured as a test to make sure the apparent value of g is correctly maintained. One evening, 100 average people move to the center of the station for a union meeting. The research technician, who has already been performing his experiment for an hour before the meeting, is disappointed that he cannot attend the meeting, and his mood sours even further by his boring experiment in which every time interval for the dropped ball is identical for the entire evening.

55. The puck in Figure 10.25 has a mass of 0.120 kg. The distance of the puck from the center of rotation is originally 40.0 cm, and the puck is sliding with a speed of 80.0 cm/s. The string is pulled downward 15.0 cm through the hole in the frictionless table. Determine the work done on the puck. (*Suggestion:* Consider the change of kinetic energy.)

56. **W** A student sits on a freely rotating stool holding two dumbbells, each of mass 3.00 kg (Fig. P10.56). When his arms are extended horizontally (Fig. P10.56a), the dumbbells are 1.00 m from the axis of rotation and the student rotates with an angular speed of 0.750 rad/s. The moment of inertia

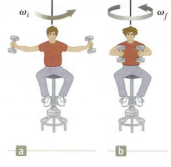

ω_i ω_f

a   b

Figure P10.56

of the student plus stool is 3.00 kg · m² and is assumed to be constant. The student pulls the dumbbells inward horizontally to a position 0.300 m from the rotation axis (Fig. P10.56b). (a) Find the new angular speed of the student. (b) Find the kinetic energy of the rotating system before and after he pulls the dumbbells inward.

57. **M** **Q|C** A 60.0-kg woman stands at the western rim of a horizontal turntable having a moment of inertia of 500 kg · m² and a radius of 2.00 m. The turntable is initially at rest and is free to rotate about a frictionless, vertical axle through its center. The woman then starts walking around the rim clockwise (as viewed from above the system) at a constant speed of 1.50 m/s relative to the Earth. Consider the woman–turntable system as motion begins. (a) Is the mechanical energy of the system constant? (b) Is the momentum of the system constant? (c) Is the angular momentum of the system constant? (d) In what direction and with what angular speed does the turntable rotate? (e) How much chemical energy does the woman's body convert into mechanical energy of the woman–turntable system as the woman sets herself and the turntable into motion?

Section 10.11 Precessional Motion of Gyroscopes

58. The angular momentum vector of a precessing gyroscope sweeps out a cone as shown in Figure P10.58. The angular speed of the tip of the angular momentum vector, called its precessional frequency, is given by $\omega_p = \tau/L$, where τ is the magnitude of the torque on the gyroscope and L is the magnitude of its angular momentum. In the motion called *precession of the equinoxes*, the Earth's axis of rotation precesses about the perpendicular to its orbital plane with a period of 2.58×10^4 yr. Model the Earth as a uniform sphere and calculate the torque on the Earth that is causing this precession.

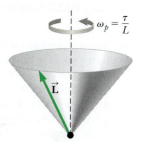

$\omega_p = \dfrac{\tau}{L}$

$\vec{L}$

Figure P10.58 A precessing angular momentum vector sweeps out a cone in space.

Section 10.12 Rolling Motion of Rigid Objects

59. **M** A cylinder of mass 10.0 kg rolls without slipping on a horizontal surface. At a certain instant, its center of mass has a speed of 10.0 m/s. Determine (a) the translational kinetic energy of its center of mass, (b) the rotational kinetic energy about its center of mass, and (c) its total energy.

60. **S** A uniform solid disk and a uniform hoop are placed side by side at the top of an incline of height h. (a) If they are released from rest and roll without slipping, which object reaches the bottom first? (b) Verify your answer by calculating their speeds when they reach the bottom in terms of h.

61. **Q|C** A metal can containing condensed mushroom soup has mass 215 g, height 10.8 cm, and diameter 6.38 cm. It is placed at rest on its side at the top of a 3.00-m-long incline that is at 25.0° to the horizontal and is then released to roll straight down. It reaches the bottom of the incline after 1.50 s. (a) Assuming mechanical energy conservation, calculate the moment of inertia of the can. (b) Which pieces of data, if any, are unnecessary for calculating the solution? (c) Why can't the moment of inertia be calculated from $I = \frac{1}{2}mr^2$ for the cylindrical can?

62. **Q|C** A tennis ball is a hollow sphere with a thin wall. It is set rolling without slipping at 4.03 m/s on a horizontal section of a track as shown in Figure P10.62. It rolls around the inside of a vertical circular loop of radius $r = 45.0$ cm. As the ball nears the bottom

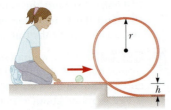

Figure P10.62

of the loop, the shape of the track deviates from a perfect circle so that the ball leaves the track at a point $h = 20.0$ cm below the horizontal section. (a) Find the ball's speed at the top of the loop. (b) Demonstrate that the ball will not fall from the track at the top of the loop. (c) Find the ball's speed as it leaves the track at the bottom. **What If?** (d) Suppose that static friction between ball and track were negligible so that the ball slid instead of rolling. Would its speed then be higher, lower, or the same at the top of the loop? (e) Explain your answer to part (d).

Section 10.13 Context Connection—Turning the Spacecraft

63. A spacecraft is in empty space. It carries on board a gyroscope with a moment of inertia of $I_g = 20.0$ kg · m² about the axis of the gyroscope. The moment of inertia of the spacecraft around the same axis is $I_s = 5.00 \times 10^5$ kg · m². Neither the spacecraft nor the gyroscope is originally rotating. The gyroscope can be powered up in a negligible period of time to an angular speed of 100 rad/s. If the orientation of the spacecraft is to be changed by 30.0°, for what time interval should the gyroscope be operated?

Additional Problems

64. **Review.** A mixing beater consists of three thin rods, each 10.0 cm long. The rods diverge from a central hub, separated from each other by 120°, and all turn in the same plane. A ball is attached to the end of each rod. Each ball has cross-sectional area 4.00 cm² and is so shaped that it has a drag coefficient of 0.600. Calculate the power input required to spin the beater at 1 000 rev/min (a) in air and (b) in water.

65. **S** A long, uniform rod of length L and mass M is pivoted about a frictionless, horizontal pin through one end. The rod is released from rest in a vertical position as shown in Figure P10.65. At the instant the rod is horizontal, find (a) its angular speed, (b) the magnitude of its angular acceleration, (c) the x and y components of the acceleration of its center of mass, and (d) the components of the reaction force at the pivot.

Figure P10.65

66. The hour hand and the minute hand of Big Ben, the Parliament tower clock in London, are 2.70 m and 4.50 m long and have masses of 60.0 kg and 100 kg, respectively (see Fig. P10.17). (a) Determine the total torque due to the weight of these hands about the axis of rotation when the time reads (i) 3:00, (ii) 5:15, (iii) 6:00, (iv) 8:20, and (v) 9:45. (You may model the hands as long, thin, uniform rods.) (b) Determine all times when the total torque about the axis of rotation is zero. Determine the times to the nearest second, solving a transcendental equation numerically.

67. **M** Two astronauts (Fig. P10.67), each having a mass of 75.0 kg, are connected by a 10.0-m rope of negligible mass. They are isolated in space, orbiting their center of mass at speeds of 5.00 m/s. Treating the astronauts as particles, calculate (a) the magnitude of the angular momentum of the two-astronaut system and (b) the rotational energy of the system. By pulling on the rope, one astronaut shortens the distance between them to 5.00 m. (c) What is the new angular momentum of the system? (d) What are the astronauts' new speeds? (e) What is the new rotational energy of the system? (f) How much chemical potential energy in the body of the astronaut was converted to mechanical energy in the system when he shortened the rope?

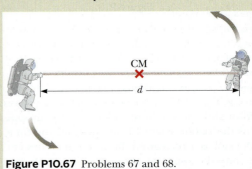

Figure P10.67 Problems 67 and 68.

68. **S** Two astronauts (Fig. P10.67), each having a mass M, are connected by a rope of length d having negligible mass. They are isolated in space, orbiting their center of mass at speeds v. Treating the astronauts as particles, calculate (a) the magnitude of the angular momentum of the two-astronaut system and (b) the rotational energy of the system. By pulling on the rope, one of the astronauts shortens the distance between them to $d/2$. (c) What is the new angular momentum of the system? (d) What are the astronauts' new speeds? (e) What is the new rotational energy of the system? (f) How much chemical potential energy in the body of the astronaut was converted to mechanical energy in the system when he shortened the rope?

69. **BIO** When a person stands on tiptoe on one foot (a strenuous position), the position of the foot is as shown in Figure P10.69a. The total gravitational force $\vec{F}_g$ on the body is supported by the normal force $\vec{n}$ exerted by the floor on the toes of one foot. A mechanical model of the situation is shown in Figure P10.69b, where $\vec{T}$ is the force exerted on the foot by the Achilles tendon and $\vec{R}$ is the force exerted on the foot by the tibia. Find the values of T, R, and θ when $F_g = 700$ N.

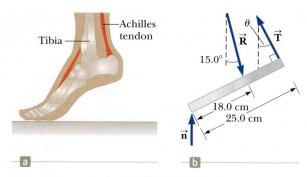

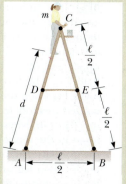

Figure P10.69

70. **S** A uniform, hollow, cylindrical spool has inside radius $R/2$, outside radius R, and mass M (Fig. P10.70). It is mounted so that it rotates on a fixed, horizontal axle. A counterweight of mass m is connected to the end of a string wound around the spool. The counterweight falls from rest at $t = 0$ to a position y at time t. Show that the torque due to the friction forces between spool and axle is

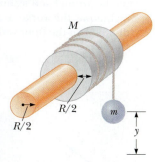

Figure P10.70

$$\tau_f = R\left[m\left(g - \frac{2y}{t^2}\right) - M\frac{5y}{4t^2}\right]$$

71. **S** The reel shown in Figure P10.71 has radius R and moment of inertia I. One end of the block of mass m is connected to a spring of force constant k, and the other end is fastened to a cord wrapped around the reel. The reel axle and the incline are frictionless. The reel is wound counterclockwise so that the spring stretches a distance d from its unstretched position and the reel is then released from rest. Find the angular speed of the reel when the spring is again unstretched.

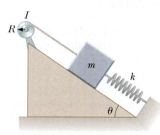

Figure P10.71

72. **W** **Review.** A block of mass $m_1 = 2.00$ kg and a block of mass $m_2 = 6.00$ kg are connected by a massless string over a pulley in the shape of a solid disk having radius $R = 0.250$ m and mass $M = 10.0$ kg. The fixed, wedge-shaped ramp makes an angle of $\theta = 30.0°$ as shown in Figure P10.72. The coefficient of kinetic friction is 0.360 for both blocks. (a) Draw force diagrams of both blocks and of the pulley. Determine (b) the acceleration of the two blocks and (c) the tensions in the string on both sides of the pulley.

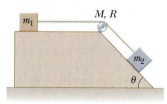

Figure P10.72

73. **M** A stepladder of negligible weight is constructed as shown in Figure P10.73, with $AC = BC = \ell = 4.00$ m. A painter of mass $m = 70.0$ kg stands on the ladder $d = 3.00$ m from

the bottom. Assuming the floor is frictionless, find (a) the tension in the horizontal bar DE connecting the two halves of the ladder, (b) the normal forces at A and B, and (c) the components of the reaction force at the single hinge C that the left half of the ladder exerts on the right half. *Suggestion:* Treat the ladder as a single object, but also treat each half of the ladder separately.

74. **S** A stepladder of negligible weight is constructed as shown in Figure P10.73, with $AC = BC = \ell$. A painter of mass m stands on the ladder a distance d from the bottom. Assuming the floor is frictionless, find (a) the tension in the horizontal bar DE connecting the two halves of the ladder, (b) the normal forces at A and B, and (c) the components of the reaction force at the single hinge C that the left half of the ladder exerts on the right half. *Suggestion:* Treat the ladder as a single object, but also treat each half of the ladder separately.

Figure P10.73 Problems 73 and 74.

75. **QC** **S** A wad of sticky clay with mass m and velocity $\vec{v}_i$ is fired at a solid cylinder of mass M and radius R (Fig. P10.75). The cylinder is initially at rest and is mounted on a fixed horizontal axle that runs through its center of mass. The line of motion of the projectile is perpendicular to the axle and at a distance $d < R$ from the center. (a) Find the angular speed of the system just after the clay strikes and sticks to the surface of the cylinder. (b) Is the mechanical energy of the clay–cylinder system constant in this process? Explain your answer. (c) Is the momentum of the clay–cylinder system constant in this process? Explain your answer.

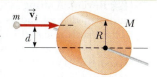

Figure P10.75

76. A common demonstration, illustrated in Figure P10.76, consists of a ball resting at one end of a uniform board of length ℓ that is hinged at the other end and elevated at an angle θ. A light cup is attached to the board at r_c so that it will catch the ball when the support stick is removed suddenly. (a) Show that the ball will lag behind the falling board when θ is less than 35.3°. (b) Assuming the board is 1.00 m long and is supported at this limiting angle, show that the cup must be 18.4 cm from the moving end.

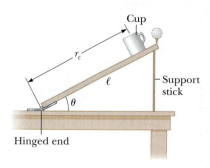

Figure P10.76

77. **BIO** The large quadriceps muscle in the upper leg terminates at its lower end in a tendon attached to the upper end of the tibia (Fig. P10.77a). The forces on the lower leg when the leg is extended are modeled as in Figure P10.77b, where $\vec{T}$ is the force in the tendon, $\vec{F}_{g,leg}$ is the gravitational force acting on the lower leg, and $\vec{F}_{g,foot}$ is the gravitational force acting on the foot. Find T when the tendon is at an angle of $\phi = 25.0°$ with the tibia, assuming $F_{g,leg} = 30.0$ N, $F_{g,foot} = 12.5$ N, and the leg is extended at an angle $\theta = 40.0°$ with respect to the vertical. Also assume the center of gravity of the tibia is at its geometric center and the tendon attaches to the lower leg at a position one-fifth of the way down the leg.

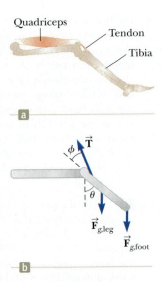

Figure P10.77

78. **S** Review. A string is wound around a uniform disk of radius R and mass M. The disk is released from rest with the string vertical and its top end tied to a fixed bar (Fig. P10.78). Show that (a) the tension in the string is one third of the weight of the disk, (b) the magnitude of the acceleration of the center of mass is $2g/3$, and (c) the speed of the center of mass is $(4gh/3)^{1/2}$ after the disk has descended through distance h. (d) Verify your answer to part (c) using the energy approach.

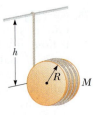

Figure P10.78

79. **BIO** **Q|C** Assume a person bends forward to lift a load "with his back" as shown in Figure P10.79a. The spine pivots mainly at the fifth lumbar vertebra, with the principal supporting force provided by the erector spinalis muscle in the back. To see the magnitude of the forces involved, consider the model shown in Figure P10.79b for a person bending forward to lift a 200-N object. The spine and upper body are represented as a uniform horizontal rod of weight 350 N, pivoted at the base of the spine. The erector spinalis muscle, attached at a point two-thirds of the way up the spine, maintains the position of the back. The angle between the spine and this muscle is $\theta = 12.0°$. Find (a) the tension T in the back muscle and (b) the compressional force in the spine. (c) Is this method a good way to lift a load? Explain your answer, using the results of parts (a) and (b). (d) Can you suggest a better method to lift a load?

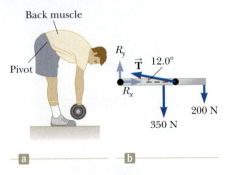

Figure P10.79

80. Why is the following situation impossible? A worker in a factory pulls a cabinet across the floor using a rope as shown in Figure P10.80a. The rope makes an angle $\theta = 37.0°$ with the floor and is tied $h_1 = 10.0$ cm from the bottom of the cabinet. The uniform rectangular cabinet has height $\ell = 100$ cm and width $w = 60.0$ cm, and it weighs 400 N. The cabinet slides with constant speed when a force $F = 300$ N is applied through the rope. The worker tires of walking backward. He fastens the rope to a point on the cabinet $h_2 = 65.0$ cm off the floor and lays the rope over his shoulder so that he can walk forward and pull as shown in Figure P10.80b. In this way, the rope again makes an angle of $\theta = 37.0°$ with the horizontal and again has a tension of 300 N. Using this technique, the worker is able to slide the cabinet over a long distance on the floor without tiring.

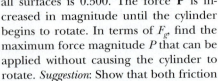

Figure P10.80

81. **GP** **S** A projectile of mass m moves to the right with a speed v_i (Fig. P10.81a). The projectile strikes and sticks to the end of a stationary rod of mass M, length d, pivoted about a frictionless axle perpendicular to the page through O (Fig. P10.81b). We wish to find the fractional change of kinetic energy in the system due to the collision. (a) What is the appropriate analysis model to describe the projectile and the rod? (b) What is the angular momentum of the system before the collision about an axis through O? (c) What is the moment of inertia of the system about an axis through O after the projectile sticks to the rod? (d) If the angular speed of the system after the collision is ω, what is the angular momentum of the system after the collision? (e) Find the angular speed ω after the collision in terms of the given quantities. (f) What is the kinetic energy of the system before the collision? (g) What is the kinetic energy of the system after the collision? (h) Determine the fractional change of kinetic energy due to the collision.

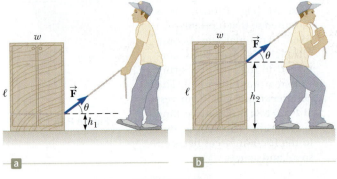

Figure P10.81

82. **Q|C** Figure P10.82 shows a vertical force applied tangentially to a uniform cylinder of weight F_g. The coefficient of static friction between the cylinder and all surfaces is 0.500. The force $\vec{P}$ is increased in magnitude until the cylinder begins to rotate. In terms of F_g, find the maximum force magnitude P that can be applied without causing the cylinder to rotate. Suggestion: Show that both friction

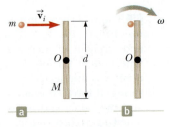

Figure P10.82

forces will be at their maximum values when the cylinder is on the verge of slipping.

83. **S** A solid sphere of mass m and radius r rolls without slipping along the track shown in Figure P10.83. It starts from rest with the lowest point of the sphere at height h above the bottom of the loop of radius R, much larger than r. (a) What is the minimum value of h (in terms of R) such that the sphere completes the loop? (b) What are the force components on the sphere at the point P if $h = 3R$?

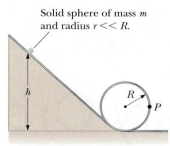

Solid sphere of mass m and radius $r \ll R$.

Figure P10.83

84. **QC** A skateboarder with his board can be modeled as a particle of mass 76.0 kg, located at his center of mass, 0.500 m above the ground. As shown in Figure P10.84, the skateboarder starts from rest in a crouching position at one lip of a half-pipe (point Ⓐ). The half-pipe forms one half of a cylinder of radius 6.80 m with its axis horizontal. On his descent, the skateboarder moves without friction and maintains his crouch so that his center of mass moves through one-quarter of a circle. (a) Find his speed at the bottom of the half-pipe (point Ⓑ). (b) Find his angular momentum about the center of curvature at this point. (c) Immediately after passing point Ⓑ, he stands up and raises his arms, lifting his center of gravity to 0.950 m above the concrete (point Ⓒ). Explain why his angular momentum is constant in this maneuver, whereas the kinetic energy of his body is not constant. (d) Find his speed immediately after he stands up. (e) How much chemical energy in the skateboarder's legs was converted into mechanical energy in the skateboarder–Earth system when he stood up?

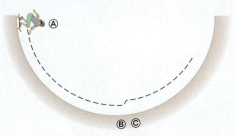

Figure P10.84

85. **BIO** **QC** When a gymnast performing on the rings executes the *iron cross*, he maintains the position at rest shown in Figure P10.85a. In this maneuver, the gymnast's feet (not shown) are off the floor. The primary muscles involved in supporting this position are the latissimus dorsi ("lats") and the pectoralis major ("pecs"). One of the rings exerts an upward force $\vec{\mathbf{F}}_h$ on a hand as shown in Figure P10.85b. The force $\vec{\mathbf{F}}_s$ is exerted by the shoulder joint on the arm. The latissimus dorsi and pectoralis major muscles exert a total force $\vec{\mathbf{F}}_m$ on the arm. (a) Using the information in the figure, find the magnitude of the force $\vec{\mathbf{F}}_m$. (b) Suppose an athlete in training cannot perform the iron cross but can hold a position similar to the figure in which the arms make a 45° angle with the horizontal rather than being horizontal. Why is this position easier for the athlete?

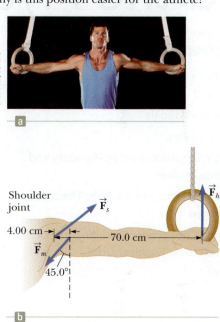

Figure P10.85

Chapter 11

Gravity, Planetary Orbits, and the Hydrogen Atom

Chapter Outline

In Chapter 1, we introduced the notion of modeling and defined four categories of models: geometric, simplification, analysis, and structural. In this chapter, we apply our analysis models to two very common *structural models:* a structural model for a large system—the Solar System—and a structural model for a small system—the hydrogen atom.

We return to Newton's law of universal gravitation—one of the fundamental force laws in nature discussed in Chapter 5—and show how it, together with our analysis models, enables us to understand the motions of planets, moons, and artificial Earth satellites.

We conclude this chapter with a discussion of Niels Bohr's model of the hydrogen atom, which represents an interesting mixture of classical and nonclassical physics. Despite the hybrid nature of the model, some of its predictions agree with experimental measurements made on hydrogen atoms. This discussion will be our first major venture into the area of *quantum physics*, which we will continue in Chapter 28.

NASA, Hubble Heritage Team, (STScI/AURA), ESA, S. Beckwith (STScI). Additional Processing: Robert Gendler

Hubble image of the Whirlpool Galaxy, M51, taken in 2005. The arms of this spiral galaxy compress hydrogen gas and create new clusters of stars. Some astronomers believe that the arms are prominent due to a close encounter with the small, yellow galaxy, NGC 5195, at the tip of one of its arms.

❰11.1 | Newton's Law of Universal Gravitation Revisited

Before 1687, a large amount of data had been collected on the motions of the Moon and the planets, but a clear understanding of the forces involved with the motions was not yet attainable. In that year, Isaac Newton provided the key that unlocked the secrets of the heavens. He knew, from the first law of motion, that a net force had to be acting on the Moon. If not, the Moon would move in a straight-line path rather than in its almost circular orbit. Newton reasoned that this force between the Moon and the Earth was an attractive force. He realized that the forces involved in the Earth–Moon attraction and in the Sun–planet attraction were not something special to those systems, but rather were particular cases of a general and universal attraction between objects.

As you should recall from Chapter 5, every particle in the Universe attracts every other particle with a force that is directly proportional to the product of their masses and inversely proportional to the square of the distance between them. If two particles have masses m_1 and m_2 and are separated by a distance r, the magnitude of the gravitational force between them is

$$F_g = G\frac{m_1 m_2}{r^2} \qquad \textbf{11.1}◀$$

where G is the **universal gravitational constant** whose value in SI units is

$$G = 6.674 \times 10^{-11}\ \text{N} \cdot \text{m}^2/\text{kg}^2 \qquad \textbf{11.2}◀$$

The force law given by Equation 11.1 is often referred to as an **inverse-square law** because the magnitude of the force varies as the inverse square of the separation of the particles. We can express this attractive force in vector form by defining a unit vector $\hat{\mathbf{r}}_{12}$ directed from m_1 toward m_2 as shown in Active Figure 11.1. The force exerted by m_1 on m_2 is

$$\vec{\mathbf{F}}_{12} = -G\frac{m_1 m_2}{r^2}\hat{\mathbf{r}}_{12} \qquad \textbf{11.3}◀$$

where the negative sign indicates that particle 2 is attracted toward particle 1. Likewise, by Newton's third law, the force exerted by m_2 on m_1, designated $\vec{\mathbf{F}}_{21}$, is equal in magnitude to $\vec{\mathbf{F}}_{12}$ and in the opposite direction. That is, these forces form an action–reaction pair, and $\vec{\mathbf{F}}_{21} = -\vec{\mathbf{F}}_{12}$.

As Newton demonstrated, the gravitational force exerted by a finite-sized, spherically symmetric mass distribution on a particle outside the distribution is the same as if the entire mass of the distribution were concentrated at its center. For example, the force on a particle of mass m at the Earth's surface has the magnitude

$$F_g = G\frac{M_E m}{R_E^2}$$

where M_E is the Earth's mass and R_E is the Earth's radius. This force is directed toward the center of the Earth.

Measurement of the Gravitational Constant

The universal gravitational constant G was first evaluated in the late 19th century, based on the results of an important experiment by Sir Henry Cavendish in 1798. The law of universal gravitation was not expressed by Newton in the form of Equation 11.1, and Newton did not mention a constant such as G. In fact, even by the time of Cavendish, a unit of force had not yet been included in the existing system of units. Cavendish's goal was to measure the density of the Earth. His results were then used by other scientists 100 years later to generate a value for G.

The apparatus he used consists of two small spheres, each of mass m, fixed to the ends of a light horizontal rod suspended by a thin wire as in Figure 11.2. Two

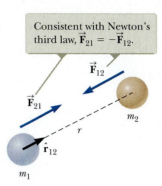

Active Figure 11.1 The gravitational force between two particles is attractive. The unit vector $\hat{\mathbf{r}}_{12}$ is directed from particle 1 toward particle 2.

Consistent with Newton's third law, $\vec{\mathbf{F}}_{21} = -\vec{\mathbf{F}}_{12}$.

Pitfall Prevention | 11.1
Be Clear on g and G
The symbol g represents the magnitude of the free-fall acceleration near a planet. At the surface of the Earth, g has an average value of $9.80\ \text{m/s}^2$. On the other hand, G is a universal constant that has the same value everywhere in the Universe.

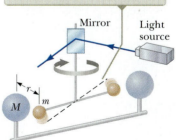

The dashed line represents the original position of the rod.

Figure 11.2 Schematic diagram of the Cavendish apparatus. As the small spheres of mass m are attracted to the large spheres of mass M, the rod rotates through a small angle. A light beam reflected from a mirror on the rotating apparatus measures the angle of rotation. (In reality, the length of wire above the mirror is much larger than that below it.)

large spheres, each of mass M, are then placed near the smaller spheres. The attractive force between the smaller and larger spheres causes the rod to rotate and twist the wire. If the system is oriented as shown in Figure 11.2, the rod rotates clockwise when viewed from the top. The angle through which it rotates is measured by the deflection of a light beam that is reflected from a mirror attached to the wire. The experiment is carefully repeated with different masses at various separations.

It is interesting that G is the least well known of the fundamental constants, with a percentage uncertainty thousands of times larger than those for other constants such as the mass m_e of the electron and the fundamental electric charge e. Several recent measurements of G vary significantly from the previous values and from one another! The search for a more precise value of G continues to be an area of active research. A 2006 experiment measured weight changes in a stationary object as a second object was brought near, resulting in a value of G of $6.674\ 3 \times 10^{-11}$ m^3/kg · s^2, with an uncertainty of $\pm 0.001\ 5\%$. A 2007 experiment measured G using a gravity gradiometer based on atom interferometry. The result of this experiment was 6.693×10^{-11} m^3/kg · s^2, with an uncertainty of $\pm 0.3\%$. The 2006 result is barely within the relatively large uncertainty range of the 2007 result!

> **QUICK QUIZ 11.1** A planet has two moons of equal mass. Moon 1 is in a circular orbit of radius r. Moon 2 is in a circular orbit of radius $2r$. What is the magnitude of the gravitational force exerted by the planet on Moon 2? (**a**) four times as large as that on Moon 1 (**b**) twice as large as that on Moon 1 (**c**) equal to that on Moon 1 (**d**) half as large as that on Moon 1 (**e**) one-fourth as large as that on Moon 1

The Gravitational Field

When Newton first published his theory of gravitation, his contemporaries found it difficult to accept the concept of a force that one object could exert on another without anything happening in the space between them. They asked how it was possible for two objects with mass to interact even though they were not in contact with each other. Although Newton himself could not answer this question, his theory was considered a success because it satisfactorily explained the motions of the planets.

▶ Gravitational field

An alternative mental representation of the gravitational force is to think of the gravitational interaction as a two-step process involving a *field,* as discussed in Section 4.1. First, one object (a *source mass*) creates a **gravitational field** $\vec{\mathbf{g}}$ throughout the space around it. Then, a second object (a *test mass*) of mass m residing in this field experiences a force $\vec{\mathbf{F}}_g = m\vec{\mathbf{g}}$. In other words, we model the *field* as exerting a force on the test mass rather than the source mass exerting the force directly. The gravitational field is defined by

$$\vec{\mathbf{g}} \equiv \frac{\vec{\mathbf{F}}_g}{m} \qquad \qquad 11.4 \blacktriangleleft$$

That is, the gravitational field at a point in space equals the gravitational force that a test mass m experiences at that point divided by the mass. Consequently, if $\vec{\mathbf{g}}$ is known at some point in space, a particle of mass m experiences a gravitational force $\vec{\mathbf{F}}_g = m\vec{\mathbf{g}}$ when placed at that point. We will also see the model of a particle in a field for electricity and magnetism in later chapters, where it plays a much larger role than it does for gravity.

As an example, consider an object of mass m near the Earth's surface. The gravitational force on the object is directed toward the center of the Earth and has a magnitude mg. Therefore, we see that the gravitational field experienced by the object at some point has a magnitude equal to the free-fall acceleration at that point. Because the gravitational force on the object has a magnitude $GM_E m/r^2$ (where M_E is the mass of the Earth), the field $\vec{g}$ at a distance r from the center of the Earth is given by

$$\vec{g} = \frac{\vec{F}_g}{m} = -\frac{GM_E}{r^2}\hat{r}$$ **11.5** ◄

where $\hat{r}$ is a unit vector pointing radially outward from the Earth and the negative sign indicates that the field vector points toward the center of the Earth as shown in Figure 11.3a. Note that the field vectors at different points surrounding the spherical mass vary in both direction and magnitude. In a small region near the Earth's surface, $\vec{g}$ is approximately constant and the downward field is uniform as indicated in Figure 11.3b. Equation 11.5 is valid at all points outside the Earth's surface, assuming that the Earth is spherical and that rotation can be neglected. At the Earth's surface, where $r = R_E$, $\vec{g}$ has a magnitude of 9.80 m/s².

The field vectors point in the direction of the acceleration a particle would experience if it were placed in the field. The magnitude of the field vector at any location is the magnitude of the free-fall acceleration at that location.

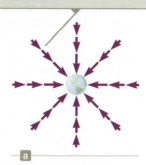

Figure 11.3 (a) The gravitational field vectors in the vicinity of a uniform spherical mass vary in both direction and magnitude. (b) The gravitational field vectors in a small region near the Earth's surface are uniform; that is, they all have the same direction and magnitude.

Example 11.1 | **The Density of the Earth**

Using the known radius of the Earth and that $g = 9.80$ m/s² at the Earth's surface, find the average density of the Earth.

SOLUTION

Conceptualize Assume the Earth is a perfect sphere. The density of material in the Earth varies, but let's adopt a simplified model in which we assume the density to be uniform throughout the Earth. The resulting density is the average density of the Earth.

Categorize This example is a relatively simple substitution problem.

Using the magnitudes in Equation 11.5 at the surface of the Earth, solve for the mass of the Earth:

$$M_E = \frac{gR_E^2}{G}$$

Substitute this mass into the definition of density (Eq. 1.1):

$$\rho_E = \frac{M_E}{V_E} = \frac{gR_E^2/G}{\frac{4}{3}\pi R_E^3} = \frac{3}{4}\frac{g}{\pi G R_E}$$

$$= \frac{3}{4}\frac{9.80 \text{ m/s}^2}{\pi(6.67 \times 10^{-11} \text{ N} \cdot \text{m}^2/\text{kg}^2)(6.37 \times 10^6 \text{ m})} = \boxed{5.51 \times 10^3 \text{ kg/m}^3}$$

What If? What if you were told that a typical density of granite at the Earth's surface is 2.75×10^3 kg/m³? What would you conclude about the density of the material in the Earth's interior?

Answer Because this value is about half the density we calculated as an average for the entire Earth, we would conclude that the inner core of the Earth has a density much higher than the average value. It is most amazing that the Cavendish experiment—which can be used to determine G and can be done today on a tabletop—combined with simple free-fall measurements of g provides information about the core of the Earth!

11.2 | Structural Models

In Chapter 1, we mentioned that we would discuss four categories of models. The fourth category is **structural models.** In these models, we propose theoretical structures in an attempt to understand the behavior of a system with which we cannot interact directly because it is far different in scale—either much smaller or much larger—from our macroscopic world.

One of the earliest structural models to be explored was that of the place of the Earth in the Universe. The movements of the planets, stars, and other celestial bodies have been observed by people for thousands of years. Early in history, scientists regarded the Earth as the center of the Universe because it appeared that objects in the sky moved around the Earth. This organization of the Earth and other objects is a structural model for the Universe called the *geocentric model*. It was elaborated and formalized by the Greek astronomer Claudius Ptolemy in the 2nd century A.D. and was accepted for the next 1400 years. In 1543, Polish astronomer Nicolaus Copernicus (1473–1543) offered a different structural model in which the Earth is part of a local Solar System, suggesting that the Earth and the other planets revolve in perfectly circular orbits about the Sun (the *heliocentric model*).

In general, a structural model contains the following features:

▶ Features of structural models

1. *A description of the physical components of the system:* In the heliocentric model, the components are the planets and the Sun.
2. *A description of where the components are located relative to one another and how they interact:* In the heliocentric model, the planets are in orbit around the Sun and they interact via the gravitational force.
3. *A description of the time evolution of the system:* The heliocentric model assumes a steady-state Solar System, with planets revolving in orbits around the Sun with fixed periods.
4. *A description of the agreement between predictions of the model and actual observations and, possibly, predictions of new effects that have not yet been observed:* The heliocentric model predicts Earth-based observations of Mars that are in agreement with historical and present measurements. The geocentric model was also able to find agreement between predictions and observations, but only at the expense of a very complicated structural model in which the planets moved in circles built on other circles. The heliocentric model, along with Newton's law of universal gravitation, predicted that a spacecraft could be sent from the Earth to Mars long before it was actually first done in the 1970s.

In Sections 11.3 and 11.4, we explore some of the details of the heliocentric model of the Solar System and supplement the description above for this structural model. In Section 11.5, we investigate a structural model of the hydrogen atom. We will use the components of structural models listed above many times throughout the book.

▌11.3 | Kepler's Laws

Danish astronomer Tycho Brahe (1546–1601) made accurate astronomical measurements over a period of 20 years and provided the basis for the currently accepted structural model of the Solar System. These precise observations, made on the planets and 777 stars, were carried out with nothing more elaborate than a large sextant and compass; the telescope had not yet been invented.

German astronomer Johannes Kepler, who was Brahe's assistant, acquired Brahe's astronomical data and spent about 16 years trying to deduce a mathematical model for the motions of the planets. After many laborious calculations, he found that Brahe's precise data on the revolution of Mars about the Sun provided the answer. Kepler's analysis first showed that the concept of circular orbits about the Sun in the heliocentric model had to be abandoned. He discovered that the orbit of Mars could be accurately described by a curve called an *ellipse*. He then generalized this analysis to include the motions of all planets. The complete analysis is summarized in three statements, known as **Kepler's laws of planetary motion,** each of which is discussed in the following sections.

Newton demonstrated that these laws are consequences of the gravitational force that exists between any two masses. Newton's law of universal gravitation, together with his laws of motion, provides the basis for a full mathematical representation of the motion of planets and satellites.

Johannes Kepler
German astronomer (1571–1630)
Kepler is best known for developing the laws of planetary motion based on the careful observations of Tycho Brahe.

iStockphoto.com/GeorgiosArt

Kepler's First Law

Kepler's first law indicates that the circular orbit of an object around a gravitational force center is a very special case and that elliptical orbits are the general situation:[1]

All planets move in elliptical orbits with the Sun at one focus.

▶ Kepler's first law

Active Figure 11.4 shows the geometry of an ellipse, which serves as our geometric model for the elliptical orbit of a planet.[2] An ellipse is mathematically defined by choosing two points, F_1 and F_2, each of which is a called a **focus,** and then drawing a curve through points for which the sum of the distances r_1 and r_2 from F_1 and F_2 is a constant. The longest distance through the center between points on the ellipse (and passing through each focus) is called the **major axis,** and this distance is $2a$. In Active Figure 11.4, the major axis is drawn along the x direction. The distance a is called the **semimajor axis.** Similarly, the shortest distance through the center between points on the ellipse is called the **minor axis** of length $2b$, where the distance b is the **semiminor axis.** Either focus of the ellipse is located at a distance c from the center of the ellipse, where $a^2 = b^2 + c^2$. In the elliptical orbit of a planet around the Sun, the Sun is at one focus of the ellipse. There is nothing at the other focus.

The **eccentricity** of an ellipse is defined as $e \equiv c/a$ and it describes the general shape of the ellipse. For a circle, $c = 0$ and the eccentricity is therefore zero. The smaller b is compared with a, the shorter the ellipse is along the y direction compared with its extent in the x direction in Active Figure 11.4. As b decreases, c increases and the eccentricity e increases. Therefore, higher values of eccentricity correspond to longer and thinner ellipses. The range of values of the eccentricity for an ellipse is $0 < e < 1$.

Eccentricities for planetary orbits vary widely in the Solar System. The eccentricity of the Earth's orbit is 0.017, which makes it nearly circular. On the other hand, the eccentricity of Mercury's orbit is 0.21, the highest of all the eight planets. Figure 11.5a (page 360) shows an ellipse with the eccentricity of that of Mercury's orbit. Notice that even this highest-eccentricity orbit is difficult to distinguish from a circle, which is one reason Kepler's first law is an admirable accomplishment. The eccentricity of the orbit of Comet Halley is 0.97, describing an orbit whose major axis is much longer than its minor axis as shown in Figure 11.5b. As a result, Comet Halley spends much of its 76-year period far from the Sun and invisible from the Earth. It is only visible to the naked eye during a small part of its orbit when it is near the Sun.

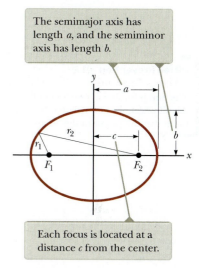

The semimajor axis has length a, and the semiminor axis has length b.

Each focus is located at a distance c from the center.

Active Figure 11.4 Plot of an ellipse.

Figure 11.5 (a) The shape of the orbit of Mercury, which has the highest eccentricity ($e = 0.21$) among the eight planets in the solar system. (b) The shape of the orbit of Comet Halley. The shape of the orbit is correct; the comet and the Sun are shown larger than in reality for clarity.

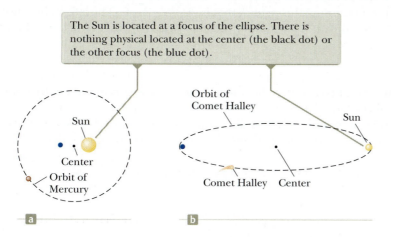

The Sun is located at a focus of the ellipse. There is nothing physical located at the center (the black dot) or the other focus (the blue dot).

---

[1]We choose a simplification model in which a body of mass m is in orbit around a body of mass M, with $M \gg m$. In this way, we model the body of mass M to be stationary. In reality, that is not true; both M and m move around the center of mass of the system of two objects. That is how we indirectly detect planets around other stars; we see the "wobbling" motion of the star as the planet and the star rotate about the center of mass.

[2]Actual orbits show perturbations due to moons in orbit around the planet and passages of the planet near other planets. We will ignore these perturbations and adopt a simplification model in which the planet follows a perfectly elliptical orbit.

Now imagine a planet in an elliptical orbit such as that shown in Active Figure 11.4 with the Sun at focus F_2. When the planet is at the far left in the diagram, the distance between the planet and the Sun is $a + c$. At this point, called the *aphelion*, the planet is at its maximum distance from the Sun. (For an object in orbit around the Earth, this point is called the *apogee*.) Conversely, when the planet is at the right end of the ellipse, the distance between the planet and the Sun is $a - c$. At this point, called the *perihelion* (for an Earth orbit, the *perigee*), the planet is at its minimum distance from the Sun.

Kepler's first law is a direct result of the inverse-square nature of the gravitational force. We have discussed circular and elliptical orbits, which are the allowed shapes of orbits for objects that are *bound* to the gravitational force center. These objects include planets, asteroids, and comets that move repeatedly around the Sun, as well as moons orbiting a planet. *Unbound* objects might also occur, such as a meteoroid from deep space that might pass by the Sun once and then never return. The gravitational force between the Sun and these objects also varies as the inverse square of the separation distance, and the allowed paths for these objects include parabolas ($e = 1$) and hyperbolas ($e > 1$).

Kepler's Second Law

Let us now look at Kepler's second law:

▶ Kepler's second law

> The radius vector drawn from the Sun to any planet sweeps out equal areas in equal time intervals.

This law can be shown to be a consequence of angular momentum conservation for an isolated system as follows. Consider a planet of mass M_p moving about the Sun in an elliptical orbit (Active Fig. 11.6a). Let us consider the planet as a system. We model the Sun to be so much more massive than the planet that the Sun does not move. The gravitational force acting on the planet is a central force, always directed along the radius vector toward the Sun. The torque on the planet due to this central force is zero because $\vec{\mathbf{F}}_g$ is parallel to $\vec{\mathbf{r}}$. That is,

$$\vec{\boldsymbol{\tau}}_{\text{ext}} \equiv \vec{\mathbf{r}} \times \vec{\mathbf{F}}_g = \vec{\mathbf{r}} \times F_g(r)\hat{\mathbf{r}} = 0$$

Recall that the external net torque on a system equals the time rate of change of angular momentum of the system; that is, $\vec{\boldsymbol{\tau}}_{\text{ext}} = d\vec{\mathbf{L}}/dt$. Therefore, because $\vec{\boldsymbol{\tau}}_{\text{ext}} = 0$ for the planet, the angular momentum $\vec{\mathbf{L}}$ of the planet is a constant of the motion:

$$\vec{\mathbf{L}} = \vec{\mathbf{r}} \times \vec{\mathbf{p}} = M_p \vec{\mathbf{r}} \times \vec{\mathbf{v}} = \text{constant}$$

We can relate this result to the following geometric consideration. In a time interval dt, the radius vector $\vec{\mathbf{r}}$ in Active Figure 11.6b sweeps out the area dA, which equals one-half the area $|\vec{\mathbf{r}} \times d\vec{\mathbf{r}}|$ of the parallelogram formed by the vectors $\vec{\mathbf{r}}$ and $d\vec{\mathbf{r}}$. Because the displacement of the planet in the time interval dt is given by $d\vec{\mathbf{r}} = \vec{\mathbf{v}}dt$, we have

$$dA = \tfrac{1}{2}|\vec{\mathbf{r}} \times d\vec{\mathbf{r}}| = \tfrac{1}{2}|\vec{\mathbf{r}} \times \vec{\mathbf{v}}dt| = \frac{L}{2M_p}dt$$

$$\frac{dA}{dt} = \frac{L}{2M_p} = \text{constant} \qquad \textbf{11.6} \blacktriangleleft$$

where L and M_p are both constants. Therefore, we conclude that the radius vector from the Sun to any planet sweeps out equal areas in equal times.

This conclusion is a consequence of the gravitational force being a central force, which in turn implies that angular momentum of the planet is constant. Therefore, the law applies to *any* situation that involves a central force, whether inverse-square or not.

Pitfall Prevention | 11.2
Where Is The Sun?
The Sun is located at one focus of the elliptical orbit of a planet. It is *not* located at the center of the ellipse.

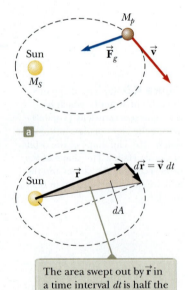

The area swept out by $\vec{\mathbf{r}}$ in a time interval dt is half the area of the parallelogram.

Active Figure 11.6 (a) The gravitational force acting on a planet acts toward the Sun, along the radius vector. (b) During a time interval dt, the vectors form a parallelogram.

> **THINKING PHYSICS 11.1**
>
> The Earth is closer to the Sun when it is winter in the Northern Hemisphere than when it is summer. July and January both have 31 days. In which month, if either, does the Earth move through a longer distance in its orbit?
>
> **Reasoning** The Earth is in a slightly elliptical orbit around the Sun. Because of angular momentum conservation, the Earth moves more rapidly when it is close to the Sun and more slowly when it is farther away. Therefore, because it is closer to the Sun in January, it is moving faster and will cover more distance in its orbit than it will in July. ◄

Kepler's Third Law

Kepler's third law reads as follows:

The square of the orbital period of any planet is proportional to the cube of the semimajor axis of the elliptical orbit.

Kepler's third law can be predicted from the inverse-square law for circular orbits. Consider a planet of mass M_p that is assumed to be moving about the Sun (mass M_S) in a circular orbit as in Active Figure 11.7. Because the gravitational force provides the centripetal acceleration of the planet as it moves in a circle, we model the planet as a particle in uniform circular motion and incorporate Newton's law of universal gravitation:

$$F_g = M_p a \quad \rightarrow \quad \frac{GM_S M_p}{r^2} = \frac{M_p v^2}{r}$$

The orbital speed of the planet is $2\pi r/T$, where T is the period; therefore, the preceding expression becomes

$$\frac{GM_S}{r^2} = \frac{(2\pi r/T)^2}{r}$$

$$T^2 = \left(\frac{4\pi^2}{GM_S}\right) r^3 = K_S r^3$$

where K_S is a constant given by

$$K_S = \frac{4\pi^2}{GM_S} = 2.97 \times 10^{-19} \text{ s}^2/\text{m}^3$$

This equation is also valid for elliptical orbits if we replace r with the length a of the semimajor axis (see Active Fig. 11.4):

$$T^2 = \left(\frac{4\pi^2}{GM_S}\right) a^3 = K_S a^3 \qquad \textbf{11.7}◄$$

► Kepler's third law

Equation 11.7 is Kepler's third law. Because the semimajor axis of a circular orbit is its radius, Equation 11.7 is valid for both circular and elliptical orbits. Note that the constant of proportionality K_S is independent of the mass of the planet. Equation 11.7 is therefore valid for *any* planet. If we were to consider the orbit of a satellite about the Earth, such as the Moon, the constant would have a different value, with the Sun's mass replaced by the Earth's mass; that is, $K_E = 4\pi^2/GM_E$.

Table 11.1 (page 362) is a collection of useful planetary data. The last column verifies that the ratio T^2/r^3 is constant. The small variations in the values in this column are because of uncertainties in the data measured for the periods and semimajor axes of the planets.

Recent astronomical work has revealed the existence of a large number of solar system objects beyond the orbit of Neptune. In general, these objects lie in the *Kuiper belt*, a region that extends from about 30 AU (the orbital radius of Neptune)

Active Figure 11.7 A planet of mass M_p moving in a circular orbit about the Sun. Kepler's third law relates the period of the orbit to the radius. The orbits of all planets except Mercury are nearly circular.

TABLE 11.1 | Useful Planetary Data

Body	Mass (kg)	Mean Radius (m)	Period of Revolution (s)	Mean Distance from the Sun (m)	$\frac{T^2}{r^3}$ (s²/m³)
Mercury	3.30×10^{23}	2.44×10^6	7.60×10^6	5.79×10^{10}	2.98×10^{-19}
Venus	4.87×10^{24}	6.05×10^6	1.94×10^7	1.08×10^{11}	2.99×10^{-19}
Earth	5.97×10^{24}	6.37×10^6	3.156×10^7	1.496×10^{11}	2.97×10^{-19}
Mars	6.42×10^{23}	3.39×10^6	5.94×10^7	2.28×10^{11}	2.98×10^{-19}
Jupiter	1.90×10^{27}	6.99×10^7	3.74×10^8	7.78×10^{11}	2.97×10^{-19}
Saturn	5.68×10^{26}	5.82×10^7	9.29×10^8	1.43×10^{12}	2.95×10^{-19}
Uranus	8.68×10^{25}	2.54×10^7	2.65×10^9	2.87×10^{12}	2.97×10^{-19}
Neptune	1.02×10^{26}	2.46×10^7	5.18×10^9	4.50×10^{12}	2.94×10^{-19}
Pluto[a]	1.25×10^{22}	1.20×10^6	7.82×10^9	5.91×10^{12}	2.96×10^{-19}
Moon	7.35×10^{22}	1.74×10^6	—	—	—
Sun	1.989×10^{30}	6.96×10^8	—	—	—

[a]In August, 2006, the International Astronomical Union adopted a definition of a planet that separates Pluto the other the eight planets. Pluto is now defined as a "dwarf planet" like the asteroid Ceres.

to 50 AU. (An AU is an *astronomical unit,* equal to the radius of the Earth's orbit.) Current estimates identify at least 70 000 objects in this region with diameters larger than 100 km. The first Kuiper belt object (KBO) is Pluto, discovered in 1930 and formerly classified as a planet. Starting in 1992, many more have been detected. Several have diameters in the 1 000-km range, such as Varuna (discovered in 2000), Ixion (2001), Quaoar (2002), Sedna (2003), Haumea (2004), Orcus (2004), and Makemake (2005). One KBO, Eris, discovered in 2005, is believed to be significantly larger than Pluto. Other KBOs do not yet have names, but they are currently indicated by their year of discovery and a code, such as 2009 YE7 and 2010 EK139.

A subset of about 1 400 KBOs are called "Plutinos" because, like Pluto, they exhibit a resonance phenomenon, orbiting the Sun two times in the same time interval as Neptune revolves three times. The contemporary application of Kepler's laws suggest the excitement of this active area of current research.

QUICK QUIZ 11.2 An asteroid is in a highly eccentric elliptical orbit around the Sun. The period of the asteroid's orbit is 90 days. Which of the following statements is true about the possibility of a collision between this asteroid and the Earth? **(a)** There is no possible danger of a collision **(b)** There is a possibility of a collision. **(c)** There is not enough information to determine whether there is a danger of a collision.

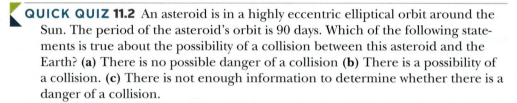

THINKING PHYSICS 11.2

The novel *Icebound,* by Dean Koontz (Bantam Books, 2000), is a story of a group of scientists trapped on a floating iceberg near the North Pole. One of the devices the scientists have with them is a transmitter with which they can fix their position with "the aid of a geosynchronous polar satellite." Can a satellite in a *polar* orbit be *geosynchronous*?

Reasoning A geosynchronous satellite is one that stays over one location on the Earth's surface at all times. Therefore, an antenna on the surface that receives signals from the satellite, such as a television dish, can stay pointed in a fixed direction toward the sky. The satellite must be in an orbit with the correct radius such that its orbital period is the same as that of the Earth's rotation.

This orbit results in the satellite appearing to have no east–west motion relative to the observer at the chosen location. Another requirement is that a geosynchronous satellite *must be in orbit over the equator.* Otherwise it would appear to undergo a north–south oscillation during one orbit. Therefore, it would be impossible to have a geosynchronous satellite in a *polar* orbit. Even if such a satellite were at the proper distance from the Earth, it would be moving rapidly in the north–south direction, resulting in the necessity of accurate tracking equipment. What's more, it would be below the horizon for long periods of time, making it useless for determining one's position. ◄

Example 11.2 | A Geosynchronous Satellite

Consider a satellite of mass m moving in a circular orbit around the Earth at a constant speed v and at an altitude h above the Earth's surface as illustrated in Figure 11.8.

(A) Determine the speed of satellite in terms of G, h, R_E (the radius of the Earth), and M_E (the mass of the Earth).

SOLUTION

Conceptualize Imagine the satellite moving around the Earth in a circular orbit under the influence of the gravitational force. This motion is similar to that of the International Space Station, the Hubble Space Telescope, and other objects in orbit around the Earth.

Categorize The satellite must have a centripetal acceleration. Therefore, we categorize the satellite as a particle under a net force and a particle in uniform circular motion.

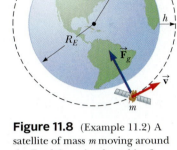

Figure 11.8 (Example 11.2) A satellite of mass m moving around the Earth in a circular orbit of radius r with constant speed v. The only force acting on the satellite is the gravitational force $\vec{F}_g$. (Not drawn to scale.)

Analyze The only external force acting on the satellite is the gravitational force, which acts toward the center of the Earth and keeps the satellite in its circular orbit.

Apply the particle under a net force and particle in uniform circular motion models to the satellite:

$$F_g = ma \quad \rightarrow \quad G\frac{M_E m}{r^2} = m\left(\frac{v^2}{r}\right)$$

Solve for v, noting that the distance r from the center of the Earth to the satellite is $r = R_E + h$:

$$(1) \quad v = \sqrt{\frac{GM_E}{r}} = \sqrt{\frac{GM_E}{R_E + h}}$$

(B) If the satellite is to be *geosynchronous* (that is, appearing to remain over a fixed position on the Earth), how fast is it moving through space?

SOLUTION

To appear to remain over a fixed position on the Earth, the period of the satellite must be 24 h = 86 400 s and the satellite must be in orbit directly over the equator.

Solve Kepler's third law (Equation 11.7, with $a = r$ and $M_S \rightarrow M_E$) for r:

$$r = \left(\frac{GM_E T^2}{4\pi^2}\right)^{1/3}$$

Substitute numerical values:

$$r = \left[\frac{(6.67 \times 10^{-11}\ \text{N} \cdot \text{m}^2/\text{kg}^2)(5.97 \times 10^{24}\ \text{kg})(86\ 400\ \text{s})^2}{4\pi^2}\right]^{1/3}$$

$$= 4.22 \times 10^7\ \text{m}$$

Use Equation (1) to find the speed of the satellite:

$$v = \sqrt{\frac{(6.67 \times 10^{-11}\ \text{N} \cdot \text{m}^2/\text{kg}^2)(5.97 \times 10^{24}\ \text{kg})}{4.22 \times 10^7\ \text{m}}}$$

$$= 3.07 \times 10^3\ \text{m/s}$$

Finalize The value of r calculated here translates to a height of the satellite above the surface of the Earth of almost 36 000 km. Therefore, geosynchronous satellites have the advantage of allowing an earthbound antenna to be aimed in a fixed direction, but there is a disadvantage in that the signals between the Earth and the satellite must travel a long distance. It is difficult to use geosynchronous satellites for optical observation of the Earth's surface because of their high altitude.

What If? What if the satellite motion in part (A) were taking place at height h above the surface of another planet more massive than the Earth but of the same radius? Would the satellite be moving at a higher speed or a lower speed than it does around the Earth?

Answer If the planet exerts a larger gravitational force on the satellite due to its larger mass, the satellite must move with a higher speed to avoid moving toward the surface. This conclusion is consistent with the predictions of Equation (1), which shows that because the speed v is proportional to the square root of the mass of the planet, the speed increases as the mass of the planet increases.

◤ 11.4 | Energy Considerations in Planetary and Satellite Motion

So far we have approached orbital mechanics from the point of view of forces and angular momentum. Let us now investigate the motion of planets in orbit from the *energy* point of view.

Consider an object of mass m moving with a speed v in the vicinity of a massive object of mass $M \gg m$. This two-object system might be a planet moving around the Sun, a satellite orbiting the Earth, or a comet making a one-time flyby past the Sun. We will model the two objects of mass m and M as an isolated system. If we assume that M is at rest in an inertial reference frame (because $M \gg m$), the total energy E of the two-object system is the sum of the kinetic energy of the object of mass m and the gravitational potential energy of the system:

$$E = K + U_g$$

Recall from Section 6.9 that the gravitational potential energy U_g associated with *any pair* of particles of masses m_1 and m_2 separated by a distance r is given by

$$U_g = -\frac{Gm_1 m_2}{r}$$

where we have defined $U_g \to 0$ as $r \to \infty$; therefore, in our case, the total energy of the system of m and M is

$$E = \tfrac{1}{2}mv^2 - \frac{GMm}{r} \qquad \textbf{11.8} \blacktriangleleft$$

Equation 11.8 shows that E may be positive, negative, or zero, depending on the value of v at a particular separation distance r. If we consider the energy diagram method of Section 6.10, we can show the potential and total energies of the system as a function of r as in Figure 11.9. A planet moving around the Sun and a satellite in orbit around the Earth are *bound* systems, such as those we discussed in Section 11.3; the Earth will always stay near the Sun and the satellite near the Earth. In Figure 11.9, these systems are represented by a total energy E that is negative. The point at which the total energy line intersects the potential energy curve is a turning point, the maximum separation distance r_{max} between the two bound objects.

A one-time meteoroid flyby represents an unbound system. The meteoroid interacts with the Sun but is not bound to it. Therefore, the meteoroid can in theory move infinitely far away from the Sun as represented in Figure 11.9 by a total energy line in the positive region of the graph. This line never intersects the potential energy curve, so all values of r are possible.

For a bound system, such as the Earth and Sun, E is necessarily less than zero because we have chosen the convention that $U_g \to 0$ as $r \to \infty$. We can easily establish that $E < 0$ for the system consisting of an object of mass m moving in a circular orbit

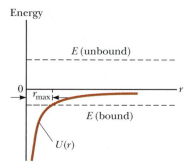

Figure 11.9 The lower total energy line represents a bound system. The separation distance r between the two gravitationally bound objects never exceeds r_{max}. The upper total energy line represents an unbound system of two objects interacting gravitationally. The separation distance r between the two objects can have any value.

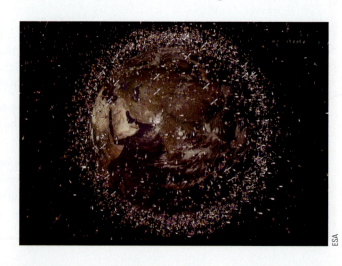

Many artificial satellites have been placed in orbit about the Earth. This diagram shows many satellites in low Earth orbits. This region of space is becoming very crowded: in 2009, a U.S. commercial Iridium satellite collided with an inactive Russian Kosmos satellite, destroying both. (The debris field shown in the image is an artist's impression based on actual data. However, the debris objects are shown at an exaggerated size to make them visible at the scale shown.)

about an object of mass $M \gg m$ (Fig. 11.8). Applying Newton's second law to the object of mass m gives

$$\sum F = ma \quad \rightarrow \quad \frac{GMm}{r^2} = \frac{mv^2}{r}$$

Multiplying both sides by r and dividing by 2 gives

$$\tfrac{1}{2}mv^2 = \frac{GMm}{2r} \qquad \qquad \textbf{11.9} \blacktriangleleft$$

Substituting this result into Equation 11.8, we obtain

$$E = \frac{GMm}{2r} - \frac{GMm}{r}$$

$$E = -\frac{GMm}{2r} \quad \text{(circular orbits)} \qquad \textbf{11.10} \blacktriangleleft$$

This result clearly shows that the total energy must be negative in the case of circular orbits. Furthermore, Equation 11.9 shows that the kinetic energy of an object in a circular orbit is equal to one-half the magnitude of the potential energy of the system (when the potential energy is chosen to be zero at infinite separation).

The total energy is also negative in the case of elliptical orbits. The expression for E for elliptical orbits is the same as Equation 11.10, with r replaced by the semimajor axis a:

$$E = -\frac{GMm}{2a} \quad \text{(elliptical orbits)} \qquad \textbf{11.11} \blacktriangleleft$$

▶ Total energy of a planet–star system

Combining this statement of energy conservation with our earlier discussion of conservation of angular momentum, we see that both the total energy and the total angular momentum of a gravitationally bound, two-object system are constants of the motion.

QUICK QUIZ 11.3 A comet moves in an elliptical orbit around the Sun. Which point in its orbit (perihelion or aphelion) represents the highest value of (**a**) the speed of the comet, (**b**) the potential energy of the comet–Sun system, (**c**) the kinetic energy of the comet, and (**d**) the total energy of the comet–Sun system?

Example 11.3 | Changing the Orbit of a Satellite

A space transportation vehicle releases a 470-kg communications satellite while in an orbit 280 km above the surface of the Earth. A rocket engine on the satellite boosts it into a geosynchronous orbit. How much energy does the engine have to provide?

SOLUTION

Conceptualize Notice that the height of 280 km is much lower than that for a geosynchronous satellite, 36 000 km, as mentioned in Example 11.2. Therefore, energy must be expended to raise the satellite to this much higher position.

Categorize This example is a substitution problem.

Find the initial radius of the satellite's orbit when it is still in the vehicle's cargo bay:

$$r_i = R_E + 280 \text{ km} = 6.65 \times 10^6 \text{ m}$$

Use Equation 11.10 to find the difference in energies for the satellite–Earth system with the satellite at the initial and final radii:

$$\Delta E = E_f - E_i = -\frac{GM_E m}{2r_f} - \left(-\frac{GM_E m}{2r_i} \right) = -\frac{GM_E m}{2} \left(\frac{1}{r_f} - \frac{1}{r_i} \right)$$

continued

11.3 *cont.*

Substitute numerical values, using $r_f = 4.22 \times 10^7$ m from Example 11.2:

$$\Delta E = -\frac{(6.67 \times 10^{-11} \text{ N} \cdot \text{m}^2/\text{kg}^2)(5.97 \times 10^{24} \text{ kg})(470 \text{ kg})}{2}$$

$$\times \left(\frac{1}{4.22 \times 10^7 \text{ m}} - \frac{1}{6.65 \times 10^6 \text{ m}} \right)$$

$$= \boxed{1.19 \times 10^{10} \text{ J}}$$

which is the energy equivalent of 89 gal of gasoline. NASA engineers must account for the changing mass of the spacecraft as it ejects burned fuel, something we have not done here. Would you expect the calculation that includes the effect of this changing mass to yield a greater or a lesser amount of energy required from the engine?

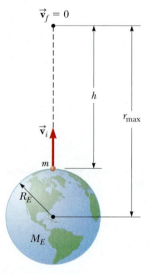

Figure 11.10 An object of mass m projected upward from the Earth's surface with an initial speed v_i reaches a maximum altitude $h = r_{max} - R_E$.

Escape Speed

Suppose an object of mass m is projected vertically upward from the Earth's surface with an initial speed v_i as in Figure 11.10. We can use energy considerations to find the minimum value of the initial speed such that the object will continue to move infinitely far away from the Earth. Equation 11.8 gives the total energy of the object–Earth system at any point when the speed of the object and its distance from the center of the Earth are known. At the surface of the Earth, $r_i = R_E$. When the object reaches its maximum altitude, $v_f = 0$ and $r_f = r_{max}$. Because the total energy of the isolated object–Earth system is conserved, substitution of these conditions into Equation 11.8 gives

$$\tfrac{1}{2}mv_i^2 - \frac{GM_E m}{R_E} = -\frac{GM_E m}{r_{max}}$$

Solving for v_i^2 gives

$$v_i^2 = 2GM_E \left(\frac{1}{R_E} - \frac{1}{r_{max}} \right) \qquad \text{11.12} \blacktriangleleft$$

For a given maximum altitude $h = r_{max} - R_E$, we can use this equation to find the required initial speed.

We are now in a position to calculate **escape speed,** which is the minimum speed the object must have at the Earth's surface to continue to move away forever. Traveling at this minimum speed, the object continues to move away from the Earth as its speed asymptotically approaches zero. Letting $r_{max} \to \infty$ in Equation 11.12 and setting $v_i = v_{esc}$, we have

$$v_{esc} = \sqrt{\frac{2GM_E}{R_E}} \qquad \text{11.13} \blacktriangleleft$$

This expression for v_{esc} is independent of the mass of the object. For example, a spacecraft has the same escape speed as a molecule. Furthermore, the result is independent of the direction of the velocity and ignores air resistance.

Note also that Equations 11.12 and 11.13 can be applied to objects projected from *any* planet. That is, in general, the escape speed from any planet of mass M and radius R is

$$v_{esc} = \sqrt{\frac{2GM}{R}} \qquad \text{11.14} \blacktriangleleft$$

A list of escape speeds for the planets, the Moon, and the Sun is given in Table 11.2. Note that the values vary from 2.3 km/s for the Moon to about 618 km/s for the Sun. These results, together with some ideas from the kinetic theory of gases (Chapter 16), explain why our atmosphere does not contain significant amounts of hydrogen, which is the most abundant element in the Universe. As we shall see later, gas molecules have an average kinetic energy that depends on the temperature of the gas. Lighter molecules in an atmosphere have translational speeds that are closer to the escape speed than more massive molecules, so they have a higher probability of escaping from the planet and the lighter molecules diffuse into space. This mechanism explains why the Earth does not retain hydrogen molecules and helium atoms in its atmosphere but does retain much heavier molecules, such as oxygen and nitrogen. On the other hand, Jupiter has a very large escape speed (60 km/s), which enables it to retain hydrogen, the primary constituent of its atmosphere.

> **Pitfall Prevention | 11.3**
> **You Can't Really Escape**
> Although Equation 11.13 provides the "escape speed" from the Earth, complete escape from the Earth's gravitational influence is impossible because the gravitational force is of infinite range. No matter how far away you are, you will always feel some gravitational force due to the Earth.

Example 11.4 | Escape Speed of a Rocket

Calculate the escape speed from the Earth for a 5 000-kg spacecraft and determine the kinetic energy it must have at the Earth's surface to move infinitely far away from the Earth.

SOLUTION

Conceptualize Imagine projecting the spacecraft from the Earth's surface so that it moves farther and farther away, traveling more and more slowly, with its speed approaching zero. Its speed will never reach zero, however, so the object will never turn around and come back.

Categorize This example is a substitution problem.

Use Equation 11.13 to find the escape speed:

$$v_{esc} = \sqrt{\frac{2GM_E}{R_E}} = \sqrt{\frac{2(6.67 \times 10^{-11} \text{ N} \cdot \text{m}^2/\text{kg}^2)(5.97 \times 10^{24} \text{ kg})}{6.37 \times 10^6 \text{ m}}}$$

$$= 1.12 \times 10^4 \text{ m/s}$$

Evaluate the kinetic energy of the spacecraft from Equation 6.16:

$$K = \tfrac{1}{2}mv_{esc}^2 = \tfrac{1}{2}(5.00 \times 10^3 \text{ kg})(1.12 \times 10^4 \text{ m/s})^2$$

$$= 3.13 \times 10^{11} \text{ J}$$

The calculated escape speed corresponds to about 25 000 mi/h. The kinetic energy of the spacecraft is equivalent to the energy released by the combustion of about 2 300 gal of gasoline.

Black Holes

In Chapter 10, we briefly described a rare event called a supernova, the catastrophic explosion of a very massive star. The material that remains in the central core of such an object continues to collapse, and the core's ultimate fate depends on its mass. If the core has a mass less than 1.4 times the mass of our Sun, it gradually cools down and ends its life as a white dwarf star. If, however, the core's mass is greater than that, it may collapse further due to gravitational forces. What remains is a neutron star, discussed in Chapter 10, in which the mass of a star is compressed to a radius of about 10 km. (On the Earth, a teaspoon of this material would weigh about 5 billion tons!)

An even more unusual star death may occur when the core has a mass greater than about three solar masses. The collapse may continue until the star becomes a very small object in space, commonly referred to as a **black hole.** In effect, black holes are the remains of stars that have collapsed under their own gravitational force. If an object such as a spacecraft comes close to a black hole, it experiences an extremely strong gravitational force and is trapped forever.

The escape speed from any spherical body depends on the mass and radius of the body. The escape speed for a black hole is very high because of the concentration of the star's mass into a sphere of very small radius. If the escape speed exceeds the speed of light c, radiation from the body (e.g. visible light) cannot escape and the body appears to be black, hence the origin of the term *black hole*. The critical radius R_S at which the escape speed is c is called the **Schwarzschild radius** (Fig. 11.11). The imaginary surface of a sphere of this radius surrounding the black hole is called the

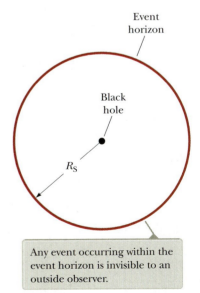

Figure 11.11 A black hole. The distance R_S equals the Schwarzschild radius.

> Any event occurring within the event horizon is invisible to an outside observer.

Figure 11.12 Hubble Space Telescope image of the galaxy M107. This galaxy contains about 800 billion stars and is 28 million light years from the Earth. Scientists believe that a supermassive black hole exists at the center of the galaxy.

event horizon, which is the limit of how close you can approach the black hole and hope to be able to escape.

Although light from a black hole cannot escape, light from events taking place near the black hole should be visible. For example, it is possible for a binary star system to consist of one normal star and one black hole. Material surrounding the ordinary star can be pulled into the black hole, forming an **accretion disk** around the black hole. Friction among particles in the accretion disk results in transformation of mechanical energy into internal energy. As a result, the orbital height of the material above the event horizon decreases and the temperature rises. This high-temperature material emits a large amount of radiation, extending well into the x-ray region of the electromagnetic spectrum. These x-rays are characteristic of a black hole. Several possible candidates for black holes have been identified by observation of these x-rays.

Figure 11.12 shows a Hubble Space Telescope photograph of M107, known as the Sombrero galaxy. Scientists have shown that the speed of revolution of the stars could not be maintained unless a mass one billion times the mass of the Sun is present at its center. This is strong evidence for a supermassive black hole at the center of the galaxy.

Black holes are of considerable interest to those searching for **gravity waves,** which are ripples in space–time caused by changes in a gravitational system. These ripples can be caused by a star collapsing into a black hole, a binary star consisting of a black hole and a visible companion, and supermassive black holes at a galaxy center. A gravity wave detector, the Laser Interferometer Gravitational Wave Observatory (LIGO), is currently being built and tested in the United States, and hopes are high for detecting gravitational waves with this instrument.

11.5 | Atomic Spectra and the Bohr Theory of Hydrogen

In the preceding sections, we described a structural model for a large-scale system, the Solar System. Let us now do the same for a very small-scale system, the hydrogen atom. We shall find that a Solar System model of the atom, with a few extra features, provides explanations for some of the experimental observations made on the hydrogen atom.

As you may have already learned in a chemistry course, the hydrogen atom is the simplest known atomic system and an especially important one to understand. Much of what is learned about the hydrogen atom (which consists of one proton and one electron) can be extended to single-electron ions such as He^+ and Li^{2+}. Furthermore, a thorough understanding of the physics underlying the hydrogen atom can then be used to describe more complex atoms and the periodic table of the elements.

In this section, we will investigate the changes in the structural model of the hydrogen atom during the second decade of the 20th century. Early in that decade, the structural model had the following components, following the format outlined in Section 11.2:

1. *A description of the physical components of the system:* In the hydrogen atom model, the physical components are the electron and a positive charge distribution.

2. *A description of where the components are located relative to one another and how they interact:* The hydrogen atom model at the time was the Rutherford model, to be discussed in Chapter 29. In this model, the positive charge is concentrated in a small region of space called the *nucleus.* The electron is in orbit around the nucleus. The particle-like nature of the positive charge and the word *proton* were not yet understood at the time, so we will avoid referring to the nucleus as a proton in this discussion. The interaction between the electron and the nucleus is the electric force.

3. *A description of the time evolution of the system:* In the hydrogen atom model of the early 20th century, the time evolution was unclear and not understood.

4. *A description of the agreement between predictions of the model and actual observations and, possibly, predictions of new effects that have not yet been observed:* Rutherford's model was unable to explain the spectral lines exhibited by hydrogen that had been observed experimentally and are discussed below. Furthermore, as we discuss in Chapter 29, the model predicted an unstable atom, clearly at odds with reality.

Atomic systems can be investigated by observing *electromagnetic waves* emitted from the atom. Our eyes are sensitive to visible light, one type of electromagnetic wave. A wave, which is a disturbance that propagates through space, will be one of our four simplification models around which we will identify analysis models, as we have done for a particle, a system, and a rigid object. Think of ocean waves as an example; they represent disturbances of the ocean surface and move across the surface toward the shore. A common form of periodic wave is the sinusoidal wave, whose shape is depicted in Figure 11.13. If this graph represents an electromagnetic wave, the vertical axis represents the magnitude of the electric field. (We will study electric fields in Chapter 19.) The horizontal axis is position in the direction of travel of the wave. The distance between two consecutive crests of the wave is called the **wavelength** λ. As the wave travels to the right with a speed v, any point on the wave travels a distance of one wavelength in a time interval of one period T (the time interval for one cycle), so the wave speed is given by $v = \lambda/T$. The inverse of the period, $1/T$, is called the **frequency** f of the wave; it represents the number of cycles per second. Therefore, the speed of the wave is often written as $v = \lambda f$. In this section, because we shall deal with electromagnetic waves—which travel at the speed of light c—the appropriate relation is

$$c = \lambda f \qquad\qquad \textbf{11.15} \blacktriangleleft$$

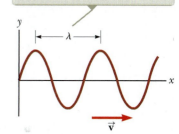

Any point on the wave moves a distance of one wavelength λ in a time interval equal to the period T of the wave.

Figure 11.13 A sinusoidal wave traveling to the right with wave speed v.

▶ Relation between wavelength, frequency, and wave speed

Suppose an evacuated glass tube is filled with hydrogen (or some other gas). If a voltage applied between metal electrodes in the tube is great enough to produce an electric current in the gas, the tube emits light with colors that are characteristic of the gas. (That is how a neon sign works.) When the emitted light is analyzed with a device called a spectroscope, in which the light passes through a narrow slit, a series of discrete **spectral lines** is observed, each line corresponding to a different wavelength, or color, of light. Such a series of spectral lines is commonly referred to as an **emission spectrum.** The wavelengths contained in a given spectrum are characteristic of the element emitting the light. Figure 11.14 (page 370) is a semigraphical representation of the spectra of various elements. It is semigraphical because the horizontal axis is linear in wavelength, but the vertical axis has no significance. Because no two elements emit the same line spectrum, this phenomenon represents a marvelous and reliable technique for identifying elements in a substance.

In addition to emitting light at specific wavelengths, an element can also absorb light at specific wavelengths. The spectral lines corresponding to this process form what is known as an **absorption spectrum.** An absorption spectrum can be obtained by passing a continuous radiation spectrum (one containing all wavelengths) through a vapor of the element being analyzed. The absorption spectrum consists of a series of dark lines superimposed on the otherwise continuous spectrum (Fig. 11.14b).

Figure 11.14 Visible spectra.
(a) Line spectra produced by
emission in the visible range for the
elements hydrogen, mercury, and
neon. (b) The absorption spectrum
for hydrogen. The dark absorption
lines occur at the same wavelengths
as the emission lines for hydrogen
shown in (a).

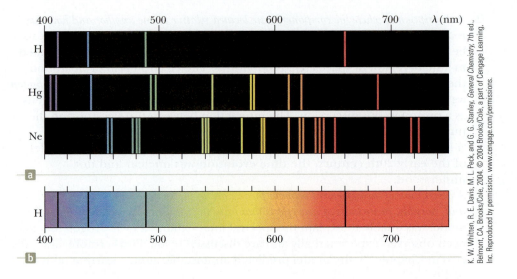

K. W. Whitten, R. E. Davis, M. L. Peck, and G. G. Stanley, *General Chemistry*, 7th ed.,
Belmont, CA, Brooks/Cole, 2004. © 2004 Brooks/Cole, a part of Cengage Learning,
Inc. Reproduced by permission. www.cengage.com/permissions.

The emission spectrum of hydrogen shown in Figure 11.15 includes four visible
lines that occur at wavelengths of 656.3 nm, 486.1 nm, 434.1 nm, and 410.2 nm. In
1885, Johann Balmer (1825–1898) found that the wavelengths of these and other
invisible lines can be described by the following simple empirical equation:

$$\lambda = 364.56 \, \frac{n^2}{n^2 - 4} \qquad n = 3, 4, 5, \ldots$$

in which n is an integer starting at 3 and the wavelengths given by this expression
are in nanometers. These spectral lines are called the **Balmer series.** The first line
in the Balmer series, at 656.3 nm, corresponds to $n = 3$, the line at 486.1 nm corresponds to $n = 4$, and so on. At the time this equation was formulated, it had no
valid theoretical basis; it simply predicted the wavelengths correctly. Therefore, this
equation is not based on a model but is simply a trial-and-error equation that happens to work. A few years later, Johannes Rydberg (1854–1919) recast the equation
in the following form:

▶ Rydberg equation

$$\frac{1}{\lambda} = R_{\text{H}}\left(\frac{1}{2^2} - \frac{1}{n^2}\right) \qquad n = 3, 4, 5, \ldots \qquad \textbf{11.16} \blacktriangleleft$$

where n may have integral values of 3, 4, 5, . . . and R_{H} is a constant, now called the
Rydberg constant, with a value of $R_{\text{H}} = 1.097\ 373\ 2 \times 10^7\ \text{m}^{-1}$. Equation 11.16 is
no more based on a model than is Balmer's equation. In this form, however, we can
compare it with the predictions of a structural model of the hydrogen atom that is
described below.

At the beginning of the 20th century, scientists were perplexed by the failure of
classical physics to explain the characteristics of atomic spectra. Why did atoms of a
given element emit only certain wavelengths of radiation so that the emission spectrum displayed discrete lines? Furthermore, why did the atoms absorb many of the
same wavelengths that they emitted? In 1913, Niels Bohr provided an explanation of
atomic spectra that includes some features of the currently accepted theory. Using
the simplest atom, hydrogen, Bohr described a structural model for the atom called
the **Bohr theory of the hydrogen atom.** His model of the hydrogen atom contains
some classical features that can be related to our analysis models as well as some
revolutionary postulates that could not be justified within the framework of classical
physics. The components of Bohr's structural model as it applies to the hydrogen
atom are as follows:

1. *A description of the physical components of the system:* In the hydrogen atom model,
   the physical components are the electron and a positive charge distribution,
   just as in Rutherford's model.

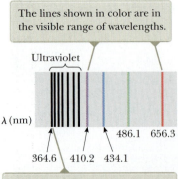

The lines shown in color are in
the visible range of wavelengths.

Ultraviolet

λ (nm)

486.1   656.3

364.6   410.2   434.1

This line is the shortest wavelength
line and is in the ultraviolet region
of the electromagnetic spectrum.

Figure 11.15 The Balmer series of
spectral lines for atomic hydrogen,
with several lines marked with the
wavelength in nanometers. (The
horizontal wavelength axis is not to
scale.)

2. *A description of where the components are located relative to one another and how they interact:* The electron moves in a circular orbit about the nucleus under the influence of the electric force of attraction as in Figure 11.16. This notion is again consistent with the Rutherford model.

3. *A description of the time evolution of the system:* Here is where Bohr's model deviates from Rutherford's. We discuss three major parts of the theory:

   (a) Bohr's model claims that only certain electron orbits are stable, and they are the only orbits in which we find the electron. In these orbits, the hydrogen atom does not emit energy in the form of radiation. Hence, the total energy of the atom remains constant, and classical mechanics can be used to describe the electron's motion. This restriction to certain orbits is a new idea that is not consistent with classical physics. As we shall see in Chapter 24, an accelerating electron should emit energy by electromagnetic radiation. Therefore, according to the conservation of energy equation, the emission of radiation from the atom should result in a decrease in the energy of the atom. Bohr's postulate boldly claims that this radiation simply does not happen.

   (b) The size of the stable electron orbits is determined by a condition imposed on the electron's orbital angular momentum. The allowed orbits are those for which the electron's orbital angular momentum about the nucleus is an integral multiple of $\hbar \equiv h/2\pi$:

$$m_e v r = n\hbar \qquad n = 1, 2, 3, \ldots \qquad \textbf{11.17}\blacktriangleleft$$

   where h is **Planck's constant** ($h = 6.63 \times 10^{-34}$ J · s; we will see Planck's constant extensively in our studies of modern physics). This new idea cannot be related to any of the models we have developed so far.

   It can be related, however, to a model that will be developed in later chapters, and we shall return to this idea at that time to see how it is predicted by the model. This concept is our first introduction to a notion from **quantum mechanics,** which describes the behavior of microscopic particles. The orbital radii are *quantized.*

   (c) Radiation is emitted by the hydrogen atom when the atom makes a transition from a more energetic initial state to a lower state. The transition cannot be visualized or treated classically. In particular, the frequency f of the radiation emitted in the transition is related to the change in the atom's energy. The frequency of the emitted radiation is found from

$$E_i - E_f = hf \qquad \textbf{11.18}\blacktriangleleft$$

   where E_i is the energy of the initial state, E_f is the energy of the final state, and $E_i > E_f$. The notion of energy being emitted only when a transition occurs is nonclassical. Given this notion, however, Equation 11.18 is simply the conservation of energy equation $\Delta E = \Sigma T \rightarrow E_f - E_i = -hf$. On the left is the change in energy of the system—the atom—and on the right is the energy transferred out of the system by electromagnetic radiation.

4. *A description of the agreement between predictions of the model and actual observations and, possibly, predictions of new effects that have not yet been observed:* In the discussion below, we see how the structural model makes predictions and agrees with some experimental results.

The electric potential energy of the system shown in Figure 11.16 is found from Equation 6.35, $U_e = -k_e e^2/r$, where k_e is the electric constant, e is the magnitude of the charge on the electron, and r is the electron–nucleus separation. Therefore,

Bain Collection, Prints & Photographs Division, Library of Congress, LC-DIG-ggbain-35303

Niels Bohr
(1885–1962)
Bohr, a Danish physicist, was an active participant in the early development of quantum mechanics and provided much of its philosophical framework. During the 1920s and 1930s, Bohr headed the Institute for Advanced Studies in Copenhagen. The institute was a magnet for many of the world's best physicists and provided a forum for the exchange of ideas. Bohr was awarded the 1922 Nobel Prize in Physics for his investigation of the structure of atoms and of the radiation emanating from them.

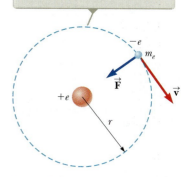

The orbiting electron is allowed to be only in specific orbits of discrete radii.

Figure 11.16 Diagram representing Bohr's model of the hydrogen atom.

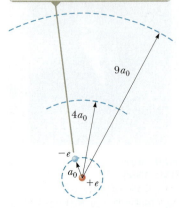

The electron is shown in the lowest-energy orbit, but it could be in any of the allowed orbits.

$9a_0$

$4a_0$

$-e$

a_0

$+e$

Active Figure 11.17 The first three circular orbits predicted by the Bohr model for hydrogen.

▶ Total energy of the hydrogen atom

▶ Radii of Bohr orbits in hydrogen

▶ The Bohr radius

▶ Energies of quantum states of the hydrogen atom

the total energy of the atom, which contains both kinetic and potential energy terms, is

$$E = K + U_e = \tfrac{1}{2} m_e v^2 - k_e \frac{e^2}{r} \qquad \textbf{11.19} \blacktriangleleft$$

According to component 3(a) of the structural model, the energy of the system remains constant; the system is isolated because the structural model does not allow for electromagnetic radiation for a given orbit.

Applying Newton's second law to this system, we see that the magnitude of the attractive electric force on the electron, $k_e e^2 / r^2$ (Eq. 5.12), is equal to the product of its mass and its centripetal acceleration ($a_c = v^2 / r$):

$$\frac{k_e e^2}{r^2} = \frac{m_e v^2}{r}$$

From this expression, the kinetic energy of the electron is found to be

$$K = \tfrac{1}{2} m_e v^2 = \frac{k_e e^2}{2r} \qquad \textbf{11.20} \blacktriangleleft$$

Substituting this value of K into Equation 11.19 gives the following expression for the total energy E of the hydrogen atom:

$$E = -\frac{k_e e^2}{2r} \qquad \textbf{11.21} \blacktriangleleft$$

Note that the total energy is negative,[3] indicating a bound electron–proton system. Therefore, energy in the amount of $k_e e^2 / 2r$ must be added to the atom just to separate the electron and proton by an infinite distance and make the total energy zero.[4] An expression for r, the radius of the allowed orbits, can be obtained by eliminating v by substitution between Equation 11.17 from component 3(b) of the structural model and Equation 11.20:

$$r_n = \frac{n^2 \hbar^2}{m_e k_e e^2} \qquad n = 1, 2, 3 \ldots \qquad \textbf{11.22} \blacktriangleleft$$

This result determines the discrete radii of the electron orbits. The integer n is called a **quantum number** and specifies the particular allowed **quantum state** of the atomic system.

The orbit for which $n = 1$ has the smallest radius; it is called the **Bohr radius** a_0 and has the value

$$a_0 = \frac{\hbar^2}{m_e k_e e^2} = 0.052\ 9 \text{ nm} \qquad \textbf{11.23} \blacktriangleleft$$

The first three Bohr orbits are shown to scale in Active Figure 11.17.

The quantization of the orbit radii immediately leads to quantization of the energy of the atom, which can be seen by substituting $r_n = n^2 a_0$ into Equation 11.21. The allowed energies of the atom are

$$E_n = -\frac{k_e e^2}{2 a_0} \left(\frac{1}{n^2} \right) \qquad n = 1, 2, 3, \ldots \qquad \textbf{11.24} \blacktriangleleft$$

Insertion of numerical values into Equation 11.24 gives

$$E_n = -\frac{13.606 \text{ eV}}{n^2} \qquad n = 1, 2, 3, \ldots \qquad \textbf{11.25} \blacktriangleleft$$

---

[3]Compare this expression with Equation 11.10 for a gravitational system.

[4]This process is called *ionizing* the atom. In theory, ionization requires separating the electron and proton by an infinite distance. In reality, however, the electron and proton are in an environment with huge numbers of other particles. Therefore, ionization means separating the electron and proton by a distance large enough so that the interaction of these particles with other entities in their environment is larger than the remaining interaction between them.

(Recall from Section 9.7 that $1 \text{ eV} = 1.60 \times 10^{-19}$ J.) The lowest quantum state, corresponding to $n = 1$, is called the **ground state** and has an energy of $E_1 = -13.606$ eV. The next state, the **first excited state,** has $n = 2$ and an energy of $E_2 = E_1/2^2 = -3.401$ eV. Active Figure 11.18 is an **energy level diagram** showing the energies of these discrete energy states and the corresponding quantum numbers. This diagram is another semigraphical representation. The vertical axis is linear in energy, but the horizontal axis has no significance. The horizontal lines correspond to the allowed energies. The atomic system cannot have any energies other than those represented by the lines. The vertical lines with arrowheads represent transitions between states, during which energy is emitted.

The upper limit of the quantized levels, corresponding to $n \rightarrow \infty$ (or $r \rightarrow \infty$) and $E \rightarrow 0$, represents the state for which the electron is removed from the atom.[5] Above this energy is a continuum of available states for the ionized atom. The minimum energy required to ionize the atom is called the **ionization energy.** As can be seen from Active Figure 11.18, the ionization energy for hydrogen, predicted by Bohr's structural model, is 13.6 eV. This finding constituted a major achievement for the Bohr model because the ionization energy for hydrogen had already been measured to be 13.6 eV!

Active Figure 11.18 also shows various transitions of the atom from one state to a lower state, as referred to in component 3(c) of the structural model. As the energy of the atom decreases in a transition, the difference in energy between the states is carried away by electromagnetic radiation as described by Equation 11.18. Those transitions ending on $n = 2$ are shown in color, corresponding to the color of the light they represent. The transitions ending on $n = 2$ form the Balmer series of spectral lines, the wavelengths of which are correctly predicted by the Rydberg equation (see Eq. 11.16). Active Figure 11.18 also shows other spectral series (the Lyman series and the Paschen series) that were found after Balmer's discovery.

Equation 11.24, together with Equation 11.18, can be used to calculate the frequency of the radiation that is emitted when the atom makes a transition[6] from a high-energy state to a low-energy state:

$$f = \frac{E_i - E_f}{h} = \frac{k_e e^2}{2 a_0 h} \left(\frac{1}{n_f^2} - \frac{1}{n_i^2} \right)$$

11.26 ◄ ► Frequency of radiation emitted from hydrogen

Because the quantity expressed in the Rydberg equation is wavelength, it is convenient to convert frequency to wavelength, using $c = f\lambda$, to obtain

$$\frac{1}{\lambda} = \frac{f}{c} = \frac{k_e e^2}{2 a_0 h c} \left(\frac{1}{n_f^2} - \frac{1}{n_i^2} \right)$$

11.27 ◄ ► Emission wavelengths of hydrogen

Notice that the *theoretical* expression, Equation 11.27, is identical to the *empirical* Rydberg equation (Equation 11.16), provided that the combination of constants $k_e e^2/2 a_0 h c$ is equal to the experimentally determined Rydberg constant and that $n_f = 2$. After Bohr demonstrated the agreement of the constants in these two equations to a precision of about 1%, it was soon recognized as the crowning achievement of his structural model of the atom.

One question remains: What is the significance of $n_f = 2$? Its importance is simply because those transitions ending on $n_f = 2$ result in radiation that happens to lie in the visible; therefore, they were easily observed! As seen in Active Figure 11.18, other series of lines end on other final states. These lines lie in regions of the spectrum not

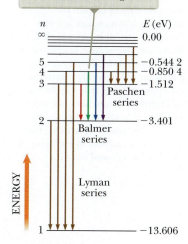

The colored arrows for the Balmer series indicate that this series results in the emission of visible light.

Active Figure 11.18 An energy-level diagram for hydrogen. The discrete allowed energies are plotted on the vertical axis. Nothing is plotted on the horizontal axis, but the horizontal extent of the diagram is made large enough to show allowed transitions. Quantum numbers are given on the left and energies (in electron volts) on the right. Vertical arrows represent the four lowest energy transitions in each of the spectral series shown.

---

[5] The phrase "the electron is removed from the atom" is very commonly used, but, of course, we realize that we mean that the electron and nucleus are separated *from each other.*

[6] The phrase "the electron makes a transition" is also commonly used, but we will use "the atom makes a transition" to emphasize that the energy belongs to the system of the atom, not just to the electron. This wording is similar to our discussion in Chapter 6 of gravitational potential energy belonging to the system of an object and the Earth, not to the object alone.

Pitfall Prevention | 11.4
The Bohr Model Is Great, But . . .
The Bohr model correctly predicts the ionization energy and general features of the spectrum for hydrogen, but it cannot account for the spectra of more complex atoms and is unable to predict many subtle spectral details of hydrogen and other simple atoms. Scattering experiments show that the electron in a hydrogen atom does not move in a flat circle around the nucleus. Instead, the atom is spherical. The ground-state angular momentum of the atom is zero and not $\hbar$.

visible to the eye, the infrared and ultraviolet. The generalized Rydberg equation for any initial and final states is

$$\frac{1}{\lambda} = R_H\left(\frac{1}{n_f^2} - \frac{1}{n_i^2}\right)$$

11.28 ◀

In this equation, different series correspond to different values of n_f and different lines within a series correspond to varying values of n_i.

Bohr immediately extended his structural model for hydrogen to other elements in which all but one electron had been removed. Ionized elements such as He^+, Li^{2+}, and Be^{3+} were suspected to exist in hot stellar atmospheres, where frequent atomic collisions occur with enough energy to completely remove one or more atomic electrons. Bohr showed that many mysterious lines observed in the Sun and several stars could not be due to hydrogen, but were correctly predicted by his theory if attributed to singly ionized helium.

QUICK QUIZ 11.4 A hydrogen atom makes a transition from the $n = 3$ level to the $n = 2$ level. It then makes a transition from the $n = 2$ level to the $n = 1$ level. Which transition results in emission of the longest-wavelength photon? (a) the first transition (b) the second transition (c) neither transition because the wavelengths are the same for both

---

Example 11.5 | Electronic Transitions in Hydrogen

The electron in a hydrogen atom makes a transition from the $n = 2$ energy level to the ground level ($n = 1$). Find the wavelength and frequency of the emitted photon.

SOLUTION

Conceptualize Imagine the electron in a circular orbit about the nucleus as in the Bohr model in Figure 11.16. When the electron makes a transition to a lower stationary state, it emits a photon with a given frequency.

Categorize We evaluate the results using equations developed in this section, so we categorize this example as a substitution problem.

Use Equation 11.28 to obtain λ, with $n_i = 2$ and $n_f = 1$:

$$\frac{1}{\lambda} = R_H\left(\frac{1}{1^2} - \frac{1}{2^2}\right) = \frac{3R_H}{4}$$

$$\lambda = \frac{4}{3R_H} = \frac{4}{3(1.097 \times 10^7 \text{ m}^{-1})} = 1.22 \times 10^{-7} \text{ m} = \boxed{122 \text{ nm}}$$

Use Equation 11.15 to find the frequency of the photon:

$$f = \frac{c}{\lambda} = \frac{3.00 \times 10^8 \text{ m/s}}{1.22 \times 10^{-7} \text{ m}} = \boxed{2.47 \times 10^{15} \text{ Hz}}$$

▶11.6 | Context Connection: Changing from a Circular to an Elliptical Orbit

In part (A) of Example 11.2, we discussed a spacecraft in a circular orbit around the Earth. From our studies of Kepler's laws in this chapter, we are also aware that an elliptical orbit is possible for our spacecraft. Let us investigate how the motion of our spacecraft can be changed from a circular to an elliptical orbit, which will set us up for the conclusion to our *Mission to Mars* Context.

Let us identify the system as the spacecraft and the Earth, *but not the portion of the fuel in the spacecraft that we use to change the orbit.* In a given orbit, the total energy of the spacecraft–Earth system is given by Equation 11.10,

$$E = -\frac{GMm}{2r}$$

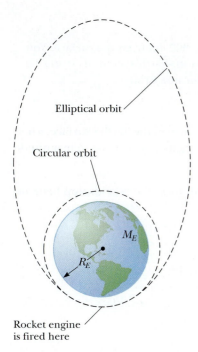

Figure 11.19 A spacecraft, originally in a circular orbit about the Earth; fires its engines and enters an elliptical orbit about the Earth.

This energy includes the kinetic energy of the spacecraft and the potential energy associated with the gravitational force between the spacecraft and the Earth. If the rocket engines are fired, the exhausted fuel can be seen as doing work on the spacecraft–Earth system because the thrust force moves through a displacement. As a result, the total energy of the spacecraft–Earth system increases.

The spacecraft has a new, higher energy but is constrained to be in an orbit that includes the original starting point. It cannot be in a higher-energy circular orbit having a larger radius because this orbit would not contain the starting point. The only possibility is that the orbit is elliptical. Figure 11.19 shows the change from the original circular orbit to the new elliptical orbit for our spacecraft.

Equation 11.11 gives the energy of the spacecraft–Earth system for an elliptical orbit. Therefore, if we know the new energy of the orbit, we can find the semimajor axis of the elliptical orbit. Conversely, if we know the semimajor axis of an elliptical orbit we would like to achieve, we can calculate how much additional energy is required from the rocket engines. This information can then be converted to a required burn time for the rockets.

Larger amounts of energy increase supplied by the rocket engines will move the spacecraft into elliptical orbits with larger semimajor axes. What happens if the burn time of the engines is so long that the total energy of the spacecraft–Earth system becomes positive? A positive energy refers to an *unbound* system. Therefore, in this case, the spacecraft will *escape* from the Earth, going into a hyperbolic path that would not bring it back to the Earth.

This process is the essence of what must be done to transfer to Mars. Our rocket engines must be fired to leave the circular parking orbit and escape the Earth. At this point, our thinking must shift to a spacecraft–Sun system rather than a spacecraft–Earth system. From this point of view, the spacecraft in orbit around the Earth can also be considered to be in a circular orbit around the Sun, moving along with the Earth, as shown in Figure 11.20. The orbit is not a perfect circle because there are perturbations corresponding to its extra motion around the Earth, but these perturbations are small compared with the radius of the orbit around the Sun. When our engines are fired to escape from the Earth, our orbit around the Sun changes from a circular orbit (ignoring the perturbations) to an elliptical one with the Sun at one focus. We shall choose the semimajor axis of our elliptical orbit so that it intersects the orbit of Mars! In the Context 2 Conclusion, we shall look at more details of this process.

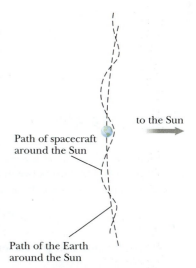

Figure 11.20 A spacecraft in orbit about the Earth can be modeled as one in a circular orbit about the Sun, with its orbit about the Earth appearing as small perturbations from the circular orbit.

Example 11.6 | How High Do We Go?

Imagine that you are in a spacecraft in circular orbit around the Earth, at a height of $h = 300$ km from the surface. You fire your rocket engines, and as a result the magnitude of the total energy of the spacecraft–Earth system decreases by 10.0%. What is the greatest height of your spacecraft above the surface of the Earth in your new orbit?

SOLUTION

Conceptualize Study Figure 11.19, which represents this situation. We wish to find the distance to the Earth's surface, when the spacecraft is at the topmost point in the figure. Notice that because the total energy is negative, a decrease in magnitude represents an increase in energy.

Categorize We don't need an analysis model for this problem. We can evaluate the result from the relationship between orbit energy and semimajor axis as exhibited by Equation 11.11.

- -

Analyze

Set up a ratio of the energies of the two orbits, using Equations 11.10 and 11.11 for circular and elliptical orbits:

$$\frac{E_{elliptical}}{E_{circular}} = \frac{\left(-\dfrac{GMm}{2a}\right)}{\left(-\dfrac{GMm}{2r}\right)} = \frac{r}{a} = f$$

The ratio f is equal to 0.900 because of the 10.0% decrease in magnitude of the total energy. Find a in terms of r:

$$a = \frac{r}{f}$$

Substitute for the orbital radius in terms of the Earth's radius and the initial height of the spacecraft above the surface:

$$a = \frac{1}{f}(R_E + h)$$

The maximum distance from the center of the Earth will occur when the spacecraft is at apogee and is given by $r_{max} = 2a - r$. Find the value of this maximum distance:

$$r_{max} = 2a - r = \frac{2}{f}(R_E + h) - (R_E + h) = \left(\frac{2}{f} - 1\right)(R_E + h)$$

Subtract the radius of the Earth from r_{max} to find the maximum height above the Earth's surface:

$$(1) \quad h_{max} = \left(\frac{2}{f} - 1\right)(R_E + h) - R_E$$

Substitute numerical values:

$$h_{max} = \left(\frac{2}{0.900} - 1\right)(6.37 \times 10^3 \text{ km} + 300 \text{ km}) - 6.37 \times 10^3 \text{ km}$$

$$= \boxed{1.78 \times 10^3 \text{ km}}$$

- -

Finalize The height above the Earth's surface has been increased by a factor of almost 6 for this fuel expenditure. Notice from Equation (1) that h_{max} increases as f decreases (representing more fuel expenditure), but not in a simple way.

▶ SUMMARY

Newton's law of universal gravitation states that the gravitational force of attraction between any two particles of masses m_1 and m_2 separated by a distance r has the magnitude

$$F_g = G\frac{m_1 m_2}{r^2} \qquad \textbf{11.1} \blacktriangleleft$$

where G is the **universal gravitational constant** whose value is 6.674×10^{-11} N · m²/kg².

Rather than considering the gravitational force as a direct interaction between two objects, we can imagine that one object sets up a **gravitational field** in space:

$$\vec{g} \equiv \frac{\vec{F}_g}{m} \qquad \textbf{11.4} \blacktriangleleft$$

A second object in this field experiences a force $\vec{F}_g = m\vec{g}$ when placed in this field.

Kepler's laws of planetary motion state the following:

1. Each planet in the Solar System moves in an elliptical orbit with the Sun at one focus.
2. The radius vector drawn from the Sun to any planet sweeps out equal areas in equal time intervals.
3. The square of the orbital period of any planet is proportional to the cube of the semimajor axis of the elliptical orbit.

Kepler's first law is a consequence of the inverse-square nature of the law of universal gravitation. The **semimajor axis**

of an ellipse is a, where $2a$ is the longest dimension of the ellipse. The **semiminor axis** of the ellipse is b, where $2b$ is the shortest dimension of the ellipse. The **eccentricity** of the ellipse is $e = c/a$, where c is the distance between the center and a focus and $a^2 = b^2 + c^2$.

Kepler's second law is a consequence of the gravitational force being a central force. For a central force, the angular momentum of the planet is conserved.

Kepler's third law is a consequence of the inverse-square nature of the universal law of gravitation. Newton's second law, together with the force law given by Equation 11.1, verifies that the period T and semimajor axis a of the orbit of a planet about the Sun are related by

$$T^2 = \left(\frac{4\pi^2}{GM_S}\right)a^3 \qquad \textbf{11.7} \blacktriangleleft$$

where M_S is the mass of the Sun.

If an isolated system consists of a particle of mass m moving with a speed v in the vicinity of a massive body of mass M, the total energy of the system is constant and is

$$E = \tfrac{1}{2}mv^2 - \frac{GMm}{r} \qquad \textbf{11.8} \blacktriangleleft$$

If m moves in an elliptical orbit of major axis $2a$ about M, where $M \gg m$, the total energy of the system is

$$E = -\frac{GMm}{2a} \qquad \textbf{11.11} \blacktriangleleft$$

The total energy is negative for any bound system, that is, one in which the orbit is closed, such as a circular or an elliptical orbit.

The Bohr model of the atom successfully describes the spectra of atomic hydrogen and hydrogen-like ions. One basic assumption of this structural model is that the electron can exist only in discrete orbits such that the angular momentum $m_e v r$ is an integral multiple of $\hbar \equiv h/2\pi$. Assuming circular orbits and a simple electrical attraction between the electron and proton, the energies of the quantum states for hydrogen are calculated to be

$$E_n = -\frac{k_e e^2}{2a_0}\left(\frac{1}{n^2}\right) \quad n = 1, 2, 3, \ldots \qquad \textbf{11.24} \blacktriangleleft$$

where k_e is the Coulomb constant, e is the fundamental electric charge, n is a positive integer called a **quantum number,** and $a_0 = 0.052\,9$ nm is the **Bohr radius.**

If the hydrogen atom makes a transition from a state whose quantum number is n_i to one whose quantum number is n_f, where $n_f < n_i$, the frequency of the radiation emitted by the atom is

$$f = \frac{k_e e^2}{2a_0 h}\left(\frac{1}{n_f^2} - \frac{1}{n_i^2}\right) \qquad \textbf{11.26} \blacktriangleleft$$

Using $E_i - E_f = hf = hc/\lambda$, one can calculate the wavelengths of the radiation for various transitions. The calculated wavelengths are in excellent agreement with those in observed atomic spectra.

OBJECTIVE QUESTIONS |

☐ denotes answer available in *Student Solutions Manual/Study Guide*

1. Imagine that nitrogen and other atmospheric gases were more soluble in water so that the atmosphere of the Earth is entirely absorbed by the oceans. Atmospheric pressure would then be zero, and outer space would start at the planet's surface. Would the Earth then have a gravitational field? (a) Yes, and at the surface it would be larger in magnitude than 9.8 N/kg. (b) Yes, and it would be essentially the same as the current value. (c) Yes, and it would be somewhat less than 9.8 N/kg. (d) Yes, and it would be much less than 9.8 N/kg. (e) No, it would not.

2. The gravitational force exerted on an astronaut on the Earth's surface is 650 N directed downward. When she is in the space station in orbit around the Earth, is the gravitational force on her (a) larger, (b) exactly the same, (c) smaller, (d) nearly but not exactly zero, or (e) exactly zero?

3. Rank the magnitudes of the following gravitational forces from largest to smallest. If two forces are equal, show their equality in your list. (a) the force exerted by a 2-kg object on a 3-kg object 1 m away (b) the force exerted by a 2-kg object on a 9-kg object 1 m away (c) the force exerted by a 2-kg object on a 9-kg object 2 m away (d) the force exerted by a 9-kg object on a 2-kg object 2 m away (e) the force exerted by a 4-kg object on another 4-kg object 2 m away

4. An object of mass m is located on the surface of a spherical planet of mass M and radius R. The escape speed from the planet does not depend on which of the following? (a) M (b) m (c) the density of the planet (d) R (e) the acceleration due to gravity on that planet

5. A system consists of five particles. How many terms appear in the expression for the total gravitational potential energy of the system? (a) 4 (b) 5 (c) 10 (d) 20 (e) 25

6. Suppose the gravitational acceleration at the surface of a certain moon A of Jupiter is 2 m/s². Moon B has twice the mass and twice the radius of moon A. What is the gravitational acceleration at its surface? Neglect the gravitational acceleration due to Jupiter. (a) 8 m/s² (b) 4 m/s² (c) 2 m/s² (d) 1 m/s² (e) 0.5 m/s²

7. A satellite originally moves in a circular orbit of radius R around the Earth. Suppose it is moved into a circular orbit of radius $4R$. (i) What does the force exerted on the satellite then become? (a) eight times larger (b) four times larger (c) one-half as large (d) one-eighth as large (e) one-sixteenth as large (ii) What happens to the satellite's speed? Choose from the same possibilities (a) through (e). (iii) What happens to its period? Choose from the same possibilities (a) through (e).

8. (a) Can a hydrogen atom in the ground state absorb a photon of energy less than 13.6 eV? (b) Can this atom absorb a photon of energy greater than 13.6 eV?

9. A satellite moves in a circular orbit at a constant speed around the Earth. Which of the following statements is true? (a) No force acts on the satellite. (b) The satellite moves at constant speed and hence doesn't accelerate. (c) The satellite has an acceleration directed away from the

Earth. (d) The satellite has an acceleration directed toward the Earth. (e) Work is done on the satellite by the gravitational force.

10. Rank the following quantities of energy from largest to the smallest. State if any are equal. (a) the absolute value of the average potential energy of the Sun–Earth system (b) the average kinetic energy of the Earth in its orbital motion relative to the Sun (c) the absolute value of the total energy of the Sun–Earth system

11. Halley's comet has a period of approximately 76 years, and it moves in an elliptical orbit in which its distance from the Sun at closest approach is a small fraction of its maximum distance. Estimate the comet's maximum distance from the Sun in astronomical units (AU, the distance from the Earth to the Sun). (a) 6 AU (b) 12 AU (c) 20 AU (d) 28 AU (e) 35 AU

12. The vernal equinox and the autumnal equinox are associated with two points 180° apart in the Earth's orbit. That is, the Earth is on precisely opposite sides of the Sun when it passes through these two points. From the vernal equinox, 185.4 days elapse before the autumnal equinox. Only 179.8 days elapse from the autumnal equinox until the next vernal equinox. Why is the interval from the March (vernal) to the September (autumnal) equinox (which contains the summer solstice) longer than the interval from the September to the March equinox rather than being equal to that interval? Choose one of the following reasons. (a) They are really the same, but the Earth spins faster during the "summer" interval, so the days are shorter. (b) Over the "summer" interval, the Earth moves slower because it is farther from the Sun. (c) Over the March-to-September interval, the Earth moves slower because it is closer to the Sun. (d) The Earth has less kinetic energy when it is warmer. (e) The Earth has less orbital angular momentum when it is warmer.

13. (i) Rank the following transitions for a hydrogen atom from the transition with the greatest gain in energy to that with the greatest loss, showing any cases of equality. (a) $n_i = 2$; $n_f = 5$; (b) $n_i = 5$; $n_f = 3$; (c) $n_i = 7$; $n_f = 4$; (d) $n_i = 4$; $n_f = 7$ (ii) Rank the same transitions as in part (i) according to the wavelength of the photon absorbed or emitted by an otherwise isolated atom from greatest wavelength to smallest.

14. Let $-E$ represent the energy of a hydrogen atom. (i) What is the kinetic energy of the electron? (a) $2E$ (b) E (c) 0 (d) $-E$ (e) $-2E$ (ii) What is the potential energy of the atom? Choose from the same possibilities (a) through (e).

CONCEPTUAL QUESTIONS

☐ denotes answer available in *Student Solutions Manual/Study Guide*

1. A satellite in low-Earth orbit is not truly traveling through a vacuum. Rather, it moves through very thin air. Does the resulting air friction cause the satellite to slow down?

2. (a) Explain why the force exerted on a particle by a uniform sphere must be directed toward the center of the sphere. (b) Would this statement be true if the mass distribution of the sphere were not spherically symmetric? Explain.

3. Why don't we put a geosynchronous weather satellite in orbit around the 45th parallel? Wouldn't such a satellite be more useful in the United States than one in orbit around the equator?

4. Explain why it takes more fuel for a spacecraft to travel from the Earth to the Moon than for the return trip. Estimate the difference.

5. (a) If a hole could be dug to the center of the Earth, would the force on an object of mass m still obey Equation 11.1 there? (b) What do you think the force on m would be at the center of the Earth?

6. You are given the mass and radius of planet X. How would you calculate the free-fall acceleration on this planet's surface?

7. (a) At what position in its elliptical orbit is the speed of a planet a maximum? (b) At what position is the speed a minimum?

8. Each *Voyager* spacecraft was accelerated toward escape speed from the Sun by the gravitational force exerted by Jupiter on the spacecraft. (a) Is the gravitational force a conservative or a nonconservative force? (b) Does the interaction of the spacecraft with Jupiter meet the definition of an elastic collision? (c) How could the spacecraft be moving faster after the collision?

9. In his 1798 experiment, Cavendish was said to have "weighed the Earth." Explain this statement.

PROBLEMS

WebAssign The problems found in this chapter may be assigned online in Enhanced WebAssign.

1. denotes straightforward problem; 2. denotes intermediate problem; 3. denotes challenging problem

1. denotes full solution available in the *Student Solutions Manual/ Study Guide*

1. denotes problems most often assigned in Enhanced WebAssign.

BIO denotes biomedical problem

GP denotes guided problem

M denotes Master It tutorial available in Enhanced WebAssign

Q|C denotes asking for quantitative and conceptual reasoning

S denotes symbolic reasoning problem

shaded denotes "paired problems" that develop reasoning with symbols and numerical values

W denotes Watch It video solution available in Enhanced WebAssign

Section 11.1 Newton's Law of Universal Gravitation

Note: Problems 35, 37, and 38 in Chapter 5 can be assigned with this section.

1. Two ocean liners, each with a mass of 40 000 metric tons, are moving on parallel courses 100 m apart. What is the magnitude of the acceleration of one of the liners toward the other due to their mutual gravitational attraction? Model the ships as particles.

2. **Q|C** During a solar eclipse, the Moon, the Earth, and the Sun all lie on the same line, with the Moon between the Earth and the Sun. (a) What force is exerted by the Sun on the Moon? (b) What force is exerted by the Earth on the Moon? (c) What force is exerted by the Sun on the Earth? (d) Compare the answers to parts (a) and (b). Why doesn't the Sun capture the Moon away from the Earth?

3. **W** A 200-kg object and a 500-kg object are separated by 4.00 m. (a) Find the net gravitational force exerted by these objects on a 50.0-kg object placed midway between them. (b) At what position (other than an infinitely remote one) can the 50.0-kg object be placed so as to experience a net force of zero from the other two objects?

4. A satellite of mass 300 kg is in a circular orbit around the Earth at an altitude equal to the Earth's mean radius. Find (a) the satellite's orbital speed, (b) the period of its revolution, and (c) the gravitational force acting on it.

5. **M** In introductory physics laboratories, a typical Cavendish balance for measuring the gravitational constant G uses lead spheres with masses of 1.50 kg and 15.0 g whose centers are separated by about 4.50 cm. Calculate the gravitational force between these spheres, treating each as a particle located at the sphere's center.

6. **Review.** A student proposes to study the gravitational force by suspending two 100.0-kg spherical objects at the lower ends of cables from the ceiling of a tall cathedral and measuring the deflection of the cables from the vertical. The 45.00-m-long cables are attached to the ceiling 1.000 m apart. The first object is suspended, and its position is carefully measured. The second object is suspended, and the two objects attract each other gravitationally. By what distance has the first object moved horizontally from its initial position due to the gravitational attraction to the other object? *Suggestion:* Keep in mind that this distance will be very small and make appropriate approximations.

7. **W** The free-fall acceleration on the surface of the Moon is about one-sixth that on the surface of the Earth. The radius of the Moon is about $0.250 R_E$ (R_E = Earth's radius = 6.37×10^6 m). Find the ratio of their average densities, ρ_{Moon}/ρ_{Earth}.

8. *Why is the following situation impossible?* The centers of two homogeneous spheres are 1.00 m apart. The spheres are each made of the same element from the periodic table. The gravitational force between the spheres is 1.00 N.

9. **Review.** Miranda, a satellite of Uranus, is shown in Figure P11.9a. It can be modeled as a sphere of radius 242 km and mass 6.68×10^{19} kg. (a) Find the free-fall acceleration on its surface. (b) A cliff on Miranda is 5.00 km high. It appears on the limb at the 11 o'clock position in Figure P11.9a and is magnified in Figure P11.9b. If a devotee of extreme sports runs horizontally off the top of the cliff at 8.50 m/s, for what time interval is he in flight? (c) How far from the base of the vertical cliff does he strike the icy surface of Miranda? (d) What will be his vector impact velocity?

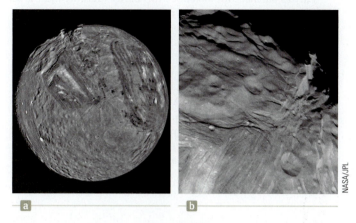

a   b

NASA/JPL

Figure P11.9

10. **W** Three uniform spheres of masses $m_1 = 2.00$ kg, $m_2 = 4.00$ kg, and $m_3 = 6.00$ kg are placed at the corners of a right triangle as shown in Figure P11.10. Calculate the resultant gravitational force on the object of mass m_2, assuming the spheres are isolated from the rest of the Universe.

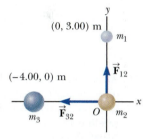

Figure P11.10

11. **W** A spacecraft in the shape of a long cylinder has a length of 100 m, and its mass with occupants is 1 000 kg. It has strayed too close to a black hole having a mass 100 times that of the Sun (Fig. P11.11). The nose of the spacecraft points toward the black hole, and the distance between the nose and the center of the black hole is 10.0 km. (a) Determine the total force on the spacecraft. (b) What is the difference in the gravitational fields acting on the occupants in the nose of the ship and on those in the rear of the ship, farthest from the black hole? (This difference in accelerations grows rapidly as the ship approaches the black hole. It puts the body of the ship under extreme tension and eventually tears it apart.)

Black hole

|← 100 m →|← //─ 10.0 km ─// →|

Figure P11.11

12. **Q|C** **S** (a) Compute the vector gravitational field at a point P on the perpendicular bisector of the line joining two objects of equal mass separated by a distance $2a$ as shown in Figure P11.12. (b) Explain physically why the field should approach zero as $r \rightarrow 0$. (c) Prove mathematically that the answer to part (a) behaves in this way. (d) Explain physically why the magnitude of the

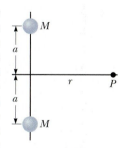

Figure P11.12

field should approach $2GM/r^2$ as $r \to \infty$. (e) Prove mathematically that the answer to part (a) behaves correctly in this limit.

13. **M** When a falling meteoroid is at a distance above the Earth's surface of 3.00 times the Earth's radius, what is its acceleration due to the Earth's gravitation?

Section 11.3 Kepler's Laws

14. Two planets X and Y travel counterclockwise in circular orbits about a star as shown in Figure P11.14. The radii of their orbits are in the ratio 3:1. At one moment, they are aligned as shown in Figure P11.14a, making a straight line with the star. During the next five years, the angular displacement of planet X is 90.0° as shown in Figure P11.14b. What is the angular displacement of planet Y at this moment?

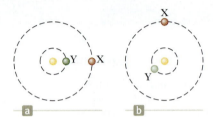

Figure P11.14

15. A communication satellite in geosynchronous orbit remains above a single point on the Earth's equator as the planet rotates on its axis. (a) Calculate the radius of its orbit. (b) The satellite relays a radio signal from a transmitter near the North Pole to a receiver, also near the North Pole. Traveling at the speed of light, how long is the radio wave in transit?

16. **Q│C** Suppose the Sun's gravity were switched off. The planets would leave their orbits and fly away in straight lines as described by Newton's first law. (a) Would Mercury ever be farther from the Sun than Pluto? (b) If so, find how long it would take Mercury to achieve this passage. If not, give a convincing argument that Pluto is always farther from the Sun than is Mercury.

17. Io, a satellite of Jupiter, has an orbital period of 1.77 days and an orbital radius of 4.22×10^5 km. From these data, determine the mass of Jupiter.

18. The *Explorer VIII* satellite, placed into orbit November 3, 1960, to investigate the ionosphere, had the following orbit parameters: perigee, 459 km; apogee, 2 289 km (both distances above the Earth's surface); period, 112.7 min. Find the ratio v_p/v_a of the speed at perigee to that at apogee.

19. **M** Plaskett's binary system consists of two stars that revolve in a circular orbit about a center of mass midway between them. This statement implies that the masses of the two stars are equal (Fig. P11.19). Assume the orbital speed of each star is $|\vec{\mathbf{v}}| = 220$ km/s and the orbital period of each is 14.4 days. Find the mass M of each star. (For comparison, the mass of our Sun is 1.99×10^{30} kg.)

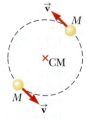

Figure P11.19

20. As thermonuclear fusion proceeds in its core, the Sun loses mass at a rate of 3.64×10^9 kg/s. During the 5 000-yr period of recorded history, by how much has the length of the year

changed due to the loss of mass from the Sun? *Suggestions:* Assume the Earth's orbit is circular. No external torque acts on the Earth–Sun system, so the angular momentum of the Earth is constant.

21. **W** Comet Halley (Fig. P11.21) approaches the Sun to within 0.570 AU, and its orbital period is 75.6 yr. (AU is the symbol for astronomical unit, where 1 AU = 1.50×10^{11} m is the mean Earth–Sun distance.) How far from the Sun will Halley's comet travel before it starts its return journey?

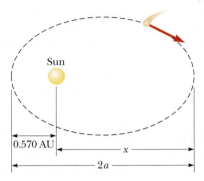

Figure P11.21 The elliptical orbit of Comet Halley. (Orbit is not drawn to scale.)

Section 11.4 Energy Considerations in Planetary and Satellite Motion

Note: Problems 50 through 53 in Chapter 6 can be assigned with this section.

22. How much work is done by the Moon's gravitational field on a 1 000-kg meteor as it comes in from outer space and impacts on the Moon's surface?

23. After the Sun exhausts its nuclear fuel, its ultimate fate will be to collapse to a *white dwarf* state. In this state, it would have approximately the same mass as it has now, but its radius would be equal to the radius of the Earth. Calculate (a) the average density of the white dwarf, (b) the surface free-fall acceleration, and (c) the gravitational potential energy associated with a 1.00-kg object at the surface of the white dwarf.

24. **S** A "treetop satellite" moves in a circular orbit just above the surface of a planet, assumed to offer no air resistance. Show that its orbital speed v and the escape speed from the planet are related by the expression $v_{esc} = \sqrt{2}v$.

25. An asteroid is on a collision course with Earth. An astronaut lands on the rock to bury explosive charges that will blow the asteroid apart. Most of the small fragments will miss the Earth, and those that fall into the atmosphere will produce only a beautiful meteor shower. The astronaut finds that the density of the spherical asteroid is equal to the average density of the Earth. To ensure its pulverization, she incorporates into the explosives the rocket fuel and oxidizer intended for her return journey. What maximum radius can the asteroid have for her to be able to leave it entirely simply by jumping straight up? On Earth she can jump to a height of 0.500 m.

26. **M** A space probe is fired as a projectile from the Earth's surface with an initial speed of 2.00×10^4 m/s. What will its speed be when it is very far from the Earth? Ignore atmospheric friction and the rotation of the Earth.

27. (a) Determine the amount of work that must be done on a 100-kg payload to elevate it to a height of 1 000 km above the Earth's surface. (b) Determine the amount of additional work that is required to put the payload into circular orbit at this elevation.

28. An object is released from rest at an altitude h above the surface of the Earth. (a) Show that its speed at a distance r from the Earth's center, where $R_E \leq r \leq R_E + h$, is

$$v = \sqrt{2GM_E\left(\frac{1}{r} - \frac{1}{R_E + h}\right)}$$

(b) Assume the release altitude is 500 km. Perform the integral

$$\Delta t = \int_i^f dt = -\int_i^f \frac{dr}{v}$$

to find the time of fall as the object moves from the release point to the Earth's surface. The negative sign appears because the object is moving opposite to the radial direction, so its speed is $v = -dr/dt$. Perform the integral numerically.

29. **W** A satellite of mass 200 kg is placed into Earth orbit at a height of 200 km above the surface. (a) Assuming a circular orbit, how long does the satellite take to complete one orbit? (b) What is the satellite's speed? (c) Starting from the satellite on the Earth's surface, what is the minimum energy input necessary to place this satellite in orbit? Ignore air resistance but include the effect of the planet's daily rotation.

30. **S** A satellite of mass m, originally on the surface of the Earth, is placed into Earth orbit at an altitude h. (a) Assuming a circular orbit, how long does the satellite take to complete one orbit? (b) What is the satellite's speed? (c) What is the minimum energy input necessary to place this satellite in orbit? Ignore air resistance but include the effect of the planet's daily rotation. Represent the mass and radius of the Earth as M_E and R_E, respectively.

31. A 500-kg satellite is in a circular orbit at an altitude of 500 km above the Earth's surface. Because of air friction, the satellite eventually falls to the Earth's surface, where it hits the ground with a speed of 2.00 km/s. How much energy was transformed into internal energy by means of air friction?

32. **S** Derive an expression for the work required to move an Earth satellite of mass m from a circular orbit of radius $2R_E$ to one of radius $3R_E$.

33. A comet of mass 1.20×10^{10} kg moves in an elliptical orbit around the Sun. Its distance from the Sun ranges between 0.500 AU and 50.0 AU. (a) What is the eccentricity of its orbit? (b) What is its period? (c) At aphelion, what is the potential energy of the comet–Sun system? *Note:* 1 AU = one astronomical unit = the average distance from the Sun to the Earth = 1.496×10^{11} m.

34. (a) What is the minimum speed, relative to the Sun, necessary for a spacecraft to escape the solar system if it starts at the Earth's orbit? (b) *Voyager 1* achieved a maximum speed of 125 000 km/h on its way to photograph Jupiter. Beyond what distance from the Sun is this speed sufficient to escape the solar system?

35. (a) A space vehicle is launched vertically upward from the Earth's surface with an initial speed of 8.76 km/s, which is less than the escape speed of 11.2 km/s. What maximum height does it attain? (b) A meteoroid falls toward the Earth. It is essentially at rest with respect to the Earth when it is at a height of 2.51×10^7 m above the Earth's surface. With what speed does the meteorite (a meteoroid that survives to impact the Earth's surface) strike the Earth?

36. **S** (a) A space vehicle is launched vertically upward from the Earth's surface with an initial speed of v_i that is comparable to but less than the escape speed v_{esc}. What maximum height does it attain? (b) A meteoroid falls toward the Earth. It is essentially at rest with respect to the Earth when it is at a height h above the Earth's surface. With what speed does the meteorite (a meteoroid that survives to impact the Earth's surface) strike the Earth? (c) **What If?** Assume a baseball is tossed up with an initial speed that is very small compared to the escape speed. Show that the result from part (a) is consistent with Equation 3.15.

Section 11.5 Atomic Spectra and the Bohr Theory of Hydrogen

37. (a) What value of n_i is associated with the 94.96-nm spectral line in the Lyman series of hydrogen? (b) **What If?** Could this wavelength be associated with the Paschen series? (c) Could this wavelength be associated with the Balmer series?

38. A hydrogen atom emits light as it undergoes a transition from the $n = 3$ state to the $n = 2$ state. Calculate (a) the energy, (b) the wavelength, and (c) the frequency of the radiation.

39. For a hydrogen atom in its ground state, compute (a) the orbital speed of the electron, (b) the kinetic energy of the electron, and (c) the electric potential energy of the atom.

40. **S** Show that the speed of the electron in the nth Bohr orbit in hydrogen is given by

$$v_n = \frac{k_e e^2}{n\hbar}$$

41. **W** A hydrogen atom is in its first excited state ($n = 2$). Calculate (a) the radius of the orbit, (b) the linear momentum of the electron, (c) the angular momentum of the electron, (d) the kinetic energy of the electron, (e) the potential energy of the system, and (f) the total energy of the system.

42. How much energy is required to ionize hydrogen (a) when it is in the ground state and (b) when it is in the state for which $n = 3$?

43. Two hydrogen atoms collide head-on and end up with zero kinetic energy. Each atom then emits light with a wavelength of 121.6 nm ($n = 2$ to $n = 1$ transition). At what speed were the atoms moving before the collision?

Section 11.6 Context Connection: Changing from a Circular to an Elliptical Orbit

44. A spacecraft of mass 1.00×10^4 kg is in a circular orbit at an altitude of 500 km above the Earth's surface. Mission Control wants to fire the engines so as to put the spacecraft in an elliptical orbit around the Earth with an apogee of 2.00×10^4 km. How much energy must be used from the fuel to achieve this orbit? (Assume that all the fuel energy goes into increasing the orbital energy. This model will give a lower limit to the required energy because some of the energy from the fuel will appear as internal energy in the hot exhaust gases and engine parts.)

45. A spacecraft is approaching Mars after a long trip from the Earth. Its velocity is such that it is traveling along a parabolic trajectory under the influence of the gravitational force from Mars. The distance of closest approach will be 300 km above the Martian surface. At this point of closest approach, the engines will be fired to slow down the spacecraft and place it in a circular orbit 300 km above the surface. (a) By what percentage must the speed of the spacecraft be reduced to achieve the desired orbit? (b) How would the answer to part (a) change if the distance of closest approach and the desired circular orbit altitude were 600 km instead of 300 km? (*Note:* The energy of the spacecraft–Mars system for a parabolic orbit is $E = 0$.)

Additional Problems

46. **S** Astronomers detect a distant meteoroid moving along a straight line that, if extended, would pass at a distance $3R_E$ from the center of the Earth, where R_E is the Earth's radius. What minimum speed must the meteoroid have if it is *not* to collide with the Earth?

47. Let Δg_M represent the difference in the gravitational fields produced by the Moon at the points on the Earth's surface nearest to and farthest from the Moon. Find the fraction $\Delta g_M/g$, where g is the Earth's gravitational field. (This difference is responsible for the occurrence of the *lunar tides* on the Earth.)

48. **S** **Review.** Two identical hard spheres, each of mass m and radius r, are released from rest in otherwise empty space with their centers separated by the distance R. They are allowed to collide under the influence of their gravitational attraction. (a) Show that the magnitude of the impulse received by each sphere before they make contact is given by $[Gm^3(1/2r - 1/R)]^{1/2}$. (b) **What If?** Find the magnitude of the impulse each receives during their contact if they collide elastically.

49. *Voyager 1* and *Voyager 2* surveyed the surface of Jupiter's moon Io and photographed active volcanoes spewing liquid sulfur to heights of 70 km above the surface of this moon. Find the speed with which the liquid sulfur left the volcano. Io's mass is 8.9×10^{22} kg, and its radius is 1 820 km.

50. **S** Two stars of masses M and m, separated by a distance d, revolve in circular orbits about their center of mass (Fig. P11.50). Show that each star has a period given by

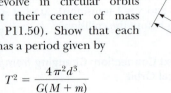

Figure P11.50

$$T^2 = \frac{4\pi^2 d^3}{G(M + m)}$$

Proceed as follows: Apply Newton's second law to each star. Note that the center-of-mass condition requires that $Mr_2 = mr_1$, where $r_1 + r_2 = d$.

51. A ring of matter is a familiar structure in planetary and stellar astronomy. Examples include Saturn's rings and a ring nebula. Consider a uniform ring of mass 2.36×10^{20} kg and

radius 1.00×10^8 m. An object of mass 1 000 kg is placed at a point A on the axis of the ring, 2.00×10^8 m from the center of the ring (Fig. P11.51). When the object is released, the attraction of the ring makes the object move along the axis toward the center of the ring (point B). (a) Calculate the gravitational potential energy of the object–ring system when the object is at A. (b) Calculate the gravitational potential energy of the system when the object is at B. (c) Calculate the speed of the object as it passes through B.

Figure P11.51

52. (a) Show that the rate of change of the free-fall acceleration with vertical position near the Earth's surface is

$$\frac{dg}{dr} = -\frac{2GM_E}{R_E^3}$$

This rate of change with position is called a *gradient*. (b) Assuming h is small in comparison to the radius of the Earth, show that the difference in free-fall acceleration between two points separated by vertical distance h is

$$|\Delta g| = \frac{2GM_E h}{R_E^3}$$

(c) Evaluate this difference for $h = 6.00$ m, a typical height for a two-story building.

53. **Review.** As an astronaut, you observe a small planet to be spherical. After landing on the planet, you set off, walking always straight ahead, and find yourself returning to your spacecraft from the opposite side after completing a lap of 25.0 km. You hold a hammer and a falcon feather at a height of 1.40 m, release them, and observe that they fall together to the surface in 29.2 s. Determine the mass of the planet.

54. **Q|C** Many people assume air resistance acting on a moving object will always make the object slow down. It can, however, actually be responsible for making the object speed up. Consider a 100-kg Earth satellite in a circular orbit at an altitude of 200 km. A small force of air resistance makes the satellite drop into a circular orbit with an altitude of 100 km. (a) Calculate the satellite's initial speed. (b) Calculate its final speed in this process. (c) Calculate the initial energy of the satellite–Earth system. (d) Calculate the final energy of the system. (e) Show that the system has lost mechanical energy and find the amount of the loss due to friction. (f) What force makes the satellite's speed increase? *Hint:* You will find a free-body diagram useful in explaining your answer.

55. The maximum distance from the Earth to the Sun (at aphelion) is 1.521×10^{11} m, and the distance of closest approach (at perihelion) is 1.471×10^{11} m. The Earth's orbital speed at perihelion is 3.027×10^4 m/s. Determine

(a) the Earth's orbital speed at aphelion and the kinetic and potential energies of the Earth–Sun system (b) at perihelion, and (c) at aphelion. (d) Is the total energy of the system constant? Explain. Ignore the effect of the Moon and other planets.

56. **GP** **S** Two spheres having masses M and $2M$ and radii R and $3R$, respectively, are simultaneously released from rest when the distance between their centers is $12R$. Assume the two spheres interact only with each other and we wish to find the speeds with which they collide. (a) What *two* isolated system models are appropriate for this system? (b) Write an equation from one of the models and solve it for $\vec{v}_1$, the velocity of the sphere of mass M at any time after release in terms of $\vec{v}_2$, the velocity of $2M$. (c) Write an equation from the other model and solve it for speed v_1 in terms of speed v_2 when the spheres collide. (d) Combine the two equations to find the two speeds v_1 and v_2 when the spheres collide.

57. **M** Two hypothetical planets of masses m_1 and m_2 and radii r_1 and r_2, respectively, are nearly at rest when they are an infinite distance apart. Because of their gravitational attraction, they head toward each other on a collision course. (a) When their center-to-center separation is d, find expressions for the speed of each planet and for their relative speed. (b) Find the kinetic energy of each planet just before they collide, taking $m_1 = 2.00 \times 10^{24}$ kg, $m_2 = 8.00 \times 10^{24}$ kg, $r_1 = 3.00 \times 10^6$ m, and $r_2 = 5.00 \times 10^6$ m. *Note:* Both the energy and momentum of the isolated two-planet system are constant.

58. *Why is the following situation impossible?* A spacecraft is launched into a circular orbit around the Earth and circles the Earth once an hour.

59. **Q|C** **Review.** Assume you are agile enough to run across a horizontal surface at 8.50 m/s, independently of the value of the gravitational field. What would be (a) the radius and (b) the mass of an airless spherical asteroid of uniform density 1.10×10^3 kg/m^3 on which you could launch yourself into orbit by running? (c) What would be your period? (d) Would your running significantly affect the rotation of the asteroid? Explain.

60. The oldest artificial satellite still in orbit is *Vanguard I*, launched March 3, 1958. Its mass is 1.60 kg. Neglecting atmospheric drag, the satellite would still be in its initial orbit, with a minimum distance from the center of the Earth of 7.02 Mm and a speed at this perigee point of 8.23 km/s. For this orbit, find (a) the total energy of the satellite–Earth system and (b) the magnitude of the angular momentum of the satellite. (c) At apogee, find the satellite's speed and its distance from the center of the Earth. (d) Find the semimajor axis of its orbit. (e) Determine its period.

61. The positron is the antiparticle to the electron. It has the same mass and a positive electric charge of the same magnitude as that of the electron. Positronium is a hydrogen-like atom consisting of a positron and an electron revolving around each other. Using the Bohr model, find (a) the allowed distances between the two particles and (b) the allowed energies of the system.

62. **S** Show that the minimum period for a satellite in orbit around a spherical planet of uniform density ρ is

$$T_{min} = \sqrt{\frac{3\pi}{G\rho}}$$

independent of the planet's radius.

63. Studies of the relationship of the Sun to our galaxy—the Milky Way—have revealed that the Sun is located near the outer edge of the galactic disc, about 30 000 ly (1 ly = 9.46×10^{15} m) from the center. The Sun has an orbital speed of approximately 250 km/s around the galactic center. (a) What is the period of the Sun's galactic motion? (b) What is the order of magnitude of the mass of the Milky Way galaxy? (c) Suppose the galaxy is made mostly of stars of which the Sun is typical. What is the order of magnitude of the number of stars in the Milky Way?

64. The Solar and Heliospheric Observatory (SOHO) spacecraft has a special orbit, located between the Earth and the Sun along the line joining them, and it is always close enough to the Earth to transmit data easily. Both objects exert gravitational forces on the observatory. It moves around the Sun in a near-circular orbit that is smaller than the Earth's circular orbit. Its period, however, is not less than 1 yr but just equal to 1 yr. Show that its distance from the Earth must be 1.48×10^9 m. In 1772, Joseph Louis Lagrange determined theoretically the special location allowing this orbit. *Suggestions:* Use data that are precise to four digits. The mass of the Earth is 5.974×10^{24} kg. You will not be able to easily solve the equation you generate; instead, use a computer to verify that 1.48×10^9 m is the correct value.

65. Consider an object of mass m, not necessarily small compared with the mass of the Earth, released at a distance of 1.20×10^7 m from the center of the Earth. Assume the Earth and the object behave as a pair of particles, isolated from the rest of the Universe. (a) Find the magnitude of the acceleration a_{rel} with which each starts to move relative to the other as a function of m. Evaluate the acceleration (b) for $m = 5.00$ kg, (c) for $m = 2\,000$ kg, and (d) for $m = 2.00 \times 10^{24}$ kg. (e) Describe the pattern of variation of a_{rel} with m.

Context 2 — CONCLUSION

A Successful Mission Plan

Now that we have explored the physics of classical mechanics, let us return to our central question for the *Mission to Mars* Context:

> **How can we undertake a successful transfer of a spacecraft from the Earth to Mars?**

We make use of the physical principles that we now understand and apply them to our journey from the Earth to Mars.

Let us start with a more modest proposition. Suppose a spacecraft is in a circular orbit around the Earth and you are a passenger on the spacecraft. If you toss a wrench in the direction of travel, tangent to the circular path, what orbital path will the wrench follow?

Let us adopt a simplification model in which the spacecraft is much more massive than the wrench. Conservation of momentum for the isolated system of the wrench and the spacecraft tells us that the spacecraft must slow down slightly once the wrench is thrown. Because of the mass difference between the wrench and spacecraft, however, we can ignore the small change in the spacecraft's speed. The wrench now enters a new orbit, from its perigee position, and the wrench–Earth system has more energy than it had when the wrench was in the circular orbit. Because the orbital energy is related to the major axis, the wrench is injected into an elliptical orbit as discussed in the Context Connection of Chapter 11 and as shown in Figure 1. Therefore, the path of the wrench is changed from a circular orbit to an elliptical orbit by providing the wrench–Earth system with extra energy. The energy is provided by the force you apply to the wrench tangent to the circular orbit because you have done work on the system. The elliptical orbit will take the wrench farther from the Earth than the circular orbit. If there were another spacecraft in a higher circular orbit than your spacecraft, you could throw the wrench so that it transfers from one spacecraft to another as shown in Figure 2. For that to occur, the elliptical orbit of the wrench must intersect with the higher spacecraft orbit. Furthermore, the wrench and the second spacecraft must arrive at the same point at the same time.

This scenario is the essence of our planned mission from the Earth to Mars. Rather than transferring a wrench between two spacecraft in orbit around the Earth, we will transfer a spacecraft between two planets in orbit around the Sun. Kinetic energy is added to the wrench–Earth system by throwing the wrench. Kinetic energy is added to the spacecraft–Sun system by firing the engines.

What if you were to throw the wrench harder and harder in the previous example? The wrench would be placed in a larger and larger elliptical orbit around the Earth. As you increased the launch velocity, you could inject the wrench into a *hyperbolic* escape orbit, relative to the Earth, and into an *elliptical* orbit around the *Sun*. This approach is the one we will take for the trip from the Earth to Mars; we will break free from a circular *parking* orbit around the Earth and

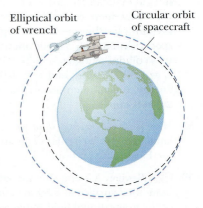

Figure 1 A wrench thrown tangent to the circular orbit of a spacecraft enters an elliptical orbit.

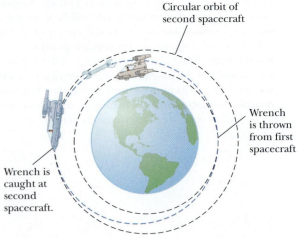

Figure 2 If a second spacecraft were in a higher circular orbit, the wrench could be carefully thrown so as to be transferred from one spacecraft to the other.

move into an elliptical *transfer* orbit around the Sun. The spacecraft will then continue on its journey to Mars, where it will enter a new parking orbit.

Now let us focus our attention on the transfer orbit part of the journey. One simple transfer orbit is called a **Hohmann transfer,** the type of transfer imparted to the wrench shown in Figure 2. The Hohmann transfer involves the least energy expenditure and therefore requires the smallest amount of fuel. As might be expected for a lowest-energy transfer, the transfer time for a Hohmann transfer is longer than for other types of orbits. We shall investigate the Hohmann transfer because of its simplicity and its general usefulness in planetary transfers.

The rocket engine on the spacecraft is fired from the parking orbit such that the spacecraft enters an elliptical orbit around the Sun at its perihelion and encounters the planet at the spacecraft's aphelion. Therefore, the spacecraft makes exactly one-half of a revolution about its elliptical path during the transfer as shown in Figure 3.

This process is energy efficient because fuel is expended only at the beginning and the end. The movement between parking orbits around the Earth and Mars is free; the spacecraft simply follows Kepler's laws while in an elliptical orbit around the Sun.

Let us perform a simple numerical calculation to see how to apply the mechanical laws to this process. We assume that the spacecraft is in a parking orbit above the Earth's surface. Notice also that the spacecraft is in orbit around the Sun, with a perturbation in its orbit caused by the Earth. Therefore, if we calculate the tangential speed of the Earth about the Sun, we can let this speed represent the average speed of the spacecraft around the Sun. This speed is calculated from Newton's second law for a particle in uniform circular motion:

$$F = ma \quad \rightarrow \quad G\frac{M_{Sun}\, m_{Earth}}{r^2} = m_{Earth}\frac{v^2}{r}$$

$$\rightarrow \quad v = \sqrt{\frac{GM_{Sun}}{r}} = \sqrt{\frac{(6.67 \times 10^{-11}\,\text{N} \cdot \text{m}^2/\text{kg}^2)(1.99 \times 10^{30}\,\text{kg})}{1.50 \times 10^{11}\,\text{m}}}$$

$$= 2.97 \times 10^4\,\text{m/s}$$

This result is the original speed of the spacecraft, to which we add a change Δv to inject the spacecraft into the transfer orbit.

The major axis of the elliptical transfer orbit is found by adding together the orbit radii of the Earth and Mars (see Fig. 3):

$$\text{Major axis} = 2a = r_{Earth} + r_{Mars}$$

$$= 1.50 \times 10^{11}\,\text{m} + 2.28 \times 10^{11}\,\text{m} = 3.78 \times 10^{11}\,\text{m}$$

Therefore, the semimajor axis is half this value:

$$a = 1.89 \times 10^{11}\,\text{m}$$

From this value, Kepler's third law is used to find the travel time, which is one-half of the period of the orbit:

$$\Delta t_{travel} = \tfrac{1}{2}T = \tfrac{1}{2}\sqrt{\frac{4\pi^2}{GM_{Sun}}\,a^3}$$

$$= \tfrac{1}{2}\sqrt{\frac{4\pi^2}{(6.67 \times 10^{-11}\,\text{N} \cdot \text{m}^2/\text{kg}^2)(1.99 \times 10^{30}\,\text{kg})}\,(1.89 \times 10^{11}\,\text{m})^3}$$

$$= 2.24 \times 10^7\,\text{s} = 0.710\,\text{yr} = 259\,\text{d}$$

Therefore, the journey to Mars will require 259 Earth days. We can also determine where in their orbits Mars and the Earth must be so that the planet will be there when the spacecraft arrives. Mars has an orbital period of 687 Earth days. During the transfer time, the angular position *change* of Mars is

$$\Delta\theta_{Mars} = \frac{259\,\text{d}}{687\,\text{d}}\,(2\pi) = 2.37\,\text{rad} = 136°$$

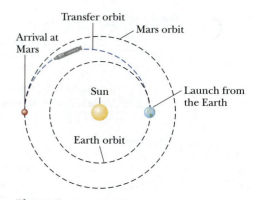

Figure 3 The Hohmann transfer orbit from the Earth to Mars. It is similar to transferring the wrench from one spacecraft to another in Figure 2, but here we are transferring a spacecraft from one planet to another.

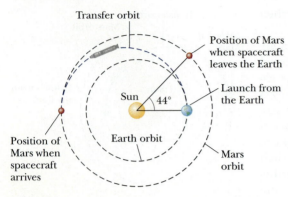

Figure 4 The spacecraft must be launched when Mars is 44° ahead of the Earth in its orbit.

BIO Effects of space travel on human health

Therefore, for the spacecraft and Mars to arrive at the same point at the same time, the spacecraft must be launched when Mars is $180° - 136° = 44°$ ahead of the Earth in its orbit. This geometry is shown in Figure 4.

With relatively simple mathematics, that is as far as we can go in describing the details of a trip to Mars. We have found the desired path, the time for the trip, and the position of Mars at launch time. Another important issue for the spacecraft captain would be that of the amount of fuel required for the trip. This question is related to the speed changes necessary to put us into a transfer orbit. These types of calculations involve energy considerations and are explored in Problem 3.

Our experiences so far with space travel have indicated a number of biological issues that will need to be addressed on the journey to Mars. In the absence of a gravitational field, the middle ear can no longer perceive a downward direction, and muscles are no longer required to maintain posture. Results include motion sickness and illusions about being right side up or upside down. The absence of gravity also results in bodily fluids distributing throughout the body, leading to symptoms similar to a cold. A serious problem in space travel is the atrophy of muscles and the loss of bone tissue due to the very different requirements of moving the body in a gravity-free environment. Bone loss occurs within 10 days of being in space because excessive amounts of calcium and phosphorus are released by the body, which could cause urinary stones and bone fracture. These issues can be addressed by spinning a spacecraft with a circular cross section around its axis so that the space travelers experience a centripetal acceleration that is equivalent to a gravitational field (see Section 9.9 on general relativity).

A difficulty that cannot be addressed by spinning the spacecraft is that of radiation. Being outside the atmosphere and magnetosphere of Earth exposes the space travelers to cosmic rays and other types of radiation. This exposure can lead to several detrimental health conditions, including cancer, cataracts, and suppression of the immune system. It is not clear at present whether protective shielding or pharmaceuticals would be sufficient to avoid these effects.

Although many considerations for a successful mission to Mars have not been addressed, we have successfully designed a transfer orbit from the Earth to Mars that is consistent with the laws of mechanics. We consequently declare success for our endeavor and bring this *Mission to Mars* Context to a close.

Questions

1. Some science fiction stories describe a twin planet to the Earth. It is exactly 180° ahead of us in the same orbit as the Earth, so we will never see it because it is on the other side of the Sun. Assuming you are in a spacecraft in orbit around the Earth, describe conceptually how you could visit this planet by altering your orbit.

2. You are in an orbiting spacecraft. Another spacecraft is in precisely the same orbit but is 1 km ahead of you, moving in the same direction around the circle. Through an oversight, your food supplies have been exhausted, but there is more than enough food in the other spacecraft. The commander of the other spacecraft is going to throw, from her spacecraft to yours, a picnic basket full of sandwiches. Give a qualitative description of how she should throw it.

Problems

1. Consider a Hohmann transfer from the Earth to Venus. (a) How long will this transfer take? (b) Should Venus be ahead of or behind the Earth in its orbit when the spacecraft leaves the Earth on its way to the rendezvous? How many degrees should Venus be ahead of or behind the Earth?

2. You are on a space station in a circular orbit 500 km above the surface of the Earth. Your passenger and guest is a large, strong, intelligent extraterrestrial. You try to teach her to play golf. Walking on the space station surface with magnetic shoes, you demonstrate a drive. The alien tees up a golf ball and hits it with incredible power, sending it off with speed Δv, relative to the space station, in a direction parallel to the instantaneous velocity vector of the space station. You notice that after you then complete precisely 2.00 orbits of the Earth, the golf ball also returns to the same location, so you reach up and catch the ball as it is passing the space station. With what speed Δv was the golf ball hit?

3. Investigate what the engine has to do to make a spacecraft follow the Hohmann transfer orbit from the Earth to Mars described in the text. Short-duration burns of our rocket engine are required to change the speed of our spacecraft whenever we alter our orbit. There are no brakes in space, so fuel is required both to increase and to decrease the speed of the spacecraft. First, ignore the gravitational attraction between the spacecraft and the planets. (a) Calculate the speed change required for switching the craft from a circular orbit around the Sun at the Earth's distance to the transfer orbit to Mars. (b) Calculate the speed change required for switching from the transfer orbit to a circular orbit around the Sun at the distance of Mars. Now consider the effects of the two planets' gravity. (c) Calculate the speed change required to carry the craft from the Earth's surface to its own independent orbit around the Sun. You may suppose the craft is launched from the Earth's equator toward the east. (d) Model the craft as falling to the surface of Mars from solar orbit. Calculate the magnitude of the speed change required to make a soft landing on Mars at the end of the fall. Mars rotates on its axis with a period of 24.6 h.

Earthquakes

Earthquakes result in massive movement of the ground, as evidenced by the accompanying photograph of severe damage caused by a magnitude 7.0 earthquake in Port-au-Prince, Haiti, in 2010. One of the most devastating events ever recorded was the magnitude 9.0 earthquake that took place on March 11, 2011, on the east coast of Japan. The earthquake triggered a devastating and widespread tsunami that killed thousands and caused major damage to buildings and several nuclear power plants.

Whereas earthquakes in Japan are relatively common, another earthquake in 2011 was far rarer. A magnitude 5.8 earthquake struck in August 2011 in the Appalachian Mountain region of Virginia in the United States. East coast earthquakes in the United States are not common. Shaking from the earthquake was felt as far north as Quebec in Canada and as far south as Atlanta, Georgia. Only minor damage was reported in towns surrounding the epicenter, although the White House and Capitol Building in Washington, D.C. were evacuated as a precaution. The National Cathedral, the Washington Monument, and the Smithsonian Castle all reported damage to structural components of the buildings.

Anyone who has experienced a serious earthquake can attest to the violent shaking it produces. In this

Figure 2 A secondary effect of some earthquakes that occur in the ocean is a tsunami. The tsunami caused by the Japanese earthquake of March 2011 caused extensive damage to the east coast of the country. This photo shows houses that have been swept off their foundations by the water as well as fires from ruptured gas lines.

Context, we shall focus on earthquakes as an application of our study of the physics of vibrations and waves.

The cause of an earthquake is a release of energy within the Earth at a point called the *focus*, or *hypocenter*, of the earthquake. The point on the Earth's surface radially above the focus is called the *epicenter*. As the energy from the focus reaches the surface, it spreads out along the surface of the Earth.

Earthquakes generally originate along a *fault*, which is a fracture or discontinuity in the rock beneath the Earth's surface. When there is sudden relative movement between the material on either side of a fault, an earthquake occurs. U.S. Geological Survey studies have shown a direct correlation between the magnitude of an earthquake and the size of nearby faults. Furthermore, such studies indicate that large-magnitude earthquakes can last as long as 2 minutes.

We might expect that the risk of damage in an earthquake decreases as one moves farther from the epicenter, and over long distances that assumption is correct. For example, structures in Kansas are not affected by earthquakes in California. In regions close to the earthquake, however, the notion of decrease in risk with

Figure 1 A day after the magnitude 7.0 earthquake in Port-au-Prince, Haiti, on January 13, 2010, a young woman climbs over the rubble of a collapsed store.

distance is not consistent. Consider, for example, the following comparisons describing local and distant effects resulting from two different earthquakes.

With regard to the magnitude-7.9 Michoacán earthquake, September 19, 1985:[1]

An earthquake rattled the coast of Mexico in the state of Michoacán, about 400 kilometers west of Mexico City. Near the coast, the shaking of the ground was mild and caused little damage. As the seismic waves raced inland, the ground shook even less, and by the time the waves were 100 kilometers from Mexico City, the shaking had nearly subsided. Nevertheless, the seismic waves induced severe shaking in the city, and some areas continued to shake for several minutes after the seismic waves had passed. Some 300 buildings collapsed and more than 20,000 people died.

A magnitude-6.3 earthquake ocurred on February 22, 2011, 10 km southeast of Christchurch, New Zealand. Flight crews from the New York Air National Guard were at Christchurch International Airport, 12 km northwest of the city, when the quake struck, and reported that they were safe and unharmed, and that the airport had water and electricity.

Consider, however, a contrasting situation much farther away, 200 km from Christchurch:[2]

The magnitude 6.3 earthquake . . . was strong enough to shake 30 million tonnes of ice loose from Tasman Glacier at Aoraki Mt Cook National Park. Passengers of two explorer boats were hit with waves of up to 3.5 metres as the ice crashed into Terminal Lake under the Tasman Glacier at the mountain.

Figure 3 Severe damage occurred in localized regions of Mexico City in 1985, even though the epicenter of the Michoacán earthquake was hundreds of kilometers away.

It is clear from these comparisons that the notion of a simple decrease in risk with distance is misleading. We will use these comparisons as motivation in our study of the physics of vibrations and waves so that we can better analyze the risk of damage to structures in an earthquake. Our study here will also be important when we investigate electromagnetic waves in Chapters 24 through 27. In this Context, we shall address the central question:

> **How can we choose locations and build structures to minimize the risk of damage in an earthquake?**

[1] *American Scientist,* November–December 1992, p. 566.

[2] *New Zealand Herald,* 22 February 2011

Chapter | 12

Oscillatory Motion

Chapter Outline

To reduce swaying in tall buildings because of the wind, tuned dampers are placed near the top of the building. These mechanisms include an object of large mass that oscillates under computer control at the same frequency as the building, reducing the swaying. The 730-ton suspended sphere in the photograph above is part of the tuned damper system of the Taipei Financial Center, at one time the world's tallest building.

Y ou are most likely familiar with several examples of *periodic* motion, such as the oscillations of an object on a spring, the motion of a pendulum, and the vibrations of a stringed musical instrument. Numerous other systems exhibit periodic behavior. For example, the molecules in a solid oscillate about their equilibrium positions; electromagnetic waves, such as light waves, radar, and radio waves, are characterized by oscillating electric and magnetic field vectors; in alternating-current circuits, such as in your household electrical service, voltage and current vary periodically with time. In this chapter, we will investigate mechanical systems that exhibit periodic motion.

We have experienced a number of situations in which the net force on a particle is constant. In these situations, the acceleration of the particle is also constant and we can describe the motion of the particle using the particle under constant acceleration model and the kinematic equations of Chapter 2. If a force acting on a particle varies in time, the acceleration of the particle also changes with time and so the kinematic equations cannot be used.

A special kind of periodic motion occurs when the force that acts on a particle is always directed toward an equilibrium position and is proportional to the position of the particle relative to the equilibrium position. We shall study this special type of varying force in this chapter. When this type of force acts on a particle, the particle exhibits *simple harmonic motion,* which will serve as an analysis model for a large class of oscillation problems.

⟨12.1 | Motion of an Object Attached to a Spring

As a model for simple harmonic motion, consider a block of mass m attached to the end of a spring, with the block free to move on a frictionless, horizontal surface (Active Fig. 12.1). When the spring is neither stretched nor compressed, the block is at rest at the position called the **equilibrium position** of the system, which we identify as $x = 0$ (Active Fig. 12.1b). We know from experience that such a system oscillates back and forth if disturbed from its equilibrium position.

We can understand the oscillating motion of the block in Active Figure 12.1 qualitatively by first recalling that when the block is displaced to a position x, the spring exerts on the block a force that is proportional to the position and given by **Hooke's law** (see Section 6.4):

$$F_s = -kx \qquad \text{12.1} \blacktriangleleft \qquad \blacktriangleright \text{Hooke's law}$$

We call F_s a **restoring force** because it is always directed toward the equilibrium position and therefore *opposite* the displacement of the block from equilibrium. That is, when the block is displaced to the right of $x = 0$ in Active Figure 12.1a, the position is positive and the restoring force is directed to the left. When the block is displaced to the left of $x = 0$ as in Figure 12.1c, the position is negative and the restoring force is directed to the right.

When the block is displaced from the equilibrium point and released, it is a particle under a net force and consequently undergoes an acceleration. Applying Newton's second law to the motion of the block, with Equation 12.1 providing the net force in the x direction, we obtain

$$-kx = ma_x$$

$$a_x = -\frac{k}{m}x \qquad \text{12.2} \blacktriangleleft$$

That is, the acceleration of the block is proportional to its position, and the direction of the acceleration is opposite the direction of the displacement of the block from equilibrium. Systems that behave in this way are said to exhibit **simple harmonic motion.** An object moves with simple harmonic motion whenever its acceleration is proportional to its position and is oppositely directed to the displacement from equilibrium.

If the block in Active Figure 12.1 is displaced to a position $x = A$ and released from rest, its *initial* acceleration is $-kA/m$. When the block passes through the equilibrium position $x = 0$, its acceleration is zero. At this instant, its speed is a maximum because the acceleration changes sign. The block then continues to travel to the left of equilibrium with a positive acceleration and finally reaches $x = -A$, at which time its acceleration is $+kA/m$ and its speed is again zero as discussed in Sections 6.4 and 6.6. The block completes a full cycle of its motion by returning to the original position, again passing through $x = 0$ with maximum speed. Therefore, the block oscillates between the turning points $x = \pm A$. In the absence of friction, this idealized motion will continue forever because the force exerted by the spring is conservative. Real systems are generally subject to friction, so they do not oscillate forever. We shall explore the details of the situation with friction in Section 12.6.

> **Pitfall Prevention | 12.1**
> **The Orientation of the Spring**
> Active Figure 12.1 shows a *horizontal* spring, with an attached block sliding on a frictionless surface. Another possibility is a block hanging from a *vertical* spring. All the results we discuss for the horizontal spring are the same for the vertical spring with one exception: when the block is placed on the vertical spring, its weight causes the spring to extend. If the resting position of the block is defined as $x = 0$, the results of this chapter also apply to this vertical system.

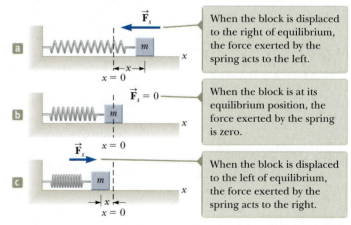

When the block is displaced to the right of equilibrium, the force exerted by the spring acts to the left.

When the block is at its equilibrium position, the force exerted by the spring is zero.

When the block is displaced to the left of equilibrium, the force exerted by the spring acts to the right.

Active Figure 12.1 A block attached to a spring moving on a frictionless surface.

QUICK QUIZ 12.1 A block on the end of a spring is pulled to position $x = A$ and released from rest. In one full cycle of its motion, through what total distance does it travel? **(a)** $A/2$ **(b)** A **(c)** $2A$ **(d)** $4A$

12.2 | Analysis Model: Particle in Simple Harmonic Motion

The motion described in the preceding section occurs so often that we identify the **particle in simple harmonic motion** model to represent such situations. To develop a mathematical representation for this model, we will generally choose x as the axis along which the oscillation occurs; hence, we will drop the subscript-x notation in this discussion. Recall that, by definition, $a = dv/dt = d^2x/dt^2$, so we can express Equation 12.2 as

$$\frac{d^2x}{dt^2} = -\frac{k}{m}x \qquad \text{12.3} \blacktriangleleft$$

If we denote the ratio k/m with the symbol ω^2 (we choose ω^2 rather than ω so as to make the solution we develop below simpler in form), then

$$\omega^2 = \frac{k}{m} \qquad \text{12.4} \blacktriangleleft$$

and Equation 12.3 can be written in the form

$$\frac{d^2x}{dt^2} = -\omega^2 x \qquad \text{12.5} \blacktriangleleft$$

Let's now find a mathematical solution to Equation 12.5, that is, a function $x(t)$ that satisfies this second-order differential equation and is a mathematical representation of the position of the particle as a function of time. We seek a function whose second derivative is the same as the original function with a negative sign and multiplied by ω^2. The trigonometric functions sine and cosine exhibit this behavior, so we can build a solution around one or both of them. The following cosine function is a solution to the differential equation:

$$x(t) = A \cos (\omega t + \phi) \qquad \text{12.6} \blacktriangleleft$$

where A, ω, and ϕ are constants. To show explicitly that this solution satisfies Equation 12.5, notice that

$$\frac{dx}{dt} = A \frac{d}{dt} \cos (\omega t + \phi) = -\omega A \sin (\omega t + \phi) \qquad \text{12.7} \blacktriangleleft$$

$$\frac{d^2x}{dt^2} = -\omega A \frac{d}{dt} \sin (\omega t + \phi) = -\omega^2 A \cos (\omega t + \phi) \qquad \text{12.8} \blacktriangleleft$$

Comparing Equations 12.6 and 12.8, we see that $d^2x/dt^2 = -\omega^2 x$ and Equation 12.5 is satisfied.

The parameters A, ω, and ϕ are constants of the motion. To give physical significance to these constants, it is convenient to form a graphical representation of the motion by plotting x as a function of t as in Active Figure 12.2a. First, A, called the **amplitude** of the motion, is simply the maximum value of the position of the particle in either the positive or negative x direction. The constant ω is called the **angular frequency,** and it has units[1] of radians per second. It is a measure of how rapidly the oscillations are occurring; the more oscillations per unit time, the higher the value of ω. From Equation 12.4, the angular frequency is

$$\omega = \sqrt{\frac{k}{m}} \qquad \text{12.9} \blacktriangleleft$$

Pitfall Prevention | 12.2
A Nonconstant Acceleration
The acceleration of a particle in simple harmonic motion is not constant. Equation 12.3 shows that its acceleration varies with position x. Therefore, we *cannot* apply the kinematic equations of Chapter 2 in this situation.

Pitfall Prevention | 12.3
Where's the Triangle?
Equation 12.6 includes a trigonometric function, a *mathematical function* that can be used whether it refers to a triangle or not. In this case, the cosine function happens to have the correct behavior for representing the position of a particle in simple harmonic motion.

▶ Position versus time for a particle in simple harmonic motion

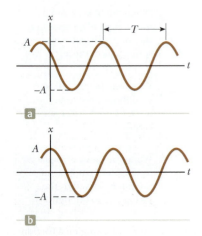

Active Figure 12.2 (a) An x–t graph for a particle undergoing simple harmonic motion. The amplitude of the motion is A, and the period (defined in Eq. 12.10) is T. (b) The x–t graph for the special case in which $x = A$ at $t = 0$ and hence $\phi = 0$.

---

[1]We have seen many examples in earlier chapters in which we evaluate a trigonometric function of an angle. The argument of a trigonometric function, such as sine or cosine, *must* be a pure number. The radian is a pure number because it is a ratio of lengths. Angles in degrees are pure numbers because the degree is an artificial "unit"; it is not related to measurements of lengths. The argument of the trigonometric function in Equation 12.6 must be a pure number. Therefore, ω *must* be expressed in radians per second (and not, for example, in revolutions per second) if t is expressed in seconds. Furthermore, other types of functions such as logarithms and exponential functions require arguments that are pure numbers.

The constant angle ϕ is called the **phase constant** (or initial phase angle) and, along with the amplitude A, is determined uniquely by the position and velocity of the particle at $t = 0$. If the particle is at its maximum position $x = A$ at $t = 0$, the phase constant is $\phi = 0$ and the graphical representation of the motion is as shown in Active Figure 12.2b. The quantity $(\omega t + \phi)$ is called the **phase** of the motion. Notice that the function $x(t)$ is periodic and its value is the same each time ωt increases by 2π radians.

Equations 12.1, 12.5, and 12.6 form the basis of the mathematical representation of the particle in simple harmonic motion model. If you are analyzing a situation and find that the force on an object modeled as a particle is of the mathematical form of Equation 12.1, you know the motion is that of a simple harmonic oscillator and the position of the particle is described by Equation 12.6. If you analyze a system and find that it is described by a differential equation of the form of Equation 12.5, the motion is that of a simple harmonic oscillator. If you analyze a situation and find that the position of a particle is described by Equation 12.6, you know the particle undergoes simple harmonic motion.

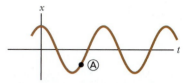

Figure 12.3 (Quick Quiz 12.2) An x–t graph for a particle undergoing simple harmonic motion. At a particular time, the particle's position is indicated by Ⓐ in the graph.

QUICK QUIZ 12.2 Consider a graphical representation (Fig. 12.3) of simple harmonic motion as described mathematically in Equation 12.6. When the particle is at point Ⓐ on the graph, what can you say about its position and velocity? (a) The position and velocity are both positive. (b) The position and velocity are both negative. (c) The position is positive, and the velocity is zero. (d) The position is negative, and the velocity is zero. (e) The position is positive, and the velocity is negative. (f) The position is negative, and the velocity is positive.

QUICK QUIZ 12.3 Figure 12.4 shows two curves representing particles undergoing simple harmonic motion. The correct description of these two motions is that the simple harmonic motion of particle B is (a) of larger angular frequency and larger amplitude than that of particle A, (b) of larger angular frequency and smaller amplitude than that of particle A, (c) of smaller angular frequency and larger amplitude than that of particle A, or (d) of smaller angular frequency and smaller amplitude than that of particle A.

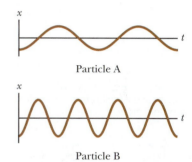

Particle A

Particle B

Figure 12.4 (Quick Quiz 12.3) Two x–t graphs for particles undergoing simple harmonic motion. The amplitudes and frequencies are different for the two particles.

Let us investigate further the mathematical description of simple harmonic motion. The **period** T of the motion is the time interval required for the particle to go through one full cycle of its motion (Active Fig. 12.2a). That is, the values of x and v for the particle at time t equal the values of x and v at time $t + T$. Because the phase increases by 2π radians in a time interval of T,

$$[\omega(t + T) + \phi] - (\omega t + \phi) = 2\pi$$

Simplifying this expression gives $\omega T = 2\pi$, or

$$T = \frac{2\pi}{\omega} \qquad \textbf{12.10}\blacktriangleleft$$

The inverse of the period is called the **frequency** f of the motion. Whereas the period is the time interval per oscillation, the frequency represents the number of oscillations the particle undergoes per unit time interval:

$$f = \frac{1}{T} = \frac{\omega}{2\pi} \qquad \textbf{12.11}\blacktriangleleft$$

The units of f are cycles per second, or **hertz** (Hz). Rearranging Equation 12.11 gives

$$\omega = 2\pi f = \frac{2\pi}{T} \qquad \textbf{12.12}\blacktriangleleft$$

Equations 12.9 through 12.11 can be used to express the period and frequency of the motion for the particle in simple harmonic motion in terms of the characteristics m and k of the system as

$$T = \frac{2\pi}{\omega} = 2\pi\sqrt{\frac{m}{k}} \qquad \textbf{12.13}\blacktriangleleft \quad \blacktriangleright \text{ Period}$$

Pitfall Prevention | 12.4
Two Kinds of Frequency
We identify two kinds of frequency for a simple harmonic oscillator: f, called simply the *frequency*, is measured in hertz, and ω, the *angular frequency*, is measured in radians per second. Be sure you are clear about which frequency is being discussed or requested in a given problem. Equations 12.11 and 12.12 show the relationship between the two frequencies.

▶ Frequency

$$f = \frac{1}{T} = \frac{1}{2\pi}\sqrt{\frac{k}{m}}$$ **12.14**◀

That is, the period and frequency depend *only* on the mass of the particle and the force constant of the spring and *not* on the parameters of the motion, such as A or ϕ. As we might expect, the frequency is larger for a stiffer spring (larger value of k) and decreases with increasing mass of the particle.

We can obtain the velocity and acceleration[2] of a particle undergoing simple harmonic motion from Equations 12.7 and 12.8:

▶ Velocity of a particle in simple harmonic motion

$$v = \frac{dx}{dt} = -\omega A \sin(\omega t + \phi)$$ **12.15**◀

▶ Acceleration of a particle in simple harmonic motion

$$a = \frac{d^2x}{dt^2} = -\omega^2 A \cos(\omega t + \phi)$$ **12.16**◀

From Equation 12.15, we see that because the sine and cosine functions oscillate between ±1, the extreme values of the velocity v are $\pm\omega A$. Likewise, Equation 12.16 shows that the extreme values of the acceleration a are $\pm\omega^2 A$. Therefore, the *maximum* values of the magnitudes of the velocity and acceleration are

▶ Maximum magnitudes of velocity and acceleration in simple harmonic motion

$$v_{\text{max}} = \omega A = \sqrt{\frac{k}{m}}\, A$$ **12.17**◀

$$a_{\text{max}} = \omega^2 A = \frac{k}{m}\, A$$ **12.18**◀

Figure 12.5a plots position versus time for an arbitrary value of the phase constant. The associated velocity–time and acceleration–time curves are illustrated in Figures 12.5b and 12.5c, respectively. They show that the phase of the velocity differs from the phase of the position by $\pi/2$ rad, or 90°. That is, when x is a maximum or a minimum, the velocity is zero. Likewise, when x is zero, the speed is a maximum. Furthermore, notice that the phase of the acceleration differs from the phase of the position by π radians, or 180°. For example, when x is a maximum, a has a maximum magnitude in the opposite direction.

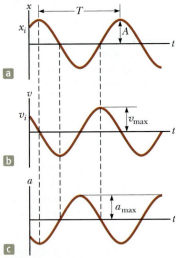

Figure 12.5 Graphical representation of simple harmonic motion. (a) Position versus time. (b) Velocity versus time. (c) Acceleration versus time. Notice that at any specified time the velocity is 90° out of phase with the position and the acceleration is 180° out of phase with the position.

QUICK QUIZ 12.4 An object of mass m is hung from a spring and set into oscillation. The period of the oscillation is measured and recorded as T. The object of mass m is removed and replaced with an object of mass $2m$. When this object is set into oscillation, what is the period of the motion? **(a)** $2T$ **(b)** $\sqrt{2}\,T$ **(c)** T **(d)** $T/\sqrt{2}$ **(e)** $T/2$

Equation 12.6 describes simple harmonic motion of a particle in general. Let's now see how to evaluate the constants of the motion. The angular frequency ω is evaluated using Equation 12.9. The constants A and ϕ are evaluated from the initial conditions, that is, the state of the oscillator at $t = 0$.

Suppose a block is set into motion by pulling it from equilibrium by a distance A and releasing it from rest at $t = 0$ as in Active Figure 12.6. We must then require our solutions for $x(t)$ and $v(t)$ (Eqs. 12.6 and 12.15) to obey the initial conditions that $x(0) = A$ and $v(0) = 0$:

$$x(0) = A \cos\phi = A$$

$$v(0) = -\omega A \sin\phi = 0$$

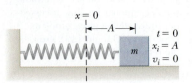

Active Figure 12.6 A block–spring system that begins its motion from rest with the block at $x = A$ at $t = 0$.

These conditions are met if $\phi = 0$, giving $x = A \cos\omega t$ as our solution. To check this solution, notice that it satisfies the condition that $x(0) = A$ because $\cos 0 = 1$.

The position, velocity, and acceleration of the block versus time are plotted in Figure 12.7a for this special case. The acceleration reaches extreme values of $\mp\omega^2 A$ when the position has extreme values of $\pm A$. Furthermore, the velocity has extreme

---

[2]Because the motion of a simple harmonic oscillator takes place in one dimension, we denote velocity as v and acceleration as a, with the direction indicated by a positive or negative sign as in Chapter 2.

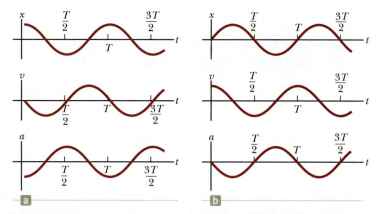

Figure 12.7 (a) Position, velocity, and acceleration versus time for the block in Active Figure 12.6 under the initial conditions that at $t = 0$, $x(0) = A$, and $v(0) = 0$. (b) Position, velocity, and acceleration versus time for the block in Active Figure 12.8 under the initial conditions that at $t = 0$, $x(0) = 0$, and $v(0) = v_i$.

values of $\pm\omega A$, which both occur at $x = 0$. Hence, the quantitative solution agrees with our qualitative description of this system.

Let's consider another possibility. Suppose the system is oscillating and we define $t = 0$ as the instant the block passes through the unstretched position of the spring while moving to the right (Active Fig. 12.8). In this case, our solutions for $x(t)$ and $v(t)$ must obey the initial conditions that $x(0) = 0$ and $v(0) = v_i$:

$$x(0) = A \cos \phi = 0$$

$$v(0) = -\omega A \sin \phi = v_i$$

The first of these conditions tells us that $\phi = \pm\pi/2$. With these choices for ϕ, the second condition tells us that $A = \mp v_i/\omega$. Because the initial velocity is positive and the amplitude must be positive, we must have $\phi = -\pi/2$. Hence, the solution is

$$x = \frac{v_i}{\omega} \cos\left(\omega t - \frac{\pi}{2}\right)$$

The graphs of position, velocity, and acceleration versus time for this choice of $t = 0$ are shown in Figure 12.7b. Notice that these curves are the same as those in Figure 12.7a, but shifted to the right by one-fourth of a cycle. This shift is described mathematically by the phase constant $\phi = -\pi/2$, which is one-fourth of a full cycle of 2π.

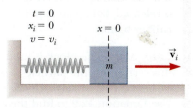

Active Figure 12.8 The block–spring system is undergoing oscillation, and $t = 0$ is defined at an instant when the block passes through the equilibrium position $x = 0$ and is moving to the right with speed v_i.

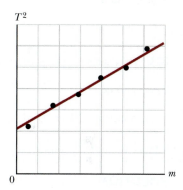

Figure 12.9 (Thinking Physics 12.1) A graph of experimental data: the square of the period versus mass of a block in a block–spring system.

> **THINKING PHYSICS 12.1**

We know that the period of oscillation of an object attached to a spring is proportional to the square root of the mass of the object (Eq. 12.13). Therefore, if we perform an experiment in which we place objects with a range of masses on the end of a spring and measure the period of oscillation of each object–spring system, a graph of the square of the period versus the mass will result in a straight line as suggested in Figure 12.9. We find, however, that the line does not go through the origin. Why not?

Reasoning The line does not go through the origin because the spring itself has mass. Therefore, the resistance to changes in motion of the system is a combination of the mass of the object on the end of the spring

and the mass of the oscillating spring coils. The entire mass of the spring is not oscillating in the same way, however. The coil of the spring attached to the object is oscillating over the same amplitude as the object, but the coil at the fixed end of the spring is not oscillating at all. For a cylindrical spring, energy arguments can be used to show that the effective additional mass representing the oscillations of the spring is one-third of the mass of the spring. The square of the period is proportional to the total oscillating mass, but the graph in Figure 12.9 shows the square of the period versus only the mass of the object on the spring. A graph of period squared versus total mass (mass of the object on the spring plus the effective oscillating mass of the spring) would pass through the origin. ◄

Example 12.1 | A Block–Spring System

A 200-g block connected to a light spring for which the force constant is 5.00 N/m is free to oscillate on a frictionless, horizontal surface. The block is displaced 5.00 cm from equilibrium and released from rest as in Active Figure 12.6.

(A) Find the period of its motion.

SOLUTION

Conceptualize Study Active Figure 12.6 and imagine the block moving back and forth in simple harmonic motion once it is released. Set up an experimental model in the vertical direction by hanging a heavy object such as a stapler from a strong rubber band.

Categorize The block is modeled as a particle in simple harmonic motion. We find values from equations developed in this section for the particle in simple harmonic motion model, so we categorize this example as a substitution problem.

Use Equation 12.9 to find the angular frequency of the block–spring system:

$$\omega = \sqrt{\frac{k}{m}} = \sqrt{\frac{5.00 \text{ N/m}}{200 \times 10^{-3} \text{ kg}}} = 5.00 \text{ rad/s}$$

Use Equation 12.13 to find the period of the system:

$$T = \frac{2\pi}{\omega} = \frac{2\pi}{5.00 \text{ rad/s}} = \boxed{1.26 \text{ s}}$$

(B) Determine the maximum speed of the block.

SOLUTION

Use Equation 12.17 to find v_{max}:

$$v_{max} = \omega A = (5.00 \text{ rad/s})(5.00 \times 10^{-2} \text{ m}) = \boxed{0.250 \text{ m/s}}$$

(C) What is the maximum acceleration of the block?

SOLUTION

Use Equation 12.18 to find a_{max}:

$$a_{max} = \omega^2 A = (5.00 \text{ rad/s})^2 (5.00 \times 10^{-2} \text{ m}) = \boxed{1.25 \text{ m/s}^2}$$

(D) Express the position, velocity, and acceleration as functions of time in SI units.

SOLUTION

Find the phase constant from the initial condition that $x = A$ at $t = 0$:

$$x(0) = A \cos \phi = A \quad \rightarrow \quad \phi = 0$$

Use Equation 12.6 to write an expression for $x(t)$:

$$x = A \cos (\omega t + \phi) = \boxed{0.050 \, 0 \cos 5.00t}$$

Use Equation 12.15 to write an expression for $v(t)$:

$$v = -\omega A \sin (\omega t + \phi) = \boxed{-0.250 \sin 5.00t}$$

Use Equation 12.16 to write an expression for $a(t)$:

$$a = -\omega^2 A \cos (\omega t + \phi) = \boxed{-1.25 \cos 5.00t}$$

Example 12.2 | Watch Out for Potholes!

A car with a mass of 1 300 kg is constructed so that its frame is supported by four springs. Each spring has a force constant of 20 000 N/m. Two people riding in the car have a combined mass of 160 kg. Find the frequency of vibration of the car after it is driven over a pothole in the road and the car oscillates vertically.

SOLUTION

Conceptualize Think about your experiences with automobiles. When you sit in a car, it moves downward a small distance because your weight is compressing the springs further. If you push down on the front bumper and release it, the front of the car oscillates a few times.

Categorize We imagine the car as being supported by a single spring and model the car as a particle in simple harmonic motion.

12.2 *cont.*

Analyze First, let's determine the effective spring constant of the four springs combined. For a given extension x of the springs, the combined force on the car is the sum of the forces from the individual springs.

Find an expression for the total force on the car:

$$F_{\text{total}} = \sum(-kx) = -\left(\sum k\right)x$$

In this expression, x has been factored from the sum because it is the same for all four springs. The effective spring constant for the combined springs is the sum of the individual spring constants.

Evaluate the effective spring constant:

$$k_{\text{eff}} = \sum k = 4 \times 20\,000 \text{ N/m} = 80\,000 \text{ N/m}$$

Use Equation 12.14 to find the frequency of vibration:

$$f = \frac{1}{2\pi}\sqrt{\frac{k_{\text{eff}}}{m}} = \frac{1}{2\pi}\sqrt{\frac{80\,000 \text{ N/m}}{1\,460 \text{ kg}}} = \boxed{1.18 \text{ Hz}}$$

Finalize The mass we used here is that of the car plus the people because that is the total mass that is oscillating. Also notice that we have explored only up-and-down motion of the car. If an oscillation is established in which the car rocks back and forth such that the front end goes up when the back end goes down, the frequency will be different.

What If? Suppose the car stops on the side of the road and the two people exit the car. One of them pushes downward on the car and releases it so that it oscillates vertically. Is the frequency of the oscillation the same as the value we just calculated?

Answer The suspension system of the car is the same, but the mass that is oscillating is smaller: it no longer includes the mass of the two people. Therefore, the frequency should be higher. Let's calculate the new frequency, taking the mass to be 1 300 kg:

$$f = \frac{1}{2\pi}\sqrt{\frac{k_{\text{eff}}}{m}} = \frac{1}{2\pi}\sqrt{\frac{80\,000 \text{ N/m}}{1\,300 \text{ kg}}} = 1.25 \text{ Hz}$$

As predicted, the new frequency is a bit higher.

12.3 | Energy of the Simple Harmonic Oscillator

Let us examine the mechanical energy of a system in which a particle undergoes simple harmonic motion, such as the block–spring system illustrated in Active Figure 12.1. Because the surface is frictionless, the system is isolated and we expect the total mechanical energy of the system to be constant. We assume a massless spring, so the kinetic energy of the system corresponds only to that of the block. We can use Equation 12.15 to express the kinetic energy of the block as

$$K = \tfrac{1}{2}mv^2 = \tfrac{1}{2}m\omega^2 A^2 \sin^2(\omega t + \phi) \qquad \textbf{12.19} \blacktriangleleft$$

▶ Kinetic energy of a simple harmonic oscillator

The elastic potential energy stored in the spring for any elongation x is given by $\tfrac{1}{2}kx^2$ (see Eq. 6.22). Using Equation 12.6 gives

$$U = \tfrac{1}{2}kx^2 = \tfrac{1}{2}kA^2 \cos^2(\omega t + \phi) \qquad \textbf{12.20} \blacktriangleleft$$

▶ Potential energy of a simple harmonic oscillator

We see that K and U are *always* positive quantities or zero. Because $\omega^2 = k/m$, we can express the total mechanical energy of the simple harmonic oscillator as

$$E = K + U = \tfrac{1}{2}kA^2[\sin^2(\omega t + \phi) + \cos^2(\omega t + \phi)]$$

From the identity $\sin^2\theta + \cos^2\theta = 1$, we see that the quantity in square brackets is unity. Therefore, this equation reduces to

$$\boxed{E = \tfrac{1}{2}kA^2} \qquad \textbf{12.21} \blacktriangleleft$$

▶ Total energy of a simple harmonic oscillator

That is, the total mechanical energy of a simple harmonic oscillator is a constant of the motion and is proportional to the square of the amplitude. The total mechanical energy is equal to the maximum potential energy stored in the spring when $x = \pm A$

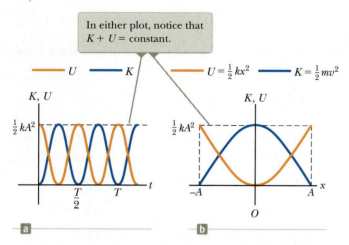

In either plot, notice that $K + U = $ constant.

Active Figure 12.10 (a) Kinetic energy and potential energy versus time for a simple harmonic oscillator with $\phi = 0$. (b) Kinetic energy and potential energy versus position for a simple harmonic oscillator.

▶ Velocity as a function of position for a simple harmonic oscillator

because $v = 0$ at these points and there is no kinetic energy. At the equilibrium position, where $U = 0$ because $x = 0$, the total energy, all in the form of kinetic energy, is again $\frac{1}{2}kA^2$.

Plots of the kinetic and potential energies versus time appear in Active Figure 12.10a, where we have taken $\phi = 0$. At all times, the sum of the kinetic and potential energies is a constant equal to $\frac{1}{2}kA^2$, the total energy of the system.

The variations of K and U with the position x of the block are plotted in Active Figure 12.10b. Energy is continuously being transformed between potential energy stored in the spring and kinetic energy of the block.

Active Figure 12.11 illustrates the position, velocity, acceleration, kinetic energy, and potential energy of the block–spring system for one full period of the motion. Most of the ideas discussed so far are incorporated in this important figure. Study it carefully.

Finally, we can obtain the velocity of the block at an arbitrary position by expressing the total energy of the system at some arbitrary position x as

$$E = K + U = \tfrac{1}{2}mv^2 + \tfrac{1}{2}kx^2 = \tfrac{1}{2}kA^2$$

and then solving for v:

$$v = \pm\sqrt{\frac{k}{m}(A^2 - x^2)} = \pm\omega\sqrt{A^2 - x^2} \qquad \textbf{12.22} ◀$$

When you check Equation 12.22 to see whether it agrees with known cases, you find that it verifies that the speed is a maximum at $x = 0$ and is zero at the turning points $x = \pm A$.

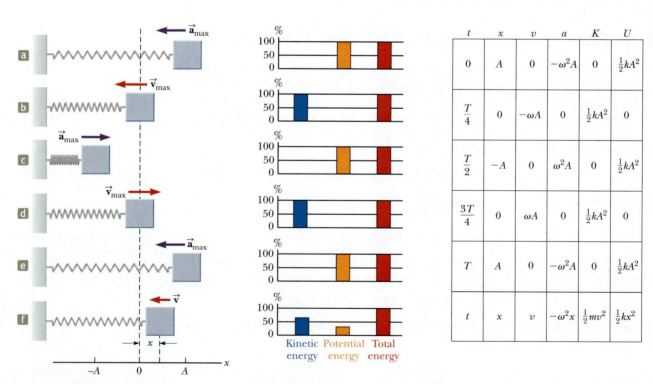

Active Figure 12.11 (a) through (e) Several instants in the simple harmonic motion for a block–spring system. Energy bar graphs show the distribution of the energy of the system at each instant. The parameters in the table at the right refer to the block–spring system, assuming at $t = 0$, $x = A$; hence, $x = A\cos\omega t$. For these five special instants, one of the types of energy is zero. (f) An arbitrary point in the motion of the oscillator. The system possesses both kinetic energy and potential energy at this instant as shown in the bar graph.

> ## THINKING PHYSICS 12.2

An object oscillating on the end of a horizontal spring slides back and forth over a frictionless surface. During one oscillation, you set an identical object at the maximum displacement point, with instant-acting glue on its surface. Just as the oscillating object reaches its largest displacement and is momentarily at rest, it adheres to the new object by means of the glue and the two objects continue the oscillation together. Does the period of the oscillation change? Does the amplitude of oscillation change? Does the energy of the oscillation change?

Reasoning The period of oscillation changes because the period depends on the mass that is oscillating (Eq. 12.13). The amplitude does not change. Because the new object was added under the special condition that the original object was at rest, the combined objects are at rest at this point also, defining the amplitude as the same as in the original oscillation. The energy does not change either. At the maximum displacement point, the energy is all potential energy stored in the spring, which depends only on the force constant and the amplitude, not on the mass of the object. The object of increased mass will pass through the equilibrium point with lower speed than in the original oscillation but with the same kinetic energy. Another approach is to think about how energy could be transferred into the oscillating system. No work was done on the system (nor did any other form of energy transfer occur), so the energy in the system cannot change. ◄

You may wonder why we are spending so much time studying simple harmonic oscillators. We do so because they are good models of a wide variety of physical phenomena. For example, recall the Lennard–Jones potential discussed in Example 6.9. This complicated function describes the forces holding atoms together. Figure 12.12a shows that for small displacements from the equilibrium position, the potential energy curve for this function approximates a parabola, which represents the potential energy function for a simple harmonic oscillator. Therefore, we can model the complex atomic binding forces as being due to tiny springs as depicted in Figure 12.12b.

The ideas presented in this chapter apply not only to block–spring systems and atoms, but also to a wide range of situations that include bungee jumping, playing a musical instrument, and viewing the light emitted by a laser. You will see more examples of simple harmonic oscillators as you work through this book.

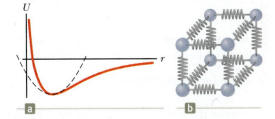

Figure 12.12 (a) If the atoms in a molecule do not move too far from their equilibrium positions, a graph of potential energy versus separation distance between atoms is similar to the graph of potential energy versus position for a simple harmonic oscillator (dashed black curve). (b) The forces between atoms in a solid can be modeled by imagining springs between neighboring atoms.

Example 12.3 | Oscillations on a Horizontal Surface

A 0.500-kg cart connected to a light spring for which the force constant is 20.0 N/m oscillates on a frictionless, horizontal air track.

(A) Calculate the maximum speed of the cart if the amplitude of the motion is 3.00 cm.

SOLUTION

Conceptualize The system oscillates in exactly the same way as the block in Active Figure 12.11, so use that figure in your mental image of the motion.

Categorize The cart is modeled as a particle in simple harmonic motion.

Analyze Use Equation 12.21 to express the total energy of the oscillator system and equate it to the kinetic energy of the system when the cart is at $x = 0$:

$$E = \tfrac{1}{2}kA^2 = \tfrac{1}{2}mv_{max}^2$$

Solve for the maximum speed and substitute numerical values:

$$v_{max} = \sqrt{\frac{k}{m}}\,A = \sqrt{\frac{20.0\ \text{N/m}}{0.500\ \text{kg}}}(0.030\ 0\ \text{m}) = \boxed{0.190\ \text{m/s}}$$

(B) What is the velocity of the cart when the position is 2.00 cm?

SOLUTION

Use Equation 12.22 to evaluate the velocity:

$$v = \pm\sqrt{\frac{k}{m}(A^2 - x^2)}$$

$$= \pm\sqrt{\frac{20.0\ \text{N/m}}{0.500\ \text{kg}}[(0.030\ 0\ \text{m})^2 - (0.020\ 0\ \text{m})^2]} = \boxed{\pm 0.141\ \text{m/s}}$$

The positive and negative signs indicate that the cart could be moving to either the right or the left at this instant.

continued

12.3 cont.

(C) Compute the kinetic and potential energies of the system when the position of the cart is 2.00 cm.

SOLUTION

Use the result of part (B) to evaluate the kinetic energy at
$x = 0.020\ 0$ m:

$$K = \tfrac{1}{2}mv^2 = \tfrac{1}{2}(0.500\ \text{kg})(0.141\ \text{m/s})^2 = \boxed{5.00 \times 10^{-3}\,\text{J}}$$

Evaluate the elastic potential energy at $x = 0.020\ 0$ m:

$$U = \tfrac{1}{2}kx^2 = \tfrac{1}{2}(20.0\ \text{N/m})(0.0200\ \text{m})^2 = \boxed{4.00 \times 10^{-3}\,\text{J}}$$

Finalize The sum of the kinetic and potential energies in part (C) is equal to the total energy, which can be found from Equation 12.21. That must be true for *any* position of the cart.

What If? The cart in this example could have been set into motion by releasing the cart from rest at $x = 3.00$ cm. What if the cart were released from the same position, but with an initial velocity of $v = -0.100$ m/s? What are the new amplitude and maximum speed of the cart?

Answer We can respond to this question by applying an energy approach

First calculate the total energy of the system at $t = 0$:

$$E = \tfrac{1}{2}mv^2 + \tfrac{1}{2}kx^2$$

$$= \tfrac{1}{2}(0.500\ \text{kg})(-0.100\ \text{m/s})^2 + \tfrac{1}{2}(20.0\ \text{N/m})(0.030\ 0\ \text{m})^2$$

$$= 1.15 \times 10^{-2}\,\text{J}$$

Equate this total energy to the potential energy of the system when the cart is at the endpoint of the motion:

$$E = \tfrac{1}{2}kA^2$$

Solve for the amplitude A:

$$A = \sqrt{\frac{2E}{k}} = \sqrt{\frac{2(1.15 \times 10^{-2}\,\text{J})}{20.0\ \text{N/m}}} = 0.033\ 9\ \text{m}$$

Equate the total energy to the kinetic energy of the system when the cart is at the equilibrium position:

$$E = \tfrac{1}{2}mv_{\text{max}}^2$$

Solve for the maximum speed:

$$v_{\text{max}} = \sqrt{\frac{2E}{m}} = \sqrt{\frac{2(1.15 \times 10^{-2}\,\text{J})}{0.500\ \text{kg}}} = 0.214\ \text{m/s}$$

The amplitude and maximum velocity are larger than the previous values because the cart was given an initial velocity at $t = 0$.

◄12.4 | The Simple Pendulum

When θ is small, a simple pendulum's motion can be modeled as simple harmonic motion about the equilibrium position $\theta = 0$.

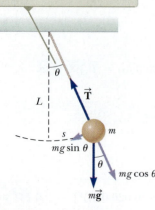

Active Figure 12.13 A simple pendulum.

The **simple pendulum** is another mechanical system that exhibits periodic motion. It consists of a particle-like bob of mass m suspended by a light string of length L that is fixed at the upper end as in Active Figure 12.13. For a real object, as long as the size of the object is small relative to the length of the string, the pendulum can be modeled as a simple pendulum, so we adopt the particle model. When the bob is pulled to the side and released, it oscillates about the lowest point, which is the equilibrium position. The motion occurs in a vertical plane and is driven by the gravitational force.

The forces acting on the bob are the force $\vec{\mathbf{T}}$ acting along the string and the gravitational force $m\vec{\mathbf{g}}$. The component vector of the gravitational force tangent to the curved path of the bob and of magnitude $mg \sin\theta$ always acts toward $\theta = 0$, opposite the displacement of the bob from the lowest position. The gravitational force is therefore a restoring force, and we can use Newton's second law to write the equation of motion in the tangential direction as

$$F_t = ma_t \quad \rightarrow \quad -mg \sin\theta = m\frac{d^2s}{dt^2}$$

where s is the position measured along the circular arc in Active Figure 12.13 and the negative sign indicates that F_t acts toward the equilibrium position. Because $s = L\theta$ (Eq. 10.1a) and L is constant, this equation reduces to

$$\frac{d^2\theta}{dt^2} = -\frac{g}{L}\sin\theta$$

Considering θ as the position, let us compare this equation to Equation 12.5, which is of a similar, but not identical, mathematical form. The right side is proportional to $\sin\theta$ rather than to θ; hence, we conclude that the motion is *not* simple harmonic motion because the equation describing the motion is not of the form of Equation 12.5. If we assume that θ is *small* (less than about 10° or 0.2 rad), however, we can use a simplification model called the **small angle approximation,** in which $\sin\theta \approx \theta$, where θ is measured in radians. Table 12.1 shows angles, in degrees and radians, and the sines of these angles. As long as θ is less than about 10°, the angle in radians and its sine are the same, at least to within an accuracy of less than 1.0%.

Therefore, for small angles, the equation of motion becomes

$$\frac{d^2\theta}{dt^2} = -\frac{g}{L}\theta \qquad \textbf{12.23} \blacktriangleleft$$

Now we have an expression with exactly the same mathematical form as Equation 12.5, with $\omega^2 = g/L$, and so we conclude that the motion is approximately simple harmonic motion for small amplitudes. Modeling the solution after Equation 12.6, θ can therefore be written as $\theta = \theta_{max}\cos(\omega t + \phi)$, where θ_{max} is the *maximum angular position* and the angular frequency ω is

$$\omega = \sqrt{\frac{g}{L}} \qquad \textbf{12.24} \blacktriangleleft$$

The period of the motion is

$$T = \frac{2\pi}{\omega} = 2\pi\sqrt{\frac{L}{g}} \qquad \textbf{12.25} \blacktriangleleft$$

▶ Angular frequency for a simple pendulum

▶ Period for a simple pendulum

TABLE 12.1 | Angles and Sines of Angles

Angle in Degrees	Angle in Radians	Sine of Angle	Percent Difference
0°	0.000 0	0.000 0	0.0%
1°	0.017 5	0.017 5	0.0%
2°	0.034 9	0.034 9	0.0%
3°	0.052 4	0.052 3	0.0%
5°	0.087 3	0.087 2	0.1%
10°	0.174 5	0.173 6	0.5%
15°	0.261 8	0.258 8	1.2%
20°	0.349 1	0.342 0	2.1%
30°	0.523 6	0.500 0	4.7%

We see that the period and frequency of a simple pendulum oscillating at small angles depend only on the length of the string and the acceleration due to gravity. Because the period is *independent* of the mass, we conclude that *all* simple pendula that are of equal length and are at the same location (so that g is constant) oscillate with the same period. Experiments show that this conclusion is correct.

Notice the importance of our modeling technique in this discussion. Equation 12.23 is a mathematical representation of the simple pendulum. This representation has exactly the same *mathematical* form as Equation 12.5 for the block on a spring, despite the fact that there are clear *physical* differences between the two systems. Despite the physical differences, because the mathematical representations are the same, we can immediately write the solution of the angular position θ for the pendulum and identify its angular frequency ω as in Equation 12.24. This is a very powerful technique, made possible by the fact that we are forming a mathematical model of the physical system.

Pitfall Prevention | 12.4
Not True Simple Harmonic Motion
The pendulum *does not* exhibit true simple harmonic motion for *any* angle. If the angle is less than about 10°, the motion is close to and can be *modeled* as simple harmonic.

QUICK QUIZ 12.5 A grandfather clock depends on the period of a pendulum to keep correct time. (**i**) Suppose a grandfather clock is calibrated correctly and then a mischievous child slides the bob of the pendulum downward on the oscillating rod. Does the grandfather clock run (**a**) slow, (**b**) fast, or (**c**) correctly? (**ii**) Suppose the grandfather clock is calibrated correctly at sea level and is then taken to the top of a very tall mountain. Does the grandfather clock run (**a**) slow, (**b**) fast, or (**c**) correctly?

> **THINKING PHYSICS 12.3**
>
> You set up two oscillating systems: a simple pendulum and a block hanging from a vertical spring. You carefully adjust the length of the pendulum so that both oscillators have the same period. You now take the two oscillators to the Moon. Will they still have the same period as each other? What happens if you observe the two oscillators in an orbiting spacecraft? (Assume that the spring is one with open space between the coils when it is unstretched, so the spring can be both stretched and compressed.)
>
> **Reasoning** The block hanging from the spring will have the same period on the Moon that it had on the Earth because the period depends on the mass of the block and the force constant of the spring, neither of which have changed. The pendulum's period on the Moon will be different from its period on the Earth because the period of the pendulum depends on the value of g. Because g is smaller on the Moon than on the Earth, the pendulum will oscillate with a longer period.
>
> In the orbiting spacecraft, the block–spring system will oscillate with the same period as on the Earth when it is set into motion because the period does not depend on gravity. The pendulum will not oscillate at all; if you pull it to the side from a direction you define as "vertical" and release it, it stays there. Because the spacecraft is in free-fall while in orbit around the Earth, the effective gravity is zero and there is no restoring force on the pendulum. ◀

Example **12.4** | **A Connection Between Length and Time**

Christian Huygens (1629–1695), the greatest clockmaker in history, suggested that an international unit of length could be defined as the length of a simple pendulum having a period of exactly 1 s. How much shorter would our length unit be if his suggestion had been followed?

SOLUTION

Conceptualize Imagine a pendulum that swings back and forth in exactly 1 second. Based on your experience in observing swinging objects, can you make an estimate of the required length? Hang a small object from a string and simulate the 1-s pendulum.

Categorize This example involves a simple pendulum, so we categorize it as an application of the concepts introduced in this section.

Analyze Solve Equation 12.25 for the length and substitute the known values:

$$L = \frac{T^2 g}{4\pi^2} = \frac{(1.00 \text{ s})^2 (9.80 \text{ m/s}^2)}{4\pi^2} = \boxed{0.248 \text{ m}}$$

Finalize The meter's length would be slightly less than one-fourth of its current length. Also, the number of significant digits depends only on how precisely we know g because the time has been defined to be exactly 1 s.

12.5 | The Physical Pendulum

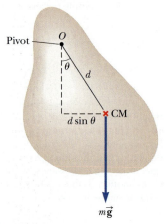

Pivot — O

$d \sin \theta$ ✖ CM

$m\vec{g}$

Figure 12.14 A physical pendulum pivoted at O.

Suppose you balance a wire coat hanger so that the hook is supported by your extended index finger. When you give the hanger a small angular displacement with your other hand and then release it, it oscillates. If a hanging object oscillates about a fixed axis that does not pass through its center of mass and the object cannot be approximated as a point mass, we cannot treat the system as a simple pendulum. In this case, the system is called a **physical pendulum.**

Consider a rigid object pivoted at a point O that is a distance d from the center of mass (Fig. 12.14). The gravitational force provides a torque about an axis through O, and the magnitude of that torque is $mgd \sin \theta$, where θ is as shown in Figure 12.14. We model the object as a rigid object under a net torque and use the rotational form of Newton's second law, $\Sigma \tau_{\text{ext}} = I\alpha$, where I is the moment of inertia of the object about the axis through O. The result is

$$-mgd \sin \theta = I \frac{d^2\theta}{dt^2}$$

The negative sign indicates that the torque about O tends to decrease θ. That is, the gravitational force produces a restoring torque. If we again assume θ is small, the approximation $\sin \theta \approx \theta$ is valid and the equation of motion reduces to

$$\frac{d^2\theta}{dt^2} = -\left(\frac{mgd}{I}\right)\theta = -\omega^2\theta \qquad \textbf{12.26} \blacktriangleleft$$

Because this equation is of the same mathematical form as Equation 12.5, its solution is that of the simple harmonic oscillator. That is, the solution of Equation 12.26 is given by $\theta = \theta_{\text{max}} \cos(\omega t + \phi)$, where θ_{max} is the maximum angular position and

$$\omega = \sqrt{\frac{mgd}{I}}$$

The period is

▶ Period of a physical pendulum

$$T = \frac{2\pi}{\omega} = 2\pi\sqrt{\frac{I}{mgd}} \qquad \textbf{12.27} \blacktriangleleft$$

This result can be used to measure the moment of inertia of a flat, rigid object. If the location of the center of mass—and hence the value of d—is known, the moment of inertia can be obtained by measuring the period. Finally, notice that Equation 12.27 reduces to the period of a simple pendulum (Eq. 12.25) when $I = md^2$, that is, when all the mass is concentrated at the center of mass.

Notice again the importance of modeling here, as discussed for the simple pendulum. Because the mathematical representation in Equation 12.26 is identical in form to that in Equation 12.5, we were able to immediately write the solution for the physical pendulum.

QUICK QUIZ 12.6 Two students, Alex and Brian, are in a museum watching the swinging of a pendulum with a large bob. Alex says, "I'm going to sneak past the fence and stick some chewing gum on the top of the pendulum bob to change its period of oscillation." Brian says, "That won't change the period. The period of a pendulum is independent of mass." Which student is correct? **(a)** Alex **(b)** Brian

Example 12.5 | A Swinging Rod

A uniform rod of mass M and length L is pivoted about one end and oscillates in a vertical plane (Fig. 12.15). Find the period of oscillation if the amplitude of the motion is small.

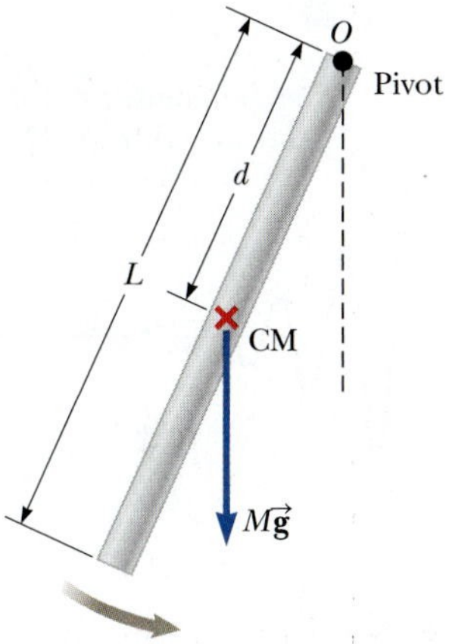

SOLUTION

Conceptualize Imagine a rod swinging back and forth when pivoted at one end. Try it with a meterstick or a scrap piece of wood.

Categorize Because the rod is not a point particle, we categorize it as a physical pendulum.

Figure 12.15 (Example 12.5) A rigid rod oscillating about a pivot through one end is a physical pendulum with $d = L/2$.

Analyze In Chapter 10, we found that the moment of inertia of a uniform rod about an axis through one end is $\frac{1}{3}ML^2$. The distance d from the pivot to the center of mass of the rod is $L/2$.

Substitute these quantities into Equation 12.27:
$$T = 2\pi\sqrt{\frac{\frac{1}{3}ML^2}{Mg(L/2)}} = 2\pi\sqrt{\frac{2L}{3g}}$$

Finalize In one of the Moon landings, an astronaut walking on the Moon's surface had a belt hanging from his space suit, and the belt oscillated as a physical pendulum. A scientist on the Earth observed this motion on television and used it to estimate the free-fall acceleration on the Moon. How did the scientist make this calculation?

12.6 | Damped Oscillations

The oscillatory motions we have considered so far have been for ideal systems, that is, systems that oscillate indefinitely under the action of only one force, a linear restoring force. In many real systems, nonconservative forces such as friction or air resistance retard the motion. Consequently, the mechanical energy of the system diminishes in time, and the motion is described as a **damped oscillation.**

Consider an object moving through a medium such as a liquid or a gas. One common type of retarding force on the object, which we discussed in Chapter 5, is proportional to the velocity of the object and acts in the direction opposite that of the object's velocity relative to the medium. This type of force is often observed when an object is oscillating slowly in air, for instance. Because the resistive force can be expressed as $\vec{R} = -b\vec{v}$, where b is a constant related to the strength of the resistive force, and the restoring force exerted on the system is $-kx$, Newton's second law gives us

$$\sum F_x = -kx - bv = ma_x$$
$$-kx - b\frac{dx}{dt} = m\frac{d^2x}{dt^2} \qquad \textbf{12.28}\blacktriangleleft$$

The solution of this differential equation requires mathematics that may not yet be familiar to you, so it will simply be stated without proof. When the parameters of the system are such that $b < \sqrt{4mk}$ so that the resistive force is small, the solution to Equation 12.28 is

$$x = (Ae^{-(b/2m)t})\cos(\omega t + \phi) \qquad \textbf{12.29}\blacktriangleleft$$

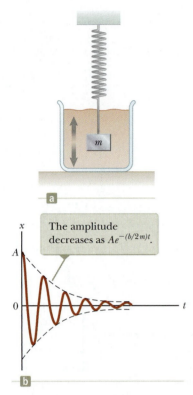

The amplitude decreases as $Ae^{-(b/2m)t}$.

a

b

Active Figure 12.16 (a) One example of a damped oscillator is an object attached to a spring and submersed in a viscous liquid. (b) Graph of position versus time for a damped oscillator.

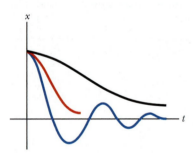

Figure 12.17 Graphs of position versus time for an underdamped oscillator (blue curve), a critically damped oscillator (red curve), and an overdamped oscillator (black curve).

where the angular frequency of the oscillation is

$$\omega = \sqrt{\frac{k}{m} - \left(\frac{b}{2m}\right)^2}$$ **12.30**◄

This result can be verified by substituting Equation 12.29 into Equation 12.28. It is convenient to express the angular frequency of a damped oscillator in the form

$$\omega = \sqrt{\omega_0^2 - \left(\frac{b}{2m}\right)^2}$$

where $\omega_0 = \sqrt{k/m}$ represents the angular frequency in the absence of a retarding force (the undamped oscillator) and is called the **natural frequency**[3] of the system.

In Active Figure 12.16a, we see one example of a damped system. The object suspended from the spring experiences both a force from the spring and a resistive force from the surrounding liquid. Active Figure 12.16b shows the position as a function of time for an object oscillating in the presence of a retarding force. When the retarding force is small, the oscillatory character of the motion is preserved but the amplitude decreases exponentially in time, with the result that the motion ultimately becomes undetectable. Any system that behaves in this way is known as a **damped oscillator.** The dashed black lines in Active Figure 12.16b, which define the *envelope* of the oscillatory curve, represent the exponential factor in Equation 12.29. This envelope shows that the amplitude decays exponentially with time. For motion with a given spring constant and object mass, the oscillations dampen more rapidly for larger values of the retarding force.

When the magnitude of the retarding force is small such that $b/2m < \omega_0$, the system is said to be **underdamped.** The resulting motion is represented by the blue curve in Figure 12.17. As the value of b increases, the amplitude of the oscillations decreases more and more rapidly. When b reaches a critical value b_c such that $b_c/2m = \omega_0$, the system does not oscillate and is said to be **critically damped.** In this case, the system, once released from rest at some nonequilibrium position, approaches but does not pass through the equilibrium position. The graph of position versus time for this case is the red curve in Figure 12.17.

If the medium is so viscous that the retarding force is large compared with the restoring force—that is, if $b/2m > \omega_0$—the system is **overdamped.** Again, the displaced system, when free to move, does not oscillate but rather simply returns to its equilibrium position. As the damping increases, the time interval required for the system to approach equilibrium also increases as indicated by the black curve in Figure 12.17. For critically damped and overdamped systems, there is no angular frequency ω and the solution in Equation 12.29 is not valid.

12.7 | Forced Oscillations

We have seen that the mechanical energy of a damped oscillator decreases in time as a result of the resistive force. It is possible to compensate for this energy decrease by applying an external force that does positive work on the system. Such an oscillator then undergoes **forced oscillations.** At any instant, energy can be transferred into the system by an applied force that acts in the direction of motion of the oscillator. For example, a child on a swing can be kept in motion by appropriately timed "pushes." The amplitude of motion remains constant if the energy input per cycle of motion exactly equals the decrease in mechanical energy in each cycle that results from resistive forces.

A common example of a forced oscillator is a damped oscillator driven by an external force $F(t)$ that varies periodically, such as $F(t) = F_0 \sin \omega t$, where ω is the angular frequency of the driving force and F_0 is a constant. In general, the frequency ω of the driving force is different from the natural frequency ω_0 of the oscillator. Newton's second law in this situation gives

$$\sum F_x = ma_x \quad \rightarrow \quad F_0 \sin \omega t - b\frac{dx}{dt} - kx = m\frac{d^2x}{dt^2}$$ **12.31**◄

---

[3]In practice, both ω_0 and $f_0 = \omega_0/2\pi$ are described as the natural frequency. The context of the discussion will help you determine which frequency is being discussed.

Again, the solution of this equation is rather lengthy and will not be presented. After the driving force on an initially stationary object begins to act, the amplitude of the oscillation will increase. After a sufficiently long time interval, when the energy input per cycle from the driving force equals the amount of mechanical energy transformed to internal energy for each cycle, a steady-state condition is reached in which the oscillations proceed with constant amplitude. In this case, Equation 12.31 has the solution

$$x = A\cos(\omega t + \phi) \qquad \textbf{12.32}$$

where

$$A = \frac{F_0/m}{\sqrt{\left(\omega^2 - \omega_0^2\right)^2 + \left(\dfrac{b\omega}{m}\right)^2}} \qquad \textbf{12.33}$$

and where $\omega_0 = \sqrt{k/m}$ is the natural frequency of the undamped oscillator ($b = 0$).

Equation 12.33 shows that the amplitude of the forced oscillator is constant for a given driving force because it is being driven in steady state by an external force. For small damping the amplitude becomes large when the frequency of the driving force is near the natural frequency of oscillation, or when $\omega \approx \omega_0$ as can be seen in Equation 12.33. The dramatic increase in amplitude near the natural frequency is called **resonance,** and the natural frequency ω_0 is called the **resonance frequency** of the system.

Figure 12.18 is a graph of amplitude as a function of frequency for the forced oscillator, with varying resistive forces. Note that the amplitude increases with decreasing damping ($b \to 0$) and that the resonance curve flattens as the damping increases. In the absence of a damping force ($b = 0$), we see from Equation 12.33 that the steady-state amplitude approaches infinity as $\omega \to \omega_0$. In other words, if there are no resistive forces in the system and we continue to drive an oscillator with a sinusoidal force at the resonance frequency, the amplitude of motion will build up without limit. This situation does not occur in practice because some damping is always present in real oscillators.

Resonance appears in many areas of physics. For example, certain electric circuits have resonance frequencies. This fact is exploited in radio tuners, which allow you to select the station you wish to hear. Vibrating strings and columns of air also have resonance frequencies, which allow them to be used for musical instruments, which we shall discuss in Chapter 14.

12.8 | Context Connection: Resonance in Structures

In the preceding section, we investigated the phenomenon of resonance in which an oscillating system exhibits its maximum response to a periodic driving force when the frequency of the driving force matches the oscillator's natural frequency. We now apply this understanding to the interaction between the shaking of the ground during an earthquake and structures attached to the ground. The structure is the oscillator. It has a set of natural frequencies, determined by its stiffness, its mass, and the details of its construction. The periodic driving force is supplied by the shaking of the ground.

A disastrous result can occur if a natural frequency of the building matches a frequency contained in the ground shaking. In this case, the resonance vibrations of the building can build to a very large amplitude, large enough to damage or destroy the building. This result can be avoided in two ways. The first involves designing the structure so that natural frequencies of the building lie outside the range of earthquake frequencies. (A typical range of earthquake frequencies is 0–15 Hz.) Such a building can be designed by varying its size or mass structure. The second method involves incorporating sufficient damping in the building. This method may not change the resonance frequency significantly, but it will lower the response to the natural frequency as in Figure 12.18. It will also flatten the resonance curve, so the building will respond to a wide range of frequencies but with relatively small amplitude at any given frequency.

We now describe two examples involving resonance excitations in bridge structures. One example of such a structural resonance occurred in 1940, when the Tacoma Narrows Bridge in Washington State was destroyed by resonant vibrations (Fig. 12.19). The winds were not particularly strong on that occasion, but the bridge

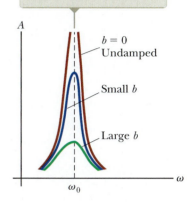

When the frequency ω of the driving force equals the natural frequency ω_0 of the oscillator, resonance occurs.

Figure 12.18 Graph of amplitude versus frequency for a damped oscillator when a periodic driving force is present. Notice that the shape of the resonance curve depends on the size of the damping coefficient b.

Figure 12.19 (a) In 1940, turbulent winds set up torsional vibrations in the Tacoma Narrows Bridge, causing it to oscillate at a frequency near one of the natural frequencies of the bridge structure. (b) Once established, this resonance condition led to the bridge's collapse. (Mathematicians and physicists are currently challenging this interpretation.)

still collapsed because vortices (turbulences) generated by the wind blowing through the bridge occurred at a frequency that matched a natural frequency of the bridge. The flapping of this wind across the roadway (similar to the flapping of a flag in a strong breeze) provided the periodic driving force that brought the bridge down into the river.

As a second example, consider soldiers marching across a bridge. They are commanded to break step while on the bridge because of resonance. If the marching frequency of the soldiers matches that of the bridge, the bridge could be set into resonance oscillation. If the amplitude becomes large enough, the bridge could actually collapse. Just such a situation occurred on April 14, 1831, when the Broughton suspension bridge in England collapsed while troops marched over it. Investigations after the accident showed that the bridge was near failure, and the resonance vibration induced by the marching soldiers caused it to fail sooner than it otherwise might have.

Resonance gives us our first clue to responding to the central question for this Context. Suppose a building is far from the epicenter of an earthquake so that the ground shaking is small. If the shaking frequency matches a natural frequency of the building, a very effective energy coupling occurs between the ground and the building. Therefore, even for relatively small shaking, the ground, by resonance, can feed energy into the building efficiently enough to cause the failure of the structure. The structure must be carefully designed so as to reduce the resonance response.

❯ SUMMARY

An object attached to the end of a spring moves with a motion called **simple harmonic motion,** and the system is called a **simple harmonic oscillator.**

The time for one complete oscillation of the object is called the **period** T of the motion. The inverse of the period is the **frequency** f of the motion, which equals the number of oscillations per second:

$$f = \frac{1}{T} = \frac{\omega}{2\pi} \qquad \text{12.11} ◀$$

The velocity and acceleration of an object in simple harmonic oscillation are

$$v = \frac{dx}{dt} = -\omega A \sin(\omega t + \phi) \qquad \text{12.15} ◀$$

$$a = \frac{d^2x}{dt^2} = -\omega^2 A \cos(\omega t + \phi) \qquad \text{12.16} ◀$$

Therefore, the maximum speed of the object is ωA and its maximum acceleration is of magnitude $\omega^2 A$. The speed is zero when the object is at its turning points, $x = \pm A$, and the speed is a maximum at the equilibrium position, $x = 0$. The magnitude of the acceleration is a maximum at the turning points and is zero at the equilibrium position.

The kinetic energy and potential energy of a simple harmonic oscillator vary with time and are given by

$$K = \tfrac{1}{2}mv^2 = \tfrac{1}{2}m\omega^2 A^2 \sin^2(\omega t + \phi) \qquad \text{12.19} ◀$$

$$U = \tfrac{1}{2}kx^2 = \tfrac{1}{2}kA^2 \cos^2(\omega t + \phi) \qquad \text{12.20} ◀$$

The **total energy** of a simple harmonic oscillator is a constant of the motion and is

$$E = \tfrac{1}{2}kA^2 \qquad \text{12.21} ◀$$

The potential energy of a simple harmonic oscillator is a maximum when the particle is at its turning points (maximum displacement from equilibrium) and is zero at the equilibrium position. The kinetic energy is zero at the turning points and is a maximum at the equilibrium position.

A **simple pendulum** of length L exhibits motion that can be modeled as simple harmonic for small angular displacements from the vertical, with a period of

$$T = 2\pi \sqrt{\frac{L}{g}} \qquad \text{12.25} ◀$$

The period of a simple pendulum is independent of the mass of the suspended object.

A **physical pendulum** exhibits motion that can be modeled as simple harmonic for small angular displacements from equilibrium about a pivot that does not go through the center of mass. The period of this motion is

$$T = 2\pi \sqrt{\frac{I}{mgd}} \qquad \text{12.27} ◀$$

where I is the moment of inertia about an axis through the pivot and d is the distance from the pivot to the center of mass.

Damped oscillations occur in a system in which a resistive force opposes the motion of the oscillating object. If such a system is set in motion and then left to itself, its mechanical energy decreases in time because of the presence of the nonconservative resistive force. It is possible to compensate for this transformation of energy by driving the system with an external periodic force. The oscillator in this case is undergoing **forced oscillations.** When the frequency of the driving force matches the natural frequency of the *undamped* oscillator, energy is efficiently transferred to the oscillator and its steady-state amplitude is a maximum. This situation is called **resonance.**

▶Analysis Model for Problem Solving

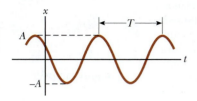

Particle in Simple Harmonic Motion If a particle is subject to a force of the form of Hooke's law $F = -kx$, the particle exhibits **simple harmonic motion.** Its position is described by

$$x(t) = A \cos (\omega t + \phi) \qquad \textbf{12.6}◀$$

where A is the **amplitude** of the motion, ω is the **angular frequency,** and ϕ is the **phase constant.** The value of ϕ depends on the initial position and initial velocity of the oscillator.

The period of the oscillation is related to the parameters of a block–spring system according to

$$T = \frac{2\pi}{\omega} = 2\pi\sqrt{\frac{m}{k}} \qquad \textbf{12.13}◀$$

▶ OBJECTIVE QUESTIONS

☐ denotes answer available in *Student Solutions Manual/Study Guide*

1. Which of the following statements is *not* true regarding a mass–spring system that moves with simple harmonic motion in the absence of friction? (a) The total energy of the system remains constant. (b) The energy of the system is continually transformed between kinetic and potential energy. (c) The total energy of the system is proportional to the square of the amplitude. (d) The potential energy stored in the system is greatest when the mass passes through the equilibrium position. (e) The velocity of the oscillating mass has its maximum value when the mass passes through the equilibrium position.

2. The position of an object moving with simple harmonic motion is given by $x = 4 \cos (6\pi t)$, where x is in meters and t is in seconds. What is the period of the oscillating system? (a) 4 s (b) $\frac{1}{6}$ s (c) $\frac{1}{3}$ s (d) 6π s (e) impossible to determine from the information given

3. A block–spring system vibrating on a frictionless, horizontal surface with an amplitude of 6.0 cm has an energy of 12 J. If the block is replaced by one whose mass is twice the mass of the original block and the amplitude of the motion is again 6.0 cm, what is the energy of the system? (a) 12 J (b) 24 J (c) 6 J (d) 48 J (e) none of those answers

4. A runaway railroad car, with mass 3.0×10^5 kg, coasts across a level track at 2.0 m/s when it collides elastically with a spring-loaded bumper at the end of the track. If the spring constant of the bumper is 2.0×10^6 N/m, what is the maximum compression of the spring during the collision? (a) 0.77 m (b) 0.58 m (c) 0.34 m (d) 1.07 m (e) 1.24 m

5. An object of mass 0.40 kg, hanging from a spring with a spring constant of 8.0 N/m, is set into an up-and-down simple harmonic motion. What is the magnitude of the acceleration of the object when it is at its maximum displacement of 0.10 m? (a) zero (b) 0.45 m/s² (c) 1.0 m/s² (d) 2.0 m/s² (e) 2.4 m/s²

6. If an object of mass m attached to a light spring is replaced by one of mass $9m$, the frequency of the vibrating system changes by what factor? (a) $\frac{1}{9}$ (b) $\frac{1}{3}$ (c) 3.0 (d) 9.0 (e) 6.0

7. If a simple pendulum oscillates with small amplitude and its length is doubled, what happens to the frequency of its motion? (a) It doubles. (b) It becomes $\sqrt{2}$ times as large. (c) It becomes half as large. (d) It becomes $1/\sqrt{2}$ times as large. (e) It remains the same.

8. An object–spring system moving with simple harmonic motion has an amplitude A. When the kinetic energy of the object equals twice the potential energy stored in the spring, what is the position x of the object? (a) A (b) $\frac{1}{3}A$ (c) $A/\sqrt{3}$ (d) 0 (e) none of those answers

9. A mass–spring system moves with simple harmonic motion along the x axis between turning points at $x_1 = 20$ cm and $x_2 = 60$ cm. For parts (i) through (iii), choose from the same five possibilities. (i) At which position does the particle have the greatest magnitude of momentum? (a) 20 cm (b) 30 cm (c) 40 cm (d) some other position (e) The greatest value occurs at multiple points. (ii) At which position does the particle have greatest kinetic energy? (iii) At which position does the particle–spring system have the greatest total energy?

10. A particle on a spring moves in simple harmonic motion along the x axis between turning points at $x_1 = 100$ cm and $x_2 = 140$ cm. (i) At which of the following positions does the particle have maximum speed? (a) 100 cm (b) 110 cm (c) 120 cm (d) at none of those positions (ii) At which position does it have maximum acceleration? Choose from the same possibilities as in part (i). (iii) At which position is the greatest net force exerted on the particle? Choose from the same possibilities as in part (i).

11. A block with mass $m = 0.1$ kg oscillates with amplitude $A = 0.1$ m at the end of a spring with force constant $k = 10$ N/m on a frictionless, horizontal surface. Rank the periods of the following situations from greatest to smallest. If any periods are equal, show their equality in your ranking. (a) The system is as described above. (b) The system is as described in situation (a) except the amplitude is 0.2 m. (c) The situation is as described in situation (a) except the mass is 0.2 kg. (d) The situation is as described in situation (a) except the spring has force constant 20 N/m. (e) A small resistive force makes the motion underdamped.

12. For a simple harmonic oscillator, answer yes or no to the following questions. (a) Can the quantities position and velocity have the same sign? (b) Can velocity and acceleration have the same sign? (c) Can position and acceleration have the same sign?

13. A simple pendulum has a period of 2.5 s. (i) What is its period if its length is made four times larger? (a) 1.25 s (b) 1.77 s (c) 2.5 s (d) 3.54 s (e) 5 s (ii) What is its period if the length is held constant at its initial value and the mass of the suspended bob is made four times larger? Choose from the same possibilities.

14. You attach a block to the bottom end of a spring hanging vertically. You slowly let the block move down and find that it hangs at rest with the spring stretched by 15.0 cm. Next, you lift the block back up to the initial position and release it from rest with the spring unstretched. What maximum distance does it move down? (a) 7.5 cm (b) 15.0 cm (c) 30.0 cm (d) 60.0 cm (e) The distance cannot be determined without knowing the mass and spring constant.

15. The top end of a spring is held fixed. A block is hung on the bottom end as in Figure OQ12.15a, and the frequency f of the oscillation of the system is measured. The block, a second identical block, and the spring are carried up in a space shuttle to Earth orbit. The two blocks are attached to the ends of the spring. The spring is compressed without making adjacent coils touch (Fig. OQ12.15b), and the system is released to oscillate while floating within the shuttle cabin (Fig. OQ12.15c). What is the frequency of oscillation for this system in terms of f? (a) $f/2$ (b) $f/\sqrt{2}$ (c) f (d) $\sqrt{2}f$ (e) $2f$

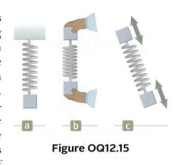

Figure OQ12.15

CONCEPTUAL QUESTIONS

denotes answer available in *Student Solutions Manual/Study Guide*

1. (a) If the coordinate of a particle varies as $x = -A \cos \omega t$, what is the phase constant in Equation 12.6? (b) At what position is the particle at $t = 0$?

2. The equations listed in Table 2.2 give position as a function of time, velocity as a function of time, and velocity as a function of position for an object moving in a straight line with constant acceleration. The quantity v_{xi} appears in every equation. (a) Do any of these equations apply to an object moving in a straight line with simple harmonic motion? (b) Using a similar format, make a table of equations describing simple harmonic motion. Include equations giving acceleration as a function of time and acceleration as a function of position. State the equations in such a form that they apply equally to a block–spring system, to a pendulum, and to other vibrating systems. (c) What quantity appears in every equation?

3. Is a bouncing ball an example of simple harmonic motion? Is the daily movement of a student from home to school and back simple harmonic motion? Why or why not?

4. Figure CQ12.4 shows graphs of the potential energy of four different systems versus the position of a particle in each system. Each particle is set into motion with a push at an arbitrarily chosen location. Describe its subsequent motion in each case (a), (b), (c), and (d).

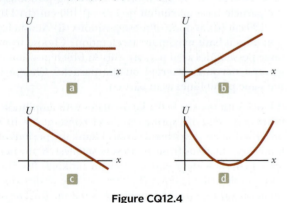

Figure CQ12.4

5. A simple pendulum can be modeled as exhibiting simple harmonic motion when θ is small. Is the motion periodic when θ is large?

6. Is it possible to have damped oscillations when a system is at resonance? Explain.

7. The mechanical energy of an undamped block–spring system is constant as kinetic energy transforms to elastic potential energy and vice versa. For comparison, explain what happens to the energy of a damped oscillator in terms of the mechanical, potential, and kinetic energies.

8. A student thinks that any real vibration must be damped. Is the student correct? If so, give convincing reasoning. If not, give an example of a real vibration that keeps constant amplitude forever if the system is isolated.

9. Will damped oscillations occur for any values of b and k? Explain.

10. If a pendulum clock keeps perfect time at the base of a mountain, will it also keep perfect time when it is moved to the top of the mountain? Explain.

11. A pendulum bob is made from a sphere filled with water. What would happen to the frequency of vibration of this pendulum if there were a hole in the sphere that allowed the water to leak out slowly?

12. You are looking at a small, leafy tree. You do not notice any breeze, and most of the leaves on the tree are motionless. One leaf, however, is fluttering back and forth wildly. After a while, that leaf stops moving and you notice a different leaf moving much more than all the others. Explain what could cause the large motion of one particular leaf.

13. Consider the simplified single-piston engine in Figure CQ12.13. Assuming the wheel rotates with constant angular speed, explain why the piston rod oscillates in simple harmonic motion.

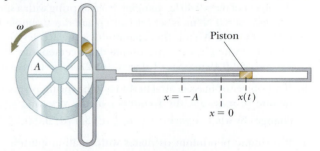

Figure CQ12.13

PROBLEMS

WebAssign The problems found in this chapter may be assigned online in Enhanced WebAssign.

1. denotes straightforward problem; 2. denotes intermediate problem; 3. denotes challenging problem

1. denotes full solution available in the *Student Solutions Manual/ Study Guide*

1. denotes problems most often assigned in Enhanced WebAssign.

BIO denotes biomedical problem

GP denotes guided problem

M denotes Master It tutorial available in Enhanced WebAssign

Q|C denotes asking for quantitative and conceptual reasoning

S denotes symbolic reasoning problem

shaded denotes "paired problems" that develop reasoning with symbols and numerical values

W denotes Watch It video solution available in Enhanced WebAssign

Note: Ignore the mass of every spring except in Problem 69.

Section 12.1 Motion of an Object Attached to a Spring

Note: Problems 16, 17, 23, and 69 in Chapter 6 can also be assigned with this section.

1. A 0.60-kg block attached to a spring with force constant 130 N/m is free to move on a frictionless, horizontal surface as in Active Figure 12.1. The block is released from rest when the spring is stretched 0.13 m. At the instant the block is released, find (a) the force on the block and (b) its acceleration.

2. When a 4.25-kg object is placed on top of a vertical spring, the spring compresses a distance of 2.62 cm. What is the force constant of the spring?

Section 12.2 Analysis Model: Particle in Simple Harmonic Motion

3. **M** The position of a particle is given by the expression $x = 4.00 \cos (3.00\pi t + \pi)$, where x is in meters and t is in seconds. Determine (a) the frequency and (b) period of the motion, (c) the amplitude of the motion, (d) the phase constant, and (e) the position of the particle at $t = 0.250$ s.

4. **Q|C** You attach an object to the bottom end of a hanging vertical spring. It hangs at rest after extending the spring 18.3 cm. You then set the object vibrating. (a) Do you have enough information to find its period? (b) Explain your answer and state whatever you can about its period.

5. **W** A 7.00-kg object is hung from the bottom end of a vertical spring fastened to an overhead beam. The object is set into vertical oscillations having a period of 2.60 s. Find the force constant of the spring.

6. **S** The initial position, velocity, and acceleration of an object moving in simple harmonic motion are x_i, v_i, and a_i; the angular frequency of oscillation is ω. (a) Show that the position and velocity of the object for all time can be written as

$$x(t) = x_i \cos \omega t + \left(\frac{v_i}{\omega}\right) \sin \omega t$$

$$v(t) = -x_i \omega \sin \omega t + v_i \cos \omega t$$

(b) Using A to represent the amplitude of the motion, show that

$$v^2 - ax = v_i^2 - a_i x_i = \omega^2 A^2$$

7. A particle moves in simple harmonic motion with a frequency of 3.00 Hz and an amplitude of 5.00 cm. (a) Through what total distance does the particle move during one cycle of its motion? (b) What is its maximum speed? Where does this maximum speed occur? (c) Find the maximum acceleration of the particle. Where in the motion does the maximum acceleration occur?

8. **Q|C** A ball dropped from a height of 4.00 m makes an elastic collision with the ground. Assuming no mechanical energy is lost due to air resistance, (a) show that the ensuing motion is periodic and (b) determine the period of the motion. (c) Is the motion simple harmonic? Explain.

9. A particle moving along the x axis in simple harmonic motion starts from its equilibrium position, the origin, at $t = 0$ and moves to the right. The amplitude of its motion is 2.00 cm, and the frequency is 1.50 Hz. (a) Find an expression for the position of the particle as a function of time. Determine (b) the maximum speed of the particle and (c) the earliest time ($t > 0$) at which the particle has this speed. Find (d) the maximum positive acceleration of the particle and (e) the earliest time ($t > 0$) at which the particle has this acceleration. (f) Find the total distance traveled by the particle between $t = 0$ and $t = 1.00$ s.

10. **M** A 1.00-kg glider attached to a spring with a force constant of 25.0 N/m oscillates on a frictionless, horizontal air track. At $t = 0$, the glider is released from rest at $x = -3.00$ cm (that is, the spring is compressed by 3.00 cm). Find (a) the period of the glider's motion, (b) the maximum values of its speed and acceleration, and (c) the position, velocity, and acceleration as functions of time.

11. **Review.** A particle moves along the x axis. It is initially at the position 0.270 m, moving with velocity 0.140 m/s and acceleration -0.320 m/s². Suppose it moves as a particle under constant acceleration for 4.50 s. Find (a) its position and (b) its velocity at the end of this time interval. Next, assume it moves as a particle in simple harmonic motion for 4.50 s and $x = 0$ is its equilibrium position. Find (c) its position and (d) its velocity at the end of this time interval.

12. **W** A simple harmonic oscillator takes 12.0 s to undergo five complete vibrations. Find (a) the period of its motion, (b) the frequency in hertz, and (c) the angular frequency in radians per second.

13. **M** A 0.500-kg object attached to a spring with a force constant of 8.00 N/m vibrates in simple harmonic motion with

an amplitude of 10.0 cm. Calculate the maximum value of its (a) speed and (b) acceleration, (c) the speed and (d) the acceleration when the object is 6.00 cm from the equilibrium position, and (e) the time interval required for the object to move from $x = 0$ to $x = 8.00$ cm.

14. **W** In an engine, a piston oscillates with simple harmonic motion so that its position varies according to the expression

$$x = 5.00 \cos\left(2t + \frac{\pi}{6}\right)$$

where x is in centimeters and t is in seconds. At $t = 0$, find (a) the position of the piston, (b) its velocity, and (c) its acceleration. Find (d) the period and (e) the amplitude of the motion.

15. A vibration sensor, used in testing a washing machine, consists of a cube of aluminum 1.50 cm on edge mounted on one end of a strip of spring steel (like a hacksaw blade) that lies in a vertical plane. The strip's mass is small compared with that of the cube, but the strip's length is large compared with the size of the cube. The other end of the strip is clamped to the frame of the washing machine that is not operating. A horizontal force of 1.43 N applied to the cube is required to hold it 2.75 cm away from its equilibrium position. If it is released, what is its frequency of vibration?

Section 12.3 Energy of the Simple Harmonic Oscillator

16. A block–spring system oscillates with an amplitude of 3.50 cm. The spring constant is 250 N/m and the mass of the block is 0.500 kg. Determine (a) the mechanical energy of the system, (b) the maximum speed of the block, and (c) the maximum acceleration.

17. A block of unknown mass is attached to a spring with a spring constant of 6.50 N/m and undergoes simple harmonic motion with an amplitude of 10.0 cm. When the block is halfway between its equilibrium position and the end point, its speed is measured to be 30.0 cm/s. Calculate (a) the mass of the block, (b) the period of the motion, and (c) the maximum acceleration of the block.

18. A 2.00-kg object is attached to a spring and placed on a frictionless, horizontal surface. A horizontal force of 20.0 N is required to hold the object at rest when it is pulled 0.200 m from its equilibrium position (the origin of the x axis). The object is now released from rest from this stretched position, and it subsequently undergoes simple harmonic oscillations. Find (a) the force constant of the spring, (b) the frequency of the oscillations, and (c) the maximum speed of the object. (d) Where does this maximum speed occur? (e) Find the maximum acceleration of the object. (f) Where does the maximum acceleration occur? (g) Find the total energy of the oscillating system. Find (h) the speed and (i) the acceleration of the object when its position is equal to one-third the maximum value.

19. **M** To test the resiliency of its bumper during low-speed collisions, a 1 000-kg automobile is driven into a brick wall. The car's bumper behaves like a spring with a force constant 5.00×10^6 N/m and compresses 3.16 cm as the car is brought to rest. What was the speed of the car before impact, assuming no mechanical energy is transformed or transferred away during impact with the wall?

20. A 200-g block is attached to a horizontal spring and executes simple harmonic motion with a period of 0.250 s. The total

energy of the system is 2.00 J. Find (a) the force constant of the spring and (b) the amplitude of the motion.

21. **W** A 50.0-g object connected to a spring with a force constant of 35.0 N/m oscillates with an amplitude of 4.00 cm on a frictionless, horizontal surface. Find (a) the total energy of the system and (b) the speed of the object when its position is 1.00 cm. Find (c) the kinetic energy and (d) the potential energy when its position is 3.00 cm.

22. **GP** **Review.** A 65.0-kg bungee jumper steps off a bridge with a light bungee cord tied to her body and to the bridge. The unstretched length of the cord is 11.0 m. The jumper reaches the bottom of her motion 36.0 m below the bridge before bouncing back. We wish to find the time interval between her leaving the bridge and her arriving at the bottom of her motion. Her overall motion can be separated into an 11.0-m free fall and a 25.0-m section of simple harmonic oscillation. (a) For the free-fall part, what is the appropriate analysis model to describe her motion? (b) For what time interval is she in free fall? (c) For the simple harmonic oscillation part of the plunge, is the system of the bungee jumper, the spring, and the Earth isolated or nonisolated? (d) From your response in part (c) find the spring constant of the bungee cord. (e) What is the location of the equilibrium point where the spring force balances the gravitational force exerted on the jumper? (f) What is the angular frequency of the oscillation? (g) What time interval is required for the cord to stretch by 25.0 m? (h) What is the total time interval for the entire 36.0-m drop?

23. A particle executes simple harmonic motion with an amplitude of 3.00 cm. At what position does its speed equal half of its maximum speed?

24. **QC** **S** A simple harmonic oscillator of amplitude A has a total energy E. Determine (a) the kinetic energy and (b) the potential energy when the position is one-third the amplitude. (c) For what values of the position does the kinetic energy equal one-half the potential energy? (d) Are there any values of the position where the kinetic energy is greater than the maximum potential energy? Explain.

25. The amplitude of a system moving in simple harmonic motion is doubled. Determine the change in (a) the total energy, (b) the maximum speed, (c) the maximum acceleration, and (d) the period.

Section 12.4 The Simple Pendulum
Section 12.5 The Physical Pendulum

Note: Problem 1.60 in Chapter 1 can also be assigned with this section.

26. A "seconds pendulum" is one that moves through its equilibrium position once each second. (The period of the pendulum is precisely 2 s.) The length of a seconds pendulum is 0.992 7 m at Tokyo, Japan, and 0.994 2 m at Cambridge, England. What is the ratio of the free-fall accelerations at these two locations?

27. **M** A physical pendulum in the form of a planar object moves in simple harmonic motion with a frequency of 0.450 Hz. The pendulum has a mass of 2.20 kg, and the pivot is located 0.350 m from the center of mass. Determine the moment of inertia of the pendulum about the pivot point.

28. **S** A physical pendulum in the form of a planar object moves in simple harmonic motion with a frequency f. The

pendulum has a mass m, and the pivot is located a distance d from the center of mass. Determine the moment of inertia of the pendulum about the pivot point.

29. The angular position of a pendulum is represented by the equation $\theta = 0.032\,0 \cos \omega t$, where θ is in radians and $\omega = 4.43$ rad/s. Determine the period and length of the pendulum.

30. A small object is attached to the end of a string to form a simple pendulum. The period of its harmonic motion is measured for small angular displacements and three lengths. For lengths of 1.000 m, 0.750 m, and 0.500 m, total time intervals for 50 oscillations of 99.8 s, 86.6 s, and 71.1 s are measured with a stopwatch. (a) Determine the period of motion for each length. (b) Determine the mean value of g obtained from these three independent measurements and compare it with the accepted value. (c) Plot T^2 versus L and obtain a value for g from the slope of your best-fit straight-line graph. (d) Compare the value found in part (c) with that obtained in part (b).

31. A very light rigid rod of length 0.500 m extends straight out from one end of a meterstick. The combination is suspended from a pivot at the upper end of the rod as shown in Figure P12.31. The combination is then pulled out by a small angle and released. (a) Determine the period of oscillation of the system. (b) By what percentage does the period differ from the period of a simple pendulum 1.00 m long?

0.500 m

Figure P12.31

32. **S** A particle of mass m slides without friction inside a hemispherical bowl of radius R. Show that if the particle starts from rest with a small displacement from equilibrium, it moves in simple harmonic motion with an angular frequency equal to that of a simple pendulum of length R. That is, $\omega = \sqrt{g/R}$.

33. **M Review.** A simple pendulum is 5.00 m long. What is the period of small oscillations for this pendulum if it is located in an elevator (a) accelerating upward at 5.00 m/s²? (b) Accelerating downward at 5.00 m/s²? (c) What is the period of this pendulum if it is placed in a truck that is accelerating horizontally at 5.00 m/s²?

34. **S** Consider the physical pendulum of Figure 12.14. (a) Represent its moment of inertia about an axis passing through its center of mass and parallel to the axis passing through its pivot point as I_{CM}. Show that its period is

$$T = 2\pi \sqrt{\frac{I_{CM} + md^2}{mgd}}$$

where d is the distance between the pivot point and the center of mass. (b) Show that the period has a minimum value when d satisfies $md^2 = I_{CM}$.

35. **Q C** A simple pendulum has a mass of 0.250 kg and a length of 1.00 m. It is displaced through an angle of 15.0° and then released. Using the analysis model of a particle in simple harmonic motion, what are (a) the maximum speed of the bob, (b) its maximum angular acceleration, and (c) the maximum restoring force on the bob? (d) **What If?** Solve parts (a) through (c) again by using analysis models introduced in earlier chapters. (e) Compare the answers.

Section 12.6 Damped Oscillations

36. **S** Show that the time rate of change of mechanical energy for a damped, undriven oscillator is given by $dE/dt = -bv^2$ and hence is always negative. To do so, differentiate the expression for the mechanical energy of an oscillator, $E = \frac{1}{2}mv^2 + \frac{1}{2}kx^2$, and use Equation 12.28.

37. **W** A pendulum with a length of 1.00 m is released from an initial angle of 15.0°. After 1 000 s, its amplitude has been reduced by friction to 5.50°. What is the value of $b/2m$?

38. **S** Show that Equation 12.29 is a solution of Equation 12.28 provided that $b^2 < 4mk$.

Section 12.7 Forced Oscillations

39. A 2.00-kg object attached to a spring moves without friction ($b = 0$) and is driven by an external force given by the expression $F = 3.00 \sin(2\pi t)$, where F is in newtons and t is in seconds. The force constant of the spring is 20.0 N/m. Find (a) the resonance angular frequency of the system, (b) the angular frequency of the driven system, and (c) the amplitude of the motion.

40. **S** Considering an undamped, forced oscillator ($b = 0$), show that Equation 12.32 is a solution of Equation 12.31, with an amplitude given by Equation 12.33.

41. As you enter a fine restaurant, you realize that you have accidentally brought a small electronic timer from home instead of your cell phone. In frustration, you drop the timer into a side pocket of your suit coat, not realizing that the timer is operating. The arm of your chair presses the light cloth of your coat against your body at one spot. Fabric with a length L hangs freely below that spot, with the timer at the bottom. At one point during your dinner, the timer goes off and a buzzer and a vibrator turn on and off with a frequency of 1.50 Hz. It makes the hanging part of your coat swing back and forth with remarkably large amplitude, drawing everyone's attention. Find the value of L.

42. A baby bounces up and down in her crib. Her mass is 12.5 kg, and the crib mattress can be modeled as a light spring with force constant 700 N/m. (a) The baby soon learns to bounce with maximum amplitude and minimum effort by bending her knees at what frequency? (b) If she were to use the mattress as a trampoline—losing contact with it for part of each cycle—what minimum amplitude of oscillation does she require?

43. **M** Damping is negligible for a 0.150-kg object hanging from a light, 6.30-N/m spring. A sinusoidal force with an amplitude of 1.70 N drives the system. At what frequency will the force make the object vibrate with an amplitude of 0.440 m?

Section 12.8 Context Connection: Resonance in Structures

44. **Q C** People who ride motorcycles and bicycles learn to look out for bumps in the road and especially for *washboarding*, a condition in which many equally spaced ridges are worn into the road. What is so bad about washboarding? A motorcycle has several springs and shock absorbers in its suspension, but you can model it as a single spring supporting a block. You can estimate the force constant by thinking about how far the spring compresses when a heavy rider sits on the seat. A motorcyclist traveling at highway speed must be particularly careful of washboard bumps that are a certain distance apart. What is the order of magnitude of their separation distance?

45. Four people, each with a mass of 72.4 kg, are in a car with a mass of 1 130 kg. An earthquake strikes. The vertical oscillations of the ground surface make the car bounce up and down on its suspension springs, but the driver manages to pull off the road and stop. When the frequency of the shaking is 1.80 Hz, the car exhibits a maximum amplitude of vibration. The earthquake ends and the four people leave the car as fast as they can. By what distance does the car's undamaged suspension lift the car's body as the people get out?

Additional Problems

46. **Q|C** (a) A hanging spring stretches by 35.0 cm when an object of mass 450 g is hung on it at rest. In this situation, we define its position as $x = 0$. The object is pulled down an additional 18.0 cm and released from rest to oscillate without friction. What is its position x at a moment 84.4 s later? (b) Find the distance traveled by the vibrating object in part (a). (c) **What If?** Another hanging spring stretches by 35.5 cm when an object of mass 440 g is hung on it at rest. We define this new position as $x = 0$. This object is also pulled down an additional 18.0 cm and released from rest to oscillate without friction. Find its position 84.4 s later. (d) Find the distance traveled by the object in part (c). (e) Why are the answers to parts (a) and (c) so different when the initial data in parts (a) and (c) are so similar and the answers to parts (b) and (d) are relatively close? Does this circumstance reveal a fundamental difficulty in calculating the future?

47. **Review.** A large block P attached to a light spring executes horizontal, simple harmonic motion as it slides across a frictionless surface with a frequency $f = 1.50$ Hz. Block B rests on it as shown in Figure P12.47, and the coefficient of static friction between the two is $\mu_s = 0.600$. What maximum amplitude of oscillation can the system have if block B is not to slip?

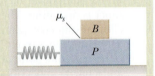

Figure P12.47 Problems 47 and 48.

48. **S** **Review.** A large block P attached to a light spring executes horizontal, simple harmonic motion as it slides across a frictionless surface with a frequency f. Block B rests on it as shown in Figure P12.47, and the coefficient of static friction between the two is μ_s. What maximum amplitude of oscillation can the system have if block B is not to slip?

49. A particle with a mass of 0.500 kg is attached to a horizontal spring with a force constant of 50.0 N/m. At the moment $t = 0$, the particle has its maximum speed of 20.0 m/s and is moving to the left. (a) Determine the particle's equation of motion, specifying its position as a function of time. (b) Where in the motion is the potential energy three times the kinetic energy? (c) Find the minimum time interval required for the particle to move from $x = 0$ to $x = 1.00$ m. (d) Find the length of a simple pendulum with the same period.

50. **BIO** To account for the walking speed of a bipedal or quadrupedal animal, model a leg that is not contacting the ground as a uniform rod of length ℓ, swinging as a physical pendulum through one-half of a cycle, in resonance. Let θ_{max} represent its amplitude. (a) Show that the animal's speed is given by the expression

$$v = \frac{\sqrt{6g\ell}\,\sin\theta_{max}}{\pi}$$

if θ_{max} is sufficiently small that the motion is nearly simple harmonic. An empirical relationship that is based on the same model and applies over a wider range of angles is

$$v = \frac{\sqrt{6g\ell\,\cos(\theta_{max}/2)}\,\sin\theta_{max}}{\pi}$$

(b) Evaluate the walking speed of a human with leg length 0.850 m and leg-swing amplitude 28.0°. (c) What leg length would give twice the speed for the same angular amplitude?

51. The mass of the deuterium molecule (D_2) is twice that of the hydrogen molecule (H_2). If the vibrational frequency of H_2 is 1.30×10^{14} Hz, what is the vibrational frequency of D_2? Assume the "spring constant" of attracting forces is the same for the two molecules.

52. *Why is the following situation impossible?* Your job involves building very small damped oscillators. One of your designs involves a spring–object oscillator with a spring of force constant $k = 10.0$ N/m and an object of mass $m = 1.00$ g. Your design objective is that the oscillator undergo many oscillations as its amplitude falls to 25.0% of its initial value in a certain time interval. Measurements on your latest design show that the amplitude falls to the 25.0% value in 23.1 ms. This time interval is too long for what is needed in your project. To shorten the time interval, you double the damping constant b for the oscillator. This doubling allows you to reach your design objective.

53. A horizontal plank of mass 5.00 kg and length 2.00 m is pivoted at one end. The plank's other end is supported by a spring of force constant 100 N/m (Fig. P12.53). The plank is displaced by a small angle θ from its horizontal equilibrium position and released. Find the angular frequency with which the plank moves with simple harmonic motion.

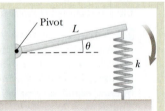

Figure P12.53 Problems 53 and 54.

54. **S** A horizontal plank of mass m and length L is pivoted at one end. The plank's other end is supported by a spring of force constant k (Fig. P12.53). The plank is displaced by a small angle θ from its horizontal equilibrium position and released. Find the angular frequency with which the plank moves with simple harmonic motion.

55. **W** A simple pendulum with a length of 2.23 m and a mass of 6.74 kg is given an initial speed of 2.06 m/s at its equilibrium position. Assume it undergoes simple harmonic motion. Determine (a) its period, (b) its total energy, and (c) its maximum angular displacement.

56. **S** A block of mass m is connected to two springs of force constants k_1 and k_2 in two ways as shown in Figure P12.56. In both cases, the block moves on a frictionless table after it is displaced from equilibrium and released. Show that in the two cases the block exhibits simple harmonic motion with periods

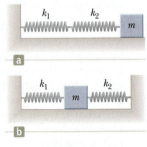

Figure P12.56

(a) $T = 2\pi\sqrt{\dfrac{m(k_1 + k_2)}{k_1 k_2}}$ and (b) $T = 2\pi\sqrt{\dfrac{m}{k_1 + k_2}}$

57. **Review.** One end of a light spring with force constant $k = 100$ N/m is attached to a vertical wall. A light string is tied to the other end of the horizontal spring. As shown in Figure P12.57, the string changes from horizontal to vertical as it passes over a pulley of mass M in the shape of a solid disk of radius $R = 2.00$ cm. The pulley is free to turn on a fixed, smooth axle. The vertical section of the string supports an object of mass $m = 200$ g. The string does not slip at its contact with the pulley. The object is pulled downward a small distance and released. (a) What is the angular frequency ω of oscillation of the object in terms of the mass M? (b) What is the highest possible value of the angular frequency of oscillation of the object? (c) What is the highest possible value of the angular frequency of oscillation of the object if the pulley radius is doubled to $R = 4.00$ cm?

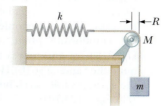

Figure P12.57

58. **S** After a thrilling plunge, bungee jumpers bounce freely on the bungee cord through many cycles. After the first few cycles, the cord does not go slack. Your little brother can make a pest of himself by figuring out the mass of each person, using a proportion that you set up by solving the following problem. An object of mass m is oscillating freely on a light vertical spring with a period T. An object of unknown mass m' on the same spring oscillates with a period T'. Determine (a) the spring constant and (b) the unknown mass.

59. **M** A small ball of mass M is attached to the end of a uniform rod of equal mass M and length L that is pivoted at the top (Fig. P12.59). Determine the tensions in the rod (a) at the pivot and (b) at the point P when the system is stationary. (c) Calculate the period of oscillation for small displacements from equilibrium and (d) determine this period for $L = 2.00$ m.

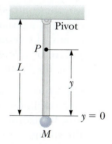

Figure P12.59

60. **S** Your thumb squeaks on a plate you have just washed. Your sneakers squeak on the gym floor. Car tires squeal when you start or stop abruptly. You can make a goblet sing by wiping your moistened finger around its rim. When chalk squeaks on a blackboard, you can see that it makes a row of regularly spaced dashes. As these examples suggest, vibration commonly results when friction acts on a moving elastic object. The oscillation is not simple harmonic motion, but is called *stick-and-slip*. This problem models stick-and-slip motion.

A block of mass m is attached to a fixed support by a horizontal spring with force constant k and negligible mass (Fig. P12.60). Hooke's law describes the spring both in extension and in compression. The block sits on a long horizontal board, with which it has coefficient of static friction μ_s and a smaller coefficient of kinetic friction μ_k. The board moves to the right at constant speed v. Assume the block spends most of its time sticking to the board and moving to the right with it, so the speed v is small in comparison to the average speed the block has as it slips back toward the left. (a) Show that the

maximum extension of the spring from its unstressed position is very nearly given by $\mu_s mg/k$. (b) Show that the block oscillates around an equilibrium position at which the spring is stretched by $\mu_k mg/k$. (c) Graph the block's position versus time. (d) Show that the amplitude of the block's motion is

$$A = \frac{(\mu_s - \mu_k)mg}{k}$$

(e) Show that the period of the block's motion is

$$T = \frac{2(\mu_s - \mu_k)mg}{vk} + \pi\sqrt{\frac{m}{k}}$$

It is the excess of static over kinetic friction that is important for the vibration. "The squeaky wheel gets the grease" because even a viscous fluid cannot exert a force of static friction.

Figure P12.60

61. **Q|C Review.** This problem extends the reasoning of Problem 48 in Chapter 8. Two gliders are set in motion on an air track. Glider 1 has mass $m_1 = 0.240$ kg and moves to the right with speed 0.740 m/s. It will have a rear-end collision with glider 2, of mass $m_2 = 0.360$ kg, which initially moves to the right with speed 0.120 m/s. A light spring of force constant 45.0 N/m is attached to the back end of glider 2 as shown in Figure P8.48. When glider 1 touches the spring, superglue instantly and permanently makes it stick to its end of the spring. (a) Find the common speed the two gliders have when the spring is at maximum compression. (b) Find the maximum spring compression distance. The motion after the gliders become attached consists of a combination of (1) the constant-velocity motion of the center of mass of the two-glider system found in part (a) and (2) simple harmonic motion of the gliders relative to the center of mass. (c) Find the energy of the center-of-mass motion. (d) Find the energy of the oscillation.

62. **S** A ball of mass m is connected to two rubber bands of length L, each under tension T as shown in Figure P12.62. The ball is displaced by a small distance y perpendicular to the length of the rubber bands. Assuming the tension does not change, show that (a) the restoring force is $-(2T/L)y$ and (b) the system exhibits simple harmonic motion with an angular frequency $\omega = \sqrt{2T/mL}$.

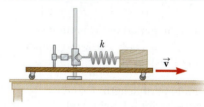

Figure P12.62

63. **Q|C Review.** A particle of mass 4.00 kg is attached to a spring with a force constant of 100 N/m. It is oscillating on a frictionless, horizontal surface with an amplitude of 2.00 m. A 6.00-kg object is dropped vertically on top of the 4.00-kg object as it passes through its equilibrium point. The two objects stick together. (a) What is the new amplitude of the vibrating system after the collision? (b) By what factor has the period of the system changed? (c) By how much does the energy of the system change as a result of the collision? (d) Account for the change in energy.

64. S A smaller disk of radius r and mass m is attached rigidly to the face of a second larger disk of radius R and mass M as shown in Figure P12.64. The center of the small disk is located at the edge of the large disk. The large disk is mounted at its center on a frictionless axle. The assembly is rotated through a small angle θ from its equilibrium position and released. (a) Show that the speed of the center of the small disk as it passes through the equilibrium position is

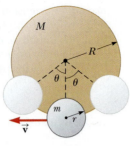

Figure P12.64

$$v = 2\left[\frac{Rg(1 - \cos \theta)}{(M/m) + (r/R)^2 + 2}\right]^{1/2}$$

(b) Show that the period of the motion is

$$T = 2\pi\left[\frac{(M + 2m)R^2 + mr^2}{2mgR}\right]^{1/2}$$

65. S A pendulum of length L and mass M has a spring of force constant k connected to it at a distance h below its point of suspension (Fig. P12.65). Find the frequency of vibration of the system for small values of the amplitude (small θ). Assume the vertical suspension rod of length L is rigid, but ignore its mass.

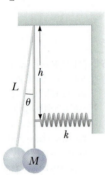

Figure P12.65

66. Consider the damped oscillator illustrated in Figure 12.16a. The mass of the object is 375 g, the spring constant is 100 N/m, and $b = 0.100$ N · s/m. (a) Over what time interval does the amplitude drop to half its initial value? (b) **What If?** Over what time interval does the mechanical energy drop to half its initial value? (c) Show that, in general, the fractional rate at which the amplitude decreases in a damped harmonic oscillator is one-half the fractional rate at which the mechanical energy decreases.

67. An object of mass $m_1 = 9.00$ kg is in equilibrium when connected to a light spring of constant $k = 100$ N/m that is

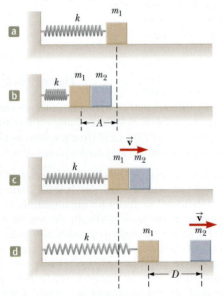

Figure P12.67

fastened to a wall as shown in Figure P12.67a. A second object, $m_2 = 7.00$ kg, is slowly pushed up against m_1, compressing the spring by the amount $A = 0.200$ m (see Fig. P12.67b). The system is then released, and both objects start moving to the right on the frictionless surface. (a) When m_1 reaches the equilibrium point, m_2 loses contact with m_1 (see Fig. P12.67c) and moves to the right with speed v. Determine the value of v. (b) How far apart are the objects when the spring is fully stretched for the first time (the distance D in Fig. P12.67d)?

68. S **Review.** *Why is the following situation impossible?* You are in the high-speed package delivery business. Your competitor in the next building gains the right-of-way to build an evacuated tunnel just above the ground all the way around the Earth. By firing packages into this tunnel at just the right speed, your competitor is able to send the packages into orbit around the Earth in this tunnel so that they arrive on the exact opposite side of the Earth in a very short time interval. You come up with a competing idea. Figuring that the distance *through* the Earth is shorter than the distance *around* the Earth, you obtain permits to build an evacuated tunnel through the center of the Earth. By simply dropping packages into this tunnel, they fall downward and arrive at the other end of your tunnel, which is in a building right next to the other end of your competitor's tunnel. Because your packages arrive on the other side of the Earth in a shorter time interval, you win the competition and your business flourishes. *Note:* An object at a distance r from the center of the Earth is pulled toward the center of the Earth only by the mass within the sphere of radius r (the reddish region in Fig. P12.68). Assume the Earth has uniform density.

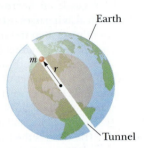

Figure P12.68

69. S A block of mass M is connected to a spring of mass m and oscillates in simple harmonic motion on a frictionless, horizontal track (Fig. P12.69). The force constant of the spring is k, and the equilibrium length is ℓ. Assume all portions of the spring oscillate in phase and the velocity of a segment of the spring of length dx is proportional to the distance x from the fixed end; that is, $v_x = (x/\ell)v$. Also, notice that the mass of a segment of the spring is $dm = (m/\ell) \, dx$. Find (a) the kinetic energy of the system when the block has a speed v and (b) the period of oscillation.

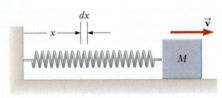

Figure P12.69

Mechanical Waves

Chapter Outline

Stefano Cellai/AGE fotostock

Three musicians play the alpenhorn in Valais, Switzerland. In this chapter, we explore the behavior of sound waves such as those coming from these large musical instruments.

Most of us experienced waves as children when we dropped pebbles into a pond. The disturbance created by a pebble manifests itself as ripples that move outward from the point at which the pebble lands in the water. If you were to carefully examine the motion of a leaf floating near the point where the pebble enters the water, you would see that the leaf moves up and down and back and forth about its original position but does not undergo any net displacement away from or toward the source of the disturbance. The *disturbance* in the water moves over a long distance, but a given small *element of the water* oscillates only over a very small distance. This behavior is the essence of wave motion.

The world is full of other kinds of waves, including sound waves, waves on strings, seismic waves, radio waves, and x-rays. Most waves can be placed in one of two categories. **Mechanical waves** are waves that disturb and propagate through a medium; the ripple in the water because of the pebble and a sound wave, for which air is the medium, are examples of mechanical waves. The opening photograph shows an example of one possible source of sound waves in air: blowing on very large pipes of different dimensions. **Electromagnetic waves** are a special class of waves that do not require a medium to propagate, as discussed with regard to the absence of the ether in Section 9.2; light waves and radio waves are two familiar examples. In this chapter, we shall confine our attention to the study of mechanical waves, deferring our study of electromagnetic waves to Chapter 24.

⟨13.1 | Propagation of a Disturbance

In the introduction, we alluded to the essence of wave motion: the transfer of a *disturbance* through space without the accompanying transfer of *matter*. The propagation of the disturbance also represents a transfer of energy; therefore, we can view waves as a means of energy transfer. In the list of energy transfer mechanisms in Section 7.1, we see two entries that depend on waves: mechanical waves and electromagnetic radiation. These entries are to be contrasted with another entry—matter transfer—in which the energy transfer is accompanied by a movement of matter through space.

All waves carry energy, but the amount of energy transmitted through a medium and the mechanism responsible for the energy transport differ from case to case. For instance, the power of ocean waves during a storm is much greater than that of sound waves generated by a musical instrument.

All mechanical waves require (1) some source of disturbance, (2) a medium that can be disturbed, and (3) some physical mechanism through which elements of the medium can influence one another. This final requirement assures that a disturbance to one element will cause a disturbance to the next so that the disturbance will indeed propagate through the medium.

One way to demonstrate wave motion is to flip the free end of a long rope that is under tension and has its opposite end fixed as in Figure 13.1. In this manner, a single **pulse** is formed and travels (to the right in Fig. 13.1) with a definite speed. The rope is the medium through which the pulse travels. Figure 13.1 represents consecutive "snapshots" of the traveling pulse. The shape of the pulse changes very little as it travels along the rope.

As the pulse travels, each rope element that is disturbed moves in a direction perpendicular to the direction of propagation. Figure 13.2 illustrates this point for a particular element, labeled *P*. Note that there is no motion of any part of the rope that is in the direction of the wave. A disturbance such as this one in which the elements of the disturbed medium move perpendicularly to the direction of propagation is called a **transverse wave.**

In another class of mechanical waves, called **longitudinal waves,** the elements of the medium undergo displacements *parallel* to the direction of propagation. Sound waves in air, for instance, are longitudinal. Their disturbance corresponds to a series of high- and low-pressure regions that may travel through air or through any material medium with a certain speed. A longitudinal pulse can be easily produced in a stretched spring as in Figure 13.3. A group of coils at the free end is pushed forward and pulled back. This action produces a pulse in the form of a compressed region of coils that travels along the spring.

So far, we have provided pictorial representations of a traveling pulse and hope you have begun to develop a mental representation of such a pulse. Let us now develop a mathematical representation for the propagation of this pulse. Consider a pulse traveling to the right with constant speed v on a long, stretched string as in Figure 13.4. The pulse moves along the x axis (the axis of the string), and the transverse (up-and-down) displacement of the elements of the string is described by means of the position y.

Figure 13.4a represents the shape and position of the pulse at time $t = 0$. At this time, the shape of the pulse, whatever it may be, can be represented by some mathematical function that we will write as $y(x, 0) = f(x)$. This function describes the vertical position y of the element of the string located at each value of x at time

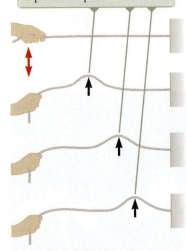

As the pulse moves along the string, new elements of the string are displaced from their equilibrium positions.

Figure 13.1 A hand moves the end of a stretched string up and down once (red arrow), causing a pulse to travel along the string.

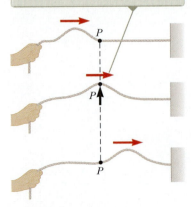

The direction of the displacement of any element at a point *P* on the string is perpendicular to the direction of propagation (red arrow).

Figure 13.2 The displacement of a particular string element for a transverse pulse traveling on a stretched string.

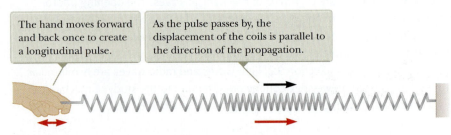

The hand moves forward and back once to create a longitudinal pulse.

As the pulse passes by, the displacement of the coils is parallel to the direction of the propagation.

Figure 13.3 A longitudinal pulse along a stretched spring.

$t = 0$. Because the speed of the pulse is v, the pulse has traveled to the right a distance vt at time t (Fig. 13.4b). We adopt a simplification model in which the shape of the pulse does not change with time.[1] Therefore, at time t, the shape of the pulse is the same as it was at time $t = 0$, as in Figure 13.4a. Consequently, an element of the string at x at this time has the same y position as an element located at $x - vt$ had at time $t = 0$:

$$y(x, t) = y(x - vt, 0)$$

In general, then, we can represent the position y for all values of x and t, measured in a stationary frame with the origin at O, as

$$y(x, t) = f(x - vt) \qquad \text{(pulse traveling to the right)} \qquad \textbf{13.1} \blacktriangleleft$$

If the pulse travels to the left, the position of an element of the string is described by

$$y(x, t) = f(x + vt) \qquad \text{(pulse traveling to the left)} \qquad \textbf{13.2} \blacktriangleleft$$

The function y, sometimes called the **wave function,** depends on the two variables x and t. For this reason, it is often written $y(x, t)$, which is read "y as a function of x and t."

It is important to understand the meaning of y. Consider a point P on the string, identified by a particular value of its x coordinate as in Figure 13.4. As the pulse passes through P, the y coordinate of this point increases, reaches a maximum, and then decreases to zero. The wave function $y(x, t)$ represents the y position of any element of string located at position x at any time t. Furthermore, if t is fixed (e.g., in the case of taking a snapshot of the pulse), the wave function y as a function of x, sometimes called the **waveform,** defines a curve representing the actual geometric shape of the pulse at that time.

QUICK QUIZ 13.1 (i) In a long line of people waiting to buy tickets, the first person leaves and a pulse of motion occurs as people step forward to fill the gap. As each person steps forward, the gap moves through the line. Is the propagation of this gap (a) transverse or (b) longitudinal? (ii) Consider "the wave" at a baseball game when people stand up and shout as the wave arrives at their location and the resultant pulse moves around the stadium. Is this wave (a) transverse or (b) longitudinal?

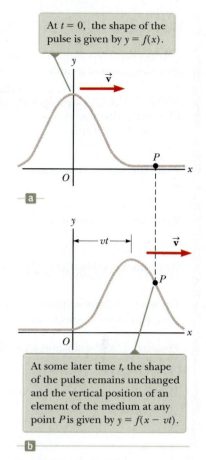

At $t = 0$, the shape of the pulse is given by $y = f(x)$.

a

At some later time t, the shape of the pulse remains unchanged and the vertical position of an element of the medium at any point P is given by $y = f(x - vt)$.

b

Figure 13.4 A one-dimensional pulse traveling to the right with a speed v.

Example 13.1 | A Pulse Moving to the Right

A pulse moving to the right along the x axis is represented by the wave function

$$y(x, t) = \frac{2}{(x - 3.0t)^2 + 1}$$

where x and y are measured in centimeters and t is measured in seconds. Find expressions for the wave function at $t = 0$, $t = 1.0$ s, and $t = 2.0$ s.

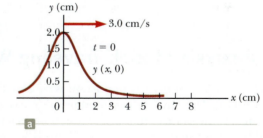

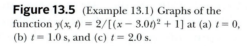

Figure 13.5 (Example 13.1) Graphs of the function $y(x, t) = 2/[(x - 3.0t)^2 + 1]$ at (a) $t = 0$, (b) $t = 1.0$ s, and (c) $t = 2.0$ s.

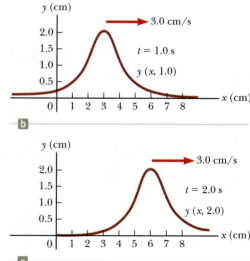

SOLUTION

Conceptualize Figure 13.5a shows the pulse represented by this wave function at $t = 0$. Imagine this pulse moving to the right and maintaining its shape as suggested by Figures 13.5b and 13.5c.

continued

---

[1] In reality, the pulse changes its shape and gradually spreads out during the motion. This effect, called *dispersion,* is common to many mechanical waves, but we adopt a simplification model that ignores this effect.

13.1 *cont.*

Categorize We categorize this example as a relatively simple analysis problem in which we interpret the mathematical representation of a pulse.

. . . .

Analyze The wave function is of the form $y = f(x - vt)$. Inspection of the expression for $y(x, t)$ and comparison to Equation 13.1 reveal that the wave speed is $v = 3.0$ cm/s. Furthermore, by letting $x - 3.0t = 0$, we find that the maximum value of y is given by $A = 2.0$ cm.

Write the wave function expression at $t = 0$:

$$y(x, 0) = \frac{2}{x^2 + 1}$$

Write the wave function expression at $t = 1.0$ s:

$$y(x, 1.0) = \frac{2}{(x - 3.0)^2 + 1}$$

Write the wave function expression at $t = 2.0$ s:

$$y(x, 2.0) = \frac{2}{(x - 6.0)^2 + 1}$$

For each of these expressions, we can substitute various values of x and plot the wave function. This procedure yields the wave functions shown in the three parts of Figure 13.5.

. . . .

Finalize These snapshots show that the pulse moves to the right without changing its shape and that it has a constant speed of 3.0 cm/s.

What If? What if the wave function were

$$y(x, t) = \frac{4}{(x + 3.0t)^2 + 1}$$

How would that change the situation?

Answer One new feature in this expression is the plus sign in the denominator rather than the minus sign. The new expression represents a pulse with a similar shape as that in Figure 13.5, but moving to the left as time progresses. Another new feature here is the numerator of 4 rather than 2. Therefore, the new expression represents a pulse with twice the height of that in Figure 13.5.

❰13.2 | Analysis Model: Traveling Wave

In this section, we introduce an important wave function whose shape is shown in Active Figure 13.6. The wave represented by this curve is called a **sinusoidal wave** because the curve is the same as that of the function $\sin \theta$ plotted against θ. A sinusoidal wave could be established on the rope in Figure 13.1 by shaking the end of the rope up and down in simple harmonic motion.

The sinusoidal wave is the simplest example of a periodic continuous wave and can be used to build more complex waves (see Section 14.6). The brown curve in Active Figure 13.6 represents a snapshot of a traveling sinusoidal wave at $t = 0$, and the blue curve represents a snapshot of the wave at some later time t. Imagine two types of motion that can occur. First, the entire waveform in Active Figure 13.6 moves to the right so that the brown curve moves toward the right and eventually reaches the position of the blue curve. This movement is the motion of the *wave*. If we focus on one element of the medium, such as the element at $x = 0$, we see that each element moves up and down along the y axis in simple harmonic motion. This movement is the motion of the *elements of the medium*. It is important to differentiate between the motion of the wave and the motion of the elements of the medium.

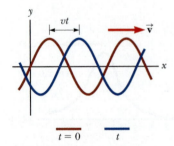

Active Figure 13.6 A one-dimensional sinusoidal wave traveling to the right with a speed v. The brown curve represents a snapshot of the wave at $t = 0$, and the blue curve represents a snapshot at some later time t.

In the early chapters of this book, we developed several analysis models based on three simplification models: the particle, the system, and the rigid object. With our introduction to waves, we can develop a new simplification model, the **wave,** that will allow us to explore more analysis models for solving problems. An ideal particle has zero size. We can build physical objects with nonzero size as combinations of particles. Therefore, the particle can be considered a basic building block. An ideal wave has a single frequency and is infinitely long; that is, the wave exists throughout the Universe. (A wave of finite length must necessarily have a mixture of frequencies.) When this concept is explored in Section 14.6, we will find that ideal waves can be combined to build complex waves, just as we combined particles.

In what follows, we will develop the principal features and mathematical representations of the analysis model of a **traveling wave.** This model is used in situations in which a wave moves through space without interacting with other waves or particles.

Active Figure 13.7a shows a snapshot of a wave moving through a medium. Active Figure 13.7b shows a graph of the position of one element of the medium as a function of time. A point in Active Figure 13.7a at which the displacement of the element from its normal position is highest is called the **crest** of the wave. The lowest point is called the **trough.** The distance from one crest to the next is called the **wavelength** λ (Greek letter lambda). More generally, the wavelength is the minimum distance between any two identical points on adjacent waves as shown in Active Figure 13.7a.

If you count the number of seconds between the arrivals of two adjacent crests at a given point in space, you measure the **period** T of the waves. In general, the period is the time interval required for two identical points of adjacent waves to pass by a point as shown in Active Figure 13.7b. The period of the wave is the same as the period of the simple harmonic oscillation of one element of the medium.

The same information is more often given by the inverse of the period, which is called the **frequency** f. In general, the frequency of a periodic wave is the number of crests (or troughs, or any other point on the wave) that pass a given point in a unit time interval. The frequency of a sinusoidal wave is related to the period by the expression

$$f = \frac{1}{T} \qquad \qquad \textbf{13.3} \blacktriangleleft$$

The frequency of the wave is the same as the frequency of the simple harmonic oscillation of one element of the medium. The most common unit for frequency, as we learned in Chapter 12, is s^{-1}, or **hertz** (Hz). The corresponding unit for T is seconds.

The maximum position of an element of the medium relative to its equilibrium position is called the **amplitude** A of the wave as indicated in Active Figure 13.7.

Waves travel with a specific speed, and this speed depends on the properties of the medium being disturbed. For instance, sound waves travel through room-temperature air with a speed of about 343 m/s (781 mi/h), whereas they travel through most solids with a speed greater than 343 m/s.

Consider the sinusoidal wave in Active Figure 13.7a, which shows the position of the wave at $t = 0$. Because the wave is sinusoidal, we expect the wave function at this instant to be expressed as $y(x, 0) = A \sin ax$, where A is the amplitude and a is a constant to be determined. At $x = 0$, we see that $y(0, 0) = A \sin a(0) = 0$, consistent with Active Figure 13.7a. The next value of x for which y is zero is $x = \lambda/2$. Therefore,

$$y\left(\frac{\lambda}{2}, 0\right) = A \sin\left(a\frac{\lambda}{2}\right) = 0$$

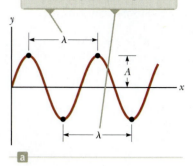

The wavelength λ of a wave is the distance between adjacent crests or adjacent troughs.

a

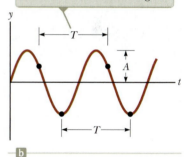

The period T of a wave is the time interval required for the element to complete one cycle of its oscillation and for the wave to travel one wavelength.

b

Active Figure 13.7 (a) A snapshot of a sinusoidal wave. (b) The position of one element of the medium as a function of time.

Pitfall Prevention | 13.1
What's the Difference Between Active Figures 13.7a and 13.7b?
Notice the visual similarity between Active Figures 13.7a and 13.7b. The shapes are the same, but (a) is a graph of vertical position versus horizontal position, whereas (b) is vertical position versus time. Active Figure 13.7a is a pictorial representation of the wave *for a series of elements of the medium;* it is what you would see at an instant of time. Active Figure 13.7b is a graphical representation of the position of *one element of the medium* as a function of time. That both figures have the identical shape represents Equation 13.1: a wave is the *same* function of both x and t.

For this equation to be true, we must have $a\lambda/2 = \pi$, or $a = 2\pi/\lambda$. Therefore, the function describing the positions of the elements of the medium through which the sinusoidal wave is traveling can be written

$$y(x, 0) = A \sin\left(\frac{2\pi}{\lambda} x\right) \qquad \text{13.4} \blacktriangleleft$$

where the constant A represents the wave amplitude and the constant λ is the wavelength. Notice that the vertical position of an element of the medium is the same whenever x is increased by an integral multiple of λ. Based on our discussion of Equation 13.1, if the wave moves to the right with a speed v, the wave function at some later time t is

$$y(x, t) = A \sin\left[\frac{2\pi}{\lambda}(x - vt)\right] \qquad \text{13.5} \blacktriangleleft$$

If the wave were traveling to the left, the quantity $x - vt$ would be replaced by $x + vt$ as we learned when we developed Equations 13.1 and 13.2.

By definition, the wave travels through a displacement Δx equal to one wavelength λ in a time interval Δt of one period T. Therefore, the wave speed, wavelength, and period are related by the expression

$$v = \frac{\Delta x}{\Delta t} = \frac{\lambda}{T} \qquad \text{13.6} \blacktriangleleft$$

Substituting this expression for v into Equation 13.5 gives

$$y = A \sin\left[2\pi\left(\frac{x}{\lambda} - \frac{t}{T}\right)\right] \qquad \text{13.7} \blacktriangleleft$$

This form of the wave function shows the *periodic* nature of y. Note that we will often use y rather than $y(x, t)$ as a shorthand notation. At any given time t, y has the *same* value at the positions x, $x + \lambda$, $x + 2\lambda$, and so on. Furthermore, at any given position x, the value of y is the same at times t, $t + T$, $t + 2T$, and so on.

We can express the wave function in a convenient form by defining two other quantities, the **angular wave number** k (usually called simply the **wave number**) and the **angular frequency** ω:

▶ Angular wave number

$$k \equiv \frac{2\pi}{\lambda} \qquad \text{13.8} \blacktriangleleft$$

▶ Angular frequency

$$\omega \equiv \frac{2\pi}{T} = 2\pi f \qquad \text{13.9} \blacktriangleleft$$

Using these definitions, Equation 13.7 can be written in the more compact form

▶ Wave function for a sinusoidal wave

$$y = A \sin(kx - \omega t) \qquad \text{13.10} \blacktriangleleft$$

Using Equations 13.3, 13.8, and 13.9, the wave speed v originally given in Equation 13.6 can be expressed in the following alternative forms:

$$v = \frac{\omega}{k} \qquad \text{13.11} \blacktriangleleft$$

▶ Speed of a sinusoidal wave

$$v = \lambda f \qquad \text{13.12} \blacktriangleleft$$

The wave function given by Equation 13.10 assumes the vertical position y of an element of the medium is zero at $x = 0$ and $t = 0$. That need not be the case. If it is not, we generally express the wave function in the form

▶ General expression for a sinusoidal wave

$$y = A \sin(kx - \omega t + \phi) \qquad \text{13.13} \blacktriangleleft$$

where ϕ is the **phase constant,** just as we learned in our study of periodic motion in Chapter 12. This constant can be determined from the initial conditions. The primary equations in the mathematical representation of the traveling wave analysis model are Equations 13.3, 13.10, and 13.12.

QUICK QUIZ 13.2 A sinusoidal wave of frequency f is traveling along a stretched string. The string is brought to rest, and a second traveling wave of frequency $2f$ is established on the string. (**i**) What is the wave speed of the second wave? (a) twice that of the first wave (b) half that of the first wave (c) the same as that of the first wave (d) impossible to determine (**ii**) From the same choices, describe the wavelength of the second wave. (**iii**) From the same choices, describe the amplitude of the second wave.

Example 13.2 | **A Traveling Sinusoidal Wave**

A sinusoidal wave traveling in the positive x direction has an amplitude of 15.0 cm, a wavelength of 40.0 cm, and a frequency of 8.00 Hz. The vertical position of an element of the medium at $t = 0$ and $x = 0$ is also 15.0 cm as shown in Figure 13.8.

(A) Find the wave number k, period T, angular frequency ω, and speed v of the wave.

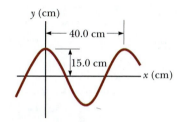

Figure 13.8 (Example 13.2) A sinusoidal wave of wavelength $\lambda = 40.0$ cm and amplitude $A = 15.0$ cm.

SOLUTION

Conceptualize Figure 13.8 shows the wave at $t = 0$. Imagine this wave moving to the right and maintaining its shape.

Categorize We will evaluate parameters of the wave using equations generated in the preceding discussion, so we categorize this example as a substitution problem.

Evaluate the wave number from Equation 13.8:

$$k = \frac{2\pi}{\lambda} = \frac{2\pi \text{ rad}}{40.0 \text{ cm}} = \boxed{15.7 \text{ rad/m}}$$

Evaluate the period of the wave from Equation 13.3:

$$T = \frac{1}{f} = \frac{1}{8.00 \text{ s}^{-1}} = \boxed{0.125 \text{ s}}$$

Evaluate the angular frequency of the wave from Equation 13.9:

$$\omega = 2\pi f = 2\pi(8.00 \text{ s}^{-1}) = \boxed{50.3 \text{ rad/s}}$$

Evaluate the wave speed from Equation 13.12:

$$v = \lambda f = (40.0 \text{ cm})(8.00 \text{ s}^{-1}) = \boxed{3.20 \text{ m/s}}$$

(B) Determine the phase constant ϕ and write a general expression for the wave function.

SOLUTION

Substitute $A = 15.0$ cm, $y = 15.0$ cm, $x = 0$, and $t = 0$ into Equation 13.13:

$$15.0 = (15.0)\sin\phi \quad\rightarrow\quad \sin\phi = 1 \quad\rightarrow\quad \phi = \frac{\pi}{2}\text{ rad}$$

Write the wave function:

$$y = A\sin\left(kx - \omega t + \frac{\pi}{2}\right) = A\cos(kx - \omega t)$$

Substitute the values for A, k, and ω in SI units into this expression:

$$y = \boxed{0.150 \cos(15.7x - 50.3t)}$$

The Linear Wave Equation

In Figure 13.1, we demonstrated how to create a pulse by jerking a taut string up and down once. To create a series of such pulses—a wave—let's replace the hand with an oscillating blade vibrating in simple harmonic motion. Active Figure 13.9 (page 422) represents snapshots of the wave created in this way at intervals of $T/4$. Because the end of the blade oscillates in simple harmonic motion, each element of the string, such as that at P, also oscillates vertically with simple harmonic motion. Therefore, every element of the string can be treated as a simple

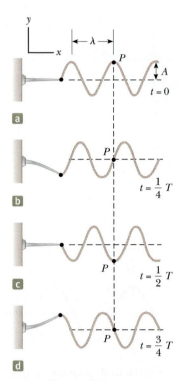

Active Figure 13.9 One method for producing a sinusoidal wave on a string. The left end of the string is connected to a blade that is set into oscillation. Every element of the string, such as that at point P, oscillates with simple harmonic motion in the vertical direction.

Pitfall Prevention | 13.2
Two Kinds of Speed/Velocity
Do not confuse v, the speed of the wave as it propagates along the string, with v_y, the transverse velocity of a point on the string. The speed v is constant for a uniform medium, whereas v_y varies sinusoidally.

harmonic oscillator vibrating with a frequency equal to the frequency of oscillation of the blade.[2] Notice that while each element oscillates in the y direction, the wave travels in the x direction with a speed v. Of course, that is the definition of a transverse wave.

If we define $t = 0$ as the time for which the configuration of the string is as shown in Active Figure 13.9a, the wave function can be written as in Equation 13.10:

$$y = A \sin (kx - \omega t)$$

We can use this expression to describe the motion of any element of the string. An element at point P (or any other element of the string) moves only vertically, and so its x coordinate remains constant. Therefore, the **transverse speed** v_y (not to be confused with the wave speed v) and the **transverse acceleration** a_y of elements of the string are

$$v_y = \frac{dy}{dt}\bigg]_{x=\text{constant}} = \frac{\partial y}{\partial t} = -\omega A \cos(kx - \omega t) \qquad \text{13.14} \blacktriangleleft$$

$$a_y = \frac{dv_y}{dt}\bigg]_{x=\text{constant}} = \frac{\partial v_y}{\partial t} = \frac{\partial^2 y}{\partial t^2} = -\omega^2 A \sin(kx - \omega t) \qquad \text{13.15} \blacktriangleleft$$

These expressions incorporate partial derivatives because y depends on both x and t. In the operation $\partial y / \partial t$, for example, we take a derivative with respect to t while holding x constant. The maximum magnitudes of the transverse speed and transverse acceleration are simply the absolute values of the coefficients of the cosine and sine functions:

$$v_{y,\text{max}} = \omega A \qquad \text{13.16} \blacktriangleleft$$

$$a_{y,\text{max}} = \omega^2 A \qquad \text{13.17} \blacktriangleleft$$

The transverse speed and transverse acceleration of elements of the string do not reach their maximum values simultaneously. The transverse speed reaches its maximum value (ωA) when $y = 0$, whereas the magnitude of the transverse acceleration reaches its maximum value ($\omega^2 A$) when $y = \pm A$. Finally, Equations 13.16 and 13.17 are identical in mathematical form to the corresponding equations for simple harmonic motion, Equations 12.17 and 12.18.

QUICK QUIZ 13.3 The amplitude of a wave is doubled, with no other changes made to the wave. As a result of this doubling, which of the following statements is correct? **(a)** The speed of the wave changes. **(b)** The frequency of the wave changes. **(c)** The maximum transverse speed of an element of the medium changes. **(d)** Statements **(a)** through **(c)** are all true. **(e)** None of statements **(a)** through **(c)** is true.

Now let's take derivatives of our wave function (Eq. 13.10) with respect to position at a fixed time, similar to the process by which we took derivatives with respect to time in Equations 13.14 and 13.15:

$$\frac{dy}{dx}\bigg]_{t=\text{constant}} = \frac{\partial y}{\partial x} = kA \cos(kx - \omega t) \qquad \text{13.18} \blacktriangleleft$$

$$\frac{d^2 y}{dx^2}\bigg]_{t=\text{constant}} = \frac{\partial^2 y}{\partial x^2} = -k^2 A \sin(kx - \omega t) \qquad \text{13.19} \blacktriangleleft$$

Comparing Equations 13.15 and 13.19, we see that

$$A \sin(kx - \omega t) = -\frac{1}{k^2}\frac{\partial^2 y}{\partial x^2} = -\frac{1}{\omega^2}\frac{\partial^2 y}{\partial t^2} \quad \rightarrow \quad \frac{\partial^2 y}{\partial x^2} = \frac{k^2}{\omega^2}\frac{\partial^2 y}{\partial t^2}$$

---

[2]In this arrangement, we are assuming that a string element always oscillates in a vertical line. The tension in the string would vary if an element were allowed to move sideways. Such motion would make the analysis very complex.

Using Equation 13.11, we can rewrite this expression as

$$\frac{\partial^2 y}{\partial x^2} = \frac{1}{v^2}\frac{\partial^2 y}{\partial t^2}$$

13.20 ◀ ▶ Linear wave equation

which is known as the **linear wave equation.** If we analyze a situation and find this kind of relationship between derivatives of a function describing the situation, wave motion is occurring. Equation 13.20 is a differential equation representation of the traveling wave model. The solutions to the equation describe **linear mechanical waves.** We have developed the linear wave equation from a sinusoidal mechanical wave traveling through a medium, but it is much more general. The linear wave equation successfully describes waves on strings, sound waves, and also electromagnetic waves.[3] What's more, although the sinusoidal wave that we have studied is a solution to Equation 13.20, the general solution to the equation is *any* function of the form $y(x, t) = f(x \pm vt)$ as discussed in Section 13.1.

Nonlinear waves are more difficult to analyze, but they are an important area of current research, especially in optics. An example of a nonlinear mechanical wave is one for which the amplitude is not small compared with the wavelength.

Example 13.3 | A Solution to the Linear Wave Equation

Verify that the wave function presented in Example 13.1 is a solution to the linear wave equation.

SOLUTION

Conceptualize Look back at Figure 13.5 for a pictorial representation of the pulse. Imagine the pulse moving to the right as suggested by the three parts of the figure.

Categorize This is not an example of the traveling wave model because the moving entity is a single pulse, with no discernible wavelength or frequency. The linear wave equation, however, applies to both waves and pulses.

Analyze Write an expression for the wave function:

$$y(x, t) = \frac{2}{(x - 3.0t)^2 + 1}$$

Take partial derivatives of this function with respect to x and to t:

$$(1) \quad \frac{\partial^2 y}{\partial x^2} = \frac{12(x - 3.0t)^2 - 4.0}{[(x - 3.0t)^2 + 1]^3}$$

$$(2) \quad \frac{\partial^2 y}{\partial t^2} = \frac{108(x - 3.0t)^2 - 36}{[(x - 3.0t)^2 + 1]^3} = 9.0\frac{[12(x - 3.0t)^2 - 4.0]}{[(x - 3.0t)^2 + 1]^3}$$

Use Equations (1) and (2) to find a relationship between the left sides of these expressions:

$$\frac{\partial^2 y}{\partial x^2} = \frac{1}{9.0}\frac{\partial^2 y}{\partial t^2}$$

Finalize Comparing this result with Equation 13.20, we see that the wave function is a solution to the linear wave equation if the speed at which the pulse moves is 3.0 cm/s. We have already determined in Example 13.1 that this speed is indeed the speed of the pulse, so we have proven what we set out to do.

13.3 | The Speed of Transverse Waves on Strings

An aspect of the behavior of linear mechanical waves is that the wave speed depends only on the properties of the medium through which the wave travels. Waves for which the amplitude A is small relative to the wavelength λ can be represented as linear waves. In this section, we determine the speed of a transverse wave traveling on a stretched string.

---

[3]In the case of electromagnetic waves, *y* is interpreted to represent an electric field, which we will study in Chapter 24.

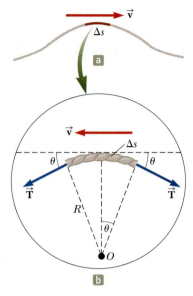

Figure 13.10 (a) In the reference frame of the Earth, a pulse moves to the right on a string with speed v. (b) In a frame of reference moving to the right with the pulse, the small element of length Δs moves to the left with speed v.

Let us use a mechanical analysis to derive the expression for the speed of a pulse traveling on a stretched string under tension T. Consider a pulse moving to the right with a uniform speed v, measured relative to a stationary (with respect to the Earth) inertial reference frame as shown in Figure 13.10a. Recall from Chapter 9 that Newton's laws are valid in any inertial reference frame. Therefore, let us view this pulse from a different inertial reference frame, one that moves along with the pulse at the same speed so that the pulse appears to be at rest in the frame as in Figure 13.10b. In this reference frame, the pulse remains fixed and each element of the string moves to the left through the pulse shape.

A short element of the string, of length Δs, forms an approximate arc of a circle of radius R as shown in the magnified view in Figure 13.10b. In our moving frame of reference, the element of the string moves to the left with speed v. As it travels through the arc, we can model the element as a particle in uniform circular motion. This element has a centripetal acceleration of v^2/R, which is supplied by components of the force $\vec{T}$ whose magnitude is the tension in the string. The force $\vec{T}$ acts on each side of the element, tangent to the arc, as in Figure 13.10b. The horizontal components of $\vec{T}$ cancel, and each vertical component $T \sin\theta$ acts downward. Hence, the magnitude of the total radial force on the element is $2T \sin\theta$. Because the element is small, θ is small and we can use the small-angle approximation $\sin\theta \approx \theta$. Therefore, the magnitude of the total radial force is

$$F_r = 2T \sin\theta \approx 2T\theta$$

The element has mass $m = \mu\,\Delta s$, where μ is the mass per unit length of the string. Because the element forms part of a circle and subtends an angle of 2θ at the center, $\Delta s = R(2\theta)$, and

$$m = \mu\,\Delta s = 2\mu R\theta$$

Applying Newton's second law to this element in the radial direction gives

$$F_r = \frac{mv^2}{R} \quad\rightarrow\quad 2T\theta = \frac{2\mu R\theta v^2}{R} \quad\rightarrow\quad T = \mu v^2$$

Solving for v gives

▶ Speed of a wave on a stretched string

$$v = \sqrt{\frac{T}{\mu}}$$

13.21◀

Notice that this derivation is based on the assumption that the pulse height is small relative to the length of the pulse. Using this assumption, we were able to use the approximation $\sin\theta \approx \theta$. Furthermore, the model assumes that the tension T is not affected by the presence of the pulse, so T is the same at all points on the pulse. Finally, this proof does *not* assume any particular shape for the pulse. We therefore conclude that a pulse of *any shape* will travel on the string with speed $v = \sqrt{T/\mu}$, without any change in pulse shape.

Pitfall Prevention | 13.3
Multiple T's
Do not confuse the T in Equation 13.21 for the tension with the symbol T used in this chapter for the period of a wave. The context of the equation should help you identify which quantity is meant. There simply aren't enough letters in the alphabet to assign a unique letter to each variable!

◣ **QUICK QUIZ 13.4** Suppose you create a pulse by moving the free end of a taut string up and down once with your hand beginning at $t = 0$. The string is attached at its other end to a distant wall. The pulse reaches the wall at time t. Which of the following actions, taken by itself, decreases the time interval required for the pulse to reach the wall? More than one choice may be correct. (a) moving your hand more quickly, but still only up and down once by the same amount (b) moving your hand more slowly, but still only up and down once by the same amount (c) moving your hand a greater distance up and down in the same amount of time (d) moving your hand a lesser distance up and down in the same amount of time (e) using a heavier string of the same length and under the same tension (f) using a lighter string of the same length and under the same tension (g) using a string of the same linear mass density but under decreased tension (h) using a string of the same linear mass density but under increased tension

> **THINKING PHYSICS 13.1**

A secret agent is trapped in a building on top of an elevator car at a lower floor. He attempts to signal a fellow agent on the roof by tapping a message in Morse code on the elevator cable so that transverse pulses move upward on the cable. As the pulses move up the cable toward the accomplice, does the speed with which they move stay the same, increase, or decrease? If the pulses are sent 1 s apart, are they received 1 s apart by the agent on the roof?

Reasoning The elevator cable can be modeled as a vertical string. The speed of waves on the cable is a function of the tension in the cable. As the waves move higher on the cable, they encounter increased tension because each higher point on the cable must support the weight of all the cable below it (and the elevator). Therefore, the speed of the pulses increases as they move higher on the cable. The frequency of the pulses will not be affected because each pulse takes the same time interval to reach the top. They will still arrive at the top of the cable at intervals of 1 s. ◄

Example 13.4 | The Speed of a Pulse on a Cord

A uniform string has a mass of 0.300 kg and a length of 6.00 m. The string passes over a pulley and supports a 2.00-kg object (Fig. 13.11). Find the speed of a pulse traveling along this string.

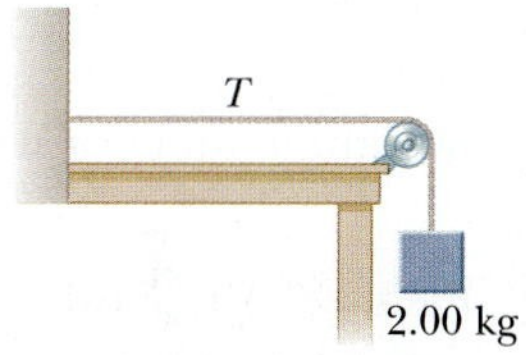

SOLUTION

Conceptualize In Figure 13.11, the hanging block establishes a tension in the horizontal string. This tension determines the speed with which waves move on the string.

Categorize To find the tension in the string, we model the hanging block as a particle in equilibrium. Then we use the tension to evaluate the wave speed on the string using Equation 13.21.

Figure 13.11 (Example 13.4) The tension T in the cord is maintained by the suspended object. The speed of any wave traveling along the cord is given by $v = \sqrt{T/\mu}$.

Analyze Apply the particle in equilibrium model to the block:

$$\sum F_y = T - m_{block}g = 0$$

Solve for the tension in the string:

$$T = m_{block}g$$

Use Equation 13.21 to find the wave speed, using $\mu = m_{string}/\ell$ for the linear mass density of the string:

$$v = \sqrt{\frac{T}{\mu}} = \sqrt{\frac{m_{block}g\,\ell}{m_{string}}}$$

Evaluate the wave speed:

$$v = \sqrt{\frac{(2.00 \text{ kg})(9.80 \text{ m/s}^2)(6.00 \text{ m})}{0.300 \text{ kg}}} = \boxed{19.8 \text{ m/s}}$$

Finalize The calculation of the tension neglects the small mass of the string. Strictly speaking, the string can never be exactly straight; therefore, the tension is not uniform.

Example 13.5 | Rescuing the Hiker

An 80.0-kg hiker is trapped on a mountain ledge following a storm. A helicopter rescues the hiker by hovering above him and lowering a cable to him. The mass of the cable is 8.00 kg, and its length is 15.0 m. A sling of mass 70.0 kg is attached to the end of the cable. The hiker attaches himself to the sling, and the helicopter then accelerates upward. Terrified by hanging from the cable in midair, the hiker tries to signal the pilot by sending transverse pulses up the cable. A pulse takes 0.250 s to travel the length of the cable. What is the acceleration of the helicopter? Assume the tension in the cable is uniform.

SOLUTION

Conceptualize Imagine the effect of the acceleration of the helicopter on the cable. The greater the upward acceleration, the larger the tension in the cable. In turn, the larger the tension, the higher the speed of pulses on the cable.

continued

> **13.5** *cont.*

Categorize This problem is a combination of one involving the speed of pulses on a string and one in which the hiker and sling are modeled as a particle under a net force.

Analyze Use the time interval for the pulse to travel from the hiker to the helicopter to find the speed of the pulses on the cable:

$$v = \frac{\Delta x}{\Delta t} = \frac{15.0 \text{ m}}{0.250 \text{ s}} = 60.0 \text{ m/s}$$

Solve Equation 13.21 for the tension in the cable:

$$v = \sqrt{\frac{T}{\mu}} \quad \rightarrow \quad T = \mu v^2$$

Model the hiker and sling as a particle under a net force, noting that the acceleration of this particle of mass m is the same as the acceleration of the helicopter:

$$\sum F = ma \quad \rightarrow \quad T - mg = ma$$

Solve for the acceleration:

$$a = \frac{T}{m} - g = \frac{\mu v^2}{m} - g = \frac{m_{cable} \, v^2}{\ell_{cable} \, m} - g$$

Substitute numerical values:

$$a = \frac{(8.00 \text{ kg})(60.0 \text{ m/s})^2}{(15.0 \text{ m})(150.0 \text{ kg})} - 9.80 \text{ m/s}^2 = \boxed{3.00 \text{ m/s}^2}$$

Finalize A real cable has stiffness in addition to tension. Stiffness tends to return a wire to its original straight-line shape even when it is not under tension. For example, a piano wire straightens if released from a curved shape; package-wrapping string does not.

Stiffness represents a restoring force in addition to tension and increases the wave speed. Consequently, for a real cable, the speed of 60.0 m/s that we determined is most likely associated with a smaller acceleration of the helicopter.

❮13.4 | Reflection and Transmission

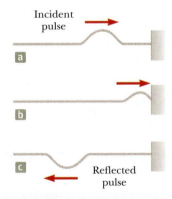

Active Figure 13.12 The reflection of a traveling pulse at the fixed end of a stretched string. The reflected pulse is inverted, but its shape is otherwise unchanged.

The traveling wave model describes waves traveling through a uniform medium without interacting with anything along the way. We now consider how a traveling wave is affected when it encounters a change in the medium. For example, consider a pulse traveling on a string that is rigidly attached to a support at one end as in Active Figure 13.12. When the pulse reaches the support, a severe change in the medium occurs: the string ends. As a result, the pulse undergoes **reflection**; that is, the pulse moves back along the string in the opposite direction.

Notice that the reflected pulse is *inverted*. This inversion can be explained as follows. When the pulse reaches the fixed end of the string, the string produces an upward force on the support. By Newton's third law, the support must exert an equal-magnitude and oppositely directed (downward) reaction force on the string. This downward force causes the pulse to invert upon reflection.

Now consider another case. This time, the pulse arrives at the end of a string that is free to move vertically as in Active Figure 13.13. The tension at the free end is maintained because the string is tied to a ring of negligible mass that is free to slide vertically on a smooth post without friction. Again, the pulse is reflected, but this time it is not inverted. When it reaches the post, the pulse exerts a force on the free end of the string, causing the ring to accelerate upward. The ring rises as high as the incoming pulse, and then the downward component of the tension force pulls the ring back down. This

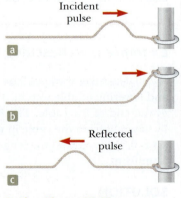

Active Figure 13.13 The reflection of a traveling pulse at the free end of a stretched string. The reflected pulse is not inverted.

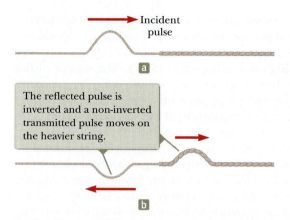

→ Incident pulse

a

The reflected pulse is inverted and a non-inverted transmitted pulse moves on the heavier string.

←

b

Active Figure 13.14 (a) A pulse traveling to the right on a light string approaches the junction with a heavier string. (b) The situation after the pulse reaches the junction.

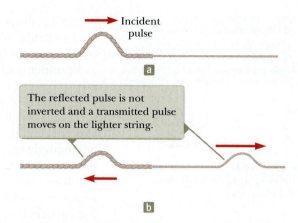

→ Incident pulse

a

The reflected pulse is not inverted and a transmitted pulse moves on the lighter string.

→

←

b

Active Figure 13.15 (a) A pulse traveling to the right on a heavy string approaches the junction with a lighter string. (b) The situation after the pulse reaches the junction.

movement of the ring produces a reflected pulse that is not inverted and that has the same amplitude as the incoming pulse.

Finally, consider a situation in which the boundary is intermediate between these two extremes. In this case, part of the energy in the incident pulse is reflected and part undergoes **transmission;** that is, some of the energy passes through the boundary. For instance, suppose a light string is attached to a heavier string as in Active Figure 13.14. When a pulse traveling on the light string reaches the boundary between the two strings, part of the pulse is reflected and inverted and part is transmitted to the heavier string. The reflected pulse is inverted for the same reasons described earlier in the case of the string rigidly attached to a support.

The reflected pulse has a smaller amplitude than the incident pulse. In Section 13.5, we show that the energy carried by a wave is related to its amplitude. According to the principle of the conservation of energy, when the pulse breaks up into a reflected pulse and a transmitted pulse at the boundary, the sum of the energies of these two pulses must equal the energy of the incident pulse. Because the reflected pulse contains only part of the energy of the incident pulse, its amplitude must be smaller.

When a pulse traveling on a heavy string strikes the boundary between the heavy string and a lighter one as in Active Figure 13.15, again part is reflected and part is transmitted. In this case, the reflected pulse is not inverted.

In either case, the relative heights of the reflected and transmitted pulses depend on the relative densities of the two strings. If the strings are identical, there is no discontinuity at the boundary and no reflection takes place.

According to Equation 13.21, the speed of a wave on a string decreases as the mass per unit length of the string increases. In other words, a wave travels more slowly on a heavy string than on a light string if both are under the same tension. The following general rules apply to reflected waves: When a wave or pulse travels from medium A to medium B and $v_A > v_B$ (that is, when B is denser than A), it is inverted upon reflection. When a wave or pulse travels from medium A to medium B and $v_A < v_B$ (that is, when A is denser than B), it is not inverted upon reflection.

13.5 | Rate of Energy Transfer by Sinusoidal Waves on Strings

Waves transport energy through a medium as they propagate. For example, suppose an object is hanging on a stretched string and a pulse is sent down the string as in Figure 13.16a. When the pulse meets the suspended object, the object is momentarily displaced upward as in Figure 13.16b. In the process, energy is transferred to the object and appears as an increase in the gravitational potential energy of the

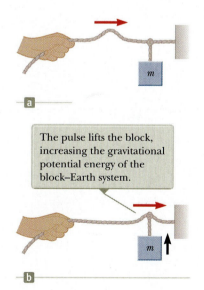

a

The pulse lifts the block, increasing the gravitational potential energy of the block–Earth system.

b

Figure 13.16 (a) A pulse travels to the right on a stretched string, carrying energy with it. (b) The energy of the pulse arrives at the hanging block.

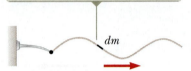

Each element of the string is a simple harmonic oscillator and therefore has kinetic energy and potential energy associated with it.

dm

Figure 13.17 A sinusoidal wave traveling along the *x* axis on a stretched string.

object–Earth system. This section examines the rate at which energy is transported along a string. We shall assume a one-dimensional sinusoidal wave in the calculation of the energy transferred.

Consider a sinusoidal wave traveling on a string (Fig. 13.17). The source of the energy is some external agent at the left end of the string. We can consider the string to be a nonisolated system. As the external agent performs work on the end of the string, moving it up and down, energy enters the system of the string and propagates along its length. Let's focus our attention on an infinitesimal element of the string of length *dx* and mass *dm*. Each such element moves vertically with simple harmonic motion. Therefore, we can model each element of the string as a simple harmonic oscillator, with the oscillation in the *y* direction. All elements have the same angular frequency ω and the same amplitude *A*. The kinetic energy *K* associated with a moving particle is $K = \frac{1}{2}mv^2$. If we apply this equation to the infinitesimal element, the kinetic energy *dK* associated with the up and down motion of this element is

$$dK = \tfrac{1}{2}(dm)\, v_y^{\,2}$$

where v_y is the transverse speed of the element. If μ is the mass per unit length of the string, the mass *dm* of the element of length *dx* is equal to $\mu\, dx$. Hence, we can express the kinetic energy of an element of the string as

$$dK = \tfrac{1}{2}(\mu\, dx)\, v_y^{\,2} \qquad \qquad \textbf{13.22}\blacktriangleleft$$

Substituting for the general transverse speed of an element of the medium using Equation 13.14 gives

$$dK = \tfrac{1}{2}\mu[-\omega A \cos(kx - \omega t)]^2\, dx = \tfrac{1}{2}\mu\omega^2 A^2 \cos^2(kx - \omega t)\, dx$$

If we take a snapshot of the wave at time $t = 0$, the kinetic energy of a given element is

$$dK = \tfrac{1}{2}\mu\omega^2\, A^2 \cos^2 kx\, dx$$

Integrating this expression over all the string elements in a wavelength of the wave gives the total kinetic energy K_λ in one wavelength:

$$K_\lambda = \int dK = \int_0^\lambda \tfrac{1}{2}\mu\omega^2 A^2 \cos^2 kx\, dx = \tfrac{1}{2}\mu\omega^2 A^2 \int_0^\lambda \cos^2 kx\, dx$$

$$= \tfrac{1}{2}\mu\omega^2 A^2 \left[\tfrac{1}{2}x + \frac{1}{4k} \sin 2kx\right]_0^\lambda = \tfrac{1}{2}\mu\omega^2 A^2 \left[\tfrac{1}{2}\lambda\right] = \tfrac{1}{4}\mu\omega^2 A^2 \lambda$$

In addition to kinetic energy, there is potential energy associated with each element of the string due to its displacement from the equilibrium position and the restoring forces from neighboring elements. A similar analysis to that above for the total potential energy U_λ in one wavelength gives exactly the same result:

$$U_\lambda = \tfrac{1}{4}\mu\omega^2 A^2 \lambda$$

The total energy in one wavelength of the wave is the sum of the potential and kinetic energies:

$$E_\lambda = U_\lambda + K_\lambda = \tfrac{1}{2}\mu\omega^2 A^2 \lambda \qquad \qquad \textbf{13.23}\blacktriangleleft$$

As the wave moves along the string, this amount of energy passes by a given point on the string during a time interval of one period of the oscillation. Therefore, the power *P*, or rate of energy transfer T_{MW} associated with the mechanical wave, is

$$P = \frac{T_{MW}}{\Delta t} = \frac{E_\lambda}{T} = \frac{\tfrac{1}{2}\mu\omega^2 A^2 \lambda}{T} = \tfrac{1}{2}\mu\omega^2 A^2 \left(\frac{\lambda}{T}\right)$$

▶ Power of a wave

$$P = \tfrac{1}{2}\mu\omega^2 A^2 v \qquad \qquad \textbf{13.24}\blacktriangleleft$$

Equation 13.24 shows that the rate of energy transfer by a sinusoidal wave on a string is proportional to (a) the square of the frequency, (b) the square of the amplitude, and (c) the wave speed. In fact, the rate of energy transfer in *any* sinusoidal wave is proportional to the square of the angular frequency and to the square of the amplitude.

QUICK QUIZ 13.5 Which of the following, taken by itself, would be most effective in increasing the rate at which energy is transferred by a wave traveling along a string? (a) reducing the linear mass density of the string by one-half (b) doubling the wavelength of the wave (c) doubling the tension in the string (d) doubling the amplitude of the wave

Example **13.6** | **Power Supplied to a Vibrating String**

A taut string for which $\mu = 5.00 \times 10^{-2}$ kg/m is under a tension of 80.0 N. How much power must be supplied to the string to generate sinusoidal waves at a frequency of 60.0 Hz and an amplitude of 6.00 cm?

SOLUTION

Conceptualize Consider Active Figure 13.9 again and notice that the vibrating blade supplies energy to the string at a certain rate. This energy then propagates to the right along the string.

Categorize We evaluate quantities from equations developed in the chapter, so we categorize this example as a substitution problem.

Use Equation 13.24 to evaluate the power:
$$P = \tfrac{1}{2}\mu\omega^2 A^2 v$$

Use Equations 13.9 and 13.21 to substitute for ω and v:
$$P = \tfrac{1}{2}\mu(2\pi f)^2 A^2\left(\sqrt{\frac{T}{\mu}}\right) = 2\pi^2 f^2 A^2 \sqrt{\mu T}$$

Substitute numerical values:
$$P = 2\pi^2(60.0\ \text{Hz})^2(0.060\ 0\ \text{m})^2\sqrt{(0.050\ 0\ \text{kg/m})(80.0\ \text{N})} = \boxed{512\ \text{W}}$$

What If? What if the string is to transfer energy at a rate of 1 000 W? What must be the required amplitude if all other parameters remain the same?

Answer Let us set up a ratio of the new and old power, reflecting only a change in the amplitude:
$$\frac{P_{\text{new}}}{P_{\text{old}}} = \frac{\tfrac{1}{2}\mu\omega^2 A_{\text{new}}^2\, v}{\tfrac{1}{2}\mu\omega^2 A_{\text{old}}^2\, v} = \frac{A_{\text{new}}^2}{A_{\text{old}}^2}$$

Solving for the new amplitude gives
$$A_{\text{new}} = A_{\text{old}}\sqrt{\frac{P_{\text{new}}}{P_{\text{old}}}} = (6.00\ \text{cm})\sqrt{\frac{1\ 000\ \text{W}}{512\ \text{W}}} = 8.39\ \text{cm}$$

13.6 | Sound Waves

Let us turn our attention from transverse waves to longitudinal waves. As stated in Section 13.1, for longitudinal waves the elements of the medium undergo displacements parallel to the direction of wave motion. Sound waves in air are the most important examples of longitudinal waves. Sound waves can travel through any material medium, however, and their speed depends on the properties of that medium. Table 13.1 (page 430) provides examples of the speed of sound in different media.

In Section 13.1, we began our investigation of waves by imagining the creation of a single pulse that traveled down a string (Figure 13.1) or a spring (Figure 13.3).

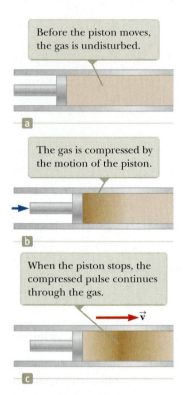

Before the piston moves, the gas is undisturbed.

a

The gas is compressed by the motion of the piston.

b

When the piston stops, the compressed pulse continues through the gas.

$\vec{v}$

c

Figure 13.18 Motion of a longitudinal pulse through a compressible gas. The compression (darker region) is produced by the moving piston.

TABLE 13.1 | Speed of Sound in Various Media

Medium	v (m/s)	Medium	v (m/s)	Medium	v (m/s)
Gases		**Liquids at 25°C**		**Solids[a]**	
Hydrogen (0°C)	1 286	Glycerol	1 904	Pyrex glass	5 640
Helium (0°C)	972	Sea water	1 533	Iron	5 950
Air (20°C)	343	Water	1 493	Aluminum	6 420
Air (0°C)	331	Mercury	1 450	Brass	4 700
Oxygen (0°C)	317	Kerosene	1 324	Copper	5 010
		Methyl alcohol	1 143	Gold	3 240
		Carbon tetrachloride	926	Lucite	2 680
				Lead	1 960
				Rubber	1 600

[a]Values given are for propagation of longitudinal waves in bulk media. Speeds for longitudinal waves in thin rods are smaller, and speeds of transverse waves in bulk are smaller yet.

Let's do something similar for sound. We describe pictorially the motion of a one-dimensional longitudinal sound pulse moving through a long tube containing a compressible gas as shown in Figure 13.18. A piston at the left end can be quickly moved to the right to compress the gas and create the pulse. Before the piston is moved, the gas is undisturbed and of uniform density as represented by the uniformly shaded region in Figure 13.18a. When the piston is pushed to the right (Fig. 13.18b), the gas just in front of it is compressed (as represented by the more heavily shaded region); the pressure and density in this region are now higher than they were before the piston moved. When the piston comes to rest (Fig. 13.18c), the compressed region of the gas continues to move to the right, corresponding to a longitudinal pulse traveling through the tube with speed v.

One can produce a one-dimensional *periodic* sound wave in the tube of gas in Figure 13.18 by causing the piston to move in simple harmonic motion. The results are shown in Active Figure 13.19. The darker parts of the colored areas in this figure represent regions in which the gas is compressed and the density and pressure are above their equilibrium values. A compressed region is formed whenever the piston is pushed into the tube. This compressed region, called a **compression,** moves through the tube, continuously compressing the region just in front of itself. When the piston is pulled back, the gas in front of it expands and the pressure and density in this region fall below their equilibrium values (represented by the lighter parts of the colored areas in Active Fig. 13.19). These low-pressure regions, called **rarefactions,** also propagate along the tube, following the compressions. Both regions move at the speed of sound in the medium.

As the piston oscillates back and forth in a sinusoidal fashion, regions of compression and rarefaction are continuously set up. The distance between two successive compressions (or two successive rarefactions) equals the wavelength λ. As these regions travel along the tube, any small element of the medium moves with simple harmonic motion parallel to the direction of the wave (in other words, longitudinally). If $s(x, t)$ is the position of a small element measured relative to its equilibrium position,[4] we can express this position function as

$$s(x, t) = s_{max} \cos(kx - \omega t) \qquad \textbf{13.25} \blacktriangleleft$$

where s_{max} is the maximum position relative to equilibrium, often called the **displacement amplitude.** Equation 13.25 represents the **displacement wave,** where k is the wave number and ω is the angular frequency of the piston. The variation ΔP in

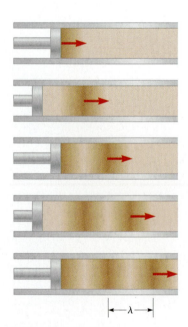

$\leftarrow \lambda \rightarrow$

Active Figure 13.19 A longitudinal wave propagating through a gas-filled tube. The source of the wave is an oscillating piston at the left.

---

[4]We use $s(x, t)$ here instead of $y(x, t)$ because the displacement of elements in the medium is not perpendicular to the x direction.

the pressure[5] of the gas measured from its equilibrium value is also sinusoidal; it is given by

$$\Delta P = \Delta P_{max} \sin(kx - \omega t) \qquad \textbf{13.26} \blacktriangleleft$$

The **pressure amplitude** ΔP_{max} is the maximum change in pressure from the equilibrium value, and Equation 13.26 represents the **pressure wave.** The pressure amplitude is proportional to the displacement amplitude s_{max}:

$$\Delta P_{max} = \rho v \omega s_{max} \qquad \textbf{13.27} \blacktriangleleft$$

where ρ is the density of the medium, v is the wave speed, and ωs_{max} is the maximum longitudinal speed of an element of the medium. It is these pressure variations in a sound wave that result in an oscillating force on the eardrum, leading to the sensation of hearing.

This discussion shows that a sound wave may be described equally well in terms of either pressure or displacement. A comparison of Equations 13.25 and 13.26 shows that the pressure wave is 90° out of phase with the displacement wave. Graphs of these functions are shown in Figure 13.20. Note that the change in pressure from equilibrium is a maximum when the displacement is zero, whereas the displacement is a maximum when the pressure change is zero.

Note that Figure 13.20 presents two graphical representations of the longitudinal wave: one for position of the elements of the medium and the other for pressure variation. They are *not* pictorial representations for longitudinal waves, however. For transverse waves, the element displacement is perpendicular to the direction of propagation and the pictorial and graphical representations look the same because the perpendicularity of the oscillations and propagation is matched by the perpendicularity of x and y axes. For longitudinal waves, the oscillations and propagation exhibit no perpendicularity, so those pictorial representations look like Active Figure 13.19.

The speed of sound depends on the temperature of the medium. For sound traveling through air, the relationship between wave speed and air temperature is

$$v = 331\sqrt{1 + \frac{T_C}{273}} \qquad \textbf{13.28} \blacktriangleleft$$

where v is in meters/second, 331 m/s is the speed of sound in air at 0°C, and T_C is the air temperature in degrees Celsius. Using this equation, one finds that at 20°C, the speed of sound in air is approximately 343 m/s.

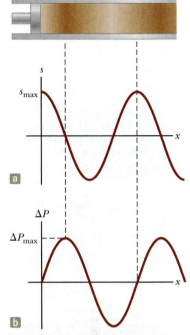

Figure 13.20 (a) Displacement amplitude versus position and (b) pressure amplitude versus position for a sinusoidal longitudinal wave. The displacement wave is 90° out of phase with the pressure wave.

THINKING PHYSICS 13.2

Why does thunder produce an extended "rolling" sound when its source, a lightning strike, occurs in a fraction of a second? How does lightning produce thunder in the first place?

Reasoning Let us assume that we are at ground level and ignore ground reflections. When cloud-to-ground lightning strikes, a channel of ionized air carries a very large electric current from the cloud to the ground. (We will study electric current in Chapter 21.) The result is a very rapid temperature increase of this channel of air as it carries the current. The temperature increase causes a sudden expansion of the air. This expansion is so sudden and so intense that a tremendous disturbance is produced in the air: thunder. The thunder rolls because the lightning channel is a long, extended source; the entire length of the channel produces the sound at essentially the same instant of time. Sound produced at the end of the channel nearest you reaches you first, but sounds from progressively farther portions of the channel reach you shortly thereafter. If the lightning channel were a perfectly straight line, the resulting sound might be a steady roar, but the zigzagged shape of the path results in the rolling variation in loudness. ◀

---

[5]We will formally introduce pressure in Chapter 15. In the case of longitudinal waves in a gas, each compressed area is a region of higher-than-average pressure and density, and each stretched region is a region of lower-than-average pressure and density.

❮13.7 | The Doppler Effect

When someone honks the horn of a vehicle as it travels along a highway, the frequency of the sound you hear is higher as the vehicle approaches you than it is as the vehicle moves away from you. This change is one example of the **Doppler effect,** named after Christian Johann Doppler (1803–1853), an Austrian physicist.

The Doppler effect for sound is experienced whenever there is relative motion between the source of sound and the observer. Motion of the source or observer toward the other results in the observer's hearing a frequency that is higher than the true frequency of the source. Motion of the source or observer away from the other results in the observer hearing a frequency that is lower than the true frequency of the source.

Although we shall restrict our attention to the Doppler effect for sound waves, it is associated with waves of all types. The Doppler effect for electromagnetic waves is used in police radar systems to measure the speeds of motor vehicles. Likewise, astronomers use the effect to determine the relative motions of stars, galaxies, and other celestial objects. In 1842, Doppler first reported the frequency shift in connection with light emitted by two stars revolving about each other in double-star systems. In the early 20th century, the Doppler effect for light from galaxies was used to argue for the expansion of the Universe, which led to the Big Bang theory, discussed in Chapter 31.

To see what causes this apparent frequency change, imagine you are in a boat lying at anchor on a gentle sea where the waves have a period of $T = 2.0$ s. Therefore, every 2.0 s a crest hits your boat. Figure 13.21a shows this situation with the water waves moving toward the left. If you start a stopwatch at $t = 0$ just as one crest hits, the stopwatch reads 2.0 s when the next crest hits, 4.0 s when the third crest hits, and so on. From these observations you conclude that the wave frequency is $f = 1/T = 0.50$ Hz. Now suppose you start your motor and head directly into the oncoming waves as shown in Figure 13.21b. Again you set your stopwatch to $t = 0$ as a crest hits the bow of your boat. This time, however, because you are moving toward the next wave crest as it moves toward you, it hits you less than 2.0 s after the first hit. In other words, the period you observe is shorter than the 2.0-s period you observed when you were stationary. Because $f = 1/T$, you observe a higher wave frequency than when you were at rest.

If you turn around and move in the same direction as the waves (Fig. 13.21c), you observe the opposite effect. You set your watch to $t = 0$ as a crest hits the stern of the boat. Because you are now moving away from the next crest, more than 2.0 s has elapsed on your watch by the time that crest catches you. Therefore, you observe a lower frequency than when you were at rest.

These effects occur because the *relative* speed between your boat and the crest of a wave depends on the direction of travel and on the speed of your boat. When you are moving toward the right in Figure 13.21b, this relative speed is higher than that of the wave speed, which leads to the observation of an increased frequency. When you turn around and move to the left, the relative speed is lower, as is the observed frequency of the water waves.

Let's now examine an analogous situation with sound waves in which we replace the water waves with sound waves, the water becomes the air, and the person on the boat becomes an observer listening to the sound. In this case, an observer O is moving with a speed of v_O and a sound source S is stationary. For simplicity, we assume that the air is also stationary and that the observer moves directly toward the source.

The circles in Active Figure 13.22 represent curves connecting the crests of sound waves moving away from the source. Therefore, the radial distance between adjacent circles is one wavelength. We shall take the frequency of the source to be f, the wavelength to be λ, and the speed of sound to be v. A stationary observer would detect a frequency f, where $f = v/\lambda$ (i.e., when the source and observer are both at rest, the observed frequency must equal the true frequency of the source). If the observer

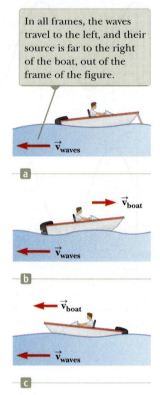

In all frames, the waves travel to the left, and their source is far to the right of the boat, out of the frame of the figure.

$\vec{v}_{\text{waves}}$

a

$\vec{v}_{\text{boat}}$

$\vec{v}_{\text{waves}}$

b

$\vec{v}_{\text{boat}}$

$\vec{v}_{\text{waves}}$

c

Figure 13.21 (a) Waves moving toward a stationary boat. (b) The boat moving toward the wave source. (c) The boat moving away from the wave source.

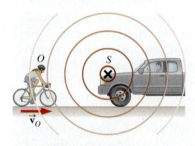

O S

$\vec{v}_O$

Active Figure 13.22 An observer O (the cyclist) moving with a speed v_O toward a stationary point source S, the horn of a parked truck. The observer hears a frequency f' that is greater than the source frequency.

moves toward the source with the speed v_O, however, the relative speed of sound experienced by the observer is higher than the speed of sound in air. Using our relative speed discussion of Section 3.6, if the sound is coming toward the observer at v and the observer is moving toward the sound at v_O, the relative speed of sound as measured by the observer is

$$v' = v + v_O$$

The frequency of sound heard by the observer is based on this apparent speed of sound:

$$f' = \frac{v'}{\lambda} = \frac{v + v_O}{\lambda} = \left(\frac{v + v_O}{v}\right)f \qquad \text{(observer moving toward source)} \qquad \text{13.29} \blacktriangleleft$$

Now consider the situation in which the source moves with a speed of v_S relative to the medium and the observer is at rest. Active Figure 13.23a shows this situation. If the source moves directly toward observer A in Active Figure 13.23a, the crests detected by the observer along a line between the source and observer are closer to one another than they would be if the source were at rest. (Active Figure 13.23b shows this effect for waves moving on the surface of water.) As a result, the wavelength λ' measured by observer A is shorter than the wavelength λ of the source. During each vibration, which lasts for a time interval T (the period), the source moves a distance $v_S T = v_S/f$ and the wavelength is *shortened* by this amount. Therefore, the observed wavelength is $\lambda' = \lambda - v_S/f$. Because $\lambda = v/f$, the frequency f' heard by observer A is

$$f' = \frac{v}{\lambda'} = \left(\frac{v}{v - v_S}\right)f \qquad \text{(source moving toward observer)} \qquad \text{13.30} \blacktriangleleft$$

That is, the frequency is *increased* when the source moves toward the observer. In a similar manner, if the source moves away from observer B at rest, the sign of v_S is reversed in Equation 13.30 and the frequency is lower.

In Equation 13.30, notice that the denominator approaches zero when the speed of the source approaches the speed of sound, resulting in the frequency f' approaching infinity. Such a situation results in waves that cannot escape from the source in the direction of motion of the source. This concentration of energy in front of the source results in a *shock wave*. Such a disturbance is noted when a jet aircraft flying at a speed equal to or greater than the speed of sound produces a *sonic boom*.

Finally, if both the source and the observer are in motion, the following general equation for the observed frequency is found:

$$f' = \left(\frac{v + v_O}{v - v_S}\right)f \qquad \text{13.31} \blacktriangleleft$$

▶ General Doppler-shift expression

In this expression, the signs for the values substituted for v_O and v_S depend on the direction of the velocity. A positive value is used for motion of the observer or the source *toward* the other, and a negative sign is used for motion of one *away from* the other.

When working with any Doppler effect problem, remember the following rule concerning signs: The word *toward* is associated with an *increase* in the observed frequency, and the words *away from* are associated with a *decrease* in the observed frequency.

Doppler Sonography BIO

The Doppler effect is used in medicine to study many different systems. For example, a technique called *Doppler sonography* is a non-invasive diagnostic procedure that can measure the speed of blood in arteries and detect turbulence in blood flow. The

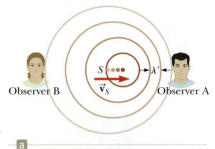

Observer B   Observer A

A point source is moving to the right with speed v_S.

Courtesy of the Educational Development Center, Newton, MA

Active Figure 13.23 (a) A source S moving with a speed v_S toward a stationary observer A and away from a stationary observer B. Observer A hears an increased frequency, and observer B hears a decreased frequency. (b) The Doppler effect in water, observed in a ripple tank. The vibrating source is moving to the right. Letters shown in the photo refer to Quick Quiz 13.6.

Pitfall Prevention | 13.4
Doppler Effect Does Not Depend on Distance
A common misconception about the Doppler effect is that it depends on the distance between the source and the observer. Although the *intensity* of a sound will vary as the distance changes, the apparent *frequency* will not; the frequency depends only on the relative speed of source and observer.

sound waves reflect from moving blood cells, undergoing a frequency shift based on the speed of the cells. The instrumentation detects the reflected sound waves and converts the frequency information to a speed of flow of the blood. By viewing the image of the heart, physicians can monitor carotid artery diseases and detect problems with heart valves. Typical diagnostic sonographic devices operate at frequencies ranging from 1 MHz to 18 MHz. Sonography is an effective method for imaging soft body tissues such as muscles, tendons, breasts, and neonatal brain at higher frequencies (7 MHz to 18 MHz). Lower frequencies (1 MHz to 6 MHz) must be used to image deeper structures in the body such as the liver and the kidney. However, the lower frequencies used in imaging these deeper structures result in lower-resolution images.

> **QUICK QUIZ 13.6** Consider detectors of water waves at three locations A, B, and C in Active Figure 13.23b. Which of the following statements is true? **(a)** The wave speed is highest at location A. **(b)** The wave speed is highest at location C. **(c)** The detected wavelength is largest at location B. **(d)** The detected wavelength is largest at location C. **(e)** The detected frequency is highest at location C. **(f)** The detected frequency is highest at location A.

> **QUICK QUIZ 13.7** You stand on a platform at a train station and listen to a train approaching the station at a constant velocity. While the train approaches, but before it arrives, what do you hear? **(a)** the intensity and the frequency of the sound both increasing **(b)** the intensity and the frequency of the sound both decreasing **(c)** the intensity increasing and the frequency decreasing **(d)** the intensity decreasing and the frequency increasing **(e)** the intensity increasing and the frequency remaining the same **(f)** the intensity decreasing and the frequency remaining the same

Example **13.7** | **Doppler Submarines**

A submarine (sub A) travels through water at a speed of 8.00 m/s, emitting a sonar wave at a frequency of 1 400 Hz. The speed of sound in the water is 1 533 m/s. A second submarine (sub B) is located such that both submarines are traveling directly toward each other. The second submarine is moving at 9.00 m/s.

(A) What frequency is detected by an observer riding on sub B as the subs approach each other?

SOLUTION

Conceptualize Even though the problem involves subs moving in water, there is a Doppler effect just like there is when you are in a moving car and listening to a sound moving through the air from another car.

Categorize Because both subs are moving, we categorize this problem as one involving the Doppler effect for both a moving source and a moving observer.

Analyze Use Equation 13.31 to find the Doppler-shifted frequency heard by the observer in sub B, being careful with the signs assigned to the source and observer speeds:

$$f' = \left(\frac{v + v_O}{v - v_S}\right) f$$

$$f' = \left[\frac{1\ 533\ \text{m/s} + (+9.00\ \text{m/s})}{1\ 533\ \text{m/s} - (+8.00\ \text{m/s})}\right](1\ 400\ \text{Hz}) = \boxed{1\ 416\ \text{Hz}}$$

(B) The subs barely miss each other and pass. What frequency is detected by an observer riding on sub B as the subs recede from each other?

SOLUTION

Use Equation 13.31 to find the Doppler-shifted frequency heard by the observer in sub B, again being careful with the signs assigned to the source and observer speeds:

$$f' = \left(\frac{v + v_O}{v - v_S}\right) f$$

$$f' = \left[\frac{1\ 533\ \text{m/s} + (-9.00\ \text{m/s})}{1\ 533\ \text{m/s} - (-8.00\ \text{m/s})}\right](1\ 400\ \text{Hz}) = \boxed{1\ 385\ \text{Hz}}$$

13.7 *cont.*

Notice that the frequency drops from 1 416 Hz to 1 385 Hz as the subs pass. This effect is similar to the drop in frequency you hear when a car passes by you while blowing its horn.

(C) While the subs are approaching each other, some of the sound from sub A reflects from sub B and returns to sub A. If this sound were to be detected by an observer on sub A, what is its frequency?

SOLUTION

The sound of apparent frequency 1 416 Hz found in part (A) is reflected from a moving source (sub B) and then detected by a moving observer (sub A). Find the frequency detected by sub A:

$$f'' = \left(\frac{v + v_O}{v - v_S}\right)f'$$

$$= \left[\frac{1\ 533\ \text{m/s} + (+8.00\ \text{m/s})}{1\ 533\ \text{m/s} - (+9.00\ \text{m/s})}\right](1\ 416\ \text{Hz}) = 1\ 432\ \text{Hz}$$

Finalize This technique is used by police officers to measure the speed of a moving car. Microwaves are emitted from the police car and reflected by the moving car. By detecting the Doppler-shifted frequency of the reflected microwaves, the police officer can determine the speed of the moving car.

13.8 | Context Connection: Seismic Waves

As mentioned in the Context introduction, the release of energy in an earthquake occurs at the **focus** or **hypocenter** of the earthquake. The **epicenter** is the point on the Earth's surface radially above the hypocenter. The released energy will propagate away from the focus of the earthquake by means of **seismic waves.** Seismic waves are like the sound waves that we have studied in the later sections of this chapter in that they are mechanical disturbances moving through a medium.

In discussing mechanical waves in this chapter, we identified two types: transverse and longitudinal. In the case of mechanical waves moving through air, we have only a longitudinal possibility. For mechanical waves moving through a solid, however, both possibilities are available because of the strong interatomic forces between elements of the solid. Therefore, in the case of seismic waves, energy propagates away from the focus both by longitudinal and transverse waves.

In the language used in earthquake studies, these two types of waves are named according to the order of their arrival at a seismograph. The longitudinal wave travels at a higher speed than the transverse wave. As a result, the longitudinal wave arrives at a seismograph first and is therefore called the **P wave,** where P stands for *primary*. The slower moving transverse wave arrives next, so it is called the **S wave,** or *secondary* wave.

Let us see why longitudinal waves travel faster than transverse waves. The speed of *all* mechanical waves follows an expression of the general form

$$v = \sqrt{\frac{\text{elastic property}}{\text{inertial property}}}$$

13.32 ◀

For a wave traveling on a string, the speed is given by Equation 13.21:

$$v = \sqrt{\frac{T}{\mu}}$$

where the elastic property is the tension in the string. It is the tension in the string that returns a displaced element of the string to equilibrium. The appropriate inertial property is the linear mass density of the string.

For a transverse wave moving in a bulk solid, the elastic property is the *shear modulus S* of the material.[6] The shear modulus is a parameter that measures the deformation of a solid to a shear force, a force in the sideways direction. For example, lay your textbook down on a table and place your hand flat on the cover. Now, move

---

[6]For details on various elastic moduli for materials see R. A. Serway and J. W. Jewett Jr., *Physics for Scientists and Engineers,* 8th ed. (Belmont, CA, Brooks-Cole: 2010), Section 12.4.

your hand in a direction away from the book spine. The book will deform so that its cross section changes from a rectangle to a parallelogram. The amount by which the book deforms under a given force from your hand is related to the shear modulus of the book. The speed of a transverse wave (an S wave) in a bulk solid is

$$v_S = \sqrt{\frac{S}{\rho}}$$
13.33◀

where ρ is the density and S is the shear modulus of the material.

For a longitudinal wave moving in a gas or liquid, the elastic property in Equation 13.32 is the *bulk modulus B* of the material. The bulk modulus is a parameter that measures the change in volume of a sample of material due to a force compressing it that is uniform over a surface area. The speed of sound in a gas is given by

$$v = \sqrt{\frac{B}{\rho}}$$
13.34◀

where B is the bulk modulus of the gas and ρ is the gas density.

Now we consider longitudinal waves moving through a bulk solid. As a wave passes through a sample of the material, the material is compressed, so the wave speed should depend on the bulk modulus. As the material is compressed along the direction of travel of the wave, however, it is also distorted in the perpendicular direction. (Imagine a partially inflated balloon that is pressed downward against a table. It spreads out in the direction parallel to the table.) The result is a shear distortion of the sample of material. Therefore, the wave speed should depend on both the bulk modulus and the shear modulus! Careful analysis shows that this wave speed is

$$v_P = \sqrt{\frac{B + \frac{4}{3}S}{\rho}}$$
13.35◀

Notice that this equation for the speed of a P wave gives a value that is larger than that for the S wave in Equation 13.33.

The wave speed for a seismic wave depends on the medium through which it travels. Typical values are 8 km/s for a P wave and 5 km/s for an S wave. Figure 13.24 shows typical seismograph traces of a distant earthquake at two seismograph stations, with the S wave clearly arriving after the P wave.

Figure 13.24 An earthquake occurs at time $t = 0$. Two seismograph traces record the arrival of seismic waves from the earthquake. The bottom trace is for a seismograph located a few hundred miles from the epicenter. The top trace shows the waves arriving at a seismograph thousands of miles from the epicenter. The time interval between arrival of the P and S waves can be used to determine the distance from the epicenter to the seismograph station.

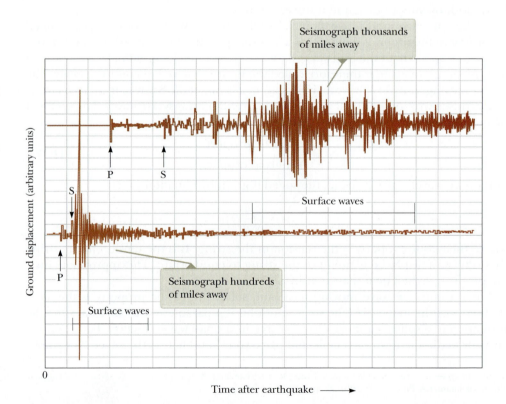

The P and S waves move through the body of the Earth. Once these waves reach the surface, the energy can propagate by additional types of waves along the surface. In a *Rayleigh wave*, the motion of the elements of the medium at the surface is a combination of longitudinal and transverse displacements so that the net motion of a point on the surface is circular or elliptical. This motion is similar to the path followed by elements of water on the ocean surface as a wave passes by, as in Active Figure 13.25. The *Love wave* is a transverse surface wave in which the transverse oscillations are parallel to the surface. Therefore, no vertical displacement of the surface occurs in a Love wave.

It is possible to use the P and S waves traveling through the body of the Earth to gain information about the structure of the Earth's interior. Measurements of a given earthquake by seismographs at various locations on the surface indicate that the Earth has an interior region that allows the passage of P waves but not S waves. This fact can be understood if this particular region is modeled as having liquid characteristics. Similar to a gas, a liquid cannot sustain a transverse force. Therefore, the transverse S waves cannot pass through this region. This information leads us to a structural model in which the Earth has a **liquid core** between radii of approximately 1.2×10^3 km and 3.5×10^3 km.

Other measurements of seismic waves allow additional interpretations of layers within the interior of the Earth, including a **solid core** at the center, a rocky region called the **mantle,** and a relatively thin outer layer called the **crust.** Figure 13.26 shows this structure. Using x-rays or ultrasound in medicine to provide information about the interior of the human body is somewhat similar to using seismic waves to provide information about the interior of the Earth.

As P and S waves propagate in the interior of the Earth, they will encounter variations in the medium. At each boundary at which the properties of the medium change, reflection and transmission occur. When the seismic wave arrives at the surface of the Earth, a small amount of the energy is transmitted into the air as low-frequency sound waves. Some of the energy spreads out along the surface in the form of Rayleigh and Love waves. The remaining wave energy is reflected back into the interior. As a result, seismic waves can travel over long distances within the Earth

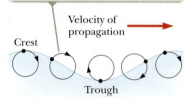

The elements at the surface move in nearly circular paths. Each element is displaced both horizontally and vertically from its equilibrium position.

Active Figure 13.25 The motion of water elements on the surface of deep water in which a wave is propagating is a combination of transverse and longitudinal displacements.

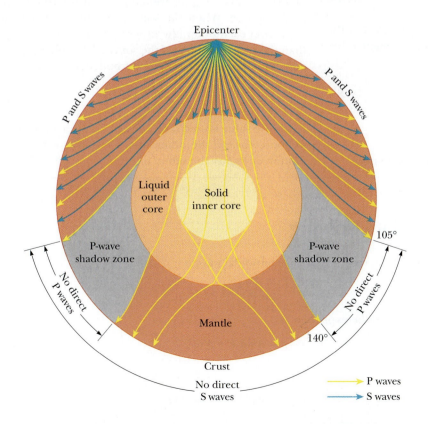

Figure 13.26 Cross section of the Earth showing paths of waves produced by an earthquake. Only P waves (yellow) can propagate in the liquid core. The S waves (blue) do not enter the liquid core. When the P waves transmit from one region to another, such as from the mantle to the liquid core, they experience *refraction*, a change in the direction of propagation. We will study refraction for light in Chapter 25. Because of the refraction for seismic waves, there is a "shadow" zone between 105° and 140° from the epicenter in which no waves following a direct path (i.e., a path with no reflections) arrive.

and can be detected at seismographs at many locations around the globe. In addition, because a relatively large fraction of the wave energy continues to be reflected at each encounter with the surface, the wave can propagate for a long time. Data are available showing seismograph activity for several hours after an earthquake, a result of the repeated reflections of seismic waves from the surface.

Another example of the reflection of seismic waves is available in the technology of oil exploration. A "thumper truck" applies large impulsive forces to the ground, resulting in low-energy seismic waves propagating into the Earth. Specialized microphones are used to detect the waves reflected from various boundaries between layers under the surface. By using computers to map out the underground structure corresponding to these layers, it is possible to detect layers likely to contain oil.

SUMMARY

A **transverse wave** is a wave in which the elements of the medium move in a direction perpendicular to the direction of the wave velocity. An example is a wave moving along a stretched string.

Longitudinal waves are waves in which the elements of the medium move back and forth parallel to the direction of the wave velocity. Sound waves in air are longitudinal.

Any one-dimensional wave traveling with a speed of v in the positive x direction can be represented by a **wave function** of the form $y = f(x - vt)$. Likewise, the wave function for a wave traveling in the negative x direction has the form $y = f(x + vt)$.

The wave function for a one-dimensional sinusoidal wave traveling to the right can be expressed as

$$y = A \sin\left[\frac{2\pi}{\lambda}(x - vt)\right] \qquad \text{13.5} \blacktriangleleft$$

where A is the **amplitude**, λ is the **wavelength**, and v is the **wave speed**. The **angular wave number** k and **angular frequency** ω are defined as follows:

$$k \equiv \frac{2\pi}{\lambda} \qquad \text{13.8} \blacktriangleleft$$

$$\omega \equiv \frac{2\pi}{T} = 2\pi f \qquad \text{13.9} \blacktriangleleft$$

where T is the **period** of the wave and f is its **frequency**.

The speed of a transverse wave traveling on a stretched string of mass per unit length μ and tension T is

$$v = \sqrt{\frac{T}{\mu}} \qquad \text{13.21} \blacktriangleleft$$

When a pulse traveling on a string meets a fixed end, the pulse is reflected and inverted. If the pulse reaches a free end, it is reflected but not inverted.

The **power** transmitted by a sinusoidal wave on a stretched string is

$$P = \tfrac{1}{2}\mu\omega^2 A^2 v \qquad \text{13.24} \blacktriangleleft$$

The change in frequency of a sound wave heard by an observer whenever there is relative motion between a wave source and the observer is called the **Doppler effect.** When the source and observer are moving toward each other, the observer hears a higher frequency than the true frequency of the source. When the source and observer are moving away from each other, the observer hears a lower frequency than the true frequency of the source. The following general equation provides the observed frequency:

$$f' = \left(\frac{v + v_O}{v - v_S}\right) f \qquad \text{13.31} \blacktriangleleft$$

A positive value is used for v_O or v_S for motion of the observer or source *toward* the other, and a negative sign is used for motion *away from* the other.

Analysis Model for Problem Solving

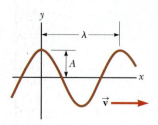

Traveling Wave. The wave speed of a sinusoidal wave is

$$v = \frac{\lambda}{T} = \lambda f \qquad \text{13.6, 13.12} \blacktriangleleft$$

A sinusoidal wave can be expressed as

$$y = A \sin(kx - \omega t) \qquad \text{13.10} \blacktriangleleft$$

OBJECTIVE QUESTIONS |

1. Which of the following statements is not necessarily true regarding mechanical waves? (a) They are formed by some source of disturbance. (b) They are sinusoidal in nature. (c) They carry energy. (d) They require a medium through which to propagate. (e) The wave speed depends on the properties of the medium in which they travel.

2. The distance between two successive peaks of a sinusoidal wave traveling along a string is 2 m. If the frequency of this wave is 4 Hz, what is the speed of the wave? (a) 4 m/s (b) 1 m/s (c) 8 m/s (d) 2 m/s (e) impossible to answer from the information given

3. Rank the waves represented by the following functions from the largest to the smallest according to (i) their amplitudes, (ii) their wavelengths, (iii) their frequencies, (iv) their periods, and (v) their speeds. If the values of a quantity are equal for two waves, show them as having equal rank. For all functions, x and y are in meters and t is in seconds. (a) $y = 4 \sin (3x - 15t)$ (b) $y = 6 \cos (3x + 15t - 2)$ (c) $y = 8 \sin (2x + 15t)$ (d) $y = 8 \cos (4x + 20t)$ (e) $y = 7 \sin (6x - 24t)$

4. If you stretch a rubber hose and pluck it, you can observe a pulse traveling up and down the hose. (i) What happens to the speed of the pulse if you stretch the hose more tightly? (a) It increases. (b) It decreases. (c) It is constant. (d) It changes unpredictably. (ii) What happens to the speed if you fill the hose with water? Choose from the same possibilities.

5. When all the strings on a guitar (Fig. OQ13.5) are stretched to the same tension, will the speed of a wave along the most massive bass string be (a) faster, (b) slower, or (c) the same as the speed of a wave on the lighter strings? Alternatively, (d) is the speed on the bass string not necessarily any of these answers?

Figure OQ13.5

Aaron Graubart/Getty Images

6. By what factor would you have to multiply the tension in a stretched string so as to double the wave speed? Assume the string does not stretch. (a) a factor of 8 (b) a factor of 4 (c) a factor of 2 (d) a factor of 0.5 (e) You could not change the speed by a predictable factor by changing the tension.

7. A sound wave can be characterized as (a) a transverse wave, (b) a longitudinal wave, (c) a transverse wave or a longitudinal wave, depending on the nature of its source, (d) one that carries no energy, or (e) a wave that does not require a medium to be transmitted from one place to the other.

8. Two sirens A and B are sounding so that the frequency from A is twice the frequency from B. Compared with the speed of sound from A, is the speed of sound from B (a) twice as fast, (b) half as fast, (c) four times as fast, (d) one-fourth as fast, or (e) the same?

9. Table 13.1 shows the speed of sound is typically an order of magnitude larger in solids than in gases. To what can this higher value be most directly attributed? (a) the difference in density between solids and gases (b) the difference in compressibility between solids and gases (c) the limited size of a solid object compared to a free gas (d) the impossibility of holding a gas under significant tension

10. A source vibrating at constant frequency generates a sinusoidal wave on a string under constant tension. If the power delivered to the string is doubled, by what factor does the amplitude change? (a) a factor of 4 (b) a factor of 2 (c) a factor of $\sqrt{2}$ (d) a factor of 0.707 (e) cannot be predicted

11. A source of sound vibrates with constant frequency. Rank the frequency of sound observed in the following cases from highest to the lowest. If two frequencies are equal, show their equality in your ranking. All the motions mentioned have the same speed, 25 m/s. (a) The source and observer are stationary. (b) The source is moving toward a stationary observer. (c) The source is moving away from a stationary observer. (d) The observer is moving toward a stationary source. (e) The observer is moving away from a stationary source.

12. (a) Can a wave on a string move with a wave speed that is greater than the maximum transverse speed $v_{y,\max}$ of an element of the string? (b) Can the wave speed be much greater than the maximum element speed? (c) Can the wave speed be equal to the maximum element speed? (d) Can the wave speed be less than $v_{y,\max}$?

13. If one end of a heavy rope is attached to one end of a lightweight rope, a wave can move from the heavy rope into the lighter one. (i) What happens to the speed of the wave? (a) It increases. (b) It decreases. (c) It is constant. (d) It changes unpredictably. (ii) What happens to the frequency? Choose from the same possibilities. (iii) What happens to the wavelength? Choose from the same possibilities.

14. If a 1.00-kHz sound source moves at a speed of 50.0 m/s toward a listener who moves at a speed of 30.0 m/s in a direction away from the source, what is the apparent frequency heard by the listener? (a) 796 Hz (b) 949 Hz (c) 1 000 Hz (d) 1 068 Hz (e) 1 273 Hz

15. As you travel down the highway in your car, an ambulance approaches you from the rear at a high speed (Fig. OQ13.15) sounding its siren at a frequency of 500 Hz. Which statement is correct? (a) You hear a frequency less than 500 Hz. (b) You hear a frequency equal to 500 Hz. (c) You hear a frequency greater than 500 Hz. (d) You hear a frequency greater than 500 Hz, whereas the ambulance driver hears a frequency lower than 500 Hz. (e) You hear a frequency less than 500 Hz, whereas the ambulance driver hears a frequency of 500 Hz.

Figure OQ13.15

© Anthony Redpath/Corbis

16. Assume a change at the source of sound reduces the wavelength of a sound wave in air by a factor of 2. (i) What happens to its frequency? (a) It increases by a factor of 4. (b) It increases by a factor of 2. (c) It is unchanged. (d) It decreases by a factor of 2. (e) It changes by an unpredictable factor. (ii) What happens to its speed? Choose from the same possibilities as in part (i).

17. Suppose an observer and a source of sound are both at rest relative to the ground and a strong wind is blowing

away from the source toward the observer. **(i)** What effect does the wind have on the observed frequency? (a) It causes an increase. (b) It causes a decrease. (c) It causes no change. **(ii)** What effect does the wind have on the observed wavelength? Choose from the same possibilities as in part (i). **(iii)** What effect does the wind have on the observed speed of the wave? Choose from the same possibilities as in part (i).

CONCEPTUAL QUESTIONS

☐ denotes answer available in *Student Solutions Manual/Study Guide*

1. *The Tunguska event.* On June 30, 1908, a meteor burned up and exploded in the atmosphere above the Tunguska River valley in Siberia. It knocked down trees over thousands of square kilometers and started a forest fire, but produced no crater and apparently caused no human casualties. A witness sitting on his doorstep outside the zone of falling trees recalled events in the following sequence. He saw a moving light in the sky, brighter than the Sun and descending at a low angle to the horizon. He felt his face become warm. He felt the ground shake. An invisible agent picked him up and immediately dropped him about a meter from where he had been seated. He heard a very loud protracted rumbling. Suggest an explanation for these observations and for the order in which they happened.

2. (a) How would you create a longitudinal wave in a stretched spring? (b) Would it be possible to create a transverse wave in a spring?

3. Why is a pulse on a string considered to be transverse?

4. Older auto-focus cameras sent out a pulse of sound and measured the time interval required for the pulse to reach an object, reflect off of it, and return to be detected. Can air temperature affect the camera's focus? New cameras use a more reliable infrared system.

5. When a pulse travels on a taut string, does it always invert upon reflection? Explain.

6. Does the vertical speed of an element of a horizontal, taut string, through which a wave is traveling, depend on the wave speed? Explain.

7. Explain how the distance to a lightning bolt (Fig. CQ13.7) can be determined by counting the seconds between the flash and the sound of thunder.

Figure CQ13.7

8. You are driving toward a cliff and honk your horn. Is there a Doppler shift of the sound when you hear the echo? If so, is it like a moving source or a moving observer? What if the reflection occurs not from a cliff, but from the forward edge of a huge alien spacecraft moving toward you as you drive?

9. If you steadily shake one end of a taut rope three times each second, what would be the period of the sinusoidal wave set up in the rope?

10. (a) If a long rope is hung from a ceiling and waves are sent up the rope from its lower end, why does the speed of the waves change as they ascend? (b) Does the speed of the ascending waves increase or decrease? Explain.

11. The radar systems used by police to detect speeders are sensitive to the Doppler shift of a pulse of microwaves. Discuss how this sensitivity can be used to measure the speed of a car.

12. How can an object move with respect to an observer so that the sound from it is not shifted in frequency?

13. In an earthquake, both S (transverse) and P (longitudinal) waves propagate from the focus of the earthquake. The focus is in the ground radially below the epicenter on the surface (Fig. CQ13.13). Assume the waves move in straight lines through uniform material. The S waves travel through the Earth more slowly than the P waves (at about 5 km/s versus 8 km/s). By detecting the time of arrival of the waves at a seismograph, (a) how can one determine the distance to the focus of the earthquake? (b) How many detection stations are necessary to locate the focus unambiguously?

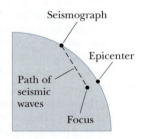

Figure CQ13.13

PROBLEMS

WebAssign The problems found in this chapter may be assigned online in Enhanced WebAssign.

1. denotes straightforward problem; **2.** denotes intermediate problem; **3.** denotes challenging problem

☐**1.** denotes full solution available in the *Student Solutions Manual/ Study Guide*

1. denotes problems most often assigned in Enhanced WebAssign.

BIO denotes biomedical problem

GP denotes guided problem

M denotes Master It tutorial available in Enhanced WebAssign

Q|C denotes asking for quantitative and conceptual reasoning

S denotes symbolic reasoning problem

shaded denotes "paired problems" that develop reasoning with symbols and numerical values

W denotes Watch It video solution available in Enhanced WebAssign

Section 13.1 Propagation of a Disturbance

1. At $t = 0$, a transverse pulse in a wire is described by the function

$$y = \frac{6.00}{x^2 + 3.00}$$

where x and y are in meters. If the pulse is traveling in the positive x direction with a speed of 4.50 m/s, write the function $y(x, t)$ that describes this pulse.

2. **QC** Ocean waves with a crest-to-crest distance of 10.0 m can be described by the wave function

$$y(x, t) = 0.800 \sin [0.628(x - vt)]$$

where x and y are in meters, t is in seconds, and $v = $ 1.20 m/s. (a) Sketch $y(x, t)$ at $t = 0$. (b) Sketch $y(x, t)$ at $t = 2.00$ s. (c) Compare the graph in part (b) with that for part (a) and explain similarities and differences. (d) How has the wave moved between graph (a) and graph (b)?

Section 13.2 Analysis Model: Traveling Wave

3. **M** The wave function for a traveling wave on a taut string is (in SI units)

$$y(x, t) = 0.350 \sin \left(10\pi t - 3\pi x + \frac{\pi}{4} \right)$$

(a) What are the speed and direction of travel of the wave? (b) What is the vertical position of an element of the string at $t = 0$, $x = 0.100$ m? What are (c) the wavelength and (d) the frequency of the wave? (e) What is the maximum transverse speed of an element of the string?

4. **S** Show that the wave function $y = e^{b(x-vt)}$ is a solution of the linear wave equation (Eq. 13.20), where b is a constant.

5. **W** The string shown in Figure P13.5 is driven at a frequency of 5.00 Hz. The amplitude of the motion is $A = 12.0$ cm, and the wave speed is $v = $ 20.0 m/s. Furthermore, the wave is such that $y = 0$ at $x = 0$ and $t = 0$. Determine (a) the angular frequency and (b) the wave number for this wave. (c) Write an expression for the wave function. Calculate (d) the maximum transverse speed and (e) the maximum transverse acceleration of an element of the string.

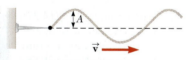

Figure P13.5

6. For a certain transverse wave, the distance between two successive crests is 1.20 m, and eight crests pass a given point along the direction of travel every 12.0 s. Calculate the wave speed.

7. (a) Write the expression for y as a function of x and t in SI units for a sinusoidal wave traveling along a rope in the negative x direction with the following characteristics: $A = 8.00$ cm, $\lambda = 80.0$ cm, $f = 3.00$ Hz, and $y(0, t) = 0$ at $t = 0$. (b) **What If?** Write the expression for y as a function of x and t for the wave in part (a) assuming $y(x, 0) = 0$ at the point $x = 10.0$ cm.

8. **M** A wave is described by $y = 0.020\ 0 \sin (kx - \omega t)$, where $k = 2.11$ rad/m, $\omega = 3.62$ rad/s, x and y are in meters, and t is in seconds. Determine (a) the amplitude, (b) the

wavelength, (c) the frequency, and (d) the speed of the wave.

9. Consider the sinusoidal wave of Example 13.2 with the wave function

$$y = 0.150 \cos (15.7x - 50.3t)$$

where x and y are in meters and t is in seconds. At a certain instant, let point A be at the origin and point B be the closest point to A along the x axis where the wave is 60.0° out of phase with A. What is the coordinate of B?

10. **W** A transverse wave on a string is described by the wave function

$$y = 0.120 \sin \left(\frac{\pi}{8}x + 4\pi t \right)$$

where x and y are in meters and t is in seconds. Determine (a) the transverse speed and (b) the transverse acceleration at $t = 0.200$ s for an element of the string located at $x = 1.60$ m. What are (c) the wavelength, (d) the period, and (e) the speed of propagation of this wave?

11. **M** A sinusoidal wave is traveling along a rope. The oscillator that generates the wave completes 40.0 vibrations in 30.0 s. A given crest of the wave travels 425 cm along the rope in 10.0 s. What is the wavelength of the wave?

12. **GP** A sinusoidal wave traveling in the negative x direction (to the left) has an amplitude of 20.0 cm, a wavelength of 35.0 cm, and a frequency of 12.0 Hz. The transverse position of an element of the medium at $t = 0$, $x = 0$ is $y = -3.00$ cm, and the element has a positive velocity here. We wish to find an expression for the wave function describing this wave. (a) Sketch the wave at $t = 0$. (b) Find the angular wave number k from the wavelength. (c) Find the period T from the frequency. Find (d) the angular frequency ω and (e) the wave speed v. (f) From the information about $t = 0$, find the phase constant ϕ. (g) Write an expression for the wave function $y(x, t)$.

13. **W** When a particular wire is vibrating with a frequency of 4.00 Hz, a transverse wave of wavelength 60.0 cm is produced. Determine the speed of waves along the wire.

14. A transverse sinusoidal wave on a string has a period $T = $ 25.0 ms and travels in the negative x direction with a speed of 30.0 m/s. At $t = 0$, an element of the string at $x = 0$ has a transverse position of 2.00 cm and is traveling downward with a speed of 2.00 m/s. (a) What is the amplitude of the wave? (b) What is the initial phase angle? (c) What is the maximum transverse speed of an element of the string? (d) Write the wave function for the wave.

Section 13.3 The Speed of Transverse Waves on Strings

15. **M** A steel wire of length 30.0 m and a copper wire of length 20.0 m, both with 1.00-mm diameters, are connected end to end and stretched to a tension of 150 N. During what time interval will a transverse wave travel the entire length of the two wires?

16. *Why is the following situation impossible?* An astronaut on the Moon is studying wave motion using the apparatus discussed in Example 13.4 and shown in Figure 13.11. He measures the time interval for pulses to travel along the horizontal wire. Assume the horizontal wire has a mass of 4.00 g and a length of 1.60 m and assume a 3.00-kg object is suspended from its extension around the pulley. The astronaut finds that a pulse requires 26.1 ms to traverse the length of the wire.

17. **W** An Ethernet cable is 4.00 m long. The cable has a mass of 0.200 kg. A transverse pulse is produced by plucking one end of the taut cable. The pulse makes four trips down and back along the cable in 0.800 s. What is the tension in the cable?

18. **Review.** A light string with a mass per unit length of 8.00 g/m has its ends tied to two walls separated by a distance equal to three-fourths the length of the string (Fig. P13.18). An object of mass m is suspended from the center of the string, putting a tension in the string. (a) Find an expression for the transverse wave speed in the string as a function of the mass of the hanging object. (b) What should be the mass of the object suspended from the string if the wave speed is to be 60.0 m/s?

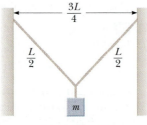

Figure P13.18

19. **M** Transverse waves travel with a speed of 20.0 m/s on a string under a tension of 6.00 N. What tension is required for a wave speed of 30.0 m/s on the same string?

20. **W** A piano string having a mass per unit length equal to 5.00×10^{-3} kg/m is under a tension of 1 350 N. Find the speed with which a wave travels on this string.

Section 13.4 Reflection and Transmission

21. A series of pulses, each of amplitude 0.150 m, are sent down a string that is attached to a post at one end. The pulses are reflected at the post and travel back along the string without loss of amplitude. When two waves are present on the same string, the net displacement of a particular element of the string is the sum of the displacements of the individual waves at that point. What is the net displacement of an element at a point on the string where two pulses are crossing (a) if the string is rigidly attached to the post and (b) if the end at which reflection occurs is free to slide up and down?

Section 13.5 Rate of Energy Transfer by Sinusoidal Waves on Strings

22. **S** A horizontal string can transmit a maximum power P_0 (without breaking) if a wave with amplitude A and angular frequency ω is traveling along it. To increase this maximum power, a student folds the string and uses this "double string" as a medium. Assuming the tension in the two strands together is the same as the original tension in the single string and the angular frequency of the wave remains the same, determine the maximum power that can be transmitted along the "double string."

23. **W** A long string carries a wave; a 6.00-m segment of the string contains four complete wavelengths and has a mass of 180 g. The string vibrates sinusoidally with a frequency of 50.0 Hz and a peak-to-valley displacement of 15.0 cm. (The "peak-to-valley" distance is the vertical distance from the farthest positive position to the farthest negative position.)

(a) Write the function that describes this wave traveling in the positive x direction. (b) Determine the power being supplied to the string.

24. **W** A taut rope has a mass of 0.180 kg and a length of 3.60 m. What power must be supplied to the rope so as to generate sinusoidal waves having an amplitude of 0.100 m and a wavelength of 0.500 m and traveling with a speed of 30.0 m/s?

25. **M** A sinusoidal wave on a string is described by the wave function

$$y = 0.15 \sin(0.80x - 50t)$$

where x and y are in meters and t is in seconds. The mass per unit length of this string is 12.0 g/m. Determine (a) the speed of the wave, (b) the wavelength, (c) the frequency, and (d) the power transmitted by the wave.

26. **M** Sinusoidal waves 5.00 cm in amplitude are to be transmitted along a string that has a linear mass density of 4.00×10^{-2} kg/m. The source can deliver a maximum power of 300 W, and the string is under a tension of 100 N. What is the highest frequency f at which the source can operate?

Section 13.6 Sound Waves

Note: Use the following values as needed unless otherwise specified. The equilibrium density of air at 20°C is $\rho = 1.20$ kg/m^3. Pressure variations ΔP are measured relative to atmospheric pressure, 1.013×10^5 N/m^2. (1 N/m^2 = 1 Pa (pascal). See Section 15.1.) The speed of sound in air is $v = 343$ m/s. Use Table 13.1 to find speeds of sound in other media.

Problem 55 in Chapter 2 can also be assigned with this section.

27. A dolphin (Fig. P13.27) in seawater at a temperature of 25°C emits a sound wave directed toward the ocean floor 150 m below. How much time passes before it hears an echo?

28. **Q|C** **W** Suppose you hear a clap of thunder 16.2 s after seeing the associated lightning strike. The speed of light in air is 3.00×10^8 m/s. (a) How far are you from the lightning strike? (b) Do you need to know the value of the speed of light to answer? Explain.

Stephen Frink/Photographer's Choice/Getty Images

Figure P13.27

29. Many artists sing very high notes in *ad lib* ornaments and cadenzas. The highest note written for a singer in a published score was F-sharp above high C, 1.480 kHz, for Zerbinetta in the original version of Richard Strauss's opera *Ariadne auf Naxos*. (a) Find the wavelength of this sound in air. (b) In response to complaints, Strauss later transposed the note down to F above high C, 1.397 kHz. By what increment did the wavelength change?

30. **BIO** A bat (Fig. P13.30) can detect very small objects, such as an insect whose length is approximately equal to one wavelength of the sound the bat makes. If a bat emits chirps at a frequency of 60.0 kHz and the speed of sound in air is 340 m/s, what is the smallest insect the bat can detect?

Figure P13.30

31. Write an expression that describes the pressure variation as a function of position and time for a sinusoidal sound wave in air. Assume the speed of sound is 343 m/s, $\lambda = 0.100$ m, and $\Delta P_{max} = 0.200$ Pa.

32. An ultrasonic tape measure uses frequencies above 20 MHz to determine dimensions of structures such as buildings. It does so by emitting a pulse of ultrasound into air and then measuring the time interval for an echo to return from a reflecting surface whose distance away is to be measured. The distance is displayed as a digital readout. For a tape measure that emits a pulse of ultrasound with a frequency of 22.0 MHz, (a) what is the distance to an object from which the echo pulse returns after 24.0 ms when the air temperature is 26°C? (b) What should be the duration of the emitted pulse if it is to include ten cycles of the ultrasonic wave? (c) What is the spatial length of such a pulse?

33. **BIO** Ultrasound is used in medicine both for diagnostic imaging (Fig. P13.33) and for therapy. For diagnosis, short pulses of ultrasound are passed through the patient's body. An echo reflected from a structure of interest is recorded, and the distance to the structure can be determined from the time delay for the echo's return. To reveal detail, the wavelength of the reflected ultrasound must be small compared to the size of the object reflecting the wave. The speed of ultrasound in human tissue is about 1 500 m/s (nearly the same as the speed of sound in water). (a) What is the wavelength of ultrasound with a frequency of 2.40 MHz? (b) In the whole set of imaging techniques, frequencies in the range 1.00 MHz to 20.0 MHz are used. What is the range of wavelengths corresponding to this range of frequencies?

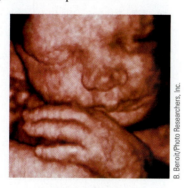

Figure P13.33

34. **M** An experimenter wishes to generate in air a sound wave that has a displacement amplitude of 5.50×10^{-6} m. The pressure amplitude is to be limited to 0.840 Pa. What is the minimum wavelength the sound wave can have?

35. **W** A sinusoidal sound wave moves through a medium and is described by the displacement wave function

$$s(x, t) = 2.00 \cos (15.7x - 858t)$$

where s is in micrometers, x is in meters, and t is in seconds. Find (a) the amplitude, (b) the wavelength, and (c) the speed of this wave. (d) Determine the instantaneous displacement from equilibrium of the elements of the medium at the position $x = 0.050\ 0$ m at $t = 3.00$ ms. (e) Determine the maximum speed of the element's oscillatory motion.

36. Calculate the pressure amplitude of a 2.00-kHz sound wave in air, assuming that the displacement amplitude is equal to 2.00×10^{-8} m.

37. **W** A sound wave in air has a pressure amplitude equal to 4.00×10^{-3} Pa. Calculate the displacement amplitude of the wave at a frequency of 10.0 kHz.

38. **W** A rescue plane flies horizontally at a constant speed searching for a disabled boat. When the plane is directly above the boat, the boat's crew blows a loud horn. By the time the plane's sound detector receives the horn's sound, the plane has traveled a distance equal to half its altitude above the ocean. Assuming it takes the sound 2.00 s to reach the plane, determine (a) the speed of the plane and (b) its altitude.

Section 13.7 The Doppler Effect

39. A driver travels northbound on a highway at a speed of 25.0 m/s. A police car, traveling southbound at a speed of 40.0 m/s, approaches with its siren producing sound at a frequency of 2 500 Hz. (a) What frequency does the driver observe as the police car approaches? (b) What frequency does the driver detect after the police car passes him? (c) Repeat parts (a) and (b) for the case when the police car is behind the driver and travels northbound.

40. **BIO** Expectant parents are thrilled to hear their unborn baby's heartbeat, revealed by an ultrasonic detector that produces beeps of audible sound in synchronization with the fetal heartbeat. Suppose the fetus's ventricular wall moves in simple harmonic motion with an amplitude of 1.80 mm and a frequency of 115 beats per minute. (a) Find the maximum linear speed of the heart wall. Suppose a source mounted on the detector in contact with the mother's abdomen produces sound at 2 000 000.0 Hz, which travels through tissue at 1.50 km/s. (b) Find the maximum change in frequency between the sound that arrives at the wall of the baby's heart and the sound emitted by the source. (c) Find the maximum change in frequency between the reflected sound received by the detector and that emitted by the source.

41. **M** Standing at a crosswalk, you hear a frequency of 560 Hz from the siren of an approaching ambulance. After the ambulance passes, the observed frequency of the siren is 480 Hz. Determine the ambulance's speed from these observations.

42. A block with a speaker bolted to it is connected to a spring having spring constant $k = 20.0$ N/m as shown in Figure P13.42 (page 444). The total mass of the block and speaker is 5.00 kg, and the amplitude of this unit's motion is 0.500 m. The speaker emits sound waves of frequency 440 Hz. Determine the highest and lowest frequencies heard by the person to the right of the speaker. Assume that the speed of sound is 343 m/s.

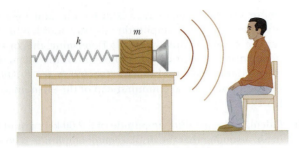

Figure P13.42

43. A siren mounted on the roof of a firehouse emits sound at a frequency of 900 Hz. A steady wind is blowing with a speed of 15.0 m/s. Taking the speed of sound in calm air to be 343 m/s, find the wavelength of the sound (a) upwind of the siren and (b) downwind of the siren. Firefighters are approaching the siren from various directions at 15.0 m/s. What frequency does a firefighter hear (c) if she is approaching from an upwind position so that she is moving in the direction in which the wind is blowing and (d) if she is approaching from a downwind position and moving against the wind?

44. *Why is the following situation impossible?* At the Summer Olympics, an athlete runs at a constant speed down a straight track while a spectator near the edge of the track blows a note on a horn with a fixed frequency. When the athlete passes the horn, she hears the frequency of the horn fall by the musical interval called a minor third. That is, the frequency she hears drops to five-sixths its original value.

45. **Review.** A tuning fork vibrating at 512 Hz falls from rest and accelerates at 9.80 m/s^2. How far below the point of release is the tuning fork when waves of frequency 485 Hz reach the release point?

46. **GP** Submarine A travels horizontally at 11.0 m/s through ocean water. It emits a sonar signal of frequency $f = 5.27 \times 10^3$ Hz in the forward direction. Submarine B is in front of submarine A and traveling at 3.00 m/s relative to the water in the same direction as submarine A. A crewman in submarine B uses his equipment to detect the sound waves ("pings") from submarine A. We wish to determine what is heard by the crewman in submarine B. (a) An observer on which submarine detects a frequency f' as described by Equation 13.31? (b) In Equation 13.31, should the sign of v_S be positive or negative? (c) In Equation 13.31, should the sign of v_O be positive or negative? (d) In Equation 13.31, what speed of sound should be used? (e) Find the frequency of the sound detected by the crewman on submarine B.

Section 13.8 Context Connection: Seismic Waves

47. **W** A seismographic station receives S and P waves from an earthquake, separated in time by 17.3 s. Assume the waves have traveled over the same path at speeds of 4.50 km/s and 7.80 km/s. Find the distance from the seismograph to the focus of the quake.

48. Two points A and B on the surface of the Earth are at the same longitude and 60.0° apart in latitude as shown in Figure P13.48. Suppose an earthquake at point A creates a P wave that reaches point B by traveling straight through the body of the Earth at a constant speed of 7.80 km/s. The earthquake also radiates a Rayleigh wave that travels at 4.50 km/s. In addition to P and S waves, Rayleigh waves are a third type of seismic wave that travels along the *surface* of the Earth rather than through the *bulk* of the Earth. (a) Which of these two seismic waves arrives at B first? (b) What is the time difference between the arrivals of these two waves at B?

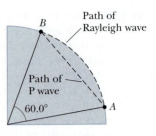

Figure P13.48

Additional Problems

49. **Review.** A block of mass $M = 0.450$ kg is attached to one end of a cord of mass 0.003 20 kg; the other end of the cord is attached to a fixed point. The block rotates with constant angular speed in a circle on a frictionless, horizontal table as shown in Figure P13.49. Through what angle does the block rotate in the time interval during which a transverse wave travels along the string from the center of the circle to the block?

Figure P13.49 Problems 49, 65, and 66.

50. **S** **Review.** A block of mass M, supported by a string, rests on a frictionless incline making an angle θ with the horizontal (Fig. P13.50). The length of the string is L, and its mass is $m \ll M$. Derive an expression for the time interval required for a transverse wave to travel from one end of the string to the other.

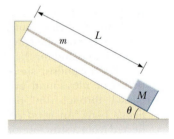

Figure P13.50

51. **Review.** A 2.00-kg block hangs from a rubber cord, being supported so that the cord is not stretched. The unstretched length of the cord is 0.500 m, and its mass is 5.00 g. The "spring constant" for the cord is 100 N/m. The block is released and stops momentarily at the lowest point. (a) Determine the tension in the cord when the block is at this lowest point. (b) What is the length of the cord in this "stretched" position? (c) If the block is held in this lowest position, find the speed of a transverse wave in the cord.

52. **S** **Review.** A block of mass M hangs from a rubber cord. The block is supported so that the cord is not stretched. The unstretched length of the cord is L_0, and its mass is m, much less than M. The "spring constant" for the cord is k. The block is released and stops momentarily at the lowest point. (a) Determine the tension in the string when the block is at this lowest point. (b) What is the length of the cord in this "stretched" position? (c) If the block is held in this lowest position, find the speed of a transverse wave in the cord.

53. A police car is traveling east at 40.0 m/s along a straight road, overtaking a car ahead of it moving east at 30.0 m/s. The police car has a malfunctioning siren that is stuck at 1 000 Hz. (a) What would be the wavelength in air of the siren sound if the police car were at rest? (b) What is the wavelength in front of the police car? (c) What is it behind the police car? (d) What is the frequency heard by the driver being chased?

54. "The wave" is a particular type of pulse that can propagate through a large crowd gathered at a sports arena (Fig. P13.54). The elements of the medium are the spectators, with zero position corresponding to their being seated and maximum position corresponding to their standing and raising their arms. When a large fraction of the spectators participates in the wave motion, a somewhat stable pulse shape can develop. The wave speed depends on people's reaction time, which is typically on the order of 0.1 s. Estimate the order of magnitude, in minutes, of the time interval required for such a pulse to make one circuit around a large sports stadium. State the quantities you measure or estimate and their values.

Figure P13.54

55. S A pulse traveling along a string of linear mass density μ is described by the wave function

$$y = [A_0 e^{-bx}] \sin (kx - \omega t)$$

where the factor in the square brackets is said to be the amplitude. (a) What is the power $P(x)$ carried by this wave at a point x? (b) What is the power $P(0)$ carried by this wave at the origin? (c) Compute the ratio $P(x)/P(0)$.

56. (a) Show that the speed of longitudinal waves along a spring of force constant k is $v = \sqrt{kL/\mu}$, where L is the unstretched length of the spring and μ is the mass per unit length. (b) A spring with a mass of 0.400 kg has an unstretched length of 2.00 m and a force constant of 100 N/m. Using the result you obtained in part (a), determine the speed of longitudinal waves along this spring.

57. W A flowerpot is knocked off a window ledge from a height $d = 20.0$ m above the sidewalk as shown in Figure P13.57. It falls toward an unsuspecting man of height $h = 1.75$ m who is standing below. Assume the man requires a time interval of $\Delta t = 0.300$ s to respond to the warning. How close to the sidewalk can the flowerpot fall before it is too late for a warning shouted from the balcony to reach the man in time?

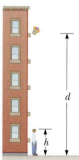

Figure P13.57
Problems 57 and 58.

58. S A flowerpot is knocked off a balcony from a height d above the sidewalk as shown in Figure P13.57. It falls toward an unsuspecting man of height h who is

standing below. Assume the man requires a time interval of Δt to respond to the warning. How close to the sidewalk can the flowerpot fall before it is too late for a warning shouted from the balcony to reach the man in time? Use the symbol v for the speed of sound.

59. Trucks carrying garbage to the town dump form a nearly steady procession on a country road, all traveling at 19.7 m/s in the same direction. Two trucks arrive at the dump every 3 min. A bicyclist is also traveling toward the dump, at 4.47 m/s. (a) With what frequency do the trucks pass the cyclist? (b) **What If?** A hill does not slow down the trucks, but makes the out-of-shape cyclist's speed drop to 1.56 m/s. How often do the trucks whiz past the cyclist now?

60. S A rope of total mass m and length L is suspended vertically. Analysis shows that for short transverse pulses, the waves above a short distance from the free end of the rope can be represented to a good approximation by the linear wave equation discussed in Section 13.2. Show that a transverse pulse travels the length of the rope in a time interval that is given approximately by $\Delta t \approx 2\sqrt{L/g}$. *Suggestion:* First find an expression for the wave speed at any point a distance x from the lower end by considering the rope's tension as resulting from the weight of the segment below that point.

61. The ocean floor is underlain by a layer of basalt that constitutes the crust, or uppermost layer, of the Earth in that region. Below this crust is found denser periodotite rock that forms the Earth's mantle. The boundary between these two layers is called the Mohorovicic discontinuity ("Moho" for short). If an explosive charge is set off at the surface of the basalt, it generates a seismic wave that is reflected back out at the Moho. If the speed of this wave in basalt is 6.50 km/s and the two-way travel time is 1.85 s, what is the thickness of this oceanic crust?

62. S Assume an object of mass M is suspended from the bottom of the rope of mass m and length L in Problem 60. (a) Show that the time interval for a transverse pulse to travel the length of the rope is

$$\Delta t = 2\sqrt{\frac{L}{mg}} \left(\sqrt{M + m} - \sqrt{M} \right)$$

(b) **What If?** Show that the expression in part (a) reduces to the result of Problem 60 when $M = 0$. (c) Show that for $m \ll M$, the expression in part (a) reduces to

$$\Delta t = \sqrt{\frac{mL}{Mg}}$$

63. M To measure her speed, a skydiver carries a buzzer emitting a steady tone at 1 800 Hz. A friend on the ground at the landing site directly below listens to the amplified sound he receives. Assume the air is calm and the speed of sound is independent of altitude. While the skydiver is falling at terminal speed, her friend on the ground receives waves of frequency 2 150 Hz. (a) What is the skydiver's speed of descent? (b) **What If?** Suppose the skydiver can hear the sound of the buzzer reflected from the ground. What frequency does she receive?

64. *Why is the following situation impossible?* Tsunamis are ocean surface waves that have enormous wavelengths (100 to 200 km), and the propagation speed for these waves is $v \approx \sqrt{gd_{avg}}$, where d_{avg} is the average depth of the water. An earthquake on the ocean floor in the Gulf of Alaska produces a tsunami that reaches Hilo, Hawaii, 4 450 km away, in a time interval of 5.88 h. (This method was used in 1856 to estimate the average depth of the Pacific Ocean long before soundings were made to give a direct determination.)

65. **Review.** A block of mass $M = 0.450$ kg is attached to one end of a cord of mass $m = 0.003\ 20$ kg; the other end of the cord is attached to a fixed point. The block rotates with constant angular speed $\omega = 10.0$ rad/s in a circle on a frictionless, horizontal table as shown in Figure P13.49. What time interval is required for a transverse wave to travel along the string from the center of the circle to the block?

66. **S Review.** A block of mass M is attached to one end of a cord of mass m; the other end of the cord is attached to a fixed point. The block rotates with constant angular speed ω in a circle on a frictionless, horizontal table as shown in Figure P13.49. What time interval is required for a transverse wave to travel along the string from the center of the circle to the block?

67. **BIO** **Q|C** A bat, moving at 5.00 m/s, is chasing a flying insect. If the bat emits a 40.0-kHz chirp and receives back an echo at 40.4 kHz, (a) what is the speed of the insect? (b) Will the bat be able to catch the insect? Explain.

68. **S** A sound wave moves down a cylinder as in Active Figure 13.19. Show that the pressure variation of the wave is described by $\Delta P = \pm \rho v \omega \sqrt{s_{max}^2 - s^2}$, where $s = s(x, t)$ is given by Equation 13.25.

69. **S** A string on a musical instrument is held under tension T and extends from the point $x = 0$ to the point $x = L$. The string is overwound with wire in such a way that its mass per unit length $\mu(x)$ increases uniformly from μ_0 at $x = 0$ to μ_L at $x = L$. (a) Find an expression for $\mu(x)$ as a function of x over the range $0 \le x \le L$. (b) Find an expression for the time interval required for a transverse pulse to travel the length of the string.

70. A train whistle ($f = 400$ Hz) sounds higher or lower in frequency depending on whether it approaches or recedes. (a) Prove that the difference in frequency between the approaching and receding train whistle is

$$\Delta f = \frac{2u/v}{1 - u^2/v^2} f$$

where u is the speed of the train and v is the speed of sound. (b) Calculate this difference for a train moving at a speed of 130 km/h. Take the speed of sound in air to be 340 m/s.

71. The Doppler equation presented in the text is valid when the motion between the observer and the source occurs on a straight line so that the source and observer are moving either directly toward or directly away from each other. If this restriction is relaxed, one must use the more general Doppler equation

$$f' = \left(\frac{v + v_O \cos \theta_O}{v - v_S \cos \theta_S} \right) f$$

where θ_O and θ_S are defined in Figure P13.71a. Use the preceding equation to solve the following problem. A train moves at a constant speed of $v = 25.0$ m/s toward the intersection shown in Figure P13.71b. A car is stopped near the crossing, 30.0 m from the tracks. The train's horn emits a frequency of 500 Hz when the train is 40.0 m from the intersection. (a) What is the frequency heard by the passengers in the car? (b) If the train emits this sound continuously and the car is stationary at this position long before the train arrives until long after it leaves, what range of frequencies do passengers in the car hear? (c) Suppose the car is foolishly trying to beat the train to the intersection and is traveling at 40.0 m/s toward the tracks. When the car is 30.0 m from the tracks and the train is 40.0 m from the intersection, what is the frequency heard by the passengers in the car now?

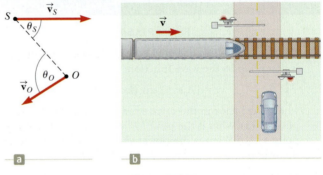

Figure P13.71

AP Photo/Danny Moloshok

Chapter

14

Superposition and Standing Waves

Chapter Outline

SUMMARY

In Chapter 13, we introduced the wave model. We have seen that waves are very different from particles. An ideal particle is of zero size, but an ideal wave is of infinite length. Another important difference between waves and particles is that we can explore the possibility of two or more waves combining at one point in the same medium. We can combine particles to form extended objects, but the particles must be at different locations. In contrast, two waves can both be present at a given location, and the ramifications of this possibility are explored in this chapter.

One ramification of the combination of waves is that only certain allowed frequencies can exist in systems with boundary conditions; that is, the frequencies are *quantized*. In Chapter 11, we learned about quantized energies of the hydrogen atom. Quantization is at the heart of quantum mechanics, a subject that is introduced formally in Chapter 28. We shall see that waves under boundary conditions explain many of the quantum phenomena. For our present purposes in this chapter, quantization enables us to understand the behavior of the wide array of musical instruments that are based on strings and air columns.

Blues master B. B. King takes advantage of standing waves on strings. He changes to higher notes on the guitar by pushing the strings against the frets on the fingerboard, shortening the lengths of the portions of the strings that vibrate.

14.1 | Analysis Model: Waves in Interference

Many interesting wave phenomena in nature cannot be described by a single traveling wave. Instead, one must analyze these phenomena in terms of a combination of traveling waves. As noted in the introduction, waves have a remarkable difference from particles in that waves can be combined at the *same* location in space. To analyze such wave combinations, we make use of the **superposition principle:**

> If two or more traveling waves are moving through a medium, the resultant value of the wave function at any point is the algebraic sum of the values of the wave functions of the individual waves.

▶ Superposition principle

When the pulses overlap, the wave function is the sum of the individual wave functions.

When the crests of the two pulses align, the amplitude is the sum of the individual amplitudes.

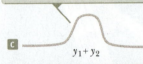

When the pulses no longer overlap, they have not been permanently affected by the interference.

Active Figure 14.1 Constructive interference. Two positive pulses travel on a stretched string in opposite directions and overlap.

▶ Constructive interference

▶ Destructive interference

Waves that obey this principle are called *linear waves*. In the case of mechanical waves, linear waves are generally characterized by having amplitudes much smaller than their wavelengths. Waves that violate the superposition principle are called *nonlinear waves* and are often characterized by large amplitudes. In this book, we deal only with linear waves.

One consequence of the superposition principle is that two traveling waves can pass through each other without being destroyed or even altered. For instance, when two pebbles are thrown into a pond and hit the surface at different locations, the expanding circular surface waves from the two locations simply pass through each other with no permanent effect. The resulting complex pattern can be viewed as two independent sets of expanding circles.

Active Figure 14.1 is a pictorial representation of the superposition of two pulses. The wave function for the pulse moving to the right is y_1, and the wave function for the pulse moving to the left is y_2. The pulses have the same speed but different shapes, and the displacement of the elements of the medium is in the positive y direction for both pulses. When the waves overlap (Active Fig. 14.1b), the wave function for the resulting complex wave is given by $y_1 + y_2$. When the crests of the pulses coincide (Active Fig. 14.1c), the resulting wave given by $y_1 + y_2$ has a larger amplitude than that of the individual pulses. The two pulses finally separate and continue moving in their original directions (Active Fig. 14.1d). Notice that the pulse shapes remain unchanged after the interaction, as if the two pulses had never met!

The combination of separate waves in the same region of space to produce a resultant wave is called **interference.** For the two pulses shown in Active Figure 14.1, the displacement of the elements of the medium is in the positive y direction for both pulses, and the resultant pulse (created when the individual pulses overlap) exhibits an amplitude greater than that of either individual pulse. Because the displacements caused by the two pulses are in the same direction, we refer to their superposition as **constructive interference.**

Now consider two pulses traveling in opposite directions on a taut string where one pulse is inverted relative to the other as illustrated in Active Figure 14.2. When these pulses begin to overlap, the resultant pulse is given by $y_1 + y_2$, but the values of the function y_2 are negative. Again, the two pulses pass through each other; because the displacements caused by the two pulses are in opposite directions, however, we refer to their superposition as **destructive interference.**

The superposition principle is the centerpiece of the analysis model called **waves in interference.** In many situations, both in acoustics and optics, waves combine according to this principle and exhibit interesting phenomena with practical applications.

▶ **QUICK QUIZ 14.1** Two symmetric pulses move in opposite directions on a string and are identical in shape except that one has positive displacements of the elements of the string and the other has negative displacements. At the moment the two pulses completely overlap on the string, what happens? (**a**) The energy associated with the pulses has disappeared. (**b**) The string is not moving. (**c**) The string forms a straight line. (**d**) The pulses have vanished and will not reappear.

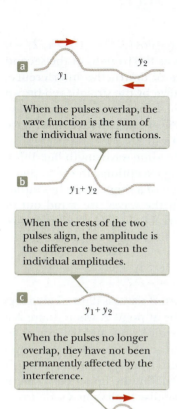

When the pulses overlap, the wave function is the sum of the individual wave functions.

When the crests of the two pulses align, the amplitude is the difference between the individual amplitudes.

When the pulses no longer overlap, they have not been permanently affected by the interference.

Active Figure 14.2 Destructive interference. Two pulses, one positive and one negative, travel on a stretched string in opposite directions and overlap.

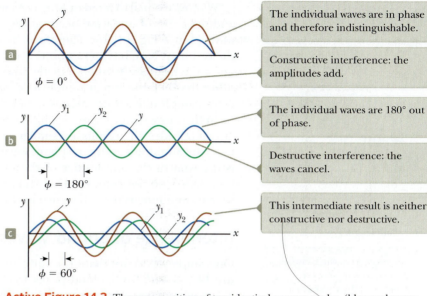

The individual waves are in phase and therefore indistinguishable.

Constructive interference: the amplitudes add.

The individual waves are 180° out of phase.

Destructive interference: the waves cancel.

This intermediate result is neither constructive nor destructive.

Active Figure 14.3 The superposition of two identical waves y_1 and y_2 (blue and green, respectively) to yield a resultant wave (red-brown).

Superposition of Sinusoidal Waves

Let us now apply the principle of superposition to two sinusoidal waves traveling in the same direction in a linear medium. If the two waves are traveling to the right and have the same frequency, wavelength, and amplitude but differ in phase, we can express their individual wave functions as

$$y_1 = A \sin (kx - \omega t) \quad y_2 = A \sin (kx - \omega t + \phi)$$

where, as usual, $k = 2\pi/\lambda$, $\omega = 2\pi f$, and ϕ is the phase constant as discussed in Section 13.2. Hence, the resultant wave function y is

$$y = y_1 + y_2 = A [\sin (kx - \omega t) + \sin (kx - \omega t + \phi)]$$

To simplify this expression, we use the trigonometric identity

$$\sin a + \sin b = 2 \cos \left(\frac{a - b}{2} \right) \sin \left(\frac{a + b}{2} \right)$$

Letting $a = kx - \omega t$ and $b = kx - \omega t + \phi$, we find that the resultant wave function y reduces to

$$y = 2A \cos \left(\frac{\phi}{2} \right) \sin \left(kx - \omega t + \frac{\phi}{2} \right) \qquad \text{14.1} \blacktriangleleft$$

◀ Resultant of two traveling sinusoidal waves

Pitfall Prevention | 14.1
Do Waves Actually *Interfere*?
In popular usage, the word *interfere* implies that an agent affects a situation in some way so as to preclude something from happening. For example, in American football, *pass interference* means that a defending player has affected the receiver so that the receiver is unable to catch the ball. This usage is very different from its use in physics, where waves pass through each other and interfere, but do not affect each other in any way. In physics, interference is similar to the notion of *combination* as described in this chapter.

This result has several important features. The resultant wave function y also is sinusoidal and has the same frequency and wavelength as the individual waves because the sine function incorporates the same values of k and ω that appear in the original wave functions. The amplitude of the resultant wave is $2A \cos (\phi/2)$, and its phase is $\phi/2$. If the phase constant ϕ equals 0, then $\cos (\phi/2) = \cos 0 = 1$ and the amplitude of the resultant wave is $2A$, twice the amplitude of either individual wave. In this case, the crests of the two waves are at the same locations in space and the waves are said to be everywhere *in phase* and therefore interfere constructively. The individual waves y_1 and y_2 combine to form the red-brown curve y of amplitude $2A$ shown in Active Figure 14.3a. Because the individual waves are in phase, they are indistinguishable in Active Figure 14.3a, where they appear as a single blue curve. In general, constructive interference occurs when $\cos (\phi/2) = \pm 1$. That is true, for example, when $\phi = 0$, 2π, 4π, ... rad, that is, when ϕ is an *even* multiple of π.

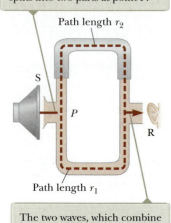

A sound wave from the speaker (S) propagates into the tube and splits into two parts at point P.

Path length r_2

S

P

R

Path length r_1

The two waves, which combine at the opposite side, are detected at the receiver (R).

Figure 14.4 An acoustical system for demonstrating interference of sound waves. The upper path length r_2 can be varied by sliding the upper section.

When ϕ is equal to π rad or to any *odd* multiple of π, then $\cos(\phi/2) = \cos(\pi/2) = 0$ and the crests of one wave occur at the same positions as the troughs of the second wave (Active Fig. 14.3b). Therefore, as a consequence of destructive interference, the resultant wave has *zero* amplitude everywhere as shown by the straight red-brown line in Active Figure 14.3b. Finally, when the phase constant has an arbitrary value other than 0 or an integer multiple of π rad (Active Fig. 14.3c), the resultant wave has an amplitude whose value is somewhere between 0 and $2A$.

In the more general case in which the waves have the same wavelength but different amplitudes, the results are similar with the following exceptions. In the in-phase case, the amplitude of the resultant wave is not twice that of a single wave, but rather is the sum of the amplitudes of the two waves. When the waves are π rad out of phase, they do not completely cancel as in Active Figure 14.3b. The result is a wave whose amplitude is the difference in the amplitudes of the individual waves.

Interference of Sound Waves

One simple device for demonstrating interference of sound waves is illustrated in Figure 14.4. Sound from a loudspeaker S is sent into a tube at point P, where there is a T-shaped junction. Half the sound energy travels in one direction, and half travels in the opposite direction. Therefore, the sound waves that reach the receiver R can travel along either of the two paths. The distance along any path from speaker to receiver is called the **path length** r. The lower path length r_1 is fixed, but the upper path length r_2 can be varied by sliding the U-shaped tube, which is similar to that on a slide trombone. When the difference in the path lengths $\Delta r = |r_2 - r_1|$ is either zero or some integer multiple of the wavelength λ (that is, $\Delta r = n\lambda$, where $n = 0, 1, 2, 3, \ldots$), the two waves reaching the receiver at any instant are in phase and interfere constructively as shown in Active Figure 14.3a. For this case, a maximum in the sound intensity is detected at the receiver. If the path length r_2 is adjusted such that the path difference $\Delta r = \lambda/2, 3\lambda/2, \ldots, n\lambda/2$ (for n odd), the two waves are exactly π rad, or 180°, out of phase at the receiver and hence cancel each other. In this case of destructive interference, no sound is detected at the receiver. This simple experiment demonstrates that a phase difference may arise between two waves generated by the same source when they travel along paths of unequal lengths. This important phenomenon will be indispensable in our investigation of the interference of light waves in Chapter 27.

> ## THINKING PHYSICS 14.1
>
> If stereo speakers are connected to the amplifier "out of phase," one speaker is moving outward when the other is moving inward. The result is a weakness in the bass notes, which can be corrected by reversing the wires on one of the speaker connections. Why are only the bass notes affected in this case and not the treble notes? For help in answering this question, note that the range of wavelengths of sound from a standard piano is from 0.082 m for the highest C to 13 m for the lowest A.
>
> **Reasoning** Imagine that you are sitting in front of the speakers, midway between them. Then, the sound from each speaker travels the same distance to you, so there is no phase difference in the sound due to a path difference. Because the speakers are connected out of phase, the sound waves are half a wavelength out of phase on leaving the speaker and, consequently, on arriving at your ear. As a result, the sound for all frequencies cancels in the simplification model of a zero-size head located exactly on the midpoint between the speakers.
>
> If the ideal head were moved off the centerline, an additional phase difference is introduced by the path length difference for the sound from the two speakers. In the case of low-frequency, long-wavelength bass notes, the path length differences are a small fraction of a wavelength, so significant cancellation still occurs. For the high-frequency, short-wavelength treble notes, a small movement of the ideal head results in a much larger fraction of a wavelength in path length difference or even multiple wavelengths. Therefore, the treble notes could be in phase with this head movement. If we now add that the head is not of zero size and that it has two ears, we can see that complete cancellation is not possible and, with even small movements of the head, one or both ears will be at or near maxima for the treble notes. The size of the head is much smaller than bass wavelengths, however, so the bass notes are significantly weakened over much of the region in front of the speakers. ◄

Example 14.1 | Two Speakers Driven by the Same Source

Two identical loudspeakers placed 3.00 m apart are driven by the same oscillator (Fig. 14.5). A listener is originally at point *O*, located 8.00 m from the center of the line connecting the two speakers. The listener then moves to point *P*, which is a perpendicular distance 0.350 m from *O*, and she experiences the *first minimum* in sound intensity. What is the frequency of the oscillator?

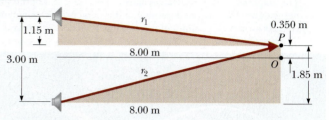

Figure 14.5 (Example 14.1) Two identical loudspeakers emit sound waves to a listener at *P*.

SOLUTION

Conceptualize In Figure 14.4, a sound wave enters a tube and is then *acoustically* split into two different paths before recombining at the other end. In this example, a signal representing the sound is *electrically* split and sent to two different loudspeakers. After leaving the speakers, the sound waves recombine at the position of the listener. Despite the difference in how the splitting occurs, the path difference discussion related to Figure 14.4 can be applied here.

Categorize Because the sound waves from two separate sources combine, we apply the waves in interference analysis model.

Analyze Figure 14.5 shows the physical arrangement of the speakers, along with two shaded right triangles that can be drawn on the basis of the lengths described in the problem. The first minimum occurs when the two waves reaching the listener at point *P* are 180° out of phase, in other words, when their path difference Δr equals $\lambda/2$.

From the shaded triangles, find the path lengths from the speakers to the listener:

$$r_1 = \sqrt{(8.00 \text{ m})^2 + (1.15 \text{ m})^2} = 8.08 \text{ m}$$
$$r_2 = \sqrt{(8.00 \text{ m})^2 + (1.85 \text{ m})^2} = 8.21 \text{ m}$$

Hence, the path difference is $r_2 - r_1 = 0.13$ m. Because this path difference must equal $\lambda/2$ for the first minimum, $\lambda = 0.26$ m.

To obtain the oscillator frequency, use Equation 13.12, $v = \lambda f$, where *v* is the speed of sound in air, 343 m/s:

$$f = \frac{v}{\lambda} = \frac{343 \text{ m/s}}{0.26 \text{ m}} = 1.3 \text{ kHz}$$

Finalize This example enables us to understand why the speaker wires in a stereo system should be connected properly. When connected the wrong way—that is, when the positive (or red) wire is connected to the negative (or black) terminal on one of the speakers and the other is correctly wired—the speakers are said to be "out of phase," with one speaker moving outward while the other moves inward, as discussed in Thinking Physics 14.1. As a consequence, the sound wave coming from one speaker destructively interferes with the wave coming from the other at point *O* in Figure 14.5. A rarefaction region due to one speaker is superposed on a compression region from the other speaker. Although the two sounds probably do not completely cancel each other (because the left and right stereo signals are usually not identical), a substantial loss of sound quality occurs at point *O*.

What If? What if the speakers were connected out of phase? What happens at point *P* in Figure 14.5?

Answer In this situation, the path difference of $\lambda/2$ combines with a phase difference of $\lambda/2$ due to the incorrect wiring to give a full phase difference of λ. As a result, the waves are in phase and there is a *maximum* intensity at point *P*.

14.2 | Standing Waves

The sound waves from the pair of loudspeakers in Example 14.1 leave the speakers in the forward direction, and we considered interference at a point in front of the speakers. Suppose we turn the speakers so that they face each other and then have them emit sound of the same frequency and amplitude. In this situation, two identical waves travel in opposite directions in the same medium as in Figure 14.6. These waves combine in accordance with the waves in interference model.

We can analyze such a situation by considering wave functions for two transverse sinusoidal waves having the same amplitude, frequency, and wavelength but traveling in opposite directions in the same medium:

$$y_1 = A \sin(kx - \omega t) \quad \text{and} \quad y_2 = A \sin(kx + \omega t)$$

Figure 14.6 Two speakers emit sound waves toward each other. Between the speakers, identical waves traveling in opposite directions combine to form standing waves.

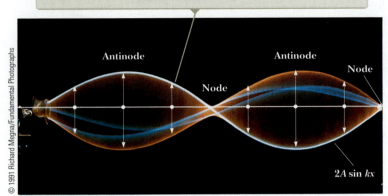

The amplitude of the vertical oscillation of any element of the string depends on the horizontal position of the element. Each element vibrates within the confines of the envelope function $2A \sin kx$.

© 1991 Richard Megna/Fundamental Photographs

Figure 14.7 Multiflash photograph of a standing wave on a string. The vertical displacement from equilibrium of an individual element of the string is proportional to $\cos \omega t$. That is, each element vibrates at an angular frequency ω.

Pitfall Prevention | 14.2
Three Types of Amplitude
We need to distinguish carefully here between the **amplitude of the individual waves,** which is A, and the **amplitude of the simple harmonic motion of the elements of the medium,** which is $2A \sin kx$. A given element in a standing wave vibrates within the constraints of the *envelope* function $2A \sin kx$, where x is that element's position in the medium. That vibration is in contrast to traveling sinusoidal waves, in which all elements oscillate with the same amplitude and the same frequency and the amplitude A of the wave is the same as the amplitude A of the simple harmonic motion of the elements. Furthermore, we can identify the **amplitude of the standing wave** as $2A$.

▶ Positions of nodes

where y_1 represents a wave traveling in the $+x$ direction and y_2 represents a wave traveling in the $-x$ direction. According to the principle of superposition, adding these two functions gives the resultant wave function y:

$$y = y_1 + y_2 = A \sin(kx - \omega t) + A \sin(kx + \omega t)$$

Using the trigonometric identity $\sin(a \pm b) = \sin a \cos b \pm \cos a \sin b$, this expression reduces to

$$y = (2A \sin kx) \cos \omega t \qquad \textbf{14.2}◀$$

Notice that this function does not look mathematically like a traveling wave because there is no function of $kx - \omega t$. Equation 14.2 represents the wave function of a **standing wave** such as that shown in Figure 14.7. A standing wave is an oscillation pattern that results from two waves traveling in opposite directions. Mathematically, this equation looks more like simple harmonic motion than wave motion for traveling waves. Every element of the medium vibrates in simple harmonic motion with the same angular frequency ω (according to the factor $\cos \omega t$). The amplitude of motion of a given element (the factor $2A \sin kx$), however, depends on its position along the medium, described by the variable x. From this result, we see that the simple harmonic motion of every element has an angular frequency of ω and a position-dependent amplitude of $2A \sin kx$.

Because the amplitude of the simple harmonic motion of an element at any value of x is equal to $2A \sin kx$, we see that the *maximum* amplitude of the simple harmonic motion has the value $2A$. This maximum amplitude is described as the amplitude of the standing wave. It occurs when the coordinate x for an element satisfies the condition $\sin kx = 1$, or when

$$kx = \frac{\pi}{2}, \frac{3\pi}{2}, \frac{5\pi}{2}, \ldots$$

Because $k = 2\pi/\lambda$, the positions of maximum amplitude, called **antinodes,** are

$$x = \frac{\lambda}{4}, \frac{3\lambda}{4}, \frac{5\lambda}{4}, \ldots = \frac{n\lambda}{4} \qquad n = 1, 3, 5, \ldots \qquad \textbf{14.3}◀$$

Note that adjacent antinodes are separated by a distance $\lambda/2$.

Similarly, the simple harmonic motion has a *minimum* amplitude of zero when x satisfies the condition $\sin kx = 0$, or when $kx = \pi, 2\pi, 3\pi, \ldots$, giving

$$x = 0, \frac{\lambda}{2}, \lambda, \frac{3\lambda}{2}, \ldots = \frac{n\lambda}{2} \qquad n = 0, 1, 2, 3, \ldots \qquad \textbf{14.4}◀$$

These points of zero amplitude, called **nodes,** are also spaced by $\lambda/2$. The distance between a node and an adjacent antinode is $\lambda/4$. The standing wave patterns produced at various times by two waves traveling in opposite directions are represented graphically in Active Figure 14.8. The upper part of each figure represents the individual traveling waves and the lower part represents the standing wave patterns. The nodes of the standing wave are labeled N and the antinodes are labeled A. At $t = 0$ (Active Fig. 14.8a), the two waves are in phase, giving a wave pattern with amplitude $2A$. One-quarter of a period later, at $t = T/4$ (Active Fig. 14.8b), the individual waves have moved one-quarter of a wavelength (one to the right and the other to the left). At this time, the waves are 180° out of phase. The individual displacements of the elements of the medium from their equilibrium positions are of equal magnitude and opposite direction for all values of x; hence, the resultant wave has zero displacement everywhere. At $t = T/2$ (Active Fig. 14.8c), the individual waves are again in phase, producing a wave pattern that is inverted relative to the $t = 0$ pattern. In the

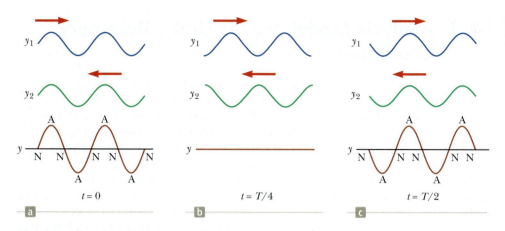

Active Figure 14.8 Standing wave patterns produced at various times by two waves of equal amplitude traveling in opposite directions. For the resultant wave y, the nodes (N) are points of zero displacement and the antinodes (A) are points of maximum displacement.

standing wave, the elements of the medium alternate in time between the extremes shown in Active Figures 14.8a and 14.8c.

> **QUICK QUIZ 14.2** Consider Active Figure 14.8 as representing a standing wave on a string. Define the velocity of elements of the string as positive if they are moving upward in the figure. (**i**) At the moment the string has the shape shown by the red-brown curve in Active Figure 14.8a, what is the instantaneous velocity of elements along the string? (**a**) zero for all elements (**b**) positive for all elements (**c**) negative for all elements (**d**) varies with the position of the element (**ii**) From the same set of choices, at the moment the string has the shape shown by the red-brown curve in Active Figure 14.8b, what is the instantaneous velocity of elements along the string?

Example 14.2 | Formation of a Standing Wave

Two waves traveling in opposite directions produce a standing wave. The individual wave functions are

$$y_1 = 4.0 \sin (3.0x - 2.0t) \qquad\qquad y_2 = 4.0 \sin (3.0x + 2.0t)$$

where x and y are measured in centimeters and t is in seconds.

(A) Find the amplitude of the simple harmonic motion of the element of the medium located at $x = 2.3$ cm.

SOLUTION

Conceptualize The waves described by the given equations are identical except for their directions of travel, so they indeed combine to form a standing wave as discussed in this section. We can represent the waves graphically by the blue and green curves in Active Figure 14.8.

Categorize We will substitute values into equations developed in this section, so we categorize this example as a substitution problem.

From the equations for the waves, we see that $A = 4.0$ cm, $k = 3.0$ rad/cm, and $\omega = 2.0$ rad/s. Use Equation 14.2 to write an expression for the standing wave:

$$y = (2A \sin kx) \cos \omega t = 8.0 \sin 3.0x \cos 2.0t$$

Find the amplitude of the simple harmonic motion of the element at the position $x = 2.3$ cm by evaluating the coefficient of the cosine function at this position:

$$y_{max} = (8.0 \text{ cm}) \sin 3.0x \big|_{x = 2.3}$$
$$= (8.0 \text{ cm}) \sin (6.9 \text{ rad}) = 4.6 \text{ cm}$$

(B) Find the positions of the nodes and antinodes if one end of the string is at $x = 0$.

SOLUTION

Find the wavelength of the traveling waves:

$$k = \frac{2\pi}{\lambda} = 3.0 \text{ rad/cm} \quad \rightarrow \quad \lambda = \frac{2\pi}{3.0} \text{ cm}$$

Use Equation 14.4 to find the locations of the nodes:

$$x = n\frac{\lambda}{2} = n\left(\frac{\pi}{3.0}\right) \text{ cm} \quad n = 0, 1, 2, 3, \ldots$$

Use Equation 14.3 to find the locations of the antinodes:

$$x = n\frac{\lambda}{4} = n\left(\frac{\pi}{6.0}\right) \text{ cm} \quad n = 1, 3, 5, 7, \ldots$$

14.3 | Analysis Model: Waves Under Boundary Conditions

In the preceding section, we discussed standing waves formed by identical waves moving in opposite directions in the same medium. One way to establish a standing wave on a string is to combine incoming and reflected waves from a rigid end. If a string is stretched between *two* rigid supports (Active Fig. 14.9a) and waves are established on the string, standing waves will be set up in the string by the continuous superposition of the waves incident on and reflected from the ends. This physical system is a model for the source of sound in any stringed instrument, such as the guitar, the violin, and the piano. The string has a number of natural patterns of oscillation, called **normal modes,** each of which has a characteristic frequency that is easily calculated.

This discussion is our first introduction to an important analysis model, the **wave under boundary conditions.** When boundary conditions are applied to a wave, we find very interesting behavior that has no analog in the physics of particles. The most prominent aspect of this behavior is **quantization.** We shall find that only certain waves—those that satisfy the boundary conditions—are allowed. The notion of quantization was introduced in Chapter 11 when we discussed the Bohr model of the atom. In that model, angular momentum was quantized. As we shall see in Chapter 29, this quantization is just an application of the wave under boundary conditions model.

In the standing wave pattern on a stretched string, the ends of the string must be nodes because these points are fixed, establishing the boundary condition on the waves. The rest of the pattern can be built from this boundary condition along with the requirement that nodes and antinodes are equally spaced and separated by one-fourth of a wavelength. The simplest pattern that satisfies these conditions has the required nodes at the ends of the string and an antinode at the center point (Active Fig. 14.9b). For this normal mode, the length of the string equals $\lambda/2$ (the distance between adjacent nodes):

$$L = \frac{\lambda_1}{2} \quad \text{or} \quad \lambda_1 = 2L$$

The next normal mode, of wavelength λ_2 (Active Fig. 14.9c), occurs when the length of the string equals one wavelength, that is, when $\lambda_2 = L$. In this mode, the two halves of the string are moving in opposite directions at a given instant, and we sometimes say that two *loops* occur. The third normal mode (Active Fig. 14.9d) corresponds to the case when the length equals $3\lambda/2$; therefore, $\lambda_3 = 2L/3$. In general, the wavelengths of the various normal modes can be conveniently expressed as

▶ Wavelengths of nomal modes

$$\lambda_n = \frac{2L}{n} \quad n = 1, 2, 3, \ldots \qquad \textbf{14.5} ◀$$

Active Figure 14.9 (a) A string of length L fixed at both ends. (b)–(d) The normal modes of vibration of the string in Active Figure 14.9a form a harmonic series. The string vibrates between the extremes shown.

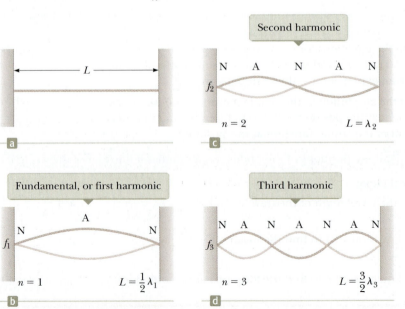

where the index n refers to the nth mode of oscillation. The natural frequencies associated with these modes are obtained from the relationship $f = v/\lambda$, where the wave speed v is determined by the tension T and linear mass density μ of the string and therefore is the same for all frequencies. Using Equation 14.5, we find that the frequencies of the normal modes are

$$f_n = \frac{v}{\lambda_n} = \frac{n}{2L}v \qquad n = 1, 2, 3, \ldots \qquad \textbf{14.6} \blacktriangleleft$$

▶ Frequencies of normal modes as functions of wave speed and length of string

Because $v = \sqrt{T/\mu}$ (Equation 13.21), we can express the natural frequencies of a stretched string as

$$f_n = \frac{n}{2L}\sqrt{\frac{T}{\mu}} \qquad n = 1, 2, 3, \ldots \qquad \textbf{14.7} \blacktriangleleft$$

▶ Frequencies of normal modes as functions of string tension and linear mass density

Equation 14.7 demonstrates the quantization that we mentioned as a feature of the wave under boundary conditions model. The frequencies are quantized because only certain frequencies of waves satisfy the boundary conditions and can exist on the string. The lowest frequency, corresponding to $n = 1$, is called the **fundamental frequency** f_1 and is

$$f_1 = \frac{1}{2L}\sqrt{\frac{T}{\mu}} \qquad \textbf{14.8} \blacktriangleleft$$

▶ Fundamental frequency of a taut string

The frequencies of the remaining normal modes are integer multiples of the fundamental frequency. Frequencies of normal modes that exhibit such an integer-multiple relationship form a **harmonic series,** and the normal modes are called **harmonics.** The fundamental frequency f_1 is the frequency of the first harmonic, the frequency $f_2 = 2f_1$ is the frequency of the second harmonic, and the frequency $f_n = nf_1$ is the frequency of the nth harmonic. Other oscillating systems, such as a drumhead, exhibit normal modes, but the frequencies are not related as integer multiples of a fundamental. Therefore, we do not use the term *harmonic* in association with those types of systems.

Let us examine further how the various harmonics are created in a string. To excite only a single harmonic, the string must be distorted into a shape that corresponds to that of the desired harmonic. After being released, the string vibrates at the frequency of that harmonic. This maneuver is difficult to perform, however, and is not how a string of a musical instrument is excited. If the string is distorted such that its shape is not that of just one harmonic, the resulting vibration includes a combination of various harmonics. Such a distortion occurs in musical instruments when the string is plucked (as in a guitar), bowed (as in a cello), or struck (as in a piano). When the string is distorted into a nonsinusoidal shape, only waves that satisfy the boundary conditions can persist on the string. These waves are the harmonics.

The frequency of a string that defines the musical note that it plays is that of the fundamental. The string's frequency can be varied by changing either the string's tension or its length. For example, the tension in guitar and violin strings is varied by a screw adjustment mechanism or by tuning pegs located on the neck of the instrument. As the tension is increased, the frequency of the normal modes increases in accordance with Equation 14.7. Once the instrument is "tuned," players vary the frequency by moving their fingers along the neck, thereby changing the length of the oscillating portion of the string. As the length is shortened, the frequency increases because, as Equation 14.7 specifies, the normal-mode frequencies are inversely proportional to string length.

Imagine that we have several strings of the same length under the same tension but varying linear mass density μ. The strings will have different wave speeds and therefore different fundamental frequencies. The linear mass density can be changed either by varying the diameter of the string or by wrapping extra mass around the string. Both of these possibilities can be seen on the guitar, on which the higher-frequency strings vary in diameter and the lower-frequency strings have additional wire wrapped around them.

> **QUICK QUIZ 14.3** When a standing wave is set up on a string fixed at both ends, which of the following statements is true? (**a**) The number of nodes is equal to the number of antinodes. (**b**) The wavelength is equal to the length of the string divided by an integer. (**c**) The frequency is equal to the number of nodes times the fundamental frequency. (**d**) The shape of the string at any instant shows a symmetry about the midpoint of the string.

Example 14.3 | Give Me a C Note!

The middle C string on a piano has a fundamental frequency of 262 Hz, and the string for the first A above middle C has a fundamental frequency of 440 Hz.

(**A**) Calculate the frequencies of the next two harmonics of the C string.

SOLUTION

Conceptualize Remember that the harmonics of a vibrating string have frequencies that are related by integer multiples of the fundamental.

Categorize This first part of the example is a simple substitution problem.

Knowing that the fundamental frequency is $f_1 = 262$ Hz, find the frequencies of the next harmonics by multiplying by integers:

$$f_2 = 2f_1 = 524 \text{ Hz}$$
$$f_3 = 3f_1 = 786 \text{ Hz}$$

(**B**) If the A and C strings have the same linear mass density μ and length L, determine the ratio of tensions in the two strings.

SOLUTION

Categorize This part of the example is more of an analysis problem than is part (A).

Analyze Use Equation 14.8 to write expressions for the fundamental frequencies of the two strings:

$$f_{1A} = \frac{1}{2L}\sqrt{\frac{T_A}{\mu}} \quad \text{and} \quad f_{1C} = \frac{1}{2L}\sqrt{\frac{T_C}{\mu}}$$

Divide the first equation by the second and solve for the ratio of tensions:

$$\frac{f_{1A}}{f_{1C}} = \sqrt{\frac{T_A}{T_C}} \rightarrow \frac{T_A}{T_C} = \left(\frac{f_{1A}}{f_{1C}}\right)^2 = \left(\frac{440}{262}\right)^2 = 2.82$$

Finalize If the frequencies of piano strings were determined solely by tension, this result suggests that the ratio of tensions from the lowest string to the highest string on the piano would be enormous. Such large tensions would make it difficult to design a frame to support the strings. In reality, the frequencies of piano strings vary due to additional parameters, including the mass per unit length and the length of the string. The What If? below explores a variation in length.

What If? If you look inside a real piano, you'll see that the assumption made in part (B) is only partially true. The strings are not likely to have the same length. The string densities for the given notes might be equal, but suppose the length of the A string is only 64% of the length of the C string. What is the ratio of their tensions?

Answer Using Equation 14.8 again, we set up the ratio of frequencies:

$$\frac{f_{1A}}{f_{1C}} = \frac{L_C}{L_A}\sqrt{\frac{T_A}{T_C}} \rightarrow \frac{T_A}{T_C} = \left(\frac{L_A}{L_C}\right)^2\left(\frac{f_{1A}}{f_{1C}}\right)^2$$

$$\frac{T_A}{T_C} = (0.64)^2\left(\frac{440}{262}\right)^2 = 1.16$$

Notice that this result represents only a 16% increase in tension, compared with the 182% increase in part (B).

14.4 | Standing Waves in Air Columns

We have discussed musical instruments that use strings, which include guitars, violins, and pianos. What about instruments classified as brasses or woodwinds? These instruments produce music using a column of air. The waves under boundary conditions model can be applied to sound waves in a column of air such as that inside an organ pipe or a clarinet. Standing waves are the result of interference between longitudinal sound waves traveling in opposite directions.

Whether a node or an antinode occurs at the end of an air column depends on whether that end is open or closed. The closed end of an air column is a **displacement node,** just as the fixed end of a vibrating string is a displacement node. Furthermore, because the pressure wave is 90° out of phase with the displacement wave (Section 13.6), the closed end of an air column corresponds to a **pressure antinode** (i.e., a point of maximum pressure variation). On the other hand, the open end of an air column is approximately a **displacement antinode** and a **pressure node.**

You may wonder how a sound wave can reflect from an open end because there may not appear to be a change in the medium at this point. It is indeed true that the medium through which the sound wave moves is air both inside and outside the pipe. Sound is a pressure wave, however, and a compression region of the sound wave is constrained by the sides of the pipe as long as the region is inside the pipe. As the compression region exits at the open end of the pipe, the constraint of the pipe is removed and the compressed air is free to expand into the atmosphere. Therefore, there is a change in the *character* of the medium between the inside of the pipe and the outside even though there is no change in the *material* of the medium. This change in character is sufficient to allow some reflection.[1]

We can determine the modes of vibration of an air column by applying the appropriate boundary condition at the end of the column, along with the requirement that nodes and antinodes be separated by one fourth of a wavelength. We shall find that the frequency for sound waves in air columns is quantized, similar to the results found for waves on strings under boundary conditions.

The first three modes of vibration of a pipe that is open at both ends are shown in Figure 14.10a. Note that the ends are displacement antinodes (approximately). In

In a pipe open at both ends, the ends are displacement antinodes and the harmonic series contains all integer multiples of the fundamental.

In a pipe closed at one end, the open end is a displacement antinode and the closed end is a node. The harmonic series contains only odd integer multiples of the fundamental.

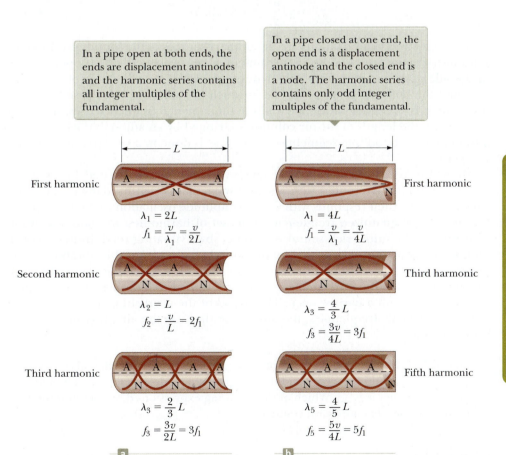

First harmonic
$\lambda_1 = 2L$
$f_1 = \frac{v}{\lambda_1} = \frac{v}{2L}$

Second harmonic
$\lambda_2 = L$
$f_2 = \frac{v}{L} = 2f_1$

Third harmonic
$\lambda_3 = \frac{2}{3}L$
$f_3 = \frac{3v}{2L} = 3f_1$

First harmonic
$\lambda_1 = 4L$
$f_1 = \frac{v}{\lambda_1} = \frac{v}{4L}$

Third harmonic
$\lambda_3 = \frac{4}{3}L$
$f_3 = \frac{3v}{4L} = 3f_1$

Fifth harmonic
$\lambda_5 = \frac{4}{5}L$
$f_5 = \frac{5v}{4L} = 5f_1$

a · b

Figure 14.10 Graphical representations of the motion of elements of air in standing longitudinal waves in (a) a column open at both ends and (b) a column closed at one end.

Pitfall Prevention | 14.3
Sound Waves In Air Are Not Transverse
Note that the standing longitudinal waves are drawn as transverse waves in Figure 14.10. It is difficult to draw longitudinal displacements because they are in the same direction as the propagation. Therefore, it is best to interpret the curves in Figure 14.10 as a graphical representation of the waves (our diagrams of string waves are pictorial representations), with the vertical axis representing horizontal position of the elements of the medium.

[1]Strictly speaking, the open end of an air column is not exactly a displacement antinode. A compression reaching an open end does not reflect until it passes beyond the end. For a tube of circular cross section, an end correction equal to approximately 0.6R, where R is the tube's radius, must be added to the length of the air column. Hence, the effective length of the air column is longer than the true length L. We ignore this end correction in this discussion.

the fundamental mode, the standing wave extends between two adjacent antinodes, which is a distance of half a wavelength. Therefore, the wavelength is twice the length of the pipe, and the frequency of the fundamental f_1 is $v/2L$. As Figure 14.10a shows, the frequencies of the higher harmonics are $2f_1, 3f_1, \ldots$. Therefore,

> in a pipe open at both ends, the natural frequencies of vibration form a harmonic series that includes all integer multiples of the fundamental frequency.

Because all harmonics are present, we can express the natural frequencies of vibration as

▶ Natural frequencies of a pipe open at both ends

$$f_n = n\frac{v}{2L} \qquad n = 1, 2, 3, \ldots \qquad \textbf{14.9} ◀$$

where v is the speed of sound in air.

If a pipe is closed at one end and open at the other, the closed end is a displacement node and the open end is a displacement antinode (Fig. 14.10b). In this case, the wavelength for the fundamental mode is four times the length of the column. Hence, the fundamental frequency f_1 is equal to $v/4L$, and the frequencies of the higher harmonics are equal to $3f_1, 5f_1, \ldots$. That is,

> in a pipe that is closed at one end, the natural frequencies of oscillation form a harmonic series that includes only odd integer multiples of the fundamental frequency.

We express this result mathematically as

▶ Natural frequencies of a pipe closed at one end and open at the other

$$f_n = n\frac{v}{4L} \qquad n = 1, 3, 5, \ldots \qquad \textbf{14.10} ◀$$

Standing waves in air columns are the primary sources of the sounds produced by wind instruments. In a woodwind instrument, a key is pressed, which opens a hole in the side of the column. This hole defines the end of the vibrating column of air (because the hole acts as an open end at which pressure can be released), so that the column is effectively shortened and the fundamental frequency rises. In a brass instrument, the length of the air column is changed by an adjustable section, as in a trombone, or by adding segments of tubing, as is done in a trumpet when a valve is pressed.

Musical instruments based on air columns are generally excited by *resonance*. The air column is presented with a sound wave that is rich in many frequencies. The air column then responds with a large-amplitude oscillation to the frequencies that match the quantized frequencies in its set of harmonics. In many woodwind instruments, the initial rich sound is provided by a vibrating reed. In brass instruments, this excitation is provided by the sound coming from the vibration of the player's lips. In a flute, the initial excitation comes from blowing over an edge at the mouthpiece of the instrument in a manner similar to blowing across the opening of a bottle with a narrow neck. The sound of the air rushing across the bottle opening has many frequencies, including one that sets the air cavity in the bottle into resonance.

QUICK QUIZ 14.4 A pipe open at both ends resonates at a fundamental frequency f_{open}. When one end is covered and the pipe is again made to resonate, the fundamental frequency is f_{closed}. Which of the following expressions describes how these two frequencies that are heard compare? (a) $f_{closed} = f_{open}$ (b) $f_{closed} = \frac{1}{2}f_{open}$ (c) $f_{closed} = 2f_{open}$ (d) $f_{closed} = \frac{3}{2}f_{open}$

QUICK QUIZ 14.5 Balboa Park in San Diego has an outdoor organ. When the air temperature increases, the fundamental frequency of one of the organ pipes (a) stays the same, (b) goes down, (c) goes up, or (d) is impossible to determine.

THINKING PHYSICS 14.2

A bugle has no valves, keys, slides, or finger holes. How can it play a song?

Reasoning Songs for the bugle are limited to harmonics of the fundamental frequency because the bugle has no control over frequencies by means of valves, keys, slides, or finger holes. The player obtains different notes by changing the tension in the lips as the bugle is played to excite different harmonics. The normal playing range of a bugle is among the third, fourth, fifth, and sixth harmonics of the fundamental. As examples, "Reveille" is played with just the three notes D (294 Hz), G (392 Hz), and B (490 Hz), and "Taps" is played with these same three notes and the D one octave above the lower D (588 Hz). Note that the frequencies of these four notes are, respectively, three, four, five, and six times the fundamental of 98 Hz. ◀

THINKING PHYSICS 14.3

If an orchestra doesn't warm up before a performance, the strings go flat and the wind instruments go sharp during the performance. Why?

Reasoning Without warming up, all the instruments will be at room temperature at the beginning of the concert. As the wind instruments are played, they fill with warm air from the player's exhalation. The increase in temperature of the air in the instrument causes an increase in the speed of sound, which raises the fundamental frequencies of the air columns. As a result, the wind instruments go sharp. The strings on the stringed instruments also increase in temperature due to the friction of rubbing with the bow. This increase in temperature results in thermal expansion, which causes a decrease in the tension in the strings. (We will study thermal expansion in Chapter 16.) With a decrease in tension, the wave speed on the strings drops and the fundamental frequencies decrease. Therefore, the stringed instruments go flat. ◀

Example 14.4 | Wind in a Culvert

A section of drainage culvert 1.23 m in length makes a howling noise when the wind blows across its open ends.

(A) Determine the frequencies of the first three harmonics of the culvert if it is cylindrical in shape and open at both ends. Take $v = 343$ m/s as the speed of sound in air.

SOLUTION

Conceptualize The sound of the wind blowing across the end of the pipe contains many frequencies, and the culvert responds to the sound by vibrating at the natural frequencies of the air column.

Categorize This example is a relatively simple substitution problem.

Find the frequency of the first harmonic of the culvert, modeling it as an air column open at both ends:

$$f_1 = \frac{v}{2L} = \frac{343 \text{ m/s}}{2(1.23 \text{ m})} = \boxed{139 \text{ Hz}}$$

Find the next harmonics by multiplying by integers:

$$f_2 = 2f_1 = \boxed{279 \text{ Hz}}$$
$$f_3 = 3f_1 = \boxed{418 \text{ Hz}}$$

(B) What are the three lowest natural frequencies of the culvert if it is blocked at one end?

SOLUTION

Find the frequency of the first harmonic of the culvert, modeling it as an air column closed at one end:

$$f_1 = \frac{v}{4L} = \frac{343 \text{ m/s}}{4(1.23 \text{ m})} = \boxed{69.7 \text{ Hz}}$$

Find the next two harmonics by multiplying by odd integers:

$$f_3 = 3f_1 = \boxed{209 \text{ Hz}}$$
$$f_5 = 5f_1 = \boxed{349 \text{ Hz}}$$

Example 14.5 | Measuring the Frequency of a Tuning Fork

A simple apparatus for demonstrating resonance in an air column is depicted in Figure 14.11. A vertical pipe open at both ends is partially submerged in water, and a tuning fork vibrating at an unknown frequency is placed near the top of the pipe. The length L of the air column can be adjusted by moving the pipe vertically. The sound waves generated by the fork are reinforced when L corresponds to one of the resonance frequencies of the pipe. For a certain pipe, the smallest value of L for which a peak occurs in the sound intensity is 9.00 cm.

(A) What is the frequency of the tuning fork?

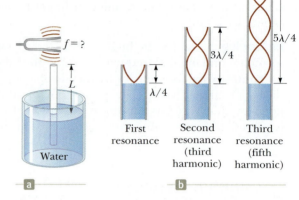

Figure 14.11 (Example 14.5) (a) Apparatus for demonstrating the resonance of sound waves in a pipe closed at one end. The length L of the air column is varied by moving the pipe vertically while it is partially submerged in water. (b) The first three normal modes of the system shown in (a).

SOLUTION

Conceptualize Consider how this problem differs from the preceding example. In the culvert, the length was fixed and the air column was presented with a mixture of very many frequencies. The pipe in this example is presented with one single frequency from the tuning fork, and the length of the pipe is varied until resonance is achieved.

Categorize This example is a simple substitution problem. Although the pipe is open at its lower end to allow the water to enter, the water's surface acts like a barrier. Therefore, this setup can be modeled as an air column closed at one end.

Use Equation 14.10 to find the fundamental frequency for $L = 0.090\ 0$ m:

$$f_1 = \frac{v}{4L} = \frac{343 \text{ m/s}}{4(0.090\ 0 \text{ m})} = 953 \text{ Hz}$$

Because the tuning fork causes the air column to resonate at this frequency, this frequency must also be that of the tuning fork.

(B) What are the values of L for the next two resonance conditions?

SOLUTION

Use Equation 13.12 to find the wavelength of the sound wave from the tuning fork:

$$\lambda = \frac{v}{f} = \frac{343 \text{ m/s}}{953 \text{ Hz}} = 0.360 \text{ m}$$

Notice from Figure 14.11b that the length of the air column for the second resonance is $3\lambda/4$:

$$L = 3\lambda/4 = 0.270 \text{ m}$$

Notice from Figure 14.11b that the length of the air column for the third resonance is $5\lambda/4$:

$$L = 5\lambda/4 = 0.450 \text{ m}$$

14.5 | Beats: Interference in Time

The interference phenomena that we have discussed so far involve the superposition of two or more waves with the same frequency. Because the resultant displacement of an element in the medium in this case depends on the position of the element, we can refer to the phenomenon as *spatial interference*. Standing waves in strings and air columns are common examples of spatial interference.

We now consider another type of interference effect, one that results from the superposition of two waves with slightly *different* frequencies. In this case, when the two waves of amplitudes A_1 and A_2 are observed at a given point, they are alternately in and out of phase. We refer to this phenomenon as *interference in time* or *temporal interference*. When the waves are in phase, the combined amplitude is $A_1 + A_2$. When they are out of phase, the combined amplitude is $|A_1 - A_2|$. The combination therefore varies between small and large amplitudes, resulting in a phenomenon called **beating**.

Although beats occur for all types of waves, they are particularly noticeable for sound waves. For example, if two tuning forks of slightly different frequencies are struck, you hear a sound of pulsating intensity.

The number of beats you hear per second, the *beat frequency*, equals the difference in frequency between the two sources. The maximum beat frequency that the human ear can detect is about 20 beats/s. When the beat frequency exceeds this value, it blends with the sounds producing the beats.

One can use beats to tune a stringed instrument, such as a piano, by beating a note against a reference tone of known frequency. The frequency of the string can then be adjusted to equal the frequency of the reference by changing the string's tension until the beats disappear; the two frequencies are then the same.

Let us look at the mathematical representation of beats. Consider two waves with equal amplitudes traveling through a medium with slightly different frequencies f_1 and f_2. We can represent the position of an element of the medium associated with each wave at a fixed point, which we choose as $x = 0$, as

$$y_1 = A \cos 2\pi f_1 t \qquad \text{and} \qquad y_2 = A \cos 2\pi f_2 t$$

Using the superposition principle, we find that the resultant position at that point is given by

$$y = y_1 + y_2 = A(\cos 2\pi f_1 t + \cos 2\pi f_2 t)$$

It is convenient to write this expression in a form that uses the trigonometric identity

$$\cos a + \cos b = 2 \cos\left(\frac{a - b}{2}\right) \cos\left(\frac{a + b}{2}\right)$$

Letting $a = 2\pi f_1 t$ and $b = 2\pi f_2 t$, we find that

$$y = \left[2A \cos 2\pi \left(\frac{f_1 - f_2}{2}\right) t\right] \cos 2\pi \left(\frac{f_1 + f_2}{2}\right) t \qquad \textbf{14.11} \blacktriangleleft$$

Graphs demonstrating the individual waves as well as the resultant wave are shown in Active Figure 14.12. From the factors in Equation 14.11, we see that the resultant wave has an effective frequency equal to the average frequency $(f_1 + f_2)/2$ and an amplitude of

$$A_{x=0} = 2A \cos 2\pi \left(\frac{f_1 - f_2}{2}\right) t \qquad \textbf{14.12} \blacktriangleleft$$

That is, the *amplitude varies in time* with a frequency of $(f_1 - f_2)/2$. When f_1 is close to f_2, this amplitude variation is slow compared with the frequency of the individual waves, as illustrated by the envelope (broken line) of the resultant wave in Active Figure 14.12b.

Note that a maximum in amplitude will be detected whenever

$$\cos 2\pi \left(\frac{f_1 - f_2}{2}\right) t = \pm 1$$

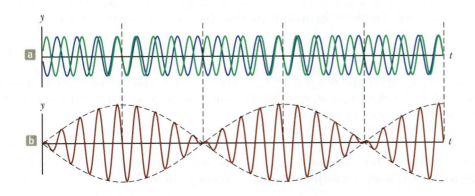

Active Figure 14.12 Beats are formed by the combination of two waves of slightly different frequencies. (a) The blue and green curves represent the individual waves. (b) The combined wave has an amplitude (dashed line) that oscillates in time.

That is, the amplitude maximizes twice in each cycle of the function on the left in the preceding expression. Therefore, the number of beats per second, or the beat frequency f_b, is twice the frequency of this function:

▶ Beat frequency

$$f_b = |f_1 - f_2| \qquad \textbf{14.13} \blacktriangleleft$$

For instance, if two tuning forks vibrate individually at frequencies of 438 Hz and 442 Hz, respectively, the resultant sound wave of the combination has a frequency of $(f_1 + f_2)/2 = 440$ Hz (the musical note A) and a beat frequency of $|f_1 - f_2| = 4$ Hz. That is, the listener hears the 440-Hz sound wave go through an intensity maximum four times every second.

QUICK QUIZ 14.6 You are tuning a guitar by comparing the sound of the string with that of a standard tuning fork. You notice a beat frequency of 5 Hz when both sounds are present. You tighten the guitar string and the beat frequency rises to 8 Hz. To tune the string exactly to the tuning fork, what should you do? (**a**) Continue to tighten the string. (**b**) Loosen the string. (**c**) It is impossible to determine.

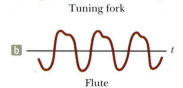

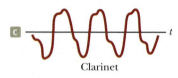

Figure 14.13 Waveforms of sound produced by (a) a tuning fork, (b) a flute, and (c) a clarinet, each at approximately the same frequency.

14.6 | Nonsinusoidal Wave Patterns

The sound wave patterns produced by most instruments are not sinusoidal. Some characteristic waveforms produced by a tuning fork, a flute, and a clarinet are shown in Figure 14.13. Although each instrument has its own characteristic pattern, Figure 14.13 shows that all three waveforms are periodic. A struck tuning fork produces primarily one harmonic (the fundamental), whereas the flute and clarinet produce many frequencies, which include the fundamental and various harmonics. The nonsinusoidal waveforms produced by a violin or clarinet, and the corresponding richness of musical tones, are the result of the superposition of various harmonics.

This phenomenon is in contrast to a percussive musical instrument, such as the drum, in which the combination of frequencies does not form a harmonic series. When frequencies that are integer multiples of a fundamental frequency are combined, the result is a *musical* sound. A listener can assign a pitch to the sound based on the fundamental frequency. Pitch is a psychological reaction to a sound that allows the listener to place the sound on a scale of low to high (bass to treble). Combinations of frequencies that are not integer multiples of a fundamental result in a *noise* rather than a musical sound. It is much harder for a listener to assign a pitch to a noise than to a musical sound.

Analysis of nonsinusoidal waveforms appears at first sight to be a formidable task. If the waveform is periodic, however, it can be represented with arbitrary precision by the combination of a sufficiently large number of sinusoidal waves that form a harmonic series. In fact, one can represent any periodic function or any function over a finite interval as a series of sine and cosine terms by using a mathematical technique based on *Fourier's theorem*. The corresponding sum of terms that represents the periodic waveform is called a **Fourier series.**

Let $y(t)$ be any function that is periodic in time, with a period of T, so that $y(t + T) = y(t)$. **Fourier's theorem** states that this function can be written

Pitfall Prevention | 14.4
Pitch Versus Frequency
A very common mistake made in speech when talking about sound is to use the term *pitch* when one means *frequency*. Frequency is the physical measurement of the number of oscillations per second, as we have defined. Pitch is a psychological reaction of humans to sound that enables a human to place the sound on a scale from high to low or from treble to bass. Therefore, frequency is the stimulus and pitch is the response. Although pitch is related mostly (but not completely) to frequency, they are not the same. A phrase such as "the pitch of the sound" is incorrect because pitch is not a physical property of the sound.

▶ Fourier's theorem

$$y(t) = \sum_n (A_n \sin 2\pi f_n t + B_n \cos 2\pi f_n t) \qquad \textbf{14.14} \blacktriangleleft$$

where the lowest frequency is $f_1 = 1/T$. The higher frequencies are integral multiples of the fundamental, so $f_n = nf_1$. The coefficients A_n and B_n represent the amplitudes of the various harmonics.

Figure 14.14 represents a harmonic analysis of the waveforms shown in Figure 14.13. Note the variation of relative intensity with harmonic content for the flute and the clarinet. In general, any musical sound contains components that are members of a harmonic series with varying relative intensities.

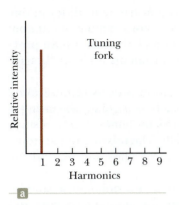 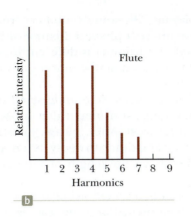 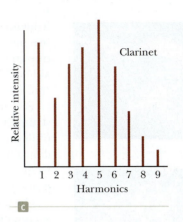

Figure 14.14 Harmonics of the waveforms shown in Figure 14.13. Note the variations in intensity of the various harmonics. Parts (a), (b), and (c) correspond to those in Figure 14.13.

We have discussed the *analysis* of a wave pattern using Fourier's theorem. The analysis involves determining the coefficients of the trigonometric functions in Equation 14.14 from a knowledge of the wave pattern. We can also perform the reverse process, *Fourier synthesis*. In this process, the various harmonics are added together to form a resultant wave pattern. As an example of Fourier synthesis, consider the building of a square wave as shown in Active Figure 14.15. The symmetry of the square wave results in only odd multiples of the fundamental combining in the synthesis. In Active Figure 14.15a, the blue curve shows the combination of *f* and 3*f*. In Active Figure 14.15b, we have added 5*f* to the combination and obtained the green curve. Notice how the general shape of the square wave is approximated, even though the upper and lower portions are not as flat as they should be.

Active Figure 14.15c shows the result of adding odd frequencies up to 9*f*, the red-brown curve. This approximation to the square wave (black curve) is better than in parts (a) and (b). To approximate the square wave as closely as possible, we would need to add all odd multiples of the fundamental frequency up to infinite frequency.

The physical mixture of harmonics can be described as the **spectrum** of the sound, with the spectrum displayed in a graphical representation such as Figure 14.14. The psychological reaction to changes in the spectrum of a sound is the detection of a change in the **timbre** or the **quality** of the sound. If a clarinet and a trumpet are both playing the same note, you will assign the same pitch to the two notes. Yet if only one of the instruments then plays the note, you will likely be able to tell which

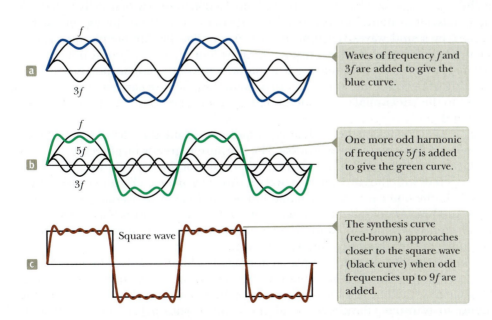

Active Figure 14.15 Fourier synthesis of a square wave represented by the sum of odd multiples of the first harmonic, which has frequency *f*.

Waves of frequency *f* and 3*f* are added to give the blue curve.

One more odd harmonic of frequency 5*f* is added to give the green curve.

The synthesis curve (red-brown) approaches closer to the square wave (black curve) when odd frequencies up to 9*f* are added.

Each musical instrument has its own characteristic sound and mixture of harmonics. Instruments shown are (a) the violin, (b) the saxophone, and (c) the trumpet.

instrument is playing. The sounds you hear from the two instruments differ in timbre because of a different physical mixture of harmonics. For example, the timbre due to the sound of a trumpet is different from that of a clarinet. You have probably developed words to describe timbres of various instruments, such as "brassy," "mellow," and "tinny."

Fourier's theorem allows us to understand the excitation process of musical instruments. In a stringed instrument that is plucked, such as a guitar, the string is pulled aside and released. After release, the string oscillates almost freely; a small damping causes the amplitude to decay to zero eventually. The mixture of harmonic frequencies depends on the length of the string, its linear mass density, and the plucking point.

On the other hand, a bowed stringed instrument, such as a violin, or a wind instrument is a forced oscillator. In the case of the violin, the alternate sticking and slipping of the bow on the string provides the periodic driving force. In the case of a wind instrument, the vibration of a reed (in a woodwind), of the lips of the player (in a brass), or the blowing of air across an edge (as in a flute) provides the periodic driving force. According to Fourier's theorem, these periodic driving forces contain a mixture of harmonic frequencies. The violin string or the air column in a wind instrument is therefore driven with a wide variety of frequencies. The frequency actually played is determined by *resonance*, which we studied in Chapter 12. The maximum response of the instrument will be to those frequencies that match or are very close to the harmonic frequencies of the instrument. The spectrum of the instrument therefore depends heavily on the strengths of the various harmonics in the initial periodic driving force.

14.7 | The Ear and Theories of Pitch Perception BIO

The human ear (Fig. 14.16) is divided into three regions: the outer ear, the middle ear, and the inner ear. The *outer ear* consists of the ear canal (which is open to the atmosphere), terminating at the eardrum (tympanum). Sound waves travel down the ear canal to the eardrum, which vibrates in response to the alternating high and low pressures of the waves. Behind the eardrum are three small bones of the *middle ear*, called the hammer, the anvil, and the stirrup because of their shapes. These bones transmit the vibration to the *inner ear*, which contains the cochlea, a snail-shaped tube about 2 cm long. The cochlea makes contact with the stirrup at the oval window and is divided along its length by the basilar membrane into an upper compartment and a lower compartment. Resting on the basilar membrane is the *organ of Corti*, which consists of about 15 000 auditory hair cells (cilia). The hair cells act as neural detectors of disturbances in the fluid filling the cochlea, caused by sound waves reaching the eardrum. The basilar membrane varies in mass per unit length and in tension along its length, and different portions of it resonate at different frequencies. The movement of the basilar membrane results in firing of nerves in the organ of Corti. Neural signals are carried by the cochlear nerve to the 8th cranial nerve and on to the brain. The brain interprets the signals as sound.

The small bones in the middle ear represent an intricate lever system that increases the force on the oval window. The pressure is greatly magnified because the surface area of the eardrum is about 20 times that of the oval window. The middle ear, together with the eardrum and oval window, acts as a matching network between the air in the outer ear and the liquid in the inner ear. The overall energy transfer between the outer ear and the inner ear is highly efficient, with pressure amplification factors of several thousand. In other words, pressure variations in the inner ear are much greater than those in the outer ear.

The ear has its own built-in protection against loud sounds. The muscles connecting the three middle-ear bones to the walls control the volume of the sound by changing the tension on the bones as sound builds up, thus hindering their ability to transmit vibrations. In addition, the eardrum becomes stiffer as the

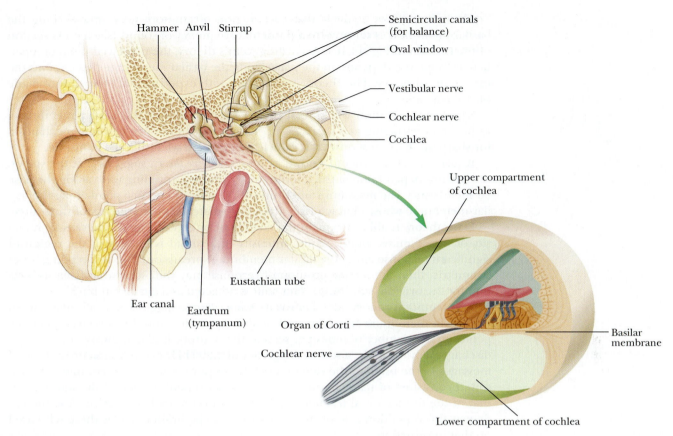

Figure 14.16 The structure of the human ear. The three tiny bones (hammer, anvil, and stirrup) that connect the eardrum to the window of the cochlea act as a double-lever system to decrease the amplitude of vibration and hence increase the pressure on the fluid in the cochlea. The right side of this figure shows a magnified cross section of the cochlea. (Magnified view from Sherwood, *Fundamentals of Human Physiology*, 4th ed., Brooks/Cole.)

sound intensity increases. These two events make the ear less sensitive to loud incoming sounds. There is a time delay between the onset of a loud sound and the ear's protective reaction, however, so a very sudden loud sound can still damage the ear.

Theories of Pitch Perception

As noted in Pitfall Prevention 14.4, *pitch* is a human psychological response to the stimulus of a sound wave. Several theories in the field of *psychoacoustics* have been proposed regarding the manner in which pitch is perceived by the human ear. We shall briefly describe the two most common and accepted theories of hearing, *place theory* and *temporal theory*.

The biggest influence on pitch perception is the frequency of the sound. The pitch can change, however, even though the frequency remains constant. For example, a large fraction of individuals perceive a change in the pitch associated with a sound when its intensity is increased rapidly while its frequency remains fixed. Any successful theory of pitch perception must address this effect. Another interesting experimental result is the *missing fundamental*. Suppose a sound consisting of a mixture of harmonics is presented to a listener. The overall pitch of the sound will be that associated with the fundamental frequency, as noted for vibrating strings in Section 14.3. Now imagine that the fundamental frequency of the sound is filtered out. Experiments show that the pitch of the sound remains the same, even though the frequency associated with that pitch is no longer present. This effect must also be predicted by a theory of pitch perception.

The place theory assumes that each region of auditory nerve fibers along the basilar membrane is sensitive to a particular frequency of sound (see a cross section of the cochlea in Fig. 14.16). According to this theory, the perceived pitch of a particular frequency depends on the place along the basilar membrane that exhibits the maximum vibration. This theory is consistent with experimental observations made on animal ears.

Significant vibration will spread over a short length of the basilar membrane, rather than occurring at a sharp point. Therefore, there must be some mechanism for sharpening the response. It is not clear at present what this mechanism is.

A variation of this theory, known as *traveling wave theory,* suggests that the stapes, the stirrup-shaped bone in the middle ear, produces a traveling wave along the basilar membrane. The maximum amplitude of the wave occurs at a point along the basilar membrane whose characteristic frequency matches the frequency of the source.

Place theory is able to explain the change of pitch with rapidly increasing intensity of the stimulus sound. As the intensity of a sound is increased, the region of significant response along the basilar membrane spreads out. If this spreading is not symmetric, then the *average* position of response may shift somewhat from its location for the original soft sound. This shift is detected as a change in pitch.

The temporal theory, also known as *timing theory,* suggests that all elements of the basilar membrane are stimulated by all frequencies, and the perceived pitch depends on the number of times per second the neurons in the auditory nerve system discharge. For example, a source frequency of 2.000 kHz causes the neurons to send messages to the brain at the rate of 2 000 times per second. For a complex musical sound consisting of many harmonics, the overall repetition rate of the waveform is that of the fundamental frequency. The temporal theory hypothesizes that the ear detects this repetition rate of the waveform and the brain decodes the pitch based on that information.

Temporal theory explains the phenomenon of the missing fundamental. Even after the fundamental frequency is filtered out of a complex sound, *the repetition rate of the combination of harmonics is still that of the fundamental.* For example, if the first harmonic of the clarinet in Figure 14.14c is filtered out, the clarinet waveform in Figure 14.13c will change shape, but it will still repeat at the same frequency, that of the fundamental. Based on this concept, the ear sends a signal to the brain related to the repetition rate of the sound and a pitch associated with the fundamental is assigned even though the fundamental is not present.

BIO Cochlear implants

One of the most amazing medical advances in recent decades is the cochlear implant, allowing some deaf individuals to hear. Deafness can occur when the hair-like sensors (cilia) in the cochlea break off over a lifetime or sometimes because of prolonged exposure to loud sounds. Because the cilia don't grow back, the ear loses sensitivity to certain frequencies of sound. The cochlear implant stimulates the nerves in the ear electronically to restore hearing loss that is due to damaged or absent cilia.

Research using modern cochlear implants suggests that the perception of pitch may depend on both location of response along the basilar membrane *and* the rate at which the neurons fire, a combination of both theories. Place theory may be dominant for frequencies above the maximum firing rate of the neurons. Research continues in this interesting area that combines physics, biology, and psychology.

⟨14.8 | Context Connection: Building on Antinodes

As an example of the application of standing waves to earthquakes, we consider the effects of standing waves in *sedimentary basins.* Many of the world's major cities are built on sedimentary basins, which are topographic depressions that over geologic time have filled with sediment. These areas provide large expanses of flat land, often surrounded by attractive mountains, as in the Los Angeles basin. Flat land for building and attractive scenery attracted early settlers and led to today's cities.

Destruction from an earthquake can increase dramatically if the natural frequencies of buildings or other structures coincide with the resonant frequencies of the

underlying basin. These resonant frequencies are associated
with three-dimensional standing waves, formed from seismic waves reflecting from the
boundaries of the basin.

To understand these standing waves, let us assume a simple model of a basin shaped
like a half-ellipsoid, similar
to an egg sliced in half along
its long diameter. Four possible normal modes associated
with ground motion in such a
basin are shown in the pictorial
representation in Active Figure 14.17. The long axis of the
ellipsoid is designated x and
the short axis is y. In Active Figure 14.17a, the entire surface
of the ground moves up and
down (that is, in and out of the
page) except at a nodal curve running around the edge of the basin.

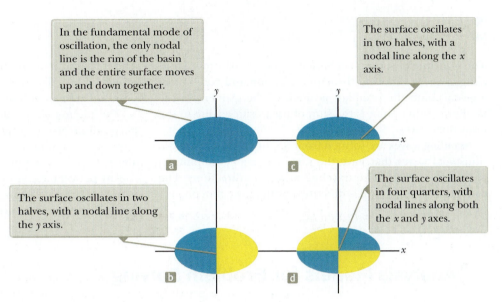

In the fundamental mode of oscillation, the only nodal line is the rim of the basin and the entire surface moves up and down together.

The surface oscillates in two halves, with a nodal line along the x axis.

The surface oscillates in two halves, with a nodal line along the y axis.

The surface oscillates in four quarters, with nodal lines along both the x and y axes.

Active Figure 14.17 Overhead views of standing waves in a basin shaped like a half-ellipsoid. In each case, if the blue element is above the plane of the page at an instant of time, the yellow element is below the plane of the page.

In Figures 14.17b and 14.17c, half the ground surface lies above and half lies below the equilibrium position, and each half oscillates up and down on either side of a nodal line. The nodal line is along the y axis in Active Figure 14.17b and along the x axis in Active Figure 14.17c. In Active Figure 14.17d, nodal lines occur along both the x and y axes and the surface oscillates in four segments, with two above the equilibrium position at any time and the other two below.

The standing wave patterns in a basin arise from seismic waves traveling horizontally between the boundaries of the basin. For structures built on sedimentary basins, the degree of seismic risk will depend on the standing wave modes excited by the interference of seismic waves trapped in the basin. It is clear that structures built on regions of maximum ground motion (i.e., the antinodes) will suffer maximum shaking, whereas structures residing near nodes will experience relatively mild ground motion. These considerations appear to have played an important role in the selective destruction that occurred in Mexico City in the Michoacán earthquake in 1985 and in the 1989 Loma Prieta earthquake, which caused the collapse of a section of the Nimitz Freeway in Oakland, California.

A similar effect occurs in bounded bodies of water, such as harbors and bays. A standing wave pattern established in such a body of water is called a **seiche**. This wave pattern can result in variations in the water level that exhibit a period of several minutes, superposed on the longer-period tidal variations. Seiches can be caused by earthquakes, tsunamis, winds, or weather disturbances. You can create a seiche in your bathtub by sliding back and forth at just the right frequency such that the water sloshes back and forth at such a large amplitude that much of it spills out onto the floor.

During the Northridge earthquake of 1994, swimming pools throughout southern California overflowed as a result of seiches set up by the shaking of the ground. Seismic events can also cause seiches very far away from the epicenter. The magnitude 8.8 Chile earthquake of February 27, 2010, caused a measureable seiche in Lake Pontchartrain, Louisiana, of height 0.15 m. A more dramatic example is the magnitude 9.0 earthquake in Japan on March 11, 2011. It caused a seiche measured at 1.8 m in Sognefjorden, the largest fjord in Norway!

We have now considered the role of standing waves in the damage caused by an earthquake. In the Context Conclusion, we will gather together the principles of vibrations and waves that we have learned to respond more fully to the central question of this Context.

SUMMARY

The **principle of superposition** states that if two or more traveling waves are moving through a medium and combine at a given point, the resultant position of the element of the medium at that point is the sum of the positions due to the individual waves.

Standing waves are formed from the superposition of two sinusoidal waves that have the same frequency, amplitude, and wavelength but are traveling in *opposite* directions. The resultant standing wave is described by the wave function

$$y = (2A \sin kx) \cos \omega t \qquad \textbf{14.2} \blacktriangleleft$$

The maximum amplitude points (called **antinodes**) are separated by a distance $\lambda/2$. Halfway between antinodes are points of zero amplitude (called **nodes**).

The phenomenon of **beats** occurs as a result of the superposition of two traveling waves of slightly different frequencies. For sound waves at a given point, one hears an alternation in sound intensity with time.

Any periodic waveform can be represented by the combination of sinusoidal waves that form a harmonic series. The process is based on **Fourier's theorem.**

Analysis Models for Problem Solving

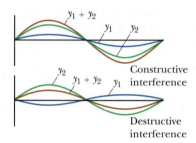

$y_1 + y_2$
y_1 y_2

y_2
$y_1 + y_2$ y_1 Constructive interference

Destructive interference

Waves in Interference. When two traveling waves having equal frequencies superimpose, the resultant wave has an amplitude that depends on the phase angle ϕ between the two waves. **Constructive interference** occurs when the two waves are in phase, corresponding to $\phi = 0, 2\pi, 4\pi, \ldots$ rad. **Destructive interference** occurs when the two waves are 180° out of phase, corresponding to $\phi = \pi, 3\pi, 5\pi, \ldots$ rad.

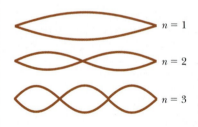

$n = 1$

$n = 2$

$n = 3$

Waves Under Boundary Conditions. When a wave is subject to boundary conditions, only certain natural frequencies are allowed; we say that the frequencies are quantized.

For waves on a string fixed at both ends, the natural frequencies are

$$f_n = \frac{n}{2L}\sqrt{\frac{T}{\mu}} \quad n = 1, 2, 3, \ldots \qquad \textbf{14.7} \blacktriangleleft$$

where T is the tension in the string and μ is its linear mass density.

For sound waves with speed v in an air column of length L open at both ends, the natural frequencies are

$$f_n = n\frac{v}{2L} \quad n = 1, 2, 3, \ldots \qquad \textbf{14.9} \blacktriangleleft$$

If an air column is open at one end and closed at the other, only odd harmonics are present and the natural frequencies are

$$f_n = n\frac{v}{4L} \quad n = 1, 3, 5, \ldots \qquad \textbf{14.10} \blacktriangleleft$$

OBJECTIVE QUESTIONS

1. A flute has a length of 58.0 cm. If the speed of sound in air is 343 m/s, what is the fundamental frequency of the flute, assuming it is a tube closed at one end and open at the other? (a) 148 Hz (b) 296 Hz (c) 444 Hz (d) 591 Hz (e) none of those answers

2. A string of length L, mass per unit length μ, and tension T is vibrating at its fundamental frequency. **(i)** If the length of the string is doubled, with all other factors held constant, what is the effect on the fundamental frequency? (a) It becomes two times larger. (b) It becomes $\sqrt{2}$ times larger. (c) It is unchanged. (d) It becomes $1/\sqrt{2}$ times as large. (e) It becomes one-half as large. **(ii)** If the mass per unit length is doubled, with all other factors held constant, what is the effect on the fundamental frequency? Choose from the same possibilities as in part (i). **(iii)** If the tension is doubled, with all other factors held constant, what is the effect

on the fundamental frequency? Choose from the same possibilities as in part (i).

3. In Figure OQ14.3, a sound wave of wavelength 0.8 m divides into two equal parts that recombine to interfere constructively, with the original difference between their path lengths being $|r_2 - r_1| = 0.8$ m. Rank the following situations according to the intensity of sound at the receiver from the highest to the lowest. Assume the tube walls absorb no sound energy. Give equal ranks to situations in which the intensity is equal. (a) From its original position, the sliding section is moved out by 0.1 m. (b) Next it slides out an additional 0.1 m. (c) It slides out still another 0.1 m. (d) It slides out 0.1 m more.

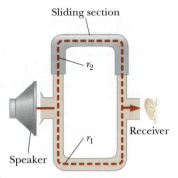

Figure OQ14.3

4. A series of pulses, each of amplitude 0.1 m, is sent down a string that is attached to a post at one end. The pulses are reflected at the post and travel back along the string without loss of amplitude. **(i)** What is the net displacement at a point on the string where two pulses are crossing? Assume the string is rigidly attached to the post. (a) 0.4 m (b) 0.3 m (c) 0.2 m (d) 0.1 m (e) 0 **(ii)** Next assume the end at which reflection occurs is free to slide up and down. Now what is the net displacement at a point on the string where two pulses are crossing? Choose your answer from the same possibilities as in part (i).

5. Suppose all six equal-length strings of an acoustic guitar are played without fingering, that is, without being pressed down at any frets. What quantities are the same for all six strings? Choose all correct answers. (a) the fundamental frequency (b) the fundamental wavelength of the string wave (c) the fundamental wavelength of the sound emitted (d) the speed of the string wave (e) the speed of the sound emitted

6. When two tuning forks are sounded at the same time, a beat frequency of 5 Hz occurs. If one of the tuning forks has a frequency of 245 Hz, what is the frequency of the other tuning fork? (a) 240 Hz (b) 242.5 Hz (c) 247.5 Hz (d) 250 Hz (e) More than one answer could be correct.

7. As oppositely moving pulses of the same shape (one upward, one downward) on a string pass through each other, at one particular instant the string shows no displacement from the equilibrium position at any point. What has happened to the energy carried by the pulses at this instant of time? (a) It was used up in producing the previous motion. (b) It is all potential energy. (c) It is all internal energy. (d) It is all kinetic energy. (e) The positive energy of one pulse adds to zero with the negative energy of the other pulse.

8. Assume two identical sinusoidal waves are moving through the same medium in the same direction. Under what condition will the amplitude of the resultant wave be greater than either of the two original waves? (a) in all cases (b) only if the waves have no difference in phase (c) only if the phase difference is less than 90° (d) only if the phase difference is less than 120° (e) only if the phase difference is less than 180°

9. An archer shoots an arrow horizontally from the center of the string of a bow held vertically. After the arrow leaves it, the string of the bow will vibrate as a superposition of what standing-wave harmonics? (a) It vibrates only in harmonic number 1, the fundamental. (b) It vibrates only in the second harmonic. (c) It vibrates only in the odd-numbered harmonics 1, 3, 5, 7, ... (d) It vibrates only in the even-numbered harmonics 2, 4, 6, 8, ... (e) It vibrates in all harmonics.

10. A tuning fork is known to vibrate with frequency 262 Hz. When it is sounded along with a mandolin string, four beats are heard every second. Next, a bit of tape is put onto each tine of the tuning fork, and the tuning fork now produces five beats per second with the same mandolin string. What is the frequency of the string? (a) 257 Hz (b) 258 Hz (c) 262 Hz (d) 266 Hz (e) 267 Hz

11. A standing wave having three nodes is set up in a string fixed at both ends. If the frequency of the wave is doubled, how many antinodes will there be? (a) 2 (b) 3 (c) 4 (d) 5 (e) 6

☐ denotes answer available in *Student Solutions Manual/Study Guide*

CONCEPTUAL QUESTIONS |

1. A crude model of the human throat is that of a pipe open at both ends with a vibrating source to introduce the sound into the pipe at one end. Assuming the vibrating source produces a range of frequencies, discuss the effect of changing the pipe's length.

2. When two waves interfere constructively or destructively, is there any gain or loss in energy in the system of the waves? Explain.

3. Explain how a musical instrument such as a piano may be tuned by using the phenomenon of beats.

4. Does the phenomenon of wave interference apply only to sinusoidal waves?

5. What limits the amplitude of motion of a real vibrating system that is driven at one of its resonant frequencies?

6. An airplane mechanic notices that the sound from a twin-engine aircraft rapidly varies in loudness when both engines

are running. What could be causing this variation from loud to soft?

7. A soft-drink bottle resonates as air is blown across its top. What happens to the resonance frequency as the level of fluid in the bottle decreases?

8. A tuning fork by itself produces a faint sound. Explain how each of the following methods can be used to obtain a louder sound from it. Explain also any effect on the time interval for which the fork vibrates audibly. (a) holding the edge of a sheet of paper against one vibrating tine (b) pressing the handle of the tuning fork against a chalkboard or a tabletop (c) holding the tuning fork above a column of air of properly chosen length as in Example 14.5 (d) holding the tuning fork close to an open slot cut in a sheet of foam plastic or cardboard (with the slot similar in size and shape to one tine of the fork and the motion of the tines perpendicular to the sheet)

▶ PROBLEMS |

Note: Unless otherwise specified, assume that the speed of sound in air is 343 m/s, its value at an air temperature at 20°C. At any other Celsius temperature T_C, the speed of sound in air is described by

$$v = 331\sqrt{1 + \frac{T_C}{273}}$$

where v is in m/s and T_C is in °C.

Section 14.1 Analysis Model: Waves in Interference

1. **W** Two waves are traveling in the same direction along a stretched string. The waves are 90.0° out of phase. Each wave has an amplitude of 4.00 cm. Find the amplitude of the resultant wave.

2. Two wave pulses A and B are moving in opposite directions, each with a speed $v = 2.00$ cm/s. The amplitude of A is twice the amplitude of B. The pulses are shown in Figure P14.2 at $t = 0$. Sketch the resultant wave at $t = 1.00$ s, 1.50 s, 2.00 s, 2.50 s, and 3.00 s.

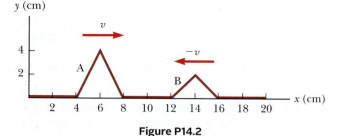

Figure P14.2

3. **M** Two sinusoidal waves in a string are defined by the wave functions

$$y_1 = 2.00 \sin (20.0x - 32.0t) \qquad y_2 = 2.00 \sin (25.0x - 40.0t)$$

where x, y_1, and y_2 are in centimeters and t is in seconds. (a) What is the phase difference between these two waves at the point $x = 5.00$ cm at $t = 2.00$ s? (b) What is the positive x value closest to the origin for which the two phases differ by $\pm\pi$ at $t = 2.00$ s? (At that location, the two waves add to zero.)

4. Two identical sinusoidal waves with wavelengths of 3.00 m travel in the same direction at a speed of 2.00 m/s. The

second wave originates from the same point as the first, but at a later time. The amplitude of the resultant wave is the same as that of each of the two initial waves. Determine the minimum possible time interval between the starting moments of the two waves.

5. **W** Two waves on one string are described by the wave functions

$$y_1 = 3.0 \cos (4.0x - 1.6t) \qquad y_2 = 4.0 \sin (5.0x - 2.0t)$$

where x and y are in centimeters and t is in seconds. Find the superposition of the waves $y_1 + y_2$ at the points (a) $x = 1.00$, $t = 1.00$; (b) $x = 1.00$, $t = 0.500$; and (c) $x = 0.500$, $t = 0$ *Note:* Remember that the arguments of the trigonometric functions are in radians.

6. Two speakers are driven by the same oscillator of frequency f. They are located a distance d from each other on a vertical pole. A man walks straight toward the lower speaker in a direction perpendicular to the pole as shown in Figure P14.6. (a) How many times will he hear a minimum in sound intensity? (b) How far is he from the pole at these moments? Let v represent the speed of sound and assume that the ground does not reflect sound.

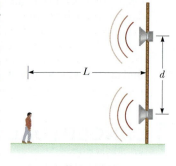

Figure P14.6

7. **W** Two pulses traveling on the same string are described by

$$y_1 = \frac{5}{(3x - 4t)^2 + 2} \qquad y_2 = \frac{-5}{(3x + 4t - 6)^2 + 2}$$

(a) In which direction does each pulse travel? (b) At what instant do the two cancel everywhere? (c) At what point do the two pulses always cancel?

8. *Why is the following situation impossible?* Two identical loudspeakers are driven by the same oscillator at frequency 200 Hz. They are located on the ground a distance $d = 4.00$ m from each other. Starting far from the speakers, a man walks straight toward the right-hand speaker as shown in

Figure P14.8. After passing through three minima in sound intensity, he walks to the next maximum and stops. Ignore any sound reflection from the ground.

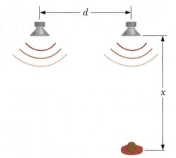

Figure P14.8

9. **M** Two traveling sinusoidal waves are described by the wave functions

$$y_1 = 5.00 \sin [\pi(4.00x - 1\,200t)]$$

$$y_2 = 5.00 \sin [\pi(4.00x - 1\,200t - 0.250)]$$

where x, y_1, and y_2 are in meters and t is in seconds. (a) What is the amplitude of the resultant wave function $y_1 + y_2$? (b) What is the frequency of the resultant wave function?

10. **W** Two identical loudspeakers are placed on a wall 2.00 m apart. A listener stands 3.00 m from the wall directly in front of one of the speakers. A single oscillator is driving the speakers at a frequency of 300 Hz. (a) What is the phase difference in radians between the waves from the speakers when they reach the observer? (b) **What If?** What is the frequency closest to 300 Hz to which the oscillator may be adjusted such that the observer hears minimal sound?

11. A tuning fork generates sound waves with a frequency of 246 Hz. The waves travel in opposite directions along a hallway, are reflected by end walls, and return. The hallway is 47.0 m long and the tuning fork is located 14.0 m from one end. What is the phase difference between the reflected waves when they meet at the tuning fork? The speed of sound in air is 343 m/s.

12. **Q|C** Two identical loudspeakers 10.0 m apart are driven by the same oscillator with a frequency of $f = 21.5$ Hz (Fig. P14.12) in an area where the speed of sound is 344 m/s. (a) Show that a receiver at point A records a minimum in sound intensity from the two speakers. (b) If the receiver is moved in the plane of the speakers, show that the path it should take so that the intensity remains at a minimum is along the hyperbola $9x^2 - 16y^2 = 144$ (shown in red-brown in Fig. P14.12). (c) Can the receiver remain at a minimum and move very far away from the two sources? If so, determine the limiting form of the path it must take. If not, explain how far it can go.

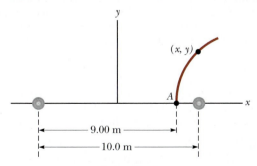

Figure P14.12

Section 14.2 Standing Waves

13. **M** Two transverse sinusoidal waves combining in a medium are described by the wave functions

$$y_1 = 3.00 \sin \pi(x + 0.600t) \qquad y_2 = 3.00 \sin \pi(x - 0.600t)$$

where x, y_1, and y_2 are in centimeters and t is in seconds. Determine the maximum transverse position of an element of the medium at (a) $x = 0.250$ cm, (b) $x = 0.500$ cm, and (c) $x = 1.50$ cm. (d) Find the three smallest values of x corresponding to antinodes.

14. **Q|C** Two waves simultaneously present on a long string have a phase difference ϕ between them so that a standing wave formed from their combination is described by

$$y(x, t) = 2A \sin\left(kx + \frac{\phi}{2}\right) \cos\left(\omega t - \frac{\phi}{2}\right)$$

(a) Despite the presence of the phase angle ϕ, is it still true that the nodes are one-half wavelength apart? Explain. (b) Are the nodes different in any way from the way they would be if ϕ were zero? Explain.

15. **W** Two sinusoidal waves traveling in opposite directions interfere to produce a standing wave with the wave function

$$y = 1.50 \sin (0.400x) \cos (200t)$$

where x and y are in meters and t is in seconds. Determine (a) the wavelength, (b) the frequency, and (c) the speed of the interfering waves.

16. Verify by direct substitution that the wave function for a standing wave given in Equation 14.2,

$$y = (2A \sin kx) \cos \omega t$$

is a solution of the general linear wave equation, Equation 13.20:

$$\frac{\partial^2 y}{\partial x^2} = \frac{1}{v^2} \frac{\partial^2 y}{\partial t^2}$$

17. **M** Two identical loudspeakers are driven in phase by a common oscillator at 800 Hz and face each other at a distance of 1.25 m. Locate the points along the line joining the two speakers where relative minima of sound pressure amplitude would be expected.

Section 14.3 Analysis Model: Waves Under Boundary Conditions

18. **W** In the arrangement shown in Figure P14.18, an object can be hung from a string (with linear mass density $\mu = 0.002\,00$ kg/m) that passes over a light pulley. The string is connected to a vibrator (of constant frequency f), and the length of the string between point P and the pulley is $L = 2.00$ m. When the mass m of the object is either 16.0 kg or 25.0 kg, standing waves are observed; no standing waves are observed with any mass between these values, however. (a) What is the frequency of the vibrator? *Note:* The greater the tension in the string, the smaller the number of nodes in the standing wave. (b) What is the largest object mass for which standing waves could be observed?

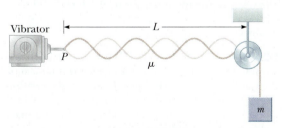

Figure P14.18

19. A string that is 30.0 cm long and has a mass per unit length of 9.00×10^{-3} kg/m is stretched to a tension of 20.0 N. Find (a) the fundamental frequency and (b) the next three frequencies that could cause standing-wave patterns on the string.

20. A violin string has a length of 0.350 m and is tuned to concert G, with $f_G = 392$ Hz. (a) How far from the end of the string must the violinist place her finger to play concert A, with $f_A = 440$ Hz? (b) If this position is to remain correct to one-half the width of a finger (that is, to within 0.600 cm), what is the maximum allowable percentage change in the string tension?

21. A string with a mass $m = 8.00$ g and a length $L = 5.00$ m has one end attached to a wall; the other end is draped over a small, fixed pulley a distance $d = 4.00$ m from the wall and attached to a hanging object with a mass $M = 4.00$ kg as in Figure P14.21. If the horizontal part of the string is plucked, what is the fundamental frequency of its vibration?

Figure P14.21

22. A standing wave is established in a 120-cm-long string fixed at both ends. The string vibrates in four segments when driven at 120 Hz. (a) Determine the wavelength. (b) What is the fundamental frequency of the string?

23. A string of length L, mass per unit length μ, and tension T is vibrating at its fundamental frequency. What effect will the following have on the fundamental frequency? (a) The length of the string is doubled, with all other factors held constant. (b) The mass per unit length is doubled, with all other factors held constant. (c) The tension is doubled, with all other factors held constant.

24. The 64.0-cm-long string of a guitar has a fundamental frequency of 330 Hz when it vibrates freely along its entire length. A fret is provided for limiting vibration to just the lower two-thirds of the string. (a) If the string is pressed down at this fret and plucked, what is the new fundamental frequency? (b) **What If?** The guitarist can play a "natural harmonic" by gently touching the string at the location of this fret and plucking the string at about one-sixth of the way along its length from the other end. What frequency will be heard then?

25. Review. A sphere of mass $M = 1.00$ kg is supported by a string that passes over a pulley at the end of a horizontal rod of length $L = 0.300$ m (Fig. P14.25). The string makes an angle $\theta = 35.0°$ with the rod. The fundamental frequency of standing waves in the portion of the string above the rod is $f = 60.0$ Hz. Find the mass of the portion of the string above the rod.

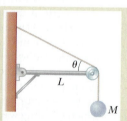

Figure P14.25
Problems 25 and 26.

26. **S** **Review.** A sphere of mass M is supported by a string that passes over a pulley at the end of a horizontal rod of length L (Fig. P14.25). The string makes an angle θ with the rod. The fundamental frequency of standing waves in the portion of the string above the rod is f. Find the mass of the portion of the string above the rod.

27. **M** The A string on a cello vibrates in its first normal mode with a frequency of 220 Hz. The vibrating segment is 70.0 cm long and has a mass of 1.20 g. (a) Find the tension in the string. (b) Determine the frequency of vibration when the string vibrates in three segments.

28. A standing-wave pattern is observed in a thin wire with a length of 3.00 m. The wave function is

$$y = 0.002\ 00 \sin(\pi x) \cos(100\pi t)$$

where x and y are in meters and t is in seconds. (a) How many loops does this pattern exhibit? (b) What is the fundamental frequency of vibration of the wire? (c) **What If?** If the original frequency is held constant and the tension in the wire is increased by a factor of 9, how many loops are present in the new pattern?

Section 14.4 Standing Waves in Air Columns

29. Calculate the length of a pipe that has a fundamental frequency of 240 Hz assuming the pipe is (a) closed at one end and (b) open at both ends.

30. **Q|C** A tunnel under a river is 2.00 km long. (a) At what frequencies can the air in the tunnel resonate? (b) Explain whether it would be good to make a rule against blowing your car horn when you are in the tunnel.

31. **BIO** The windpipe of one typical whooping crane is 5.00 feet long. What is the fundamental resonant frequency of the bird's trachea, modeled as a narrow pipe closed at one end? Assume a temperature of 37°C.

32. **W** The overall length of a piccolo is 32.0 cm. The resonating air column is open at both ends. (a) Find the frequency of the lowest note a piccolo can sound. (b) Opening holes in the side of a piccolo effectively shortens the length of the resonant column. Assume the highest note a piccolo can sound is 4 000 Hz. Find the distance between adjacent antinodes for this mode of vibration.

33. The fundamental frequency of an open organ pipe corresponds to middle C (261.6 Hz on the chromatic musical scale). The third resonance of a closed organ pipe has the same frequency. What is the length of (a) the open pipe and (b) the closed pipe?

34. A shower stall has dimensions 86.0 cm × 86.0 cm × 210 cm. Assume the stall acts as a pipe closed at both ends, with nodes at opposite sides. Assume singing voices range from 130 Hz to 2 000 Hz and let the speed of sound in the hot air be 355 m/s. For someone singing in this shower, which frequencies would sound the richest (because of resonance)?

35. **M** Two adjacent natural frequencies of an organ pipe are determined to be 550 Hz and 650 Hz. Calculate (a) the fundamental frequency and (b) the length of this pipe.

36. **BIO** Do not stick anything into your ear! Estimate the length of your ear canal, from its opening at the external ear to the eardrum. If you regard the canal as a narrow tube that is open at one end and closed at the other, at approximately what fundamental frequency would you expect your hearing to be most sensitive? Explain why you can hear especially soft sounds just around this frequency.

37. As shown in Figure P14.37, water is pumped into a tall, vertical cylinder at a volume flow rate $R = 1.00 \, \text{L/min}$. The radius of the cylinder is $r = 5.00$ cm, and at the open top of the cylinder a tuning fork is vibrating with a frequency $f = 512$ Hz. As the water rises, what time interval elapses between successive resonances?

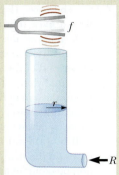

Figure P14.37
Problems 37 and 38.

38. **S** As shown in Figure P14.37, water is pumped into a tall, vertical cylinder at a volume flow rate R. The radius of the cylinder is r, and at the open top of the cylinder a tuning fork is vibrating with a frequency f. As the water rises, what time interval elapses between successive resonances?

39. A glass tube (open at both ends) of length L is positioned near an audio speaker of frequency $f = 680$ Hz. For what values of L will the tube resonate with the speaker?

40. **W** A tuning fork with a frequency of $f = 512$ Hz is placed near the top of the tube shown in Figure P14.40. The water level is lowered so that the length L slowly increases from an initial value of 20.0 cm. Determine the next two values of L that correspond to resonant modes.

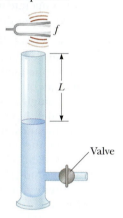

Figure P14.40

41. With a particular fingering, a flute produces a note with frequency 880 Hz at 20.0°C. The flute is open at both ends. (a) Find the air column length. (b) At the beginning of the halftime performance at a late-season football game, the ambient temperature is −5.00°C and the flutist has not had a chance to warm up her instrument. Find the frequency the flute produces under these conditions.

42. *Why is the following situation impossible?* A student is listening to the sounds from an air column that is 0.730 m long. He doesn't know if the column is open at both ends or open at only one end. He hears resonance from the air column at frequencies 235 Hz and 587 Hz.

43. An air column in a glass tube is open at one end and closed at the other by a movable piston. The air in the tube is warmed above room temperature, and a 384-Hz tuning fork is held at the open end. Resonance is heard when the piston is at a distance $d_1 = 22.8$ cm from the open end and again when it is at a distance $d_2 = 68.3$ cm from the open end. (a) What speed of sound is implied by these data? (b) How far from the open end will the piston be when the next resonance is heard?

Section 14.5 Beats: Interference in Time

44. **W** While attempting to tune the note C at 523 Hz, a piano tuner hears 2.00 beats/s between a reference oscillator and the string. (a) What are the possible frequencies of the string? (b) When she tightens the string slightly, she hears 3.00 beats/s. What is the frequency of the string now? (c) By what percentage should the piano tuner now change the tension in the string to bring it into tune?

45. **M** In certain ranges of a piano keyboard, more than one string is tuned to the same note to provide extra loudness. For example, the note at 110 Hz has two strings at this frequency. If one string slips from its normal tension of 600 N to 540 N, what beat frequency is heard when the hammer strikes the two strings simultaneously?

46. **M** **Review.** A student holds a tuning fork oscillating at 256 Hz. He walks toward a wall at a constant speed of 1.33 m/s. (a) What beat frequency does he observe between the tuning fork and its echo? (b) How fast must he walk away from the wall to observe a beat frequency of 5.00 Hz?

Section 14.6 Nonsinusoidal Wave Patterns

47. An A-major chord consists of the notes called A, C#, and E. It can be played on a piano by simultaneously striking strings with fundamental frequencies of 440.00 Hz, 554.37 Hz, and 659.26 Hz. The rich consonance of the chord is associated with near equality of the frequencies of some of the higher harmonics of the three tones. Consider the first five harmonics of each string and determine which harmonics show near equality.

48. Suppose a flutist plays a 523-Hz C note with first harmonic displacement amplitude $A_1 = 100$ nm. From Figure 14.14b read, by proportion, the displacement amplitudes of harmonics 2 through 7. Take these as the values A_2 through A_7 in the Fourier analysis of the sound and assume $B_1 = B_2 = \ldots = B_7 = 0$. Construct a graph of the waveform of the sound. Your waveform will not look exactly like the flute waveform in Figure 14.13b because you simplify by ignoring cosine terms; nevertheless, it produces the same sensation to human hearing.

Section 14.7 The Ear and Theories of Pitch Perception

49. **BIO** Some studies suggest that the upper frequency limit of hearing is determined by the diameter of the eardrum. The wavelength of the sound wave and the diameter of the eardrum are approximately equal at this upper limit. If the relationship holds exactly, what is the diameter of the eardrum of a person capable of hearing 20 000 Hz? (Assume a body temperature of 37.0°C.)

50. **BIO** If a human ear canal can be thought of as resembling an organ pipe, closed at one end, that resonates at a fundamental frequency of 3 000 Hz, what is the length of the canal? Use a normal body temperature of 37°C for your determination of the speed of sound in the canal.

Section 14.8 Context Connection: Building on Antinodes

51. An earthquake can produce a *seiche* in a lake in which the water sloshes back and forth from end to end with remarkably large amplitude and long period. Consider a seiche produced in a farm pond. Suppose the pond is 9.15 m long and assume it has a uniform width and depth. You measure that a pulse produced at one end reaches the other end in 2.50 s. (a) What is the wave speed? (b) What should be the frequency of the ground motion during the earthquake to produce a seiche that is a standing wave with antinodes at each end of the pond and one node at the center?

52. **Q|C** The Bay of Fundy, Nova Scotia, has the highest tides in the world. Assume in midocean and at the mouth of the bay the Moon's gravity gradient and the Earth's rotation make the water surface oscillate with an amplitude of a few centimeters and a period of 12 h 24 min. At the head of the bay, the amplitude is several meters. Assume the bay has a length of 210 km and a uniform depth of 36.1 m. The speed of long-wavelength water waves is given by $v = \sqrt{gd}$, where d is the water's depth. Argue for or against the proposition that the tide is magnified by standing-wave resonance.

Additional Problems

53. On a marimba (Fig. P14.53), the wooden bar that sounds a tone when struck vibrates in a transverse standing wave having three antinodes and two nodes. The lowest-frequency note is 87.0 Hz, produced by a bar 40.0 cm long. (a) Find the speed of transverse waves on the bar. (b) A resonant pipe suspended vertically below the center of the bar enhances the loudness of the emitted sound. If the pipe is open at the top end only, what length of the pipe is required to resonate with the bar in part (a)?

Figure P14.53

54. High-frequency sound can be used to produce standing-wave vibrations in a wine glass. A standing-wave vibration in a wine glass is observed to have four nodes and four antinodes equally spaced around the 20.0-cm circumference of the rim of the glass. If transverse waves move around the glass at 900 m/s, an opera singer would have to produce a high harmonic with what frequency to shatter the glass with a resonant vibration as shown in Figure P14.54?

Figure P14.54

55. A student uses an audio oscillator of adjustable frequency to measure the depth of a water well. The student reports hearing two successive resonances at 51.87 Hz and 59.85 Hz. (a) How deep is the well? (b) How many antinodes are in the standing wave at 51.87 Hz?

56. A nylon string has mass 5.50 g and length $L = 86.0$ cm. The lower end is tied to the floor, and the upper end is tied to a small set of wheels through a slot in a track on which the

wheels move (Fig. P14.56). The wheels have a mass that is negligible compared with that of the string, and they roll without friction on the track so that the upper end of the string is essentially free. At equilibrium, the string is vertical and motionless. When it is carrying a small-amplitude wave, you may assume the string is always under uniform tension 1.30 N. (a) Find the speed of transverse waves on the string. (b) The string's vibration possibilities are a set of standing-wave states, each with a node at the fixed bottom end and an antinode at the free top end. Find the node–antinode distances for each of the three simplest states. (c) Find the frequency of each of these states.

Figure P14.56

57. **Review.** Two train whistles have identical frequencies of 180 Hz. When one train is at rest in the station and the other is moving nearby, a commuter standing on the station platform hears beats with a frequency of 2.00 beats/s when the whistles operate together. What are the two possible speeds and directions the moving train can have?

58. **Review.** A loudspeaker at the front of a room and an identical loudspeaker at the rear of the room are being driven by the same oscillator at 456 Hz. A student walks at a uniform rate of 1.50 m/s along the length of the room. She hears a single tone repeatedly becoming louder and softer. (a) Model these variations as beats between the Doppler-shifted sounds the student receives. Calculate the number of beats the student hears each second. (b) Model the two speakers as producing a standing wave in the room and the student as walking between antinodes. Calculate the number of intensity maxima the student hears each second.

59. Two wires are welded together end to end. The wires are made of the same material, but the diameter of one is twice that of the other. They are subjected to a tension of 4.60 N. The thin wire has a length of 40.0 cm and a linear mass density of 2.00 g/m. The combination is fixed at both ends and vibrated in such a way that two antinodes are present, with the node between them being right at the weld. (a) What is the frequency of vibration? (b) What is the length of the thick wire?

60. **GP Review.** For the arrangement shown in Figure P14.60, the inclined plane and the small pulley are frictionless; the string supports the object of mass M at the bottom of the plane; and the string has mass m. The system is in equilibrium, and the vertical part of the string has a length h. We wish to study standing waves set up in the vertical section of the string. (a) What analysis model describes the object of mass M? (b) What analysis model describes the waves on the vertical part of the string? (c) Find the tension in the string. (d) Model the shape of the string as one leg and the hypotenuse of a right triangle. Find the whole length of the string. (e) Find the mass per unit length of the string. (f) Find the speed of waves on the string. (g) Find the lowest

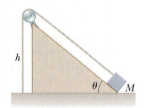

Figure P14.60

frequency for a standing wave on the vertical section of the string. (h) Evaluate this result for $M = 1.50$ kg, $m = 0.750$ g, $h = 0.500$ m, and $\theta = 30.0°$. (i) Find the numerical value for the lowest frequency for a standing wave on the sloped section of the string.

61. A 0.010 0-kg wire, 2.00 m long, is fixed at both ends and vibrates in its simplest mode under a tension of 200 N. When a vibrating tuning fork is placed near the wire, a beat frequency of 5.00 Hz is heard. (a) What could be the frequency of the tuning fork? (b) What should the tension in the wire be if the beats are to disappear?

62. **S** A standing wave is set up in a string of variable length and tension by a vibrator of variable frequency. Both ends of the string are fixed. When the vibrator has a frequency f, in a string of length L and under tension T, n antinodes are set up in the string. (a) If the length of the string is doubled, by what factor should the frequency be changed so that the same number of antinodes is produced? (b) If the frequency and length are held constant, what tension will produce $n + 1$ antinodes? (c) If the frequency is tripled and the length of the string is halved, by what factor should the tension be changed so that twice as many antinodes are produced?

63. Review. A 12.0-kg object hangs in equilibrium from a string with a total length of $L = 5.00$ m and a linear mass density of $\mu = 0.001\ 00$ kg/m. The string is wrapped around two light, frictionless pulleys that are separated by a distance of $d = 2.00$ m (Fig. P14.63a). (a) Determine the tension in the string. (b) At what frequency must the string between the pulleys vibrate to form the standing-wave pattern shown in Figure P14.63b?

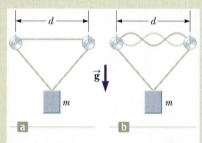

Figure P14.63 Problems 63 and 64.

64. **S** **Review.** An object of mass m hangs in equilibrium from a string with a total length L and a linear mass density μ. The string is wrapped around two light, frictionless pulleys that are separated by a distance d (Fig. P14.63a). (a) Determine the tension in the string. (b) At what frequency must the string between the pulleys vibrate to form the standing-wave pattern shown in Figure P14.63b?

65. A quartz watch contains a crystal oscillator in the form of a block of quartz that vibrates by contracting and expanding. An electric circuit feeds in energy to maintain the oscillation and also counts the voltage pulses to keep time. Two opposite faces of the block, 7.05 mm apart, are antinodes, moving alternately toward each other and away from each other. The plane halfway between these two faces is a node of the vibration. The speed of sound in quartz is equal to 3.70×10^3 m/s. Find the frequency of the vibration.

66. Two waves are described by the wave functions

$$y_1(x,\ t) = 5.00 \sin (2.00x - 10.0t)$$

$$y_2(x,\ t) = 10.0 \cos (2.00x - 10.0t)$$

where x, y_1, and y_2 are in meters and t is in seconds. (a) Show that the wave resulting from their superposition can be expressed as a single sine function. (b) Determine the amplitude and phase angle for this sinusoidal wave.

67. A string is 0.400 m long and has a mass per unit length of 9.00×10^{-3} kg/m. What must be the tension in the string if its second harmonic has the same frequency as the second resonance mode of a 1.75-m-long pipe open at one end?

68. **S** **Review.** Consider the apparatus shown in Figure P14.68a, where the hanging object has mass M and the string is vibrating in its second harmonic. The vibrating blade at the left maintains a constant frequency. The wind begins to blow to the right, applying a constant horizontal force $\vec{F}$ on the hanging object. What is the magnitude of the force the wind must apply to the hanging object so that the string vibrates in its first harmonic as shown in Figure 14.68b?

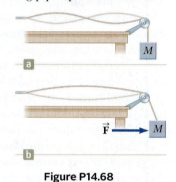

Figure P14.68

69. **Q|C** A string fixed at both ends and having a mass of 4.80 g, a length of 2.00 m, and a tension of 48.0 N vibrates in its second ($n = 2$) normal mode. (a) Is the wavelength in air of the sound emitted by this vibrating string larger or smaller than the wavelength of the wave on the string? (b) What is the ratio of the wavelength in air of the sound emitted by this vibrating string and the wavelength of the wave on the string?

Minimizing the Risk

We have explored the physics of vibrations and waves. Let us now return to our central question for this *Earthquakes* Context:

> **How can we choose locations and build structures to minimize the risk of damage in an earthquake?**

To answer this question, we shall use the physical principles that we now understand more clearly and apply them to our choices of locations and structural design.

In our discussion of simple harmonic oscillation, we learned about resonance. Resonance is one of the most important considerations in designing buildings with regard to earthquake safety. Designers of structures in earthquake-prone areas need to pay careful attention to resonance vibrations from shaking of the ground. The design features to be considered include ensuring that the resonance frequencies of the building do not match typical earthquake frequencies. In addition, the structural details should include sufficient damping to ensure that the amplitude of resonance vibration does not destroy the structure.

Resonance is a prime consideration for the design of a structure; what about, as suggested by our central question, the *location* of the structure? In Chapter 13, we discussed the role of the medium in the propagation of a wave. For seismic waves

Figure 1 Portions of the double-decked Nimitz Freeway in Oakland, California, collapsed during the Loma Prieta earthquake of 1989.

moving across the surface of the Earth, the soil on the surface is the medium. Because soil varies from one location to another, the speed of seismic waves will vary at different locations. A particularly dangerous situation exists for structures built on loose soil or mudfill. In these types of media, the interparticle forces are much weaker than in a more solid foundation such as granite bedrock. As a result, the wave speed is less in loose soil than in bedrock.

Consider Equation 13.24, which provides an expression for the rate of energy transfer by waves. This equation was derived for waves on strings, but the proportionality to the square of the amplitude and the speed is general. Because of conservation of energy, the rate of energy transfer for a wave must remain constant regardless of the medium. Therefore, according to Equation 13.24, if the wave speed decreases, as it does for seismic waves moving from rock into loose soil, the amplitude must increase. As a result, the shaking of structures built on loose soil is of larger magnitude than for those built on solid bedrock.

This factor contributed to the collapse of the Nimitz Freeway during the Loma Prieta earthquake, near San Francisco, in 1989. Figure 1 shows the results of the earthquake on the freeway. The portion of the freeway that collapsed was built on mudfill, but the surviving portion was built on bedrock. The amplitude of oscillation in the portion built on mudfill was more than five times as large as the amplitude of other portions.

Another danger for structures on loose soil is the possibility of **liquefaction** of the soil. When soil is shaken, the elements of soil can move with respect to one another and the soil tends to act like a liquid rather than a solid. It is possible for the structure to sink into the soil during an earthquake. If the liquefaction is not uniform over the foundation of the structure, the structure can deviate from its vertical orientation, as seen in the case of the Japanese police station in Figure 2. In some cases, buildings can tip over completely, as happened to some apartment buildings during a Japanese earthquake in 1964. As a result, even if the earthquake vibrations are not sufficient to damage the structure, it will be unusable in its leaning orientation.

As discussed in Section 14.8, constructing buildings or other structures where standing seismic waves can be established is dangerous. Such construction was a factor in the Michoacán earthquake of 1985. The shape of the bedrock under Mexico City resulted in standing waves, with severe damage to buildings located at antinodes.

In summary, to minimize risk of damage in an earthquake, architects and engineers must design structures to prevent destructive resonances, avoid building on loose soil, and pay attention to the underground rock formations so as to be aware of possible standing wave patterns. Other precautions can also be taken. For example, buildings can be constructed with **seismic isolation** from the ground. This method involves mounting the structure on **isolation dampers,** heavy-duty bearings that dampen the oscillations of the building, resulting in reduced amplitude of vibration. Figure 3 shows the results of the 2011 earthquake in Christchurch, New Zealand, on a building that did not take advantage of isolation dampers. Many older buildings have been retrofitted with dampers, including several in California (Los Angeles City Hall, San Francisco City Hall, Oakland City Hall) as well as in other parts of the world, such as the New Zealand Parliament Buildings. Additional measures include tuned dampers, such as that in the opening photograph of Chapter 12, shear trusses, external bracing, and other techniques.

Figure 2 A police station leans to one side due to liquefaction of the underlying soil during the Japanese earthquake of March 2011.

Figure 3 Damage to a garage building in Christchurch, New Zealand, after the magnitude 6.3 earthquake on February 22, 2011. The garage did not have isolation dampers installed to isolate it from the ground.

We have not addressed many other considerations for earthquake safety in structures, but we have been able to apply many of our concepts from oscillations and waves so as to understand some aspects of logical choices in locating and designing structures.

Problems

1. For seismic waves spreading out from a point (the epicenter) on the surface of the Earth, the intensity of the waves decreases with distance according to an inverse proportionality to distance. That is, the wave intensity is proportional to $1/r$, where r is the distance from the epicenter to the observation point. This rule applies if the medium is uniform. The intensity of the wave is proportional to the rate of energy transfer for the wave. Furthermore, we have shown that the energy of vibration of an oscillator is proportional to the square of the amplitude of the vibration. Assume that a particular earthquake causes ground shaking with an amplitude of 5.0 cm at a distance of 10 km from the epicenter. If the medium is uniform, what is the amplitude of the ground shaking at a point 20 km from the epicenter?

2. As mentioned in the text, the amplitude of oscillation during the Loma Prieta earthquake of 1989 was five times greater in areas of mudfill than in areas of bedrock. From this information, find the factor by which the seismic wave speed changed as the waves moved from the bedrock to the mudfill. Ignore any reflection of wave energy and any change in density between the two media.

3. Figure 4 is a graphical representation of the travel time for P and S waves from the epicenter of an earthquake to a seismograph as a function of the distance of travel. The following table shows the measured times of day for arrival of P waves from a particular earthquake at three seismograph locations. In the last column, fill in the times of day for the arrival of the S waves at the three seismograph locations.

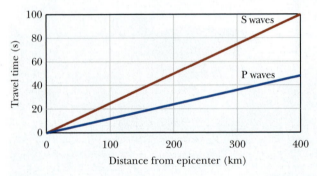

Figure 4 A graph of travel time versus distance from the epicenter for P and S waves.

Seismograph Station	Distance from Epicenter (km)	P Wave Arrival Time	S Wave Arrival Time
#1	200	15:46:06	
#2	160	15:46:01	
#3	105	15:45:54	

Heart Attacks

During an average lifetime, the human heart beats over three billion times without rest, pumping over one million barrels of blood (there are 42 gallons, or 159 liters, in a barrel). This rhythm of life, however, is sometimes interrupted by a heart attack, or a *myocardial infarction* (as it is known medically), one of the leading causes of death in the world. A heart attack occurs when there is an interruption of blood flow to the heart, often resulting in permanent damage to this vital organ. The term *cardiovascular disease* (CVD) refers to diseases affecting the heart and the blood vessels. Figure 1 shows the prevalence of deaths attributed to cardiovascular disease and total deaths per year per 100 000 men of age 35 to 74 in several developed countries. The percentage

of all deaths due to CVD ranges from a low of 19.7% in France to 48% for the Russian Federation. Cardiovascular disease accounts for 31% of all deaths in the United States each year for men of age 35 to 74. The corresponding rate for women in the same age bracket is 25%.

The human *cardiovascular system,* or *circulatory system,* has been the subject of scientific interest for millennia. The Ebers Papyrus from the 16th century BC proposed a connection between the heart and the arteries. In the second century, Galen, a prominent Greek physician famous for attempting cataract surgeries, identified the roles of the blood carried by arteries and veins. Ibn Al-Nafis, a 13th-century Arab physician, correctly

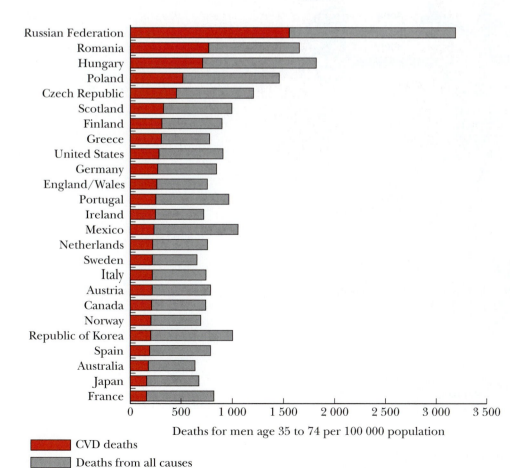

Figure 1 Annual deaths per 100 000 men from cardiovascular disease (red) and from all causes (gray) in selected countries. The largest percentage due to cardiovascular disease compared to all deaths occurs in the Russian Federation. (Graph derived from Table 2–3, p. e42, International Death Rates (Revised 2008): Death Rates (Per 100 000 Population) for Total Cardiovascular Disease, Coronary Heart Disease, Stroke, and Total Deaths in Selected Countries (Most Recent Year Available) by WRITING GROUP MEMBERS et al. for American Heart Association Statistics Committee and Stroke Statistics Subcommittee, "Heart Disease and Stroke Statistics—2009 Update: A Report From the American Heart Association Statistics Committee and Stroke Statistics Subcommittee" Circulation 119 (3): e21–e181.)

Deaths for men age 35 to 74 per 100 000 population

■ CVD deaths
■ Deaths from all causes

described the pulmonary circulation system, the portion of the cardiovascular system that delivers blood from the heart to the lungs and back. Building on the work of his predecessors, William Harvey is credited with the discovery and nearly complete description of the circulatory system in a 1628 publication, as well as with the realization that the heart was responsible for pumping blood throughout the body. Harvey correctly explained the roles of pulmonary circulation in oxygenating the blood and disposing of the carbon dioxide produced by cell metabolism, and of systemic circulation (see Fig. 2) in carrying oxygenated blood to vital organs.

While pumping oxygenated blood through its chambers, the heart itself relies on a network of vessels and capillaries surrounding its outer surface for its own oxygen supply, and is the third largest consumer of oxygen in the human body (roughly 12% of the total oxygen intake), after the liver (20%) and the brain (18%). Figure 3 shows the heart's surface and the network of blood vessels that provide oxygen to the heart.

We can identify several examples of systems that depend on the flow of fluids for proper operation. For example, if a water pipe in a home ruptures, the water supply to sinks, showers, and washing machines is

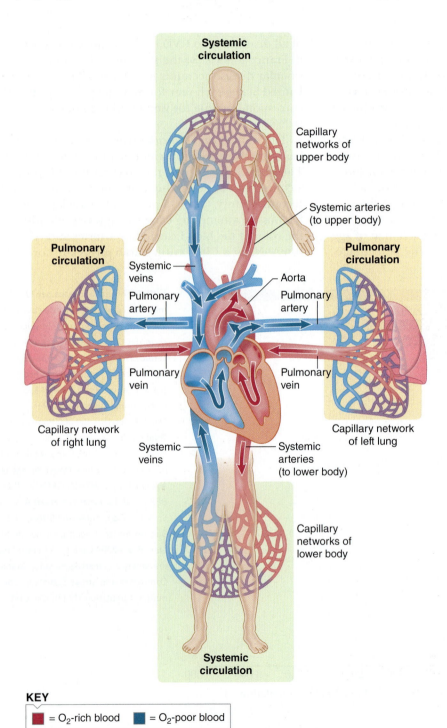

Figure 2 The human circulatory system consists of two separate loops. The pulmonary circulation system exchanges blood between the heart and the lungs. The systemic circulation system exchanges blood between the heart and the other organs of the body. (From Sherwood, *Fundamentals of Human Physiology*, 4th ed., 2012, Brooks/Cole, Figure 9.1, p. 230)

KEY

■ = O_2-rich blood   ■ = O_2-poor blood

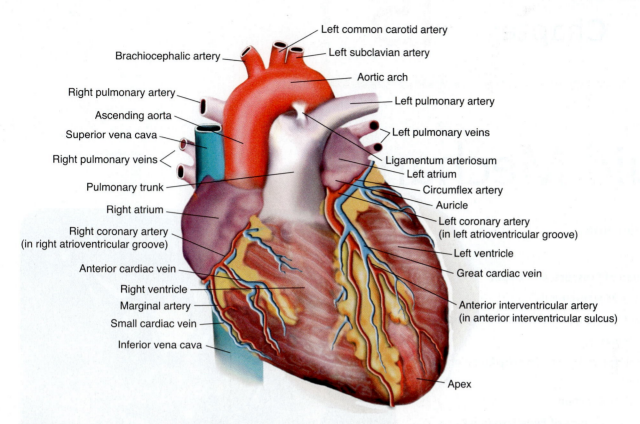

Figure 3 The human heart. In this diagram, we see sections of the major blood vessels carrying blood to the rest of the body as well as the system of vessels supplying blood to the heart itself. [from Des Jardins, *Cardiopulmonary Anatomy and Physiology: Essentials for Respiratory Care*, 5th ed., 2008, Delmar, p. 189 (Fig. 5-2, part (A))]

affected. If a hydraulic line of an automobile's brake system ruptures, the brakes may fail to operate. In a similar way, a defect in the blood vessels that affects the flow of blood to the heart can cause a number of dangerous medical conditions, including heart attacks.

Although heart attacks are sudden occurrences, in most cases they are the consequence of years of plaque buildup in arteries. During a heart attack, plaque in the arterial network of the heart ruptures, resulting in the formation of a blood clot that can cause the disruption or stoppage of blood flow to a portion of the heart, depriving it of oxygen. If oxygen deprivation lasts too long, the cells in the affected portion of the heart die, resulting in permanent heart damage.

Even if a patient survives a heart attack, it is a life-changing event. A number of lifestyle changes are necessary to reduce the risk of a subsequent heart attack, including incorporation of exercise in one's daily schedule, alterations in the diet, cessation of smoking, and a variety of medications. In addition, careful blood pressure management is necessary—again suggesting the importance of the flow of fluid within the circulatory system.

Having seen this introduction to the circulatory system and some of the impacts of heart disease and heart attacks on human lives, we will explore the physics of fluids in this Context. We will apply the principles that we learn to this question:

> **How can the principles of physics be applied in medicine to help prevent heart attacks?**

Chapter | 15

Fluid Mechanics

Chapter Outline

Ed Robinson/Pacific Stock/Photolibrary

Fish congregate around a reef in Hawaii searching for food. How do fish such as the yellow butterflyfish in the front control their movements up and down in the water? We'll find out in this chapter.

Matter is normally classified as being in one of three states: solid, liquid, or gas. Everyday experience tells us that a solid has a definite volume and shape. A brick maintains its familiar shape and size over a long time. We also know that a liquid has a definite volume but no definite shape. For example, a cup of liquid water has a fixed volume but assumes the shape of its container. Finally, an unconfined gas has neither definite volume nor definite shape. For example, if there is a leak in the natural gas supply in your home, the escaping gas continues to expand into the surrounding atmosphere. These definitions help us picture the states of matter, but they are somewhat artificial. For example, asphalt, glass, and plastics are normally considered solids, but over a long time interval they tend to flow like liquids. Likewise, most substances can be a solid, liquid, or gas (or combinations of these states), depending on the temperature and pressure. In general, the time interval required for a particular substance to change its shape in response to an external force determines whether we treat the substance as a solid, liquid, or gas.

A **fluid** is a collection of molecules that are randomly arranged and held together by weak cohesive forces between molecules and forces exerted by the walls of a container. Both liquids and gases are fluids. In our treatment of the mechanics of fluids, we shall see that no new physical principles are needed to explain such effects as the buoyant force on a submerged object and a curve ball in baseball. In this chapter, we shall apply a number of familiar analysis models to the physics of fluids.

15.1 | Pressure

Our first task in understanding the physics of fluids is to define a new quantity to describe fluids. Imagine applying a force to the surface of an object, with the force having components both parallel to and perpendicular to the surface.

If the object is a solid at rest on a table, the force component perpendicular to the surface may cause the object to flatten, depending on how hard the object is. Assuming that the object does not slide on the table, the component of the force parallel to the surface of the object will cause the object to distort. As an example, suppose you place your physics book flat on a table and apply a force with your hand parallel to the front cover and perpendicular to the spine. The book will distort, with the bottom pages staying fixed at their original location and the top pages shifting horizontally by some distance. The cross section of the book changes from a rectangle to a parallelogram. This kind of force parallel to the surface is called a *shearing force*.

We shall adopt a simplification model in which the fluids we study will be nonviscous; that is, no friction exists between adjacent layers of the fluid. Nonviscous fluids and static fluids do not sustain shearing forces. If you imagine placing your hand on a water surface and pushing parallel to the surface, your hand simply slides over the water; you cannot distort the water as you did the book. This phenomenon occurs because the interatomic forces in a fluid are not strong enough to lock atoms in place with respect to one another. The fluid cannot be modeled as a rigid object as in Chapter 10. If we try to apply a shearing force, the molecules of the fluid simply slide past one another.

Therefore, the only type of force that can exist in a fluid is one that is perpendicular to a surface. For example, the forces exerted by the fluid on the object in Figure 15.1 are everywhere perpendicular to the surfaces of the object.

The force that a fluid exerts on a surface originates in the collisions of molecules of the fluid with the surface. Each collision results in the reversal of the component of the velocity vector of the molecule perpendicular to the surface. By the impulse–momentum theorem and Newton's third law, each collision results in a force on the surface. A huge number of these impulsive forces occur every second, resulting in a constant macroscopic force on the surface. This force is spread out over the area of the surface and is related to a new quantity called *pressure*.

The pressure at a specific point in a fluid can be measured with the device pictured in Figure 15.2. The device consists of an evacuated cylinder enclosing a light piston connected to a spring. As the device is submerged in a fluid, the fluid presses in on the top of the piston and compresses the spring until the inward force of the fluid is balanced by the outward force of the spring. The force exerted on the piston by the fluid can be measured if the spring is calibrated in advance.

If *F* is the magnitude of the force exerted by the fluid on the piston and *A* is the surface area of the piston, the **pressure** *P* of the fluid at the level to which the device has been submerged is defined as the ratio of force to area:

$$P \equiv \frac{F}{A}$$

15.1 ◀   ▶ Definition of pressure

Although we have defined pressure in terms of our device in Figure 15.2, the definition is general. Because pressure is force per unit area, it has units of newtons per square meter in the SI system. Another name for the SI unit of pressure is the **pascal** (Pa):

$$1 \text{ Pa} \equiv 1 \text{ N/m}^2$$

15.2 ◀   ▶ The pascal

Notice that pressure and force are different quantities. We can have a very large pressure from a relatively small force by making the area over which the force is applied small. Such is the case with hypodermic needles. The area of the tip of the needle is very small, so a small force pushing on the needle is sufficient to cause a pressure large enough to puncture the skin. We can also create a small pressure from a large force by enlarging the area over which the force acts. Such is the principle behind the design of snowshoes. If a person were to walk on deep snow with regular shoes, it is possible for his or her feet to break through the snow and sink. Snowshoes, however, allow the force on the snow due to the weight of the person to spread out over a larger area, reducing the pressure enough so that the snow surface is not broken (Fig. 15.3, page 484).

The atmosphere exerts a pressure on the surface of the Earth and all objects at the surface. This pressure is responsible for the action of suction cups, drinking

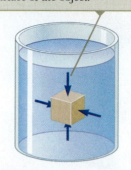

At any point on the surface of the object, the force exerted by the fluid is perpendicular to the surface of the object.

Figure 15.1 The forces exerted by a fluid on the surfaces of a submerged object. (The forces on the front and back sides of the object are not shown.)

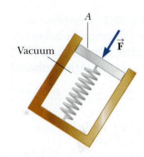

Figure 15.2 A simple device for measuring the pressure exerted by a fluid.

Pitfall Prevention | 15.1
Force and Pressure
Equation 15.1 makes a clear distinction between force and pressure. Another important distinction is that *force is a vector* and *pressure is a scalar*. There is no direction associated with pressure, but the direction of the force associated with the pressure is perpendicular to the surface on which the pressure acts.

Figure 15.3 Snowshoes keep you from sinking into soft snow because they spread the downward force you exert on the snow over a large area, reducing the pressure on the snow surface.

TABLE 15.1 | Densities of Some Common Substances at Standard Temperature (0°C) and Pressure (Atmospheric)

Substance	ρ (kg/m³)
Air	1.29
Air (at 20°C and atmospheric pressure)	1.20
Aluminum	2.70×10^3
Benzene	0.879×10^3
Brass	8.4×10^3
Copper	8.92×10^3
Ethyl alcohol	0.806×10^3
Fresh water	1.00×10^3
Glycerin	1.26×10^3
Gold	19.3×10^3
Helium gas	1.79×10^{-1}
Hydrogen gas	8.99×10^{-2}
Ice	0.917×10^3
Iron	7.86×10^3
Lead	11.3×10^3
Mercury	13.6×10^3
Nitrogen gas	1.25
Oak	0.710×10^3
Osmium	22.6×10^3
Oxygen gas	1.43
Pine	0.373×10^3
Platinum	21.4×10^3
Seawater	1.03×10^3
Silver	10.5×10^3
Tin	7.30×10^3
Uranium	19.1×10^3

straws, vacuum cleaners, and many other devices. In our calculations and end-of-chapter problems, we usually take atmospheric pressure to be

$$P_0 = 1.00 \text{ atm} \approx 1.013 \times 10^5 \text{ Pa} \qquad \text{15.3} \blacktriangleleft$$

Pressures higher than atmospheric are used in *hyperbaric medicine*, or *hyperbaric oxygen therapy* (HBOT). This type of therapy was initially developed for the treatment of disorders associated with diving accidents, such as decompression sickness and air embolisms. Today it is used for a wider range of medical situations.

To receive hyperbaric oxygen therapy, the patient reclines in a special chamber. Modern chambers are transparent, allowing the patient to see the therapist outside. Patients may read a book, listen to music, watch a movie, or simply rest during the procedure. The pressure in the chamber is slowly increased, up to as much as three times atmospheric pressure. The patient experiences the increased pressure for a time interval determined by the therapist, and then the pressure is decreased, the entire session requiring one to two hours.

Many cancer patients undergo radiation treatments. Radiation applied to the pelvic region can cause *radiation cystitis*, resulting in bladder infections, sometimes occurring years after the radiation therapy. Since 1985, hyperbaric oxygen therapy has been used to treat this condition. The therapy stimulates *angiogenesis*, the growth of new blood vessels. This growth reverses the vascular changes induced by the radiation, thereby healing the radiation-induced bladder injury.

Another area in which HBOT is used is for problem wounds, such as those associated with diabetes or amputations. The increased pressure assists with tissue oxygenation in the wound and stimulates angiogenesis in the damaged tissue. It has also been shown that the increase in pressure helps to kill various types of bacteria in the wound area.

QUICK QUIZ 15.1 Suppose you are standing directly behind someone who steps back and accidentally stomps on your foot with the heel of one shoe. Would you be better off if that person were **(a)** a large male professional basketball player wearing sneakers or **(b)** a petite woman wearing spike-heeled shoes?

THINKING PHYSICS 15.1

Suction cups can be used to hold objects onto surfaces. Why don't astronauts use suction cups to hold onto the outside surface of an orbiting spacecraft?

Reasoning A suction cup works because air is pushed out from under the cup when it is pressed against a surface. When the cup is released, it tends to spring back a bit, causing the trapped air under the cup to expand. This expansion causes a reduced pressure inside the cup. Therefore, the difference between the atmospheric pressure on the outside of the cup and the reduced pressure inside provides a net force pushing the cup against the surface. For astronauts in orbit around the Earth, almost no air exists outside the surface of the spacecraft. Therefore, if a suction cup were to be pressed against the outside surface of the spacecraft, the pressure differential needed to press the cup to the surface is not present. ◀

15.2 | Variation of Pressure with Depth

The study of fluid mechanics involves the density of a substance, defined in Equation 1.1 as the mass per unit volume for the substance. Table 15.1 lists the densities of various substances. These values vary slightly with temperature because the volume of a substance is temperature-dependent (as we shall see in Chapter 16). Note that under standard conditions (0°C and atmospheric pressure) the densities of gases are on the order of 1/1 000 the densities of solids and liquids. This difference implies that the average molecular spacing in a gas under these conditions is about ten times greater in each dimension than in a solid or liquid.

As divers know well, the pressure in the sea or a lake increases as they dive to greater depths. Likewise, atmospheric pressure decreases with increasing altitude. For this reason, aircraft flying at high altitudes must have pressurized cabins to provide sufficient oxygen for the passengers.

We now show mathematically how the pressure in a liquid increases with depth. Consider a liquid of density ρ at rest as in Figure 15.4. Let us select a sample of the liquid contained within an imaginary cylinder of cross-sectional area A extending from depth d to depth $d + h$. This sample of liquid is in equilibrium and at rest. Therefore, according to the particle in equilibrium model, the net force on the sample must be equal to zero. We will investigate the forces on the sample related to the pressure on it.

The liquid external to our sample exerts forces at all points on the sample's surface, perpendicular to it. On the sides of the sample of liquid in Figure 15.4, forces due to the pressure act horizontally and cancel in pairs on opposite sides of the sample for a net horizontal force of zero. The pressure exerted by the liquid on the sample's bottom face is P and the pressure on the top face is P_0. Therefore, from Equation 15.1, the magnitude of the upward force exerted by the liquid on the bottom of the sample is PA, and the magnitude of the downward force exerted by the liquid on the top is P_0A. In addition, a gravitational force is exerted on the sample. Because the sample is in equilibrium, the net force in the vertical direction must be zero:

$$\Sigma F_y = 0 \quad \rightarrow \quad PA - P_0A - Mg = 0$$

Because the mass of liquid in the sample is $M = \rho V = \rho Ah$, the gravitational force on the liquid in the sample is $Mg = \rho gAh$. Therefore,

$$PA = P_0A + \rho g\, Ah$$

or

$$\boxed{P = P_0 + \rho gh} \qquad \text{15.4} \blacktriangleleft$$

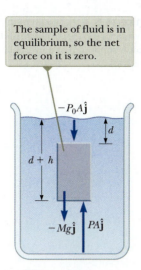

The sample of fluid is in equilibrium, so the net force on it is zero.

Figure 15.4 A sample of fluid in a larger volume of fluid is singled out.

▶ Variation of pressure with depth in a liquid

If the top surface of our sample is at $d = 0$ so that it is open to the atmosphere, P_0 is atmospheric pressure. Equation 15.4 indicates that the pressure in a liquid depends only on the depth h within the liquid. The pressure is therefore the same at all points having the same depth, independent of the shape of the container.

In view of Equation 15.4, any increase in pressure at the surface must be transmitted to every point in the liquid. This behavior was first recognized by French scientist Blaise Pascal (1623–1662) and is called **Pascal's law:**

A change in the pressure applied to an enclosed fluid is transmitted undiminished to every point of the fluid and to the walls of the container.

▶ Pascal's law

You use Pascal's law when you squeeze the sides of your toothpaste tube. The increase in pressure on the sides of the tube increases the pressure everywhere, which pushes a stream of toothpaste out of the opening.

An important application of Pascal's law is the hydraulic press illustrated by Figure 15.5 (page 486). A force $\vec{F}_1$ is applied to a small piston of area A_1. The pressure is transmitted through a liquid to a larger piston of area A_2, and force $\vec{F}_2$ is exerted by the liquid on this piston. Because the pressure is the same at both pistons, we see that $P = F_1/A_1 = F_2/A_2$. The force magnitude F_2 is therefore larger than F_1 by the multiplying factor A_2/A_1. Hydraulic brakes, car lifts, hydraulic jacks, and forklifts all make use of this principle.

QUICK QUIZ 15.2 The pressure at the bottom of a filled glass of water ($\rho = 1\,000$ kg/m^3) is P. The water is poured out, and the glass is filled with ethyl alcohol ($\rho = 806$ kg/m^3). What is the pressure at the bottom of the glass? **(a)** smaller than P **(b)** equal to P **(c)** larger than P **(d)** indeterminate

Figure 15.5 (a) Diagram of a hydraulic press. (b) A vehicle under repair is supported by a hydraulic lift in a garage.

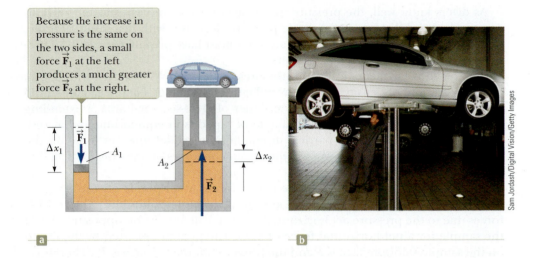

Because the increase in pressure is the same on the two sides, a small force $\vec{F}_1$ at the left produces a much greater force $\vec{F}_2$ at the right.

 THINKING PHYSICS 15.2   **BIO** Measuring blood pressure

Blood pressure is normally measured with the cuff of the sphygmomanometer around the arm. Suppose the blood pressure were measured with the cuff around the calf of the leg of a standing person. Would the reading of the blood pressure be the same here as it is for the arm?

Reasoning The blood pressure measured at the calf would be higher than that measured at the arm. If we imagine the vascular system of the body to be a vessel containing a liquid (blood), the pressure in the liquid will increase with depth. The blood at the calf is deeper in the liquid than that at the arm and is at a higher pressure.

Blood pressures are normally taken at the arm because it is at approximately the same height as the heart. If blood pressures at the calf were used as a standard, adjustments would need to be made for the height of the person and the blood pressure would be different if the person were lying down. ◀

Example **15.1** | **The Car Lift**

In a car lift used in a service station (Fig. 15.5), compressed air exerts a force on a small piston that has a circular cross section and a radius of 5.00 cm. This pressure is transmitted by a liquid to a piston that has a radius of 15.0 cm.

(A) What force must the compressed air exert to lift a car weighing 13 300 N?

SOLUTION

Conceptualize Review the material just discussed about Pascal's law to understand the operation of a car lift.

Categorize This example is a substitution problem.

Solve $F_1/A_1 = F_2/A_2$ for F_1:
$$F_1 = \left(\frac{A_1}{A_2}\right)F_2 = \frac{\pi(5.00 \times 10^{-2}\,\text{m})^2}{\pi(15.0 \times 10^{-2}\,\text{m})^2}(1.33 \times 10^4\,\text{N})$$
$$= 1.48 \times 10^3\,\text{N}$$

(B) What air pressure produces this force?

SOLUTION

Use Equation 15.1 to find the air pressure that produces this force:
$$P = \frac{F_1}{A_1} = \frac{1.48 \times 10^3\,\text{N}}{\pi(5.00 \times 10^{-2}\,\text{m})^2}$$
$$= 1.88 \times 10^5\,\text{Pa}$$

This pressure is approximately twice atmospheric pressure.

(C) Consider the lift as a nonisolated system and show that the input energy transfer is equal in magnitude to the output energy transfer.

15.1 *cont.*

SOLUTION

The energy input and output are by means of work done by the forces as the pistons move. To determine the work done, we must find the magnitude of the displacement through which each force acts. Because the liquid is modeled to be incompressible, the volume of the cylinder through which the input piston moves must equal that through which the output piston moves. The lengths of these cylinders are the magnitudes Δx_1 and Δx_2 of the displacements of the forces (see Fig. 15.5a).

Set the volumes through which the pistons move equal:

$$V_1 = V_2 \;\rightarrow\; A_1 \Delta x_1 = A_2 \Delta x_2$$

$$\frac{A_1}{A_2} = \frac{\Delta x_2}{\Delta x_1}$$

Evaluate the ratio of the input work to the output work:

$$\frac{W_1}{W_2} = \frac{F_1 \, \Delta x_1}{F_2 \, \Delta x_2} = \left(\frac{F_1}{F_2}\right)\left(\frac{\Delta x_1}{\Delta x_2}\right) = \left(\frac{A_1}{A_2}\right)\left(\frac{A_2}{A_1}\right) = 1$$

This result verifies that the work input and output are the same, as they must be to conserve energy.

Example 15.2 | The Force on a Dam

Water is filled to a height H behind a dam of width w (Fig. 15.6). Determine the resultant force exerted by the water on the dam.

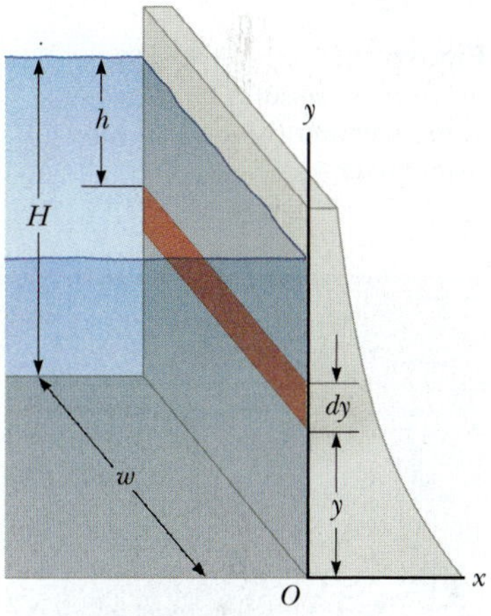

SOLUTION

Conceptualize Because pressure varies with depth, we cannot calculate the force simply by multiplying the area by the pressure. As the pressure in the water increases with depth, the force on the adjacent portion of the dam also increases.

Categorize Because of the variation of pressure with depth, we must use integration to solve this example, so we categorize it as an analysis problem.

Analyze Let's imagine a vertical y axis, with $y = 0$ at the bottom of the dam. We divide the face of the dam into narrow horizontal strips at a distance y above the bottom, such as the red strip in Figure 15.6. The pressure on each such strip is due only to the water; atmospheric pressure acts on both sides of the dam.

Figure 15.6 (Example 15.2) Water exerts a force on a dam.

Use Equation 15.4 to calculate the pressure due to the water at the depth h:

$$P = \rho g h = \rho g (H - y)$$

Use Equation 15.1 to find the force exerted on the shaded strip of area $dA = w \, dy$:

$$dF = P \, dA = \rho g (H - y) w \, dy$$

Integrate to find the total force on the dam:

$$F = \int P \, dA = \int_0^H \rho g (H - y) w \, dy = \tfrac{1}{2}\rho g w H^2$$

Finalize Notice that the thickness of the dam shown in Figure 15.6 increases with depth. This design accounts for the greater force the water exerts on the dam at greater depths.

What If? What if you were asked to find this force without using calculus? How could you determine its value?

Answer We know from Equation 15.4 that pressure varies linearly with depth. Therefore, the average pressure due to the water over the face of the dam is the average of the pressure at the top and the pressure at the bottom:

$$P_{\text{avg}} = \frac{P_{\text{top}} + P_{\text{bottom}}}{2} = \frac{0 + \rho g H}{2} = \tfrac{1}{2}\rho g H$$

The total force on the dam is equal to the product of the average pressure and the area of the face of the dam:

$$F = P_{\text{avg}} A = (\tfrac{1}{2}\rho g H)(Hw) = \tfrac{1}{2}\rho g w H^2$$

which is the same result we obtained using calculus.

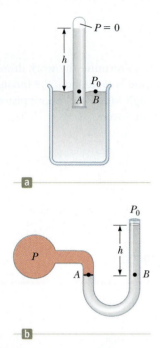

Figure 15.7 Two devices for measuring pressure: (a) a mercury barometer and (b) an open-tube manometer.

Archimedes
Greek Mathematician, Physicist, and Engineer (c. 287–212 BC)
Archimedes was perhaps the greatest scientist of antiquity. He was the first to compute accurately the ratio of a circle's circumference to its diameter, and he also showed how to calculate the volume and surface area of spheres, cylinders, and other geometric shapes. He is well known for discovering the nature of the buoyant force and was also a gifted inventor. One of his practical inventions, still in use today, is Archimedes's screw, an inclined, rotating, coiled tube used originally to lift water from the holds of ships. He also invented the catapult and devised systems of levers, pulleys, and weights for raising heavy loads. Such inventions were successfully used to defend his native city, Syracuse, during a two-year siege by Romans.

15.3 | Pressure Measurements

During the weather report on a television news program, the *barometric pressure* is often provided. Barometric pressure is the current pressure of the atmosphere, which varies over a small range from the standard value provided in Equation 15.3. How is this pressure measured?

One instrument used to measure atmospheric pressure is the common barometer, invented by Evangelista Torricelli (1608–1647). A long tube closed at one end is filled with mercury and then inverted into a dish of mercury (Fig. 15.7a). The closed end of the tube is nearly a vacuum, so the pressure at the top of the mercury column can be taken as zero. In Figure 15.7a, the pressure at point A due to the column of mercury must equal the pressure at point B due to the atmosphere. If that were not the case, a net force would move mercury from one point to the other until equilibrium was established. It therefore follows that $P_0 = \rho_{Hg}gh$, where ρ_{Hg} is the density of the mercury and h is the height of the mercury column. As atmospheric pressure varies, the height of the mercury column varies, so the height can be calibrated to measure atmospheric pressure. Let us determine the height of a mercury column for one atmosphere of pressure, $P_0 = 1$ atm $= 1.013 \times 10^5$ Pa:

$$P_0 = \rho_{Hg}gh \rightarrow h = \frac{P_0}{\rho_{Hg}g} = \frac{1.013 \times 10^5 \text{ Pa}}{(13.6 \times 10^3 \text{ kg/m}^3)(9.80 \text{ m/s}^2)} = 0.760 \text{ m}$$

Based on a calculation such as this one, one atmosphere of pressure is defined as the pressure equivalent of a column of mercury that is exactly 0.760 0 m in height at 0°C.

The open-tube manometer illustrated in Figure 15.7b is a device for measuring the pressure of a gas contained in a vessel. One end of a U-shaped tube containing a liquid is open to the atmosphere, and the other end is connected to a system of unknown pressure P. The pressures at points A and B must be the same (otherwise, the curved portion of the liquid would experience a net force and would accelerate), and the pressure at A is the unknown pressure of the gas. Therefore, equating the unknown pressure P to the pressure at point B, we see that $P = P_0 + \rho gh$. The difference in pressure $P - P_0$ is equal to ρgh. Pressure P is called the **absolute pressure,** and the difference $P - P_0$ is called the **gauge pressure.** For example, the pressure you measure in your bicycle tire is gauge pressure.

15.4 | Buoyant Forces and Archimedes's Principle

Have you ever tried to push a beach ball down under water (Fig. 15.8a)? It is extremely difficult to do because of the large upward force exerted by the water on the ball. The upward force exerted by a fluid on any immersed object is called a **buoyant force.** We can determine the magnitude of a buoyant force by applying some logic. Imagine a beach ball–sized parcel of water beneath the water surface as in Figure 15.8b. Because this parcel is in equilibrium, there must be an upward force that balances the downward gravitational force on the parcel. This upward force is

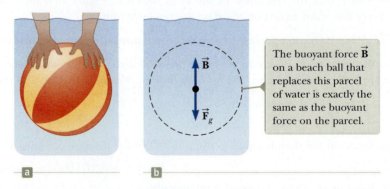

The buoyant force $\vec{B}$ on a beach ball that replaces this parcel of water is exactly the same as the buoyant force on the parcel.

Figure 15.8 (a) A swimmer pushes a beach ball under water. (b) The forces on a beach ball–sized parcel of water.

the buoyant force, and its magnitude is equal to the weight of the water in the parcel. The buoyant force is the resultant force on the parcel due to all forces applied by the fluid surrounding the parcel.

Now imagine replacing the beach ball–sized parcel of water with a beach ball of the same size. The net force applied by the fluid surrounding the beach ball is the same, regardless of whether it is applied to a beach ball or to a parcel of water. Consequently, **the magnitude of the buoyant force on an object always equals the weight of the fluid displaced by the object.** This statement is known as **Archimedes's principle.**

With the beach ball under water, the buoyant force, equal to the weight of a beach ball–sized parcel of water, is much larger than the weight of the beach ball. Therefore, there is a large net upward force, which explains why it is so hard to hold the beach ball under the water. Note that Archimedes's principle does not refer to the makeup of the object experiencing the buoyant force. The object's composition is not a factor in the buoyant force because the buoyant force is exerted by the surrounding fluid.

To better understand the origin of the buoyant force, consider a cube of solid material immersed in a liquid as in Figure 15.9. According to Equation 15.4, the pressure P_{bot} at the bottom of the cube is greater than the pressure P_{top} at the top by an amount $\rho_{fluid}gh$, where h is the height of the cube and ρ_{fluid} is the density of the fluid. The pressure at the bottom of the cube causes an *upward* force equal to $P_{bot}A$, where A is the area of the bottom face. The pressure at the top of the cube causes a *downward* force equal to $P_{top}A$. The resultant of these two forces is the buoyant force $\vec{B}$ with magnitude

$$B = (P_{bot} - P_{top})A = (\rho_{fluid}gh)A$$

$$B = \rho_{fluid}gV_{disp} \qquad \text{15.5} \blacktriangleleft$$

where $V_{disp} = Ah$ is the volume of the fluid displaced by the cube. Because the product $\rho_{fluid}V_{disp}$ is equal to the mass of fluid displaced by the object,

$$B = Mg$$

where Mg is the weight of the fluid displaced by the cube. This result is consistent with our initial statement about Archimedes's principle above, based on the discussion of the beach ball.

Before proceeding with a few examples, it is instructive to compare two common cases: the buoyant force acting on a totally submerged object and that acting on a floating object.

Case I: A Totally Submerged Object

When an object is totally submerged in a fluid of density ρ_{fluid}, the volume V_{disp} of the displaced fluid is equal to the volume V_{obj} of the object; so, from Equation 15.5, the magnitude of the upward buoyant force is $B = \rho_{fluid}gV_{obj}$. If the object has a mass M and density ρ_{obj}, its weight is equal to $F_g = Mg = \rho_{obj}gV_{obj}$, and the net force on the object is $B - F_g = (\rho_{fluid} - \rho_{obj})gV_{obj}$. Hence, if the density of the object is less than the density of the fluid, the downward gravitational force is less than the buoyant force and the unsupported object accelerates upward (Active Fig. 15.10a). If the density of the object is greater than the density of the fluid, the upward buoyant force is less than the downward gravitational force and the unsupported object sinks (Active Fig. 15.10b). If the density of the submerged object equals the density of the fluid, the net force on the object is zero and the object remains in equilibrium. Therefore, the direction of motion of an object submerged in a fluid is determined *only* by the densities of the object and the fluid.

The same behavior is exhibited by an object immersed in a gas, such as the air in the atmosphere.[1] If the object is less dense than air, like a helium-filled balloon, the object floats upward. If it is denser, like a rock, it falls downward.

---

[1]The general behavior is the same, but the buoyant force varies with height in the atmosphere due to the variation in density of the air.

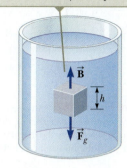

The buoyant force on the cube is the resultant of the forces exerted on its top and bottom faces by the liquid.

Figure 15.9 The external forces acting on an immersed cube are the gravitational force $\vec{F}_g$ and the buoyant force $\vec{B}$.

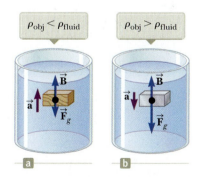

Active Figure 15.10 (a) A totally submerged object that is less dense than the fluid in which it is submerged experiences a net upward force and rises to the surface after it is released. (b) A totally submerged object that is denser than the fluid experiences a net downward force and sinks.

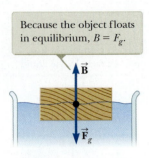

Because the object floats in equilibrium, $B = F_g$.

Active Figure 15.11 An object floating on the surface of a liquid experiences two forces, the gravitational force $\vec{F}_g$ and the buoyant force $\vec{B}$.

These hot-air balloons float on air because they are filled with air at high temperature. The buoyant force on a balloon due to the surrounding air is equal to the weight of the balloon, resulting in a net force of zero.

Case II: A Floating Object

Now consider an object of volume V_{obj} and density $\rho_{obj} < \rho_{fluid}$ in static equilibrium floating on the surface of a fluid, that is, an object that is only *partially* submerged (Active Fig. 15.11). In this case, the upward buoyant force is balanced by the downward gravitational force acting on the object. If V_{disp} is the volume of the fluid displaced by the object (this volume is the same as the volume of that part of the object beneath the surface of the fluid), the buoyant force has a magnitude $B = \rho_{fluid}gV_{disp}$. Because the weight of the object is $F_g = Mg = \rho_{obj}gV_{obj}$ and because $F_g = B$, we see that $\rho_{fluid}gV_{disp} = \rho_{obj}gV_{obj}$, or

$$\frac{V_{disp}}{V_{obj}} = \frac{\rho_{obj}}{\rho_{fluid}}$$ 15.6◀

Therefore, the fraction of the volume of the object under the liquid surface is equal to the ratio of the object density to the liquid density.

Let us consider examples of both cases. Under normal conditions, the average density of a fish in the opening photograph of this chapter is slightly greater than the density of water. That being the case, a fish would sink if it did not have some mechanism to counteract the net downward force. The fish does so by internally regulating the size of its swim bladder, a gas-filled cavity within the fish's body. Increasing its size increases the amount of water displaced, which increases the buoyant force. In this manner, fish are able to swim to various depths. Because the fish is totally submerged in the water, this example illustrates Case I.

As an example of Case II, imagine a large cargo ship. When the ship is at rest, the upward buoyant force from the water balances the weight so that the ship is in equilibrium. Only part of the volume of the ship is under water. If the ship is loaded with heavy cargo, it sinks deeper into the water. The increased weight of the ship due to the cargo is balanced by the extra buoyant force related to the extra volume of the ship that is now beneath the water surface.

QUICK QUIZ 15.3 An apple is held completely submerged just below the surface of a container of water. The apple is then moved to a deeper point in the water. Compared with the force needed to hold the apple just below the surface, what is the force needed to hold it at a deeper point? **(a)** larger **(b)** the same **(c)** smaller **(d)** impossible to determine

QUICK QUIZ 15.4 You are shipwrecked and floating in the middle of the ocean on a raft. Your cargo on the raft includes a treasure chest full of gold that you found before your ship sank and the raft is just barely afloat. To keep you floating as high as possible in the water, should you **(a)** leave the treasure chest on top of the raft, **(b)** secure the treasure chest to the underside of the raft, or **(c)** hang the treasure chest in the water with a rope attached to the raft? (Assume that throwing the treasure chest overboard is not an option you wish to consider!)

THINKING PHYSICS 15.3

A florist delivery person is delivering a flower basket to a home. The basket includes an attached helium-filled balloon, which suddenly comes loose from the basket and begins to accelerate upward toward the sky. Startled by the release of the balloon, the delivery person drops the flower basket. As the basket falls, the basket–Earth system experiences an increase in kinetic energy and a decrease in gravitational potential energy, consistent with conservation of mechanical energy. The balloon–Earth system, however, experiences an increase in *both* gravitational potential energy and kinetic energy. Is that consistent with the principle of conservation of mechanical energy? If not, from where is the extra energy coming?

Reasoning In the case of the system of the flower basket and the Earth, a good approximation to the motion of the basket can be made by ignoring the effects of the air. Therefore, the basket–Earth system can be analyzed with the isolated system model and mechanical energy is conserved. For the balloon–Earth system, we cannot ignore the effects of the air because it is the buoyant force of the air that causes the balloon to rise. Therefore, the balloon–Earth system is analyzed with the nonisolated system model. The buoyant force of the air does work across the boundary of the system, and that work results in an increase in both the kinetic and gravitational potential energies of the system. ◀

Example 15.3 | Eureka!

Archimedes supposedly was asked to determine whether a crown made for the king consisted of pure gold. According to legend, he solved this problem by weighing the crown first in air and then in water as shown in Figure 15.12. Suppose the scale read 7.84 N when the crown was in air and 6.84 N when it was in water. What should Archimedes have told the king?

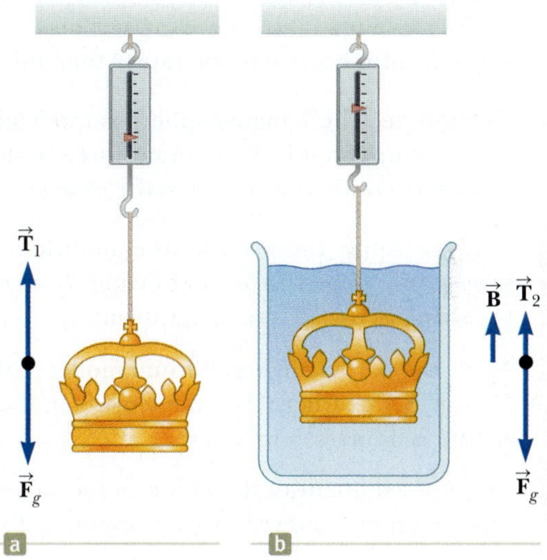

SOLUTION

Conceptualize Figure 15.12 helps us imagine what is happening in this example. Because of the buoyant force, the scale reading is smaller in Figure 15.12b than in Figure 15.12a.

Categorize This problem is an example of Case 1 discussed earlier because the crown is completely submerged. The scale reading is a measure of one of the forces on the crown, and the crown is stationary. Therefore, we can categorize the crown as a particle in equilibrium.

Figure 15.12 (Example 15.3) (a) When the crown is suspended in air, the scale reads its true weight because $T_1 = F_g$ (the buoyancy of air is negligible). (b) When the crown is immersed in water, the buoyant force $\vec{B}$ changes the scale reading to a lower value $T_2 = F_g - B$.

Analyze When the crown is suspended in air, the scale reads the true weight $T_1 = F_g$ (neglecting the small buoyant force due to the surrounding air). When the crown is immersed in water, the buoyant force $\vec{B}$ reduces the scale reading to an *apparent* weight of $T_2 = F_g - B$.

Apply the particle in equilibrium model to the crown in water:

$$\sum F = B + T_2 - F_g = 0$$

Solve for B and substitute the known values:

$$B = F_g - T_2 = 7.84 \text{ N} - 6.84 \text{ N} = 1.00 \text{ N}$$

Because this buoyant force is equal in magnitude to the weight of the displaced water, $B = \rho_w g V_{\text{disp}}$, where V_{disp} is the volume of the displaced water and ρ_w is its density. Also, the volume of the crown V_c is equal to the volume of the displaced water because the crown is completely submerged, so $B = \rho_w g V_c$.

Find the density of the crown from Equation 1.1:

$$\rho_c = \frac{m_c}{V_c} = \frac{m_c g}{V_c g} = \frac{m_c g}{(B/\rho_w)} = \frac{m_c g \rho_w}{B}$$

Substitute numerical values:

$$\rho_c = \frac{(7.84 \text{ N})(1\,000 \text{ kg/m}^3)}{1.00 \text{ N}} = 7.84 \times 10^3 \text{ kg/m}^3$$

Finalize From Table 15.1, we see that the density of gold is $19.3 \times 10^3 \text{ kg/m}^3$. Therefore, Archimedes should have reported that the king had been cheated. Either the crown was hollow, or it was not made of pure gold.

What If? Suppose the crown has the same weight but is indeed pure gold and not hollow. What would the scale reading be when the crown is immersed in water?

Answer Find the buoyant force on the crown:

$$B = \rho_w g V_w = \rho_w g V_c = \rho_w g \left(\frac{m_c}{\rho_c}\right) = \rho_w \left(\frac{m_c g}{\rho_c}\right)$$

Substitute numerical values:

$$B = (1.00 \times 10^3 \text{ kg/m}^3)\frac{7.84 \text{ N}}{19.3 \times 10^3 \text{ kg/m}^3} = 0.406 \text{ N}$$

Find the tension in the string hanging from the scale:

$$T_2 = F_g - B = 7.84 \text{ N} - 0.406 \text{ N} = 7.43 \text{ N}$$

Example 15.4 | **Changing String Vibration with Water**

One end of a horizontal string is attached to a vibrating blade, and the other end passes over a pulley as in Figure 15.13a. A sphere of mass 2.00 kg hangs on the end of the string. The string is vibrating in its second harmonic. A container of water is raised under the sphere so that the sphere is completely submerged. In this configuration, the string vibrates in its fifth harmonic as shown in Figure 15.13b. What is the radius of the sphere?

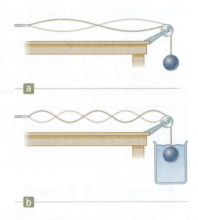

SOLUTION

Conceptualize Imagine what happens when the sphere is immersed in the water. The buoyant force acts upward on the sphere, reducing the tension in the string. The change in tension causes a change in the speed of waves on the string, which in turn causes a change in the wavelength. This altered wavelength results in the string vibrating in its fifth normal mode rather than the second.

Categorize The hanging sphere is modeled as a particle in equilibrium. One of the forces acting on it is the buoyant force from the water. We also apply the waves under boundary conditions model to the string.

Figure 15.13 (Example 15.4)
(a) When the sphere hangs in air, the string vibrates in its second harmonic. (b) When the sphere is immersed in water, the string vibrates in its fifth harmonic.

Analyze Apply the particle in equilibrium model to the sphere in Figure 15.13a, identifying T_1 as the tension in the string as the sphere hangs in air:

$$\sum F = T_1 - mg = 0$$

$$T_1 = mg$$

Apply the particle in equilibrium model to the sphere in Figure 15.13b, where T_2 is the tension in the string as the sphere is immersed in water:

$$T_2 + B - mg = 0$$

$$(1) \quad B = mg - T_2$$

The desired quantity, the radius of the sphere, will appear in the expression for the buoyant force B. Before proceeding in this direction, however, we must evaluate T_2 from the information about the standing wave.

Write the equation for the frequency of a standing wave on a string (Eq. 14.7) twice, once before the sphere is immersed and once after. Notice that the frequency f is the same in both cases because it is determined by the vibrating blade. In addition, the linear mass density μ and the length L of the vibrating portion of the string are the same in both cases. Divide the equations:

$$f = \frac{n_1}{2L}\sqrt{\frac{T_1}{\mu}} \quad \rightarrow \quad 1 = \frac{n_1}{n_2}\sqrt{\frac{T_1}{T_2}}$$

$$f = \frac{n_2}{2L}\sqrt{\frac{T_2}{\mu}}$$

Solve for T_2:

$$T_2 = \left(\frac{n_1}{n_2}\right)^2 T_1 = \left(\frac{n_1}{n_2}\right)^2 mg$$

Substitute this result into Equation (1):

$$(2) \quad B = mg - \left(\frac{n_1}{n_2}\right)^2 mg = mg\left[1 - \left(\frac{n_1}{n_2}\right)^2\right]$$

Using Equation 15.5, express the buoyant force in terms of the radius of the sphere:

$$B = \rho_{water} g V_{sphere} = \rho_{water} g \left(\tfrac{4}{3}\pi r^3\right)$$

Solve for the radius of the sphere and substitute from Equation (2):

$$r = \left(\frac{3B}{4\pi \rho_{water} g}\right)^{1/3} = \left\{\frac{3m}{4\pi \rho_{water}}\left[1 - \left(\frac{n_1}{n_2}\right)^2\right]\right\}^{1/3}$$

Substitute numerical values:

$$r = \left\{\frac{3(2.00 \text{ kg})}{4\pi(1\,000 \text{ kg/m}^3)}\left[1 - \left(\frac{2}{5}\right)^2\right]\right\}^{1/3}$$

$$= 0.073\,7 \text{ m} = \boxed{7.37 \text{ cm}}$$

Finalize Notice that only certain radii of the sphere will result in the string vibrating in a normal mode; the speed of waves on the string must be changed to a value such that the length of the string is an integer multiple of half wavelengths. This limitation is a feature of the *quantization* that was introduced in Chapters 11 and 14: the sphere radii that cause the string to vibrate in a normal mode are *quantized*.

15.5 | Fluid Dynamics

Thus far, our study of fluids has been restricted to fluids at rest, or **fluid statics.** We now turn our attention to **fluid dynamics,** the study of fluids in motion. Instead of trying to study the motion of each particle of the fluid as a function of time, we describe the properties of the fluid as a whole.

Flow Characteristics

When fluid is in motion, its flow is of one of two main types. The flow is said to be **steady,** or **laminar,** if each particle of the fluid follows a smooth path so that the paths of different particles never cross each other as in Figure 15.14. Therefore, in steady flow, the velocity of the fluid at any point remains constant in time.

Above a certain critical speed, fluid flow becomes **turbulent.** Turbulent flow is an irregular flow characterized by small, whirlpool-like regions as in Figure 15.15. As an example, the flow of water in a river becomes turbulent in regions where rocks and other obstructions are encountered, often forming "white-water" rapids.

The term **viscosity** is commonly used in fluid flow to characterize the degree of internal friction in the fluid. This internal friction, or viscous force, is associated with the resistance of two adjacent layers of the fluid against moving relative to each other. Because viscosity represents a nonconservative force, part of a fluid's kinetic energy is converted to internal energy when layers of fluid slide past one another. This conversion is similar to the mechanism by which an object sliding on a rough horizontal surface experiences a transformation of kinetic energy to internal energy.

Because the motion of a real fluid is very complex and not yet fully understood, we adopt a simplification model. As we shall see, many features of real fluids in motion can be understood by considering the behavior of an ideal fluid. In our simplification model, we make the following four assumptions:

1. *Nonviscous fluid.* In a nonviscous fluid, internal friction is ignored. An object moving through the fluid experiences no viscous force.
2. *Incompressible fluid.* The density of the fluid is assumed to remain constant regardless of the pressure in the fluid.
3. *Steady flow.* In steady flow, we assume that the velocity of the fluid at each point remains constant in time.
4. *Irrotational flow.* Fluid flow is irrotational if the fluid has no angular momentum about any point. If a small paddle wheel placed anywhere in the fluid does not rotate about the wheel's center of mass, the flow is irrotational. (If the wheel were to rotate, as it would if turbulence were present, the flow would be rotational.)

The first two assumptions in our simplification model are properties of our ideal fluid. The last two are descriptions of the way that the fluid flows.

Figure 15.14 An illustration of steady flow around an automobile in a test wind tunnel. The streamlines in the airflow are made visible by smoke particles.

Figure 15.15 Hot gases from a cigarette made visible by smoke particles. The smoke first moves in laminar flow at the bottom and then in turbulent flow above.

15.6 | Streamlines and the Continuity Equation for Fluids

If you are watering your garden and your garden hose is too short, you might do one of two things to help you reach the garden with the water (before you look for a longer hose!). You might attach a nozzle to the end of the hose, or, in the absence of a nozzle, you might place your thumb over the end of the hose, allowing the water to come out of a narrower opening. Why does either of these techniques cause the water to come out faster so that it can be projected over a longer range? We shall see the answer to this question in this section.

The path taken by a fluid particle under steady flow is called a **streamline.** The velocity of the particle is always tangent to the streamline as shown in Figure 15.16. A set of streamlines like the ones shown in Figure 15.16 form a *tube of flow.* Fluid

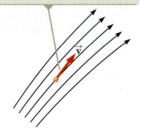

At each point along its path, the particle's velocity is tangent to the streamline.

$\vec{v}$

Figure 15.16 A particle in laminar flow follows a streamline.

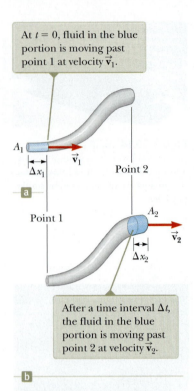

At $t = 0$, fluid in the blue portion is moving past point 1 at velocity $\vec{v}_1$.

Point 2

a

Point 1

After a time interval Δt, the fluid in the blue portion is moving past point 2 at velocity $\vec{v}_2$.

b

Figure 15.17 A fluid moving with steady flow through a pipe of varying cross-sectional area. (a) At $t = 0$, the small blue-colored portion of the fluid at the left is moving through area A_1. (b) After a time interval Δt, the blue-colored portion is that fluid that has moved through area A_2.

Figure 15.18 The speed of water spraying from the end of a garden hose increases as the size of the opening is decreased with the thumb.

particles cannot flow into or out of the sides of this tube; if they could, the streamlines would cross one another.

Consider ideal fluid flow through a pipe of nonuniform size as illustrated in Figure 15.17. Let's focus our attention on a segment of fluid in the pipe. Figure 15.17a shows the segment at time $t = 0$ consisting of the gray portion between point 1 and point 2 and the short blue portion to the left of point 1. At this time, the fluid in the short blue portion is flowing through a cross section of area A_1 at speed v_1. During the time interval Δt, the small length Δx_1 of fluid in the blue portion moves past point 1. During the same time, fluid moves past point 2 at the other end of the pipe. Figure 15.17b shows the situation at the end of the time interval Δt. The blue portion at the right end represents the fluid that has moved past point 2 through an area A_2 at a speed v_2.

The mass of fluid contained in the blue portion in Figure 15.17a is given by $m_1 = \rho A_1 \Delta x_1 = \rho A_1 v_1 \Delta t$, where ρ is the (unchanging) density of the ideal fluid. Similarly, the fluid in the blue portion in Figure 15.17b has a mass $m_2 = \rho A_2 \Delta x_2 = \rho A_2 v_2 \Delta t$. Because the fluid is incompressible and the flow is steady, however, the mass of fluid that passes point 1 in a time interval Δt must equal the mass that passes point 2 in the same time interval. That is, $m_1 = m_2$ or $\rho A_1 v_1 \Delta t = \rho A_2 v_2 \Delta t$, which means that

$$A_1 v_1 = A_2 v_2 = \text{constant} \qquad \text{15.7} \blacktriangleleft$$

This expression is called the **continuity equation for fluids.** It states that the product of the area and the fluid speed at all points along a pipe is constant for an incompressible fluid. Equation 15.7 shows that the speed is high where the tube is constricted (small A) and low where the tube is wide (large A). The product Av, which has the dimensions of volume per unit time, is called either the *volume flux* or the *flow rate*. The condition $Av = $ constant is equivalent to the statement that the volume of fluid that enters one end of a tube in a given time interval equals the volume leaving the other end of the tube in the same time interval if no leaks are present.

You demonstrate the continuity equation each time you water your garden with your thumb over the end of a garden hose as in Figure 15.18. By partially blocking the opening with your thumb, you reduce the cross-sectional area through which the water passes. As a result, the speed of the water increases as it exits the hose, and the water can be sprayed over a long distance.

QUICK QUIZ 15.5 You tape two different soda straws together end to end to make a longer straw with no leaks. The two straws have radii of 3 mm and 5 mm. You drink a soda through your combination straw. In which straw is the speed of the liquid higher? **(a)** It is higher in whichever one is nearest your mouth. **(b)** It is higher in the one of radius 3 mm. **(c)** It is higher in the one of radius 5 mm. **(d)** Neither, because the speed is the same in both straws.

Example 15.5 | **Watering a Garden**

A gardener uses a water hose 2.50 cm in diameter to fill a 30.0-L bucket. The gardener notes that it takes 1.00 min to fill the bucket. A nozzle with an opening of cross-sectional area 0.500 cm² is then attached to the hose. The nozzle is held so that water is projected horizontally from a point 1.00 m above the ground. Over what horizontal distance can the water be projected?

SOLUTION

Conceptualize Imagine any past experience you have with projecting water from a horizontal hose or a pipe. The faster the water is traveling as it leaves the hose, the farther it will land on the ground from the end of the hose.

Categorize Once the water leaves the hose, it is in free fall. Therefore, we categorize a given element of the water as a projectile. The element is modeled as a particle under constant acceleration (due to gravity) in the vertical direction and

15.5 *cont.*

a particle under constant velocity in the horizontal direction. The horizontal distance over which the element is projected depends on the speed with which it is projected. This example involves a change in area for the pipe, so we also categorize it as one in which we use the continuity equation for fluids.

· ·

Analyze We first find the speed of the water in the hose from the bucket-filling information.

Find the cross-sectional area of the hose:

$$A = \pi r^2 = \pi \frac{d^2}{4} = \pi \left[\frac{(2.50 \text{ cm})^2}{4} \right] = 4.91 \text{ cm}^2$$

Evaluate the volume flow rate:

$$Av_1 = 30.0 \text{ L/min} = \frac{30.0 \times 10^3 \text{ cm}^3}{60.0 \text{ s}} = 500 \text{ cm}^3/\text{s}$$

Solve for the speed of the water in the hose:

$$v_1 = \frac{500 \text{ cm}^3/\text{s}}{A} = \frac{500 \text{ cm}^3/\text{s}}{4.91 \text{ cm}^2} = 102 \text{ cm/s} = 1.02 \text{ m/s}$$

We have labeled this speed v_1 because we identify point 1 within the hose. We identify point 2 in the air just outside the nozzle. We must find the speed $v_2 = v_{xi}$ with which the water exits the nozzle. The subscript i anticipates that it will be the *initial* velocity component of the water projected from the hose, and the subscript x indicates that the initial velocity vector of the projected water is horizontal.

Solve the continuity equation for fluids for v_2:

$$v_2 = v_{xi} = \frac{A_1}{A_2} v_1$$

Substitute numerical values:

$$v_{xi} = \frac{4.91 \text{ cm}^2}{0.500 \text{ cm}^2} (1.02 \text{ m/s}) = 10.0 \text{ m/s}$$

We now shift our thinking away from fluids and to projectile motion. In the vertical direction, an element of the water starts from rest and falls through a vertical distance of 1.00 m.

Write Equation 2.13 for the vertical position of an element of water, modeled as a particle under constant acceleration:

$$y_f = y_i + v_{yi}t - \tfrac{1}{2}gt^2$$

Substitute numerical values:

$$-1.00 \text{ m} = 0 + 0 - \tfrac{1}{2}(9.80 \text{ m/s}^2) t^2$$

Solve for the time at which the element of water lands on the ground:

$$t = \sqrt{\frac{2(1.00 \text{ m})}{9.80 \text{ m/s}^2}} = 0.452 \text{ s}$$

Use Equation 2.5 to find the horizontal position of the element at this time, modeled as a particle under constant velocity:

$$x_f = x_i + v_{xi}t = 0 + (10.0 \text{ m/s})(0.452 \text{ s}) = \boxed{4.52 \text{ m}}$$

· ·

Finalize The time interval for the element of water to fall to the ground is unchanged if the projection speed is changed because the projection is horizontal. Increasing the projection speed results in the water hitting the ground farther from the end of the hose, but requires the same time interval to strike the ground.

15.7 | Bernoulli's Equation

You have probably experienced driving on a highway and having a large truck pass you at high speed. In this situation, you may have had the frightening feeling that your car was being pulled in toward the truck as it passed. We will investigate the origin of this effect in this section.

As a fluid moves through a region where its speed or elevation above the Earth's surface changes, the pressure in the fluid varies with these changes. The relationship between fluid speed, pressure, and elevation was first derived in 1738 by Swiss physicist Daniel Bernoulli. Consider the flow of a segment of an ideal fluid through a nonuniform pipe in a time interval Δt as illustrated in Figure 15.19.

Daniel Bernoulli
Swiss physicist (1700–1782)
Bernoulli made important discoveries in fluid dynamics. Bernoulli's most famous work, *Hydrodynamica,* was published in 1738; it is both a theoretical and a practical study of equilibrium, pressure, and speed in fluids. He showed that as the speed of a fluid increases, its pressure decreases. Referred to as "Bernoulli's principle," Bernoulli's work is used to produce a partial vacuum in chemical laboratories by connecting a vessel to a tube through which water is running rapidly.

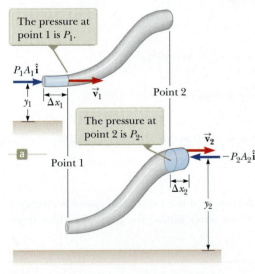

Figure 15.19 A fluid in laminar flow through a pipe. (a) A segment of the fluid at time $t = 0$. A small portion of the blue-colored fluid is at height y_1 above a reference position. (b) After a time interval Δt, the entire segment has moved to the right. The blue-colored portion of the fluid is at height y_2.

This figure is very similar to Figure 15.17, which we used to develop the continuity equation. We have added two features: the forces on the outer ends of the blue portions of fluid and the heights of these portions above the reference position $y = 0$.

The force exerted by the fluid to the left of the blue portion in Figure 15.19a has a magnitude P_1A_1. The work done by this force on the segment in a time interval Δt is $W_1 = F_1 \, \Delta x_1 = P_1 A_1 \, \Delta x_1 = P_1 V$, where V is the volume of the blue portion of fluid passing point 1 in Figure 15.19a. In a similar manner, the work done by the fluid to the right of the segment in the same time interval Δt is $W_2 = -P_2 A_2 \, \Delta x_2 = -P_2 V$, where V is the volume of the blue portion of fluid passing point 2 in Figure 15.19b. (The volumes of the blue portions of fluid in Figures 15.19a and 15.19b are equal because the fluid is incompressible.) This work is negative because the force on the segment of fluid is to the left and the displacement of the point of application of the force is to the right. Therefore, the net work done on the segment by these forces in the time interval Δt is

$$W = (P_1 - P_2) V$$

Part of this work goes into changing the kinetic energy of the segment of fluid, and part goes into changing the gravitational potential energy of the segment–Earth system. Because we are assuming streamline flow, the kinetic energy K_{gray} of the gray portion of the segment is the same in both parts of Figure 15.19. Therefore, the change in the kinetic energy of the segment of fluid is

$$\Delta K = \left(\tfrac{1}{2}mv_2^2 + K_{gray}\right) - \left(\tfrac{1}{2}mv_1^2 + K_{gray}\right) = \tfrac{1}{2}mv_2^2 - \tfrac{1}{2}mv_1^2$$

where m is the mass of the blue portions of fluid in both parts of Figure 15.19. (Because the volumes of both portions are the same, they also have the same mass.)

Considering the gravitational potential energy of the segment–Earth system, once again there is no change during the time interval for the gravitational potential energy U_{gray} associated with the gray portion of the fluid. Consequently, the change in gravitational potential energy of the system is

$$\Delta U = (mgy_2 + U_{gray}) - (mgy_1 + U_{gray}) = mgy_2 - mgy_1$$

From Equation 7.2, the total work done on the segment–Earth system by the fluid outside the segment is equal to the change in mechanical energy of the system: $W = \Delta K + \Delta U$. Substituting for each of these terms gives

$$(P_1 - P_2) V = \tfrac{1}{2}mv_2^2 - \tfrac{1}{2}mv_1^2 + mgy_2 - mgy_1 \qquad \textbf{15.8} \blacktriangleleft$$

If we divide each term by the portion volume V and recall that $\rho = m/V$, this expression reduces to

$$P_1 - P_2 = \tfrac{1}{2}\rho v_2^2 - \tfrac{1}{2}\rho v_1^2 + \rho g y_2 - \rho g y_1$$

Rearranging terms, we obtain

$$P_1 + \tfrac{1}{2}\rho v_1^2 + \rho g y_1 = P_2 + \tfrac{1}{2}\rho v_2^2 + \rho g y_2 \qquad \textbf{15.9} \blacktriangleleft$$

which is **Bernoulli's equation** applied to an ideal fluid. It is often expressed as

$$P + \tfrac{1}{2}\rho v^2 + \rho g y = \text{constant} \qquad \textbf{15.10} \blacktriangleleft$$

Bernoulli's equation says that the sum of the pressure P, the kinetic energy per unit volume $\tfrac{1}{2}\rho v^2$, and gravitational potential energy per unit volume $\rho g y$ has the same value at all points along a streamline.

When the fluid is at rest, $v_1 = v_2 = 0$ and Equation 15.9 becomes

$$P_1 - P_2 = \rho g (y_2 - y_1) = \rho g h$$

which agrees with Equation 15.4.

Although Equation 15.10 was derived for an incompressible fluid, the general behavior of pressure with speed is true even for gases: as the speed increases, the pressure decreases. This *Bernoulli effect* explains the experience with the truck on the highway at the opening of this section. As air passes between your car and the truck, it must pass through a relatively narrow channel. According to the continuity equation, the speed of the air is higher. According to the Bernoulli effect, this higher-speed air exerts less pressure on your car than the slower-moving air on the other side of your car. Therefore, there is a net force pushing you toward the truck.

QUICK QUIZ 15.6 You observe two helium balloons floating next to each other at the ends of strings secured to a table. The facing surfaces of the balloons are separated by 1–2 cm. You blow through the small space between the balloons. What happens to the balloons? **(a)** They move toward each other. **(b)** They move away from each other. **(c)** They are unaffected.

Example 15.6 | Sinking the Cruise Ship

A scuba diver is hunting for fish with a spear gun. He accidentally fires the gun so that a spear punctures the side of a cruise ship. The hole is located at a depth of 10.0 m below the water surface. With what speed does the water enter the cruise ship through the hole?

SOLUTION

Conceptualize Imagine the water streaming in through the hole. The deeper the hole is below the surface, the higher the pressure, and therefore, the higher the speed of the incoming water.

Categorize We will evaluate the result directly from Bernoulli's equation, so this is a substitution problem.

We identify point 2 as the water surface outside the ship, which we will assign as $y = 0$. At this point, the water is static, so $v_2 = 0$. We identify point 1 as a point just inside the hole in the interior of the ship because that is the point at which we wish to evaluate the speed of the water. This point is at a depth $y = -h = -10.0$ m below the water surface. We use Bernoulli's equation to compare these two points. At both points, the water is open to atmospheric pressure, so $P_1 = P_2 = P_0$.

Based on this argument, solve Bernoulli's equation for the speed v_1 of the water entering the ship and evaluate:

$$P_0 + \tfrac{1}{2}\rho(0)^2 + \rho g(0) = P_0 + \tfrac{1}{2}\rho v_1^2 + \rho g(-h) \rightarrow$$
$$v_1 = \sqrt{2gh} = \sqrt{2(9.80 \text{ m/s}^2)(10.0 \text{ m})} = 14 \text{ m/s}$$

Example 15.7 | Torricelli's Law

An enclosed tank containing a liquid of density ρ has a hole in its side at a distance y_1 from the tank's bottom (Fig. 15.20). The hole is open to the atmosphere, and its diameter is much smaller than the diameter of the tank. The air above the liquid is maintained at a pressure P. Determine the speed of the liquid as it leaves the hole when the liquid's level is a distance h above the hole.

SOLUTION

Conceptualize Imagine that the tank is a fire extinguisher. When the hole is opened, liquid leaves the hole with a certain speed. If the pressure P at the top of the liquid is increased, the liquid leaves with a higher speed. If the pressure P falls too low, the liquid leaves with a low speed and the extinguisher must be replaced.

Categorize Looking at Figure 15.20, we know the pressure at two points and the velocity at one of those points. We wish to find the velocity at the second point. Therefore, we can categorize this example as one in which we can apply Bernoulli's equation.

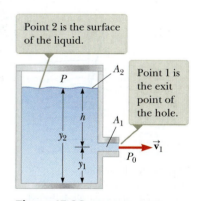

Figure 15.20 (Example 15.7) A liquid leaves a hole in a tank at speed v_1.

Analyze Because $A_2 \gg A_1$, the liquid is approximately at rest at the top of the tank, where the pressure is P. At the hole, P_1 is equal to atmospheric pressure P_0.

continued

15.7 *cont.*

Apply Bernoulli's equation between points 1 and 2 and solve for v_1, noting that $y_2 - y_1 = h$:

$$P_0 + \tfrac{1}{2}\rho v_1^2 + \rho g y_1 = P + \rho g y_2 \quad \rightarrow \quad v_1 = \sqrt{\dfrac{2(P - P_0)}{\rho} + 2gh}$$

Finalize When P is much greater than P_0 (so that the term $2gh$ can be neglected), the exit speed of the water is mainly a function of P. If the tank is open to the atmosphere, then $P = P_0$ and $v_1 = \sqrt{2gh}$, as in Example 15.6. In other words, for an open tank, the speed of the liquid leaving a hole a distance h below the surface is equal to that acquired by an object falling freely through a vertical distance h. This phenomenon is known as **Torricelli's law.**

What If? What if the position of the hole in Figure 15.20 could be adjusted vertically? If the tank is open to the atmosphere and sitting on a table, what position of the hole would cause the water to land on the table at the farthest distance from the tank?

Answer Model a parcel of water exiting the hole as a projectile. Find the time at which the parcel strikes the table from a hole at an arbitrary position y_1:

$$y_f = y_i + v_{yi}t - \tfrac{1}{2}gt^2 \quad \rightarrow \quad 0 = y_1 + 0 - \tfrac{1}{2}gt^2$$

$$t = \sqrt{\dfrac{2y_1}{g}}$$

Find the horizontal position of the parcel at the time it strikes the table:

$$x_f = x_i + v_{xi}t = 0 + \sqrt{2g(y_2 - y_1)}\,\sqrt{\dfrac{2y_1}{g}} = 2\sqrt{(y_2 y_1 - y_1^2)}$$

Maximize the horizontal position by taking the derivative of x_f with respect to y_1 (because y_1, the height of the hole, is the variable that can be adjusted) and setting it equal to zero. Solve for y_1:

$$\dfrac{dx_f}{dy_1} = \tfrac{1}{2}(2)(y_2 y_1 - y_1^2)^{-1/2}(y_2 - 2y_1) = 0 \quad \rightarrow \quad y_1 = \tfrac{1}{2}y_2$$

Therefore, to maximize the horizontal distance, the hole should be halfway between the bottom of the tank and the upper surface of the water. Below this location, the water is projected at a higher speed but falls for a short time interval, reducing the horizontal range. Above this point, the water is in the air for a longer time interval but is projected with a smaller horizontal speed.

❰15.8 | Other Applications of Fluid Dynamics

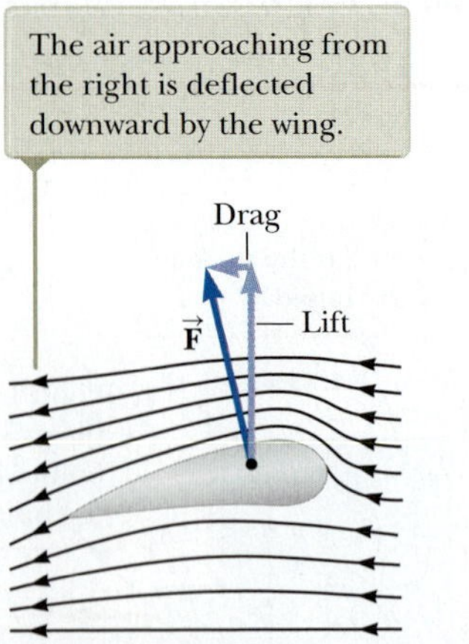

Consider the streamlines that flow around an airplane wing as shown in Figure 15.21. Let us assume that the airstream approaches the wing horizontally from the right. The tilt of the wing causes the airstream to be deflected downward. Because the airstream is deflected by the wing, the wing must exert a force on the airstream. According to Newton's third law, the airstream must exert an equal and opposite force $\vec{\mathbf{F}}$ on the wing. This force has a vertical component called the **lift** (or aerodynamic lift) and a horizontal component called **drag.** The lift depends on several factors, such as the speed of the airplane, the area of the wing, its curvature, and the angle between the wing and the horizontal. As this angle increases, turbulent flow can set in above the wing to reduce the lift.

In general, an object experiences lift by any effect that causes the fluid to change its direction as it flows past the object. Some factors that influence lift are the shape of the object, its orientation with respect to the fluid flow, spinning motion (for example, a curve ball thrown in a baseball game due to the spinning of the baseball), and the texture of the object's surface.

A number of devices operate in a manner similar to the *atomizer* in Figure 15.22. A stream of air passing over an open tube reduces the pressure above the tube. This reduction in pressure causes the liquid to rise into the air stream. The liquid is then dispersed into a fine spray of droplets. This type of system is used in perfume bottles and paint sprayers.

The air approaching from the right is deflected downward by the wing.

Drag

$\vec{\mathbf{F}}$ — Lift

Figure 15.21 Streamline flow around a moving airplane wing. By Newton's third law, the air deflected by the wing results in an upward force on the wing from the air: lift. Because of air resistance, there is also a force opposite the velocity of the wing: drag.

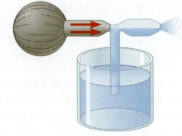

Figure 15.22 A stream of air passing over a tube dipped into a liquid causes the liquid to rise in the tube.

15.9 | Context Connection: Turbulent Flow of Blood

Fluids play a dominant role in the transport of nutrients and other materials in the human body. The circulatory system transports nutrients to cells and removes waste products, the respiratory system provides the oxygen needed for cells to consume nutrients and removes the carbon dioxide produced in these reactions, and the gastrointestinal system takes in food and removes waste from the body. Each of these systems represents a complex fluid dynamic system with unique properties.

With each heartbeat, the blood in our bodies is moved along arteries, veins, and the vast network of capillaries that comprise the blood circulation system. Blood flow in straight, healthy portions of arteries would be simple to analyze if the flow could be modeled as being *laminar*, as discussed in Section 15.5. However, such a simplification model is incorrect for at least two reasons. First, the flow of blood is unsteady because the beating heart causes a time-varying pressure differential in the arteries. Second, turbulent eddies are created as the flowing blood interacts with the walls of the arteries and smaller vessels.

Each liter of blood contains 4×10^{12} to 6×10^{12} red blood cells having diameters ranging from 6 mm to 8 mm. These cells are sufficiently large so that each side of the pancake-shaped cells experiences a different force from the surrounding fluid. The resulting torque on the cells causes them to spin, creating turbulence in the fluid. As the velocity of the blood increases, the gradients in the flow velocity between the walls and the center of the vessel become larger. These larger gradients cause the red blood cells to spin faster, stirring the blood and causing it to become even more turbulent.

The flowing blood also interacts chemically with the cells lining the walls of blood vessels. The interior surfaces of the heart and blood vessels are covered with a one-cell-thick layer of *endothelial cells,* which reduce friction between the cells and the vessel walls. These cells also have a significant role in the extraction of minerals from the blood, the passage of white blood cells into and out of the bloodstream, and in the formation of blood clots. In laminar flow, endothelial cells are football-shaped and align themselves along the direction of blood flow, creating a protective layer over the vessel walls. Blood vessels continually expand and contract because of environmental factors such as changes in temperature of the surroundings. Inherited blood vessel defects, disturbances in the nerves controlling vessel contraction, injuries, and drugs can cause blood vessels to contract. According to the continuity equation for fluids (Eq. 15.7), this will result in an increased flow velocity of the blood. When the blood flows faster and becomes more turbulent, endothelial cells respond by becoming rounded in shape, and dividing much faster than normal. The division of endothelial cells creates gaps in the coating of the blood vessel, allowing blood platelets and cholesterol-carrying lipoproteins to attach themselves to the vessel wall, and plaque begins to form in the blood vessel. The smooth walls of the blood vessel become rough, further disrupting the blood flow, which in turn affects endothelial cells downstream from the rough patches, creating even more plaque. As plaque builds up in the blood vessel, the flow channel further constricts, increasing the flow velocity and affecting large numbers of endothelial cells (Fig. 15.23).

The gradual buildup of plaque, called *atherosclerosis,* can become catastrophic when the plaque becomes unstable and ruptures. In this case, the blood is exposed to the collagen in the tissue cap of the plaque buildup. This exposure causes the blood to clot at the point of rupture, forming what is called a *thrombus.* The thrombus may continue to grow until it completely blocks the blood vessel. On the other hand, the thrombus may break loose from the site of the plaque and flow with the blood until it blocks a

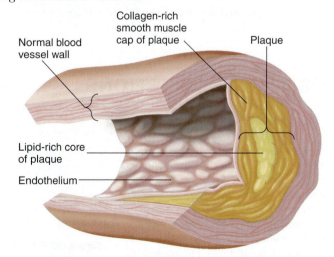

Figure 15.23 The left side of this cutaway section of a coronary vessel is normal. At the right side, atherosclerosis has led to the development of a plaque, bulging into the interior of the vessel and affecting the flow of blood. (from Sherwood, *Fundamentals of Human Physiology,* 4th ed., 2012, Brooks/ Cole, top half of Fig. 9–24, p. 253, © 2012 Brooks/Cole, a part of Cengage Learning, Inc. Reproduced by permission. www.cengage.com/permissions.)

smaller vessel somewhere downstream. Blocked blood vessels in the arms, legs, or the pelvis can result in numbness, pain, and the onset of infections such as *gangrene*. A severe blockage in a coronary blood vessel, however, can lead to a heart attack, and a blockage in a carotid artery can lead to stroke. The severity of damage to cardiac muscle during a heart attack depends on the location of the blockage. The left coronary artery (see Fig. 3 in the Context Introduction) supplies blood to 85% of the cardiac tissue, so that a blockage in this vessel at a point high on the heart near the pulmonary veins could cause great damage.

In this Context Connection, we investigated the role of turbulent flow and used the continuity equation for fluids to clarify the process of cardiovascular plaque buildup. In the Context Conclusion, we will explore an application of Bernoulli's principle to the diagnosis and prevention of cardiovascular disease and heart attacks.

SUMMARY

The **pressure** P in a fluid is the force per unit area that the fluid exerts on a surface:

$$P \equiv \frac{F}{A} \qquad \text{15.1}\blacktriangleleft$$

In the SI system, pressure has units of newtons per square meter, and $1 \text{ N/m}^2 = 1$ pascal (Pa).

The pressure in a liquid varies with depth h according to the expression

$$P = P_0 + \rho g h \qquad \text{15.4}\blacktriangleleft$$

where P_0 is the pressure at the surface of the liquid and ρ is the density of the liquid, assumed uniform.

Pascal's law states that when a change in pressure is applied to a fluid, the change in pressure is transmitted undiminished to every point in the fluid and to every point on the walls of the container.

When an object is partially or fully submerged in a fluid, the fluid exerts an upward force on the object called the **buoyant force**. According to **Archimedes's principle,** the buoyant force is equal to the weight of the fluid displaced by the object:

$$B = \rho_{\text{fluid}} g V_{\text{disp}} \qquad \text{15.5}\blacktriangleleft$$

Various aspects of fluid dynamics can be understood by adopting a simplification model in which the fluid is nonviscous and incompressible and the fluid motion is a steady flow with no turbulence.

Using this model, two important results regarding fluid flow through a pipe of nonuniform size can be obtained:

1. The flow rate through the pipe is a constant, which is equivalent to stating that the product of the cross-sectional area A and the speed v at any point is a constant. This behavior is described by the **continuity equation for fluids:**

$$A_1 v_1 = A_2 v_2 = \text{constant} \qquad \text{15.7}\blacktriangleleft$$

2. The sum of the pressure, kinetic energy per unit volume, and gravitational potential energy per unit volume has the same value at all points along a streamline. This behavior is described by **Bernoulli's equation:**

$$P_1 + \tfrac{1}{2}\rho v_1^2 + \rho g y_1 = P_2 + \tfrac{1}{2}\rho v_2^2 + \rho g y_2 \qquad \text{15.9}\blacktriangleleft$$

OBJECTIVE QUESTIONS

☐ denotes answer available in *Student Solutions Manual/Study Guide*

1. A wooden block floats in water, and a steel object is attached to the bottom of the block by a string as in Figure OQ15.1. If the block remains floating, which of the following statements are valid? (Choose all correct statements.) (a) The buoyant force on the steel object is equal to its weight. (b) The buoyant force on the block is equal to its weight. (c) The tension in the string is equal to the weight of the steel object. (d) The tension in the string is less than the weight of the steel object. (e) The buoyant force on the block is equal to the volume of water it displaces.

Figure OQ15.1

2. A beach ball filled with air is pushed about 1 m below the surface of a swimming pool and released from rest. Which of the following statements are valid, assuming the size of the ball remains the same? (Choose all correct statements.) (a) As the ball rises in the pool, the buoyant force on it increases. (b) When the ball is released, the buoyant force exceeds the gravitational force, and the ball accelerates upward. (c) The buoyant force on the ball decreases as the ball approaches the surface of the pool. (d) The buoyant force on the ball equals its weight and remains constant as the ball rises. (e) The buoyant force on the ball while it is submerged is approximately equal to the weight of a volume of water that could fill the ball.

3. Figure OQ15.3 shows aerial views from directly above two dams. Both dams are equally wide (the vertical dimension in the diagram) and equally high (into the page in the diagram). The dam on the left holds back a very large lake, and the dam on the right holds back a narrow river. Which dam has to be built more strongly? (a) the dam on the left (b) the dam on the right (c) both the same (d) cannot be predicted

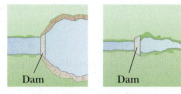

Figure OQ15.3

4. A boat develops a leak and, after its passengers are rescued, eventually sinks to the bottom of a lake. When the boat is at the bottom, what is the force of the lake bottom on the boat? (a) greater than the weight of the boat (b) equal to the weight of the boat (c) less than the weight of the boat (d) equal to the weight of the displaced water (e) equal to the buoyant force on the boat

5. A solid iron sphere and a solid lead sphere of the same size are each suspended by strings and are submerged in a tank of water. (Note that the density of lead is greater than that of iron.) Which of the following statements are valid? (Choose all correct statements.) (a) The buoyant force on each is the same. (b) The buoyant force on the lead sphere is greater than the buoyant force on the iron sphere because lead has the greater density. (c) The tension in the string supporting the lead sphere is greater than the tension in the string supporting the iron sphere. (d) The buoyant force on the iron sphere is greater than the buoyant force on the lead sphere because lead displaces more water. (e) None of those statements is true.

6. Rank the buoyant forces exerted on the following five objects of equal volume from the largest to the smallest. Assume the objects have been dropped into a swimming pool and allowed to come to mechanical equilibrium. If any buoyant forces are equal, state that in your ranking. (a) a block of solid oak (b) an aluminum block (c) a beach ball made of thin plastic and inflated with air (d) an iron block (e) a thin-walled, sealed bottle of water.

7. A person in a boat floating in a small pond throws an anchor overboard. What happens to the level of the pond? (a) It rises. (b) It falls. (c) It remains the same.

8. Three vessels of different shapes are filled to the same level with water as in Figure OQ15.8. The area of the base is the same for all three vessels. Which of the following statements are valid? (Choose all correct statements.) (a) The pressure at the top surface of vessel A is greatest because it has the largest surface area. (b) The pressure at the bottom of vessel A is greatest because it contains the most water. (c) The pressure at the bottom of each vessel is the same. (d) The force on the bottom of each vessel is not the same. (e) At a given depth below the surface of each vessel, the pressure on the side of vessel A is greatest because of its slope.

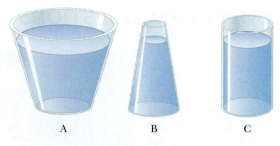

A   B   C

Figure OQ15.8

9. An ideal fluid flows through a horizontal pipe whose diameter varies along its length. Measurements would indicate that the sum of the kinetic energy per unit volume and pressure at different sections of the pipe would (a) decrease as the pipe diameter increases, (b) increase as the pipe diameter increases, (c) increase as the pipe diameter decreases, (d) decrease as the pipe diameter decreases, or (e) remain the same as the pipe diameter changes.

10. A beach ball is made of thin plastic. It has been inflated with air, but the plastic is not stretched. By swimming with fins on, you manage to take the ball from the surface of a pool to the bottom. Once the ball is completely submerged, what happens to the buoyant force exerted on the beach ball as you take it deeper? (a) It increases. (b) It remains constant. (c) It decreases. (d) It is impossible to determine.

11. One of the predicted problems due to global warming is that ice in the polar ice caps will melt and raise sea levels everywhere in the world. Is that more of a worry for ice (a) at the north pole, where most of the ice floats on water; (b) at the south pole, where most of the ice sits on land; (c) both at the north and south pole equally; or (d) at neither pole?

12. A small piece of steel is tied to a block of wood. When the wood is placed in a tub of water with the steel on top, half of the block is submerged. Now the block is inverted so that the steel is under water. (i) Does the amount of the block submerged (a) increase, (b) decrease, or (c) remain the same? (ii) What happens to the water level in the tub when the block is inverted? (a) It rises. (b) It falls. (c) It remains the same.

13. A piece of unpainted porous wood barely floats in an open container partly filled with water. The container is then sealed and pressurized above atmospheric pressure. What happens to the wood? (a) It rises in the water. (b) It sinks lower in the water. (c) It remains at the same level.

14. A glass of water contains floating ice cubes. When the ice melts, does the water level in the glass (a) go up, (b) go down, or (c) remain the same?

15. A water supply maintains a constant rate of flow for water in a hose. You want to change the opening of the nozzle so that water leaving the nozzle will reach a height that is four times the current maximum height the water reaches with the nozzle vertical. To do so, should you (a) decrease the area of the opening by a factor of 16, (b) decrease the area by a factor of 8, (c) decrease the area by a factor of 4, (d) decrease the area by a factor of 2, or (e) give up because it cannot be done?

CONCEPTUAL QUESTIONS

1. A typical silo on a farm has many metal bands wrapped around its perimeter for support as shown in Figure CQ15.1. Why is the spacing between successive bands smaller for the lower portions of the silo on the left, and why are double bands used at lower portions of the silo on the right?

Figure CQ15.1

2. **BIO** Because atmospheric pressure is about 10^5 N/m^2 and the area of a person's chest is about 0.13 m^2, the force of the atmosphere on one's chest is around 13 000 N. In view of this enormous force, why don't our bodies collapse?

3. Two thin-walled drinking glasses having equal base areas but different shapes, with very different cross-sectional areas above the base, are filled to the same level with water. According to the expression $P = P_0 + \rho gh$, the pressure is the same at the bottom of both glasses. In view of this equality, why does one weigh more than the other?

4. Why do airplane pilots prefer to take off with the airplane facing into the wind?

5. Prairie dogs ventilate their burrows by building a mound around one entrance, which is open to a stream of air when wind blows from any direction. A second entrance at ground level is open to almost stagnant air. How does this construction create an airflow through the burrow?

6. A fish rests on the bottom of a bucket of water while the bucket is being weighed on a scale. When the fish begins to swim around, does the scale reading change? Explain your answer.

7. When an object is immersed in a liquid at rest, why is the net force on the object in the horizontal direction equal to zero?

8. In Figure CQ15.8, an airstream moves from right to left through a tube that is constricted at the middle. Three table-tennis balls are levitated in equilibrium above the vertical columns through which the air escapes. (a) Why is the ball at the right higher than the one in the middle? (b) Why is the ball at the left lower than the ball at the right even though the horizontal tube has the same dimensions at these two points?

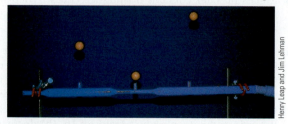

Figure CQ15.8

9. You are a passenger on a spacecraft. For your survival and comfort, the interior contains air just like that at the surface of the Earth. The craft is coasting through a very empty region of space. That is, a nearly perfect vacuum exists just outside the wall. Suddenly, a meteoroid pokes a hole, about the size of a large coin, right through the wall next to your seat. (a) What happens? (b) Is there anything you can or should do about it?

10. Does a ship float higher in the water of an inland lake or in the ocean? Why?

11. A water tower is a common sight in many communities. Figure CQ15.11 shows a collection of colorful water towers in Kuwait City, Kuwait. Notice that the large weight of the water results in the center of mass of the system being high above the ground. Why is it desirable for a water tower to have this highly unstable shape rather than being shaped as a tall cylinder?

Figure CQ15.11

12. If the airstream from a hair dryer is directed over a table-tennis ball, the ball can be levitated. Explain.

13. (a) Is the buoyant force a conservative force? (b) Is a potential energy associated with the buoyant force? (c) Explain your answers to parts (a) and (b).

14. An empty metal soap dish barely floats in water. A bar of Ivory soap floats in water. When the soap is stuck in the soap dish, the combination sinks. Explain why.

15. If you release a ball while inside a freely falling elevator, the ball remains in front of you rather than falling to the floor because the ball, the elevator, and you all experience the same downward gravitational acceleration. What happens if you repeat this experiment with a helium-filled balloon?

16. How would you determine the density of an irregularly shaped rock?

17. The water supply for a city is often provided from reservoirs built on high ground. Water flows from the reservoir, through pipes, and into your home when you turn the tap on your faucet. Why does water flow more rapidly out of a faucet on the first floor of a building than in an apartment on a higher floor?

18. Place two cans of soft drinks, one regular and one diet, in a container of water. You will find that the diet drink floats while the regular one sinks. Use Archimedes's principle to devise an explanation.

19. When ski jumpers are airborne (Fig. CQ15.19), they bend their bodies forward and keep their hands at their sides. Why?

Figure CQ15.19

PROBLEMS

ENHANCED WebAssign The problems found in this chapter may be assigned online in Enhanced WebAssign.

1. denotes straightforward problem; **2.** denotes intermediate problem; **3.** denotes challenging problem

1. denotes full solution available in the *Student Solutions Manual/Study Guide*

1. denotes problems most often assigned in Enhanced WebAssign.

BIO denotes biomedical problem

GP denotes guided problem

M denotes Master It tutorial available in Enhanced WebAssign

QIC denotes asking for quantitative and conceptual reasoning

S denotes symbolic reasoning problem

shaded denotes "paired problems" that develop reasoning with symbols and numerical values

W denotes Watch It video solution available in Enhanced WebAssign

Section 15.1 Pressure

1. The four tires of an automobile are inflated to a gauge pressure of 200 kPa. Each tire has an area of 0.024 0 m^2 in contact with the ground. Determine the weight of the automobile.

2. QIC W A 50.0-kg woman wearing high-heeled shoes is invited into a home in which the kitchen has vinyl floor covering. The heel on each shoe is circular and has a radius of 0.500 cm. (a) If the woman balances on one heel, what pressure does she exert on the floor? (b) Should the homeowner be concerned? Explain your answer.

3. M Calculate the mass of a solid gold rectangular bar that has dimensions of 4.50 cm × 11.0 cm × 26.0 cm.

4. Estimate the total mass of the Earth's atmosphere. (The radius of the Earth is 6.37 × 10^6 m, and atmospheric pressure at the surface is 1.013 × 10^5 Pa.)

Section 15.2 Variation of Pressure with Depth

5. M The spring of the pressure gauge shown in Figure P15.5 has a force constant of 1 250 N/m, and the piston has a diameter of 1.20 cm. As the gauge is lowered into water in a lake, what change in depth causes the piston to move in by 0.750 cm?

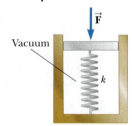

Figure P15.5

6. W The small piston of a hydraulic lift (Fig. P15.6) has a cross-sectional area of 3.00 cm^2, and its large piston has a cross-sectional area of 200 cm^2. What downward force of magnitude F_1 must be applied to the small piston for the lift to raise a load whose weight is F_g = 15.0 kN?

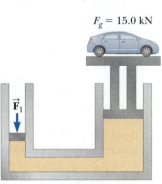

Figure P15.6

7. A container is filled to a depth of 20.0 cm with water. On top of the water floats a 30.0-cm-thick layer of oil with specific gravity 0.700. What is the absolute pressure at the bottom of the container?

8. W A swimming pool has dimensions 30.0 m × 10.0 m and a flat bottom. When the pool is filled to a depth of 2.00 m with fresh water, what is the force exerted by the water on (a) the bottom? (b) On each end? (c) On each side?

9. W (a) Calculate the absolute pressure at an ocean depth of 1 000 m. Assume the density of seawater is 1 030 kg/m^3 and the air above exerts a pressure of 101.3 kPa. (b) At this depth, what is the buoyant force on a spherical submarine having a diameter of 5.00 m?

10. (a) A very powerful vacuum cleaner has a hose 2.86 cm in diameter. With the end of the hose placed perpendicularly on the flat face of a brick, what is the weight of the heaviest brick that the cleaner can lift? (b) **What If?** An octopus uses one sucker of diameter 2.86 cm on each of the two shells of a clam in an attempt to pull the shells apart. Find the greatest force the octopus can exert on a clamshell in salt water 32.3 m deep.

11. M What must be the contact area between a suction cup (completely evacuated) and a ceiling if the cup is to support the weight of an 80.0-kg student?

12. *Why is the following situation impossible?* Figure P15.12 shows Superman attempting to drink cold water through a straw of length ℓ = 12.0 m. The walls of the tubular straw are very strong and do not collapse. With his great strength, he achieves maximum possible suction and enjoys drinking the cold water.

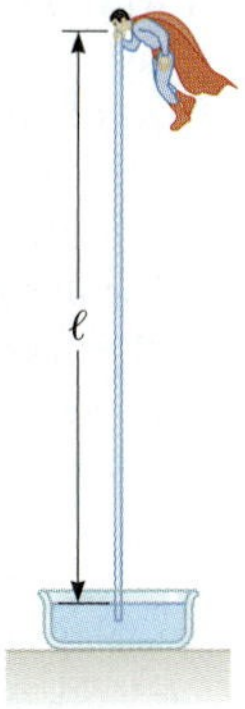

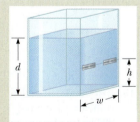

Figure P15.12

13. **Review.** The tank in Figure P15.13 is filled with water of depth d = 2.00 m. At the bottom of one sidewall is a rectangular hatch of height h = 1.00 m and width w = 2.00 m that is hinged at the top of the hatch. (a) Determine the magnitude of the force the water exerts on the hatch. (b) Find the magnitude of the torque exerted by the water about the hinges.

Figure P15.13 Problems 13 and 14.

14. **Review.** S The tank in Figure P15.13 is filled with water of depth d. At the bottom of one sidewall is a rectangular hatch of height h and width w that is hinged at the top of the hatch. (a) Determine the magnitude of the force the water exerts on the hatch. (b) Find the magnitude of the torque exerted by the water about the hinges.

15. Review. Piston ① in Figure P15.15 has a diameter of 0.250 in. Piston ② has a diameter of 1.50 in. Determine the magnitude F of the force necessary to support the 500 lb load in the absence of friction.

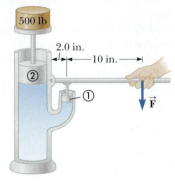

Figure P15.15

16. For the cellar of a new house, a hole is dug in the ground, with vertical sides going down 2.40 m. A concrete foundation wall is built all the way across the 9.60-m width of the excavation. This foundation wall is 0.183 m away from the front of the cellar hole. During a rainstorm, drainage from the street fills up the space in front of the concrete wall, but not the cellar behind the wall. The water does not soak into the clay soil. Find the force the water causes on the foundation wall. For comparison, the weight of the water is given by 2.40 m × 9.60 m × 0.183 m × 1 000 kg/m³ × 9.80 m/s² = 41.3 kN.

Section 15.3 Pressure Measurements

17. **W** Mercury is poured into a U-tube as shown in Figure P15.17a. The left arm of the tube has cross-sectional area A_1 of 10.0 cm², and the right arm has a cross-sectional area A_2 of 5.00 cm². One hundred grams of water are then poured into the right arm as shown in Figure P15.17b. (a) Determine the length of the water column in the right arm of the U-tube. (b) Given that the density of mercury is 13.6 g/cm³, what distance h does the mercury rise in the left arm?

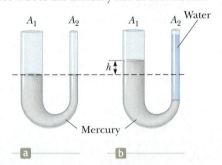

Figure P15.17

18. **Q C** Blaise Pascal duplicated Torricelli's barometer using a red Bordeaux wine, of density 984 kg/m³, as the working liquid (Fig. P15.18). (a) What was the height h of the wine column for normal atmospheric pressure? (b) Would you expect the vacuum above the column to be as good as for mercury?

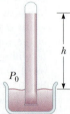

Figure P15.18

19. A backyard swimming pool with a circular base of diameter 6.00 m is filled to depth 1.50 m. (a) Find the absolute pressure at the bottom of the pool. (b) Two persons with combined mass 150 kg enter the pool and float quietly there. No water overflows. Find the pressure increase at the bottom of the pool after they enter the pool and float.

20. **S** A tank with a flat bottom of area A and vertical sides is filled to a depth h with water. The pressure is P_0 at the top surface. (a) What is the absolute pressure at the bottom of the tank? (b) Suppose an object of mass M and density less than the density of water is placed into the tank and floats. No water overflows. What is the resulting increase in pressure at the bottom of the tank?

21. Normal atmospheric pressure is 1.013×10^5 Pa. The approach of a storm causes the height of a mercury barometer to drop by 20.0 mm from the normal height. What is the atmospheric pressure?

Section 15.4 Buoyant Forces and Archimedes's Principle

22. **S** A Styrofoam slab has thickness h and density ρ_s. When a swimmer of mass m is resting on it, the slab floats in fresh water with its top at the same level as the water surface. Find the area of the slab.

23. A table-tennis ball has a diameter of 3.80 cm and average density of 0.084 0 g/cm³. What force is required to hold it completely submerged under water?

24. The gravitational force exerted on a solid object is 5.00 N. When the object is suspended from a spring scale and submerged in water, the scale reads 3.50 N (Fig. P15.24). Find the density of the object.

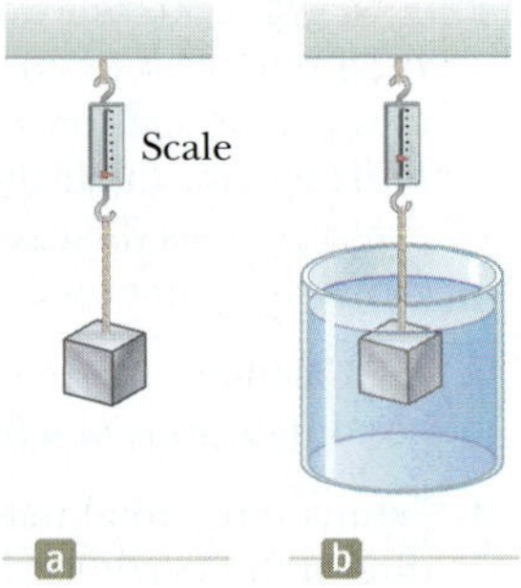

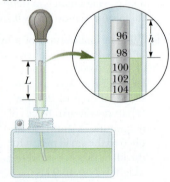

Figure P15.24 Problems 24 and 25.

25. A 10.0-kg block of metal measuring 12.0 cm by 10.0 cm by 10.0 cm is suspended from a scale and immersed in water as shown in Figure P15.24b. The 12.0-cm dimension is vertical, and the top of the block is 5.00 cm below the surface of the water. (a) What are the magnitudes of the forces acting on the top and on the bottom of the block due to the surrounding water? (b) What is the reading of the spring scale? (c) Show that the buoyant force equals the difference between the forces at the top and bottom of the block.

26. **S** A *hydrometer* is an instrument used to determine liquid density. A simple one is sketched in Figure P15.26. The bulb of a syringe is squeezed and released to let the atmosphere lift a sample of the liquid of interest into a tube containing a calibrated rod of known density. The rod, of length L and average density ρ_0, floats partially immersed in the liquid of density ρ. A length h of the rod protrudes above the surface of the liquid. Show that the density of the liquid is given by

$$\rho = \frac{\rho_0 L}{L - h}$$

Figure P15.26 Problems 26 and 28.

27. **M** A cube of wood having an edge dimension of 20.0 cm and a density of 650 kg/m³ floats on water. (a) What is the distance from the horizontal top surface of the cube to the water level? (b) What mass of lead should be placed on the cube so that the top of the cube will be just level with the water surface?

28. **Q|C** Refer to Problem 26 and Figure P15.26. A hydrometer is to be constructed with a cylindrical floating rod. Nine fiduciary marks are to be placed along the rod to indicate densities of 0.98 g/cm³, 1.00 g/cm³, 1.02 g/cm³, 1.04 g/cm³, . . . , 1.14 g/cm³. The row of marks is to start 0.200 cm from the top end of the rod and end 1.80 cm from the top end. (a) What is the required length of the rod? (b) What must be its average density? (c) Should the marks be equally spaced? Explain your answer.

29. **M** How many cubic meters of helium are required to lift a balloon with a 400-kg payload to a height of 8 000 m? Take $\rho_{He} = 0.179$ kg/m³. Assume the balloon maintains a constant volume and the density of air decreases with the altitude z according to the expression $\rho_{air} = \rho_0 e^{-z/8\,000}$, where z is in meters and $\rho_0 = 1.20$ kg/m³ is the density of air at sea level.

30. **S** **Review.** A long, cylindrical rod of radius r is weighted on one end so that it floats upright in a fluid having a density ρ. It is pushed down a distance x from its equilibrium position and released. Show that the rod will execute simple harmonic motion if the resistive effects of the fluid are negligible and determine the period of the oscillations.

31. **M** A plastic sphere floats in water with 50.0% of its volume submerged. This same sphere floats in glycerin with 40.0% of its volume submerged. Determine the densities of (a) the glycerin and (b) the sphere.

32. **Q|C** The weight of a rectangular block of low-density material is 15.0 N. With a thin string, the center of the horizontal bottom face of the block is tied to the bottom of a beaker partly filled with water. When 25.0% of the block's volume is submerged, the tension in the string is 10.0 N. (a) Find the buoyant force on the block. (b) Oil of density 800 kg/m³ is now steadily added to the beaker, forming a layer above the water and surrounding the block. The oil exerts forces on each of the four sidewalls of the block that the oil touches. What are the directions of these forces? (c) What happens to the string tension as the oil is added? Explain how the oil has this effect on the string tension. (d) The string breaks when its tension reaches 60.0 N. At this moment, 25.0% of the block's volume is still below the water line. What additional fraction of the block's volume is below the top surface of the oil?

33. **BIO** Decades ago, it was thought that huge herbivorous dinosaurs such as *Apatosaurus* and *Brachiosaurus* habitually walked on the bottom of lakes, extending their long necks up to the surface to breathe. *Brachiosaurus* had its nostrils on the top of its head. In 1977, Knut Schmidt-Nielsen pointed out that breathing would be too much work for such a creature. For a simple model, consider a sample consisting of 10.0 L of air at absolute pressure 2.00 atm, with density 2.40 kg/m³, located at the surface of a freshwater lake. Find the work required to transport it to a depth of 10.3 m, with its temperature, volume, and pressure remaining constant. This energy investment is greater than the energy that can be obtained by metabolism of food with the oxygen in that quantity of air.

34. To an order of magnitude, how many helium-filled toy balloons would be required to lift you? Because helium is an irreplaceable resource, develop a theoretical answer rather than an experimental one. In your solution, state what physical quantities you take as data and the values you measure or estimate for them.

35. A spherical vessel used for deep-sea exploration has a radius of 1.50 m and a mass of 1.20×10^4 kg. To dive, the vessel takes on mass in the form of seawater. Determine the mass the vessel must take on if it is to descend at a constant speed of 1.20 m/s, when the resistive force on it is 1 100 N in the upward direction. The density of seawater is equal to 1.03×10^3 kg/m³.

36. **W** A light balloon is filled with 400 m³ of helium at atmospheric pressure. (a) At 0°C, the balloon can lift a payload of what mass? (b) **What If?** In Table 15.1, observe that the density of hydrogen is nearly half the density of helium. What load can the balloon lift if filled with hydrogen?

Section 15.5 Fluid Dynamics

Section 15.6 Streamlines and the Continuity Equation for Fluids

Section 15.7 Bernoulli's Equation

37. A horizontal pipe 10.0 cm in diameter has a smooth reduction to a pipe 5.00 cm in diameter. If the pressure of the water in the larger pipe is 8.00×10^4 Pa and the pressure in the smaller pipe is 6.00×10^4 Pa, at what rate does water flow through the pipes?

38. Water falls over a dam of height h with a mass flow rate of R, in units of kilograms per second. (a) Show that the power available from the water is

$$P = Rgh$$

where g is the free-fall acceleration. (b) Each hydroelectric unit at the Grand Coulee Dam takes in water at a rate of 8.50×10^5 kg/s from a height of 87.0 m. The power developed by the falling water is converted to electric power with an efficiency of 85.0%. How much electric power does each hydroelectric unit produce?

39. A large storage tank with an open top is filled to a height h_0. The tank is punctured at a height h above the bottom of the tank (Fig. P15.39). Find an expression for how far from the tank the exiting stream lands.

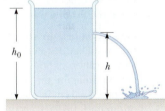

Figure P15.39

40. **Q|C** **Review.** Old Faithful Geyser in Yellowstone National Park erupts at approximately one-hour intervals, and the height of the water column reaches 40.0 m (Fig. P15.40). (a) Model the rising stream as a series of separate droplets. Analyze the free-fall motion of one of the droplets to determine the speed at which the water leaves the ground. (b) **What If?** Model the rising stream as an ideal fluid in streamline flow. Use Bernoulli's equation to determine the speed of the water as it leaves ground level. (c) How does the answer from part (a) compare with the answer from part (b)? (d) What is

Figure P15.40

the pressure (above atmospheric) in the heated underground chamber if its depth is 175 m? Assume the chamber is large compared with the geyser's vent.

41. (a) A water hose 2.00 cm in diameter is used to fill a 20.0-L bucket. If it takes 1.00 min to fill the bucket, what is the speed v at which water moves through the hose? (*Note:* 1 L = 1 000 cm³.) (b) The hose has a nozzle 1.00 cm in diameter. Find the speed of the water at the nozzle.

42. Water flows through a fire hose of diameter 6.35 cm at a rate of 0.012 0 m³/s. The fire hose ends in a nozzle of inner diameter 2.20 cm. What is the speed with which the water exits the nozzle?

43. **M** A large storage tank, open at the top and filled with water, develops a small hole in its side at a point 16.0 m below the water level. The rate of flow from the leak is found to be 2.50×10^{-3} m³/min. Determine (a) the speed at which the water leaves the hole and (b) the diameter of the hole.

44. A legendary Dutch boy saved Holland by plugging a hole of diameter 1.20 cm in a dike with his finger. If the hole was 2.00 m below the surface of the North Sea (density 1 030 kg/m³), (a) what was the force on his finger? (b) If he pulled his finger out of the hole, during what time interval would the released water fill 1 acre of land to a depth of 1 ft? Assume the hole remained constant in size.

45. A village maintains a large tank with an open top, containing water for emergencies. The water can drain from the tank through a hose of diameter 6.60 cm. The hose ends with a nozzle of diameter 2.20 cm. A rubber stopper is inserted into the nozzle. The water level in the tank is kept 7.50 m above the nozzle. (a) Calculate the friction force exerted on the stopper by the nozzle. (b) The stopper is removed. What mass of water flows from the nozzle in 2.00 h? (c) Calculate the gauge pressure of the flowing water in the hose just behind the nozzle.

46. Water is pumped up from the Colorado River to supply Grand Canyon Village, located on the rim of the canyon. The river is at an elevation of 564 m, and the village is at an elevation of 2 096 m. Imagine that the water is pumped through a single long pipe 15.0 cm in diameter, driven by a single pump at the bottom end. (a) What is the minimum pressure at which the water must be pumped if it is to arrive at the village? (b) If 4 500 m³ of water are pumped per day, what is the speed of the water in the pipe? *Note:* Assume the free-fall acceleration and the density of air are constant over this range of elevations. The pressures you calculate are too high for an ordinary pipe. The water is actually lifted in stages by several pumps through shorter pipes.

47. Figure P15.47 shows a stream of water in steady flow from a kitchen faucet. At the faucet, the diameter of the stream is 0.960 cm. The stream fills a 125-cm³ container in 16.3 s. Find the diameter of the stream 13.0 cm below the opening of the faucet.

© Cengage Learning/George Semple

Figure P15.47

48. An airplane is cruising at altitude 10 km. The pressure outside the craft is 0.287 atm; within the passenger compartment, the pressure is 1.00 atm and the temperature is 20°C. A small leak occurs in one of the window seals in the passenger compartment. Model the air as an ideal fluid to estimate the speed of the airstream flowing through the leak.

Section 15.8 Other Applications of Fluid Dynamics

49. The Bernoulli effect can have important consequences for the design of buildings. For example, wind can blow around a skyscraper at remarkably high speed, creating low pressure. The higher atmospheric pressure in the still air inside the buildings can cause windows to pop out. As originally constructed, the John Hancock Building in Boston popped windowpanes that fell many stories to the sidewalk below. (a) Suppose a horizontal wind blows with a speed of 11.2 m/s outside a large pane of plate glass with dimensions 4.00 m × 1.50 m. Assume the density of the air to be constant at 1.20 kg/m³. The air inside the building is at atmospheric pressure. What is the total force exerted by air on the windowpane? (b) **What If?** If a second skyscraper is built nearby, the airspeed can be especially high where wind passes through the narrow separation between the buildings. Solve part (a) again with a wind speed of 22.4 m/s, twice as high.

50. **Q|C** An airplane has a mass of 1.60×10^4 kg, and each wing has an area of 40.0 m². During level flight, the pressure on the lower wing surface is 7.00×10^4 Pa. (a) Suppose the lift on the airplane were due to a pressure difference alone. Determine the pressure on the upper wing surface. (b) More realistically, a significant part of the lift is due to deflection of air downward by the wing. Does the inclusion of this force mean that the pressure in part (a) is higher or lower? Explain.

51. A siphon is used to drain water from a tank as illustrated in Figure P15.51. Assume steady flow without friction. (a) If $h = 1.00$ m, find the speed of outflow at the end of the siphon. (b) **What If?** What is the limitation on the height of the top of the siphon above the end of the

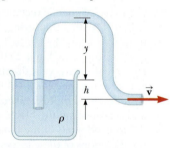

Figure P15.51

siphon? *Note:* For the flow of the liquid to be continuous, its pressure must not drop below its vapor pressure. Assume the water is at 20.0°C, at which the vapor pressure is 2.3 kPa.

Section 15.9 Context Connection: Turbulent Flow of Blood

52. **BIO** **Q|C** A common parameter that can be used to predict turbulence in fluid flow is called the *Reynolds number*. The Reynolds number for fluid flow in a pipe is a dimensionless quantity defined as

$$\text{Re} = \frac{\rho v d}{\mu}$$

where ρ is the density of the fluid, v is its speed, d is the inner diameter of the pipe, and μ is the viscosity of the fluid. Viscosity is a measure of the internal resistance of a liquid to flow and has units of Pa · s. The criteria for the type of flow are as follows:

- If Re < 2 300, the flow is laminar.
- If 2 300 < Re < 4 000, the flow is in a transition region between laminar and turbulent.
- If Re > 4 000, the flow is turbulent.

(a) Let's model blood of density 1.06×10^3 kg/m³ and viscosity 3.00×10^{-3} Pa · s as a pure liquid, that is, ignore the fact that it contains red blood cells. Suppose it is flowing in a large artery of radius 1.50 cm with a speed of 0.067 0 m/s. Show that the flow is laminar. (b) Imagine that the artery ends in a *single* capillary so that the radius of the artery reduces to a much smaller value. What is the radius of the capillary that would cause the flow to become turbulent? (c) Actual capillaries have radii of about 5–10 micrometers, much smaller than the value in part (b). Why doesn't the flow in actual capillaries become turbulent?

Additional Problems

53. Water is forced out of a fire extinguisher by air pressure as shown in Figure P15.53. How much gauge air pressure in the tank is required for the water jet to have a speed of 30.0 m/s when the water level is 0.500 m below the nozzle?

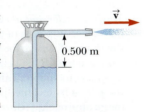

Figure P15.53

54. **S** The true weight of an object can be measured in a vacuum, where buoyant forces are absent. A measurement in air, however, is disturbed by buoyant forces. An object of volume V is weighed in air on an equal-arm balance with the use of counterweights of density ρ. Representing the density of air as ρ_{air} and the balance reading as F'_g, show that the true weight F_g is

$$F_g = F'_g + \left(V - \frac{F'_g}{\rho g}\right)\rho_{air}g$$

55. A light spring of constant $k = 90.0$ N/m is attached vertically to a table (Fig. P15.55a). A 2.00-g balloon is filled with helium (density $= 0.180$ kg/m³) to a volume of 5.00 m³ and is then connected to the spring, causing it to stretch as shown in Figure P15.55b. Determine the extension distance L when the balloon is in equilibrium.

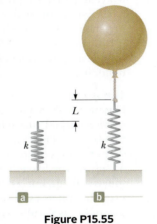

Figure P15.55

56. **S** The hull of an experimental boat is to be lifted above the water by a hydrofoil mounted below its keel as shown in Figure P15.56. The hydrofoil has a shape like that of an airplane wing. Its area projected onto a horizontal surface is A. When the boat is towed at sufficiently high speed, water of density ρ

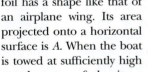

Figure P15.56

moves in streamline flow so that its average speed at the top of the hydrofoil is n times larger than its speed v_b below the hydrofoil. (a) Ignoring the buoyant force, show that the upward lift force exerted by the water on the hydrofoil has a magnitude

$$F \approx \tfrac{1}{2}(n^2 - 1)\rho v_b^2 A$$

(b) The boat has mass M. Show that the liftoff speed is given by

$$v \approx \sqrt{\frac{2Mg}{(n^2 - 1)A\rho}}$$

57. Evangelista Torricelli was the first person to realize that we live at the bottom of an ocean of air. He correctly surmised that the pressure of our atmosphere is attributable to the weight of the air. The density of air at 0°C at the Earth's surface is 1.29 kg/m³. The density decreases with increasing altitude (as the atmosphere thins). On the other hand, if we assume the density is constant at 1.29 kg/m³ up to some altitude h and is zero above that altitude, then h would represent the depth of the ocean of air. (a) Use this model to determine the value of h that gives a pressure of 1.00 atm at the surface of the Earth. (b) Would the peak of Mount Everest rise above the surface of such an atmosphere?

58. **GP** A helium-filled balloon (whose envelope has a mass of $m_b = 0.250$ kg) is tied to a uniform string of length $\ell = 2.00$ m and mass $m = 0.050$ 0 kg. The balloon is spherical with a radius of $r = 0.400$ m. When released in air of temperature 20°C and density $\rho_{air} = 1.20$ kg/m³, it lifts a length h of string and then remains stationary as shown in Figure P15.58. We wish to find the length of string lifted by the balloon. (a) When the balloon remains stationary, what is the appropriate analysis model to describe it? (b) Write a force equation for the balloon from this model in terms of the buoyant force B, the weight F_b of the balloon, the weight F_{He} of the helium, and the weight F_s of the segment of string of length h. (c) Make an appropriate substitution for each of these forces and solve symbolically for the mass m_s of the segment of string of length h in terms of m_b, r, ρ_{air}, and the density of helium ρ_{He}. (d) Find the numerical value of the mass m_s. (e) Find the length h numerically.

Figure P15.58

59. **Review.** A copper cylinder hangs at the bottom of a steel wire of negligible mass. The top end of the wire is fixed. When the wire is struck, it emits sound with a fundamental frequency of 300 Hz. The copper cylinder is then submerged in water so that half its volume is below the waterline. Determine the new fundamental frequency.

60. **S Review.** With reference to the dam studied in Example 15.2 and shown in Figure 15.6, (a) show that the total torque exerted by the water behind the dam about a horizontal axis through O is $\tfrac{1}{6}\rho gwH^3$. (b) Show that the effective line of action of the total force exerted by the water is at a distance $\tfrac{1}{3}H$ above O.

61. An incompressible, nonviscous fluid is initially at rest in the vertical portion of the pipe shown in Figure P15.61a, where $L = 2.00$ m. When the valve is opened, the fluid flows into the horizontal section of the pipe. What is the fluid's speed when all the fluid is in the horizontal section as shown in Figure P15.61b? Assume the cross-sectional area of the entire pipe is constant.

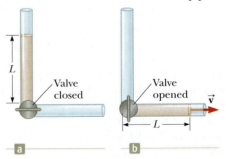

Figure P15.61

62. **S** In about 1657, Otto von Guericke, inventor of the air pump, evacuated a sphere made of two brass hemispheres

(Fig. P15.62). Two teams of eight horses each could pull the hemispheres apart only on some trials and then "with greatest difficulty," with the resulting sound likened

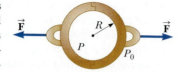

Figure P15.62

to a cannon firing. Find the force F required to pull the thin-walled evacuated hemispheres apart in terms of R, the radius of the hemispheres; P, the pressure inside the hemispheres; and atmospheric pressure P_0.

63. A 1.00-kg beaker containing 2.00 kg of oil (density = 916.0 kg/m³) rests on a scale. A 2.00-kg block of iron suspended from a spring scale is completely submerged in the oil as shown in Figure P15.63. Determine the equilibrium readings of both scales.

64. **S** A beaker of mass m_b containing oil of mass m_o and density ρ_o rests on a scale. A block of iron of mass m_{Fe} suspended from a spring scale is completely submerged in the oil as shown in Figure P15.63. Determine the equilibrium readings of both scales.

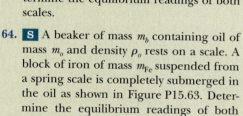

Figure P15.63
Problems 63 and 64.

65. An ice cube whose edges measure 20.0 mm is floating in a glass of ice-cold water, and one of the ice cube's faces is parallel to the water's surface. (a) How far below the water surface is the bottom face of the block? (b) Ice-cold ethyl alcohol is gently poured onto the water surface to form a layer 5.00 mm thick above the water. The alcohol does not mix with the water. When the ice cube again attains hydrostatic equilibrium, what is the distance from the top of the water to the bottom face of the block? (c) Additional cold ethyl alcohol is poured onto the water's surface until the top surface of the alcohol coincides with the top surface of the ice cube (in hydrostatic equilibrium). How thick is the required layer of ethyl alcohol?

66. **S** Show that the variation of atmospheric pressure with altitude is given by $P = P_0 e^{-\alpha y}$, where $\alpha = \rho_0 g / P_0$, P_0 is atmospheric pressure at some reference level $y = 0$, and ρ_0 is the atmospheric density at this level. Assume the decrease in atmospheric pressure over an infinitesimal change in altitude (so that the density is approximately uniform over the infinitesimal change) can be expressed from Equation 15.4 as $dP = -\rho g \, dy$. Also assume the density of air is proportional to the pressure, which, as we will see in Chapter 16, is equivalent to assuming the temperature of the air is the same at all altitudes.

67. A U-tube open at both ends is partially filled with water (Fig. P15.67a). Oil having a density 750 kg/m³ is then poured into the right arm and forms a column $L = 5.00$ cm high (Fig. P15.67b). (a) Determine the difference h in the heights of the two liquid surfaces. (b) The right arm is then shielded from any air motion while air is blown across the top of the left arm until the surfaces of the two liquids are at the same height (Fig. P15.67c). Determine the speed of the air being blown across the left arm. Take the density of air as constant at 1.20 kg/m³.

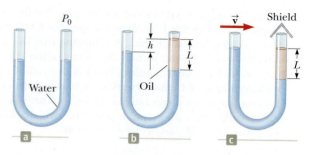

Figure P15.67

68. *Why is the following situation impossible?* A barge is carrying a load of small pieces of iron along a river. The iron pile is in the shape of a cone for which the radius r of the base of the cone is equal to the central height h of the cone. The barge is square in shape, with vertical sides of length $2r$, so that the pile of iron comes just up to the edges of the barge. The barge approaches a low bridge, and the captain realizes that the top of the pile of iron is not going to make it under the bridge. The captain orders the crew to shovel iron pieces from the pile into the water to reduce the height of the pile. As iron is shoveled from the pile, the pile always has the shape of a cone whose diameter is equal to the side length of the barge. After a certain volume of iron is removed from the barge, it makes it under the bridge without the top of the pile striking the bridge.

69. The water supply of a building is fed through a main pipe 6.00 cm in diameter. A 2.00-cm-diameter faucet tap, located 2.00 m above the main pipe, is observed to fill a 25.0-L container in 30.0 s. (a) What is the speed at which the water leaves the faucet? (b) What is the gauge pressure in the 6-cm main pipe? Assume the faucet is the only "leak" in the building.

70. The *spirit-in-glass thermometer*, invented in Florence, Italy, around 1654, consists of a tube of liquid (the spirit) containing a number of submerged glass spheres with slightly different masses (Fig. P15.70). At sufficiently low temperatures, all the spheres float, but as the temperature rises, the spheres sink one after another. The device is a crude but interesting tool for measuring temperature. Suppose the tube is filled with ethyl alcohol, whose density is 0.789 45 g/cm³ at 20.0°C and decreases to 0.780 97 g/cm³ at 30.0°C. (a) Assuming that one of the spheres has a radius of 1.000 cm and is in equilibrium halfway up the tube at 20.0°C, determine its mass. (b) When the temperature increases to 30.0°C, what mass must a second sphere of the same radius have to be in equilibrium at the halfway point? (c) At 30.0°C, the first sphere has fallen to the bottom of the tube. What upward force does the bottom of the tube exert on this sphere?

Figure P15.70

Context | 4   **CONCLUSION**

Detecting Atherosclerosis and Preventing Heart Attacks

We have explored the physics of fluids and can return to our central question for this *Heart Attacks* Context:

> **How can the principles of physics be applied in medicine to help prevent heart attacks?**

We shall apply our understanding of fluid dynamics to explore research into the causes of cardiovascular disease and heart attacks.

Traditionally, the prevention and treatment of cardiovascular disease and heart attacks has focused on a regimen of exercise, a healthy diet aimed at reducing cholesterol and preventing high blood pressure, cessation of smoking, reducing stress, and medications to reduce the patient's cholesterol and blood pressure, and to prevent blood clots from forming. In severe cases, *angioplasty*, a procedure to widen constricted arteries, or the placement of a *stent*, a mesh tube that acts as a scaffold to keep the arteries open, may also be used. This procedure is shown in Figure 1, which illustrates the use of a balloon to open a constricted artery and the placement of a stent.

Although the causes of atherosclerosis (Section 15.9) are still not known, fluid dynamics has helped illuminate many of the factors that can result in the hardening of arteries and plaque buildup that characterize this condition. As discussed in Section 15.9, recent research has focused on the response of endothelial cells lining arterial walls to turbulent flows and the buildup of plaque in constricted blood vessels.

Let us further examine the physics of an arterial constriction, or *stenosis*. Blood flow in the healthy blood vessel is laminar, and research has shown that plaque does not build up under these circumstances. The situation is very different in the constricted artery; in cases with severe narrowing, the arterial cross-section can be reduced by as much as 75%, to one-fourth of its original area. From the continuity equation for fluids (Eq. 15.7), with $A_2 = \frac{1}{4}A_1$, we obtain

$$A_1 v_1 = \left(\tfrac{1}{4}A_1\right) v_2 \quad \rightarrow \quad v_2 = 4 v_1 \qquad \blacktriangleleft \text{1}$$

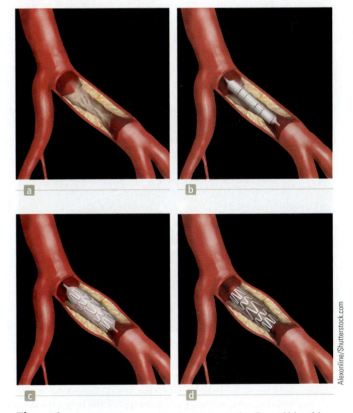

Figure 1 The placement of a stent to improve the flow of blood in a vessel. (a) An artery has a plaque buildup that restricts the flow of blood. (b) A balloon catheter with a deflated balloon and collapsed stent is inserted into the artery and guided to the site of the plaque. (c) The balloon is inflated, expanding the stent and the walls of the vessel. (d) The balloon is deflated and the catheter is removed. The stent keeps the artery open.

Alexonline/Shutterstock.com

We see that the blood flows four times faster in the constricted region of the artery. Similar to the situation shown in Figure 15.15, as a fluid speeds up, its flow becomes turbulent, so that the blood flow immediately downstream of the stenosis exhibits whirlpool-like regions. As discussed in Section 15.9, this turbulence can lead to further problems.

If we assume streamline flow for the moment, we can relate the pressure in the constricted portion 2 of the blood vessel to that in an open portion 1 by using Bernoulli's equation for a horizontal segment of the artery (Eq. 15.9):

$$P_1 + \tfrac{1}{2}\rho v_1{}^2 = P_2 + \tfrac{1}{2}\rho v_2{}^2$$

Solving for the pressure difference between these two locations,

$$\Delta P = P_2 - P_1 = \tfrac{1}{2}\rho v_1{}^2 - \tfrac{1}{2}\rho v_2{}^2 = \tfrac{1}{2}\rho(v_1{}^2 - v_2{}^2)$$

Substituting for v_2 from Equation (1),

$$\Delta P = \tfrac{1}{2}\rho\left[v_1{}^2 - (4v_1)^2\right] = -\tfrac{15}{2}\rho v_1{}^2 \qquad \textbf{2}\blacktriangleleft$$

The average *systolic* blood pressure (the pressure during the contraction of the heart) is 120 mm of mercury, or 15.7 kPa (note that this is a gauge pressure, not the absolute pressure in the blood vessel). Also, on average, blood ($\rho = 1.05 \times 10^3\,\text{kg/m}^3$) flows at a speed of $v_1 = 0.40\,\text{m/s}$. Evaluating the pressure difference in Equation (2) numerically,

$$\Delta P = -\tfrac{15}{2}(1.05 \times 10^3\,\text{kg/m}^3)(0.40\,\text{m/s})^2 = -1.3 \times 10^3\,\text{Pa}$$

Comparing this result to the initial pressure, we find

$$\frac{\Delta P}{P_0} = \frac{-1.3 \times 10^3\ \text{Pa}}{15.7 \times 10^3\ \text{Pa}} = -8.0\%$$

This 8% drop in pressure can be enough that the pressure difference between the tissue outside the vessel and that in the constriction can cause the vessel to collapse, causing a momentary interruption in blood flow. At this point, the speed of the blood goes to zero, its pressure rises again, and the vessel reopens. As the blood rushes through the constricted artery, the internal pressure drops and again the artery closes. Such a phenomenon is called *vascular flutter*. It can be heard by a physician with a stethoscope and is an indication of advanced atherosclerotic disease. Furthermore, the continued opening and closing of the artery can contribute even more to turbulence of the blood and its effects. Therefore, vascular flutter should be taken very seriously and recognized as a strong indication that medical care is needed to avoid a heart attack.

The relationship between turbulent flow occurring downstream of arterial constrictions and the build-up of plaque has been demonstrated by a number of recent studies that combine medical research, physics, and engineering to find the physical and biochemical causes of atherosclerosis and cardiovascular disease. In these studies, platelet-rich animal blood tagged with radioactive indium-111 was circulated through flow tubes with various constriction geometries. The location and amount of platelet deposits were then recorded using a device for measuring the gamma radiation emitted by the radioactive platelets. (We will study gamma radiation in Chapters 24 and 30.) This allowed researchers to determine the locations of maximum platelet deposition and their relationship to the blood flow in the artery.[1]

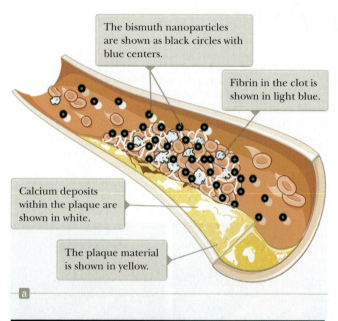

The bismuth nanoparticles are shown as black circles with blue centers.

Fibrin in the clot is shown in light blue.

Calcium deposits within the plaque are shown in white.

The plaque material is shown in yellow.

a

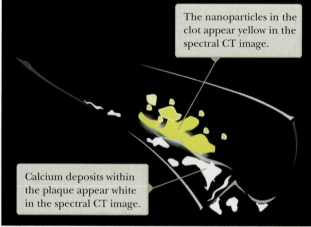

The nanoparticles in the clot appear yellow in the spectral CT image.

Calcium deposits within the plaque appear white in the spectral CT image.

Wiley-VCH Verlag GmbH & Co. KGaA. Reproduced with permission

b

Figure 2 (a) A blood vessel with ruptured atherosclerotic plaque and calcium deposits is developing a blood clot. A traditional CT image would not differentiate between the blood clot and the calcium in the plaque, making it unclear whether the image shows a clot that should be treated. To help make this differentiation, bismuth nanoparticles are targeted to a protein in the blood clot called fibrin. (b) A spectral CT image shows the nanoparticles targeted to fibrin in yellow, differentiating it from calcium, still shown in white, in the plaque.

[1]Schoephoerster, R. T., et al., "Effects of local geometry and fluid dynamics on regional platelet deposition on artificial surfaces," *Arterioscler. Thromb. Vasc. Biol.*, 1993, 13, 1806–1813.

Similar studies have used fluid dynamics to examine the deposition of plaque on artificial heart valves and helped design more hydrodynamic valves that are also less prone to plaque build-up.

More recently, researchers have begun using radioactive nanoparticles that attach themselves to plaque particles forming in the bloodstream, allowing for the earlier detection of plaque formation than with current techniques (Fig. 2). Using magnetic resonance imaging (MRI), researchers can also follow the flow of these nanoparticles to obtain unprecedented details of the flow of blood in the complicated geometry of the human cardiovascular system. (We will discuss MRIs in more detail in Context 7.) By combining nanoparticles with adult stem cells, arterial plaque in pig hearts has actually been burned away when the nanoparticles are illuminated with laser light.[2]

At the same time, advances in computing speed and memory capability have allowed computational fluid dynamic models to simulate the complex fluid interactions taking place in blood vessels (Fig. 3). One advantage of computer models is that very complicated geometries can be examined, with the ultimate goal of simulating the entire human cardiovascular system from the heart to the tiniest of capillaries.

The application of fluid dynamics to the human body, the use of magnetic resonance imaging, and the use of lasers in exploring the cardiovascular system have resulted in close cooperation between physics and medicine. This collaboration has already yielded significant results, and is certain to drive important advances in medicine in years to come.

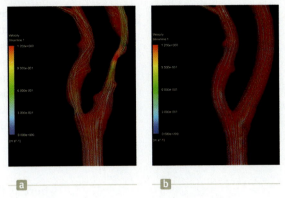

Figure 3 A computational fluid dynamic (CFD) simulation of blood flow through constricted (left) and healthy (right) arteries. (From Ding, S, Tu, J and Cheung, C 2007, "Geometric model generation for CFD simulation of blood and air flows," p. 1335–1338 in Proceedings of the 1st International Conference on Bioinformatics and Biomedical Engineering, Wuhan, China, 6–8 July 2007.)

Problems

1. **BIO** Three arterial geometries are shown in Figure 4, including a healthy blood vessel (Figure 4, top), a constricted blood vessel (Figure 4, middle), and a blood vessel with an aneurysm (a balloon-like bulge, Figure 4, bottom). The speed of the blood at point 1 in all three vessels is the same. (a) In which of the three blood vessels would the blood be flowing at a higher speed at point 2? (b) What would be the ratio of the speeds of the blood at point 2 in blood vessels (ii) and (iii)?

2. **BIO** There are many situations in physics that can be described by an equation of the form,

   (1) (driving influence) = (resistance)(result of influence)

In many cases, the driving influence can be expressed as the difference between two values of a variable at different locations in space. For example, when we study thermodynamics we will find that the rate P of energy transfer by heat through a material of cross sectional area A and length L is related to the temperature difference ΔT between the two ends of the material as follows:

$$(2) \quad \Delta T = \left(\frac{L}{kA}\right)P$$

where k is the *thermal conductivity* of the material (see Section 17.10). The temperature difference ΔT is the driving influence. The result is the rate of energy transfer P. The quantity L/kA represents resistance to the transfer of energy by heat. The ratio L/k is called the *R-value* (representing thermal resistance) for materials used to provide thermal insulation in homes and buildings.

When we study electricity, we will find that the *potential difference* ΔV between two ends of a material is a driving influence that results in a *current I* in the material as given by

$$(3) \quad \Delta V = \left(\frac{\ell}{\sigma A}\right)I$$

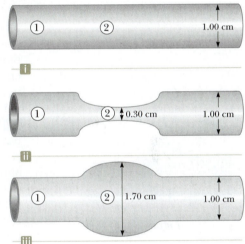

Figure 4

[2]American Heart Association. "Nanoparticles plus adult stem cells demolish plaque, study finds." *Science Daily* 21 July 2010.

The current exists in a piece of material of length ℓ and cross-sectional area A, with electrical conductivity σ (see Section 21.2). The combination in parentheses represents the resistance to the influence, which in this case is called *electrical resistance*.

Let's now think about the flow of blood in an artery. What drives the flow? What resists the flow? The flow of blood is driven by a pressure difference ΔP along an artery, just like the flow of water is driven by a pressure difference along a pipe. The following equation describes the flow of any liquid in a pipe:

$$(4) \quad \Delta P = \left(8\pi\mu \frac{L}{A} \right) v$$

where v, the speed of the fluid, is the result of the pressure difference and the resistance is related to the *viscosity* μ of the fluid. Viscosity is a measure of the internal resistance of a liquid to flow and has units of Pa · s. Honey, for example, is a more viscous fluid than water. Peanut butter is a tremendously viscous fluid. Notice the striking resemblance of Equation (4) to Equations (2) and (3).

A pulmonary artery carries deoxygenated blood from the heart to a lung. (See Figure 2 in the *Heart Attacks* Context Introduction.) Suppose a pulmonary artery has a length of 9.00 cm and a radius of 3.00 mm. A pressure difference of 400 Pa exists along the length of the artery and between its ends. (a) Find the speed of the flow of blood through this artery if the blood has a viscosity of 3.00×10^{-3} Pa · s. (b) Blood is not a simple liquid. It contains red blood cells as well as other cells. The percentage of blood volume occupied by red blood cells is called *hematocrit*. Some illnesses, for example, *polycythemia*, are characterized by increased hematocrit levels. The increased percentage of red blood cells can increase the viscosity of the blood. Suppose a patient with an increased hematocrit level has blood with a viscosity 1.80 times that of the blood in (a). What difference in pressure is required across the length of the pulmonary artery to provide the same blood speed?

Figure 5

3. **BIO** **Q|C** The human brain and spinal cord are immersed in the cerebrospinal fluid. The fluid is normally continuous between the cranial and spinal cavities and exerts a pressure of 100 to 200 mm of H_2O above the prevailing atmospheric pressure. In medical work, pressures are often measured in units of millimeters of H_2O because body fluids, including the cerebrospinal fluid, typically have the same density as water. The pressure of the cerebrospinal fluid can be measured by means of a *spinal tap* as illustrated in Figure 5. A hollow tube is inserted into the spinal column, and the height to which the fluid rises is observed. If the fluid rises to a height of 160 mm, we write its gauge pressure as 160 mm H_2O. (a) Express this pressure in pascals, in atmospheres, and in millimeters of mercury. (b) Some conditions that block or inhibit the flow of cerebrospinal fluid can be investigated by means of *Queckenstedt's test*. In this procedure, the veins in the patient's neck are compressed to make the blood pressure rise in the brain, which in turn should be transmitted to the cerebrospinal fluid. Explain how the level of fluid in the spinal tap can be used as a diagnostic tool for the condition of the patient's spine.

4. **BIO** **M** A hypodermic syringe contains a medicine with the density of water (Figure 6). The barrel of the syringe has a cross-sectional area $A = 2.50 \times 10^{-5}$ m^2, and the needle has a cross-sectional area $a = 1.00 \times 10^{-8}$ m^2. In the absence of a force on the plunger, the pressure everywhere is 1.00 atm. A force $\vec{F}$ of magnitude 2.00 N acts on the plunger, making medicine squirt horizontally from the needle. Determine the speed of the medicine as it leaves the needle's tip.

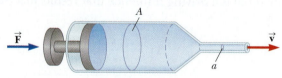

Figure 6

Appendix A

Tables

TABLE A.1 | Conversion Factors

Length

	m	cm	km	in.	ft	mi
1 meter	1	10^2	10^{-3}	39.37	3.281	6.214×10^{-4}
1 centimeter	10^{-2}	1	10^{-5}	0.393 7	3.281×10^{-2}	6.214×10^{-6}
1 kilometer	10^3	10^5	1	3.937×10^4	3.281×10^3	0.621 4
1 inch	2.540×10^{-2}	2.540	2.540×10^{-5}	1	8.333×10^{-2}	1.578×10^{-5}
1 foot	0.304 8	30.48	3.048×10^{-4}	12	1	1.894×10^{-4}
1 mile	1 609	1.609×10^5	1.609	6.336×10^4	5 280	1

Mass

	kg	g	slug	u
1 kilogram	1	10^3	6.852×10^{-2}	6.024×10^{26}
1 gram	10^{-3}	1	6.852×10^{-5}	6.024×10^{23}
1 slug	14.59	1.459×10^4	1	8.789×10^{27}
1 atomic mass unit	1.660×10^{-27}	1.660×10^{-24}	1.137×10^{-28}	1

Note: 1 metric ton = 1 000 kg.

Time

	s	min	h	day	yr
1 second	1	1.667×10^{-2}	2.778×10^{-4}	1.157×10^{-5}	3.169×10^{-8}
1 minute	60	1	1.667×10^{-2}	6.994×10^{-4}	1.901×10^{-6}
1 hour	3 600	60	1	4.167×10^{-2}	1.141×10^{-4}
1 day	8.640×10^4	1 440	24	1	2.738×10^{-5}
1 year	3.156×10^7	5.259×10^5	8.766×10^3	365.2	1

Speed

	m/s	cm/s	ft/s	mi/h
1 meter per second	1	10^2	3.281	2.237
1 centimeter per second	10^{-2}	1	3.281×10^{-2}	2.237×10^{-2}
1 foot per second	0.304 8	30.48	1	0.681 8
1 mile per hour	0.447 0	44.70	1.467	1

Note: 1 mi/min = 60 mi/h = 88 ft/s.

Force

	N	lb
1 newton	1	0.224 8
1 pound	4.448	1

(Continued)

TABLE A.1 | Conversion Factors (*continued*)

Energy, Energy Transfer

	J	ft · lb	eV
1 joule	1	0.737 6	6.242×10^{18}
1 foot-pound	1.356	1	8.464×10^{18}
1 electron volt	1.602×10^{-19}	1.182×10^{-19}	1
1 calorie	4.186	3.087	2.613×10^{19}
1 British thermal unit	1.055×10^{3}	7.779×10^{2}	6.585×10^{21}
1 kilowatt-hour	3.600×10^{6}	2.655×10^{6}	2.247×10^{25}

	cal	Btu	kWh
1 joule	0.238 9	9.481×10^{-4}	2.778×10^{-7}
1 foot-pound	0.323 9	1.285×10^{-3}	3.766×10^{-7}
1 electron volt	3.827×10^{-20}	1.519×10^{-22}	4.450×10^{-26}
1 calorie	1	3.968×10^{-3}	1.163×10^{-6}
1 British thermal unit	2.520×10^{2}	1	2.930×10^{-4}
1 kilowatt-hour	8.601×10^{5}	3.413×10^{2}	1

Pressure

	Pa	atm
1 pascal	1	9.869×10^{-6}
1 atmosphere	1.013×10^{5}	1
1 centimeter mercury[a]	1.333×10^{3}	1.316×10^{-2}
1 pound per square inch	6.895×10^{3}	6.805×10^{-2}
1 pound per square foot	47.88	4.725×10^{-4}

	cm Hg	lb/in.²	lb/ft²
1 pascal	7.501×10^{-4}	1.450×10^{-4}	2.089×10^{-2}
1 atmosphere	76	14.70	2.116×10^{3}
1 centimeter mercury[a]	1	0.194 3	27.85
1 pound per square inch	5.171	1	144
1 pound per square foot	3.591×10^{-2}	6.944×10^{-3}	1

[a]At 0°C and at a location where the free-fall acceleration has its "standard" value, 9.806 65 m/s².

TABLE A.2 | Symbols, Dimensions, and Units of Physical Quantities

Quantity	Common Symbol	Unit[a]	Dimensions[b]	Unit in Terms of Base SI Units
Acceleration	$\vec{a}$	m/s²	L/T^2	m/s²
Amount of substance	n	MOLE		mol
Angle	θ, ϕ	radian (rad)	1	
Angular acceleration	$\vec{\alpha}$	rad/s²	T^{-2}	s^{-2}
Angular frequency	ω	rad/s	T^{-1}	s^{-1}
Angular momentum	$\vec{L}$	kg · m²/s	ML^2/T	kg · m²/s
Angular velocity	$\vec{\omega}$	rad/s	T^{-1}	s^{-1}
Area	A	m²	L^2	m²
Atomic number	Z			
Capacitance	C	farad (F)	Q^2T^2/ML^2	A² · s⁴/kg · m²
Charge	q, Q, e	coulomb (C)	Q	A · s

(*Continued*)

TABLE A.2 | Symbols, Dimensions, and Units of Physical Quantities (*continued*)

Quantity	Common Symbol	Unit[a]	Dimensions[b]	Unit in Terms of Base SI Units
Charge density				
Line	λ	C/m	Q/L	$A \cdot s/m$
Surface	σ	C/m^2	Q/L^2	$A \cdot s/m^2$
Volume	ρ	C/m^3	Q/L^3	$A \cdot s/m^3$
Conductivity	σ	$1/\Omega \cdot m$	Q^2T/ML^3	$A^2 \cdot s^3/kg \cdot m^3$
Current	I	AMPERE	Q/T	A
Current density	J	A/m^2	Q/TL^2	A/m^2
Density	ρ	kg/m^3	M/L^3	kg/m^3
Dielectric constant	κ			
Electric dipole moment	$\vec{\mathbf{p}}$	$C \cdot m$	QL	$A \cdot s \cdot m$
Electric field	$\vec{\mathbf{E}}$	V/m	ML/QT^2	$kg \cdot m/A \cdot s^3$
Electric flux	Φ_E	$V \cdot m$	ML^3/QT^2	$kg \cdot m^3/A \cdot s^3$
Electromotive force	$\mathcal{E}$	volt (V)	ML^2/QT^2	$kg \cdot m^2/A \cdot s^3$
Energy	E, U, K	joule (J)	ML^2/T^2	$kg \cdot m^2/s^2$
Entropy	S	J/K	ML^2/T^2K	$kg \cdot m^2/s^2 \cdot K$
Force	$\vec{\mathbf{F}}$	newton (N)	ML/T^2	$kg \cdot m/s^2$
Frequency	f	hertz (Hz)	T^{-1}	s^{-1}
Heat	Q	joule (J)	ML^2/T^2	$kg \cdot m^2/s^2$
Inductance	L	henry (H)	ML^2/Q^2	$kg \cdot m^2/A^2 \cdot s^2$
Length	ℓ, L	METER	L	m
Displacement	$\Delta x, \Delta\vec{\mathbf{r}}$			
Distance	d, h			
Position	$x, y, z, \vec{\mathbf{r}}$			
Magnetic dipole moment	$\vec{\boldsymbol{\mu}}$	$N \cdot m/T$	QL^2/T	$A \cdot m^2$
Magnetic field	$\vec{\mathbf{B}}$	tesla (T) ($= Wb/m^2$)	M/QT	$kg/A \cdot s^2$
Magnetic flux	Φ_B	weber (Wb)	ML^2/QT	$kg \cdot m^2/A \cdot s^2$
Mass	m, M	KILOGRAM	M	kg
Molar specific heat	C	$J/mol \cdot K$		$kg \cdot m^2/s^2 \cdot mol \cdot K$
Moment of inertia	I	$kg \cdot m^2$	ML^2	$kg \cdot m^2$
Momentum	$\vec{\mathbf{p}}$	$kg \cdot m/s$	ML/T	$kg \cdot m/s$
Period	T	s	T	s
Permeability of free space	μ_0	N/A^2 ($= H/m$)	ML/Q^2	$kg \cdot m/A^2 \cdot s^2$
Permittivity of free space	ϵ_0	$C^2/N \cdot m^2$ ($= F/m$)	Q^2T^2/ML^3	$A^2 \cdot s^4/kg \cdot m^3$
Potential	V	volt (V) ($= J/C$)	ML^2/QT^2	$kg \cdot m^2/A \cdot s^3$
Power	P	watt (W) ($= J/s$)	ML^2/T^3	$kg \cdot m^2/s^3$
Pressure	P	pascal (Pa) ($= N/m^2$)	M/LT^2	$kg/m \cdot s^2$
Resistance	R	ohm (Ω) ($= V/A$)	ML^2/Q^2T	$kg \cdot m^2/A^2 \cdot s^3$
Specific heat	c	$J/kg \cdot K$	L^2/T^2K	$m^2/s^2 \cdot K$
Speed	v	m/s	L/T	m/s
Temperature	T	KELVIN	K	K
Time	t	SECOND	T	s
Torque	$\vec{\boldsymbol{\tau}}$	$N \cdot m$	ML^2/T^2	$kg \cdot m^2/s^2$
Velocity	$\vec{\mathbf{v}}$	m/s	L/T	m/s
Volume	V	m^3	L^3	m^3
Wavelength	λ	m	L	m
Work	W	joule (J) ($= N \cdot m$)	ML^2/T^2	$kg \cdot m^2/s^2$

[a]The base SI units are given in uppercase letters.

[b]The symbols M, L, T, K, and Q denote mass, length, time, temperature, and charge, respectively.

TABLE A.3 | Chemical and Nuclear Information for Selected Isotopes

Atomic Number Z	Element	Chemical Symbol	Mass Number A (* means radioactive)	Mass of Neutral Atom (u)	Percent Abundance	Half-life, if Radioactive $T_{1/2}$
−1	electron	e−	0	0.000 549		
0	neutron	n	1*	1.008 665		614 s
1	hydrogen	^{1}H = p	1	1.007 825	99.988 5	
	[deuterium	^{2}H = D]	2	2.014 102	0.011 5	
	[tritium	^{3}H = T]	3*	3.016 049		12.33 yr
2	helium	He	3	3.016 029	0.000 137	
	[alpha particle	$\alpha = {}^4$He]	4	4.002 603	99.999 863	
			6*	6.018 889		0.81 s
3	lithium	Li	6	6.015 123	7.5	
			7	7.016 005	92.5	
4	beryllium	Be	7*	7.016 930		53.3 d
			8*	8.005 305		10^{-17} s
			9	9.012 182	100	
5	boron	B	10	10.012 937	19.9	
			11	11.009 305	80.1	
6	carbon	C	11*	11.011 434		20.4 min
			12	12.000 000	98.93	
			13	13.003 355	1.07	
			14*	14.003 242		5 730 yr
7	nitrogen	N	13*	13.005 739		9.96 min
			14	14.003 074	99.632	
			15	15.000 109	0.368	
8	oxygen	O	14*	14.008 596		70.6 s
			15*	15.003 066		122 s
			16	15.994 915	99.757	
			17	16.999 132	0.038	
			18	17.999 161	0.205	
9	fluorine	F	18*	18.000 938		109.8 min
			19	18.998 403	100	
10	neon	Ne	20	19.992 440	90.48	
11	sodium	Na	23	22.989 769	100	
12	magnesium	Mg	23*	22.994 124		11.3 s
			24	23.985 042	78.99	
13	aluminum	Al	27	26.981 539	100	
14	silicon	Si	27*	26.986 705		4.2 s
15	phosphorus	P	30*	29.978 314		2.50 min
			31	30.973 762	100	
			32*	31.973 907		14.26 d
16	sulfur	S	32	31.972 071	94.93	
19	potassium	K	39	38.963 707	93.258 1	
			40*	39.963 998	0.011 7	1.28×10^9 yr
20	calcium	Ca	40	39.962 591	96.941	
			42	41.958 618	0.647	
			43	42.958 767	0.135	
25	manganese	Mn	55	54.938 045	100	
26	iron	Fe	56	55.934 938	91.754	
			57	56.935 394	2.119	

(*Continued*)

TABLE A.3 | Chemical and Nuclear Information for Selected Isotopes (*continued*)

Atomic Number Z	Element	Chemical Symbol	Mass Number A (* means radioactive)	Mass of Neutral Atom (u)	Percent Abundance	Half-life, if Radioactive $T_{1/2}$
27	cobalt	Co	57*	56.936 291		272 d
			59	58.933 195	100	
			60*	59.933 817		5.27 yr
28	nickel	Ni	58	57.935 343	68.076 9	
			60	59.930 786	26.223 1	
29	copper	Cu	63	62.929 598	69.17	
			64*	63.929 764		12.7 h
			65	64.927 789	30.83	
30	zinc	Zn	64	63.929 142	48.63	
37	rubidium	Rb	87*	86.909 181	27.83	
38	strontium	Sr	87	86.908 877	7.00	
			88	87.905 612	82.58	
			90*	89.907 738		29.1 yr
41	niobium	Nb	93	92.906 378	100	
42	molybdenum	Mo	94	93.905 088	9.25	
44	ruthenium	Ru	98	97.905 287	1.87	
54	xenon	Xe	136*	135.907 219		2.4×10^{21} yr
55	cesium	Cs	137*	136.907 090		30 yr
56	barium	Ba	137	136.905 827	11.232	
58	cerium	Ce	140	139.905 439	88.450	
59	praseodymium	Pr	141	140.907 653	100	
60	neodymium	Nd	144*	143.910 087	23.8	2.3×10^{15} yr
61	promethium	Pm	145*	144.912 749		17.7 yr
79	gold	Au	197	196.966 569	100	
80	mercury	Hg	198	197.966 769	9.97	
			202	201.970 643	29.86	
82	lead	Pb	206	205.974 465	24.1	
			207	206.975 897	22.1	
			208	207.976 652	52.4	
			214*	213.999 805		26.8 min
83	bismuth	Bi	209	208.980 399	100	
84	polonium	Po	210*	209.982 874		138.38 d
			216*	216.001 915		0.145 s
			218*	218.008 973		3.10 min
86	radon	Rn	220*	220.011 394		55.6 s
			222*	222.017 578		3.823 d
88	radium	Ra	226*	226.025 410		1 600 yr
90	thorium	Th	232*	232.038 055	100	1.40×10^{10} yr
			234*	234.043 601		24.1 d
92	uranium	U	234*	234.040 952		2.45×10^{5} yr
			235*	235.043 930	0.720 0	7.04×10^{8} yr
			236*	236.045 568		2.34×10^{7} yr
			238*	238.050 788	99.274 5	4.47×10^{9} yr
93	neptunium	Np	236*	236.046 570		1.15×10^{5} yr
			237*	237.048 173		2.14×10^{6} yr
94	plutonium	Pu	239*	239.052 163		24 120 yr

Source: G. Audi, A. H. Wapstra, and C. Thibault, "The AME2003 Atomic Mass Evaluation," *Nuclear Physics A* **729**: 337–676, 2003.

Appendix B

Mathematics Review

This appendix in mathematics is intended as a brief review of operations and methods. Early in this course, you should be totally familiar with basic algebraic techniques, analytic geometry, and trigonometry. The sections on differential and integral calculus are more detailed and are intended for students who have difficulty applying calculus concepts to physical situations.

◀ B.1 | Scientific Notation

Many quantities used by scientists often have very large or very small values. The speed of light, for example, is about 300 000 000 m/s, and the ink required to make the dot over an i in this textbook has a mass of about 0.000 000 001 kg. Obviously, it is very cumbersome to read, write, and keep track of such numbers. We avoid this problem by using a method incorporating powers of the number 10:

$$10^0 = 1$$
$$10^1 = 10$$
$$10^2 = 10 \times 10 = 100$$
$$10^3 = 10 \times 10 \times 10 = 1\,000$$
$$10^4 = 10 \times 10 \times 10 \times 10 = 10\,000$$
$$10^5 = 10 \times 10 \times 10 \times 10 \times 10 = 100\,000$$

and so on. The number of zeros corresponds to the power to which ten is raised, called the **exponent** of ten. For example, the speed of light, 300 000 000 m/s, can be expressed as 3.00×10^8 m/s.

In this method, some representative numbers smaller than unity are the following:

$$10^{-1} = \frac{1}{10} = 0.1$$

$$10^{-2} = \frac{1}{10 \times 10} = 0.01$$

$$10^{-3} = \frac{1}{10 \times 10 \times 10} = 0.001$$

$$10^{-4} = \frac{1}{10 \times 10 \times 10 \times 10} = 0.000\,1$$

$$10^{-5} = \frac{1}{10 \times 10 \times 10 \times 10 \times 10} = 0.000\,01$$

In these cases, the number of places the decimal point is to the left of the digit 1 equals the value of the (negative) exponent. Numbers expressed as some power of ten multiplied by another number between one and ten are said to be in **scientific notation.** For example, the scientific notation for 5 943 000 000 is 5.943×10^9 and that for 0.000 083 2 is 8.32×10^{-5}.

When numbers expressed in scientific notation are being multiplied, the following general rule is very useful:

$$10^n \times 10^m = 10^{n+m} \qquad \text{B.1} ◀$$

where n and m can be *any* numbers (not necessarily integers). For example, $10^2 \times 10^5 = 10^7$. The rule also applies if one of the exponents is negative: $10^3 \times 10^{-8} = 10^{-5}$.

When dividing numbers expressed in scientific notation, note that

$$\frac{10^n}{10^m} = 10^n \times 10^{-m} = 10^{n-m}$$

B.2◄

Exercises

With help from the preceding rules, verify the answers to the following equations:

1. $86\,400 = 8.64 \times 10^4$
2. $9\,816\,762.5 = 9.816\,762\,5 \times 10^6$
3. $0.000\,000\,039\,8 = 3.98 \times 10^{-8}$
4. $(4.0 \times 10^8)(9.0 \times 10^9) = 3.6 \times 10^{18}$
5. $(3.0 \times 10^7)(6.0 \times 10^{-12}) = 1.8 \times 10^{-4}$
6. $\dfrac{75 \times 10^{-11}}{5.0 \times 10^{-3}} = 1.5 \times 10^{-7}$
7. $\dfrac{(3 \times 10^6)(8 \times 10^{-2})}{(2 \times 10^{17})(6 \times 10^5)} = 2 \times 10^{-18}$

◣B.2 | Algebra

Some Basic Rules

When algebraic operations are performed, the laws of arithmetic apply. Symbols such as x, y, and z are usually used to represent unspecified quantities, called the **unknowns.**

First, consider the equation

$$8x = 32$$

If we wish to solve for x, we can divide (or multiply) each side of the equation by the same factor without destroying the equality. In this case, if we divide both sides by 8, we have

$$\frac{8x}{8} = \frac{32}{8}$$
$$x = 4$$

Next consider the equation

$$x + 2 = 8$$

In this type of expression, we can add or subtract the same quantity from each side. If we subtract 2 from each side, we have

$$x + 2 - 2 = 8 - 2$$
$$x = 6$$

In general, if $x + a = b$, then $x = b - a$.

Now consider the equation

$$\frac{x}{5} = 9$$

If we multiply each side by 5, we are left with x on the left by itself and 45 on the right:

$$\left(\frac{x}{5}\right)(5) = 9 \times 5$$
$$x = 45$$

In all cases, *whatever operation is performed on the left side of the equality must also be performed on the right side.*

The following rules for multiplying, dividing, adding, and subtracting fractions should be recalled, where *a*, *b*, *c*, and *d* are four numbers:

	Rule	Example
Multiplying	$\left(\dfrac{a}{b}\right)\left(\dfrac{c}{d}\right) = \dfrac{ac}{bd}$	$\left(\dfrac{2}{3}\right)\left(\dfrac{4}{5}\right) = \dfrac{8}{15}$
Dividing	$\dfrac{(a/b)}{(c/d)} = \dfrac{ad}{bc}$	$\dfrac{2/3}{4/5} = \dfrac{(2)(5)}{(4)(3)} = \dfrac{10}{12}$
Adding	$\dfrac{a}{b} \pm \dfrac{c}{d} = \dfrac{ad \pm bc}{bd}$	$\dfrac{2}{3} - \dfrac{4}{5} = \dfrac{(2)(5) - (4)(3)}{(3)(5)} = -\dfrac{2}{15}$

Exercises

In the following exercises, solve for *x*.

Answers

1. $a = \dfrac{1}{1 + x}$ $x = \dfrac{1 - a}{a}$

2. $3x - 5 = 13$ $x = 6$

3. $ax - 5 = bx + 2$ $x = \dfrac{7}{a - b}$

4. $\dfrac{5}{2x + 6} = \dfrac{3}{4x + 8}$ $x = -\dfrac{11}{7}$

Powers

When powers of a given quantity *x* are multiplied, the following rule applies:

$$x^n x^m = x^{n+m} \qquad \text{B.3} \blacktriangleleft$$

For example, $x^2 x^4 = x^{2+4} = x^6$.

When dividing the powers of a given quantity, the rule is

$$\frac{x^n}{x^m} = x^{n-m} \qquad \text{B.4} \blacktriangleleft$$

For example, $x^8/x^2 = x^{8-2} = x^6$.

A power that is a fraction, such as $\frac{1}{3}$, corresponds to a root as follows:

$$x^{1/n} = \sqrt[n]{x} \qquad \text{B.5} \blacktriangleleft$$

TABLE B.1 | Rules of Exponents

$$x^0 = 1$$
$$x^1 = x$$
$$x^n x^m = x^{n+m}$$
$$x^n/x^m = x^{n-m}$$
$$x^{1/n} = \sqrt[n]{x}$$
$$(x^n)^m = x^{nm}$$

For example, $4^{1/3} = \sqrt[3]{4} = 1.587\,4$. (A scientific calculator is useful for such calculations.)

Finally, any quantity x^n raised to the *m*th power is

$$(x^n)^m = x^{nm} \qquad \text{B.6} \blacktriangleleft$$

Table B.1 summarizes the rules of exponents.

Exercises

Verify the following equations:

1. $3^2 \times 3^3 = 243$
2. $x^5 x^{-8} = x^{-3}$
3. $x^{10}/x^{-5} = x^{15}$
4. $5^{1/3} = 1.709\,976$ (Use your calculator.)
5. $60^{1/4} = 2.783\,158$ (Use your calculator.)
6. $(x^4)^3 = x^{12}$

Factoring

Some useful formulas for factoring an equation are the following:

$$ax + ay + az = a(x + y + z)$$ common factor

$$a^2 + 2ab + b^2 = (a + b)^2$$ perfect square

$$a^2 - b^2 = (a + b)(a - b)$$ differences of squares

Quadratic Equations

The general form of a quadratic equation is

$$ax^2 + bx + c = 0$$ B.7 ◄

where x is the unknown quantity and a, b, and c are numerical factors referred to as **coefficients** of the equation. This equation has two roots, given by

$$x = \frac{-b \pm \sqrt{b^2 - 4ac}}{2a}$$ B.8 ◄

If $b^2 \geq 4ac$, the roots are real.

Example B.1

The equation $x^2 + 5x + 4 = 0$ has the following roots corresponding to the two signs of the square-root term:

$$x = \frac{-5 \pm \sqrt{5^2 - (4)(1)(4)}}{2(1)} = \frac{-5 \pm \sqrt{9}}{2} = \frac{-5 \pm 3}{2}$$

$$x_+ = \frac{-5 + 3}{2} = -1 \quad x_- = \frac{-5 - 3}{2} = -4$$

where x_+ refers to the root corresponding to the positive sign and x_- refers to the root corresponding to the negative sign.

Exercises

Solve the following quadratic equations:

Answers

1. $x^2 + 2x - 3 = 0$ $x_+ = 1$ $x_- = -3$
2. $2x^2 - 5x + 2 = 0$ $x_+ = 2$ $x_- = \frac{1}{2}$
3. $2x^2 - 4x - 9 = 0$ $x_+ = 1 + \sqrt{22}/2$ $x_- = 1 - \sqrt{22}/2$

Linear Equations

A linear equation has the general form

$$y = mx + b$$ B.9 ◄

where m and b are constants. This equation is referred to as linear because the graph of y versus x is a straight line as shown in Figure B.1. The constant b, called the **y-intercept**, represents the value of y at which the straight line intersects the y axis. The constant m is equal to the **slope** of the straight line. If any two points on the straight line are specified by the coordinates (x_1, y_1) and (x_2, y_2) as in Figure B.1, the slope of the straight line can be expressed as

$$\text{Slope} = \frac{y_2 - y_1}{x_2 - x_1} = \frac{\Delta y}{\Delta x}$$ B.10 ◄

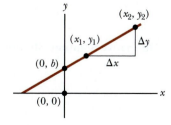

Figure B.1 A straight line graphed on an xy coordinate system. The slope of the line is the ratio of Δy to Δx.

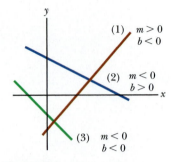

Figure B.2 The brown line has a positive slope and a negative y-intercept. The blue line has a negative slope and a positive y-intercept. The green line has a negative slope and a negative y-intercept.

Note that m and b can have either positive or negative values. If $m > 0$, the straight line has a *positive* slope as in Figure B.1. If $m < 0$, the straight line has a *negative* slope. In Figure B.1, both m and b are positive. Three other possible situations are shown in Figure B.2.

Exercises

1. Draw graphs of the following straight lines: (a) $y = 5x + 3$ (b) $y = -2x + 4$ (c) $y = -3x - 6$
2. Find the slopes of the straight lines described in Exercise 1.

Answers (a) 5 (b) -2 (c) -3

3. Find the slopes of the straight lines that pass through the following sets of points: (a) $(0, -4)$ and $(4, 2)$ (b) $(0, 0)$ and $(2, -5)$ (c) $(-5, 2)$ and $(4, -2)$

Answers (a) $\frac{3}{2}$ (b) $-\frac{5}{2}$ (c) $-\frac{4}{9}$

Solving Simultaneous Linear Equations

Consider the equation $3x + 5y = 15$, which has two unknowns, x and y. Such an equation does not have a unique solution. For example, $(x = 0, y = 3)$, $(x = 5, y = 0)$, and $(x = 2, y = \frac{9}{5})$ are all solutions to this equation.

If a problem has two unknowns, a unique solution is possible only if we have *two* pieces of information. In most common cases, those two pieces of information are equations. In general, if a problem has n unknowns, its solution requires n equations. To solve two simultaneous equations involving two unknowns, x and y, we solve one of the equations for x in terms of y and substitute this expression into the other equation.

In some cases, the two pieces of information may be (1) one equation and (2) a condition on the solutions. For example, suppose we have the equation $m = 3n$ and the condition that m and n must be the smallest positive nonzero integers possible. Then, the single equation does not allow a unique solution, but the addition of the condition gives us that $n = 1$ and $m = 3$.

Example B.2

Solve the two simultaneous equations.

$$(1) \ \ 5x + y = -8$$

$$(2) \ \ 2x - 2y = 4$$

SOLUTION

From Equation (2), $x = y + 2$. Substitution of this equation into Equation (1) gives

$$5(y + 2) + y = -8$$

$$6y = -18$$

$$y = \boxed{-3}$$

$$x = y + 2 = \boxed{-1}$$

Alternative Solution Multiply each term in Equation (1) by the factor 2 and add the result to Equation (2):

$$10x + 2y = -16$$

$$\underline{2x - 2y = 4}$$

$$12x \qquad = -12$$

$$x = \boxed{-1}$$

$$y = x - 2 = \boxed{-3}$$

Two linear equations containing two unknowns can also be solved by a graphical method. If the straight lines corresponding to the two equations are plotted in a conventional coordinate system, the intersection of the two lines represents the solution. For example, consider the two equations

$$x - y = 2$$

$$x - 2y = -1$$

These equations are plotted in Figure B.3. The intersection of the two lines has the coordinates $x = 5$ and $y = 3$, which represents the solution to the equations. You should check this solution by the analytical technique discussed earlier.

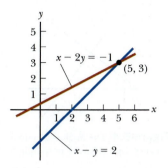

Figure B.3 A graphical solution for two linear equations.

Exercises

Solve the following pairs of simultaneous equations involving two unknowns:

	Answers
1. $x + y = 8$ $x - y = 2$	$x = 5, y = 3$
2. $98 - T = 10a$ $T - 49 = 5a$	$T = 65, a = 3.27$
3. $6x + 2y = 6$ $8x - 4y = 28$	$x = 2, y = -3$

Logarithms

Suppose a quantity x is expressed as a power of some quantity a:

$$x = a^y \qquad \text{B.11} \blacktriangleleft$$

The number a is called the **base** number. The **logarithm** of x with respect to the base a is equal to the exponent to which the base must be raised to satisfy the expression $x = a^y$:

$$y = \log_a x \qquad \text{B.12} \blacktriangleleft$$

Conversely, the **antilogarithm** of y is the number x:

$$x = \text{antilog}_a y \qquad \text{B.13} \blacktriangleleft$$

In practice, the two bases most often used are base 10, called the *common* logarithm base, and base $e = 2.718\ 282$, called Euler's constant or the *natural* logarithm base. When common logarithms are used,

$$y = \log_{10} x \quad (\text{or } x = 10^y) \qquad \text{B.14} \blacktriangleleft$$

When natural logarithms are used,

$$y = \ln x \quad (\text{or } x = e^y) \qquad \text{B.15} \blacktriangleleft$$

For example, $\log_{10} 52 = 1.716$, so antilog$_{10}$ $1.716 = 10^{1.716} = 52$. Likewise, $\ln 52 = 3.951$, so antiln $3.951 = e^{3.951} = 52$.

In general, note you can convert between base 10 and base e with the equality

$$\ln x = (2.302\ 585) \log_{10} x \qquad \text{B.16} \blacktriangleleft$$

Finally, some useful properties of logarithms are the following:

$$\left. \begin{aligned} \log(ab) &= \log a + \log b \\ \log(a/b) &= \log a - \log b \\ \log(a^n) &= n \log a \end{aligned} \right\} \quad \text{any base}$$

$$\ln e = 1$$

$$\ln e^a = a$$

$$\ln \left(\frac{1}{a} \right) = -\ln a$$

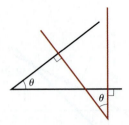

Figure B.4 The angles are equal because their sides are perpendicular.

Figure B.5 The angle θ in radians is the ratio of the arc length s to the radius r of the circle.

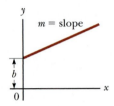

Figure B.6 A straight line with a slope of m and a y-intercept of b.

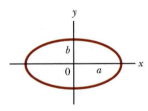

Figure B.7 An ellipse with semi-major axis a and semiminor axis b.

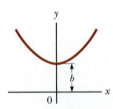

Figure B.8 A parabola with its vertex at $y = b$.

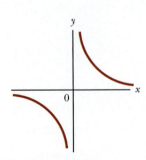

Figure B.9 A hyperbola.

◖B.3 | Geometry

The **distance** d between two points having coordinates (x_1, y_1) and (x_2, y_2) is

$$d = \sqrt{(x_2 - x_1)^2 + (y_2 - y_1)^2}$$
       B.17◄

Two angles are equal if their sides are perpendicular, right side to right side and left side to left side. For example, the two angles marked θ in Figure B.4 are the same because of the perpendicularity of the sides of the angles. To distinguish the left and right sides of an angle, imagine standing at the angle's apex and facing into the angle.

Radian measure: The arc length s of a circular arc (Fig. B.5) is proportional to the radius r for a fixed value of θ (in radians):

$$s = r\theta$$
$$\theta = \frac{s}{r}$$
       B.18◄

Table B.2 gives the **areas** and **volumes** for several geometric shapes used throughout this text.

The equation of a **straight line** (Fig. B.6) is

$$y = mx + b$$
       B.19◄

where b is the y-intercept and m is the slope of the line.

The equation of a **circle** of radius R centered at the origin is

$$x^2 + y^2 = R^2$$
       B.20◄

The equation of an **ellipse** having the origin at its center (Fig. B.7) is

$$\frac{x^2}{a^2} + \frac{y^2}{b^2} = 1$$
       B.21◄

where a is the length of the semimajor axis (the longer one) and b is the length of the semiminor axis (the shorter one).

The equation of a **parabola** the vertex of which is at $y = b$ (Fig. B.8) is

$$y = ax^2 + b$$
       B.22◄

The equation of a **rectangular hyperbola** (Fig. B.9) is

$$xy = \text{constant}$$
       B.23◄

◖TABLE B.2 | Useful Information for Geometry

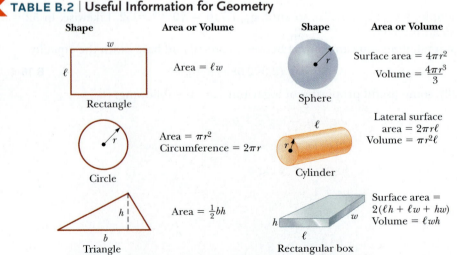

Shape	Area or Volume	Shape	Area or Volume
Rectangle	Area $= \ell w$	Sphere	Surface area $= 4\pi r^2$ Volume $= \dfrac{4\pi r^3}{3}$
Circle	Area $= \pi r^2$ Circumference $= 2\pi r$	Cylinder	Lateral surface area $= 2\pi r\ell$ Volume $= \pi r^2 \ell$
Triangle	Area $= \frac{1}{2}bh$	Rectangular box	Surface area $= 2(\ell h + \ell w + hw)$ Volume $= \ell wh$

B.4 | Trigonometry

That portion of mathematics based on the special properties of the right triangle is called trigonometry. By definition, a right triangle is a triangle containing a 90° angle. Consider the right triangle shown in Figure B.10, where side a is opposite the angle θ, side b is adjacent to the angle θ, and side c is the hypotenuse of the triangle. The three basic trigonometric functions defined by such a triangle are the sine (sin), cosine (cos), and tangent (tan). In terms of the angle θ, these functions are defined as follows:

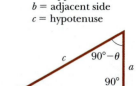

a = opposite side
b = adjacent side
c = hypotenuse

Figure B.10 A right triangle, used to define the basic functions of trigonometry.

$$\sin \theta = \frac{\text{side opposite } \theta}{\text{hypotenuse}} = \frac{a}{c} \qquad \text{B.24} \blacktriangleleft$$

$$\cos \theta = \frac{\text{side adjacent to } \theta}{\text{hypotenuse}} = \frac{b}{c} \qquad \text{B.25} \blacktriangleleft$$

$$\tan \theta = \frac{\text{side opposite } \theta}{\text{side adjacent to } \theta} = \frac{a}{b} \qquad \text{B.26} \blacktriangleleft$$

The Pythagorean theorem provides the following relationship among the sides of a right triangle:

$$c^2 = a^2 + b^2 \qquad \text{B.27} \blacktriangleleft$$

From the preceding definitions and the Pythagorean theorem, it follows that

$$\sin^2 \theta + \cos^2 \theta = 1$$

$$\tan \theta = \frac{\sin \theta}{\cos \theta}$$

The cosecant, secant, and cotangent functions are defined by

$$\csc \theta = \frac{1}{\sin \theta} \qquad \sec \theta = \frac{1}{\cos \theta} \qquad \cot \theta = \frac{1}{\tan \theta}$$

The following relationships are derived directly from the right triangle shown in Figure B.10:

$$\sin \theta = \cos (90° - \theta)$$

$$\cos \theta = \sin (90° - \theta)$$

$$\cot \theta = \tan (90° - \theta)$$

Some properties of trigonometric functions are the following:

$$\sin (-\theta) = -\sin \theta$$

$$\cos (-\theta) = \cos \theta$$

$$\tan (-\theta) = -\tan \theta$$

The following relationships apply to *any* triangle as shown in Figure B.11:

$$\alpha + \beta + \gamma = 180°$$

$$\text{Law of cosines} \begin{cases} a^2 = b^2 + c^2 - 2bc \cos \alpha \\ b^2 = a^2 + c^2 - 2ac \cos \beta \\ c^2 = a^2 + b^2 - 2ab \cos \gamma \end{cases}$$

$$\text{Law of sines} \qquad \frac{a}{\sin \alpha} = \frac{b}{\sin \beta} = \frac{c}{\sin \gamma}$$

Table B.3 (page A-14) lists a number of useful trigonometric identities.

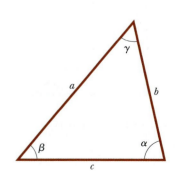

Figure B.11 An arbitrary, nonright triangle.

TABLE B.3 | **Some Trigonometric Identities**

$$\sin^2 \theta + \cos^2 \theta = 1 \qquad\qquad\qquad \csc^2 \theta = 1 + \cot^2 \theta$$

$$\sec^2 \theta = 1 + \tan^2 \theta \qquad\qquad\qquad \sin^2 \frac{\theta}{2} = \tfrac{1}{2}(1 - \cos \theta)$$

$$\sin 2\theta = 2 \sin \theta \cos \theta \qquad\qquad\qquad \cos^2 \frac{\theta}{2} = \tfrac{1}{2}(1 + \cos \theta)$$

$$\cos 2\theta = \cos^2 \theta - \sin^2 \theta \qquad\qquad\qquad 1 - \cos \theta = 2 \sin^2 \frac{\theta}{2}$$

$$\tan 2\theta = \frac{2 \tan \theta}{1 - \tan^2 \theta} \qquad\qquad\qquad \tan \frac{\theta}{2} = \sqrt{\frac{1 - \cos \theta}{1 + \cos \theta}}$$

$$\sin (A \pm B) = \sin A \cos B \pm \cos A \sin B$$

$$\cos (A \pm B) = \cos A \cos B \mp \sin A \sin B$$

$$\sin A \pm \sin B = 2 \sin \left[\tfrac{1}{2}(A \pm B) \right] \cos \left[\tfrac{1}{2}(A \mp B) \right]$$

$$\cos A + \cos B = 2 \cos \left[\tfrac{1}{2}(A + B) \right] \cos \left[\tfrac{1}{2}(A - B) \right]$$

$$\cos A - \cos B = 2 \sin \left[\tfrac{1}{2}(A + B) \right] \sin \left[\tfrac{1}{2}(B - A) \right]$$

Example **B.3**

Consider the right triangle in Figure B.12 in which $a = 2.00$, $b = 5.00$, and c is unknown. From the Pythagorean theorem, we have

$$c^2 = a^2 + b^2 = 2.00^2 + 5.00^2 = 4.00 + 25.0 = 29.0$$

$$c = \sqrt{29.0} = \boxed{5.39}$$

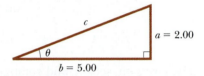

Figure B.12 (Example B.3)

To find the angle θ, note that

$$\tan \theta = \frac{a}{b} = \frac{2.00}{5.00} = 0.400$$

Using a calculator, we find that

$$\theta = \tan^{-1} (0.400) = \boxed{21.8°}$$

where $\tan^{-1} (0.400)$ is the notation for "angle whose tangent is 0.400," sometimes written as arctan (0.400).

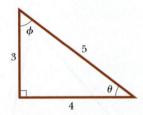

Figure B.13 (Exercise 1)

Exercises

1. In Figure B.13, identify (a) the side opposite θ (b) the side adjacent to ϕ and then find (c) $\cos \theta$, (d) $\sin \phi$, and (e) $\tan \phi$.

   Answers   (a) 3   (b) 3   (c) $\frac{4}{5}$ (d) $\frac{4}{5}$ (e) $\frac{4}{3}$

2. In a certain right triangle, the two sides that are perpendicular to each other are 5.00 m and 7.00 m long. What is the length of the third side?

   Answer   8.60 m

3. A right triangle has a hypotenuse of length 3.0 m, and one of its angles is 30°. (a) What is the length of the side opposite the 30° angle? (b) What is the side adjacent to the 30° angle?

   Answers   (a) 1.5 m   (b) 2.6 m

B.5 | Series Expansions

$$(a + b)^n = a^n + \frac{n}{1!} a^{n-1} b + \frac{n(n-1)}{2!} a^{n-2} b^2 + \cdots$$

$$(1 + x)^n = 1 + nx + \frac{n(n-1)}{2!} x^2 + \cdots$$

$$e^x = 1 + x + \frac{x^2}{2!} + \frac{x^3}{3!} + \cdots$$

$$\ln (1 \pm x) = \pm x - \frac{1}{2}x^2 \pm \frac{1}{3}x^3 - \cdots$$

$$\sin x = x - \frac{x^3}{3!} + \frac{x^5}{5!} - \cdots$$

$$\cos x = 1 - \frac{x^2}{2!} + \frac{x^4}{4!} - \cdots$$

$$\left. \tan x = x + \frac{x^3}{3} + \frac{2x^5}{15} + \cdots \; |x| < \frac{\pi}{2} \right\}$$ x in radians

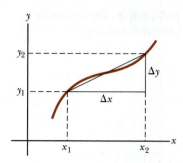

Figure B.14 The lengths Δx and Δy are used to define the derivative of this function at a point.

For $x \ll 1$, the following approximations can be used:[1]

$$(1 + x)^n \approx 1 + nx \qquad \sin x \approx x$$

$$e^x \approx 1 + x \qquad \cos x \approx 1$$

$$\ln (1 \pm x) \approx \pm x \qquad \tan x \approx x$$

B.6 | Differential Calculus

In various branches of science, it is sometimes necessary to use the basic tools of calculus, invented by Newton, to describe physical phenomena. The use of calculus is fundamental in the treatment of various problems in Newtonian mechanics, electricity, and magnetism. In this section, we simply state some basic properties and "rules of thumb" that should be a useful review to the student.

First, a **function** must be specified that relates one variable to another (e.g., a coordinate as a function of time). Suppose one of the variables is called y (the dependent variable), and the other x (the independent variable). We might have a function relationship such as

$$y(x) = ax^3 + bx^2 + cx + d$$

If a, b, c, and d are specified constants, y can be calculated for any value of x. We usually deal with continuous functions, that is, those for which y varies "smoothly" with x.

The **derivative** of y with respect to x is defined as the limit as Δx approaches zero of the slopes of chords drawn between two points on the y versus x curve. Mathematically, we write this definition as

$$\frac{dy}{dx} = \lim_{\Delta x \to 0} \frac{\Delta y}{\Delta x} = \lim_{\Delta x \to 0} \frac{y(x + \Delta x) - y(x)}{\Delta x} \qquad \text{B.28} \blacktriangleleft$$

where Δy and Δx are defined as $\Delta x = x_2 - x_1$ and $\Delta y = y_2 - y_1$ (Fig. B.14). Note that dy/dx *does not* mean dy divided by dx, but rather is simply a notation of the limiting process of the derivative as defined by Equation B.28.

A useful expression to remember when $y(x) = ax^n$, where a is a *constant* and n is *any* positive or negative number (integer or fraction), is

$$\frac{dy}{dx} = nax^{n-1} \qquad \text{B.29} \blacktriangleleft$$

---

[1] The approximations for the functions sin x, cos x, and tan x are for $x \le 0.1$ rad.

TABLE B.4 | Derivative for Several Functions

$$\frac{d}{dx}(a) = 0$$

$$\frac{d}{dx}(ax^n) = nax^{n-1}$$

$$\frac{d}{dx}(e^{ax}) = ae^{ax}$$

$$\frac{d}{dx}(\sin ax) = a \cos ax$$

$$\frac{d}{dx}(\cos ax) = -a \sin ax$$

$$\frac{d}{dx}(\tan ax) = a \sec^2 ax$$

$$\frac{d}{dx}(\cot ax) = -a \csc^2 ax$$

$$\frac{d}{dx}(\sec x) = \tan x \sec x$$

$$\frac{d}{dx}(\csc x) = -\cot x \csc x$$

$$\frac{d}{dx}(\ln ax) = \frac{1}{x}$$

$$\frac{d}{dx}(\sin^{-1} ax) = \frac{a}{\sqrt{1 - a^2 x^2}}$$

$$\frac{d}{dx}(\cos^{-1} ax) = \frac{-a}{\sqrt{1 - a^2 x^2}}$$

$$\frac{d}{dx}(\tan^{-1} ax) = \frac{a}{\sqrt{1 + a^2 x^2}}$$

Note: The symbols a and n represent constants.

If $y(x)$ is a polynomial or algebraic function of x, we apply Equation B.29 to *each* term in the polynomial and take $d[\text{constant}]/dx = 0$. In Examples B.4 through B.7, we evaluate the derivatives of several functions.

Special Properties of the Derivative

A. Derivative of the product of two functions   If a function $f(x)$ is given by the product of two functions—say, $g(x)$ and $h(x)$—the derivative of $f(x)$ is defined as

$$\frac{d}{dx}f(x) = \frac{d}{dx}[g(x)h(x)] = g\frac{dh}{dx} + h\frac{dg}{dx} \qquad \text{B.30} \blacktriangleleft$$

B. Derivative of the sum of two functions   If a function $f(x)$ is equal to the sum of two functions, the derivative of the sum is equal to the sum of the derivatives:

$$\frac{d}{dx}f(x) = \frac{d}{dx}[g(x) + h(x)] = \frac{dg}{dx} + \frac{dh}{dx} \qquad \text{B.31} \blacktriangleleft$$

C. Chain rule of differential calculus   If $y = f(x)$ and $x = g(z)$, then dy/dz can be written as the product of two derivatives:

$$\frac{dy}{dz} = \frac{dy}{dx}\frac{dx}{dz} \qquad \text{B.32} \blacktriangleleft$$

D. The second derivative   The second derivative of y with respect to x is defined as the derivative of the function dy/dx (the derivative of the derivative). It is usually written as

$$\frac{d^2y}{dx^2} = \frac{d}{dx}\left(\frac{dy}{dx}\right) \qquad \text{B.33} \blacktriangleleft$$

Some of the more commonly used derivatives of functions are listed in Table B.4.

Example B.4

Suppose $y(x)$ (that is, y as a function of x) is given by

$$y(x) = ax^3 + bx + c$$

where a and b are constants. It follows that

$$y(x + \Delta x) = a(x + \Delta x)^3 + b(x + \Delta x) + c$$

$$= a(x^3 + 3x^2\,\Delta x + 3x\,\Delta x^2 + \Delta x^3) + b(x + \Delta x) + c$$

so

$$\Delta y = y(x + \Delta x) - y(x) = a(3x^2\,\Delta x + 3x\,\Delta x^2 + \Delta x^3) + b\,\Delta x$$

Substituting this into Equation B.28 gives

$$\frac{dy}{dx} = \lim_{\Delta x \to 0} \frac{\Delta y}{\Delta x} = \lim_{\Delta x \to 0} [3ax^2 + 3ax\,\Delta x + a\,\Delta x^2] + b$$

$$\frac{dy}{dx} = \boxed{3ax^2 + b}$$

Example B.5

Find the derivative of

$$y(x) = 8x^5 + 4x^3 + 2x + 7$$

SOLUTION

Applying Equation B.29 to each term independently and remembering that d/dx (constant) $= 0$, we have

$$\frac{dy}{dx} = 8(5)x^4 + 4(3)x^2 + 2(1)x^0 + 0$$

$$\frac{dy}{dx} = 40x^4 + 12x^2 + 2$$

Example B.6

Find the derivative of $y(x) = x^3/(x+1)^2$ with respect to x.

SOLUTION

We can rewrite this function as $y(x) = x^3(x+1)^{-2}$ and apply Equation B.30:

$$\frac{dy}{dx} = (x+1)^{-2}\frac{d}{dx}(x^3) + x^3\frac{d}{dx}(x+1)^{-2}$$

$$= (x+1)^{-2}\,3x^2 + x^3(-2)(x+1)^{-3}$$

$$\frac{dy}{dx} = \frac{3x^2}{(x+1)^2} - \frac{2x^3}{(x+1)^3} = \frac{x^2(x+3)}{(x+1)^3}$$

Example B.7

A useful formula that follows from Equation B.30 is the derivative of the quotient of two functions. Show that

$$\frac{d}{dx}\left[\frac{g(x)}{h(x)}\right] = \frac{h\dfrac{dg}{dx} - g\dfrac{dh}{dx}}{h^2}$$

SOLUTION

We can write the quotient as gh^{-1} and then apply Equations B.29 and B.30:

$$\frac{d}{dx}\left(\frac{g}{h}\right) = \frac{d}{dx}(gh^{-1}) = g\frac{d}{dx}(h^{-1}) + h^{-1}\frac{d}{dx}(g)$$

$$= -gh^{-2}\frac{dh}{dx} + h^{-1}\frac{dg}{dx}$$

$$= \frac{h\dfrac{dg}{dx} - g\dfrac{dh}{dx}}{h^2}$$

❮B.7 | Integral Calculus

We think of integration as the inverse of differentiation. As an example, consider the expression

$$f(x) = \frac{dy}{dx} = 3ax^2 + b \qquad \text{B.34}◀$$

which was the result of differentiating the function

$$y(x) = ax^3 + bx + c$$

in Example B.4. We can write Equation B.34 as $dy = f(x)\,dx = (3ax^2 + b)\,dx$ and obtain $y(x)$ by "summing" over all values of x. Mathematically, we write this inverse operation as

$$y(x) = \int f(x)\,dx$$

For the function $f(x)$ given by Equation B.34, we have

$$y(x) = \int (3ax^2 + b)\,dx = ax^3 + bx + c$$

where c is a constant of the integration. This type of integral is called an *indefinite integral* because its value depends on the choice of c.

A general **indefinite integral** $I(x)$ is defined as

$$I(x) = \int f(x)\,dx \qquad \text{B.35}◀$$

where $f(x)$ is called the *integrand* and $f(x) = dI(x)/dx$.

For a *general continuous* function $f(x)$, the integral can be described as the area under the curve bounded by $f(x)$ and the x axis, between two specified values of x, say, x_1 and x_2, as in Figure B.15.

The area of the blue element in Figure B.15 is approximately $f(x_i)\,\Delta x_i$. If we sum all these area elements between x_1 and x_2 and take the limit of this sum as $\Delta x_i \to 0$, we obtain the *true* area under the curve bounded by $f(x)$ and the x axis, between the limits x_1 and x_2:

$$\text{Area} = \lim_{\Delta x_i \to 0} \sum_i f(x_i)\Delta x_i = \int_{x_1}^{x_2} f(x)\,dx \qquad \text{B.36}◀$$

Integrals of the type defined by Equation B.36 are called **definite integrals.**

One common integral that arises in practical situations has the form

$$\int x^n\,dx = \frac{x^{n+1}}{n+1} + c \quad (n \neq -1) \qquad \text{B.37}◀$$

This result is obvious, being that differentiation of the right-hand side with respect to x gives $f(x) = x^n$ directly. If the limits of the integration are known, this integral becomes a *definite integral* and is written

$$\int_{x_1}^{x_2} x^n\,dx = \frac{x^{n+1}}{n+1}\bigg|_{x_1}^{x_2} = \frac{x_2^{\,n+1} - x_1^{\,n+1}}{n+1} \quad (n \neq -1) \qquad \text{B.38}◀$$

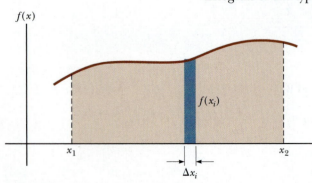

Figure B.15 The definite integral of a function is the area under the curve of the function between the limits x_1 and x_2.

Examples |

1. $\displaystyle \int_0^a x^2 \, dx = \left.\frac{x^3}{3}\right|_0^a = \frac{a^3}{3}$

3. $\displaystyle \int_3^5 x \, dx = \left.\frac{x^2}{2}\right|_3^5 = \frac{5^2 - 3^2}{2} = 8$

2. $\displaystyle \int_0^b x^{3/2} \, dx = \left.\frac{x^{5/2}}{5/2}\right|_0^b = \frac{2}{5} b^{5/2}$

Partial Integration

Sometimes it is useful to apply the method of *partial integration* (also called "integrating by parts") to evaluate certain integrals. This method uses the property

$$\int u \, dv = uv - \int v \, du \qquad\qquad \text{B.39} \blacktriangleleft$$

where u and v are *carefully* chosen so as to reduce a complex integral to a simpler one. In many cases, several reductions have to be made. Consider the function

$$I(x) = \int x^2 e^x \, dx$$

which can be evaluated by integrating by parts twice. First, if we choose $u = x^2$, $v = e^x$, we obtain

$$\int x^2 e^x \, dx = \int x^2 \, d(e^x) = x^2 e^x - 2 \int e^x x \, dx + c_1$$

Now, in the second term, choose $u = x$, $v = e^x$, which gives

$$\int x^2 e^x \, dx = x^2 e^x - 2x e^x + 2 \int e^x \, dx + c_1$$

or

$$\int x^2 e^x \, dx = x^2 e^x - 2xe^x + 2e^x + c_2$$

The Perfect Differential

Another useful method to remember is that of the *perfect differential,* in which we look for a change of variable such that the differential of the function is the differential of the independent variable appearing in the integrand. For example, consider the integral

$$I(x) = \int \cos^2 x \sin x \, dx$$

This integral becomes easy to evaluate if we rewrite the differential as $d(\cos x) = -\sin x \, dx$. The integral then becomes

$$\int \cos^2 x \sin x \, dx = -\int \cos^2 x \, d(\cos x)$$

If we now change variables, letting $y = \cos x$, we obtain

$$\int \cos^2 x \sin x \, dx = -\int y^2 \, dy = -\frac{y^3}{3} + c = -\frac{\cos^3 x}{3} + c$$

Table B.5 lists some useful indefinite integrals. Table B.6 gives Gauss's probability integral and other definite integrals. A more complete list can be found in various handbooks, such as *The Handbook of Chemistry and Physics* (Boca Raton, FL: CRC Press, published annually).

TABLE B.5 | **Some Indefinite Integrals (An arbitrary constant should be added to each of these integrals.)**

$$\int x^n \, dx = \frac{x^{n+1}}{n+1} \text{ (provided } n \neq 1)$$

$$\int \ln ax \, dx = (x \ln ax) - x$$

$$\int \frac{dx}{x} = \int x^{-1} \, dx = \ln x$$

$$\int xe^{ax} \, dx = \frac{e^{ax}}{a^2}(ax - 1)$$

$$\int \frac{dx}{a + bx} = \frac{1}{b} \ln (a + bx)$$

$$\int \frac{dx}{a + be^{cx}} = \frac{x}{a} - \frac{1}{ac} \ln (a + be^{cx})$$

$$\int \frac{x \, dx}{a + bx} = \frac{x}{b} - \frac{a}{b^2} \ln (a + bx)$$

$$\int \sin ax \, dx = -\frac{1}{a} \cos ax$$

$$\int \frac{dx}{x(x + a)} = -\frac{1}{a} \ln \frac{x + a}{x}$$

$$\int \cos ax \, dx = \frac{1}{a} \sin ax$$

$$\int \frac{dx}{(a + bx)^2} = -\frac{1}{b(a + bx)}$$

$$\int \tan ax \, dx = -\frac{1}{a} \ln (\cos ax) = \frac{1}{a} \ln (\sec ax)$$

$$\int \frac{dx}{a^2 + x^2} = \frac{1}{a} \tan^{-1} \frac{x}{a}$$

$$\int \cot ax \, dx = \frac{1}{a} \ln (\sin ax)$$

$$\int \frac{dx}{a^2 - x^2} = \frac{1}{2a} \ln \frac{a + x}{a - x} \left(a^2 - x^2 > 0\right)$$

$$\int \sec ax \, dx = \frac{1}{a} \ln (\sec ax + \tan ax) = \frac{1}{a} \ln \left[\tan \left(\frac{ax}{2} + \frac{\pi}{4} \right) \right]$$

$$\int \frac{dx}{x^2 - a^2} = \frac{1}{2a} \ln \frac{x - a}{x + a} \, (x^2 - a^2 > 0)$$

$$\int \csc ax \, dx = \frac{1}{a} \ln (\csc ax - \cot ax) = \frac{1}{a} \ln \left(\tan \frac{ax}{2} \right)$$

$$\int \frac{x \, dx}{a^2 \pm x^2} = \pm \frac{1}{2} \ln (a^2 \pm x^2)$$

$$\int \sin^2 ax \, dx = \frac{x}{2} - \frac{\sin 2ax}{4a}$$

$$\int \frac{dx}{\sqrt{a^2 - x^2}} = \sin^{-1} \frac{x}{a} = -\cos^{-1} \frac{x}{a} \, (a^2 - x^2 > 0)$$

$$\int \cos^2 ax \, dx = \frac{x}{2} + \frac{\sin 2ax}{4a}$$

$$\int \frac{dx}{\sqrt{x^2 \pm a^2}} = \ln (x + \sqrt{x^2 \pm a^2})$$

$$\int \frac{dx}{\sin^2 ax} = -\frac{1}{a} \cot ax$$

$$\int \frac{x \, dx}{\sqrt{a^2 - x^2}} = -\sqrt{a^2 - x^2}$$

$$\int \frac{dx}{\cos^2 ax} = \frac{1}{a} \tan ax$$

$$\int \frac{x \, dx}{\sqrt{x^2 \pm a^2}} = \sqrt{x^2 \pm a^2}$$

$$\int \tan^2 ax \, dx = \frac{1}{a} (\tan ax) - x$$

(Continued)

◀ **TABLE B.5** | Some Indefinite Integrals (*continued*)

$$\int \sqrt{a^2 - x^2} \, dx = \frac{1}{2}\left(x\sqrt{a^2 - x^2} + a^2 \sin^{-1}\frac{x}{|a|}\right)$$

$$\int \cot^2 ax \, dx = -\frac{1}{a}(\cot ax) - x$$

$$\int x\sqrt{a^2 - x^2} \, dx = -\frac{1}{3}(a^2 - x^2)^{3/2}$$

$$\int \sin^{-1} ax \, dx = x(\sin^{-1} ax) + \frac{\sqrt{1 - a^2x^2}}{a}$$

$$\int \sqrt{x^2 \pm a^2} \, dx = \frac{1}{2}\left[x\sqrt{x^2 \pm a^2} \pm a^2 \ln(x + \sqrt{x^2 \pm a^2})\right]$$

$$\int \cos^{-1} ax \, dx = x(\cos^{-1} ax) - \frac{\sqrt{1 - a^2x^2}}{a}$$

$$\int x(\sqrt{x^2 \pm a^2}) \, dx = \frac{1}{3}(x^2 \pm a^2)^{3/2}$$

$$\int \frac{dx}{(x^2 + a^2)^{3/2}} = \frac{x}{a^2\sqrt{x^2 + a^2}}$$

$$\int e^{ax} \, dx = \frac{1}{a}e^{ax}$$

$$\int \frac{x \, dx}{(x^2 + a^2)^{3/2}} = -\frac{1}{\sqrt{x^2 + a^2}}$$

◀ **TABLE B.6** | Gauss's Probability Integral and Other Definite Integrals

$$\int_0^\infty x^n e^{-ax} \, dx = \frac{n!}{a^{n+1}}$$

$$I_0 = \int_0^\infty e^{-ax^2} \, dx = \frac{1}{2}\sqrt{\frac{\pi}{a}} \quad \text{(Gauss's probability integral)}$$

$$I_1 = \int_0^\infty xe^{-ax^2} \, dx = \frac{1}{2a}$$

$$I_2 = \int_0^\infty x^2 e^{-ax^2} \, dx = -\frac{dI_0}{da} = \frac{1}{4}\sqrt{\frac{\pi}{a^3}}$$

$$I_3 = \int_0^\infty x^3 e^{-ax^2} \, dx = -\frac{dI_1}{da} = \frac{1}{2a^2}$$

$$I_4 = \int_0^\infty x^4 e^{-ax^2} \, dx = \frac{d^2 I_0}{da^2} = \frac{3}{8}\sqrt{\frac{\pi}{a^5}}$$

$$I_5 = \int_0^\infty x^5 e^{-ax^2} \, dx = \frac{d^2 I_1}{da^2} = \frac{1}{a^3}$$

$$\vdots$$

$$I_{2n} = (-1)^n \frac{d^n}{da^n} I_0$$

$$I_{2n+1} = (-1)^n \frac{d^n}{da^n} I_1$$

◀B.8 | Propagation of Uncertainty

In laboratory experiments, a common activity is to take measurements that act as raw data. These measurements are of several types—length, time interval, temperature, voltage, and so on—and are taken by a variety of instruments. Regardless of the measurement and the quality of the instrumentation, **there is always uncertainty associated with a physical measurement.** This uncertainty is a combination of that associated with the instrument and that related to the system being measured. An example of the former is the inability to exactly determine the position of a length measurement between the lines on a meterstick. An example of uncertainty related to the system being measured is the variation of temperature within a sample of water so that a single temperature for the sample is difficult to determine.

Uncertainties can be expressed in two ways. **Absolute uncertainty** refers to an uncertainty expressed in the same units as the measurement. Therefore, the length of a computer disk label might be expressed as (5.5 ± 0.1) cm. The uncertainty of ± 0.1 cm by itself is not descriptive enough for some purposes, however. This uncertainty is large if the measurement is 1.0 cm, but it is small if the measurement is 100 m. To give a more descriptive account of the uncertainty, **fractional uncertainty** or **percent uncertainty** is used. In this type of description, the uncertainty is divided by the actual measurement. Therefore, the length of the computer disk label could be expressed as

$$\ell = 5.5 \text{ cm} \pm \frac{0.1 \text{ cm}}{5.5 \text{ cm}} = 5.5 \text{ cm} \pm 0.018 \text{ (fractional uncertainty)}$$

or as

$$\ell = 5.5 \text{ cm} \pm 1.8\% \text{ (percent uncertainty)}$$

When combining measurements in a calculation, the percent uncertainty in the final result is generally larger than the uncertainty in the individual measurements. This is called **propagation of uncertainty** and is one of the challenges of experimental physics.

Some simple rules can provide a reasonable estimate of the uncertainty in a calculated result:

Multiplication and division: When measurements with uncertainties are multiplied or divided, add the *percent uncertainties* to obtain the percent uncertainty in the result.

Example: The Area of a Rectangular Plate

$$A = \ell w = (5.5 \text{ cm} \pm 1.8\%) \times (6.4 \text{ cm} \pm 1.6\%) = 35 \text{ cm}^2 \pm 3.4\%$$

$$= (35 \pm 1) \text{ cm}^2$$

Addition and subtraction: When measurements with uncertainties are added or subtracted, add the *absolute uncertainties* to obtain the absolute uncertainty in the result.

Example: A Change in Temperature

$$\Delta T = T_2 - T_1 = (99.2 \pm 1.5)°\text{C} - (27.6 \pm 1.5)°\text{C} = (71.6 \pm 3.0)°\text{C}$$

$$= 71.6°\text{C} \pm 4.2\%$$

Powers: If a measurement is taken to a power, the percent uncertainty is multiplied by that power to obtain the percent uncertainty in the result.

Example: The Volume of a Sphere

$$V = \tfrac{4}{3}\pi r^3 = \tfrac{4}{3}\pi(6.20 \text{ cm} \pm 2.0\%)^3 = 998 \text{ cm}^3 \pm 6.0\%$$

$$= (998 \pm 60) \text{ cm}^3$$

For complicated calculations, many uncertainties are added together, which can cause the uncertainty in the final result to be undesirably large. Experiments should be designed such that calculations are as simple as possible.

Notice that uncertainties in a calculation always add. As a result, an experiment involving a subtraction should be avoided if possible, especially if the measurements being subtracted are close together. The result of such a calculation is a small difference in the measurements and uncertainties that add together. It is possible that the uncertainty in the result could be larger than the result itself!

Periodic Table of the Elements

Group I	Group II	Transition elements						
H 1 1.007 9 $1s$								
Li 3 6.941 $2s^1$	**Be** 4 9.0122 $2s^2$							
Na 11 22.990 $3s^1$	**Mg** 12 24.305 $3s^2$							

Symbol — **Ca** 20 — Atomic number
Atomic mass† — 40.078
$4s^2$ — Electron configuration

K 19 39.098 $4s^1$	**Ca** 20 40.078 $4s^2$	**Sc** 21 44.956 $3d^14s^2$	**Ti** 22 47.867 $3d^24s^2$	**V** 23 50.942 $3d^34s^2$	**Cr** 24 51.996 $3d^54s^1$	**Mn** 25 54.938 $3d^54s^2$	**Fe** 26 55.845 $3d^64s^2$	**Co** 27 58.933 $3d^74s^2$
Rb 37 85.468 $5s^1$	**Sr** 38 87.62 $5s^2$	**Y** 39 88.906 $4d^15s^2$	**Zr** 40 91.224 $4d^25s^2$	**Nb** 41 92.906 $4d^45s^1$	**Mo** 42 95.94 $4d^55s^1$	**Tc** 43 (98) $4d^55s^2$	**Ru** 44 101.07 $4d^75s^1$	**Rh** 45 102.91 $4d^85s^1$
Cs 55 132.91 $6s^1$	**Ba** 56 137.33 $6s^2$	57–71*	**Hf** 72 178.49 $5d^26s^2$	**Ta** 73 180.95 $5d^36s^2$	**W** 74 183.84 $5d^46s^2$	**Re** 75 186.21 $5d^56s^2$	**Os** 76 190.23 $5d^66s^2$	**Ir** 77 192.2 $5d^76s^2$
Fr 87 (223) $7s^1$	**Ra** 88 (226) $7s^2$	89–103**	**Rf** 104 (261) $6d^27s^2$	**Db** 105 (262) $6d^37s^2$	**Sg** 106 (266)	**Bh** 107 (264)	**Hs** 108 (277)	**Mt** 109 (268)

*Lanthanide series

La 57 138.91 $5d^16s^2$	**Ce** 58 140.12 $5d^14f^16s^2$	**Pr** 59 140.91 $4f^36s^2$	**Nd** 60 144.24 $4f^46s^2$	**Pm** 61 (145) $4f^56s^2$	**Sm** 62 150.36 $4f^66s^2$

**Actinide series

Ac 89 (227) $6d^17s^2$	**Th** 90 232.04 $6d^27s^2$	**Pa** 91 231.04 $5f^26d^17s^2$	**U** 92 238.03 $5f^36d^17s^2$	**Np** 93 (237) $5f^46d^17s^2$	**Pu** 94 (244) $5f^67s^2$

Note: Atomic mass values given are averaged over isotopes in the percentages in which they exist in nature.
†For an unstable element, mass number of the most stable known isotope is given in parentheses.
††Elements 114 and 116 have not yet been officially named.
Note: For a description of the atomic data, visit *physics.nist.gov/PhysRefData/Elements/per_text.html*.

Group III	Group IV	Group V	Group VI	Group VII	Group 0
				H 1	**He** 2
				1.007 9	4.002 6
				$1s^1$	$1s^2$
B 5	**C** 6	**N** 7	**O** 8	**F** 9	**Ne** 10
10.811	12.011	14.007	15.999	18.998	20.180
$2p^1$	$2p^2$	$2p^3$	$2p^4$	$2p^5$	$2p^6$
Al 13	**Si** 14	**P** 15	**S** 16	**Cl** 17	**Ar** 18
26.982	28.086	30.974	32.066	35.453	39.948
$3p^1$	$3p^2$	$3p^3$	$3p^4$	$3p^5$	$3p^6$

Ni 28	**Cu** 29	**Zn** 30	**Ga** 31	**Ge** 32	**As** 33	**Se** 34	**Br** 35	**Kr** 36
58.693	63.546	65.41	69.723	72.64	74.922	78.96	79.904	83.80
$3d^8 4s^2$	$3d^{10} 4s^1$	$3d^{10} 4s^2$	$4p^1$	$4p^2$	$4p^3$	$4p^4$	$4p^5$	$4p^6$
Pd 46	**Ag** 47	**Cd** 48	**In** 49	**Sn** 50	**Sb** 51	**Te** 52	**I** 53	**Xe** 54
106.42	107.87	112.41	114.82	118.71	121.76	127.60	126.90	131.29
$4d^{10}$	$4d^{10} 5s^1$	$4d^{10} 5s^2$	$5p^1$	$5p^2$	$5p^3$	$5p^4$	$5p^5$	$5p^6$
Pt 78	**Au** 79	**Hg** 80	**Tl** 81	**Pb** 82	**Bi** 83	**Po** 84	**At** 85	**Rn** 86
195.08	196.97	200.59	204.38	207.2	208.98	(209)	(210)	(222)
$5d^9 6s^1$	$5d^{10} 6s^1$	$5d^{10} 6s^2$	$6p^1$	$6p^2$	$6p^3$	$6p^4$	$6p^5$	$6p^6$
Ds 110	**Rg** 111	**Cn** 112		$114^{\dagger\dagger}$		$116^{\dagger\dagger}$		
(271)	(272)	(285)		(289)		(292)		

Eu 63	**Gd** 64	**Tb** 65	**Dy** 66	**Ho** 67	**Er** 68	**Tm** 69	**Yb** 70	**Lu** 71
151.96	157.25	158.93	162.50	164.93	167.26	168.93	173.04	174.97
$4f^7 6s^2$	$4f^7 5d^1 6s^2$	$4f^8 5d^1 6s^2$	$4f^{10} 6s^2$	$4f^{11} 6s^2$	$4f^{12} 6s^2$	$4f^{13} 6s^2$	$4f^{14} 6s^2$	$4f^{14} 5d^1 6s^2$
Am 95	**Cm** 96	**Bk** 97	**Cf** 98	**Es** 99	**Fm** 100	**Md** 101	**No** 102	**Lr** 103
(243)	(247)	(247)	(251)	(252)	(257)	(258)	(259)	(262)
$5f^7 7s^2$	$5f^7 6d^1 7s^2$	$5f^8 6d^1 7s^2$	$5f^{10} 7s^2$	$5f^{11} 7s^2$	$5f^{12} 7s^2$	$5f^{13} 7s^2$	$5f^{14} 7s^2$	$5f^{14} 6d^1 7s^2$

Appendix D

SI Units

TABLE D.1 | SI Units

Base Quantity	SI Base Unit	
	Name	Symbol
Length	meter	m
Mass	kilogram	kg
Time	second	s
Electric current	ampere	A
Temperature	kelvin	K
Amount of substance	mole	mol
Luminous intensity	candela	cd

TABLE D.2 | Some Derived SI Units

Quantity	Name	Symbol	Expression in Terms of Base Units	Expression in Terms of Other SI Units
Plane angle	radian	rad	m/m	
Frequency	hertz	Hz	s^{-1}	
Force	newton	N	$kg \cdot m/s^2$	J/m
Pressure	pascal	Pa	$kg/m \cdot s^2$	N/m^2
Energy	joule	J	$kg \cdot m^2/s^2$	$N \cdot m$
Power	watt	W	$kg \cdot m^2/s^3$	J/s
Electric charge	coulomb	C	$A \cdot s$	
Electric potential	volt	V	$kg \cdot m^2/A \cdot s^3$	W/A
Capacitance	farad	F	$A^2 \cdot s^4/kg \cdot m^2$	C/V
Electric resistance	ohm	Ω	$kg \cdot m^2/A^2 \cdot s^3$	V/A
Magnetic flux	weber	Wb	$kg \cdot m^2/A \cdot s^2$	$V \cdot s$
Magnetic field	tesla	T	$kg/A \cdot s^2$	
Inductance	henry	H	$kg \cdot m^2/A^2 \cdot s^2$	$T \cdot m^2/A$

Answers to Quick Quizzes and Odd-Numbered Problems

CHAPTER 1

Answers to Quick Quizzes

1. False.
2. (b)
3. Scalars: (a), (d), (e). Vectors: (b), (c).
4. (c)
5. (a)
6. (b)
7. (b)
8. (d)

Answers to Odd-Numbered Problems

1. 7.69 cm
3. 23 kg
5. (b) only
7. No
9. 151 μm
11. 667 lb/s
13. 2.86 cm
15. (a) 7.14×10^{-2} gal/s (b) 2.70×10^{-4} m^3/s (c) 1.03 h
17. $\sim 10^6$ balls in a room 4 m by 4 m by 3 m
19. $\sim 10^2$ piano tuners
21. 31 556 926.0 s
23. (a) 3 (b) 4 (c) 3 (d) 2
25. 5.2 m^3, 3%
27. 1.38×10^3 m
29. 288°; 108°
31. $(-2.75, -4.76)$ m
33. (a) 2.24 m (b) 2.24 m at 26.6°
35. This situation can *never* be true because the distance is an arc of a circle between two points, whereas the magnitude of the displacement vector is a straight-line chord of the circle between the same points.
37. approximately 420 ft at $-3°$
39. (a) $a = 5.00$ and $b = 7.00$ (b) For vectors to be equal, all their components must be equal. A vector equation contains more information than a scalar equation.
41. 47.2 units at 122°
43. (a) $8.00\hat{\mathbf{i}} + 12.0\hat{\mathbf{j}} - 4.00\hat{\mathbf{k}}$ (b) $2.00\hat{\mathbf{i}} + 3.00\hat{\mathbf{j}} - 1.00\hat{\mathbf{k}}$ (c) $-24.0\hat{\mathbf{i}} - 36.0\hat{\mathbf{j}} + 12.0\hat{\mathbf{k}}$
45. 196 cm at 345°
47. (a) $2.00\hat{\mathbf{i}} - 6.00\hat{\mathbf{j}}$ (b) $4.00\hat{\mathbf{i}} + 2.00\hat{\mathbf{j}}$ (c) 6.32 (d) 4.47 (e) 288°; 26.6°
49. (a) 10.4 cm (b) $\theta = 35.5°$
51. 240 m at 237°
53. 0.141 nm
55. 70.0 m
57. 316 m
59. 106°
61. 1.15°
63. 0.449%

65. (a) 0.529 cm/s (b) 11.5 cm/s
67. (a) 185 N at 77.8° from the positive x axis (b) $(-39.3\hat{\mathbf{i}} - 181\hat{\mathbf{j}})$ N
69. (a) (10.0 m, 16.0 m) (b) This center of mass of the tree distribution is the same location whatever order we take the trees in.
71. (a) $\vec{\mathbf{R}}_1 = a\hat{\mathbf{i}} + b\hat{\mathbf{j}}$ (b) $R_1 = (a^2 + b^2)^{1/2}$ (c) $\vec{\mathbf{R}}_2 = a\hat{\mathbf{i}} + b\hat{\mathbf{j}} + c\hat{\mathbf{k}}$

CHAPTER 2

Answers to Quick Quizzes

1. (c)
2. (b)
3. (a)-(e), (b)-(d), (c)-(f)
4. (b)
5. (c)
6. (e)

Answers to Odd-Numbered Problems

1. (a) 5 m/s (b) 1.2 m/s (c) -2.5 m/s (d) -3.3 m/s (e) 0
3. (a) 2.30 m/s (b) 16.1 m/s (c) 11.5 m/s
5. (a) -2.4 m/s (b) -3.8 m/s (c) 4.0 s
7. (a) 5 m/s (b) -2.5 m/s (c) 0 (d) $+5$ m/s
9. (a) 5.00 m (b) 4.88×10^3 s
11. (a) 2.00 m (b) -3.00 m/s (c) -2.00 m/s^2
13. (a) 20 m/s, 5 m/s (b) 263 m
15. (a) 1.3 m/s^2 (b) $t = 3$ s, $a = 2$ m/s^2 (c) $t = 6$ s, $t > 10$ s (d) $a = -1.5$ m/s^2, $t = 8$ s
17. -16.0 cm/s^2
19. (a) 6.61 m/s (b) -0.448 m/s^2
21. (a) 4.98×10^{-9} s (b) 1.20×10^{15} m/s^2
23. 3.10 m/s
25. (a) 35.0 s (b) 15.7 m/s
27. (a)

(b) Particle under constant acceleration.
(c) $v_f^2 = v_i^2 + 2a(x_f - x_i)$ (Equation 2.14)
(d) $a = \dfrac{v_f^2 - v_i^2}{2\Delta x}$ (e) 1.25 m/s^2 (f) 8.00 s
29. David will be unsuccessful. The average human reaction time is about 0.2 s (research on the Internet) and a dollar bill is about 15.5 cm long, so David's fingers are about 8 cm from the end of the bill before it is dropped. The bill will fall about 20 cm before he can close his fingers.
31. (a) 7.82 m (b) 0.782 s
33. (a) 10.0 m/s up (b) 4.68 m/s down
35. 1.79 s
37. (a) 5.25 m/s^2 (b) 168 m (c) 52.5 m/s
39. (a) 70.0 mi/h · s = 31.3 m/s^2 = $3.19g$ (b) 321 ft = 97.8 m

41. (a) -202 m/s^2 (b) 198 m
43. (a) 3.00 m/s (b) 6.00 s (c) -0.300 m/s^2 (d) 2.05 m/s
45. (a) 25.4 s (b) 15.0 km/h
47. 1.60 m/s^2
49. (a) 41.0 s (b) 1.73 km (c) -184 m/s
51. (a) 5.32 m/s^2 for Laura and 3.75 m/s^2 for Healan
    (b) 10.6 m/s for Laura and 11.2 m/s for Healan
    (c) Laura, by 2.63 m
    (d) 4.47 m at $t = 2.84$ s
53. (a) 3.00 s (b) -15.3 m/s
    (c) 31.4 m/s down and 34.8 m/s down
55. (a) 26.4 m (b) 6.89%
57. $0.577v$

CHAPTER 3
Answers to Quick Quizzes

1. (a)
2. (i) (b) (ii) (a)
3. 15°, 30°, 45°, 60°, 75°
4. (c)
5. (i) (b) (ii) (d)

Answers to Odd-Numbered Problems

1. (a) 4.87 km at 209° from E (b) 23.3 m/s (c) 13.5 m/s at 209°
3. (a) $5.00t\hat{\mathbf{i}} + 1.50t^2\hat{\mathbf{j}}$ (b) $5.00\hat{\mathbf{i}} + 3.00t\hat{\mathbf{j}}$
    (c) 10.0 m, 6.00 m (d) 7.81 m/s
5. (a) $(0.800\hat{\mathbf{i}} - 0.300\hat{\mathbf{j}})$ m/s^2 (b) 339°
    (c) $(360\hat{\mathbf{i}} - 72.7\hat{\mathbf{j}})$ m, $-15.2°$
7. 12.0 m/s
9. 22.4° or 89.4°
11. (a) 2.81 m/s horizontal (b) 60.2° below the horizontal
13. (a) The ball clears by 0.89 m. (b) while descending
15. 67.8°
17. 9.91 m/s
19. (a) $(0, 50.0$ m$)$ (b) $v_{xi} = 18.0$ m/s; $v_{yi} = 0$ (c) Particle under constant acceleration (d) Particle under constant velocity (e) $v_{xf} = v_{xi}$; $v_{yf} = -gt$ (f) $x_f = v_{xi}t$; $y_f = y_i - \frac{1}{2}gt^2$ (g) 3.19 s (h) 36.1 m/s, $-60.1°$
21. (a) 18.1 m/s (b) 1.13 m (c) 2.79 m
23. 377 m/s^2
25. 0.281 rev/s
27. 7.58×10^3 m/s, 5.80×10^3 s
29. 1.48 m/s^2 inward and 29.9° backward
31. (a) 13.0 m/s^2 (b) 5.70 m/s (c) 7.50 m/s^2
33. 153 km/h at 11.3° north of west
35. (a) 2.02×10^3 s (b) 1.67×10^3 s (c) Swimming with the current does not compensate for the time lost swimming against the current.
37. (a) 57.7 km/h at 60.0° west of vertical (b) 28.9 km/h downward
39. 15.3 m
41. 22.6 m/s
43. The relationship between the height h and the walking speed is $h = (4.16 \times 10^{-3})v_x^2$, where h is in meters and v_x is in meters per second. At a typical walking speed of 4 to 5 km/h, the ball would have to be dropped from a height of about 1 cm, clearly much too low for a person's hand. Even at Olympic-record speed for the 100-m run (confirm on the Internet), this situation would only occur if the ball is dropped from about 0.4 m, which is also below the hand of a normally proportioned person.
45. (a) 101 m/s (b) 3.27×10^4 ft (c) 20.6 s
47. (a) 9.80 m/s^2, downward (b) 10.7 m/s
49. 54.4 m/s^2
51. The initial height of the ball when struck is 3.94 m, which is too high for the batter to hit the ball.
53. (a) 6.80 km (b) 3.00 km vertically above the impact point
    (c) 66.2°
55. (a) 20.0 m/s (b) 5.00 s (c) $(16.0\hat{\mathbf{i}} - 27.1\hat{\mathbf{j}})$ m/s (d) 6.53 s
    (e) $24.5\hat{\mathbf{i}}$ m
57. (a) 22.9 m/s (b) 360 m (c) 114 m/s, -44.3 m/s
59. (a) 1.52 km (b) 36.1 s (c) 4.05 km
61. (a) 4.00 km/h (b) 4.00 km/h

CHAPTER 4
Answers to Quick Quizzes

1. (d)
2. (a)
3. (d)
4. (b)
5. (i) (c) (ii) (a)
6. (c)
7. (c)

Answers to Odd-Numbered Problems

1. (a) $\frac{1}{3}$ (b) 0.750 m/s^2
3. (a) $(2.50\hat{\mathbf{i}} + 5.00\hat{\mathbf{j}})$N (b) 5.59 N
5. (a) $(6.00\hat{\mathbf{i}} + 15.0\hat{\mathbf{j}})$N (b) 16.2 N
7. (a) 5.00 m/s^2 at 36.9° (b) 6.08 m/s^2 at 25.3°
9. (a) 534 N (b) 54.5 kg
11. (a) 3.64×10^{-18} N (b) 8.93×10^{-30} N is 408 billion times smaller.
13. 2.58 N
15. (a) -4.47×10^{15} m/s^2 (b) $+2.09 \times 10^{-10}$ N
17. (a) 15.0 lb up (b) 5.00 lb up (c) 0
19. 100 N and 204 N
21. (a) 3.43 kN (b) 0.967 m/s horizontally forward
23. $T_1 = 253$ N, $T_2 = 165$ N, $T_3 = 325$ N
25. (a) 706 N (b) 814 N (c) 706 N (d) 648 N
27. (a) $a = g \tan \theta$ (b) 4.16 m/s^2
29. 8.66 N east
31. 3.73 m
33. (a) $T_1 = 79.8$ N, $T_2 = 39.9$ N (b) 2.34 m/s^2
35. 950 N
37. (a) $F_x > 19.6$ N (b) $F_x \le -78.4$ N
    (c)

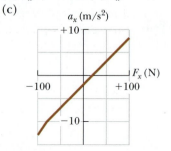

39. (a) 0.529 m below its initial level (b) 7.40 m/s upward
41. (a) Removing mass (b) 13.7 mi/h · s
43. (a)

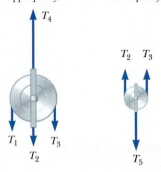

Upper pulley: *Lower pulley:*

(b) $Mg/2$, $Mg/2$, $Mg/2$, $3Mg/2$, Mg (c) $Mg/2$
45. (a)

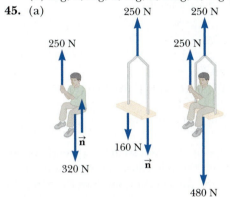

(b) 0.408 m/s^2 (c) 83.3 N
47. (a) 2.20 m/s^2 (b) 27.4 N
49. (a)

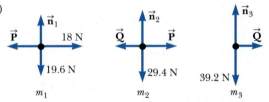

(b) 2.00 m/s^2 to the right (c) 4.00 N on m_1, 6.00 N right on m_2, 8.00 N right on m_3 (d) 14.0 N between m_1 and m_2, 8.00 N between m_2 and m_3 (e) The m_2 block models the heavy block of wood. The contact force on your back is modeled by the force between the m_2 and the m_3 blocks, which is much less than the force F. The difference between F and this contact force is the net force causing the acceleration of the 5-kg pair of objects. The acceleration is real and nonzero, but it lasts for so short a time that it is never associated with a large velocity. The frame of the building and your legs exert forces, small in magnitude relative to the hammer blow, to bring the partition, block, and you to rest again over a time large relative to the hammer blow.
51. $(M + m_1 + m_2)(m_1 g/m_2)$
53. (a) 4.90 m/s^2 (b) 3.13 m/s at 30.0° below the horizontal (c) 1.35 m (d) 1.14 s (e) The mass of the block makes no difference.
55. 1.16 cm

57. (a) 30.7° (b) 0.843 N
59. $\vec{F} = mg \cos \theta \sin \theta \hat{i} + (M + m \cos^2 \theta) g \hat{j}$

CHAPTER 5
Answers to Quick Quizzes

1. (b)
2. (b)
3. (b)
4. (i) (a) (ii) (b)
5. (a)
6. (a) Because the speed is constant, the only direction the force can have is that of the centripetal acceleration. The force is larger at Ⓒ than at Ⓐ because the radius at Ⓒ is smaller. There is no force at Ⓑ because the wire is straight. (b) In addition to the forces in the centripetal direction in (a), there are now tangential forces to provide the tangential acceleration. The tangential force is the same at all three points because the tangential acceleration is constant.

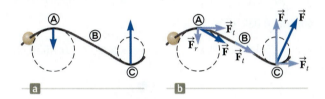

7. (c)

Answers to Odd-Numbered Problems

1. $\mu_s = 0.727$, $\mu_k = 0.577$
3. (a) 0.306 (b) 0.245
5. (a) 1.11 s (b) 0.875 s
7. 37.8 N
9. (a) 1.78 m/s^2 (b) 0.368 (c) 9.37 N (d) 2.67 m/s
11. 6.84 m
13. (a)

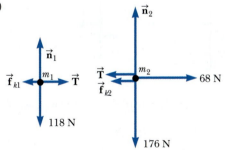

(b) 1.29 m/s^2 to the right (c) 27.2 N
15. The situation is impossible because maximum static friction cannot provide the acceleration necessary to keep the book stationary on the seat.
17. any speed up to 8.08 m/s
19. $v \leq 14.3$ m/s
21. (a) $(68.6\hat{i} + 784\hat{j})$N (b) $a = 0.857$ m/s^2
23. (a) 1.15×10^4 N up (b) 14.1 m/s
25. No. The archeologist needs a vine of tensile strength equal to or greater than 1.38 kN to make it across.

27. (a) $v = 4.81$ m/s (b) 700 N
29. (a) 1.47 N · s/m (b) 2.04×10^{-3} s (c) 2.94×10^{-2} N
31. (a) $B = \dfrac{9.80 \text{ m/s}^2}{0.300 \text{ m/s}} = 32.7 \text{ s}^{-1}$ (b) 9.80 m/s² down
    (c) 4.90 m/s² down
33. (a) $0.034\,7$ s⁻¹ (b) 2.50 m/s (c) $a = -cv$
35. 0.613 m/s² toward the Earth
37. 2.97 nN
39. 0.212 m/s², opposite the velocity vector
41. 0.835 rev/s
43. (a) $M = 3m \sin\theta$ (b) $T_1 = 2mg \sin\theta$, $T_2 = 3mg \sin\theta$
    (c) $a = \dfrac{g \sin\theta}{1 + 2\sin\theta}$
    (d) $T_1 = 4mg \sin\theta \left(\dfrac{1 + \sin\theta}{1 + 2\sin\theta}\right)$
    $T_2 = 6mg \sin\theta \left(\dfrac{1 + \sin\theta}{1 + 2\sin\theta}\right)$
    (e) $M_{\text{max}} = 3m(\sin\theta + \mu_s \cos\theta)$
    (f) $M_{\text{min}} = 3m(\sin\theta - \mu_s \cos\theta)$
    (g) $T_{2,\text{max}} - T_{2,\text{min}} = (M_{\text{max}} - M_{\text{min}})g = 6\mu_s mg \cos\theta$
45. (a) 0.087 1 (b) 27.4 N
47. (a) 5.19 m/s (b) 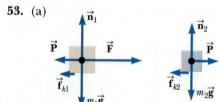 (c) 555 N

49. 2.14 rev/min
51. $F = 394$ N, $\theta = 84.7°$ with respect to the positive x axis.
53. (a)

(b) F (c) $F - P$ (d) P (e) m_1: $F - P = m_1 a$; m_2: $P = m_2 a$
(f) $a = \dfrac{F - \mu_1 m_1 g - \mu_2 m_2 g}{m_1 + m_2}$
(g) $P = \dfrac{m_2}{m_1 + m_2}[F + m_1(\mu_2 - \mu_1)g]$
55. (a) 2.13 s (b) 1.66 m
57. 12.8 N
59. (a) $\theta = 70.4°$ and $\theta = 0°$ (b) $\theta = 0°$ (c) The period is too large. (d) Zero is always a solution for the angle. (e) There are never more than two solutions.
61. (a) 0.013 2 m/s (b) 1.03 m/s (c) 6.87 m/s
63. (a) 735 N (b) 732 N (c) The gravitational force is larger. The normal force is smaller, just like it is when going over the top of a Ferris wheel.

CHAPTER 6
Answers to Quick Quizzes

1. (a)
2. (c), (a), (d), (b)
3. (d)
4. (a)
5. (b)
6. (c)
7. **(i)** (c) **(ii)** (a)
8. (d)

Answers to Odd-Numbered Problems

1. (a) 472 J (b) 2.76 kN
3. (a) 31.9 J (b) 0 (c) 0 (d) 31.9 J
5. -4.70×10^3 J
7. 28.9
9. (a) 16.0 J (b) 36.9°
11. 16.0
13. $\vec{\mathbf{A}} = 7.05$ m at 28.4°
15. (a) 7.50 J (b) 15.0 J (c) 7.50 J (d) 30.0 J
17. (a) 0.938 cm (b) 1.25 J
19. (a) 2.04×10^{-2} m (b) 720 N/m
21. 7.37 N/m
23. (a) Design the spring constant so that the weight of one tray removed from the pile causes an extension of the springs equal to the thickness of one tray. (b) 316 N/m (c) We do not need to know the length and width of the tray.
25. 50.0 J
27. 0.299 m/s
29. (a) 1.13 kN/m (b) 0.518 m = 51.8 cm
31. (a) 60.0 J (b) 60.0 J
33. (a) 1.20 J (b) 5.00 m/s (c) 6.30 J
35. 878 kN up
37. (a) 97.8 J (b) $(-4.31\hat{\mathbf{i}} + 31.6\hat{\mathbf{j}})$ N (c) 8.73 m/s
39. (a) 2.5 J (b) -9.8 J (c) -12 J
41. (a) $U_A = 2.59 \times 10^5$ J, $U_B = 0$, $\Delta U = -2.59 \times 10^5$ J
    (b) $U_A = 0$, $U_B = -2.59 \times 10^5$ J, $\Delta U = -2.59 \times 10^5$ J
43. (a) 125 J (b) 50.0 J (c) 66.7 J (d) nonconservative (e) The work done on the particle depends on the path followed by the particle.
45. (a) 30.0 J (b) 51.2 J (c) 42.4 J (d) Friction is a nonconservative force.
47. (a) 40.0 J (b) -40.0 J (c) 62.5 J
49. The book hits the ground with 20.0 J of kinetic energy. The book–Earth system now has zero gravitational potential energy, for a total energy of 20.0 J, which is the energy put into the system by the librarian.
51. (a) -4.77×10^9 J (b) 569 N (c) 569 N up
53. 2.52×10^7 m
55.

Stable   Unstable   Neutral

57. 0.27 MJ/kg for a battery. 17 MJ/kg for hay is 63 times larger. 44 MJ/kg for gasoline is 2.6 times larger still. 142 MJ/kg for hydrogen is 3.2 times larger than that.

59. 90.0 J

61. 0.131 m

63. (a) $U(x) = 1 + 4e^{-2x}$ (b) The force must be conservative because the work the force does on the particle on which it acts depends only on the original and final positions of the particle, not on the path between them.

65. (a) $x = 3.62m/(4.30 - 23.4m)$, where x is in meters and m is in kilograms (b) 0.095 1 m (c) 0.492 m (d) 6.85 m (e) The situation is impossible. (f) The extension is directly proportional to m when m is only a few grams. Then it grows faster and faster, diverging to infinity for $m = 0.184$ kg.

67. (a) $\vec{F}_1 = (20.5\hat{i} + 14.3\hat{j})$ N, $\vec{F}_2 = (-36.4\hat{i} + 21.0\hat{j})$ N
(b) $\Sigma\vec{F} = (-15.9\hat{i} + 35.3\hat{j})$ N
(c) $\vec{a} = (-3.18\hat{i} + 7.07\hat{j})$ m/s^2
(d) $\vec{v} = (-5.54\hat{i} + 23.7\hat{j})$ m/s
(e) $\vec{r} = (-2.30\hat{i} + 39.3\hat{j})$ m (f) 1.48 kJ (g) 1.48 kJ
(h) The work–kinetic energy theorem is consistent with Newton's second law.

69. (a)

(b) The slope of the line is 116 N/m. (c) We use all the points listed and also the origin. There is no visible evidence for a bend in the graph or nonlinearity near either end. (d) 116 N/m (e) 12.2 N

CHAPTER 7
Answers to Quick Quizzes

1. (a) For the television set, energy enters by electrical transmission (through the power cord). Energy leaves by heat (from hot surfaces into the air), mechanical waves (sound from the speaker), and electromagnetic radiation (from the screen). (b) For the gasoline-powered lawn mower, energy enters by matter transfer (gasoline). Energy leaves by work (on the blades of grass), mechanical waves (sound), and heat (from hot surfaces into the air). (c) For the hand-cranked pencil sharpener, energy enters by work (from your hand turning the crank). Energy leaves by work (done on the pencil), mechanical waves (sound), and heat due to the temperature increase from friction.

2. (i) (b) (ii) (b) (iii) (a)

3. (a)

4. $v_1 = v_2 = v_3$

5. (c)

Answers to Odd-Numbered Problems

1. (a) $\Delta E_{int} = Q + T_{ET} + T_{ER}$ (b) $\Delta K + \Delta U + \Delta E_{int} = W + Q + T_{MW} + T_{MT}$ (c) $\Delta U = Q + T_{MT}$
(d) $0 = Q + T_{MT} + T_{ET} + T_{ER}$

3. (a) $v = (3gR)^{1/2}$ (b) 0.098 0 N down

5. 10.2 m

7. (a) 4.43 m/s (b) 5.00 m

9. $\sqrt{\dfrac{8gh}{15}}$

11. 5.49 m/s

13. 2.04 m

15. (a) 0.791 m/s (b) 0.531 m/s

17. (a) 5.60 J (b) 2.29 rev

19. 168 J

21. (a) -160 J (b) 73.5 J (c) 28.8 N (d) 0.679

23. (a) 4.12 m (b) 3.35 m

25. (a) 1.40 m/s (b) 4.60 cm after release (c) 1.79 m/s

27. (a) Isolated. The only external influence on the system is the normal force from the slide, but this force is always perpendicular to its displacement so it performs no work on the system. (b) No, the slide is frictionless. (c) $E_{system} = mgh$
(d) $E_{system} = \frac{1}{5}mgh + \frac{1}{2}mv_1^2$ (e) $E_{system} = mgy_{max} + \frac{1}{2}mv_{xi}^2$
(f) $v_i = \sqrt{\dfrac{8gh}{5}}$ (g) $y_{max} = h(1 - \frac{4}{5}\cos^2\theta)$ If friction is present, mechanical energy of the system would *not* be conserved, so the child's kinetic energy at all points after leaving the top of the waterslide would be reduced when compared with the frictionless case. Consequently, her launch speed, maximum height reached, and final speed would be reduced as well.

29. 1.23 kW

31. $\sim 10^4$ W

33. 5×10^3 N

35. $145

37. 236 s or 3.93 min

39. (a) 423 mi/gal (b) 776 mi/gal

41. (a) 10.2 kW (b) 10.6 kW (c) 5.82 MJ

43. 830 N

45. (a) 0.588 J (b) 0.588 J (c) 2.42 m/s (d) $K = 0.196$ J, $U = 0.392$ J

47. Her arms would need to be 1.36 m long to perform this task. This is significantly longer than the human arm.

49. (a) 11.1 m/s (b) 1.00×10^3 J (c) 1.35 m

51. (a) $K = 2 + 24t^2 + 72t^4$, where t is in seconds and K is in joules (b) $a = 12t$ and $F = 48t$, where t is in seconds, a is in m/s^2, and F is in newtons (c) $P = 48t + 288t^3$, where t is in seconds and P is in watts (d) 1.25×10^3 J

53. (a) -6.08×10^3 J (b) -4.59×10^3 J (c) 4.59×10^3 J

55. (a) 1.53 J at $x = 6.00$ cm, 0 J at $x = 0$ (b) 1.75 m/s (c) 1.51 m/s (d) The answer to part (c) is not half the answer to part (b) because the equation for the speed of an oscillator is not linear in position.

57. (a) $-\mu_k gx/L$ (b) $(\mu_k gL)^{1/2}$

59. 2.92 m/s

61. (a) 2.17 kW (b) 58.6 kW

63. $\sim 10^3$ W peak or $\sim 10^2$ W sustainable

65. (a) $x = -4.0$ mm (b) -1.0 cm

67. 33.4 kW

69. (a) 100 J (b) 0.410 m (c) 2.84 m/s (d) -9.80 mm (e) 2.85 m/s

71. 0.328

73. $v = 1.24$ m/s

75. (a) No. The change in the kinetic energy of the plane is equal to the *net* work done by all forces doing work on it. In this case, there are two such forces, the thrust due to the engine and a resistive force due to the air. Since the work done by the air resistance force is negative, the net work done (and hence the change in kinetic energy) is less than the positive work done by the engine thrust. Also, because the thrust from the engine and the air resistance force are nonconservative forces, mechanical energy is not conserved. (b) 77 m/s

77. (a) 14.1 m/s (b) 800 N (c) 771 N (d) 1.57 kN up

79. (a) 0.400 m (b) 4.10 m/s (c) The block stays on the track.

81. (a) 6.15 m/s (b) 9.87 m/s

83. (a) 21.0 m/s (b) 16.1 m/s

Context 1 Conclusion

1. (a) 315 kJ (b) 220 kJ (c) 187 kJ (d) 127 kJ (e) 14.0 m/s (f) 40.5% (g) 187 kJ

2. (a) Conventional car = 581 MJ; Hybrid car = 220 MJ (b) Conventional car = 11.4%; Hybrid car = 30.0%

CHAPTER 8
Answers to Quick Quizzes

1. (d)

2. (b), (c), (a)

3. (i) (c), (e) (ii) (b), (d)

4. (b)

5. (b)

6. (i) (a) (ii) (b)

Answers to Odd-Numbered Problems

1. 40.5 g

3. (a) $(9.00\hat{\mathbf{i}} - 12.0\hat{\mathbf{j}})$ kg · m/s (b) 15.0 kg · m/s at 307°

5. (a) $v_{pi} = -0.346$ m/s (b) $v_{gi} = 1.15$ m/s

7. (a) $(-6.00\hat{\mathbf{i}})$ m/s (b) 8.40 J (c) The original energy is in the spring. (d) A force had to be exerted over a displacement to compress the spring, transferring energy into it by work. The cord exerts force, but over no displacement. (e) System momentum is conserved with the value zero. (f) The forces on the two blocks are internal forces, which cannot change the momentum of the system; the system is isolated. (g) Even though there is motion afterward, the final momenta are of equal magnitude in opposite directions, so the final momentum of the system is still zero.

9. 260 N normal to the wall

11. (a) 13.5 N · s (b) 9.00 kN

13. 15.0 N in the direction of the initial velocity of the exiting water stream.

15. (a) 2.50 m/s (b) 37.5 kJ

17. 91.2 m/s

19. 0.556 m

21. (a) 0.284 (b) 1.15×10^{-13} J and 4.54×10^{-14} J

23. (a) 2.24 m/s (b) Coupling order makes no difference.

25. $(3.00\hat{\mathbf{i}} - 1.20\hat{\mathbf{j}})$ m/s

27. $v_O = 3.99$ m/s and $v_Y = 3.01$ m/s

29. 2.50 m/s at $-60.0°$

31. (a) $(-9.33\hat{\mathbf{i}} - 8.33\hat{\mathbf{j}})$ Mm/s (b) 439 fJ

33. (a) 1.07 m/s at $-29.7°$ (b) $\dfrac{\Delta K}{K_i} = -0.318$

35. $\vec{\mathbf{r}}_{CM} = (0\hat{\mathbf{i}} + 1.00\hat{\mathbf{j}})$ m

37. 3.57×10^8 J

39. (a) $(1.40\hat{\mathbf{i}} + 2.40\hat{\mathbf{j}})$ m/s (b) $(7.00\hat{\mathbf{i}} + 12.0\hat{\mathbf{j}})$ kg · m/s

41. 0.700 m

43. 4.41 kg

45. (a) 3.90×10^7 N (b) 3.20 m/s^2

47. (a) 787 m/s (b) 138 m/s

49. (a) $1.33\hat{\mathbf{i}}$ m/s (b) $-235\hat{\mathbf{i}}$ N (c) 0.680 s (d) $-160\hat{\mathbf{i}}$ N · s and $+160\hat{\mathbf{i}}$ N · s (e) 1.81 m (f) 0.454 m (g) -427 J (h) $+107$ J (i) Equal friction forces act through different distances on person and cart, to do different amounts of work on them. The total work on both together, -320 J, becomes $+320$ J of extra internal energy in this perfectly inelastic collision.

51. (a) 2.07 m/s^2 (b) 3.88 m/s

53. (a) Momentum of the bullet–block system is conserved in the collision, so you can relate the speed of the block and bullet immediately after the collision to the initial speed of the bullet. Then, you can use conservation of mechanical energy for the bullet–block–Earth system to relate the speed after the collision to the maximum height. (b) 521 m/s upward

55. (a) -0.667 m/s (b) 0.952 m

57. (a) 100 m/s (b) 374 J

59. (a) $-0.256\hat{\mathbf{i}}$ m/s and $0.128\hat{\mathbf{i}}$ m/s (b) $-0.064\,2\hat{\mathbf{i}}$ m/s and 0 (c) 0 and 0

61. $2v_i$ for the particle with mass m and 0 for the particle with mass $3m$.

63. 0.403

65. $\vec{\mathbf{F}} = \left(\dfrac{3Mgx}{L}\right)\hat{\mathbf{j}}$

CHAPTER 9
Answers to Quick Quizzes

1. (d)

2. (a)

3. (a)

4. (c)

5. (c), (d)

6. (a) $m_3 > m_2 = m_1$ (b) $K_3 = K_2 > K_1$ (c) $u_2 > u_3 = u_1$

Answers to Odd-Numbered Problems

3. $0.866c$

5. $0.866c$

7. (a) 25.0 yr (b) 15.0 yr (c) 12.0 ly

9. 1.55 ns

11. $0.800c$

13. (a) 20.0 m (b) 19.0 m (c) $0.312c$

15. (a) 17.4 m (b) 3.30°

17. (a) 2.50×10^8 m/s (b) 4.98 m (c) -1.33×10^{-8} s

19. $0.357c$

21. $0.960c$

23. (a) 2.73×10^{-24} kg · m/s (b) 1.58×10^{-22} kg · m/s
(c) 5.64×10^{-22} kg · m/s

25. 4.51×10^{-14}

27. $0.285c$

29. (a) 0.582 MeV (b) 2.45 MeV

31. (a) 3.07 MeV (b) $0.986c$

33. (a) $0.979c$ (b) $0.065\,2c$ (c) 15.0
(d) $0.999\,999\,97c$; $0.948c$; 1.06

35. (a) 938 MeV (b) 3.00 GeV (c) 2.07 GeV

37. (a) 5.37×10^{-11} J = 335 MeV
(b) 1.33×10^{-9} J = 8.31 GeV

39. (a) 4.08 MeV (b) 29.6 MeV

41. 4.28×10^{9} kg/s

43. (a) 2.66×10^{7} m (b) 3.87 km/s (c) -8.35×10^{-11}
(d) 5.29×10^{-10} (e) $+4.46 \times 10^{-10}$

45. (a) $(1 - 1.12 \times 10^{-10})c$ (b) 6.00×10^{27} J (c) \$$1.83 \times 10^{20}$

47. 0.712%

49. 2.97×10^{-26} kg

51. (a) $\sim 10^{2}$ or 10^{3} s (b) $\sim 10^{8}$ km

53. (a) $0.946c$ (b) 0.160 ly (c) 0.114 yr (d) 7.49×10^{22} J

55. The trackside observer measures the length to be 31.2 m, so the supertrain is measured to fit in the tunnel, with 19.8 m to spare.

57. (a) 229 s (b) 174 s

59. (a) 76.0 minutes (b) 52.1 minutes

61. 1.02 MeV

63. (a) $0.800c$ (b) 7.51×10^{3} s (c) 1.44×10^{12} m (d) $0.385c$
(e) 4.88×10^{3} s

65. (a) Tau Ceti exploded 16.0 years before the Sun.
(b) The two stars blew up simultaneously.

CHAPTER 10

Answers to Quick Quizzes

1. (i) (c) (ii) (b)

2. (b)

3. (i) (b) (ii) (a)

4. (a)

5. (i) (b) (ii) (a)

6. (b)

7. (b)

8. (a)

9. (i) (b) (ii) (c) (iii) (a)

Answers to Odd-Numbered Problems

1. (a) 5.00 rad, 10.0 rad/s, 4.00 rad/s²
(b) 53.0 rad, 22.0 rad/s, 4.00 rad/s²

3. (a) 0.209 rad/s² (b) yes

5. 50.0 rev

7. (a) 5.24 s (b) 27.4 rad

9. 13.7 rad/s²

11. (a) 126 rad/s (b) 3.77 m/s (c) 1.26 km/s² (d) 20.1 m

13. 0.572

15. (a) 56.5 rad/s (b) 22.4 rad/s (c) -7.63×10^{-3} rad/s²
(d) 1.77×10^{5} rad (e) 5.81×10^{3} m

17. 1.03×10^{-3} J

19. (a) 24.5 m/s (b) no (c) no (d) no (e) no (f) yes

21. (a) 1.95 s (b) If the pulley were massless, the acceleration would be larger by a factor 35/32.5 and the time shorter by the square root of the factor 32.5/35. That is, the time would be reduced by 3.64%.

23. -3.55 N · m

25. $\hat{\mathbf{i}} + 8.00\hat{\mathbf{j}} + 22.0\hat{\mathbf{k}}$

27. $\vec{\boldsymbol{\tau}} = (2.00\hat{\mathbf{k}})$ N · m

29. 0.896 m

31. (a) 1.04 kN at 60.0° upward and to the right
(b) $(370\hat{\mathbf{i}} + 910\hat{\mathbf{j}})$ N

33. (a) $f_s = 268$ N, $n = 1\,300$ N (b) 0.324

35. $F_t = 724$ N, $F_s = 716$ N

37. 21.5 N

39. (a) 21.6 kg · m² (b) 3.60 N · m (c) 52.5 rev

41. 0.312

43. (a) Particle under a net force (b) Rigid object under a net torque (c) 118 N (d) 156 N (e) $\dfrac{r^2}{a}(T_2 - T_1)$
(f) 1.17 kg · m²

45. (a) 11.4 N (b) 7.57 m/s² (c) 9.53 m/s (d) 9.53 m/s

47. $60.0\hat{\mathbf{k}}$ kg · m²/s

49. 1.20 kg · m²/s

51. (a) 0.433 kg · m²/s (b) 1.73 kg · m²/s

53. (a) 7.20×10^{-3} kg · m²/s (b) 9.47 rad/s

55. 5.99×10^{-2} J

57. (a) The mechanical energy of the system is not constant. Some chemical energy is converted into mechanical energy. (b) The momentum of the system is not constant. The turntable bearing exerts an external northward force on the axle. (c) The angular momentum of the system is constant. (d) 0.360 rad/s counterclockwise (e) 99.9 J

59. (a) 500 J (b) 250 J (c) 750 J

61. (a) 1.21×10^{-4} kg · m² (b) Knowing the height of the can is unnecessary. (c) The mass is not uniformly distributed; the density of the metal can is larger than that of the soup.

63. 131 s

65. (a) $(3g/L)^{1/2}$ (b) $3g/2L$ (c) $-\frac{3}{2}g\hat{\mathbf{i}} - \frac{3}{4}g\hat{\mathbf{j}}$
(d) $-\frac{3}{2}Mg\hat{\mathbf{i}} + \frac{1}{4}Mg\hat{\mathbf{j}}$

67. (a) $3\,750$ kg · m²/s (b) 1.88 kJ (c) $3\,750$ kg · m²/s
(d) 10.0 m/s (e) 7.50 kJ (f) 5.62 kJ

69. $T = 1.68$ kN; $R = 2.34$ kN; $\theta = 21.2°$

71. $\omega = \sqrt{\dfrac{2mgd \sin\theta + kd^2}{I + mR^2}}$

73. (a) $T = 133$ N (b) $n_A = 429$ N, $n_B = 257$ N
(c) $R_x = 133$ N, to the right; $R_y = 257$ N, downward

75. (a) $\omega = 2mv_i d/[M + 2m]R^2$ (b) No; some mechanical energy of the system changes into internal energy. (c) The momentum of the system is not constant. The axle exerts a backward force on the cylinder when the clay strikes.

77. 209 N

79. (a) 2.71 kN (b) 2.65 kN (c) You should lift "with your knees" rather than "with your back." (d) In this situation, you can make the compressional force in your spine about ten times smaller by bending your knees and lifting with your back as straight as possible.

81. (a) isolated system (angular momentum) (b) $mv_i d/2$

(c) $\left(\frac{1}{12}M + \frac{1}{4}m\right)d^2$ (d) $\left(\frac{1}{12}M + \frac{1}{4}m\right)d^2\omega$ (e) $\dfrac{6mv_i}{(M + 3m)d}$

(f) $\frac{1}{2}mv_i^2$ (g) $\dfrac{3m^2v_i^2}{2(M + 3m)}$ (h) $-\dfrac{M}{M + 3m}$

83. (a) $2.70R$ (b) $F_x = -20mg/7$, $F_y = -mg$

85. (a) 9.28 kN (b) The moment arm of the force $\vec{F}_h$ is no longer 70 cm from the shoulder joint but only 49.5 cm, therefore reducing $\vec{F}_m$ to 6.56 kN.

CHAPTER 11

Answers to Quick Quizzes

1. (e)
2. (a)
3. (a) Perihelion (b) Aphelion (c) Perihelion
   (d) All points
4. (a)

Answers to Odd-Numbered Problems

1. 2.67×10^{-7} m/s^2
3. (a) 2.50×10^{-5} N toward the 500-kg object (b) between the objects and 2.45 m from the 500-kg object
5. 7.41×10^{-10} N
7. 2/3
9. (a) 7.61 cm/s^2 (b) 363 s (c) 3.08 km
   (d) 28.9 m/s at 72.9° below the horizontal
11. (a) 1.31×10^{17} N (b) 2.62×10^{12} N/kg
13. 0.614 m/s^2, toward the Earth
15. (a) 4.22×10^7 m (b) 0.285 s
17. 1.90×10^{27} kg
19. 1.26×10^{32} kg
21. 35.1 AU
23. (a) 1.84×10^9 kg/m^3 (b) 3.27×10^6 m/s^2
   (c) -2.08×10^{13} J
25. 1.78 km
27. (a) 850 MJ (b) 2.71×10^9 J
29. (a) 5.30×10^3 s (b) 7.79 km/s (c) 6.43×10^9 J
31. 1.58×10^{10} J
33. (a) 0.980 (b) 127 yr (c) -2.13×10^{17} J
35. (a) 1.00×10^7 m (b) 1.00×10^4 m/s
37. (a) 5 (b) no (c) no
39. (a) 2.19×10^6 m/s (b) 13.6 eV (c) -27.2 eV
41. (a) 0.212 nm (b) 9.95×10^{-25} kg · m/s
   (c) 2.11×10^{-34} kg · m^2/s (d) 3.40 eV (e) -6.80 eV
   (f) -3.40 eV
43. 4.42×10^4 m/s
45. (a) 29.3% (b) no change
47. 2.25×10^{-7}
49. 492 m/s
51. (a) -7.04×10^4 J
   (b) -1.57×10^5 J (c) 13.2 m/s
53. 7.79×10^{14} kg

55. (a) 2.93×10^4 m/s
   (b) $K = 2.74 \times 10^{33}$ J, $U = -5.39 \times 10^{33}$ J
   (c) $K = 2.56 \times 10^{33}$ J, $U = -5.21 \times 10^{33}$ J
   (d) Yes; $E = -2.65 \times 10^{33}$ J at both aphelion and perihelion.
57. (a) $v_1 = m_2\sqrt{\dfrac{2G}{d(m_1 + m_2)}}$, $v_2 = m_1\sqrt{\dfrac{2G}{d(m_1 + m_2)}}$,
   $v_{\text{rel}} = \sqrt{\dfrac{2G(m_1 + m_2)}{d}}$
   (b) 1.07×10^{32} J and 2.67×10^{31} J
59. (a) 15.3 km (b) 1.66×10^{16} kg (c) 1.13×10^4 s
   (d) No; its mass is so large compared with yours that you would have a negligible effect on its rotation.
61. (a) $r_n = 0.106n^2$, where r_n is in nanometers and $n = 1, 2, 3, \ldots$
   (b) $E_n = -\dfrac{6.80}{n^2}$, where E_n is in electron volts and $n = 1, 2, 3, \ldots$
63. (a) 2×10^8 yr (b) $\sim10^{41}$ kg (c) 10^{11}
65. (a) $(2.77 \text{ m/s}^2)\left(1 + \dfrac{m}{5.97 \times 10^{24} \text{ kg}}\right)$ (b) 2.77 m/s^2
   (c) 2.77 m/s^2 (d) 3.70 m/s^2 (e) Any object with mass small compared to the Earth starts to fall with acceleration 2.77 m/s^2. As m increases to become comparable to the mass of the Earth, the acceleration increases and can become arbitrarily large. It approaches a direct proportionality to m.

Context 2 Conclusion

1. (1) 146 d (b) Venus 53.9° behind the Earth
2. 1.30×10^3 m/s
3. (a) 2.95 km/s (b) 2.65 km/s (c) 10.7 km/s
   (d) 4.80 km/s

CHAPTER 12

Answers to Quick Quizzes

1. (d)
2. (f)
3. (a)
4. (b)
5. (i) (a) (ii) (a)
6. (a)

Answers to Odd-Numbered Problems

1. (a) 17 N to the left (b) 28 m/s^2 to the left
3. (a) 1.50 Hz (b) 0.667 s (c) 4.00 m (d) π rad (e) 2.83 m
5. 40.9 N/m
7. (a) 20 cm (b) 94.2 cm/s as the particle passes through equilibrium (c) 17.8 m/s^2 at maximum excursion from equilibrium
9. (a) $x = 2.00 \cos(3.00\pi t - 90°)$ or $x = 2.00 \sin(3.00\pi t)$ where x is in centimeters and t is in seconds (b) 18.8 cm/s
   (c) 0.333 s (d) 178 cm/s^2 (e) 0.500 s (f) 12.0 cm
11. (a) -2.34 m (b) -1.30 m/s (c) $-0.076\,3$ m (d) 0.315 m/s

13. (a) 40.0 cm/s (b) 160 cm/s^2 (c) 32.0 cm/s
 (d) -96.0 cm/s^2 (e) 0.232 s
15. 12.0 Hz
17. (a) 0.542 kg (b) 1.81 s (c) 1.20 m/s^2
19. 2.23 m/s
21. (a) 28.0 mJ (b) 1.02 m/s (c) 12.2 mJ (d) 15.8 mJ
23. 2.60 cm and -2.60 cm
25. (a) E increases by a factor of 4. (b) v_{max} is doubled.
 (c) a_{max} is doubled. (d) period is unchanged.
27. 0.944 kg $\cdot$ m^2
29. 1.42 s, 0.499 m
31. (a) 2.09 s (b) 4.08%
33. (a) 3.65 s (b) 6.41 s (c) 4.24 s
35. (a) 0.820 m/s (b) 2.57 rad/s^2 (c) 0.641 N
 (d) $v_{max} = 0.817$ m/s, $\alpha_{max} = 2.54$ rad/s^2, $F_{max} = 0.634$ N
 (e) The answers are close but not exactly the same. The answers computed from conservation of energy and from Newton's second law are more precise.
37. 1.00×10^{-3} s^{-1}
39. (a) 3.16 s^{-1} (b) 6.28 s^{-1} (c) 5.09 cm
41. 11.0 cm
43. 0.641 Hz or 1.31 Hz
45. 1.56×10^{-2} m
47. 6.62 cm
49. (a) $x = 2\cos\left(10t + \dfrac{\pi}{2}\right)$ (b) ± 1.73 m
 (c) 0.052 4 s = 52.4 ms
 (d) 0.098 0 m
51. 9.19×10^{13} Hz
53. 7.75 s^{-1}
55. (a) 3.00 s (b) 14.3 J (c) $\theta = 25.5°$
57. (a) $\omega = \sqrt{\dfrac{200}{0.400 + M}}$, where ω is in s^{-1} and M is in kilograms
 (b) 22.4 s^{-1} (c) 22.4 s^{-1}
59. (a) $2Mg$ (b) $Mg\left(1 + \dfrac{y}{L}\right)$ (c) $\dfrac{4\pi}{3}\sqrt{\dfrac{2L}{g}}$ (d) 2.68 s
61. (a) 0.368 m/s (b) 3.51 cm (c) 40.6 mJ (d) 27.7 mJ
63. (a) 1.26 m (b) 1.58 (c) The energy decreases by 120 J.
 (d) Mechanical energy is transformed into internal energy in the perfectly inelastic collision.
65. $\dfrac{1}{2\pi L}\sqrt{gL + \dfrac{kh^2}{M}}$
67. (a) 0.500 m/s (b) 8.56 cm
69. (a) $\frac{1}{2}\left(M + \frac{1}{3}m\right)v^2$ (b) $2\pi\sqrt{\dfrac{M + \frac{1}{3}m}{k}}$

CHAPTER 13

Answers to Quick Quizzes

1. (i) (b) (ii) (a)
2. (i) (c) (ii) (b) (iii) (d)
3. (c)
4. (f) and (h)
5. (d)
6. (e)
7. (e)

Answers to Odd-Numbered Problems

1. $y = \dfrac{6.00}{(x - 4.50t)^2 + 3.00}$, where x and y are in meters and t is in seconds.
3. (a) $3.33\hat{\mathbf{i}}$ m/s (b) -5.48 cm (c) 0.667 m (d) 5.00 Hz
 (e) 11.0 m/s
5. (a) 31.4 rad/s (b) 1.57 rad/m
 (c) $y = 0.120 \sin(1.57x - 31.4t)$, where x and y are in meters and t is in seconds (d) 3.77 m/s (e) 118 m/s^2
7. (a) $y = 0.080\ 0 \sin(2.5\pi x + 6\pi t)$
 (b) $y = 0.080\ 0 \sin(2.5\pi x + 6\pi t - 0.25\pi)$
9. ± 6.67 cm
11. 0.319 m
13. 2.40 m/s
15. 0.329 s
17. 80.0 N
19. 13.5 N
21. (a) zero (b) 0.300 m
23. (a) $y = 0.075 \sin(4.19x - 314t)$ (b) 625 W
25. (a) 62.5 m/s (b) 7.85 m (c) 7.96 Hz (d) 21.1 W
27. 0.196 s
29. (a) 23.2 cm (b) 1.38 cm
31. $\Delta P = 0.200 \sin(20\pi x - 6\ 860\pi t)$, where ΔP is in pascals, x is in meters, and t is in seconds.
33. (a) 0.625 mm (b) 1.50 mm to 75.0 μm
35. (a) 2.00 μm (b) 40.0 cm (c) 54.6 m/s (d) -0.433 μm
 (e) 1.72 mm/s
37. 1.55×10^{-10} m
39. (a) 3.04 kHz (b) 2.08 kHz (c) 2.62 kHz; 2.40 kHz
41. 26.4 m/s
43. (a) 0.364 m (b) 0.398 m (c) 941 Hz (d) 938 Hz
45. 19.7 m
47. 184 km
49. 0.084 3 rad
51. (a) 39.2 N (b) 0.892 m (c) 83.6 m/s
53. (a) 0.343 m (b) 0.303 m (c) 0.383 m (d) 1.03 kHz
55. (a) $\dfrac{\mu\omega^3}{2k} A_0^2 e^{-2bx}$ (b) $\dfrac{\mu\omega^3}{2k} A_0^2$ (c) e^{-2bx}
57. 7.82 m
59. (a) 0.515 trucks per minute (b) 0.614 trucks per minute
61. 6.01 km
63. (a) 55.8 m/s (b) 2 500 Hz
65. 8.43×10^{-3} s
67. (a) 3.29 m/s (b) The bat will be able to catch the insect because the bat is traveling at a higher speed in the same direction as the insect.
69. (a) $\mu(x) = \dfrac{(\mu_L - \mu_0)x}{L} + \mu_0$
 (b) $\Delta t = \dfrac{2L}{3\sqrt{T}(\mu_L - \mu_0)}(\mu_L^{3/2} - \mu_0^{3/2})$
71. (a) 531 Hz (b) 466 Hz to 539 Hz (c) 568 Hz

CHAPTER 14

Answers to Quick Quizzes

1. (c)
2. (i) (a) (ii) (d)

3. (d)
4. (b)
5. (c)
6. (b)

Answers to Odd-Numbered Problems

1. 5.66 cm
3. (a) 156° (b) 0.058 4 cm
5. (a) −1.65 cm (b) −6.02 cm (c) 1.15 cm
7. (a) y_1: positive x direction; y_2: negative x direction
   (b) 0.750 s (c) 1.00 m
9. (a) 9.24 m (b) 600 Hz
11. 91.3°
13. (a) 4.24 cm (b) 6.00 cm (c) 6.00 cm
    (d) 0.500 cm, 1.50 cm, 2.50 cm
15. (a) 15.7 m (b) 31.8 Hz (c) 500 m/s
17. at 0.089 1 m, 0.303 m, 0.518 m, 0.732 m, 0.947 m, and 1.16 m from one speaker
19. (a) 78.6 Hz (b) 157 Hz, 236 Hz, 314 Hz
21. 19.6 Hz
23. (a) reduced by $\frac{1}{2}$ (b) reduced by $1/\sqrt{2}$
    (c) increased by $\sqrt{2}$
25. 1.86 g
27. (a) 163 N (b) 660 Hz
29. (a) 0.357 m (b) 0.715 m
31. 57.9 Hz
33. (a) 0.656 m (b) 1.64 m
35. (a) 50.0 Hz (b) 1.72 m
37. 158 s
39. $n(0.252\ \text{m})$ with $n = 1, 2, 3, \ldots$
41. (a) 0.195 m (b) 841 Hz
43. (a) 349 m/s (b) 1.14 m
45. 5.64 beats/s
47. The second harmonic of E is close to the third harmonic of A, and the fourth harmonic of C# is close to the fifth harmonic of A.
49. 1.76 cm
51. (a) 3.66 m/s (b) 0.200 Hz
53. (a) 34.8 m/s (b) 0.986 m
55. (a) 21.5 m (b) seven
57. 3.85 m/s away from the station or 3.77 m/s toward the station
59. (a) 59.9 Hz (b) 20.0 cm
61. (a) 45.0 or 55.0 Hz (b) 162 or 242 N
63. (a) 78.9 N (b) 211 Hz
65. 262 kHz
67. 31.1 N
69. (a) larger (b) 2.43

Context 3 Conclusion

1. 3.5 cm
2. The speed decreases by a factor of 25
3. Station 1: 15:46:32
   Station 2: 15:46:24
   Station 3: 15:46:09

CHAPTER 15

Answers to Quick Quizzes

1. (a)
2. (a)
3. (b)
4. (b) or (c)
5. (b)
6. (a)

Answers to Odd-Numbered Problems

1. 1.92×10^4 N
3. 24.8 kg
5. 8.46 m
7. 1.05×10^7 Pa
9. (a) 1.02×10^7 Pa (b) 6.61×10^5 N
11. 7.74×10^{-3} m^2
13. (a) 2.94×10^4 N (b) 1.63×10^4 N · m
15. 2.31 lb
17. (a) 20.0 cm (b) 0.490 cm
19. (a) 116 kPa (b) 52.0 Pa
21. 98.6 kPa
23. 0.258 N down
25. (a) 4.9 N down, 16.7 N up (b) 86.2 N (c) By either method of evaluation, the buoyant force is 11.8 N up.
27. (a) 7.00 cm (b) 2.80 kg
29. $V = 1.52 \times 10^3$ m^3
31. (a) 1 250 kg/m^3 (b) 500 kg/m^3
33. 1.01 kJ
35. 2.67×10^3 kg
37. 12.8 kg/s
39. $2\sqrt{h(h_0 - h)}$
41. (a) 106 cm/s = 1.06 m/s (b) 424 cm/s = 4.24 m/s
43. (a) 17.7 m/s (b) 1.73 mm
45. (a) 27.9 N (b) 3.32×10^4 kg (c) 7.26×10^4 Pa
47. 0.247 cm
49. (a) 452 N outward (b) 1.81 kN outward
51. (a) 4.43 m/s (b) 10.1 m
53. 347 m/s
55. 0.604 m
57. (a) 8.01 km (b) yes
59. 291 Hz
61. 4.43 m/s
63. upper scale: 17.3 N; lower scale: 31.7 N
65. (a) 18.3 mm (b) 14.3 mm (c) 8.56 mm
67. (a) 1.25 cm (b) 14.3 m/s
69. (a) 2.65 m/s (b) 2.31×10^4 Pa

Context 4 Conclusion

1. (a) The blood in vessel (ii) would have the highest speed at point 2. (b) $v_{ii} = 32 v_{iii}$.
2. (a) 1.67 m/s (b) 720 Pa
3. (a) 1.57 kPa, 0.015 5 atm, 11.8 mm (b) Blockage of the fluid within the spinal column or between the skull and the spinal column would prevent the fluid level from rising.
4. 12.6 m/s

Index

Standard Abbreviations and Symbols for Units

Symbol	Unit	Symbol	Unit
A	ampere	K	kelvin
u	atomic mass unit	kg	kilogram
atm	atmosphere	kmol	kilomole
Btu	British thermal unit	L	liter
C	coulomb	lb	pound
°C	degree Celsius	ly	light-year
cal	calorie	m	meter
d	day	min	minute
eV	electron volt	mol	mole
°F	degree Fahrenheit	N	newton
F	farad	Pa	pascal
ft	foot	rad	radian
G	gauss	rev	revolution
g	gram	s	second
H	henry	T	tesla
h	hour	V	volt
hp	horsepower	W	watt
Hz	hertz	Wb	weber
in.	inch	yr	year
J	joule	Ω	ohm

Mathematical Symbols Used in the Text and Their Meaning

Symbol	Meaning
$=$	is equal to
$\equiv$	is defined as
$\neq$	is not equal to
$\propto$	is proportional to
$\sim$	is on the order of
$>$	is greater than
$<$	is less than
$\gg (\ll)$	is much greater (less) than
$\approx$	is approximately equal to
Δx	the change in x
$\displaystyle\sum_{i=1}^{N} x_i$	the sum of all quantities x_i from $i = 1$ to $i = N$
$\lvert x \rvert$	the absolute value of x (always a nonnegative quantity)
$\Delta x \to 0$	Δx approaches zero
$\dfrac{dx}{dt}$	the derivative of x with respect to t
$\dfrac{\partial x}{\partial t}$	the partial derivative of x with respect to t
$\displaystyle\int$	integral